Handbook of Ecological and Ecosystem Engineering

Handbook of Ecological and Ecosystem Engineering

Edited by

Majeti Narasimha Vara Prasad
School of Life Sciences
University of Hyderabad (an Institution of Eminence)
Hyderabad, Telangana, India

WILEY

Registered Offices
John Wiley & Sons, Inc., 111 River Street, Hoboken, NJ 07030, USA
John Wiley & Sons Ltd, The Atrium, Southern Gate, Chichester, West Sussex, PO19 8SQ, UK

Editorial Office
The Atrium, Southern Gate, Chichester, West Sussex, PO19 8SQ, UK

For details of our global editorial offices, customer services, and more information about Wiley products visit us at www.wiley.com.

Wiley also publishes its books in a variety of electronic formats and by print-on-demand. Some content that appears in standard print versions of this book may not be available in other formats.

For general information on our other products and services or for technical support, please contact our Customer Care Department within the United States at (800) 762-2974, outside the United States at (317) 572-3993 or fax (317) 572-4002.

Wiley also publishes its books in a variety of electronic formats. Some content that appears in print may not be available in electronic formats. For more information about Wiley products, visit our web site at www.wiley.com.

Library of Congress Cataloging-in-Publication Data

Names: Prasad, M. N. V. (Majeti Narasimha Vara), 1953– editor.
Title: Handbook of ecological and ecosystem engineering / M.N.V. Prasad, University of Hyderabad, School of Life Sciences
Description: First edition. | Hoboken : Wiley, 2021. | Includes bibliographical references and index.
Identifiers: LCCN 2020051312 (print) | LCCN 2020051313 (ebook) | ISBN 9781119678533 (hardback) | ISBN 9781119678502 (adobe pdf) | ISBN 9781119678601 (epub)
Subjects: LCSH: Ecological engineering. | Sustainable engineering.
Classification: LCC GE350 .P73 2021 (print) | LCC GE350 (ebook) | DDC 628–dc23
LC record available at https://lccn.loc.gov/2020051312
LC ebook record available at https://lccn.loc.gov/2020051313

Cover Design: Wiley
Cover Image: Courtesy of Maibam Meiti Dhanaraj

Set in 9.5/12.5pt STIXTwoText by Straive, Pondicherry, India
Printed and bound by CPI Group (UK) Ltd, Croydon, CR0 4YY

C9781119678533_180521

Contents

List of Contributors

Nadeesh M. Adassooriya
Department of Food Science & Technology, Wayamba University of Sri Lanka, Makandura, Sri Lanka

Madhoolika Agrawal
Department of Botany, Institute of Science, Banaras Hindu University, Varanasi, Uttar Pradesh, India

S.B. Agrawal
Department of Botany, Institute of Science, Banaras Hindu University, Varanasi, Uttar Pradesh, India

Jitendra Ahirwal
Restoration Ecology Laboratory, Department of Environmental Science and Engineering, Indian Institute of Technology (Indian School of Mines), Dhanbad, Jharkhand, India

J.R.B. Alfred
Nature Environment & Wildlife Society (NEWS), Kolkata, West Bengal, India

H.A.D.B. Amarasiri
Department of Pharmacy, The Open University of Sri Lanka, Nugegoda, Sri Lanka

Luís P. Azevedo
Materials Engineering Department, Federal University of Ouro Preto, Ouro Preto, Minas Gerais, Brazil

Bruno Barbosa
MEtRICS, Departamento de Ciências e Tecnologia da Biomassa, Faculdade de Ciências e Tecnologia, FCT, Universidade Nova de Lisboa, Caparica, Portugal; Universidade Feevale, Novo Hamburgo – RS, Brazil

S.P. Biswas
Department of Life Sciences, Dibrugarh University, Dibrugarh, Assam, India

E. Bocharnikova
Department of Functional Ecology, Institute Basic Biological Problems, Pushchino, Russia

Cláudia Cseko Nolasco de Carvalho
Department of Biology, State University of Alagoas, Santana do Ipanema, Alagoas, Brazil

Udo Censkowsky
Bluesensus – Sustainability & Seafood Consulting, Weilheim, Germany

S.B. Chaphekar
Homi Bhabha State University, Mumbai, Maharashtra, India

Smita Chaudhry
Institute of Environmental Studies, Kurukshetra University, Kurukshetra, Haryana, India

Biswajit Roy Chowdhury
Nature Environment & Wildlife Society (NEWS), Kolkata, West Bengal, India

Ronan Courtney
Department of Biological Science, Faculty of Science & Engineering, Bernal Institute, University of Limerick, Limerick, Ireland

Sijagurumayum Geetanjali Devi
Department of Forestry and Environmental Science, Manipur University, Imphal, Manipur, India

Ajanta Dey
Nature Environment & Wildlife Society (NEWS), Kolkata, West Bengal, India

Bhupinder Dhir
School of Sciences, Indira Gandhi National Open University, New Delhi, India

Denise C.R. Espinosa
Polytechnic School of University of São Paulo, Chemical Engineering Department, São Paulo, Brazil

Mihail Eva
Department of Geography, Faculty of Geography and Geology, "Alexandru Ioan Cuza" University, Iasi, Romania

Ana Luísa Fernando
MEtRICS, Departamento de Ciências e Tecnologia da Biomassa, Faculdade de Ciências e Tecnologia, FCT, Universidade Nova de Lisboa, Caparica, Portugal

Seemaa S. Ghate
Women Scientist 'B', Department of Science and Technology, New Delhi, India; Knowhow Foundation, Pune, Maharashtra, India

Anu Gopinath
Department of Aquatic Environment Management, Kerala University of Fisheries and Ocean Studies, Kochi, Kerala, India

Anna Grobelak
Faculty of Infrastructure and Environment, Institute of Environmental Engineering, Czestochowa University of Technology, Czestochowa, Poland

Marta Jaskulak
Faculty of Infrastructure and Environment, Institute of Environmental Engineering, Czestochowa University of Technology, Czestochowa, Poland; LGCgE, Laboratoire de Génie Civil et géo-Environnement University of Lille, IMT Lille Douai, Univ. Artois, Yncrea Hauts-de-France, Lille, France

Lander de Jesus Alves
Postgraduate Program in Biology and Biotechnology of Microorganisms, State University of Santa Cruz (UESC), Ilhéus, Bahia, Brazil

Anu Joy
Department of Chemistry and Research Centre, St Albert's College, Ernakulam, Kerala, India

Atya Kapley
Director's Research Cell, National Environmental Engineering Research Institute, (CSIR: NEERI), Nagpur, Maharashtra, India

Rupinder Kaur
Soil Ecosystems and Restoration Ecology Lab, Department of Botany, Panjab University, Chandigarh, India

M.L. Khan
Department of Botany, Dr. Harisingh Gour Vishwavidyalaya (A Central University), Sagar, Madhya Pradesh, India

Nazmun Ara Khanom
Department of Chemistry, Jorhat Institute of Science & Technology (JIST), Jorhat, Assam, India

Amit Kumar
Department of Geoinformatics, Central University of Jharkhand, Ranchi, Jharkhand, India

Rima Kumari
Department of Environmental Science and Engineering, Indian Institute of Technology (Indian School of Mines), Dhanbad, Jharkhand, India

Carlos A.F. Lagarinhos
Polytechnic School of University of São Paulo, Metallurgical and Materials Department, São Paulo, Brazil

Suvha Lama
Director's Research Cell, National Environmental Engineering Research Institute (CSIR: NEERI), Nagpur, Maharashtra, India

R.P. Madav
Terracon Ecotech Pvt. Ltd., Mumbai, Maharashtra, India

D.N. Magana-Arachchi
National Institute of Fundamental Studies, Kandy, Sri Lanka

Subodh Kumar Maiti
Restoration Ecology Laboratory, Department of Environmental Science and Engineering, Indian Institute of Technology (Indian School of Mines), Dhanbad, Jharkhand, India

Thaís de Marchi Soares
Department of Soil Science, University of Sao Paulo, Piracicaba, São Paulo, Brazil

V. Matichenkov
Department of Ecology and Physiology of Autotrophic Organisms, Institute Basic Biological Problems, Pushchino, Russia

Maibam Dhanaraj Meitei
Department of Forestry and Environmental Science, Manipur University, Imphal, Manipur, India

Florin-Constantin Mihai
CERNESIM Center, Department of Sciences, Interdisciplinary Research Institute, "Alexandru Ioan Cuza" University, Iasi, Romania

Chanchal Kumar Mitra
Department of Biochemistry, University of Hyderabad, Hyderabad, Telangana, India

Akash Nag
Department of Environmental Sciences, Central University of Jharkhand, Ranchi, Jharkhand, India

Fábio Carvalho Nunes
Academic Department, Federal Institute Baiano (IFBAIANO), Santa Inês, Bahia, Brazil

Rashmi Paliwal
Institute of Environmental Studies, Kurukshetra University, Kurukshetra, Haryana, India

Sara Pelaez Sanchez
Department of Biological Science, Faculty of Science & Engineering, Bernal Institute, University of Limerick, Limerick, Ireland

Marcin Pietrzykowski
Department of Forest Ecology and Silviculture, Faculty of Forestry, University of Agriculture in Krakow, Krakow, Poland

Majeti Narasimha Vara Prasad
School of Life Sciences, University of Hyderabad, Hyderabad, Telangana, India

Richa Rai
Plant Ecology and Climate Change Science Division, CSIR National Botanical Research Institute, Lucknow, Uttar Pradesh, India

Sanchayita Rajkhowa
Department of Chemistry, Jorhat Institute of Science & Technology (JIST), Jorhat, Assam, India

Purabi Saikia
Department of Environmental Sciences, Central University of Jharkhand, Ranchi, Jharkhand, India

Jyotirmoy Sarma
Department of Chemistry, Kaziranga University, Jorhat, Assam, India

Olaf Schmidt
School of Agriculture and Food Science, University College Dublin, Dublin, Ireland

Petra Schneider
Magdeburg-Stendal University of Applied Sciences, Magdeburg, Germany

L.R. Sendanayake
Postgraduate Institute of Science, University of Peradeniya, Peradeniya, Sri Lanka

Fernando G. da Silva Araújo
Materials Engineering Department, Federal University of Ouro Preto, Ouro Preto, Minas Gerais, Brazil

Anand Narain Singh
Soil Ecosystems and Restoration Ecology Lab, Department of Botany, Panjab University, Chandigarh, India

Gagan Preet Singh Sidhu
Centre for Applied Biology in Environment Sciences, Kurukshetra University, Kurukshetra, Haryana, India

José Rodrigues de Souza Filho
Academic Department, Federal Institute Baiano (IFBAIANO), Catu, Bahia, Brazil

Jorge A.S. Tenório
Polytechnic School of University of São Paulo, Chemical Engineering Department, São Paulo, Brazil

Niteshwori Thongam
Department of Forestry and Environmental Science, Manipur University, Imphal, Manipur, India

R.P. Wanigatunge
Department of Plant and Molecular Biology, Faculty of Science, University of Kelaniya, Kelaniya, Sri Lanka

Shailendra Yadav
Director's Research Cell, National Environmental Engineering Research Institute, (CSIR: NEERI), Nagpur, Maharashtra, India

Preface

There is a growing global concern about environment and sustainable development. The 2030 Agenda for Sustainable Development was launched in 2015 to end poverty and set the world on a path of peace, prosperity and opportunity for all on a healthy planet (UN Sustainable development goals 2015). The 17 Sustainable Development Goals (SDGs) demand nothing short of a transformation of the financial, economic and political systems that govern our societies today to guarantee the human rights of all. They require immense political will and ambitious action by all stakeholders. Global efforts to date have been insufficient to deliver the desired to current and future generations. Due to COVID-19, globally visible and an unprecedented health emergency, economic and social crisis, threatening lives and livelihoods jeopardizing the SDG's Agenda's promise to the current and future generations.

In this regard an integrated approach of management and restoration of ecosystems is a clear global concern. With increasing population explosion and exploitation of natural capital pressure on the planet earth is steeply increasing. In the last half century, the world human population has doubled, the consumption of natural resources has been multiplied by six, biodiversity has drastically decreased and the degradation of many ecosystems and global environmental modifications are glaring.

Odum H.T., used the term "ecological engineering" in 1960s, in several publications. Mitsch and Jøørgensen highlighted the necessity of ecological engineering in 1989.

Ecology and Ecosystems study is the only scientific field capable of providing the scientific bases necessary to implement ecosystem manipulations e.g. to mitigate the effect of global climatic changes. Such ecosystem manipulations could potentially have very negative impacts if not well designed.

Conversely, modern scientific ecology needs concrete examples of applied management Practitioners, stakeholders, and policy makers for ecosystem engineering are indeed good forcing processes to encompass the range of the studied ecological variables in terms of heterogeneity, limits and the occurring interactions between these variables.

In the same vein, the full development of ecological engineering requires stakeholders, scientists from engineering sciences, socio-economical sciences, and all environmental sciences to work together both to build a common general framework and to ground their work on concrete common projects.

Ecological engineering is also particularly relevant for developing countries. Indeed, the need for sustainable exploitation of ecosystems is very high in these countries where populations are still quickly growing and where individuals and institutions have often difficulties to develop practices based on high input rates (energy, water, mineral nutrients).

In this context this book assumes considerable significance. Thus, this book aimed at integrating multiple ecosystem services of the “Earth Ecosystem” resources (biotic and abiotic) keeping in view the UN 17 Sustainable Development Goals. **Phytobiome** and **microbiome** services that humans derive are focussed. Essentially Ecological and Ecosystem Engineering principles of the global ecosystem (Earth) for sustainable development will be dealt in this work

Goals of Ecological Engineering are

1) the restoration of ecosystems that have been substantially disturbed by human activities such as environmental pollution or land disturbance
2) the development of new sustainable ecosystems that have both human and ecosystems, and ecological value.
3) Ecological restoration - the return of an ecosystem to a close approximation of its condition prior to disturbance

Terms that are synonyms, subdisciplines, or fields similar to the crux of this books are: Bioengineering, biomanipulation, biospherics, ecohydrology, ecological engineering ecosystem rehabilitation, ecotechnology, engineering ecology, habitat reconstruction, nature cure, nature engineering, reclamation ecology, restoration ecology, river and lake restoration, solar aquatics, sustainable development, agroecology, synthetic ecology and wetland restoration.

Ecological Engineering - the design of sustainable ecosystems that integrate human ecosystems with its natural environment for the benefit of both. An attempt has been made and designed for the benefit of Environmental Engineers, Pollution Engineers and Energy Engineers, Researchers working in Water, Energy footprints, Carbon foot print, Environmental Biotechnologists etc. Researchers and academicians, undergraduate and graduate students in the broad area of Environmental Sciences, Ago-forestry and agronomy, Land scape engineers etc.

65 Authors from 9 countries viz., Brazil, Germany, India, Ireland, Poland, Portugal, Romania, Russia and Sri Lanka participated and contributed to various topics listed below:

1) Ecological engineering and ecosystem services - theory and practice
2) Ecological and ecosystem engineering for economic and environmental revitalization
3) Environmental issues and priority areas for ecological engineering initiatives
4) The role of Soil meso and macrofauna indicators for restoration assessment success in rehabilitated mine site
5) Urban Environmental issues – solutions by ecological and ecosystem engineering
6) Mitigation of Urban Environmental issues by applying ecological and ecosystem engineering
7) Soil fertility restoration, theory and practice
8) Extracellular soil enzymes act as moderator to restore carbon in the soil habitats
9) Ecological engineering for solid waste reduction and resource recovery
10) Urban floods and mitigation by applying ecological and ecosystem engineering
11) Ecological engineering and restoration of mine ecosystems
12) Restoration of abandoned mines
13) Wetland and watershed Lake restoration
14) Restoration of Riverine Health: An Ecohydrological Approach - Flow Regimes and Aquatic Biodiversity
15) Ecosystem services of phoomdi islands of Loktak, a dying Ramsar site in north-east India
16) The application of reefs in shoreline protection

17) Mangroves, as shore engineers, are nature-based solutions for ensuring coastal protection
18) Forests degradation prevention through nature-based solution: an Indian perspective
19) Restoring Ecosystem Services of Degraded Forest Ecosystems in a Changing Climate
20) Forest degradation prevention
21) Use of plants for air quality improvement
22) Phylloremediation for Mitigating Air Pollution
23) Air quality improvement using green cover
24) Air quality improvement using phytodiversity and plant architecture
25) Information explosion in digital science and ecosystems
26) Nanotechnology in Ecological and Ecosystem Engineering

Editor
22-10-2020

1

Ecological Engineering and Ecosystem Services – Theory and Practice

Fábio Carvalho Nunes[1], Thaís de Marchi Soares[2], Lander de Jesus Alves[3], José Rodrigues de Souza Filho[4], Cláudia Cseko Nolasco de Carvalho[5], and Majeti Narasimha Vara Prasad[6]

[1] *Academic Department, Federal Institute Baiano (IFBAIANO), Santa Inês, Bahia, Brazil*
[2] *Department of Soil Science, University of Sao Paulo, Piracicaba, São Paulo, Brazil*
[3] *Postgraduate Program in Biology and Biotechnology of Microorganisms, State University of Santa Cruz (UESC), Ilhéus, Bahia, Brazil*
[4] *Academic Department, Federal Institute Baiano (IFBAIANO), Catu, Bahia, Brazil*
[5] *Department of Biology, State University of Alagoas, Santana do Ipanema, Alagoas, Brazil*
[6] *School of Life Sciences, University of Hyderabad, Hyderabad, Telangana, India*

1.1 Introduction

In vain we try to snatch the veil from Nature;
She remains mysterious even in bright daylight,
Neither the machine nor any mechanism would reveal
The secrets she hid from us.

Source: Goethe, J. Faust, Part One, Trans. © 1987, Oxford University Press.

Humanity has witnessed magnificent technical-scientific advancements in recent decades. Robots feel the texture of asteroids and smell the atmosphere of Mars, whereas shining dipods dissect the soul of DNA. However, there is a strong existential and ethical imbalance, so much so that some individuals already compare themselves to Olympian gods and feel supralunary,[1] so their relationship with the inner (body–mind) and outer (planet Earth) "house" has profoundly changed; they believe that the total translation of the genetic code, human cloning, and interstellar travels are just a matter of time and would represent the last frontier to transcend time–space, as they would no longer be bound to the same body and planet Earth [2].

Society experiences a double *hybris* associated with our way of seeing and being in the world; it happens because we believe that everything is knowable since mystery and paradox are just what is not yet known, that everything can be manipulated and resources are unlimited [3], and that the wings of humanity can reach the sun: Icarus syndrome.[2] When human pride is confronted with the complexity of the *Cosmos*, it reveals its flight of Icarus, from which perplexity emerges; frustration stems from one's inability to face the abyss imposed by practical life.

Handbook of Ecological and Ecosystem Engineering, First Edition. Edited by Majeti Narasimha Vara Prasad.

The current culture becomes increasingly blind to the mysterious, to the impenetrable in life [4]. This blindness has made systems produced or changed by humanity – called *Black Swans* by Taleb [5] – increasingly fragile and susceptible to unpredictability. Complex systems created by humanity tend to develop uncontrolled cascades and chains of reactions, which decrease predictability and generate oversizing – the world becomes increasingly technological, although, paradoxically, that makes things even more unpredictable [4].

In contemporary times, it is believed that science can unveil everything, as observed in the advent of Modernity. However, Nature remains mysterious, and it is necessary to humbly acknowledge that "what we observe is not nature itself, but nature exposed to our method of questioning" [6] or, according to Johann von Goethe [7], that Nature "has so much freedom that we, despite our knowledge and science, cannot fully penetrate it."

Indeed, "in vain we try to snatch the veil from Nature," but it does not mean we should not nourish our understanding of the world. The pursuit of knowledge is a source of light; it helps us better understand the world and ourselves, but we must not let ourselves be blinded to the point of feeling like Moirae[3] – who control the thread of life – or like lords of Nature. We are more than its dependents: we are Nature becoming aware of itself.

Ecological engineering has emerged and developed in the fruitful context of technological advancements and environmental questioning of the post-World War II period. It has developed methods capable of analyzing matter and energy flows in different ecosystems to highlight economies' dependence on natural sources [8], make sustainability feasible, and enable the cultural transformation of human-Nature relationships.

The concept of ecosystem service has emerged in the wake of ecological engineering development to systematically express, qualify, and quantify the benefits of natural ecosystems for human societies by considering material aspects such as food and raw material supply, as well as immaterial aspects such as culture and religion.

Human beings inhabit the world to nourish the material and immaterial dimensions. We remove natural resources such as water, food, and raw materials from the environment to nourish the material dimension, a process whose direct and indirect adverse impacts vary in magnitudes and range. On the other hand, we do not always need to extract natural resources to nourish the immaterial dimension; our interaction with soil, wind, or mountains is satisfying.

In light of the preceding, it is evident that ecological engineering must go beyond analyzing the dynamics of ecosystems and their contribution to the economy (materials and services) by considering the immaterial dimensions – which are little explored in studies – in a more productive way. The subtlety of the immaterial dimensions makes them difficult to analyze, but that does not mean we should not try.

There is still a long way to go when it comes to measuring ecosystem services, primarily immaterial goods. However, it is worth emphasizing the importance of reflecting about, and improving, the concept of ecosystem service to enable the daily approach and praxis that take into consideration the benefits of ecosystems to other ecosystems that are at different spatial and temporal scales and that end up directly or indirectly affecting systems created by humanity.

Nothing is ready or finished, and the role of science is to point out problems, reveal contradictions, refine meanings, and suggest solutions. Ecological engineering is a relatively new, comprehensive, interdisciplinary science that proposes a different way of seeing our planet. Earth comprises large systems that continuously interact with each other and enable the dynamics of the biosphere, which comprises the anthroposphere and its subsystems (Figure 1.1). Thus, ecological engineering has the challenge of proposing less interventionist engineering strategies that do not weaken systems and favor antifragility (or that do so to a lesser extent).[4]

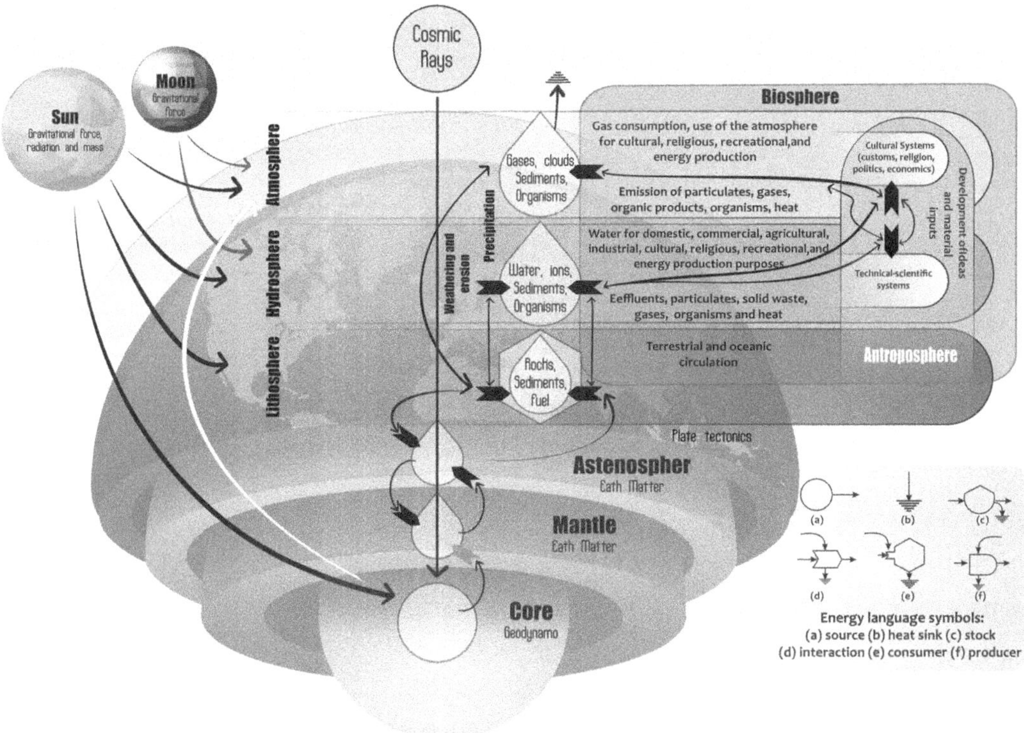

Figure 1.1 Global systems and their interactions. *See colour plate section for colour representation of this figure.*

Ecological engineering must not forget that natural systems are complex, interdependent, and full of nonlinear responses. It must keep in mind that interventions in complex systems do not necessarily need to be complicated and full of intricacies. Perhaps the primary challenges faced by ecological engineering lie in denying the idea that Nature should be controlled and in showing that heuristic actions can be more effective since are less opaque and, therefore, less unpredictable.

The aims of this chapter are to address issues associated with the aforementioned aspects, primarily the concepts of ecological engineering and ecosystem services, and present and discuss examples of qualitative and quantitative measurements of services supplied by specific ecosystems and practical examples applied in different contexts. The chapter reflects on the need to reframe our perspective and actions and reinvent the human-Nature relationship.

1.2 Ecological Engineering: History and Definition

The basic concepts of ecological engineering were elaborated in the mid-twentieth century, primarily through the contribution of ecologist Howard Thomas Odum. He defined it as the science that analyzes matter and energy flow in ecosystems dominated by humans to show the dependence of human production systems on natural sources, based on quantitative indices. Subsequently, Mitsch and Jorgensen [9] and Mitsch [10] have expanded the concept and defined ecological engineering as the science that accounts for the design of ecosystems and human organization to benefit both, without taking into consideration quantitative indices as determinant factors.

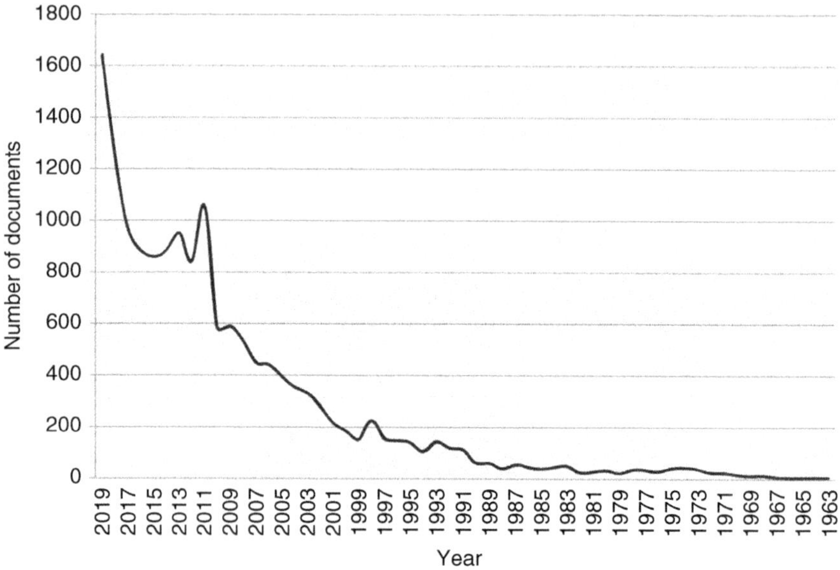

Figure 1.2 Searches on the Scopus database using the term *ecological engineering* show the evolution (1948–2019) of the number of articles, reviews, and books published in international journals that used the term. *Source:* Scopus, © 2020, Elsevier.

Although the initial concepts of ecological engineering were developed in the mid-twentieth century, studies associated with this science only started to gain room in the 1990s, as shown in Figure 1.2. The primary reasons for the increased number of studies about ecological engineering

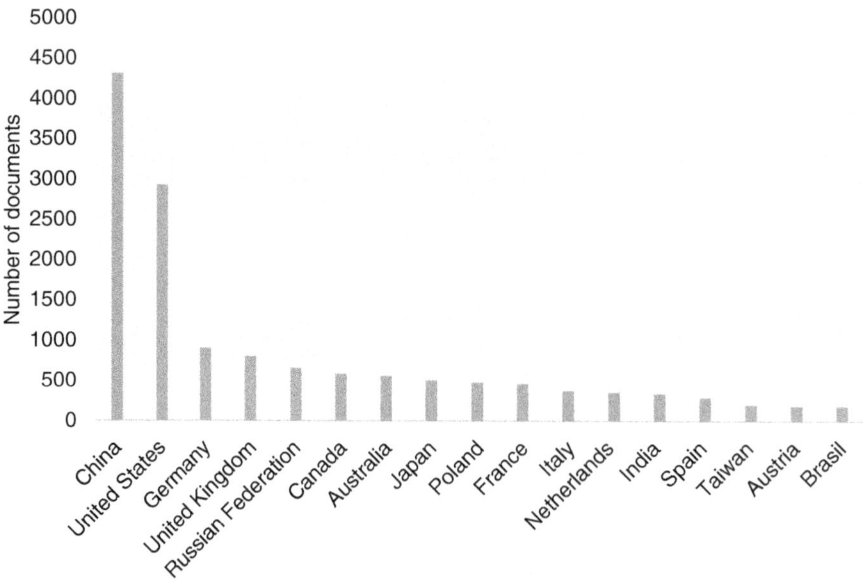

Figure 1.3 Searches on the Scopus database using the term *ecological engineering* show the evolution (1948–2019), by country or territory, of the number of articles, reviews, and books published in international journals that used that expression. *Source:* Scopus, © 2020, Elsevier.

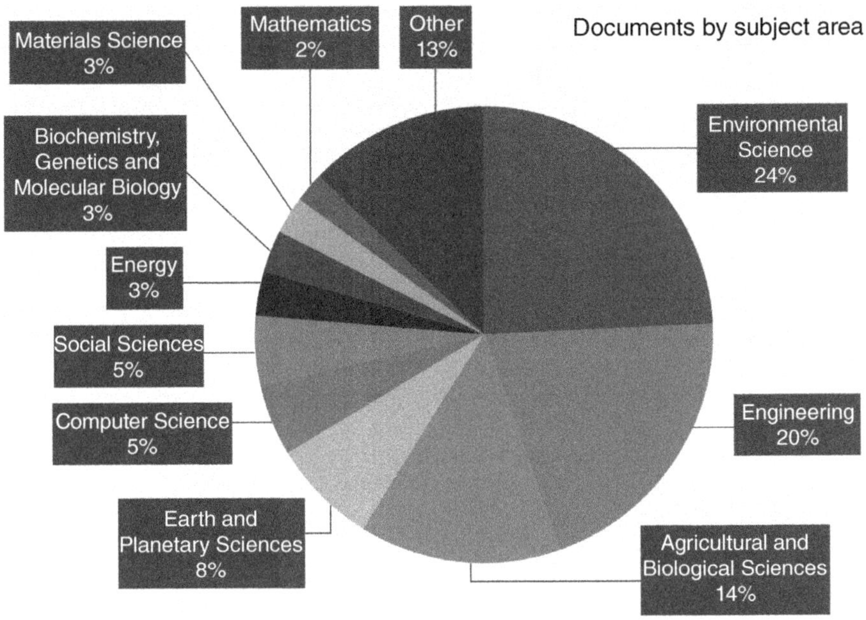

Figure 1.4 Searches on the Scopus database using the term *ecological engineering* and shows the evolution (1948–2019), by subarea of knowledge, of the number of articles, reviews, and books published in international journals that used the term. *Source:* Scopus, © 2020, Elsevier.

lie in the expansion of international commissions, conferences, and meetings aimed at sustainable development, such as the World Commission on Environment and Development – which published the report "Our Common Future" - [11] – and the Second United Nations Conference on Environment and Development (ECO-92).

From the 1990s to the present day, studies about ecological engineering have increased in number, gained global range, and reached new fields of knowledge. However, China, the United States, Germany, the United Kingdom, and Australia stand out in the number of published studies in this field, whereas environmental sciences, agriculture, engineering, earth sciences, and social sciences are the subfields mostly dedicated to developing and publishing studies on this topic (Figures 1.3 and 1.4).

Although studies about ecological engineering have expanded, the view that natural ecosystems must be controlled under the basic premise of meeting human needs remains. Mitsch and Jorgensen [9] have emphasized that true ecological engineering can only be practiced based on the belief that humanity exists on the premises of Nature, which is in contrast to the current attitude that sees natural ecosystems based on conditions of human society. Human actions must move toward Nature rather than against it.

The basic ecological principles of ecological engineering were formulated as 12 guidelines or commandments [9, 12] that must be considered when planning any activity that falls under the scope of ecological engineering. These principles are described in Table 1.1.

The effective implementation of these principles is challenging, primarily in developing countries, where research resources are scarce or insufficient. In addition, production systems designed to meet global demands and balance the trade balance of these countries – such as food and input production – do not respect the slower pace required by ecological engineering, which contributes to the expansion of degradation processes in ecosystems and in services provided by them.

Table 1.1 Basic ecological principles of ecological engineering.

Principles	Description
Forcing functions.	Ecosystem structures and functions are determined by the forcing functions of the system. According to this principle, ecological engineering interventions should stimulate the system's forces to improve the ecosystem's performance by strengthening its antifragility, i.e. by improving its structure and functions.
Relationships between biological function and chemical composition	Homeostasis of ecosystems requires accordance between biological function and chemical composition.
Recycling management.	It is necessary to match recycling pathways and rates to ecosystems in environmental management to reduce the effects of pollution.
Ecosystems are self-designing systems.	The more one works with the self-designing ability of Nature, the lower the costs of energy to maintain that system.
Ecosystem-specific space–time scales.	Processes of ecosystems have characteristic time and space scales that should be accounted for in environmental management.
Chemical and biological diversity	Chemical and biological diversity contribute to the spectrum of buffering capacities and the self-designing ability of ecosystems. A wide variety of chemical and biological components should be introduced or maintained for the ecosystem's self-designing ability to choose from. Thereby, a broad spectrum of buffer capacities is available to meet the impacts of anthropogenic pollution.
Transition zones and agricultural management.	Transition zones are as important for ecosystems as membranes are for cells. Agricultural management should therefore consider the importance of transition zones.
Coupling between ecosystems and agricultural management.	Coupling between ecosystems should be used to the benefit of ecosystems in the application of ecotechnology and environmental management of agricultural systems.
Ecotechnology, environmental management, and components of ecosystems.	The application of ecotechnology and environmental management must consider that the components of an ecosystem are interconnected and interrelated and form a network, which implies that direct, as well as indirect, effects are relevant.
Ecosystem history.	It is important to realize that, overall, an ecosystem has a history in the application of ecotechnology and environmental management.
Ecosystems and geographical edges.	Ecosystems are most vulnerable at the geographical edges; therefore, ecological management should take advantage of ecosystems and their biota in their optimal geographical range.
Ecosystem hierarchy.	Ecosystems are hierarchical systems, and all the components forming the various levels of the hierarchy make up a structure that is important for ecosystem functioning.

1.3 Ecosystem Services: History, Concepts, and Dimensions

> Close your eyes, prick your ears, and from the softest sound to the wildest noise, from the simplest tone to the highest harmony, from the most violent, passionate scream to the gentlest words of sweet reason, it is by Nature who speaks, revealing her being, her power, her life, and her relatedness so that a blind person, to whom the infinitely visible world is denied, can grasp an infinite vitality in what can be heard.
>
> *Johann Wolfgang von Goethe*

Ancient thinkers emphasized the first concepts of services provided by Nature to human beings, nurturing what we today call *provision services* and enjoying sounds, auras, and odors that nurture body and soul. The book *On Airs, Waters and Places* by Hippocrates (460–370 BCE) – a Greek philosopher considered the father of medicine in the Western world – correlates the environment to different nosological frameworks and population features and includes aspects of daily activities and environmental effects on individuals' health in its analysis.

Services provided by ecosystems to human beings have been described in mythological, philosophical, scientific, and poetic texts. However, it was naturalist and geographer Alexander von Humboldt who bequeathed us our current vision of Nature. Humboldt – who was influenced by the poet Goethe, the philosopher Kant, English romantic writers, and American transcendentalists – offered the idea that Nature exists only insofar as we perceive it within us; according to him, the outer world, ideas, and feelings were associated with each other, so he wrote: "Nature must be known through feelings" [13]. In addition, Humboldt sought to unravel God's designs through Nature.

Alexander von Humboldt's vision was lost over time; however, it was recovered from the 1970s onwards primarily due to the academic elaboration of the concept of Nature services by botanist Walter Westman. Certainly our current vision of services provided by ecosystems has deep roots stemming from Humboldt's contribution, such as the concept of cultural, leisure, and religious services, which play a fundamental role in nourishing the immaterial dimension of human beings.

When Walter Westman formulated the concept of Nature services, he pointed out that benefits deriving from ecosystems should be listed to increase public interest in biodiversity conservation and ensure that decisions are made based on robust information. Later, in 1981, Paul and Anne Ehrlich [14] suggested the term *ecosystem services*, which is used today.

The contribution of ecosystems to human wellbeing has been described based on the concept of ecosystem services, which was introduced in the 1970s [15] – as previously highlighted – although it was only consolidated in the 1990s due to the economic valuation of 17 services by Costanza et al. [1]. The diffusion of this concept beyond the academic environment became evident after the Millennium Ecosystem Assessment [16], when studies associated with this concept began to gain attention (Figure 1.5). Its importance was acknowledged by various fields of knowledge (Figure 1.6); by professionals from all over the world (Figure 1.7), primarily from the United States, United Kingdom, China, Germany, and Australia; and by the business sector [17].

Several definitions of *ecosystem services* are available in the literature. MEA, for example, has defined ecosystem services as the benefits humanity gets from ecosystems [16], whereas Constanza et al. [1] defined them as the direct and indirect human benefits resulting from ecosystem functions. Thus, *functions* are biophysical ecosystem features essential to their own functioning, and *services* are the product of two or more ecosystem functions; a single ecosystem function can contribute to two or more ecosystem services.

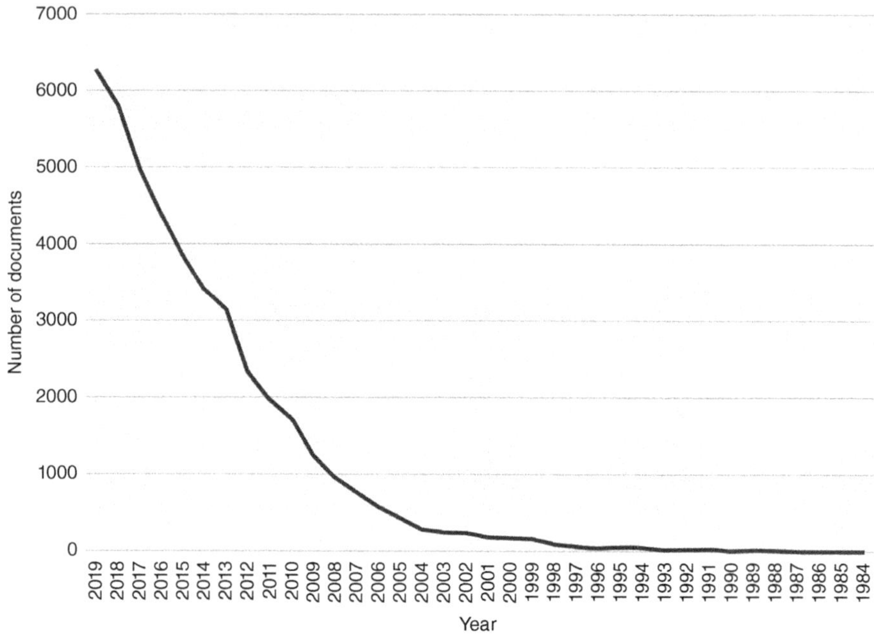

Figure 1.5 Articles, reviews, and books published on ecosystem services. *Source:* Data from 1984 to 2019, Scopus 2020.

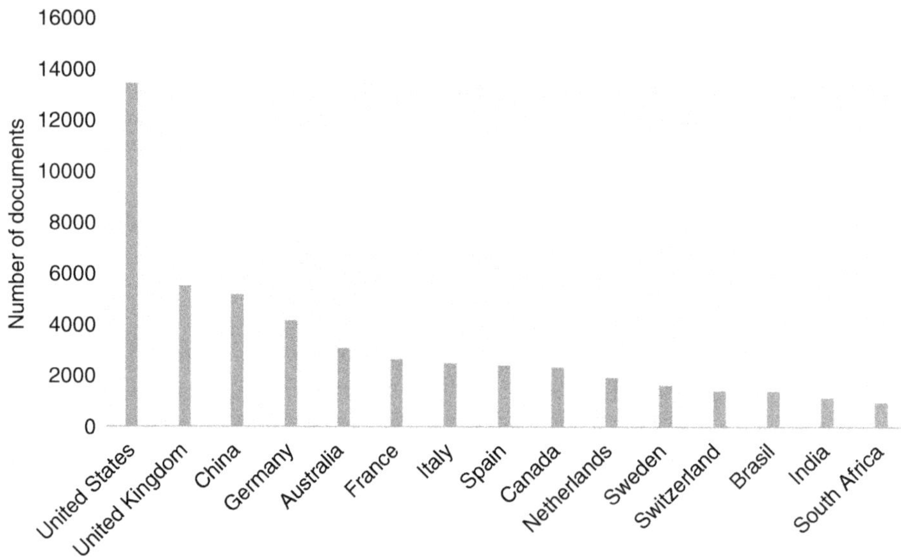

Figure 1.6 Articles, reviews, and books published on ecosystem services by country or territory. *Source:* Data from 1984 to 2019, Scopus 2020.

The primary idea observed in the relevant literature on this topic is that the concept of ecosystem services is inherently anthropocentric – as seen in Costanza et al. [1], De Groot et al. [18], Daily and Farley [19], MEA [16], Boyd and Banzhaf [20], and Fisher et al. [21] – since human presence is a valorization agent that allows us to view basic ecological structures and processes as entities full of value, as advocated by De Groot et al. [18].

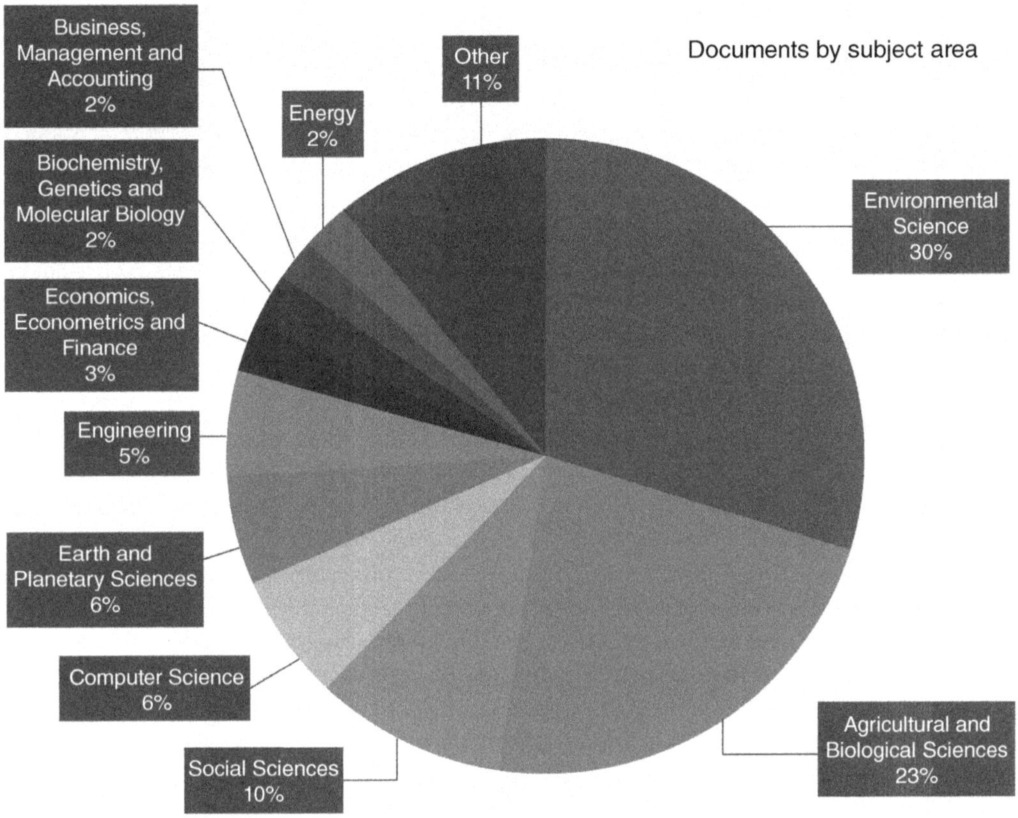

Figure 1.7 Articles, reviews, and books published on ecosystem services by area. *Source:* Data from 1984 to 2019, Scopus 2020.

There are several criticisms of the concept of ecosystem services, as well as their valuation proposals, primarily quantitative methodologies. The main criticisms emphasize that the concept is eminently anthropocentric, focuses on economic evaluation, promotes nature-commodification relationships, and, therefore, favors exploitative human-Nature relationships; moreover, it goes against biodiversity conservation goals. In addition, it is worth emphasizing that definitions and classifications are inaccurate and that all ecosystem process results are desirable and strengthen the network of relationships in the ecosystem.

Supporters of the concept of ecosystem services refute the critics and explain that this concept goes beyond its anthropocentric nature and instrumental values since it leads to the interest of actors involved in biodiversity conservation and encourages the reconnection, complementarity, and ethical commitment of individuals and society to ecosystems. In addition, the inaccuracy of definitions and classifications improves transdisciplinary collaboration; most environmental services are not linked to the market.

Despite the arguments for and counterarguments against the concept of ecosystem services, it has evolved over the last 30 years from a general heuristic model that highlights the importance of Nature for human wellbeing to a model that highlights how the appropriation of specific ecological structures or processes affects human wellbeing [22]. However, it is important to understand how ecosystems benefit other ecosystems that end up directly or indirectly influencing humanity.

1.3.1 Sizing Ecosystem Services

The first step toward a comprehensive assessment of ecosystem services lies in translating ecological complexity (structures and processes) into a more limited number of ecosystem functions that provide goods and services that are sized and valued by humans [18]. According to the aforementioned authors, understanding the functions of a given ecosystem and the nature and magnitude of its value for humans provides an empirical basis for the classification of useful aspects for humanity.

Many scholars have tried to measure the value of services provided by natural environments. Thus, as a science, economics has developed several forms of analysis over the years, which can be divided into three stages: natural resource economics, environmental economics, and ecological economics [23]. Numerous studies focused on standardizing these investigations were published during this period. Among them, one finds Constanza [24–26], De Groot et al. [18], Hein et al. [27], and Andrade and Romeiro [28].

Constanza et al. (1987) surveyed the economic value of 17 ecosystem services in 16 biomes; they only considered renewable resources and excluded non-renewable fuels and the atmosphere. This study has shown the lack of standardization in the categorization of ecosystem services, as well as the difficulty of presenting the real valuation of these services.

Overall, the valuation of ecosystem services depends on how much individuals would be willing to pay for a specific service, which makes this type of assessment relative since value is specific to objectives of choice (moment) and individuals evaluate things in different ways. A single person can assign two different values to the same service depending on the person's momentary state and whether they have greater or lesser need of it at that moment [29]. In addition, these services are often indispensable; for many individuals, no institution or technology can make them expendable [19].

Several scholars have tried to improve methodologies in the last two decades to achieve well-founded valuation parameters with applicability in different fields. Accordingly, De Groot et al. [18] have shown that data on ecosystem services often appears on incompatible scales. According to them, even if a given ecosystem service is sometimes valued at the price individuals are willing to pay for it, in some cases this service is so necessary for their survival that it is more logical to measure how much individuals would be willing to pay to avoid losing it.

According to Merico [30], it is necessary to differentiate value from intrinsic value among different economic value types associated with natural resources. *Use value* derives from the way the environment is used, such as mineral resource extraction or birdwatching. On the other hand, *intrinsic value* is the value of a given good – even its potential value – such as a certain plant species growing in a specific area or a particular insect species.

Based on the assumption that every ecosystem service derives from a given ecosystem function that has value for human beings, De Groot et al. [18] have systematized a wide variety of ecosystem functions by categorizing ecosystem services provided by such functions, either individually or jointly. The authors mentioned above divided ecosystem functions into four different groups: (i) regulation function (gas, climate, and water, among others); (ii) habitat function (places, such as refuges and estuaries, which provide habitat for animal and plant species); (iii) production function (food production), and; (iv) information function (cultural, recreational, historical, spiritual, scientific, among others); this division enabled the same groups in different environments. It is important to highlight that both regulation and habitat functions provide support to and maintain natural components; thus, they contribute to the provision of other functions [28].

The biggest challenge today in the study of ecosystem services is standardizing terminologies and methodologies, which is often an obstacle to more consistent analyses [31]; this issue can be associated with a lack of collaborative studies among scientists from different fields [29]. However, significant efforts have been made to establish standardization in these studies, such as those carried out by De Groot et al. [18], the Millennium Ecosystem Assessment (MA, 2003) [32], Santos and Silva [33], and Souza et al. [34].

Santos and Silva [33] conducted a survey of ecosystem services in a coastal zone; they traveled to the study site to make an inventory based on environmental and geological indicators. This strategy was also adopted by Seppelt et al. [35] and reformulated by Crossman et al. [36], who suggested that the evaluated service should be described based on indicators such as quantification unit (i.e. area, time) and scale (local, regional, global).

Souza et al. [34] have adapted the methodologies suggested by the authors mentioned above (see Tables 1.2–1.4) and used them to value ecosystem services provided by beaches in Ilhéus County, Bahia State, Brazil (Figures 1.8 and 1.9). They concluded that the greater the concentration of technical systems, the smaller the variety and the poorer the quality of provided services, highlighting the importance of conducting diagnosis and ecological valuation studies to substantiate environmental management processes.

Table 1.2 Indicators for the valuation of regulating and supporting ecosystem services.

Regulating and supporting service	Low (1)	Medium (2)	High (3)
Natural sediment retention	Absence of vegetation in the backshore or along the beach ridge	Occurrence of vegetation in the backshore or along the beach ridge over at least 50% of the shoreline	Occurrence of vegetation in the backshore or along the beach ridge over more than 50% of the shoreline
Aquifer recharge	Absence of sandy terraces or terraces with waterproofed surface	Occurrence of sandy terraces in at least 50% of the shoreline	Occurrence of sandy terraces in over 50% of the shoreline
Water control and storage	Absence of wetlands or mangrove forests	Occurrence of wetlands or mangrove forests in less than 50% of the shoreline	Occurrence of wetlands or mangrove forests in over 50% of the shoreline
Pollutant assimilation and recycling	Absence of wetlands or mangrove forests	Occurrence of wetlands or mangrove forests in less than 50% of the shoreline	Occurrence of wetlands or mangrove forests in over 50% of the shoreline
Wave energy dissipation	Absence of surf zone	Surf zone with up to three breakers	Surf zone with more than three breakers
Foreshore zone natural protection	Absence of coral reefs and/or sandstone banks	Occurrence of coral reefs and/or sandstone banks in less than 50% of the shoreline	Occurrence of coral reefs and/or sandstone banks in less than 50% of the shoreline
Backshore zone natural protection	Absence of beach ridge	Occurrence of beach ridge in less than 50% of the shoreline	Occurrence of beach ridge in over 50% of the shoreline
Marine refuge and/or nursery	Absence of estuaries, coral reefs, or sea turtle nesting areas	Occurrence of at least one refuge/nursery area (estuaries, coral reefs, or sea turtle nesting areas)	Occurrence of more than one refuge/nursery area (estuaries, coral reefs, or sea turtle nesting areas)
Terrestrial or transitional refuge and/or nursery	Absence of mangrove forests, restinga, or Atlantic rainforest	Occurrence of at least one refuge/nursery area (mangrove forests, restinga, Atlantic rainforest)	Occurrence of more than one refuge/nursery area (mangrove forests, restinga, Atlantic rainforest)

Table 1.3 Indicators for the valuation of provisioning ecosystem services.

Provisioning Service	Low (1)	Medium (2)	High (3)
Natural food production	Absence of activities such as fisheries, shellfish gathering, or gathering wild plants	Occurrence ofat least one activity (i.e. fisheries, shellfish gathering or gathering wild plants)	Occurrence of more than one activity (i.e. fisheries, shellfish gathering or gathering wild plants)
Food production in farmed areas	Absence of activities such as crops, animal breeding, fish farming, etc.	Occurrence of at least one activity (i.e. crops, animal breeding, fish farming, etc.)	Occurrence of more than one activity (i.e. crops, animal breeding, fish farming, etc.)
Water resources	Absence of surface water bodies or aquifers	Occurrence of at least one source of water (i.e. surface water bodies or aquifers)	Occurrence of more than one source of water (i.e. surface water bodies or aquifers)
Ornamental resources	Absence of ornamental resources (i.e. deadwood, oysters, plants, fish, rocks, minerals)	Occurrence of at least one ornamental resource (i.e. deadwood, oysters, plants, fish, rocks, minerals)	Occurrence of more than one ornamental resource (i.e. deadwood, oysters, plants, fish, rocks, minerals)
Genetic resources	Occurrence of anthropized areas, pastures, or monocultures	Occurrence of restinga or agroforestry systems	Occurrence of forests, coral reefs, estuaries, or mangrove forests

Table 1.4 Indicators for the valuation of information and cultural ecosystem services.

Information and cultural service	Low (1)	Medium (2)	High (3)
Ecotourism	Absence of locations with quality for ecotourism, such as hiking and diving	Occurrence of at least one location with quality for ecotourism, such as hiking and diving	Occurrence of more than one location with quality for ecotourism, such as hiking and diving
Historical/ cultural tourism	Absence of buildings or areas with known historical value	Occurrence of at least one building or area with known historical value	Occurrence of more than one building or area with known historical value
Recreation and leisure	Low recreational quality	Medium recreational quality	High recreational quality
Scenic quality	Absence of natural attractions (e.g. cliffs)	Occurrence of at least one natural attraction (e.g. cliffs)	Occurrence of more than one natural attraction (e.g. cliffs)

Primary information collected by methodologies such as the ones applied by Santos and Silva [33], Seppelt et al. [35], and Souza et al. [34] is essential to enable researchers to perform environmental mapping and modeling. Chan and Ruckelshaus [37] reported efforts to map and model ecosystem services, such as the Multiscale Integrated Models of Ecosystem Services (MIMES), Artificial Intelligence for Ecosystem Services (ARIES), and Integrated Valuation of Ecosystem Services and Tradeoffs (InVEST). However, these models are considered recent. Studies about them remain scarce in the literature [31]. The applicability and effectiveness of the adopted tools remain significantly questionable.

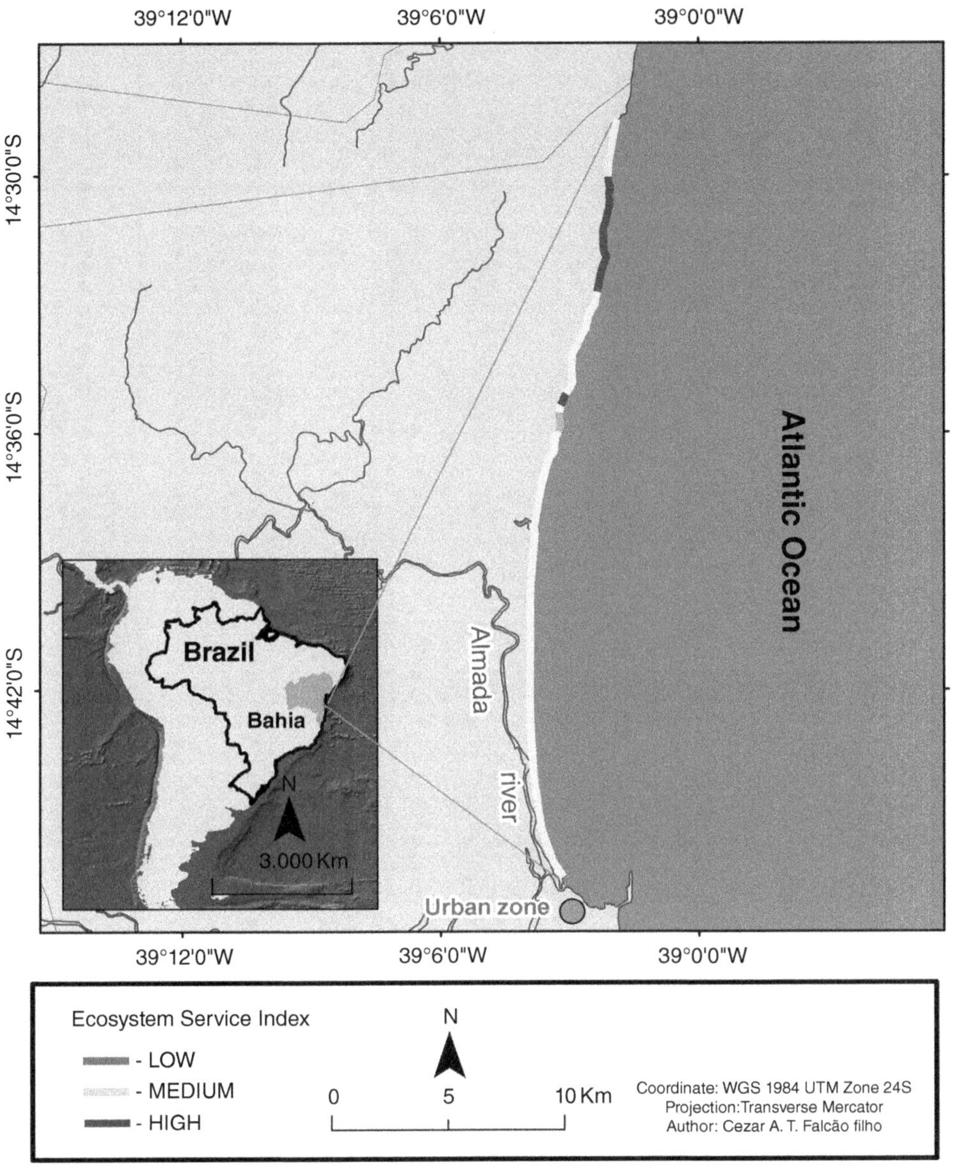

Figure 1.8 Ecosystem service index obtained for the northern shoreline of the municipality of Ilhéus, state of Bahia, Brazil.

It is important to emphasize that the analysis of ecosystem services applied to management and decision-making processes does not need to include economic assessments [31], as shown in the study by Ruckelshaus et al. [38], who evaluated 20 pilot studies about the application of ecosystem services and concluded that many decision-makers wished to examine the consequences of their actions for traditional market commodities in monetary terms, as well as the non-economic benefits, including cultural aspects and biodiversity values. The misconception that the assessment of ecosystem services requires economic assessment can be a barrier to scientific development.

ECOSYSTEM SERVICE	VALUE ATTRIBUTED TO EACH SERVICE EVALUATED																					
	OS	S1	S2	PR	L	C	M1	M2	M3	PT1	PT2	V	B	PA	JA1	JA2	MS	J	FO	SD1	SD2	SM
REGULATING AND SUPPORTING SERVICES																						
Natural sediment retention	3	3	2	3	3	3	3	2	3	3	2	3	3	3	3	3	3	3	3	3	2	2
Aquifer recharge	3	3	3	3	3	3	3	3	3	3	3	3	3	3	3	3	3	3	3	3	2	2
Water control and storage	2	2	2	2	2	2	2	2	3	2	1	1	1	2	2	2	2	2	2	1	2	2
Pollutant assimilation and recycling	2	2	2	2	2	2	2	2	3	2	1	1	1	2	2	2	2	3	2	1	2	2
Wave energy dissipation	2	2	2	2	2	2	2	2	2	2	2	2	2	2	2	2	2	2	2	2	2	2
Foreshore zone natural protection	1	1	1	1	1	1	1	1	1	1	1	1	1	1	1	1	1	1	1	1	1	1
Backshore zone natural protection	3	2	2	2	2	2	3	1	3	3	1	2	2	3	2	2	3	2	3	3	1	1
Marine refuge and/or nursery	2	2	3	3	2	2	2	1	2	2	1	1	1	2	2	1	2	2	1	1	1	1
Terrestrial or transitional refuge and/or nursery	2	2	3	3	1	2	2	2	3	3	1	2	2	2	3	2	3	3	2	3	2	2
PROVISIONING SERVICES																						
Natural food production	2	2	3	3	2	2	2	2	3	3	2	2	2	2	2	2	2	2	2	2	3	3
Food production in farmed areas	2	2	2	3	2	2	1	2	2	2	2	2	2	2	2	2	2	2	2	3	1	1
Water resources	2	3	3	3	3	3	2	2	3	2	1	2	2	3	3	2	3	3	2	2	2	2
Ornamental resources	2	2	2	2	1	1	1	2	2	2	1	2	2	2	2	1	2	2	3	2	3	3
Genetic resources	3	2	3	3	2	2	2	2	3	2	1	2	2	2	2	2	3	3	2	2	1	1
INFORMATION AND CULTURAL SERVICES																						
Ecotourism	2	2	3	2	1	1	1	1	1	1	1	1	1	1	1	1	1	1	1	1	1	1
Historical/ cultural tourism	1	1	1	1	1	1	1	1	1	1	1	1	1	1	1	1	1	1	1	1	1	1
Recreation and leisure	2	2	2	2	2	2	2	2	2	2	2	2	2	2	3	2	2	2	2	2	2	2
Scenic quality	3	1	3	3	1	1	1	1	2	2	1	1	1	2	2	1	2	2	1	1	1	1
TOTAL (Ecosystem services index)	39	36	42	43	33	34	33	31	42	38	25	31	31	37	38	32	39	39	35	34	30	31

Figure 1.9 Qualitative assessment of ecosystem services by beaches in the municipality of Ilhéus, state of Bahia, Brazil. Northern shoreline beaches – PS: Pé de Serra; S1: Sargi – trecho 1; S2: Sargi – trecho 2; PR: Ponta do Ramo; L: Luzimares; C: Coqueiros; M1: Mamoã – trecho 1; M2: Mamoã – trecho 2; M3: Mamoã – trecho 3; PT1: Ponta da Tulha – trecho 1; PT2: Ponta da Tulha – trecho 2; V: Verdesmares; B: Barramares; PA: Paraíso do Atlântico; JA1: Jóia do Atlântico – trecho 1; JA2: Jóia do Atlântico – trecho 2; MS: Mar e Sol; J: Japará; FO: Fazenda de Osmar; SD1: São Domingos – trecho 1; SD2: São Domingos – trecho 2; SM: São Miguel.

As the concept of ecosystem service increasingly becomes an operational tool, it is necessary to explain the complexity of the relationships among production, ecosystem service appropriation [22], and impacts on different time and space scales. So far, there has been considerable focus on spatial patterns of ecosystem service provision and appropriation; however, the temporal dynamics have been little explored, as highlighted by Rau et al. [22], who proposed a new way of categorizing ecosystem services based on their temporal dynamics by differentiating linear from nonlinear dynamics in the provision and appropriation of these services.

The authors above have suggested that temporal dynamics can be better integrated into studies about ecosystem services through four stages: (i) defining the appropriate time limits of the system; (ii) identifying the primary types of ecosystem dynamics; (iii) evaluating the spatial scale at which the dynamics develop in the system; and (iv) developing measures to assess such dynamics.

Considering the temporal dynamics of ecosystem services by following the stages suggested by Rau et al. [22] is a way to better plan the management of ecosystem services.

1.3.2 Agriculture and Ecosystem Services

Agriculture is the most widely managed ecosystem globally since it covers approximately one-third of the earth's surface. Therefore, given its importance, this item addresses agricultural systems, ecosystem services provided by these systems, and their issues, possibilities, and challenges.

Agricultural ecosystems provide food, fibers, fuels, and a range of products essential to human life; thus, they are classified as ecosystem provisioning services. However, agricultural ecosystems depend on many other support and regulation services – such as soil fertility, pollinating agents, and water supplies – to provide these services [16, 39].

There are so-called *ecosystem disservices* to agriculture if one considers the most widespread definition of ecosystem services, according to which Nature provides services essential to human survival. All natural ecosystem processes that may reduce productivity or increase production costs – such as herbivory and plants' competition for space, nutrients, and water – are classified as disservices to agriculture [40].

The intensity of the impact that ecosystem services or disservices have on agricultural systems depends on the diversity, composition, and functioning of the landscape surrounding them [41]. Management of ecosystem processes began after humans stopped being hunter-gatherers and became farmers, based on the degree of service or disservice to agricultural production increase. Management, or manipulation, of processes reached unprecedented levels due to the *Green Revolution*,[5] after which agriculture primarily started to follow a latifundiary monoculture model.

The idea that ecosystem processes must provide products necessary for human survival is dangerous since it makes room for individuals to understand that one can enjoy and interfere in processes that took thousands of years to exist in harmony on our planet. Conservationists and ecologists may see ecosystem management for agricultural purposes as an irreversible interference in natural processes. On the other hand, ecological engineers see changes caused by agriculture in the ecosystem and its services as a means of preserving the natural system and, at the same time, providing provisioning services. Thus, the core question appears to be: can there be a middle ground between the two approaches? In other words, is it possible to manage ecosystems so that all their ecological processes – if well investigated, interpreted, and understood – can be considered ecosystem services (excluding the idea that there are ecological processes acting as disservices to agriculture)?

In light of the current scenario, according to which food production accounts for approximately 69% of greenhouse gas emissions, 70% of freshwater use, and the annual loss of approximately 6 million hectares of agricultural areas due to soil desertification processes [42, 43], it is necessary to manage ecosystem services to provide food of the right amount and quality to the population without generating irreversible environmental burdens. From the moment when a given human activity starts imposing negative "feedbacks" on the population, this activity becomes unstable in the long term and may eventually destroy the population itself [44].

Following the trend of manipulating natural elements to enable increasingly productive agricultural systems, the introduction of genetically modified organisms – so-called *transgenics* – is considered the most significant technological revolution in the recent history of agriculture [45]. An organism is considered genetically modified when a given material is subjected to genetic changes that would happen neither naturally nor through crossing or recombination processes [46].

Transgenic organisms have spread worldwide due to an annual increase in their use in agricultural sites, as well as to their import and export in the form of processed foods and animal feed. The trade of these organisms started with promises such as increasing productivity without the need to convert more land into agricultural fields, reducing the use of pesticides, enabling the efficient use of water through drought-resistant varieties, greater efficiency in fertilizer use, less need to prepare the soil, and, consequently, lower carbon emissions and erosion rates, among others. However, many of these environmental benefits are not observed in all regions and remain scientifically uncertain [47]. Part of the uncertainty results from rapid technology adoption in a world thirsty for profits and higher productivity to guarantee human food, as well as for political reasons and due to the interests of large corporations in accelerating the introduction of these products in the market.

Although the aim of this chapter is not to address how transgenic organisms have been placed in the global market, under what interests or research methodologies, we would like to raise some questions of ecological importance to reason about the difficulties in maintaining ecosystem services in full operation in the face of some techniques used in agriculture. Based on our viewpoint, biotechnology may enable the formation of entirely new ecosystems in the future. But will these ecosystems be efficient and desirable? Will ecosystem services be equally present as they are in natural ecosystems? These questions are certainly far from being accurately answered, but scientists and society must consider them.

One cannot forget that arable lands became part of the ecosystem from the moment human beings started to produce their own food. Thus, when one cultivates genetically modified species, these new organisms are introduced into the ecosystem. Because ecosystems are a complex set of organisms in constant interaction, together they perform a variety of ecological processes; therefore, it is expected that the introduction of a new organism in these ecosystems will change these processes. Thus, it is essential to evaluate the effects of transgenic cultures as a whole, rather than just the results of a single organism in a single environment and in a single period of time. Individual analyses cannot reflect the environmental effects accumulating in ecosystems as a whole.

Below, we list just a few examples of organisms that transgenic plants come into ecological contact with when they are grown [48]:

- Other plants belonging to the same, or different, species
- Herbivores feeding on plants' shoot or roots
- Natural enemies such as plants competing for space, nutrients, water, and light through allelopathy

- Pollinating insects visiting transgenic plants' flowers
- Symbiotic organisms, such as mycorrhizae and nitrogen-fixing bacteria, which can live in plants' shoot and roots
- Detritivores and decomposers feeding on dead plant parts and promoting nutrient cycling

Most ecosystem processes in both natural and agricultural systems have the soil as a critical and dynamic regulatory core [49]. Soil accounts for environmental services that go far beyond being a substrate for plant development; soil contributes to all four ecosystem service dimensions. It is accountable for maintaining water springs, storing carbon, maintaining biodiversity, providing essential nutrients for plant species' growth, structuring the soil and controlling pests, and housing most of Earth's biodiversity.

Soil is an open and constantly changing system that is part of the biosphere and interfaces with the atmosphere, hydrosphere, and lithosphere systems; it performs uninterrupted matter and energy exchanges among these spheres, which affects their dynamics. Thus, soil is a strategic point to help mitigate environmental impacts and as the core of Earth's *critical zone*, i.e. the layer supporting life on Earth that goes from the top of trees to the bottom of aquifers.

Agricultural systems are mostly self-regulated due to high biodiversity in the soil. Thus, biodiversity loss reduces the resilience of agricultural systems and leads to increasing dependence on input addition. Given its wide physical and chemical heterogeneity, soil hosts a complex and varying biological community of organisms that help maintain an extremely large diversity of processes at a small or large scale [50, 51]. It may be why we are so far from acquiring knowledge about soil biodiversity and its contribution to the function of ecosystems presenting physical and chemical features similar to the one we have [52–54].

Soil conservation is the basic requirement to achieve sustainable agriculture with high ecosystem service functionality. Conservation measures in tropical environments with low natural soil fertility and high rainfall rates most of the year are of paramount importance to stop cultivated soils from reaching irreversible chemical and physical degradation levels that could compromise the provision of environmental services.

It is necessary to keep in mind that where a rural property is located can influence the type of crops to be grown and the price of the land, to have an idea about the importance of ecosystem services for agriculture. The greater the physical, chemical, and biological quality, the easier the access to water; the more the climate favors the culture, the greater the economic value of the land. It is also necessary to keep in mind that ecosystem services and disservices are not limited to rural property; they are also in the landscape surrounding the property and can move between habitats [40]. Therefore, the more natural and inserted in the landscape agriculture is, the less an ecosystem process will be considered a disservice to agriculture. If we take an ecological process that happens naturally and in harmony with other processes as a disservice to agriculture, it is likely because we are not using the ideal agricultural system.

Agriculture, regardless of type (conventional, ecological, conservationist), always has some impact on ecosystems since it is incorporated into a set of countless ecological niches. Thus, our role is to reduce environmental impacts as much as possible and use technology – which may contribute to the loss of essential ecological processes to the proper functioning of the ecosystem as a whole – as little as possible.

Jogersen and Nielsen [44] have indicated some guidelines to be taken into consideration for agriculture development purposes:

1) Closing cycles
2) Enabling direct energy and material flows in increasingly smaller cycles
3) Increasing diversity in agricultural systems by using, for example, integrated ecological agriculture
4) Minimizing pesticide and fertilizer use, or finding the proper balance between economics and ecology, by using best management practices based on the application of environmental management models
5) Increasing agricultural pattern complexity in time and space by using a wide variety of crops and domestic animals, small fields, fallows, wetlands, and so on

Decision-making in agricultural management processes must be coordinated with the surrounding ecosystem. Thus, principles advocated by agroecology can be great allies in these processes. *Agroecology* is defined as the application of social and ecological concepts in sustainable food production systems; thus, it can be used as a scientific and methodological tool to achieve agricultural multifunctionality [55–57]. Organic, conservationist, biodynamic, ecological, syntrophic, and natural agriculture, as well as permaculture and forestry-livestock integration systems, are examples of agricultural systems based on agroecological principles. See Table 1.5 to better understand agroecology, the application of its principles, and how they influence ecosystem services.

Table 1.5 Articles on agroecology and ecosystem services.

Salient summary	References
Agroforestry boosts soil health in the humid and sub-humid tropics.	Muchane et al. [58]
Contribution of agroecological farming systems to ecosystem services	Boeraeve et al. [59]
How can integrated valuation of ecosystem services help to understand and steer agroecological transitions?	Dendoncker et al. [60]
Examining multifunctionality for crop yield and ecosystem services in five systems of agroecological intensification	Garbach et al. [61]
Intersection between biodiversity conservation, agroecology, and ecosystem services	Liere et al. [57]
How to implement biodiversity-based agriculture to enhance ecosystem services: a review	Duru et al. [62]
A social–ecological analysis of ecosystem services in two different farming systems	Andersson et al. [63]
Quantifying the impacts of ecological restoration on biodiversity and ecosystem services in agroecosystems: a global meta-analysis	Barral et al. [64]
Integrated crop–livestock systems: strategies to achieve synergy between agricultural production and environmental quality	Lemaire et al. [65]
Conservation agriculture and ecosystem services: an overview	Palm et al. [66]
Agroecological practices for sustainable agriculture: a review.	Wezel et al. [67]
Framework for systematic indicator selection to assess the effects of land management on ecosystem services	van Oudenhoven et al. [68]
The value of producing food, energy, and ecosystem services within an agroecosystem	Porter et al. [69]
Measures of the effects of agricultural practices on ecosystem services	Dale and Polasky [70]
Agroecology: the ecology of food systems	Francis et al. [71]

Ecological engineering must be used to engineer, i.e. to create, agricultural models capable of producing food in sufficient quantities to meet the demands of a growing population while conserving ecosystem services. The practice of ecological engineering focused on encouraging society to adapt to Nature (and not the other way around) can be a great ally in developing productive models aimed at reducing the impacts of agriculture on ecosystem services and exploiting their benefits for both humans and nature.

Ecological engineering must always be thought of as a multidisciplinary science. Efforts from specialists aimed at a common goal can help strengthen the integrative approach necessary for the maintenance and strengthening of services provided by different ecosystems.

1.4 Final Considerations: Challenges for the Future

Ecological engineering is a promising field of knowledge [72], as well as the only academic discipline capable of merging ecology and engineering based on the idea that ecological principles and management can be combined in all ecosystem types, from wildlands to cities, current ecosystems, and the analog ecosystems of the future [73]. Yet there are many challenges to be faced, such as those addressed by Jones [74]: ethical, relational, and intellectual.

The ethical challenge requires explaining the meaning of the statement that ecological engineering integrates human society with its natural environment to benefit both, which encompasses its practices [74]. This perspective must also contemplate the reasoning about, and the application of, ecosystem services, which are primarily anthropocentric. As expressed in this chapter, why not try a new concept of nature services, which also assesses the benefits of ecosystems for other ecosystems rather than just for humanity? This new concept would relate well to the aspirations proposed by ecological engineering since its origin – i.e. integrating humanity into its natural environment for reciprocal benefits.

The relational challenge refers to developing and strengthening relationships with other academic disciplines [74], ancestral knowledge, longings, and viewpoints of civil society. Environmental sciences, agriculture, engineering, and earth and social sciences are the primary scientific partners of ecological engineering today; however, it is possible to expand its dialogue with computer sciences, materials sciences, energy, chemistry, biochemistry, and countless others. On the other hand, companies and professionals in the business, education, and political fields are the primary social partners of ecological engineering, although partnerships with health sciences, philanthropic entities, and traditional communities can be easily added and expanded.

Ecological engineering was founded by intellectuals who had different viewpoints about ecosystems due to ecological and engineering perspectives deriving from their academic lineage. The intellectual approximation of perspectives about ecosystems to unify the key principles of this science in a way that is useful, coherent, understandable, and accessible to all is a challenge that requires significant efforts from the academic community [74].

Facing these challenges must be thought of in an integrated way; it requires the assumption of partnerships among professionals from different fields and fruitful dialogue with constituted powers, communities, and social actors. The task is arduous, but it can be implemented if every individual is willing to communicate and redefine their knowledge, based on their experience, to incorporate new perspectives and build a stronger knowledge base.

Notes

1 Reference to the Aristotelian concept of the world, which was improved by Ptolemy (second century), according to which the *Cosmos* is divided into *sublunary* (imperfect world of human beings) and *supralunary* (perfect world of eternal and absolute beings) regions.
2 Reference to the Greek myth of Icarus, son of Daedalus, who did not listen to the advice of his astute father and flew with wax wings beyond where he should have. Daedalus represents the experience, thoughtfulness and moderation of the sage, whereas Icarus represents the impetus and immaturity of youth, which can lead to misfortune and even to death.
3 Reference to the Greek myth of the Moirae (Fates), three sisters who wove and cut the threads determining the fate of gods and human beings.
4 Concept proposed and developed by Nassin Nicholas Taleb in the book *Antifragile: Things That Gain from Disorder*, Random House, 2012. Antifragility is the opposite of fragility; it is a property of complex natural or social systems that benefit from stressors, volatility, and randomness. Over-intervening in natural systems that have survived billions of years by depriving them of stressful, random, and volatile interactions can weaken them and/or trigger unpredictable consequences due to multiplicative chains of unexpected effects.
5 It refers to the development and dissemination of new agricultural practices, such as the use of agrochemicals, pesticides, and seed processing, which have enabled significant production increase since the 1960s.

References

1 Costanza, R., dArge, R., de Groot, R. et al. (1997). The value of the world's ecosystem services and natural capital. *Nature* 387: 253–260.
2 Nunes, F.C. (2016). Por uma nova Geografia Física: razão-ação-emoção. *Caderno de Geografia* 26, número especial 2: 380–396. https://doi.org/10.5752/p.2318-2962.2016v26nesp2p380.
3 Unger, N. M., 2000. O encantamento do humano: Ecologia e espiritualidade. Edições Loyola, São Paulo, Brasil. 2ª edição. 94p.
4 Taleb, N.N., 2012. Antifragile: Things That Gain from Disorder. Ed.: Random House. 544p.
5 Taleb, N.N. (2010). The Black Swan: The Impact of the Highly Improbable. Ed.: Random House, 2ed., 444p.
6 Heisenberg, W. (1958). *Physics and Philosophy: The Revolution in Modern Science*. Harper & Brothers.
7 von Goethe, J. (1997). *Máximas e reflexões. Tradução de Afonso Teixeira da Mota*. Lisboa: Guimarães Editores.
8 Odum, H.T. et al. (1988). *Environmental Systems and Public Policies*. EUA: University of Florida Press.
9 Mitsch, W.J. and Jorgensen, S.E. (1989). *Ecological Engineering, an Introduction to Ecotechnology*, 472. New York, NY: John Wiley.
10 Mitsch, W.J. (1995). Ecological engineering: from Gainesville to Beijing – a comparison of approaches in the United States and China. In: *Maximum Power. The Ideas and Applications of H. T. Odum* (ed. C.A.S. Hall). USA: University Press of Colorado.
11 United Nations (UN). (1987). Report of the World Commission on Environment and Development: Our Common Future.
12 Straskraba, M. (1993). Ecotechnology as a new means for environmental management. *Ecological Engineering* 2: 311–331.

13 Wulf, A., 2016. A invenção da natureza: A vida e as descobertas de Alexander von Humboldt. São Paulo, EditoraPlaneta, 1 ed.

14 Ehrlich, P.R. and Ehrlich, A. (1981). Extinction: *The Causes and Consequences of the Disappearance of Species*. New York: Random House.

15 Nahlik, A.M., Kentula, M.E., Fennessy, M.S. et al. (2012). Where is the consensus? A proposed foundation for moving ecosystem service concepts into practice. *Ecological Economics* 77: 27–35.

16 Millennium Ecosystem Assessment (MA) (2005). *Ecosystems and Human Well-being: Synthesis.* Washington, DC: Island Press.

17 World Business Council for Sustainable Development (WBCSD). (2013). Ecosystem services and biodiversity tools to support business decision-making. Eco4Biz-Interactive-Ecosystem-services-and-biodiversity-tools-Abr-2013.pdf (cebds.org).

18 De Groot, R.S., Wilson, M.A., and Boumans, R.M.J. (2002). A typology for the classification, description and valuation of ecosystem functions, goods and services. *Ecological Economics* 41: 393–408.

19 Daly, H. and Farley, J. (2004). *Economia Ecológica: Princípios e Aplicações.* Lisboa: Instituto Paget.

20 Boyd, J. and Banzhafa, S. (2007). What are ecosystem services? The need for standardized environmental accounting units. *Ecological Economics* 63: 616–626.

21 Fisher, B., Turner, R.K., and Morling, P. (2009). Defining and classifying ecosystem services for decision making. *Ecological Economics* 68: 643–653.

22 Rau, A.-L., von Wehrden, H., and Abson, D.J. (2018). Temporal dynamics of ecosystem services. *Ecological Economics*, Elsevier 151 (C): 122–130.

23 Mattos, K.M.d.C., Mattos, K.M.d.C., and Mattos, A. (2005). Valoração econômica do meio ambiente dentro do contexto do desenvolvimento sustentável. *Revista Gestão Industrial* 1 (2): 105–117.

24 Constanza, R. (2000). Social goals and the valuation of ecosystem services. *Ecosystems* 3: 4–10. https://doi.org/10.1007/s100210000002.

25 Constanza, R. (1989). What is ecological economics? *Ecological Economics* 1: 1–7. https://doi.org/10.1016/0921-8009(89)90020-7.

26 Constanza, R., D'Arge, R., De Groot, R.S. et al. (1987). The value of the world's ecosystem services and natural capital. *Nature* (387): 253–260. https://doi.org/10.1038/387253a0.

27 Hein, L., Van Koppen, K., De Groot, R., and Van Ierland, E.C. (2006). Spatial scales, stakeholders and the valuation of ecosystem service. *Ecological Economics* 57: 209–228. https://doi.org/10.1016/j.ecolecon.2005.04.005.

28 Andrade, D. C.; Romeiro, A. R, 2009. Serviçosecossistêmicos e suaimportância para o sistemaeconômico e o bem-estarhumano. Texto para Discussão. Texto para Discussão. IE/UNICAMP. n. 155.

29 Bockstael, N.E. et al. (2000). On measuring economic values for nature. *Environmental Science & Technology* 34: 1384–1389.

30 Merico, L.F.K. (1996). Introdução à economiaecológica. Blumenau: Edifurb.

31 Elliff, C.I. and Kikuchi, R.K.P. (2015). The ecosystem service approach and its application as a tool for integrated coastal management. *Natureza & Conservação* 13: 105–111.

32 Millennium Ecosystem Assessment (2003). *Ecosystem and Human Well-Being: A Framework for Assessment*. Washington, D.C: Island Press.

33 Santos, R.C. and Silva, I.R. (2012). Serviços ecossistêmicos oferecidos pelas praias do município de Camaçari, Litoral Norte do estado da Bahia, Brasil. *Caderno Geociências* 9: 47–56.

34 Souza Filho, J.R., Silva, I.R., and Nunes, F.C. (2019). Avaliação qualitativa dos serviços ecossistêmicos oferecidos pelas praias da APA Lagoa Encantada/Rio Almada, Bahia, Brasil. *Caminhos de Geografia, Uberlândia* 20 (72): 15–32.

35 Seppelt, R. et al. (2012). Form follows function? Proposing a blueprint for ecosystem service assessments based on reviews and case studies. *Ecological Indicators* 21: 145–154.

36 Crossman, N.D. et al. (2013). A blueprint for mapping and modelling ecosystem services. *Ecosystem Services* 4: 4–14.

37 Chan, K.M.A. and Ruckelshaus, M. (2010). Characterizing changes in marine ecosystem services. *F1000 Biology Reports* 2 https://doi.org/10.3410/B2-54.

38 Ruckelshaus, M. et al. (2013). Notes from the field: lessons learned from using ecosystem service approaches to inform real-world decisions. *Ecological Economics* https://doi.org/10.1016/j.ecolecon.2013.07.009.

39 National Research Council (NCR) (2005). *Valuing Ecosystem Services: Toward Better Environmental Decision-Making*. Washington, DC: National Academy Press.

40 Zhang, W., Ricketts, T.H., Kremen, C. et al. (2007). Ecosystem services and dis-services to agriculture. *Ecological Economics* 64: 253–260.

41 Tilman, D. (1999). Global environmental impacts of agricultural expansion: the need for sustainable and efficient practices. *Proceedings of the National Academy of Sciences of the United States of America* 96: 5995–6000.

42 Empresa Brasileira de Pesquisa Agropecuária – EMBRAPA, 2019. O desafio do uso da água na agricultura brasileira. Disponível em: <https://www.embrapa.br/agua-na-agricultura/sobre-o-tema> Acesso em: 01 abril 2019.

43 Organização das Nações Unidas (ONU). (1997). Convenção das Nações Unidas de combate à desertificação nos países afetados por seca grave e/ou desertificação, particularmente na África. MMA.

44 Jørgensen, S.E. and Nielsen, S.N. (1996). Application of ecological engineering principles in agriculture. *Ecological Engineering* 7: 373–381.

45 Ervin, D.E., Welsh, R., Batie, S.S. et al. (2003). Towards an ecological systems approach in public research for environmental regulation of transgenic crops. *Agriculture, Ecosystems & Environment* 99: 1–14.

46 European Union (EU). (2001). 2001/18/EC=Directive 2001/18/EC of the European Parliament and of the Council (2001) on the deliberate release into the environment of genetically modified organisms and repealing Council Directive 90/220/EEC.

47 Ervin, D., Welsh, R., Batie, S., and Carpentier, C.L. (2001). Public research for environmental regulation of transgenic crops. Environmental Sciences and Resources Program, Portland State University, Portland, OR.

48 Lovei, G. (2001). Ecological risks and benefits of transgenic plants. *New Zealand Plant Protection* 54: 93–100.

49 Barrios, E. (2007). Soil biota, ecosystem services and land productivity. *Ecological Economics* 64: 269–285.

50 Ettema, C.H. and Wardle, D.A. (2002). Spatial soil ecology. *Trends in Ecology and Evolution* 17: 177–183.

51 Tiedje, J.M., Cho, J.C., Murray, A. et al. (2001). Soil teeming with life: new frontiers to soil science. In: *Sustainable Management of Soil Organic Matter* (eds. R.M. Rees, B.C. Ball, C.D. Campbell and C.A. Watson), 393–412. Wallingford: CAB International.

52 Brussaard, L., Kuyper, T.W., Didden, W.A.M. et al. (2004). Biological soil quality from biomass to biodiversity— importance and resilience to management stress and disturbance. In: *Managing Soil Quality: Challenges in Modern Agriculture* (eds. P. Schønning, S. Emholt and B.T. Christensen), 139–161. Wallingford: CAB International.

53 Swift, M.J., Izac, A.M.N., and van Noordwijk, M. (2004). Biodiversity and ecosystem services in agricultural landscapes — are we asking the right questions? *Agriculture, Ecosystems & Environment* 104: 113–134.

54 Wall, D.H. and Virginia, R.A. (2000). The world beneath our feet: soil biodiversity and ecosystem functioning. In: *Nature and Human Society: The Quest for a Sustainable World. Committee for the Second Forum on Biodiversity* (eds. P.H. Raven and T. Williams), 225–241. Washington D.C.: National Academy of Sciences and National Research Council.

55 Altieri, M.A. and Nicholls, C.I. (2012). Agroecology scaling up for food sovereignty and resiliency. In: *Sustainable Agriculture Reviews* (ed. E. Lichtfouse), 1–29. Dordrecht: Springer Netherlands.

56 Gliessman, S. (2007). *Agroecology: The Ecology of Sustainable Food Systems*. Boca Raton, FL: CRC Press.

57 Liere, H., Jha, S., and Philpott, S.M. (2017). Intersection between biodiversity conservation, agroecology, and ecosystem services. *Agroecology and Sustainable Food Systems* 41: 723–760.

58 Muchane, M.N., Sileshi, G.W., Gripenberg, S. et al. (2020). Agroforestry boosts soil health in the humid and sub-humid tropics: a meta-analysis. *Agriculture, Ecosystems and Environment* 295: 106899.

59 Boeraeve, F., Dendoncker, N., Cornélis, J.-T. et al. (2020). Contribution of agroecological farming systems to the delivery of ecosystem services. *Journal of Environmental Management* 260: 109576.

60 Dendoncker, N., Boeraeve, F., Crouzat, E. et al. (2018). How can integrated valuation of ecosystem services help understanding and steering agroecological transitions? *Ecology and Society* 23.

61 Garbach, K., Milder, J.C., DeClerck, F.A.J. et al. (2017). Examining multi-functionality for crop yield and ecosystem services in five systems of agroecological intensification. *International Journal of Agricultural Sustainability* 15: 11–28.

62 Duru, M., Therond, O., Martin, G. et al. (2015). How to implement biodiversity-based agriculture to enhance ecosystem services: a review. *Agronomy for Sustainable Development* 35: 1259–1281.

63 Andersson, E., Nykvist, B., Malinga, R. et al. (2015). A social–ecological analysis of ecosystem services in two different farming systems. *Ambio* 44 (S1): 102–112.

64 Barral, M.P., Rey Benayas, J.M., Meli, P. et al. (2015). Quantifying the impacts of ecological restoration on biodiversity and ecosystem services in agroecosystems: a global meta-analysis. *Agriculture, Ecosystems and Environment* 202: 223–231.

65 Lemaire, G., Franzluebbers, A., Carvalho, P.C.F. et al. (2014). Integrated crop–livestock systems: strategies to achieve synergy between agricultural production and environmental quality. *Agriculture, Ecosystems & Environment* 190: 4–8.

66 Palm, C., Blanco-Canqui, H., DeClerck, F. et al. (2014). Conservation agriculture and ecosystem services: an overview. *Agriculture, Ecosystems & Environment* 187: 87–105.

67 Wezel, A., Casagrande, M., Celette, F. et al. (2014). Agroecological practices for sustainable agriculture. A review. *Agronomy for Sustainable Development* 34: 1–20.

68 van Oudenhoven, A.P.E., Petz, K., Alkemade, R. et al. (2012). Framework for systematic indicator selection to assess effects of land management on ecosystem services. *Ecological Indicators* 21: 110–122.

69 Porter, J., Costanza, R., Sandhu, H. et al. (2009). The value of producing food, energy, and ecosystem services within an agro-ecosystem. *Ambio: A Journal of the Human Environment* 38 (4): 186–193.

70 Dale, V.H. and Polasky, S. (2007). Measures of the effects of agricultural practices on ecosystem services. *Ecological Economics* 64: 286–296.

71 Francis, C., Lieblein, G., Gliessman, S. et al. (2003). Agroecology: the ecology of food systems. *Journal of Sustainable Agriculture* 22 (3): 99–118.

72 Mitsch, W.J. and Jørgensen, S.E. (2003). Ecological engineering: a field whose time has come. *Ecological Engineering* 20: 363–377.

73 Williams, J.W. and Jackson, S.T. (2007). Novel climates, no-analog communities, and ecological surprises. *Frontiers in Ecology and the Environment* 5: 475–482.

74 Jones, C.G. (2012). Grand challenges for the future of ecological engineering. *Ecological Engineering* 45: 80–84.

2

Ecological and Ecosystem Engineering for Economic-Environmental Revitalization

Bruno Barbosa[1,2] *and Ana Luísa Fernando*[1]

[1] *MEtRICs, Departamento de Ciências e Tecnologia da Biomassa, Faculdade de Ciências e Tecnologia, FCT, Universidade Nova de Lisboa, Caparica, Portugal*
[2] *Universidade Feevale, Novo Hamburgo – RS, Brazil*

2.1 Introduction

A variety of factors can confer marginality to a particular soil [1–3]. This fact is reflected in the number of scientific publications related to ecological and ecosystem engineering. In a brief search of the literature, it is simple to find many published articles containing results from soils contaminated with heavy metals or organic compounds, or describing other chemical, physical, and biological properties that are unfavorable to the proper functioning and balance of soil ecosystems [4–8]. In a particular degraded soil, multiple disturbances in physical, environmental, and biological factors can be found. Indeed, soils may be marginal due to factors such as temperature, waterlogging, unfavorable soil texture and stoniness, dryness, shallow rooting depth, poor chemical properties of the soil (salinity and sodicity), steep slopes, soil layer restrictions, low biological production, acidity and alkalinity, the presence of organic and inorganic contaminants, and so on [9, 10]. Emergent contaminants, such as microplastics, can also confer marginality to soils [11, 12]. Likewise, socioeconomic factors may influence the marginality of a given soil [13, 14].

Physical factors such as soil compaction can interfere with soil drainage, the depth of plant roots, soil porosity, soil aeration, and gas transport properties, factors that together can also induce changes in the soil microbial composition and increase the emissions of greenhouse gases such as N_2O [15]. Such physical disturbance of soil can lead to a scarcity of oxygen at the root level, limiting plant growth and nutrition and creating conditions that may lead to the development of soil pathogens [10]. These changes in soil may also cause low rates of water infiltration and low hydraulic conductivity, which in turn will lead to lower gas diffusivities and air permeability, increasing the risk of developing anaerobic conditions in compacted soils [16, 17]. Such events and consequences induced by altering a single soil factor show the complex level of disturbance that can occur when more than one soil factor is changed. The same thing can occur with other physical factors at the chemical and biological/environmental levels.

Inorganic and organic compounds can alter soils' chemical conditions, constituting chemical factors that provoke marginality in soil environments to different degrees. Organic and inorganic

Handbook of Ecological and Ecosystem Engineering, First Edition. Edited by Majeti Narasimha Vara Prasad.

soil contaminants pose numerous risks for the soil environment, producing ecological dysfunctions, disorders in the structure and functioning of soil microbial communities and soil biota in general, and animal and human health [18]. For example, heavy metals in soils can lead to toxicosis in both animals and plants [19, 20]. These soil contaminants can impact biochemical processes in plants related to the production of enzymes and antioxidants involved in protein mobilization and in the photosynthetic process [21], also impacting plant growth parameters, yields, and water and nutrient balances [22, 23]. Indeed, heavy metals can cause disturbances and dysfunctions in plant organisms not only in their biochemical processes, but also in their physiological and metabolic processes, including the absorption and assimilation of nitrogen [24]. Soil microorganisms play different roles in the biogeochemical process of the soil, but they can be severely affected by the presence of heavy metals in soils [25, 26]. Long-term exposure to heavy metals can alter the richness, diversity, and structure of soils' microbial communities. Factors like soil pH, which can be changed by heavy metals in the soil environment, can also alter the fungal community [27]. These are just some examples of how physical and chemical factors can affect soils. Most of the time, these factors act simultaneously on soil biota and soil properties: multiple restricting factors act synergistically, either positively (leading to soil revitalization) or negatively (increasing marginality) [10]. For example, despite having recorded losses in biodiversity caused by heavy metals, other authors have not registered any negative effect on the soil's microbial functional stability [28]. So, how can we control the path and direction of these synergies? Can engineering interventions predict and induce soil ecosystems in the direction of revitalization? Have we any control over the impacts resulting from our interventions?

Ecological and environmental engineering approaches that encompass physical and chemical technologies and operations to improve soil properties together with ecosystem engineering technologies – meaning the application of bioremediation, dendroremediation, phytoremediation, mycoremediation, and associated processes – with a goal of modifying, restoring, maintaining, and/or (re-)creating functional ecosystems should be included in a remediation strategy and plan, as appropriate for the edaphic and climatic conditions of a given marginal soil and the specific properties that make it marginal [29]. Directly or indirectly, specific approaches and technologies should synergistically act with the physical, biological, and environmental factors of degraded land to provide socioeconomic and environmental revitalization [14, 30, 31]. However, despite proposing the remediation of the properties of a particular soil, ecological and ecosystem engineering do not always act in harmony. There is not always a dialogue between research areas and groups to propose broader remediation strategies.

In recent years, industrial crops have been considered for the simultaneous purpose of ecological restoration and biomass production for bioenergy, biofuels, biomaterials, etc. [32–40]. It is believed that by crossing specific plant features and mechanisms (agronomic, physiological, bioenergetic, tolerance to toxic elements and compounds, and phytoremediation potentials) with the appropriate consortium of technologies used by ecological and ecosystem engineering, and also with the site-specific features of a particular marginal soil, it will be possible not only to decrease environmental risks but also to provide economic benefits in such limited conditions [41]. Indeed, industrial crops that are suitable for use in phytoremediation, together with other biological and physical-chemical soil-restoration technologies, can promote ecosystem and economic revitalization in different types of marginal lands, since ecosystem functions can be restored and a sustainable chain and source of biomass for bioeconomy markets is generated [13, 42–46].

In this chapter, the aim is to show how phytoremediation technology can dialogue with the physical and chemical technologies of soil remediation, widely used by ecological engineering, as well as how such technology may be in concert with biological technologies used by the ecosystem engineering to revitalize marginal soils. For each factor that confers a degree of marginality in a particular degraded soil, it will be assessed how remediation approaches may contribute to this revitalization goal. Here, *revitalization* should be understood as the process by which certain ecological and socio-economical functions of a particular marginal soil ecosystem can be re-established through an intervention and restoration process.

2.2 Revitalization of Physical/Environmental Factors

Many soils present different stages of degradation due to a variety of physical factors acting alone or in synergy. Physical factors such as low temperature, limited soil drainage, unfavorable texture and stoniness, shallow rooting depth, salinity, sodicity, and acidity can induce marginality in soils or reflect a stage of marginality [10]. Depending on the degree of soil marginality, more than one of these factors (and others not described here) may be acting in an ecosystem. In many cases, some factors induce others, causing more pronounced consequences in the balance of a soil ecosystem. This section will discuss physical factors that can be overcome by implementing phytoremediation strategies and, whenever possible, possible links with other biological and physical-chemical remediation technologies.

2.2.1 Low Temperature

Description of the marginality factor: Low temperature is a physical factor that inhibits, restricts, or avoids the development and growth of almost all plant species since cellular homeostasis is interrupted, enzyme activity is inhibited, and photosynthesis is damaged [10, 47]. Cold affects crop growth and yield [48]. The stress caused by cold induces the disintegration of the cell membrane, reducing the growth and development of plants, e.g. some perennial crops. This is due to the accumulation of malondialdehyde (MDA) as a result of lipid peroxidation in the membranes [49]. Wheat cultivation exposed to 3 °C for 48 and 72 hours showed decreased chlorophyll levels and CO_2 assimilation [50]. The effect on transpiration and photosynthesis rates caused reduced activity of ATP synthase, which in turn further restricted the regeneration of RuBisCo and limited photophosphorylation [49]. Cold stress can cause oxidative stress in crops, negatively affecting membrane properties and enzymatic activities that lead to necrosis of plants and tissues [51]. Examples of these adverse effects on growth and development were reported in crops like *Musa* spp., cotton, sugar cane, *Nicotiana tabacum*, and *Brassica juncea* [49].

Revitalization by ecosystem engineering: Limitations due to soil located in a cold region or subjected to freezing can be overcome by plants that are adapted to these conditions. Freeze-tolerant plant species have several strategies to reduce the possibility of freezing their cells and tissues, including maintaining high intracellular concentrations of solutes and nucleating ice outside cells [51]. Interestingly, these plants generally have xerophytic adaptations to reduce conditions of limited water availability in soil. Plants such as winter wheat have the genetic capacity to acclimate, strengthen, and acquire tolerance to freezing down to −20 °C [49]. Cold tolerance can be more easily observed in plant species with enhanced antioxidant enzyme activities [52]. The tolerance to cold stress in alfalfa varieties during the germination phase seems to be associated with increased activity of antioxidant enzymes such as superoxide dismutase (SOD), ascorbate peroxidase (APX), and catalase (CAT) in aerial tissues [53]. Also, exposing soybean seedlings to very low temperatures (1 °C) resulted in the increased activity of antioxidant enzymes [54].

Acclimatization or adaptation to cold can also occur through the synthesis of cryoprotective molecules such as soluble sugars (sucrose, raffinose, stachyose, trehalose), sugar alcohols (sorbitol, ribitol, inositol), and nitrogen compounds of low molecular weight (proline, glycine betaine) [55]. Other plants increase the activity of antioxidant enzymes to protect against cold and maintain cellular redox homeostasis. This acclimatization to cold temperature conditions affects the cellular lipid composition, increasing the proportion of unsaturated fatty acids that make up the phospholipids necessary to maintain the functions of the plasma membrane [49].

Plants that show tolerance and adaptability to cold tend to grow slowly. Such plants display C3 photosynthesis and store sugars in underground tissues. These species have developed a very efficient respiratory system, which allows them to mobilize reserves extremely quickly, usually in short growing seasons. The development factors and physiological responses of these plants are under genetic control [55]. Genetic manipulation of plants with these characteristics may confer adaptation to soils in colder climatic regions, contributing to the revitalization of local markets through the production of biomass to feed the bioeconomy. Transgenic plants, whose reproductive organs accumulate glycinabetaine, exhibit greater tolerance to abiotic stresses [56].

The vast majority of C3 species are temperate crops, while C4 species are mainly tropical. Despite the vast majority of C4s being adapted to warmer climates, *Miscanthus* spp. has some tolerance to cold, evidenced by its cultivation in temperate climates in northern Europe, including Germany and Denmark [57, 58]. *Miscanthus* spp. appears in late spring and goes into senescence in early autumn in Northern Europe, where it has a shorter canopy period compared to C3 grasses. Recently, some attempts have been made to identify the characteristics of the most cold-tolerant varieties of *Miscanthus* spp., and, at the moment, some studies are trying to improve cold-tolerant *Miscanthus* cultivars, and specifically ice-tolerance in the year of establishment, which is expected to produce higher biomass yields across Europe [59]. The productivity of many of these cold-resistant cultivars will also improve if increasing temperatures (due to climate change) in these latitudes are confirmed. *Arundo donax* also has a wide capacity for climatic acclimatization and resistance to cold, growing in cold regions as Denmark, Germany, and the United Kingdom [60, 61]. The plants *Pennisetum purpureum* and *A. donax* were indicated by Jensen and Eller [62] as biomass crops, candidates for development in temperate European paludiculture due to their high growth and biomass production rates, ease of harvest, and resistance to temporary flooding. In this study, the hybrid C4 of the species *P. purpureum* showed higher yields and photosynthetic rates than the giant reed (C3), but the latter showed more stable results on its physiological parameters [62]. Genetic characteristics seem to be related to the cold tolerance of other biomass crops such as switchgrass. Switchgrass rhizomes can survive in soils exposed to air temperatures as low as −20 °C [63]. A cold-resistant variety of switchgrass produced 17.54 Mg dry matter ha^{-1} in the third year after planting in a sandy soil in the Netherlands at 52° latitude [61]. Annual crops such as cardoon and rapeseed (*Brassica* spp.) show good tolerance to low temperatures [64].

The type of revitalization discussed in this point has a more economic character: the possibility of introducing those cold-resistant crop species in an area not previously cultivated may provide further income for farmers. Moreover, introducing a vegetative cover may increase organic matter in the soil, with benefits for biodiversity, soil structure, and erosion control.

2.2.2 Limited Soil Drainage and Shallow Rooting Depth

Description of the marginality factor: Part of the growth and survival strategy of plants is related to the depth and distribution of their roots, especially in ecosystems where water is a limiting factor [65]. Here, roots play a predominant role for hydrological and biogeochemical cycling [66]. The depth of rooting is governed by the depth of water infiltration and is strongly influenced by the soil's properties, the moment and magnitude of water intake, and evaporative demand [66]. Plant species exposed to water stress, such as slight soil moisture deficiencies, have deeper root profiles [66]. In general, roots tend to be deeper in ecosystems limited by water, colder regions, humid climates, or places where the rainy season occurs during the winter. The deepest roots are found in arid environments where a long dry season occurs and water is limited, and where the penetration of the roots seems to be related to the depth of infiltration and the intensity of evaporative demand [66]. Also, soil compaction can lead to limited drainage, shallow rooting depths, negative effects on soil porosity and gas transport properties, reduction of critical levels of aerobic microbial activity, and promotion of N_2O emission [10, 15–17, 67].

Revitalization by ecological engineering: Microbially induced calcite precipitation has been referred to as a method that uses bacteria and cementation solutions to effectively reduce oil cracking [68]. Other physical methods and approaches to reduce the cracking of clayey soil include compaction control, geogrid, and plant root reinforcement [69]. Compaction control and similar technologies have high associated costs and low long-term maintenance [68].

Revitalization by ecosystem engineering: Introducing perennial and annual plants capable of enduring these soil conditions may be a solution. Mechanical stresses and chemical stimuli exerted by soil conditions may affect the ability of plant roots to penetrate the soil. The physical force that the soil exerts on the root of a plant that tries to penetrate along the soil profile is influenced by the soil's density, porosity, and, above all, moisture [70]. Several alternatives have been used, such as the plows and chisels, cover crops, and species with strong root systems and high biomass production, among other management techniques [70]. Other plant species have developed adaptive responses to these conditions by increasing or decreasing the rate of elongation of the apical region, swelling or shrinking its diameter [71]. In Europe, it is easier to find shallow soils in mountainous slope areas. A threshold of 30–35 cm relative to the rooting depth constitutes a severe restriction on the cultivation of perennial crops (or other crops) such as giant reed and switchgrass [61]. Shallow soil implies limited nutrients and water resources, as well as very limited soil workability [61].

Biomass crops other than switchgrass and giant reed can be adapted to such soil conditions or modify their marginal aspects. Cardoon fields, for example, do not suffer from soil compaction [64, 72]. The roots of perennial plants can also accelerate the decomposition of organic matter in the soil in the long term through the action of biological and physical mechanisms [73]. *Miscanthus* spp. can be used to rehabilitate ecosystem functions and physical properties of marginal soils such as mine soils [74]. However, the mechanization associated with its cultivation can also cause the reverse effect. Switchgrass can also be well-adapted to shallow soil depth and light soils with reduced fertility [75]. Introducing those industrial crops in these soil types can ameliorate the soils through the increment of organic matter and structure [76]. When explored for biomaterials, bioproducts, biochemicals, and bioenergy can help revitalize these areas economically.

2.2.3 Unfavorable Texture and Stoniness

Description of the marginality factor: Soil aggregates play an essential role in accumulating and stabilizing soil organic matter since they provide physical protection to soils [77]. Also, the soil's physical structure is essential for organic carbon storage [78]. So, topsoil's heavy clay texture affects plant growth and crop parameters and is directly related to plant nutrient supply, soil moisture conditions, and rooting conditions [10]. Soil texture affects the balance between water and gases, which is very stable over time independent of soil management [79]. Soil texture plays an important role in plant tolerance to drought at the germination stage [61]. Stoniness and dryness may influence water and oxygen availability in the root zone and soil stability [10].

Revitalization by ecosystem engineering: Soil ecosystems dominated by fungal communities show higher carbon retention than soil communities dominated by bacteria. Generally, afforestation is a suitable technique to improve carbon sequestration on marginal lands [80]. Switchgrass and giant reed are well adapted to a broad range of soils with different relative abundances of clay, silt, sand, organic matter, and coarse material [61]. Most mine soils are marginal soils with no topsoil, consisting of mine spoils whose properties can range from loose, coarse-textured material with many rock fragments to highly compacted clayey material [81]. Industrial crops like *Miscanthus* spp., giant reed, switchgrass, *P. purpureum*, and others can tolerate mine tailings and produce quality biomass for energy and other purposes such that soil properties are restored [82–85]. Also, soybeans can be easily adapted to a variety of soil textures [86].

2.2.4 Sloping Areas

Description of the marginality factor: Sloping areas change in elevation with respect to planimetric distance [61]. Steep rocky slopes are marginal lands: they are composed of unstable and unbalanced ecosystems that are not favorable to germination, growth, and plant development and are characterized by high soil loss, high evapotranspiration, and low mechanical stability for root development [87]. Such steeper slopes tend to occur in higher areas of the landscape and usually induce more pronounced apparent dryness, often associated with shallow soils, leading to heavy runoff, mechanization difficulties, minimal cultivation opportunities and land management, a greater risk of soil degradation (erosion), and landslides [10].

Revitalization by ecological engineering: Ecological engineering and the use of super-absorbent polymers can improve the environment through the hydrologic effect of aboveground biomass, improving the adsorption and complexing capacity of hydrophilic functional groups, promoting the formation of a water-blocking layer between soil particles and thus inhibiting moisture from moving from the soil surface to the atmosphere or to the slope's rock layer, enhancing water uptake and the utilization of water for plant growth, decreasing evapotranspiration, increasing germination rates in grass and woody plant species, and increasing plant survival [87]. This technology can also be used to improve and optimize the use of irrigation water in regions affected by drought, since water and nutrients can be stored and released when needed, enhancing the yields of crops [88].

Revitalization by ecosystem engineering: Shallow soils are often combined with slopes, and plants like *Miscanthus* spp., giant reed, and switchgrass can easily adapt to such conditions [61, 89]. Bioengineering techniques for slope stabilization and landslide mitigation are already using many plant species that are being investigated for bioenergy purposes. The roots of giant reed and *P. purpureum*, for example, can act in sloping areas by a variety of mechanical mechanisms: (i) they reinforce the soil, increasing shear resistance, like jute nets, jute reinforcement grids, and geotextiles; (ii) they provide support to slope soils through buttress and arching, since they can anchor deeply in firm strata, similar to shear wrenches and screws or pretensioned rock anchors; (iii) they help compact the soil, increasing the densification and solidification of soil materials, similar to wooden, concrete, or lime piles; and (iv) they unite surface soil particles, increasing cohesive and adhesive forces and reducing susceptibility to erosion and landslides [90, 91].

2.2.5 Dryness

Description of the marginality factor: Dryness is a severe water stress that induces changes in the maturity time of plants and losses in their biomass yields, causing leaf senescence and increasing the stem/leaf ratio as well the translocation of nutrients to the roots [92, 93]. Dryness induces the closure of stomata, affecting the photosynthetic process and leading to less water loss through transpiration [94, 95]. Dryness can also be seen as the ratio between annual precipitation and potential evapotranspiration in a particular area, causing marginality [61]. When dry conditions develop in soil, its surface becomes dry and hard, making root penetration difficult. However, in some soils, the deepest layers can contain enough moisture for plant growth. Even if some roots manage to penetrate these deeper layers, deleterious effects should be expected to occur in the more superficial soil layers [96].

Revitalization by ecosystem engineering: In dry conditions, cytokinin concentrations on plants decrease, stoma closes, and growth is redirected from the aerial components of the plants to the underground ones. The inoculation of plants with cytokinin-producing bacteria in water deficit conditions can be an efficient approach to overcome dryness [97]. Drought tolerance of the bioenergy crop giant reed (*A. donax* L.) is high, and it is well-suited to warm, temperate, and semiarid environments with high temperatures and prolonged summer dryness [98]. The same authors pointed out *Miscanthus* spp., switchgrass, and other fast-growing, fast-yielding energy crops exhibited high tolerance to drought [65]. However, during the establishment phase of these perennials, it is essential to avoid drought. From the second year onward, drought tolerance increases since rhizomes are already established, reaching water sources from deeper soil horizons [61]. Despite this, the productivity of these crops can be affected by drought.

Techniques such as mutagenesis have been identified as valid strategies for increasing the genetic diversity of giant reed, increasing its performance under dry soil conditions [99]. Moreover, giant reed can maintain its photosynthetic capacity and leaf growth to minimize the stress caused by drought [100]. In fact, giant reed can change its water sources depending on water availability, maximize water uptake efficiency, and satisfy evapotranspirative demands [101, 102]. When water is scarce, the giant reed species can improve the efficiency of water use and maintain high productivity. This fact indicates that the species can grow under conditions of moderate irrigation during periods of drought.

Saccharum spontaneum spp. aegyptiacum is another potential bioenergy crop species that produces high biomass yields while tolerating environments characterized by drought stress, high temperatures, and high vapor pressure deficit [94]. It is a C4 plant with a rapid growth rate, few or no natural enemies, and very active assimilation rates during drought-stress periods, and it can use water very efficiently [103]. *Saccharum spontaneum spp. aegyptiacum* is so well adapted to drought conditions in Southern Europe that it displays biomass yields even higher than *Miscanthus* spp. and *A. donax* grown in the same experimental conditions and under well-watered, rainfed conditions [104].

Reed canary grass shows high plasticity and is one of the best candidates for most marginal land revitalization. It can be distributed in and dominate various habitats, including wetlands and riparian zones; and in a cool temperature climate, it can withstand flooding for long periods and exhibits excellent drought tolerance [98]. The energy crops perennial ryegrass (*Lolium perenne* L.), *Panicum* spp., and *Sorghum* spp. are also well adapted and resistant to drought conditions [105, 106].

2.2.6 Waterlogging

Description of the marginality factor: *Waterlogging* means excess water in the root zone. Excessive soil moisture adversely affects crop growth and development by causing restricting rooting conditions and low-oxygen stress and reducing soil strength for workability and trafficability [10, 107]. Flooding induces changes in soil pH and redox potential [108]. Today this factor affects up to 10% of global land area [109].

Revitalization by ecosystem engineering: Flooding tolerance found in halophytes and glycophytes is often associated with the production of adventitious roots containing aerenchyma, resulting in an internal supply of O_2 and preventing anoxia in tissue [110]. Flooding causes a reduction in the photosynthetic capacity of many plants, such as *L. perenne* [108]. Waterlogged marginal land can be revitalized through the adaptation and growth of plants for biomass production.

Flooded soils are characterized by the absence of oxygen and reduced chemical conditions. The adaptation of a plant species to these conditions depends on the growth capacity of its roots and its aeration. The effectiveness of aeration within the roots of plant species such as *Vetiveria zizanoides*, wetland grass species, sorghum, maize, and sugarcane is determined by: (i) anatomical characteristics such as the length of the aerenchyma, dimensions of the stele, cortical cell layout pattern, and existence of barriers to radial O_2 losses; (ii) morphological characteristics such as root diameter and the number of lateral roots; and (iii) physiological processes, such as respiratory demand for oxygen [111]. Sunflowers and maize also display adaptations to flooded conditions/low levels of oxygen in the soil, such as elongation of the roots and inhibition/reduction of the growth aerial plant components [112]. Aquaporin genes of *Sorghum bicolor* L. seem to be related to the answer to waterlogging stress [113]. Liu and Jiang [114] reported that waterlogging and submergence stress decreased the root activity of SOD, catalase, peroxidase, and ascorbate peroxidase in perennial ryegrasses, indicating that maintaining antioxidant activity and carbohydrate and minimizing lipid peroxidation could contribute to better waterlogging and submergence tolerance in that bioenergy crop. Giant reed quickly adapts to flooding conditions with the absence of oxygen. After prolonged oxygen scarcity, the species can readjust and restore its physiological conditions. Depending on the severity of the stress, it can adapt its uptake of photosynthetic CO_2, related to a combination of stomatal and mesophyll diffusional constraints [115]. Japanese sugarcane and *S. bicolor* are pointed out as bioenergy crops tolerant to multiple environmental stresses in marginal land, tolerating waterlogging conditions [116].

2.3 Revitalization of Chemical Factors

2.3.1 Acidity

Description of the marginality factor: Soil surface acidity can reduce nutrient availability in soil solutions, increasing the concentration of heavy metals (especially aluminum) and their toxic effects on decomposition and nitrogen fixation microbial communities [61, 117, 118]. The occurrence of toxicity caused by aluminum depends on the interaction of this element with other limiting factors such as nitrogen, phosphorus, pH, and iron [119].

Revitalization strategy: According to Zhao and collaborators [119], three proposals can be applied to acidic soils to improve plant growth: (i) combined application of lime and nutrients, depending on the type of soil and plants; (ii) identification and development of tolerant plant species of multiple stressors; and (iii) inoculation of plants with beneficial microorganisms.

Revitalization by ecosystem engineering: Plants can reduce the toxicity of aluminum ions by exuding compounds that can chelate aluminum or sequester aluminum in plant cells through complexes with organic acids [120]. Other plants also develop associations with arbuscular symbiotic mycorrhizal fungi, increasing the plants' resistance to soil acidity and phytotoxic levels caused by aluminum in soil [121]. Bioenergy crops like *A. donax* can adapt to a sub-severe pH threshold

(5–5.5): performance is optimal between pH 6 and 7, moderate between pH 7 and 8, and poor between pH 4 and 5 [61]. Switchgrass tolerates acidic soils in all the growth and development phases, from seedling to adult; its roots can tolerate a pH of 3.7 and show excellent growth in soils with a pH between 6 and 7 [61]. Other bioenergy crops tolerant of acid soils include soybean [122, 123] and *Jatropha curcas* [124]. Adjusting the pH and adding fertilizers to the soil can also improve phytoremediation efficiency technology, as discussed in the following subsection [125].

2.3.2 Heavy Metals and Organic Contaminants

Description of the marginality factor: Disposal sites and industrial and agricultural soils are contaminated with multiple pollutants. Soils are often contaminated with inorganic and organic pollutants (heavy metals, metalloids, endocrine disruptors, pesticides, oil, and derivatives), posing risks in the food chain and threats to human health and the functioning of natural ecosystems [126]. Multiple pollutants in soil tend to interact, making the remediation process more complex. For example, the interaction between organic pollutants and heavy metals changes their speciation, solubility, and bioavailability, inhibiting or promoting the efficiency of mutual remediation [127]. There may be competition between pollutants for adsorbent and enzyme binding sites, as well as the double pressure effect on plants and soil microorganisms [128]. Heavy metals inhibit microbial metabolism and enzyme activity, reducing the degradation efficiency of organic pollutants [129]. Thus, promoting remediation in a synchronized manner is not a simple task.

Revitalization by ecological engineering: Physical and chemical technologies for soil remediation, such as washing, electrokinetic remediation, chemical oxidation, nanomaterials remediation, adsorption, filtration, granular activated carbon, and photocatalysis, can remove heavy metals from soils but affect different ecological functions of the soil by changing either the pH, moisture content, or oxidation potential [130, 131]. For example, studies conducted by Ma et al. [127] showed that electroremediation technology combined with activated bamboo charcoal was effective in the simultaneous removal of 2,4-dichlorophenol and Cd from a soil; however, the application of such technology seems to decrease the water content in soil significantly. Likewise, the studies of Villa et al. [130] showed that the Fenton process, despite being very efficient for removing DDT (dichlorodiphenyltrichloroethane) and diesel from contaminated soil, creates oxidizing conditions that contribute to the dissolution of organic matter initially present in the soil and increase the water solubility of hydrophobic compounds (like DDT), accelerating the possible contamination of ground and surface waters. Nevertheless, in some situations, the conventional Fenton process can be effective by treating the soil with lower organic content and pH and can be integrated with biological remediation, which is more economical and ecological [132]. Other technologies include soil replacement, consisting of diluting the contaminated soil with a large amount of clean soil, covering it, and mixing it with contaminated soil. This approach very effectively reduces the toxic effect of the heavy metals in soil but requires large volumes of work and earthworks that disturb the functioning of these ecosystems [133].

Revitalization by ecosystem engineering: Bioremediation technology minimizes the toxic effect of certain heavy metals in less toxic forms, revitalizing natural ecosystems through the use of microorganisms and enzymes that purify the contaminated environment [131]. Other variants include the remediation of heavy metal pollution in soil using a combination of microbial immobilization and carbon microspheres [134]. This approach leads to the formation of stable bacterial colonies and reduces heavy metal pollution. However, in the case of soils contaminated with heavy metals, this technology has some limitations: (i) heavy metals are not biodegradable; (ii) microorganisms produce toxic metabolites, which are eliminated during the process; and (iii) the microorganism's response to heavy metals depends on the concentration and availability of the heavy metals [131]. It is more efficient when associated with phytoremediation technology through symbiotic associations with plants [23].

Nanobioremediation is an emerging technology that uses biosynthesized nanoparticles for the decontamination (adsorption) of heavy metals from contaminated soils. The biological synthesis of nanoparticles using plant extracts or microorganisms can be an ecological and economical alternative to chemical and physical methods [135].

Biological remediation technologies like phytoremediation use plants' abilities and mechanisms (phytodegradation, phytostabilization, phytoextraction, and phytovolatilization), together with their microbial consortia, to effectively reduce the bioavailability of toxic heavy metals and degrade organic pollutants by metabolic decomposition [136–139]. These technologies have lower remediation costs and can be applied in situ and on a large scale, maintaining ecological sustainability through the use of plant organisms capable of restoring soil functions [140–142].

Recently, bioenergy crops have been considered for the simultaneous purpose of ecological restoration and biomass production [36–40, 143]. For example, switchgrass can be used to decompose (phytodegradation) and remove contaminants from the soil, including herbicides, trinitrotoluene, polychlorinated biphenyls, polynuclear aromatic hydrocarbons, chromium, and radionuclides [144]. Castor bean (*Ricinus communis* L.) is a primary colonizer suitable for dealing with toxic conditions in soils, probably because arbuscular fungi may be involved in root metal sequestration. Although it is not a metal accumulator, it can be useful in stabilizing mine tailings, promoting the reduction of the heavy metal's bioavailability, dispersion, and concomitant risks to human health [145]. In addition to extracting most toxic metals (Cd, Pb, Ni, As, Cu, etc.) as well as organic contaminants (pesticides), the plant can increase soil fertility and reduce soil erosion and is also important for generating biodiesel and employment/social development in local communities [146]. *Miscanthus* spp. can also tolerate soils that are poor in various elements, such as nitrogen, and contaminated with heavy metals and hydrocarbons, making it suitable for phytostabilization [147]. Aided phytostabilization technology combines soil amendments with the plant consortium and associated microorganisms and is well-suited to stabilize mine tailings and acid mine drainage [148]. *A. donax* and *Miscanthus* spp. can also be used in the phytoextraction of Zn from soils and the phytostabilization of heavy metal contamination and preventing leaching/groundwater contamination [149]. The energy crops *P. purpureum* and *Pennisetum thyphoideum* can also be used successfully to aid the phytostabilization of mine tailings rich in the heavy metals Cd, Zn, and Pb [82].

This ability of certain energy crops to tolerate heavy metal contamination has a genetic cause: in giant reed, for example, the transport and detoxification of heavy metals seems to be related to the expression of glutathione reductase genes, and antioxidant activities and gene expression are specific to a particular toxic metal [150]. This knowledge can be important in genetic engineering. In addition, the inoculation of plant growth-promoting bacteria may also increase efficiency in aided phytostabilization [83]. Also, mycological assisted phytoremediation enhancement of bioenergy crops can be a suitable approach to increase the efficiency of this technology. The inoculation of fungi in energy crops such as corn and sunflower and favoring their phytoremediation efficiency in soils contaminated with Cr, Cd, Pb, and Cu also increased their respective biomass productivity and oil production in sunflower [151].

2.3.3 Salinity and Sodicity

Description of the marginality factor: The salinization process is focused on arid and semi-arid regions and depends on the soil texture and relative water and salt content [152, 153]. It consists of changes in the patterns of precipitation, evapotranspiration, and landscape hydrology caused by the presence of chlorides, sulfites, nitrates, and bicarbonates of sodium, calcium, magnesium, and potassium that affect the microbial activity of soils and vegetation zonation and cover [153]. Since natural biological, biochemical, hydrological, and erosional earth cycles can be affected by the salinization process, environmental and agricultural services can also be lost [154].

Revitalization by ecosystem engineering: The species *A. donax* L. can be classified as "moderately salt-tolerant" [155]. At 42 dS/m, the giant cane's overall growth is affected, but with zero mortality [155]. Gas exchange of giant reed leaves and their overall tolerance to salt is greater than in many food crops (like corn and rice) and energy crops such as Miscanthus, Populus, and Salix that are indigenous in regions where giant reed is now an invasive species [155]. Although *A. donax* L. is naturally tolerant, some research has tried to improve this tolerance to salts and select which varieties are best adapted to these stress conditions. Both salinity and water stress affect stomatal closure and eventually decrease the photosynthetic rate of this crop, but that some giant reed ecotypes are more tolerant to salinity than to water stress [156]. Also, the inoculation of arbuscular mycorrhizal fungi (*Rhizophagus intraradices* and *Funneliformis mosseae*) in giant reed can improve phosphorous extraction by the roots and the plant's overall tolerance to salinity. The plants' nutritional status has been improved along with the efficiency of nutrient use and response to saline stress conditions [157].

Miscanthus tolerance to saline conditions seems to be lower than that of giant reed. Biomass yields of *Miscanthus* spp. were reduced by 50% at 10.65 dS/m NaCl [158]. In greenhouse experiments, upland switchgrass showed low tolerance to 5 dS/m (with a reduction of 77% in the aboveground biomass, dry weight), while the lowland ecotype showed higher tolerance (with a reduction of 20% in the aboveground biomass, dry weight) [61].

2.4 Economic Revitalization of Degraded Soil Ecosystems

The sociotechnical paradigm related to bioeconomy topics is based on the assumption that biomass resources can be used to reduce dependence on fossil-based resources, revitalize rural economies, ensure energy security, mitigate climate change, and provide environmental services [159]. In the previous sections, we saw how industrial crops can tolerate and provide remediation for a variety of soil/environmental conditions. Once produced in such environmental/physical conditions, biomass can be used for the economic revitalization of marginal lands [160].

Lignocellulosic crops (e.g. miscanthus, giant reed, switchgrass) represent a source of fermentable sugars, cellulose, hemicellulose, starch, and lignin that can be used for the production of either bioenergy and biofuels or biomaterials and bioproducts [161]. The biomass of bioenergy crops like *Miscanthus* spp., jatropha, and castor bean can be used for bioethanol production, and biomass from *A. donax*, *Cannabis sativa*, *Cynara cardunculus,* and kenaf can be used for bioenergy and fiber purposes [29]. The lignocellulosic biomass can also be used to produce medium-density fiberboards, insulation mats, composites, nanocellulose, pulp for paper, etc. [33, 162–164]. Other industrial crops can provide feedstock (e.g. gums, resins, rubber, terpenes, phenolic substances, triglycerides, alkaloids, among others) to produce high-value-added bio-based products and green chemicals (food flavors, pharmaceuticals, cosmetics, nutraceutical products, bioplastics, lubricants, etc.) [165–173] along with second-generation biofuels through an integrated biorefinery approach. Figure 2.1 shows a schematic image of processing options and multiple bio-based products derived from industrial crops.

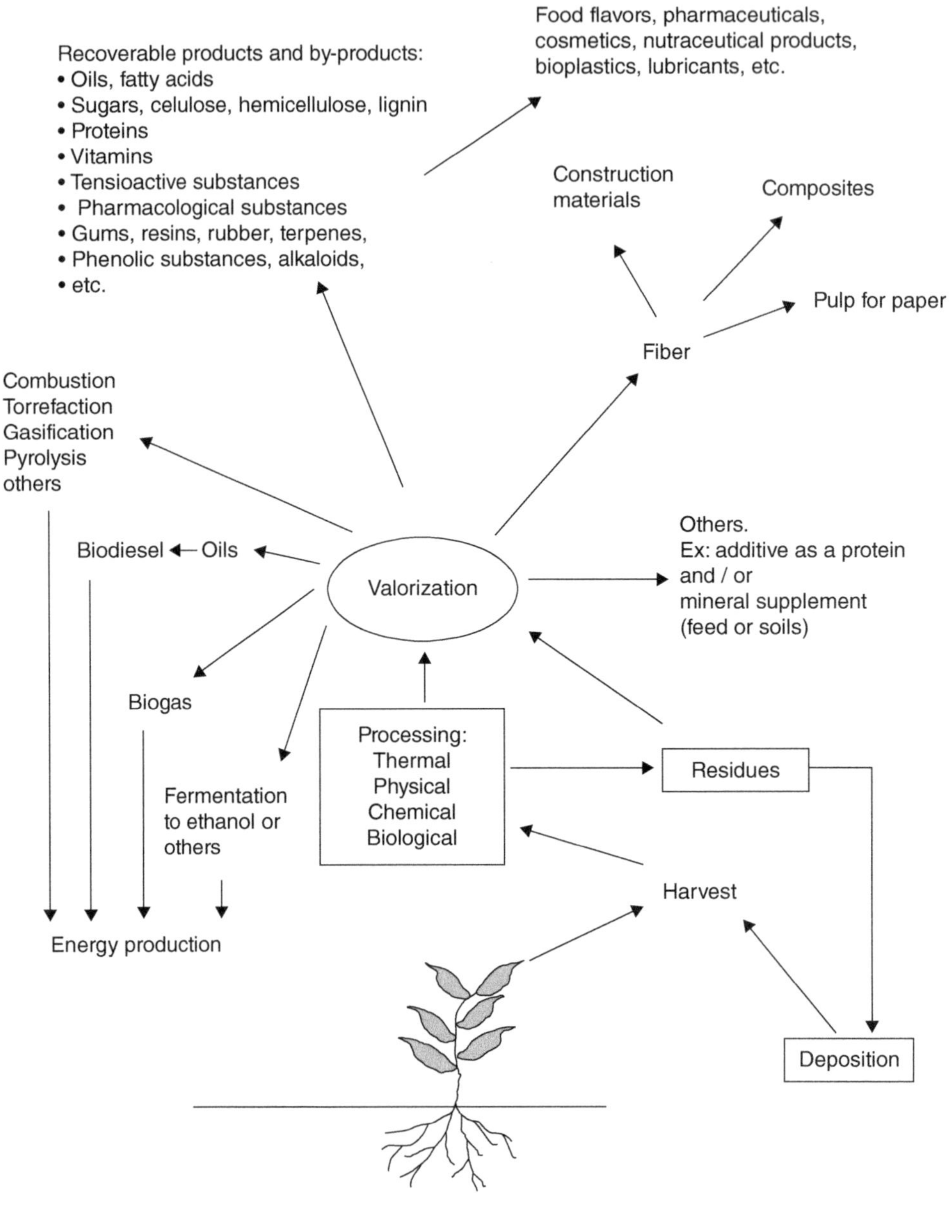

Figure 2.1 Examples of processing options and multiple bio-based products derived from industrial crops.

2.5 Conclusions

Ecological engineering encompasses a variety of physical and chemical remediation technologies and, together with ecosystem engineering technologies such as bioremediation, dendroremediation, mycoremediation, and phytoremediation (and associated technologies), can be successfully applied to modify, restore, maintain, and/or (re-)create functional ecosystems on marginal and degraded land. The joint application of approaches can result in more effective land recovery and restoration than using them separately.

Given soil cracking and low soil drainage, ecological engineering techniques such as microbially induced calcite precipitation, compaction control, surcharge loading, geogrids, and plant root reinforcement can reduce soil cracking. Super-absorbent polymers can promote water uptake and utilization in sloping areas, which can optimize irrigation water schemes in regions affected by drought, erosion, and desertification. Extracted washing, electrokinetic remediation, chemical oxidation, nanomaterial remediation, adsorption, filtration, granular activated carbon, and photocatalysis are physical and chemical technologies that can effectively remove and stabilize heavy metals from marginal and degraded soils by changing the pH, moisture content, or oxidation potential. All these technologies can be applied efficiently to improve marginal lands; but when applied alone, they usually affect and can even stop soil functions.

Together with ecological engineering approaches, ecosystem engineering and, specifically, phytoremediation can further reduce soil constraints and promote the development of ecological functions. Under soil-cracking conditions, introducing perennial energy crops can accelerate the decomposition of organic matter available in soil (incorporated, for example, via biosolids and other organic amendments), increasing soil aeration and fertility. These perennial crops can stabilize the soil's physical structure and promote soil organic carbon storage. They can also adapt to a variety of clay textures and promote the balance between water and gases in different soil types, including stony soil. This can be an essential aspect of soil management and restoration, especially for soils prone to drought and affected by salinization. Other industrial crops like *V. zizanoides*, sorghum, giant reed, and perennial ryegrass can also promote soil remediation given waterlogging or cold conditions, showing high tolerance to a huge range of environmental conditions. Many of the perennial crops used for bioenergy purposes produce high yields and fast growth at the same time, which can stimulate fungal and microorganism communities that help restore physical and chemical constraints in soils. In sloping areas, perennial crops like giant reed, switchgrass, *Miscanthus* spp., and *P. purpureum* can reinforce soil aggregates and provide support, increasing cohesive and adhesive forces and reducing susceptibility to erosion and landslides. The same crops can remove and immobilize the toxicological effect of heavy metals, other salts, and inorganic and emergent contaminants that reach the soil environment, providing a very efficient and cheap means to restore soil and improve soil conditions for the development of soil microorganisms, fungi, and other soil biota while also generating biomass for the bioeconomy (bioenergy, fiber, and other bio-products).

References

1 Shortall, O.K. (2013). "Marginal land" for energy crops: exploring definitions and embedded assumptions. *Energ Policy* 62: 19–27.

2 Tve, A.M., Robinson, D.A., and Lark, R.M. (2013). Gradual and anthropogenic soil change for fertility and carbon on marginal sandy soils. *Geoderma* 207–208: 35–48.

3 Nsanganwimana, F., Pourrut, B., Mench, M., and Douay, F. (2014). Suitability of Miscanthus species for managing inorganic and organic contaminated land and restoring ecosystem services. A review. *J. Environ. Manage.* 143: 123–134.

4 Sheoran, V., Sheoran, A.S., and Poonam, P. (2008). Remediation techniques for contaminated soils. *Environ. Eng. Manag. J.* 7: 379–387.

5 Changfeng, L., Kehai, Z., Wenqiang, Q. et al. (2019). A review on heavy metals contamination in soil: effects, sources, and remediation techniques. *Soil Sediment Contam.* 28: 380–394.

6 Guo, J., Muhammad, H., Lv, X. et al. (2020). Prospects and applications of plant growth promoting rhizobacteria to mitigate soil metal contamination: a review. *Chemosphere* 246: 125823.

7 Qayyum, S., Khan, I., Meng, K. et al. (2020). A review on remediation technologies for heavy metals contaminated soil. *Cent. Asian J. Environ. Sci. Technol. Innov.* 1: 21–29.

8 Wan, X., Lei, M., and Chen, T. (2020). Review on remediation technologies for arsenic-contaminated soil. *Front. Environ. Sci. Eng.* 14: 24.

9 Kang, S., Post, W., Wang, D. et al. (2013). Hierarchical marginal land assessment for land use planning. *Land Use Policy* 30: 106–113.

10 Terres, J.M., Hagyo, A., Wania, A. (2014). Scientific contribution on combining biophysical criteria underpinning the delineation of agricultural areas affected by specific constraints: Methodology and factsheets for plausiblecriteria combinations. In "JRC Scientific and Technical Reports", Publications Office of the European Union, Brussels, Belgium.

11 Awet, T.T., Kohl, Y., Meier, F. et al. (2018). Effects of polystyrene nanoparticles on the microbiota and functional diversity of enzymes in soil. *Environ. Sci. Eur.* 30: 11.

12 Rillig, M.C., Lehmann, A., Souza-Machado, A.A., and Yang, G. (2019). Microplastic effects on plants. *New Phytol.* 223: 1066–1070.

13 Lewandowski, I. (2015). Securing a sustainable biomass supply in a growing bioeconomy. *Glob. Food Secur.* 6: 34–42.

14 Mehemood, M.A., Ibrahim, M., Rashid, U. et al. (2016). Biomass production for bioenergy using marginal lands. *Sustain. Prod. Consum.* 9: 3–21.

15 Berisso, F.E., Schjønning, P., Keller, T. et al. (2012). Persistent effects of subsoil compaction on pore size distribution and gas transport in a loamy soil. *Soil Tillage Res.* 122: 42–51.

16 Ahmadi, S.H., Plauborg, F., Andersen, M.N. et al. (2011). Effects of irrigation strategies and soils on field grown potatoes: root distribution. *Agric. Water Manag.* 98: 1280–1290.

17 Colombi, T., Braun, S., Keller, T., and Walter, A. (2017). Artificial macropores attract crop roots and enhance plant productivity on compacted soils. *Sci. Total Environ.* 574: 1283–1293.

18 Tang, J., Zhang, J., Ren, L. et al. (2019). Diagnosis of soil contamination using microbiological indices: a review on heavy metal pollution. *J. Environ. Manag.* 242: 121–130.

19 Amuno, S., Bedos, L., Kodzhahinchev, V. et al. (2020). Comparative study of arsenic toxicosis and ocular pathology in wild muskrats (*Ondatra zibethicus*). and red squirrels (*Tamiasciurus hudsonicus*). breeding in arsenic contaminated areas of Yellowknife, Northwest Territories (Canada). *Chemosphere* 248: 126011.

20 Amuno, S., Shekh, K., Kodzhahinchev, V., and Niyogi, S. (2020). Neuropathological changes in wild muskrats (*Ondatra zibethicus*). and red squirrels (*Tamiasciurus hudsonicus*). breeding in arsenic endemic areas of Yellowknife, Northwest Territories (Canada).: arsenic and cadmium accumulation in the brain and biomarkers of oxidative stress. *Sci. Total Environ.* 704: 135426.

21 Seneviratne, M., Rajakaruna, N., Rizwan, M. et al. (2019). Heavy metal-induced oxidative stress on seed germination and seedling development: a critical review. *Environ. Geochem. Health* 41: 1813–1831.

22 Athar, R. and Ahmad, M. (2002). Heavy metal toxicity: effect on plant growth and metal uptake by wheat, and on free living Azotobacter. *Water Air Soil Pollut.* 138: 165–180.

23 Chibuike, G.U. and Obiora, S.C. (2014). Heavy metal polluted soils: effect on plants and bioremediation methods. *Appl. Environ. Soil Sci.* 2014: 752708.

24 Hussain, S., Khaliq, A., Noor, M.A. et al. (2019). Metal toxicity and nitrogen metabolism in plants: An overview. In: *Carbon and Nitrogen Cycling in Soil* (eds. R. Datta, R. Meena, S. Pathan and M. Ceccherini), 221–248. Singapore: Springer.

25 Lemmel, F., Maunoury-Danger, F., Fanesi, A. et al. (2019). Soil properties and multi-pollution affect taxonomic and functional bacterial diversity in a range of French soils displaying an anthropisation gradient. *Microb. Ecol.* 77: 993–1013.
26 Wise, B.R., Roane, T.M., and Mosier, A.C. (2019). Community composition of nitrite reductase gene sequences in an acid mine drainage environment. *Environ. Microbiol.* 79: 562–575.
27 Pan, X., Zhang, S., Zhong, Q. et al. (2020). Effects of soil chemical properties and fractions of Pb, Cd, and Zn on bacterial and fungal communities. *Sci. Total Environ.* 715: 136904.
28 Jiang, R., Wang, M., Chen, W. et al. (2020). Changes in the integrated functional stability of microbial community under chemical stresses and the impacting factors in field soils. *Ecol. Indic.* 110: 105919.
29 Pandey, V.C., Bajpai, O., and Singh, N. (2016). Energy crops in sustainable phytoremediation. *Renew. Sust. Energ. Rev.* 54: 58–73.
30 McIntyre, T. (2003). Phytoremediation of heavy metals from soils. *Adv. Biochem. Eng. Biotec.* 78: 97–123.
31 Dale, V.H., Kline, K.L., Buford, M.A. et al. (2016). Incorporating bioenergy into sustainable designs. *Renew. Sust. Energ. Rev.* 56: 1158–1171.
32 Oliveira, J.S., Duarte, M.P., Christian, D.G. et al. (2001). Environmental aspects of Miscanthus production. In: *Miscanthus for Energy and Fibre* (eds. M.B. Jones and M. Walsh), 172–178. London, UK: James & James (Science Publishers). Ltd.
33 Fernando, A.L., Godovikova, V., and Oliveira, J.F.S. (2004). *Miscanthus* x *giganteus*: contribution to a sustainable agriculture of a future/present – oriented biomaterial. *Mater. Sci. Forum* 455–456: 437–441.
34 Fernando, A.L. (2013). Environmental aspects of Kenaf production and use. In: *Kenaf: A Multi-Purpose Crop for Several Industrial Applications*, vol. 117 (eds. A. Monti and E. Alexopoulou) Green Energy and Technology, 83–104. London: Springer.
35 Alexopoulou, E., Cosentino, S.L., Danalatos, N. et al. (2013). New insights from the BIOKENAF project. In: *Kenaf: A Multi-Purpose Crop for Several Industrial Applications*, vol. 117 (eds. A. Monti and E. Alexopoulou), Green Energy and Technology, 177–203. London: Springer.
36 Fernando, A.L., Costa, J., Barbosa, B. et al. (2018). Environmental impact assessment of perennial crops cultivation on marginal soils in the Mediterranean region. *Biomass Bioenergy* 111: 174–186.
37 Abdelsalam, I.M., Elshobary, M., Eladawy, M.M., and Nagah, M. (2019). Utilization of multi-tasking non-edible plants for phytoremediation and bioenergy source – a review. *Phyton-Int. J. Exp. Bot.* 88: 69–90.
38 Chen, X., Kumari, D., Cao, C.J. et al. (2019). A review on remediation technologies for nickel contaminated soil. *Hum. Ecol. Risk. Assess.* 26: 571–585.
39 Kumar, V. and Kumar, P. (2019). A review on feasibility of phytoremediation technology for heavy metals removal. *Arch. Agric. Environ. Sci.* 4: 326–341.
40 Naila, A., Meerdink, G., Jayasena, V. et al. (2019). A review on global metal accumulators – mechanism, enhancement, commercial application, and research trend. *Environ. Sci. Pollut. Res.* 26: 26449–26471.
41 Barbosa, B., Costa, J., Boléo, S. et al. (2016). Phytoremediation of inorganic compounds. In: *Electrokinetics Across Disciplines and Continents – New Strategies for Sustainable Development* (eds. A.B. Ribeiro, E.P. Mateus and N. Couto), 373–400. Switzerland: Springer International Publishing.
42 Testa, R., Foderà, M., Di Trapani, A.M. et al. (2016). Giant reed as energy crop for Southern Italy: An economic feasibility study. *Renew. Sust. Energ. Rev.* 58: 558–564.
43 Bracco, S., Calicioglu, O., San Juan, M.G., and Flammini, A. (2018). Assessing the contribution of bioeconomy to the total economy: a review of national frameworks. *Sustainability* 10: 1698.

44 Stephanie, S., An, D.S., Margot, V. et al. (2019). Phytomining to re-establish phosphorus-poor soil conditions for nature restoration on former agricultural land. *Plant Soil* 440: 233–246.

45 Sidella, S., Barbosa, B., Costa, J. et al. (2016). Screening of giant reed clones for phytoremediation of lead contaminated soils. In: *Perennial Biomass Crops for a Resource Constrained World* (eds. S. Barth, D. Murphy-Bokern, O. Kalinina, et al.), 191–197. Switzerland: Springer International Publishing.

46 Barbosa, B., Costa, J., and Fernando, A.L. (2018). Production of energy crops in heavy metals contaminated land: opportunities and risks. In: *Land Allocation for Biomass* (eds. R. Li and A. Monti), 83–102. Cham, Switzerland: Springer.

47 Porter, J.R. and Semenov, M.A. (2005). Crop responses to climatic variation. *Philos. Trans. R. Soc. B* 360: 2021–2035.

48 Hasanuzzaman, M., Nahar, K., and Fujita, M. (2013). Extreme temperatures, oxidative stress and antioxidant defense in plants. In: *Abiotic Stress – Plant Responses and Applications in Agriculture* (eds. K. Vahdati and C. Leslie), 169–205. London, UK: IntechOpen Limited.

49 Awasthi, R., Bhandari, K., and Nayyar, H. (2015). Temperature stress and redox homeostasis in agricultural crops. *Front. Environ. Sci.* 3: 1–24.

50 Yordanova, R. and Popova, L. (2007). Effect of exogenous treatment with salicylic acid on photosynthetic activity and antioxidant capacity of chilled wheat plants. *Gen. Appl. Plant Physiol.* 33: 155–170.

51 Allen, D.J. and Ort, D.R. (2001). Impacts of chilling temperatures on photosynthesis in warm-climate plants. *Trends Plant Sci.* 6: 36–42.

52 Fahimirad, S., Karimzadeh, G., and Ghanati, F. (2013). Cold-induced changes of antioxidant enzymes activity and lipid peroxidation in two canola (*Brassica napus* L.). Cultivars. *J. Plant Physiol. Breed.* 3: 1–11.

53 Wang, W.B., Kim, Y.H., Lee, H.S. et al. (2009). Differential antioxidation activities in two alfalfa cultivars under chilling stress. *Plant Biotechnol. Rep.* 3: 301–307.

54 Posmyk, M.M., Corbineau, F., Vinel, D. et al. (2001). Osmoconditioning reduces physiological and biochemical damage induced by chilling in soybean seeds. *Physiol. Plant.* 111: 473–482.

55 Janská, A., Maršík, P., Zelenková, S., and Ovesná, J. (2010). Cold stress and acclimation – what is important for metabolic adjustment? *Plant Biol.* 12: 395–405.

56 Chen, T.H.H. and Murata, N. (2008). Glycinebetaine: an effective protectant against abiotic stress in plants. *Trends Plant Sci.* 13: 499–505.

57 Larsen, S., Jaiswal, D., Bentsen, N.S. et al. (2016). Comparing predicted yield and yield stability of willow and Miscanthus across Denmark. *Glob. Change Biol. Bioenergy* 6: 1061–1070.

58 McCalmont, J.P., Hastings, A., McNamara, N.P. et al. (2017). Environmental costs and benefits of growing *Miscanthus* for bioenergy in the UK. *Glob. Change Biol. Bioenergy* 9: 489–507.

59 Kandel, T.P., Hastings, A., Jørgensen, U., and Olesen, J.E. (2016). Simulation of biomass yield of regular and chilling tolerant Miscanthus cultivars and reed canary grass in different climates of Europe. *Ind. Crop. Prod.* 86: 329–333.

60 El Bassam, N. (2010). *Handbook of Bioenergy Crops – a Complete Reference to Species, Development and Applications*. London, United Kingdom: Earthscan, Ltd.

61 Parenti, A., Lambertini, C., and Monti, A. (2018). Areas with natural constraints to agriculture: possibilities and limitations for the cultivation of Switchgrass (*Panicum virgatum* L.) and Giant Reed (*Arundo donax* L.) in Europe. In: *Land Allocation for Biomass* (eds. R. Li and A. Monti), 39–63. Cham, Switzerland: Springer.

62 Jensen, A.B. and Eller, F. (2020). Hybrid Napier grass (*Pennisetum purpureum* Schumach × *P. americanum* (L.). Leeke cv. Pakchong 1). and Giant reed (*Arundo donax* L.). as

candidate species in temperate European paludiculture: growth and gas exchange responses to suboptimal temperatures. *Aquat. Bot.* 160: 103165.
63 Poudel, H.P., Sanciangco, M.D., Kaeppler, S.M. et al. (2019). Quantitative trait loci for freezing tolerance in a lowland x upland switchgrass population. *Front. Plant Sci.* 10: 372.
64 Paschalidou, A., Tsatiris, M., and Kitikidou, K. (2019). Perennial vs annual energy crops-SWOT analysis (case study: Greece). *Int. Refereed J. Eng. Sci.* 7: 1–24.
65 Barbosa, B., Costa, J., Fernando, A.L., and Papazoglou, E.G. (2015). Wastewater reuse for fiber crops cultivation as a strategy to mitigate desertification. *Ind. Crop. Prod.* 68: 17–23.
66 Collins, D.B.G. and Bras, R.L. (2007). Plant rooting strategies in water-limited ecosystems. *Water Resour. Res.* 43: 1–10.
67 Pietola, L., Horn, R., and Yli-Halla, M. (2005). Effects of trampling by cattle on the hydraulic and mechanical properties of soil. *Soil Tillage Res.* 82: 99–108.
68 Liu, B., Zhu, C., Tang, C.S. et al. (2020). Bio-remediation of desiccation cracking in clayey soils through microbially induced calcite precipitation (MICP). *Eng. Geol.* 264: 105389.
69 Butt, W.A., Mir, B.A., and Jha, J.N. (2016). Strength behavior of clayey soil reinforced with human hair as a natural fibre. *Geotech. Geol. Eng.* 34: 411–417.
70 Bartzen, B.T., Hoelscher, G.L., Ribeiro, L.L.O., and Seidel, E.P. (2019). How the soil resistance to penetration affects the development of agricultural crops? *J. Exp. Agric. Int.* 30: 1–17.
71 Calusi, B., Tramacere, F., Gualtieri, S. et al. (2020). Plant root penetration and growth as a mechanical inclusion problem. *Int. J. Non Linear Mech.* 120: 103344.
72 Grammelis, P., Malliopoulou, A., Basinas, P., and Danalatos, N.G. (2008). Cultivation and characterization of *Cynara cardunculus* for solid biofuels production in the Mediterranean region. *Int. J. Mol. Sci.* 9: 1241–1258.
73 Lu, J., Dijkstra, F.A., Wang, P., and Cheng, W. (2019). Roots of non-woody perennials accelerated long-term soil organic matter decomposition through biological and physical mechanisms. *Soil Biol. Biochem.* 134: 42–53.
74 Guzman, J.G., Ussiri, D.A.N., and Lal, R. (2019). Soil physical properties following conversion of a reclaimed minesoil to bioenergy crop production. *Catena* 176: 289–295.
75 Alexopoulou, E., Zanetti, F., Papazoglou, E.G. et al. (2017). Long-term studies on switchgrass grown on a marginal area in Greece under different varieties and nitrogen fertilization rates. *Ind. Crop. Prod.* 107: 446–452.
76 Fernando, A.L., Boléo, S., Barbosa, B. et al. (2015). Perennial grass production opportunities on marginal Mediterranean land. *Bioenergy Res.* 8: 1523–1537.
77 O'Brien, S.L. and Jastrow, J.D. (2013). Physical and chemical protection in hierarchical soil aggregates regulates soil carbon and nitrogen recovery in restored perennial grasslands. *Soil Biol. Biochem.* 61: 1–13.
78 Zhong, X., Li, J., Li, X. et al. (2017). Physical protection by soil aggregates stabilizes soil organic carbon under simulated N deposition in a subtropical forest of China. *Geoderma* 285: 323–332.
79 Kv, U., Km, R., and Naik, D. (2019). Role of soil physical, chemical and biological properties for soil health improvement and sustainable agriculture. *J. Pharmacogn. Phytochem.* 8: 1256–1267.
80 Niu, X. and Duiker, S.W. (2006). Carbon sequestration potential by afforestation of marginal agricultural land in the Midwestern U.S. *For. Ecol. Manag.* 223: 415–427.
81 Ussiri, D.A.N., Guzman, J.G., Lal, R., and Somireddy, U. (2019). Bioenergy crop production on reclaimed mine land in the North Appalachian region, USA. *Biomass Bioenergy* 125: 188–195.
82 Gao, B., Zhang, X., Tian, C. et al. (2019). Effects of amendments and aided phytostabilization of an energy crop on the metal availability and leaching in mine tailings using a pot test. *Environ. Sci. Pollut. Res.* 27: 2745–2759.

83 Hamidpour, M., Nemati, H., Abbaszadeh-Dahaji, P., and Roosta, H.R. (2019). Effects of plant growth-promoting bacteria on EDTA-assisted phytostabilization of heavy metals in a contaminated calcareous soil. *Environ. Geochem. Health*: 3. https://doi.org/10.1007/s10653-019-00422-3.

84 Von Cossel, M., Lewandowski, I., Elbersen, B. et al. (2019). Marginal agricultural land low-input systems for biomass production. *Energies* 12: 3123.

85 Von Cossel, M., Wagner, M., Lask, J. et al. (2019). Prospects of bioenergy cropping systems for a more social-ecologically sound bioeconomy. *Agronomy* 9: 605.

86 Cordeiro, C.F.S. and Echer, F.R. (2019). Interactive effects of nitrogen-fixing bacteria inoculation and nitrogen fertilization on soybean yield in unfavorable edaphoclimatic environments. *Sci. Rep.* 9: 1–11.

87 Yang, L., Yang, Y., Chen, Z. et al. (2014). Influence of super absorbent polymer on soil water retention, seed germination and plant survivals for rocky slopes eco-engineering. *Ecol. Eng.* 62: 27–32.

88 Khodadadi-Dehkordi, D. (2016). The effects of superabsorbent polymers on soils and plants. *Pertanika J. Trop. Agric. Sci.* 39: 267–298.

89 Cosentino, S.L., Copani, V., Scalici, G. et al. (2015). Soil erosion mitigation by perennial species under Mediterranean environment. *Bioenergy Res.* 8: 1538–1547.

90 Singh, A.K. (2010). Bioengineering techniques of slope stabilization and landslide mitigation. *Disaster Prev. Manag.* 19: 384–397.

91 Cantalice, J.R.B., Nunes, E.O.S., Cavalcante, D.M. et al. (2019). Vegetative-hydraulic parameters generated by agricultural crops for laminar flows under a semi-arid environment of Pernambuco, Brazil. *Ecol. Indic.* 106: 105496.

92 Buxton, D.R. (1996). Quality-related characteristics of forages as influenced by plant environment and agronomic factors. *Anim. Feed Sci. Technol.* 59: 37–49.

93 Deléglise, C., Meisser, M., Mosimann, E. et al. (2015). Drought-induced shifts in plants traits, yields and nutritive value under realistic grazing and mowing managements in a mountain grassland. *Agric. Ecosyst. Environ.* 213: 94–104.

94 Scordia, D., Testa, G., Cosentino, S.L. et al. (2015). Soil water effect on crop growth, leaf gas exchange, water and radiation use efficiency of *Saccharum spontaneum* L. Ssp. *aegyptiacum* (willd.). hackel in semi-arid Mediterranean environment. *Ital. J. Agron.* 10: 185–191.

95 Wilmowicz, E., Kućko, A., Burchardt, S., and Przywieczerski, T. (2019). Molecular and hormonal aspects of drought-triggered flower shedding in yellow lupine. *Int. J. Mol. Sci.* 20: 3731.

96 Blum, A., Johnson, J.W., Ramseur, E.L., and Tollner, E.W. (1991). The effect of a drying top soil and a possible non-hydraulic root signal on wheat growth and yield. *J. Exp. Bot.* 42: 1225–1231.

97 Arkhipova, T.N., Prinsen, E., Veselov, S.U. et al. (2007). Cytokinin producing bacteria enhance plant growth in drying soil. *Plant Soil* 292: 305–315.

98 Scordia, D. and Cosentino, S.L. (2019). Perennial energy grasses: resilient crops in a changing European agriculture. *Agriculture* 9: 169.

99 Zegada-Lizarazu, W., Salvi, S., and Monti, A. (2020). Assessment of mutagenized giant reed clones for yield, drought resistance and biomass quality. *Biomass Bioenergy* 134: 105501.

100 Haworth, M., Marino, G., Riggi, E. et al. (2019). The effect of summer drought on the yield of *Arundo donax* is reduced by the retention of photosynthetic capacity and leaf growth later in the growing season. *Ann. Bot.* 124: 567–579.

101 Cosentino, S.L., Scordia, D., Sanzone, E. et al. (2014). Response of giant reed (*Arundo donax* L.) to nitrogen fertilization and soil water availability in semi-arid Mediterranean environment. *Eur. J. Agron.* 60: 22–32.

102 Zegada-Lizarazu, W. and Monti, A. (2019). Deep root growth, ABA adjustments and root water uptake response to soil water deficit in giant reed. *Ann. Bot.* 124: 605–615.

103 Cosentino, S.L., Copani, V., Testa, G., and Scordia, D. (2015). *Saccharum spontaneum* L. ssp. *aegyptiacum* (Willd.). Hack. a potential perennial grass for biomass production in marginal land in semi-arid Mediterranean environment. *Ind. Crop. Prod.* 75: 93–102.

104 Scordia, D., Testa, G., Copani, V. et al. (2017). Lignocellulosic biomass production of Mediterranean wild accessions (*Oryzopsis miliacea*, *Cymbopogon hirtus*, *Sorghum halepense* and *Saccharum spontaneum*) in a semi-arid environment. *Field Crops Res.* 214: 56–65.

105 Dąbrowski, P., Baczewska-Dąbrowska, A.H., Kalaji, H.M. et al. (2019). Exploration of chlorophyll a fluorescence and plant gas exchange parameters as indicators of drought tolerance in perennial ryegrass. *Sensors* 19: 2736.

106 Taylor, G., Donnison, I.S., Murphy-Bokern, D. et al. (2019). Sustainable bioenergy for climate mitigation: developing drought-tolerant trees and grasses. *Ann. Bot.* 124: 513–520.

107 Zhang, Y., Chen, Y., Lu, H. et al. (2016). Growth, lint yield and changes in physiological attributes of cotton under temporal waterlogging. *Field Crops Res.* 194: 83–93.

108 Muhammad, A.A. (2012). Waterlogging stress in plants: a review. *Afr. J. Agric. Res.* 7: 1976–1981.

109 Arguello, M.N., Mason, R.E., Roberts, T.L. et al. (2016). Performance of soft red winter wheat subjected to field soil waterlogging: grain yield and yield components. *Field Crops Res.* 194: 57–64.

110 Colmer, T.D. and Flowers, T.J. (2008). Flooding tolerance in halophytes. *New Phytol.* 179: 964–974.

111 McDonald, M.P., Galwey, N.W., and Colmer, T.D. (2002). Similarity and diversity in adventitious root anatomy as related to root aeration among a range of wetland and dryland grass species. *Plant Cell Environ.* 25: 441–451.

112 Akhtar, I. and Nazir, N. (2013). Effect of waterlogging and drought stress in plants. *Int. J. Water Res. Environ. Sci.* 2: 34–40.

113 Kadam, S., Abril, A., Dhanapal, A.P. et al. (2017). Characterization and regulation of aquaporin genes of Sorghum [*Sorghum bicolor* (L.). Moench] in response to waterlogging stress. *Front. Plant Sci.* 8: 1–14.

114 Liu, M. and Jiang, Y. (2015). Genotypic variation in growth and metabolic responses of perennial ryegrass exposed to short-term waterlogging and submergence stress. *Plant Physiol. Biochem.* 95: 57–64.

115 Pompeiano, A., Reyes, T.H., Moles, T.M. et al. (2019). Photosynthetic and growth responses of *Arundo donax* L. plantlets under different oxygen deficiency stresses and reoxygenation. *Front. Plant Sci.* 10: 408.

116 Quinn, L.D., Straker, K.C., Guo, J. et al. (2015). Stress-tolerant feedstocks for sustainable bioenergy production on marginal land. *Bioenergy Res.* 8: 1081–1100.

117 Martins, A.P., Denardin, L.G.O., Tiecher, T. et al. (2020). Nine-year impact of grazing management on soil acidity and aluminum speciation and fractionation in a long-term no-till integrated crop-livestock system in the subtropics. *Geoderma* 359: 113986.

118 Niu, H., Leng, Y., Ran, S. et al. (2020). Toxicity of soil labile aluminum fractions and aluminum species in soil water extracts on the rhizosphere bacterial community of tall fescue. *Ecotoxicol. Environ. Saf.* 187: 109828.

119 Zhao, X.Q., Chen, R.F., and Shen, R.F. (2014). Coadaptation of plants to multiple stresses in acidic soils. *Soil Sci.* 179: 503–513.

120 Ashraf, S., Dixit, S., Ramteke, P.W., and Rizvi, A.Z. (2019). Interactive role of brassinosteroids and calcium ameliorates in response to the aluminium toxicity in plants. *Int. J. Trend Sci. Res. Dev.* 3: 183–203.

121 Aguilera, P., Borie, F., Seguel, A., and Cornejo, P. (2019). How does the use of non-host plants affect arbuscular mycorrhizal communities and levels and nature of glomalin in crop rotation systems established in acid andisols? In: *Mycorrhizal Fungi in South America* (eds. M. Pagano and M. Lugo), Fungal Biology, 147–158. Cham, Switzerland: Springer.

122 Zheng, S.J. (2010). Crop production on acidic soils: overcoming aluminium toxicity and phosphorus deficiency. *Ann. Bot.* 106: 183–184.

123 Kuswantoro, H. and Zen, S. (2013). Performance of acid-tolerant soybean promising lines in two planting seasons. *Int. J. Biol.* 5: 49–56.

124 Konaka, T., Ishimoto, Y., Yamada, M. et al. (2019). Tolerance evaluation of *Jatropha curcas* and *Acacia burkei* to acidic and copper/nickel-contaminated soil. *J. Environ. Biol.* 40: 1109–1114.

125 Awa, S.H. and Hadibarata, T. (2020). Removal of heavy metals in contaminated soil by phytoremediation mechanism: a review. *Water Air Soil Pollut.* 231: 1–15.

126 Ye, S., Zeng, G., Wu, H. et al. (2017). Biological technologies for the remediation of co-contaminated soil. *Crit. Rev. Biotechnol.* 37: 1062–1076.

127 Ma, J.W., Wang, F.Y., Huang, Z.H., and Wang, H. (2010). Simultaneous removal of 2,4-dichlorophenol and Cd from soils by electrokinetic remediation combined with activated bamboo charcoal. *J. Hazard. Mater.* 176: 715–720.

128 Ye, S., Zeng, G., Wu, H. et al. (2017). Co-occurrence and interactions of pollutants, and their impacts on soil remediation – a review. *Crit. Rev. Environ. Sci. Technol.* 47: 1528–1553.

129 Ojuederie, O.B. and Babalola, O.O. (2017). Microbial and plant-assisted bioremediation of heavy metal polluted environments: a review. *Int. J. Environ. Res. Public Health* 14: 1504.

130 Villa, R.D., Trovó, A.G., and Nogueira, R.F.P. (2008). Environmental implications of soil remediation using the Fenton process. *Chemosphere* 71: 43–50.

131 Akhtar, F.Z., Archana, K.M., Krishnaswamy, V.G., and Rajagopal, R. (2020). Remediation of heavy metals (Cr, Zn). Using physical, chemical and biological methods: a novel approach. *SN Appl. Sci.* 2: 267.

132 Cheng, M., Zeng, G., Huang, D. et al. (2016). Hydroxyl radicals based advanced oxidation processes (AOPs). For remediation of soils contaminated with organic compounds: a review. *Chem. Eng. J.* 284: 582–598.

133 Yao, Z., Li, J., Xie, H., and Yu, C. (2012). Review on remediation technologies of soil contaminated by heavy metals. *Procedia Environ. Sci.* 16: 722–729.

134 Sun, X., Meng, J., Huo, S. et al. (2020). Remediation of heavy metal pollution in soil by microbial immobilization with carbon microspheres. *Int. J. Environ. Sci. Dev.* 11: 43–47.

135 Yadav, K.K., Singh, J.K., Gupta, N., and Kumar, V. (2017). A review of nanobioremediation technologies for environmental cleanup: a novel biological approach. *J. Mater. Environ. Sci.* 8: 740–757.

136 Pauwels, M., Willems, G., Roosens, N. et al. (2008). Merging methods in molecular and ecological genetics to study the adaptation of plants to anthropogenic metal-polluted sites: implications for phytoremediation. *Mol. Ecol.* 17: 109–119.

137 Coppa, E., Astolfi, S., Beni, C. et al. (2020). Evaluating the potential use of Cu-contaminated soils for giant reed (*Arundo donax*, L.). cultivation as a biomass crop. *Environ. Sci. Pollut. Res.* 27: 8662–8672.

138 Manoj, S.R., Karthik, C., Kadirvelu, K. et al. (2020). Understanding the molecular mechanisms for the enhanced phytoremediation of heavy metals through plant growth promoting rhizobacteria: a review. *J. Environ. Manag.* 254: 109779.

139 Patra, D.K., Pradhan, C., and Patra, H.K. (2020). Toxic metal decontamination by phytoremediation approach: concept, challenges, opportunities and future perspectives. *Environ. Technol. Innov.* 18: 100672.

140 Gomes, H.I. (2012). Phytoremediation for bioenergy: challenges and opportunities. *Environ. Technol. Rev.* 1: 59–66.

141 Yang, Y., Zhou, X., Tie, B. et al. (2017). Comparison of three types of oil crop rotation systems for effective use and remediation of heavy metal contaminated agricultural soil. *Chemosphere* 188: 148–156.

142 Zhou, J., Chen, L.H., Peng, L. et al. (2020). Phytoremediation of heavy metals under an oil crop rotation and treatment of biochar from contaminated biomass for safe use. *Chemosphere* 247: 125856.

143 Papazoglou, E.G. and Fernando, A.L. (2017). Preliminary studies on the growth, tolerance and phytoremediation ability of sugarbeet (*Beta vulgaris* L.). grown on heavy metal contaminated soil. *Ind. Crop. Prod.* 107: 463–471.

144 Parrish, D.J. and Fike, J.H. (2005). The biology and agronomy of switchgrass for biofuels. *Crit. Rev. Plant Sci.* 24: 423–459.

145 Ruiz-Olivares, A., Carrillo-González, R., González-Chávez, M.C.A., and Soto-Hernández, R.M. (2013). Potential of castor bean (*Ricinus communis* L.). for phytoremediation of mine tailings and oil production. *J. Environ. Manag.* 114: 316–323.

146 Bauddh, K., Singh, K., Singh, B., and Singh, R.P. (2015). *Ricinus communis*: a robust plant for bio-energy and phytoremediation of toxic metals from contaminated soil. *Ecol. Eng.* 84: 640–652.

147 Pidlisnyuk, V., Stefanovska, T., Lewis, E.E. et al. (2014). Miscanthus as a productive biofuel crop for phytoremediation. *Crit. Rev. Plant Sci.* 33: 1–19.

148 Barbosa, B. and Fernando, A.L. (2018). Aided phytostabilization of mine waste. In: *Bio-Geotechnologies for Mine Site Rehabilitation* (eds. M.N.V. Prasad, P.J.C. Favas and S.K. Maiti), 147–157. UK: Elsevier Inc.

149 Barbosa, B., Boléo, S., Sidella, S. et al. (2015). Phytoremediation of heavy metal-contaminated soils using the perennial energy crops *Miscanthus* spp. and *Arundo donax* L. *Bioenergy Res.* 8: 1500–1511.

150 Shaheen, S., Ahmad, R., Mahmood, Q. et al. (2019). Gene expression and biochemical response of giant reed under Ni and Cu stress. *Int. J. Phytoremediation* 21: 1474–1485.

151 Iram, S., Basri, R., Ahmad, K.S., and Jaffri, S.B. (2019). Mycological assisted phytoremediation enhancement of bioenergy crops *Zea mays* and *Helianthus annuus* in heavy metal contaminated lithospheric zone. *Soil Sediment Contam.* 28: 411–430.

152 Rengasamy, P. (2006). World salinization with emphasis on Australia. *J. Exp. Bot.* 57: 1017–1023.

153 Bui, E.N. (2013). Soil salinity: a neglected factor in plant ecology and biogeography. *J. Arid Environ.* 92: 14–25.

154 Dahlhaus, P.G., Cox, J.W., Simmons, C.T., and Smitt, C.M. (2008). Beyond hydrogeologic evidence: challenging the current assumptions about salinity processes in the Corangamite region, Australia. *Hydrogeol. J.* 16: 1283.

155 Nackley, L.L. and Kim, S.H. (2015). A salt on the bioenergy and biological invasions debate: salinity tolerance of the invasive biomass feedstock *Arundo donax*. *Glob. Change Biol. Bioenergy* 7: 752–762.

156 Sánchez, E., Scordia, D., Lino, G. et al. (2015). Salinity and water stress effects on biomass production in different *Arundo donax* L. clones. *Bioenergy Res.* 8: 1461–1479.

157 Romero-Munar, A., Baraza, E., Gulías, J., and Cabot, C. (2019). Arbuscular mycorrhizal fungi confer salt tolerance in giant reed (*Arundo donax* l.) plants grown under low phosphorus by reducing leaf NA+ concentration and improving phosphorus use efficiency. *Front. Plant Sci.* 10: 843.

158 Stavridou, E., Hastings, A., Webster, R.J., and Robson, P. (2017). The impact of soil salinity on the yield, composition and physiology of the bioenergy grass *Miscanthus* × *giganteus*. *Glob. Change Biol. Bioenergy* 9: 92–104.

159 Burnham, M., Eaton, W., Selfa, T. et al. (2017). The politics of imaginaries and bioenergy sub-niches in the emerging Northeast U.S. bioenergy economy. *Geoforum* 82: 66–76.

160 Dickinson, N.M., Baker, A.J.M., Doronila, A. et al. (2009). Phytoremediation of inorganics: realism and synergies. *Int. J. Phytoremediation* 11: 97–114.

161 Fernando, A.L., Rettenmaier, N., Soldatos, P., and Panoutsou, C. (2018). Sustainability of perennial crops production for bioenergy and bioproducts. In: *Perennial Grasses for Bioenergy and Bioproducts* (ed. E. Alexopoulou), 245–283. UK: Academic Press, Elsevier Inc.

162 Pires, J.R.A., Souza, V.L., and Fernando, A.L. (2019). Valorization of energy crops as a source for nanocellulose production–current knowledge and future prospects. *Ind. Crop. Prod.* 140: 111642.

163 Fernando, A.L., Barbosa, B., Costa, J., and Papazoglou, E.G. (2016). Giant reed (*Arundo donax* L.).: a multipurpose crop bridging phytoremediation with sustainable bio-economy. In: *Bioremediation and Bioeconomy* (ed. M.N.V. Prasad), 77–95. UK: Elsevier Inc.

164 Fernando, A.L., Duarte, M.P., Vatsanidou, A., and Alexopoulou, E. (2015). Environmental aspects of fiber crops cultivation and use. *Ind. Crop. Prod.* 68: 105–115.

165 Pascoal, A., Quirantes-Piné, R., Fernando, A.L. et al. (2015). Phenolic composition and antioxidant activity of kenaf leaves. *Ind. Crop. Prod.* 78: 116–123.

166 Souza, V.G.L., Fernando, A.L., Pires, J.R.A. et al. (2017). Physical properties of chitosan films incorporated with natural antioxidants. *Ind. Crop. Prod.* 107: 565–572.

167 Souza, V.G.L., Rodrigues, P.F., Duarte, M.P., and Fernando, A.L. (2018). Antioxidant migration studies in chitosan films incorporated with plant extracts. *J. Renew. Mater.* 6: 548–558.

168 Pires, J.R.A., Souza, V.G.L., and Fernando, A.L. (2018). Chitosan/montmorillonite bionanocomposites incorporated with rosemary and ginger essential oil as packaging for fresh poultry meat. *Food Packag. Shelf Life* 17: 142–149.

169 Souza, V.G.L., Rodrigues, C., Ferreira, L. et al. (2019). in vitro; bioactivity of novel chitosan bionanocomposites incorporated with different essential oils. *Ind. Crop. Prod.* 140: 111563.

170 Zanetti, F., Monti, A., and Berti, M.T. (2013). Challenges and opportunities for new industrial oilseed crops in EU-27: a review. *Ind. Crop. Prod.* 50: 580–595.

171 Righini, D., Zanetti, F., Martínez-Force, E. et al. (2019). Shifting sowing of camelina from spring to autumn enhances the oil quality for bio-based applications in response to temperature and seed carbon stock. *Ind. Crop. Prod.* 137: 66–73.

172 Hemida, A. and Abdelrahman, M. (2020). Monitoring separation tendency of partial asphalt replacement by crumb rubber modifier and guayule resin. *Constr. Build. Mater.* 251: 118967.

173 Ren, X. and Cornish, K. (2019). Eggshell improves dynamic properties of durable guayule rubber composites co-reinforced with silanized silica. *Ind. Crop. Prod.* 138: 111440.

3

Environmental Issues and Priority Areas for Ecological Engineering Initiatives

Sanchayita Rajkhowa[1], *Nazmun Ara Khanom*[1], *and Jyotirmoy Sarma*[2]

[1] *Department of Chemistry, Jorhat Institute of Science & Technology (JIST), Jorhat, Assam, India*
[2] *Department of Chemistry, Kaziranga University, Jorhat, Assam, India*

3.1 Introduction

What's the use of a fine house if you haven't got a tolerable planet to put it on?

Henry David Thoreau.
Familiar Letters of Henry David Thoreau, *ed. Franklin Benjamin Sanborn,*
©1894, Houghton Mifflin Harcourt.

The term *ecological engineering* was coined by Howard T. Odum as early as the 1960s, and thus he is regarded as the father of the field [37]. He suggested ecological engineering for "those cases in which the energy supplied by man is small relative to the natural sources, but sufficient to produce large effects in the resulting patterns and processes" [37]. Since then, an array of definitions of ecological engineering has been proposed by many researchers. Odum has proposed several definitions, such as "the management of nature. . ., an endeavour with singular aspects supplementary to those of traditional engineering" in 1971 [38] and "the engineering of new ecosystem designs. . .that use systems that are mainly self-organizing" in 1983 [39]. His contemporary, Ma Shijun, is famous as the "father of ecological engineering in China" for presenting his concept and ideology regarding ecological engineering [26, 27]. Teal also presented a definition of ecological engineering: "to use ecological processes within natural or constructed imitations of natural systems to achieve engineering goals" [53]. In another report, ecological engineering is defined as "the design of sustainable ecosystems that integrate human society with its natural environment for the benefit of both" [34]. "The practice of joining the economy of society to the environment symbiotically by fitting technological design with ecological *self design*" is another definition offered by Odum [40].

The basic idea of all these definitions is to create new designs to restore the ecosystem with sustainable development and help achieve harmony and integration of human society with Nature. The principles of ecological engineering combine the themes of basic and applied science to implement designing, contrasting, and restoring the ecosystem [49]. Engineering is a discipline that studies and solves technological problems. On the other hand, environmental engineering helps

Handbook of Ecological and Ecosystem Engineering, First Edition. Edited by Majeti Narasimha Vara Prasad.

associate human society with the environment. This technology comprehends half of the interface while the other half is contributed by the self-organizing ecosystem to complete the environment. Merging these two interfaces leads to ecological engineering, which uses natural resources with simultaneous economic growth for societal advancement. Figure 3.1 represents the exchange of materials and services in the ecological engineering domain.

The International Ecological Engineering Society (IEES) was founded in 1991 to conduct ecological engineering activities worldwide to promote the exchange of ecological engineering problems between scientific and educational organizations, private enterprises, non-governmental and

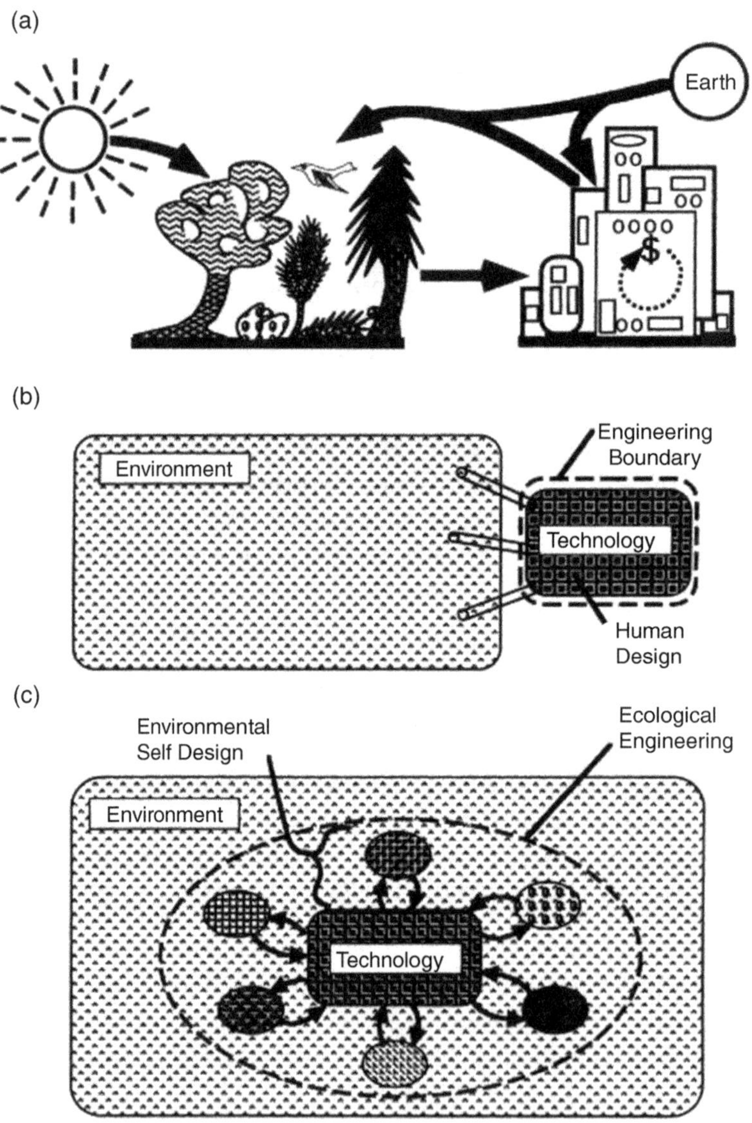

Figure 3.1 Sketches of the exchange of materials and services in the ecological engineering domain. (a) Unified system of environment and technology; (b) traditional boundary of environmental engineering design; (c) boundary of ecological engineering design. *Source:* Odum and Odum [40]. © 2003, Elsevier.

governmental organizations; and to facilitate cooperation among engineers, ecologists, and other scientists. *Ecotechnology* is used as a synonym for *ecological engineering* in the eastern part of Europe, defining it as a technological means to understand ecosystem management–based ecology with minimal cost and limited harm to the environment [50]. A number of sub-disciplines, synonyms, and amalgamations of the two fields, as well as fields similar to ecological engineering are found in literature, such as industrial ecology, engineering ecology, agroecology, restoration ecology, reclamation ecology, synthetic ecology, bioengineering, ecotechnology, and nature engineering; aquatic, terrestrial, and wetland restoration; biomanipulation; habitat reconstruction and rehabilitation; and so on.

The common aim of research by various groups with different backgrounds is to attain and promote the restoration[1] and management of the ecosystem. Advances in engineering and environmental sciences during the twentieth century have made an immense contribution to ecological principles. A holistic approach toward the restoration and management of the biosphere with advanced engineering problem-solving skills has led to the synergistic co-existence of Nature and developed societies. In the 1960s, contradictory statements were raised in scientific communities on introducing a similar discipline, especially when environmental engineering was already well established. This new discipline integrates science and technology, deals with sustainability and self-designed ecosystem; and serving the purpose of remediation and prevention of environmental issues and also addressing alarming global problems such as pollution and protecting the environment.

Ecological engineering envisages the practice of engineering methods and designs for ecological sciences, bridging these two disciplines with the interpretation of complex, uncertain, and variable natural systems. This approach can result in a unique archetype for designing engineering methods. Human beings are integral components of the ecosystem, providing the initial elements and composition and influencing the environment significantly. The self-organizing ecosystem eventually takes over these changes and looks for the best route to adjust to artificial modulations imposed on it.

The approach of ecological engineering can be briefly summarized as (i) improvement of new or existing infrastructure with modern complex building models but using environmentally benign materials and ecofriendly designs, referred to as a “hard” approach[2]; (ii) incorporation of a “soft” approach, i.e. the replacement of buildings and infrastructure with natural habitats such as salt marshes, mangroves, or oyster reefs; (iii) designing hybrid ecological engineering where natural habitats or vegetation are made to coexist with built infrastructure[10]. The approach of ecological engineering may differ with the environment, geographic locations, resources, requirements etc., but the common goal of all the approaches is to restore the natural habitats with the best possible infrastructural design for the benefit of the environment with the progress of development.

The field of ecological engineering has developed immensely within the research and academic dimensions to achieve higher endeavors for environmental sustainability and mankind. Several essential steps are proposed by researchers as part of ecological engineering activities [2, 31, 33, 35, 50, 60]:

- Propose principles.
- Design processes and techniques.
- Implement the techniques.
- Organize research archives.
- Create international societies.
- Include ecological engineering in academic curricula.

Proposing principles is considered the most significant step as the foundation of ecological engineering will be established upon it. These principles must connect ecological theory with ecological practices on real ground, which has been a challenge to researchers for years. The primary principles of ecological engineering proposed by Ma [28] were designed to formulate species symbiosis, cycling, and regeneration, and harmonize the ecological structure with its function. Later, 12 commandants or guidelines were formulated as principles of ecological engineering [17]:

1) Ecosystem structure and functions are determined by the forcing functions of the system.
2) Homeostasis of ecosystems requires accordance between biological function and chemical composition.
3) It is necessary in environmental management to match recycling pathways and rates to ecosystems to reduce the effect of pollution.
4) Ecosystems are self-designing systems. The more one works with the self-designing ability of Nature, the lower the cost of the energy to maintain that system.
5) Processes of ecosystems have characteristic time and space scales that should be accounted for in environmental management.
6) Chemical and biological diversity contribute to the spectrum of buffering capacities and the self-designing ability of ecosystems.
7) Ecotones (transition zones) are as important for ecosystems as membranes are for cells.
8) The coupling between ecosystems should be utilized to the benefit of the ecosystems in the application of ecotechnology and in environmental management of agricultural systems.
9) The components of an ecosystem are interconnected and interrelated and form a network, which implies that direct as well as indirect effects are important.
10) It is essential to realize that an ecosystem has a history when applying ecotechnology and environmental management in general.
11) Ecosystems are most vulnerable at their geographical edges.
12) Ecosystems are hierarchical systems, and all the components forming the various levels of the hierarchy make up a structure that is important for the function of the ecosystem.

Based on these principles, a few ecotechnology rules can be derived for proper management of the environment and ecosystem. These rules are summarized in Table 3.1. The terms *ecological engineering* and *ecotechnology* are used interchangeably, although the former defines the creation and restoration of the ecosystem, while the latter describes ecosystem management.

3.2 Basic Concepts of Ecological Engineering

In addition to the 12 principles, some basic concepts of ecological engineering are involved in solving environmental issues. The following concepts thus individually or collectively explain the benefits with an engineering approach:

1) It is based on the self-designing capacity of the ecosystem.
2) It can be used as an acid test of ecological theories.
3) It depends on system approaches.
4) It conserves non-renewable sources of energy.
5) It supports biological conservation.

Table 3.1 Rules of ecotechnology.

Rule no.	Rule	Brief explanation
1)	Minimize energy waste.	Limit fossil energy utilization, thus reducing global warming; find new energy sources with promising efficiency.
2)	Recycle.	Recycle and reuse materials, synthetic or natural; reduce environmental degradation.
3)	Retain all kinds of structures.	Natural structures and biodiversity persevere.
4)	Consider long-term horizons.	Sustainable life will be realized with long-time evaluations.
5)	Do not forget that humankind is dependent on many organisms and that their loss may lessen our ability to survive in the changing environment.	Continuously replace food sources and medications.
6)	Consider economic dynamics.	Incorporate dynamic ecosystem models in conventional practices.
7)	Nature is a teaching ground for handling complex systems. In particular, the ability to survive by adapting to changing conditions can be learned only from Nature.	The preparation of new technological models should be adopted from principles of Nature and based on adapting to continuous change in the environment.
8)	People, as part of the ecosystem, are directly dependent on solar energy and indirectly dependent on energy stored in coal and oil as well as vegetable and animal food.	The need for energy for growth and reproduction is inevitable: we are dependent on the environment.
9)	People are sensitive to external inputs, particularly solar radiation and the supply of minerals from the earth.	Environmental impacts can be both beneficial and detrimental. Extraction and use of minerals from the earth's crust is for human welfare, while exposure to harmful UV radiation is the root cause of diseases such as cancer.
10)	Manage the environment as an interconnected system, not as isolated subsystems.	Every great discovery made by humans has come with a price to be paid by the environment sooner or later. A systematic and detailed investigation must be carried out with products having multi-mediated feedbacks so that negative environmental impacts can be minimized.
11)	Evaluate the available management options simultaneously.	The various routes for solving environmental issues must be thoroughly investigated and assessed.
12)	Include secondary effects elsewhere.	A global system of evaluation is necessary to evaluate the output measures of environmental management practices.
13)	Do not exceed the ecosystem's homeostatic (assimilation) capacity.	The environment should not be deteriorated to a level from which it cannot recover or be restored to its original state.

(*Continued*)

Table 3.1 (Continued)

Rule no.	Rule	Brief explanation
14	Consider ecosystem self-adaptation to management strategies.	Resistance to chemical treatment by organisms (flies, parasites, super-bugs) is a well-known problem. Introducing new species (plants or animals) to a region can be helpful. However, the new species' ability to adapt to the environment must be considered.
15	Evaluate the socioeconomic environment.	Economic welfare bodies and environmentalists should work together to achieve common goals.
16	Evaluate all possible human uses of the environment.	While designing a model or project, one must consider the maximum utilization of environmental resources with as little impact as possible.
17	Base measures on ecosystem principles.	Ecosystem principles play a vital role in the successful launch of environmental management models.

These concepts have been discussed in detail by Mitsch and Jørgensen [34]. Further, four philosophical Aristotelian causes have been suggested by Ulanowicz [58, 59] to argue the change-causing activities of engineering, as follows:

- Formal cause – the essence or nature of a thing
- Material cause – the matter
- Efficient cause –the source of change
- Final cause – the goal or purpose

In general, the efficient causes of traditionally practiced engineering models are based on Newtonian-mechanical approaches in the form of machine, design, and method. Societal as well as natural demands influence these approaches by material and formal causes. Material causes arise due to dependence on natural resources in terms of the extraction of raw materials and energy, while formal causes are based on the actual extraction of those materials from nature to meet societal demands. In this way, anthropocentric engineering design reflects societal norms: safety, welfare, and quality of life. Although the final cause may be a replication of societal overlap with nature, it primarily focuses on protecting the environment with nature-preserving services and designs. Incorporating both organized simplicity and disorganized complexity in the four Aristotelian causes make traditional engineering designs advantageous. On the other hand, the Newtonian worldview covers the controlled development of elements, design of systems, and maintaining controlled conditions to serve and preserve societal needs. The ecological engineering worldview, design, system, and modeling have been successfully addressed by the Aristotelian four causes' scheme. However, the Newtonian approach has failed to reveal the *nature of nature*, indicating that it is an insufficient tool for designing the ecosystem [19].

Nature represents a self-organizing, self-sustaining system up to an exploitation threshold. Thus, any new approach must formulate a scheme to minimize external influence and also fulfill societal demands. Newtonian knowledge was found to be insufficient, and engineering science proved to be sufficiently efficient in developing the field of ecological engineering in terms of design concepts and management of the ecosystem. Gattie and Foutz [18] recognized the amalgamation of systems with engineering ecology as "the development, synthesis, and insight-building of theory and principles from the fields of general systems science, systems ecology, and engineering, as they apply to making complex, holistic ecosystems, tractable study units of nature." The foundation of traditional engineering has been built on the principles of engineering sciences; therefore, these principles must also be rigorously followed in the discipline of ecological engineering.

Before applying engineering concepts and theory to the ecosystem, the distinctive perceptions of ecosystem ecology and system ecology should be considered. The former is a discipline of measuring, quantifying, and tracing the flow of energy and materials via biotic as well as abiotic media of an ecosystem, while the latter defines the science of exploring the properties and functions of the ecosystem, which are known to mediate the energy and material flow within the entire system. Based on the principles and characteristics already mentioned, the following four proposed domains are relevant to the development of ecological engineering in terms of these two ecologies [19]:

1) Core system ecology theory
2) Ecological network analysis
3) Ecological modeling
4) System science

3.3 Practice and Implication of Ecological Engineering

A Newtonian approach alone may not be sufficient in formulating the design and principles of ecological engineering while framing a sustainable system. A balance between the domains of ecology, engineering, and design proposes a sustainable system model including an open living ecosystem that coexists with societal development. Traditional ecological knowledge[3] (TEK) and practices may prove to be revolutionary alternatives in generating thought-provoking ideas for managing and restoring the environment by designing sustainable ecological engineering methods [2]. TEK has been practiced by indigenous and local populations to meet socioeconomic requirements since ancient times. With the integration of new techniques and concern about the environment, these practices have evolved over the years to ensure ecosystem sustainability and conservation. A World Commission on Environment and Development [62] report stated, "Local communities are the repositories of vast accumulations of traditional knowledge and experience that links humanity with its ancient origins. Their disappearance is a loss for the larger society, which could learn a great deal from their traditional skills in sustainably managing very complex ecological systems." For more complex ecosystems with living and non-living sections, a Newtonian worldview should embrace an organic approach to extend its boundaries. Traditional engineering, however, may not be successful in a natural system that depends on controlled distribution imposed on the environment and emergent behavior of the system during ecosystem management and design.

The practice of ecological engineering has been observed in various ecosystems with self-design models and sustainability. Several large projects around the world have been designed based on these concepts: in the eastern part of the United States, restoration of salt marshes on Delaware Bay [45, 54]; river restoration in Jutland, Denmark, with farm and civilization protection [44]; reduction of nitrogen loading in the Gulf of Mexico [36]; restoration of Southeast Asian tropical mangroves [3]; and Mississippi River delta restoration [11, 12]. Mitsch [32] showed that well-designed ecological engineering following oil spills in the Gulf of Mexico is more important than a traditional engineering approach that resulted only in wasting money and resulted in great harm to the ecosystem.

Practical approaches to ecological engineering are supported by programs and academic discussions. North American universities have developed several such programs: a separate environmental and ecological engineering department has been set up at Purdue University, Oregon State University has renamed its agricultural engineering department to biological and ecological engineering, and the University of Florida offers graduate-level courses on ecological engineering. Other universities, such as the University of California Davis, Ohio State University, and McGill University, have special programs designed for students in this discipline.

3.4 Priority Areas for Ecological Engineering

The world population is growing rapidly and is expected to reach 10 billion by 2050 [5, 25]. With this accelerated population growth, exploitation of natural resources, fragmentation of land and water bodies, and extensive damage to ecosystems are inevitable consequences of urbanization and infrastructure development as a part of modern civilization. The prevailing processes of using non-renewable natural resources and allochthonous energy inputs for long periods of time impose a future threat to the environment as a consequence of the exhaustion of these resources. Such resources are considered successful providers of energy and services due to their high calorific value.[4] Urbanization, on the other hand, is linked to artificial floods, drying up of water bodies, soil erosion, and pollution of all strata of the ecosystem; it ultimately leads to imbalances in the ecosystem, agricultural outputs, and ability to maintain the biodiversity of the planet. A balance between the natural ecosystem and ecosystem engineering services must be retained for the vitality of mankind. It is vital to identify key areas of conservation priorities and meticulously design ecological engineering services with ecological principles and rules to develop and manage structures that effectively enhance industrialization with a reduced impact on the environment.

Xiao and Xiao [63] mathematically identified potential areas and hotspots/coldspots based on the amount of change in ecosystem services (i.e. provisioning, regulating, cultural and supporting services to human obtained from aquatic ecosystems, forest ecosystems, grassland ecosystems and agro ecosystems), assuming 100% conversion of all non-forest land into forest. The ecological engineers, planners, and managers establish a monitoring program that objectifies decisions, ensures target fulfillment, and addresses social needs. For any restoration project to be successful, there are three primary stages of an effective adaptive management plan: (i) a clear vision and goal statement, (ii) a conceptualized design or model, and (iii) a definite framework [55]. The target or goal statement is a means to evaluate the system's practical applicability that can be assessed with the project's performance criteria. A properly designed model helps in devising a target-oriented project based on ecological science. A project framework is based on a system-development matrix by incorporating the knowledge gained through the previous two ingredients required for a project. During the process of implementing ecological engineering models and then developing and restoring a particular region, plausible outcomes must be foreseen and required measures undertaken to correct problems.

The term *restoration* is defined as any process that focuses on returning a system to its original, pre-existing situation. The Society for Ecological Restoration has defined *ecological restoration* as "the process of assisting the recovery of an ecosystem that has been degraded, damaged, or destroyed" [48]. The aim of ecological restoration is to replicate structures with reference to natural ecosystems as a basic model, maintaining biodiversity, functionality, and other parameters.

Ecological engineering involves creating and then restoring sustainable ecosystems, thus benefiting both humans and Nature. Some of the major areas that require the attention of ecological engineering projects for development and restoration are discussed next.

3.4.1 Coastal Ecosystem Restoration

Protection and preservation of coastal regions were traditionally meant to safeguard against flooding. However, incomplete knowledge about the state of the system with this traditional approach often impacted the local ecology negatively and caused unpredicted outcomes that eventually affected the ecosystem more significantly. Incorporating ecological services into coastal ecosystem protection and restoration, due to the need for novel, cost-effective, sustainable designs for solving problems like rising sea levels due to climate change and minimizing the anthropogenic impacts on these ecosystems, has attracted the attention of many researchers [13].

To fulfill the objectives of coastal ecology restoration, the methods used should be selected based on the paradigm of a few species (such as mussel beds, oyster beds, and vegetation) that can participate in the enhancements with their ability to modify/save the physical structure of the environment; the construction of dams or dikes also enhances ecosystem' biodiversity. In addition, the unique ability of these species to trap and stabilize sediments in intertidal areas leads to soil elevation and attenuation of tides; they are also suitable for construction of low-cost, sustainable dams. For example, marram grass (*Ammophilaarenaria*) can trap sand, hence building protective sand dunes that invigorate dikes in sandy coastal regions. Dams or dikes constructed with reef-forming shellfish species, coral reefs, wetlands, or mangrove forests are self-sustaining and require low maintenance, minimum reinforcement, and less financial assistance.

Complex engineering designs often find it impossible to incorporate ecosystem engineering species into a system's local demographic conditions. In such cases, traditional designs can be revised to enhance inhabitability by simply modifying ecological structures with better ecological values for coastal protection purposes. This may not result in a cost-effective approach, but mitigation of engineering impacts on ecology is definitely realized with acceptance from the local community.

The primary goal of coastal protection is to protect the land from seasonal floods and erosion of shorelines. An occasional or annual flood threatens a large geographic scale with mostly temporary damage, while the erosion of shorelines is very site-specific and has a greater impact on the entire coastline [61]. Groyns and revetments are traditional protective measures that minimize the threats of shoreline erosion; dams are constructed to prevent floods and protect the mainland on a larger scale, whereas breakwaters are effective against both floods and erosion on a smaller scale. A comprehensive and integrated assessment and evaluation of the coastal ecosystem for management, maintenance, stabilization, and protection has become the primary focus of recent researchers seeking to develop innovative methodologies [51]. Weight of evidence (WOE) [1], ecological risk assessment (ERA) [9], and the ecosystem services approach (ESA) [57] are among several well-designed assessment models that emphasize pollution rather than the validation of ecosystem integrity; a few integrative approaches primarily focus on the assessment of ecosystem health. Nevertheless, an integrated ecosystem health assessment approach (IEHA) in combination with ecological indexes results in better evaluation of community structures at a macro level as well as biodiversity measures of ecological systems [14, 29].

According to a report by Brown and McLachlan [7], more than 80% of sedimentary shorelines are lost due to construction of artificial structures in coastal areas. With the replacement of coastal marshes by brick walls, benthos are immensely effected on either side of the wall [47], although the exact cause for the adverse effects on adjacent habitats is difficult to pinpoint. As an example, natural rocky shores of Sydney Harbor are affected by the disintegration of the shore due to the infrastructure of intertidal seawalls and urbanization [20]. A case study by Burt and Bartholomew [8]discussed coastal development and modification in the Arabian Gulf by encouraging colonization of fish and corals to enhance ecological benefits, integrity, and sustainability of modern urban development. Similarly, a case study in the Jiangsu coastal area of China based on a dimensionless polygon area method reflected the structure, health, and function of an ecosystem that greatly depends on temporal[5] and spatial[6] changes. The results of this assessment may be applicable for policy makers as a reference for the planning and development of coastal areas [52].

3.4.2 Mangrove Restoration

Mangrove forests are integral parts of coastal ecosystems of ecological importance; 69 species of plants are referred to as mangroves. Mangroves cover 146 530 km of the world's tropical shorelines [16]. Reports witnessed a significant decline in the area of mangroves from 198 000 km in 1980 to 157 630 km in 1990, and it continues to decrease at a noticeable rate [16]. This loss is taking place primarily in countries like the Philippines, Thailand, Vietnam, Malaysia, and the US (Florida and Puerto Rico)[24]. A large re forestation program for the restoration of anthropogenic degradation caused to mangrove forests by planting *Sonneratia Apetala* species over an area of 1600 km^2 on the accreted muddy shores of Bangladesh was successfully implemented with the aim of providing sufficiently raised land for accretion of sediments to support agriculture and vegetation. It has been estimated that about 800 km^2 of land has been recovered as a result of these plantings after a loss of almost 50% caused by cyclones and insect outbreaks [4]. However, several similar attempts have failed to restore the mangroves in other countries. Nevertheless, a thorough study of the habitants and meticulous projection of ecological engineering projects will lead to successful restoration of mangroves, ensuring the coexistence of modern civilization with Nature.

Mangroves are intertidal forests along the shorelines of tropical regions, submerged under seawater most of the time and physiologically adaptable to saline water. The trees have special adaptations in their roots that let them survive in soft muddy soil and transport oxygen from the atmosphere to their roots in anaerobic conditions. Reproduction is supported by the unique nature of their seeds, which can float on water and are transported to other sites.

Mangrove forests naturally stabilize shorelines by reducing tidal currents and wind energy hitting the shore. They provide food and shelter to the coastal fish and shellfish colonies; habitat for various species of sea birds as well as wading birds; and timber, medical products, and useful chemicals (Figure 3.2). In spite of being home to a wide range of flora and fauna, mangroves are often considered worthless swamps and have been destroyed by hydrological impacts over the years. A major reason for mangrove die-backs is annual hypersalinity variations caused by rainwater and freshwater flow. Although mangrove forests are known to undergo self-repairing or secondary succession every 15–30 years; they can perform this self-sustenance only if the tidal hydrology is not disturbed and the flow of floating seeds is not interrupted.

The need to restore mangroves was only been realized toward the end of the twentieth century. Traditional ecological restoration of these forests initially aimed at rehabilitation or management of natural resources such as planting timber-producing trees. However, this approach is not the same as an ecological restoration process. For ecological restoration of mangrove forests, several aspects must be taken into consideration: a cohesive ecosystem, alternate stable state of the ecosystem, and development of a new ecosystem once the ecological threshold is crossed. Although a complete restoration of mangroves may not be practically viable, the objective is to minimize the difference between the undisturbed (may be naturally disturbed) and restored states of the

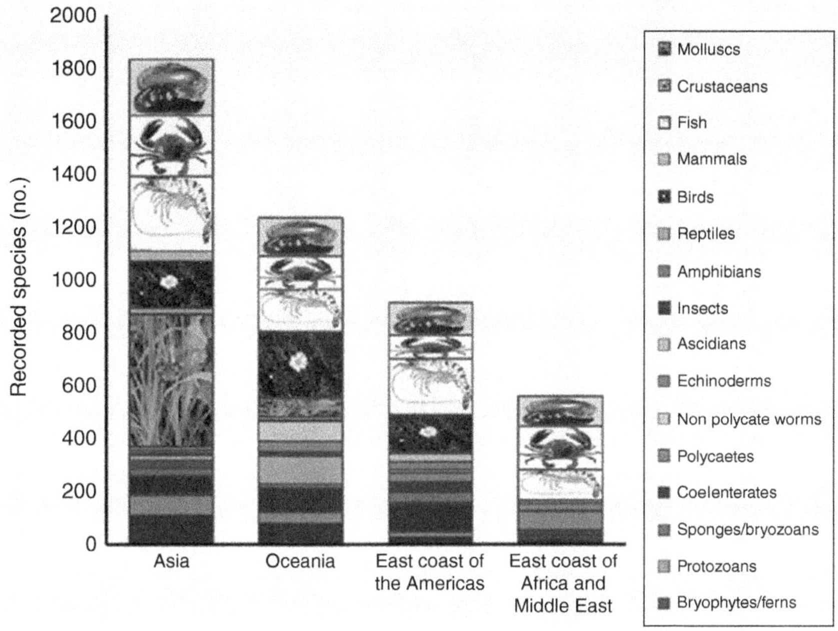

Figure 3.2 Schematic representation of the geographical distribution of mangrove biodiversity. *Source:* Adapted from Biswas et al. [3]. © 2009, Springer Nature. *See colour plate section for colour representation of this figure.*

ecosystem that correlate to the principle of [3, 48]. Common manipulation practices exercised during restoration to revitalize the ecosystem primarily include the following:

- *Structural manipulation*: Tree planting and hydrological engineering designs
- *Compositional manipulation*: Seeding and planting numerous species, enhancing diversity and habitat recovery

In addition to trees, land, and water, humans are considered the fourth component of the mangrove ecosystem due to their undeniable interference in most parts of Southeast Asian countries. Therefore, ecological engineering principles can be further manipulated by incorporating significant social and economic issues related to coastal environments and ensuring active participation by local communities along with supporting their livelihoods.

3.4.3 River and Wetland Restoration

Rivers and other bodies of water (ponds, lakes, streams, waterfalls, etc.) are source of freshwater and support human and animal life on earth. A clean, self-sustaining natural water system plays a vital role in ecological and socioeconomic development. With increasing population and thereby an increased pollution load on the environment, the sustainability of these natural services is of great concern. Thus, restoration of rivers and streams has attracted the attention of ecologists worldwide. This field is a booming enterprise and has received huge investments; it is reported that the US has spent billions of dollars on river restoration [30, 41].

Even though the importance of river restoration is understood, there is a gap between this consensus and the successful implementation of projects, which becomes evident from the rapid deterioration of freshwater sources. The restoration method selected must be based on several conditions regarding the particular river, such as geographic location, length/size, area covered, extent and nature of existing damage, stakeholder requirements, and the goal of the process. Palmer et al. [42] have proposed five standard criteria for monitoring the successfulness of ecological river restoration:

- A dynamic ecological endpoint is identified in advance and used to guide the restoration.
- Ecological conditions of the river are measurably enhanced.
- The river ecosystem is more self-sustaining than prior to restoration.
- Implementing the restoration does not inflict irreparable harm.
- Some level of both pre and post-project assessment is conducted and information made available.

Questions may arise, such as what kinds of indicators to select that will be significant, repeatable, and reasonable for successful project evaluations. However, indicators may vary from the re-establishment of a species to multispecies community restoration, depending on the predefined goals. There are two perspectives for selecting suitable indicators: one focuses on improving water quality as compared to the pre-restoration state (i.e. moving away from the degraded state), while the other demonstrates how close the water quality is to that of a reference site (i.e. approaching the predetermined desired state) [42]. In this context, the chosen indicators must be easy to measure; be susceptible to changes in the system; display clear, predictable responses to such changes; and also be integrative. Jansson et al. [22] discussed the outcomes of implementation and practice of river restoration programs at the international and national levels to evaluate the understanding, accomplishments, and advancement of river restoration services.

Recently, a paradigm shift has occurred, such that river restoration design and construction imitate natural structures such as channelized streams without interfering with biomes,[7] i.e. biomic restoration is greatly harnessed by biogeomorphic agents as *nature's river restorers* [64]. This approach is referred to as *biomic river restoration*. Biomic rivers are known to be resilient to adverse effects and are true examples of sustainable management or restoration solutions that reflect the future of river restoration projects. Figure 3.3 shows a section of Nippersink Creek (Illinois, US) being filled in to allow efficient farming by excavating a channel, contouring the banks, anchoring mats to control erosion, adding a mulching blanket and seeds to encourage native vegetation, and planting aquatic plants. The project is a perfect example of river restoration with multiple goals such as reduction of flooding downstream; improved water quality; habitat for fish, wading birds; and waterfowl; and involvement of youngsters in the process.

Wetlands are shallow-water ecosystems that include marshes, bogs, vernal pools, and seagrass beds. These are also described as functions and services with broad ecological interest that serve as melting pots for public and private services promoting both consumptive and non-consumptive assistance. Wetlands are considered a perfect example that integrates the approaches of two diverse fields – natural science and social science – thus enabling the economic validation of policy development and research [56]. For economic validation, however, ecosystems are subdivided into small sets that can deliver functions and services since it may not be possible to include the entire ecosystem during validation with standard methods.

Wetland restoration focuses on the construction of water conservation and protection of riverine wetlands, as they provide shelter to aquatic animals and plants. Water quality improvement, biodiversity maintenance, flood abatement, and socioeconomic uplift are among the additions to the ecosystem [6]. Another important future aspect of wetland restoration is its contribution to controlling global climate change. About 6% of the world's land is covered by wetlands, while approximately 12% of the global carbon pool is known to exist in these wetlands; thus they play a

Figure 3.3 A 2.6 km section of Nippersink Creek (Illinois, US) is filled in to make farming more efficient. *Source:* Reproduced with permission from Zedler, Joy B. (2000). "Progress in wetland restoration ecology." *Trends in Ecology & Evolution* 15: 402–407. ©2000, Elsevier.

vital role in the global carbon cycle [15]. Climate change must also be addressed during the implementation and practice of restoration projects, with future ecosystem management policies as an inevitable domain of climate change and mitigation processes.

On the other hand, there has been a decline in several ecosystem services resulting in severe alteration or complete disappearance of wetlands in recent years. For instance, alteration of habitat by a multistage dam constructed on the Min River in China led to a rapid decrease in the population of *S. Prenanti*[8] [46]. A very common example is floodplains, which can include one or more types of wetlands, primarily including submerged grasslands, wet meadows, forests, woodlands, and shrublands. Degradation of these ecosystems can be correlated with the deterioration of freshwater biodiversity due to habitat alteration, species migration, pollution, and flood-control mechanisms. The disappearance of natural floodplains and rapid extinction of aquatic and riparian species pose a worldwide threat to the biodiversity of these systems. Perseverance of existing floodplains and restoration of hydrological ecosystems, riparian vegetation, and sediment transportation require urgent attention to retain ecological integrity and sustainability.

3.4.4 Ecological Engineering in Soil Restoration and Agriculture

Agriculture is influenced by an array of biotic and abiotic stresses; a strategic technique is required to address plant health and issues that affect productivity. The integrity and conservation of the agro-atmosphere is critical for sustainable agriculture. Extensive exploitation of ecosystems to enhance productivity can affect agro-ecosystems via erosion of soil, contamination of water, loss of biodiversity, pest issues, and disruption in the usual flow of the surrounding ecology in order to allow safe food manufacturing. The role of ecological engineering is to assimilate soil and pest management schemes with conventional agricultural practices that benefit both the ecosystem and productivity. It is a combination of agricultural knowledge, crop economics, and ecology applied to development and restoration to retain an overall sustainable agricultural ecosystem that has been disturbed by human activities.

The prime focus of management schemes is to reduce the use of chemical pesticides and overuse of fertilizers and agricultural chemicals that cause environmental pollution that further can enter the food chain; and also to minimize drug resistance and resurgence in pests and bugs over time. Replacing these detrimental chemicals with biofertilizers such as mycorrhiza improves the soil health, helps plants to uptake micro/macro nutrients from the soil, and supplies these nutrients at an optimum level that inhibits the growth of pests. Thus ecological engineering uses traditional vegetation management practices to control the pests without damaging soil health. A collective eco-friendly approach adopted by the farming community can successfully implement this technology to manage the pest population and build up the biota by enriching the organic matter content in the soil.

To support a good agricultural system, the soil ecosystem must support the growth and sustainability of the land. However, soil erosion interrupts agriculture in many parts of the world. Although soil erosion is a naturally occurring process, it is accelerated by anthropogenic interventions into Nature and landscapes. Improper management of agricultural land also contributes to soil erosion.

Land consolidation is a tool for creating sustainable rural agriculture and improving the efficiency of cultivation, which ultimately helps develop the rural cultivation culture in countries like the Czech Republic (CR) [43]. Land consolidation is a process of systematically reallocating or redistributing fragmented agricultural land and properly arranging the shape and size of these lands in rural areas [23]. At the same time, land consolidation contributes to the improvement of environmental status, protects soil, provides ecological stability, helps with water management, and controls flooding and erosion. Soil degradation and the urgent need to conserve soil have been the most-discussed topics in the last few decades. Land consolidation can improve soil stability when proper management policies are implemented. However, soil management practices (in terms of agricultural soil) such as till age[9] may have negative consequences on the soil biota since they can lead to soil decomposition and the mineralization of organically bound nutrients. Therefore, these practices must be devised wisely to minimize the adverse impact on soil quality while providing agricultural benefits. In this context, tillage is often combined with other practices like mulches and cover crops, whereas enrichment with organic wastes, compost, and biochar favors soil fertility and biota. In addition, conservation of soil's natural and biological efficiency improves and maintains the soil's organic matter content, thus facilitating soil preservation.

At this point, ecological engineering can play a pivotal role that combines management practices with soil biota enhancement designs to achieve maximum crop yield, improve overall service, and maintain sustainable biodiversity with maximum yield (Figure 3.4). A compromise between agricultural yield and sustainability can be achieved by appropriate manipulation of ecosystem processes. Figure 3.4 depicts a direct correlation between land-use intensity and yield. It suggests that yield is directly proportional to the magnitude of land-use intensity. The yield is maximum under intensive management when the external source inputs are highest. Soil ecological engineering in harmony with naturally occurring biological processes is known to replace external inputs by either maintaining yields with fewer external inputs or increasing yields without simultaneous increment in external inputs.

Several other fields have developed ecological engineering practices, such as desert forest restoration, ocean and aquatic life restoration, forest restoration, urban reconciliation of living roofs and walls, all kinds of riparian ecosystem restoration, arctic megafaunal restoration, etc. With the aim of mitigating past adverse impacts and creating novel projects to benefit sustainable ecological growth in the near future, improved environmental policies using ecological engineering approaches can create awareness at the local, regional, and international level regardless of the field where these policies have been adopted.

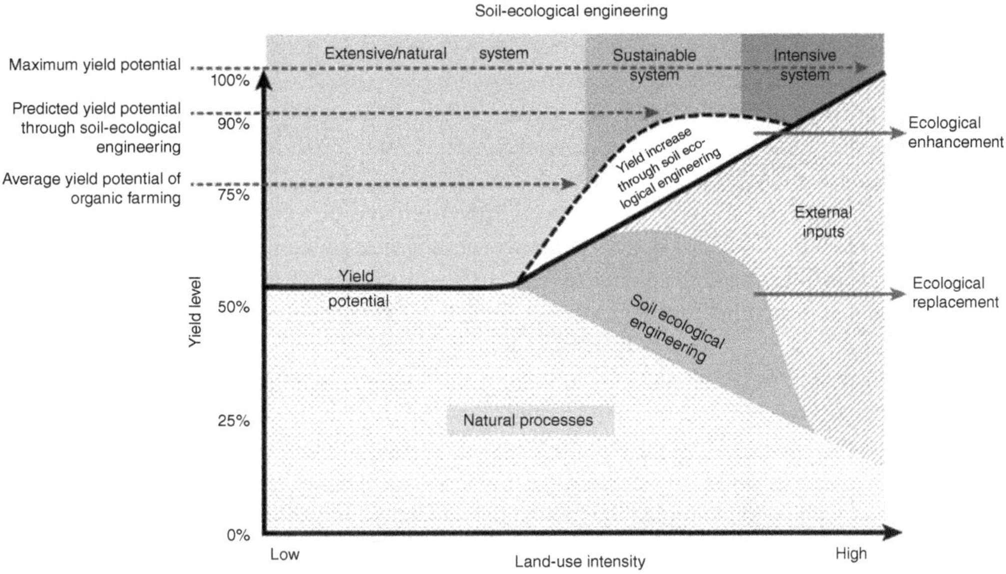

Figure 3.4 Conceptual model showing the contribution of external resource inputs and natural biological processes to an ecosystem function (yield), depending on land-use intensity. *Source:* Bender, S.F., Wagg, C., and van der Heijden, M.G.A. (2016). "An underground revolution: biodiversity and soil ecological engineering for agricultural sustainability." *Trends in Ecology & Evolution* 31 (6): 440–452. © 2016, Elsevier.

3.5 Conclusion

The contribution of ecological engineering in designing, building, and operating a new ecosystem is noteworthy. The approach of ecological engineering designs is not as simple as it appears since such designs are proposed to apply to species that are known to evolve within a newly created ecosystem. Although new ecosystems are called by various names such as *domestic ecosystems*, *interface ecosystems*, and *living machines*, they are ultimately the outcome of creative designs incorporated with self-organized properties of the systems themselves. This approach brings about the selection of species naturally within the framework of ecological engineering designs reflecting the manmade designs through the response of natural choices. This feature makes this field a unique and intellectually motivating branch of applied engineering.

With the aim of solving environmental problems, ecological engineering addresses a subset of issues that have been created anthropogenically and then must be resolved by ecological designs. Among these designs, pollution control or treatment is the most frequent in all environmental areas, where polluted materials are considered resources. The designs are formulated in such a way that pollutants are either stabilized or broken down into useful by-products with the natural development of the ecosystem, thereby converting the problems into solutions for the successful implementation of ecological engineering principles.

In this chapter, we have discussed hard, soft, and hybrid approaches for improving new or existing infrastructure with modern, complex building models. The primary principles of ecological engineering proposed by Ma [28] were designed to formulate species symbiosis, cycling, and regeneration and harmonize the ecological structure with its function. Later, 12 commandants or guidelines were formulated as principles of ecological engineering. Based on these basic principles, rules of ecotechnology can be derived for proper management of the environment and ecosystem.

The terms *ecological engineering* and *ecotechnology* are used interchangeably, although the former defines the creation and restoration of ecosystem while the later describes ecosystem management. Nature represents a self-organizing, self-sustaining system up to an exploitation threshold. Thus, any new approach must include a scheme to minimize external influence and also fulfill societal demands. A balance between the natural ecosystem and ecosystem engineering services must be achieved for the vitality of mankind. However, it is vital to identify key conservation priorities and carefully design ecological engineering services with ecological principles and rules to develop and manage structures that effectively enhance industrialization with a reduced impact on the environment.

This chapter discussed in great detail some of the major areas those require the attention of ecological engineering projects for development and restoration. For instance, we have explained coastal development and restoration and proposed ways that ecological engineering could be applied productively to improve the ecological benefits of infrastructure and natural habitat for corals and fishes and also encourage sustainability with the growth of urbanization. A common ecological engineering approach adopted for successful restoration of mangroves practices analytical thought processes with minimum exploitation of the mangrove ecosystem. Various nonstructural measurements are applied in the development of flood hazard plans, including locally grown flood-fighting crops that help with risk management as well as controlled river engineering by ecological restoration. Likewise, we can rely more on internal regulatory methods rather than traditional agricultural practices.

Soil ecology restoration is the next step toward attaining an overall sustainable agricultural ecosystem. One of the major challenges in the near future is sustaining the human population while maintaining the integrity of the environment. Earth is a closed system, and thus it will be wise to apply human potential toward achieving this goal. Therefore, the expansion of restoration projects in terms of expectations and goals must be realized to achieve all the plausible ecological aspects. Ecological restoration and management should make both explicit and implicit efforts to predict and recommend the ecological future of a location. These attempts need to be extrapolated beyond the prediction within the boundary of scientific limitations during the implementation of ecological restoration and management projects.

Notes

1 Management of ecosystems for productivity and harmony with the economy is called *restoration*, which implies the state of the ecosystem before conditions were changed by economic development.

2 Hard approaches include novel designs for infrastructure, or modifications to an existing wall, revetment, groyne, etc. that cannot be removed and where vegetation or soft sediments cannot be added.

3 *Traditional ecological knowledge* (TEK) is ecological knowledge held by indigenous and local cultures rooted in long-term, intimate involvement with local ecosystems.

4 The amount of energy produced by the complete combustion of a material or fuel and measured in units of energy per amount of material, e.g. $kJ\ kg^{-1}$.

5 Seasonal and annual.

6 Northern, central, and southern regions.

7 The term *biome* refers to a fully integrated climatic, geologic, and ecologic system, typically at or in excess of the river catchment scale.

8 *Schizothoraxprenanti* (*S. prenanti*) is a species of ray-finned fish in the genus *Schizothorax* from the middle and upper parts of the Yangtze basin in China.
9 *Tillage* is the agricultural preparation of soil by mechanical agitation of various types, e.g. digging, stirring, and turning over.

References

1 Bebianno, M.J., Pereira, C.G., Rey, F. et al. (2015). Integrated approach to assess ecosystem health in harbor areas. *Sci. Total Environ.* 514: 92–107.
2 Bergen, S.D., Bolton, S.M., and Fridley, J.L. (2001). Design principles for ecological engineering. *Ecol. Eng.* 18: 201–210.
3 Biswas, S.R., Mallik, A.U., Choudhury, J.K., and Nishat, A. (2009). A unified framework for the restoration of Southeast Asian mangroves – bridging ecology, society and economics. *Wetlands Ecol. Manage.* 17: 365–383.
4 Blasco, F., Aizpuru, M., and Gers, C. (2001). Depletion of the mangroves of continental Asia. *Wetlands Ecol. Manage.* 9 (3): 245–256.
5 Bloom, D.E. and Bloom, D.E. (2011). 7 Billion and counting. *Science* 333: 562–569.
6 Brander, L., Brouwer, R., and Wagtendonk, A. (2013). Economic valuation of regulating services provided by wetlands in agricultural landscapes: a meta-analysis. *Ecol. Eng.* 56: 89–96.
7 Brown, A.C. and McLachlan, A. (2002). Sandy shore ecosystems and the threats facing them: some predictions for the year 2025. *Environ. Conserv.* 29: 62–77.
8 Burt, J.A. and Bartholomew, A. (2019). Towards more sustainable coastal development in the Arabian Gulf: opportunities for ecological engineering in an urbanized seascape. *Mar. Pollut. Bull.* 142: 93–102.
9 Caeiro, S., Vaz-Fernandes, P., Martinho, A.P. et al. (2017). Environmental risk assessment in a contaminated estuary: an integrated weight of evidence approach as a decision support tool. *Ocean Coast. Manage.* 143: 51–62.
10 Chapman, M.G. and Underwood, A.J. (2011). Evaluation of ecological engineering of "armoured" shorelines to improve their value as habitat. *J. Exp. Mar. Biol. Ecol.* 400: 302–313.
11 Costanza, R., Mitsch, W.J., and Day, J.W. (2006). A new vision for New Orleans and the Mississippi delta: applying ecological economics and ecological engineering. *Front. Ecol. Environ.* 4: 465–472.
12 Day, J.W. Jr., Hall, C.A.S., Yánez-Arancibia, A. et al. (2009). Ecology in times of scarcity. *BioScience* 59: 321–331.
13 Day, J.W., Psuty, N.P., and Perez, B.C. (2000). The role of pulsing events in the functioning of coastal barriers and wetlands: implications for human impact, management and the response to sea level rise. In: Concepts and Controversies in Salt Marsh Ecology (eds. M.P. Weinstein and D. Dreeger), 633–659. Dordrecht, The Netherlands: Kluwer Academic Publishers.
14 Donohue, I., Petchey, O.L., Montoya, J.M. et al. (2013). On the dimensionality of ecological stability. *Ecol. Lett.* 16 (4): 421.
15 Ferrati, R., Canziani, G.A., and Moreno, D.R. (2005). Estero del Ibera: hydrometeorological and hydrological characterization. *Ecol. Model.* 186: 3–15.
16 Food and Agricultural Organization (FAO).(2003). New global mangrove estimate.
17 Fu-liu, X. and Shu, T. (2000). Conceptual understandings of the ecological engineering issues in China and Abroad. *J. Environ. Sci.* 12: 203–208.

18 Gattie, D.K. and Foutz, T.L. (2007). Systems and engineering ecology: developing formal foundations for ecological engineering. *Int. J. Eng. Educ.* 23 (4): 683–690.

19 Gattie, D.K., Kellam, N.N., and Turk, H.J. (2007). Informing ecological engineering through ecological network analysis, ecological modelling, and concepts of systems and engineering ecology. *Ecol. Model.* 208: 25–40.

20 Goodsell, P.J., Chapman, M.G., and Underwood, A.J. (2007). Differences between biota in anthropogenically fragmented habitats and in naturally patchy habitats. *Mar. Ecol. Prog. Ser.* 351: 15–23.

22 Jansson, R., Backx, H., Boulton, A.J. et al. (2005). Stating mechanisms and refining criteria for ecologically successful river restoration: a comment on Palmer et al. (2005). *J. Appl. Ecol.* 42: 218–222.

23 Leń, P. and Noga, K. (2018). Prioritization of land consolidation interventions in the villages of Central Poland. *J. Ecol. Eng.* 19 (2): 246–254.

24 Lewis, R.R. III (2005). Ecological engineering for successful management and restoration of mangrove forests. *Ecol. Eng.* 24: 403–418.

25 Lutz, W. and Kc, S. (2010). Dimensions of global population projections: what do we know about future population trends and structures? *Science* 365: 2779–2791.

26 Ma, S. (1985). Ecological engineering:application of ecosystem principles. *Environ. Conserv.* 12: 331–335.

27 Ma, S. and Yan, J. (1989a). Ecological engineering for treatment and utilization of wastewater. New York: Wiley. pp. 185-217.

28 Ma, S. and Yan, J. (1989b). Ecological Engineering: An Introduction to Technology [M] (eds. W.J. Mitsch and S.E. Jørgensen), 365–380. New York: Wiley.

29 Mace, G.M., Norris, K., and Fitter, A.H. (2012). Biodiversity and ecosystem services: a multilayered relationship. *Trends Ecol. Evol.* 27 (1): 19–26.

30 Malakoff, D. (2004). The river doctor. *Science* 305: 937–939.

31 Martin, J.F., Roy, E.D., Diemont, S.A.W., and Ferguson, B.G. (2010). Traditional Ecological Knowledge (TEK): ideas, inspiration, and designs for ecological engineering. *Ecol. Eng.* 36: 839–849.

32 Mitsch, W.J. (2010). The 2010 Gulf of Mexico oil spill: what would Mother Nature do? *Ecol. Eng.* 36: 1607–1610.

33 Mitsch, W.J. and Jørgensen, S.E. (eds.) (1989). Ecological Engineering: An Introduction to Ecotechnology. New York: Wiley.

34 Mitsch, W.J. and Jørgensen, S.E. (2003). Ecological engineering: afield whose time has come. *Ecol. Eng.* 20: 363–377.

35 Mitsch, W.J. and Jørgensen, S.E. (2004). Ecological Engineering and Ecosystem Resturation. New York: Wiley.

36 Mitsch, W.J., Day, J.W. Jr., Wendell Gilliam, J. et al. (2001). Reducing nitrogen loading to the Gulf of Mexico from the Mississippi River Basin: strategies to counter a persistent ecological problem. *BioScience* 51: 373–388.

37 Odum, H.T. (1962). Ecological tools and their use: man and the ecosystem. Lockwood Conference on the Suburban Forest and Ecology.

38 Odum, H.T. (1971). Environment, Power and Society. New York: Wiley.

39 Odum, H.T. (1983). System Ecology. New York: Wiley.

40 Odum, H.T. and Odum, B. (2003). Concepts and methods of ecological engineering. *Ecol. Eng.* 20: 339–361.

41 Palmer, M.A., Hart, D.D., Allan, J.D., and the National River Restoration Science Synthesis Working Group (2003). Bridging engineering, ecological, and geomorphic science to enhance riverine restoration: local and national efforts. In: Proceedings of the National Symposium on Urban and Rural Stream Protection and Restoration (eds. P. Bizier and P. DeBarry). Philadelphia: EWRI World Water and Environmental Congress. Reston, VA: American Society of Civil Engineers.

42 Palmer, M.A., Bernhardt, E.S., Allan, J.D. et al. (2005). Standards for ecologically successful river restoration. *J. Appl. Ecol.* 42: 208–217.

43 Pašakarnis, G. and Maliene, V. (2010). Towards sustainable development in central and Eastern Europe: applying land consolidation. *Land Use Policy* 27 (2): 545–549.

44 Pedersen, M.L., Andersen, J.M., Nielsen, K., and Linnemann, M. (2007). Restoration of the Skjern River and its valley: project description and general ecological changes in the project area. *Ecol. Eng.* 30: 131–144.

45 Peterson, S.B., Teal, J.M., and Mitsch, W.J. (eds.) (2005). Delaware Bay salt marsh restoration. *Ecol. Eng.* 25 (special issue): 199–314.

46 Qin, L.L., Li, K.F., Li, Y. et al. (2016). A habitat similarity model based on vague sets to assess Schizothorax prenanti spawning habitat. *Ecol. Eng.* 96: 86–93. https://doi.org/10.1016/j.ecoleng.2016.04.007.

47 Seitz, R.D. and Lawless, A.S. (2006). Landscape-level impacts of shoreline development on Chesapeake Bay benthos and their predators. In: Management, Policy, Science, and Engineering of Nonstructural Erosion Control in the Chesapeake Bay (eds. S.Y. Erdle, J.L.D. Davis and K.G. Sellner), 63–70. Chesapeake Bay: CRC Publ. No. 08-164.

48 Society for Ecological Restoration (SER).(2002). The SER Primer on Ecological Restoration. SER International Science and Policy Working Group. www.ser.org/content/ecological restorationprimer.asp.

49 Sree Latha, E. and Jesu Rajan, S. (2018). Ecological engineering for sustainable agriculture: simple concept with greater impact. *Int. J. Sci. Res. Publ.* 8 (2): 123–125.

50 Straškraba, M. (1993). Ecotechnology as a new means for environmental management. *Ecol. Eng.* 2: 311–332.

51 Tang, D., Zou, X., Liu, X. et al. (2015). Integrated ecosystem health assessment based on eco-exergy theory: a case study of the Jiangsu coastal area. *Ecol. Indic.* 48: 107–119.

52 Tang, D., Liu, X., and Zou, X. (2018). An improved method for integrated ecosystem health assessments based on the structure and function of coastal ecosystems: a case study of the Jiangsu coastal area, China. *Ecol. Indic.* 84: 82–95.

53 Teal, J.M. (1991). Contributions of marshes and saltmarshes to ecological engineering. In: Ecological Engineering for Wastewater Treatment (eds. C. Etnier and B. Guterstam), 55–62. Gothenburg, Sweden: Bokskogen.

54 Teal, J.M. and Weinstein, M.P. (2002). Ecological engineering, design, and construction considerations for marsh restorations in Delaware Bay, USA. *Ecol. Eng.* 18: 607–618.

55 Thom, R.M. (2000). Adaptive management of coastal ecosystem restoration projects. *Ecol. Eng.* 15: 365–372.

56 Turner, R.K., van den Bergh, J.C.J.M., Soderquist, T. et al. (2000). Ecological–economic analysis of wetlands: scientific integration for management and policy. *Ecol. Econ.* 35 (1): 7–23.

57 Uehara, T. and Mineo, K. (2017). Regional sustainability assessment framework for integrated coastal zone management: satoumi, ecosystem services approach, and inclusive wealth. *Ecol. Indic.* 73: 716–725.

58 Ulanowicz, R.E. (1990). Aristotelian causalities in ecosystem development. *Oikos* 57: 42–48.

59 Ulanowicz, R.E. (1999). Life after Newton: an ecological metaphysic. *Biosystems* 50: 127–142.

60 Van der Ryn, S. and Cowan, S. (1996). Ecological Design. Washington, DC: Island Press.

61 Wolters, M., Bakker, J.P., Bertness, M.D. et al. (2005). Saltmarsh erosion and restoration in south-east England: squeezing the evidence requires realignment. *J. Appl. Ecol.* 42: 844–851.

62 World Commission on Environment and Development (1987). Our Common Future, 98. Oxford, UK: Oxford University Press.

63 Xiao, Y. and Xiao, Q. (2018). Identifying key areas of ecosystem services potential to improve ecological management in Chongqing City, southwest China. *Environ. Monit. Assess.* 190: 258.

64 Johnson, M.F., Thorne, C.R., Castro, J.M. et al. (2020). Biomic river restoration: A new focus for river management. *River Res. App.* 36 (1): 3–12.

4

Soil Meso- and Macrofauna Indicators of Restoration Success in Rehabilitated Mine Sites

Sara Pelaez Sanchez[1], *Ronan Courtney*[1], *and Olaf Schmidt*[2]

[1] *Department of Biological Science, Faculty of Science & Engineering, Bernal Institute, University of Limerick, Limerick, Ireland*
[2] *School of Agriculture and Food Science, University College Dublin, Dublin, Ireland*

4.1 Introduction

Soil supports life on Earth and is a significant contributor to ecosystem services [1, 2]. The Status of the World's Soil Resources report identified soil pollution as one of the main threats affecting soils and ecosystem services globally [3]. The problem is expected to worsen as the global population continues to grow and, with it, demand for raw materials such as metals and minerals [4]. Notably, the mining of raw materials is among the worst anthropogenic perturbations as almost all the components and attributes of the original ecosystem are destroyed [5]. Today, soil pollution is a global concern due to the potentially hazardous impacts of these pollutants on soil quality, crop yield and quality [6], and food safety and public health [7].

4.2 Restoration to Combat Land Degradation

Restoration allows the reestablishment of some attributes of a former ecosystem; it offers an opportunity for ecosystem habitat enhancement and an increase in the ecological value by creating new ecosystem services [8]. In recent years, the science of restoration ecology has turned away from accurate restoration to past conditions, instead of stressing future trajectories and the resilience of (socio-)ecological systems [9, 10]. Ecosystem restoration is increasingly taking place worldwide [11] as an important contribution to urgently needed responses to environmental degradation, biodiversity loss, and anthropogenic climate change.

Restoration has evolved significantly over time and has multiple possible connotations [12, 13]. *Rehabilitation* is a generic term used for ecological repair activities that aim to restore ecosystem functioning rather than the biodiversity and integrity of a designated native reference ecosystem. When rehabilitation is used for mined lands or post-industrial sites, it is sometimes called *reclamation* [13]. For ecosystem restoration in post-mining sites, the core focus is the restoration of soil processes [8]. For the purposes of this chapter, we use *restoration* (terminology in McDonald [14]), *remediation*, *rehabilitation*, *reclamation*, and *ecosystem restoration* as synonyms in the literature search. The mining restoration literature encompasses an extensive collection of terminologies for defining restoration practices [15–17].

Handbook of Ecological and Ecosystem Engineering, First Edition. Edited by Majeti Narasimha Vara Prasad.

Herein, we highlight the current gaps in rehabilitation assessment, especially the role of invertebrates as indicators of restoration success, and present future research needs in metalliferous mine tailings.

4.3 Mine Rehabilitation

4.3.1 Mine Tailings

Metalliferous mining produces waste of two distinct physical forms. Very coarse waste rock (usually 2–20 cm in diameter) consists mainly of unmineralized overburden rock, excavated to uncover the ore-body, and rock from areas within the ore-body that contain metals at concentrations below the cut-off grade. Mine tailings, on the other hand, are fine-grained (<2 mm) deposits from the final stage separators. An estimated 10+ billion tons of mine tailings are produced per year worldwide [18, 19]. Tailings management facilities are commonly constructed to store tailings. Tailings sites can be susceptible to dusting and erosion, present a pollution risk, and require some form of reclamation.

4.3.2 Rehabilitation of Mine Tailings

Historically, abandoned mine tailing sites have been prone to dusting or water erosion and are therefore a potential source of contamination to the surrounding environment [20]. The absence of environmental legislation (pre-1990s) limited rehabilitation after the termination of mining operations and post-closure monitoring programs [21]. Today in Europe, mine waste facilities and tailings disposal areas in most jurisdictions are operated under license from regulators that often necessitate the implementation of closure strategies and rehabilitation (Council Directive 2006//21/EC [22]).

Phytoremediation may be an effective means for reducing metal loading in land reclaimed after surface coal mining, such as in the UK [23]. According to Xie and Van Zyl [24], phytostabilization stands out as the best potential phytoremediation technique for reducing metal(loid) mobility in sulfidic tailings. Phytoremediation has also been proven effective for bauxite residue [25] and Pb/Zn tailings in Ireland [20].

4.3.3 The Challenge of Metal Mine Rehabilitation

Degradation and landform modifications from mining activities produce some of the most difficult landscapes to restore [16]. Self-regulation is the goal of post-mining ecological land rehabilitation, where natural processes take over the control and management of the land and govern the development of a self-sustaining system [26]. To achieve such endpoints, other challenges need to be overcome; in particular, mine spoils and mine tailings pose very harsh conditions for plant establishment and growth due to low organic matter, heavy metal contents, and other unfavorable physicochemical characteristics [27–29].

Herein, we understand the rehabilitated tailings from metal mining as novel ecosystems, and we follow the definition proposed by Macdonald and King [30]: i.e. novel ecosystems are assemblages of organisms and abiotic elements that form self-organizing systems that may not have historical precedent. Rehabilitated mine tailings that have a residue as substrate – even when it is a type of redeveloped soil or artificial soil (revegetated and amended) – cannot replicate the pre-disturbance soil that develops slowly over long periods in natural systems [31]. We can accept that a certain ecological abiotic or biotic threshold has been passed and can no longer be restored to a pre-disturbance state.

One of the primary challenges for practitioners when working with such sites is the uncertainty of long-term resilience and the evolutionary and ecological implications [32, 33]. It requires a better understanding of when and why an irreversible ecological threshold has caused an ecosystem to shift to a new stable state [34]. At the same time, mine waste requiring rehabilitation represents an exciting research field and an opportunity to determine limiting factors for soil fauna and vegetation colonization and establishment. Ultimately, we want to increase understanding of how rehabilitated sites can support keystone soil fauna species that, in turn, can accelerate soil processes and ecosystem development.

Post-mining rehabilitated sites present hostile conditions, such as lack of suitable food, adverse physicochemical conditions, unfavorable moisture conditions, and large surface temperature fluctuations [35, 36]. However, primary succession can occur [37, 38], and soil fauna communities gradually develop even in these harsh conditions [39]. In naturally and artificially revegetated grassland and wetland soils from historical Irish mine sites, Good [40] observed that some tailings sites had well-developed soil conditions to support Staphylinid assemblages. However, these areas need careful study before definitive statements can be made regarding the sustainability of residue rehabilitation strategies. More information is needed, and long-term field trials should be designed to target soil faunal colonization and activity in mine wastes and their effects on ecosystem function [41].

On top of the challenges emerging when working with novel ecosystems are tailings cover systems, which can be considered manufactured soils or so-called *Technosols* [42]. Pedogenic processes occurring in Technosols are similar to those occurring in natural soils [43]. However, the ingredients used during their construction can significantly influence their formation, the resulting soil properties, and the soil's capacity to provide ecosystem services. For mine tailings deposited decades previously, some soil-forming processes may have initiated, leading to a certain level of soil development [44]. It seems clear that in post-mining scenarios, assembling an entire geoecological system by applying the rehabilitation engineer approach (principally using machinery and fertilizer) cannot replicate the pre-disturbance geosystem landscape [26]. For successful ecosystem reconstruction, it is necessary to further understand the physical, chemical, and biological characteristics of these Technosols. Research should move from simply evaluating vegetation establishment techniques to a more comprehensive assessment of the ecological processes and developments, including below ground [45, 46]. As part of this, improved knowledge of soil fauna recolonization in post-mining sites is now considered crucial for meeting restoration objectives after mining [16] and forming self-regulating ecosystems [45, 47].

4.4 Restoration Success Assessment: Monitoring Diversity, Vegetation, and Ecological Processes

Evaluation of whether goals and objectives are achieved and assessment of the effects (biotic and abiotic responses to management treatments) of restoration activities are currently missing from many ecological restoration projects [13]. Such information will enhance conservation outcomes, aid policy implementation, and boost further investment in restoration science [48]. As part of a long-established practice, the mining industry has focused on compensation standards set for mitigation, which are often based only loosely on ecological criteria. Traditionally, assessment studies in restoration projects can be categorized according to three majors attributes they focus on: (i) diversity, (ii) vegetation, and (iii) ecological processes [48], which also correspond to the foundation of the majority of management and restoration decisions in mine site rehabilitation worldwide [15, 16, 20, 25, 45].

4.4.1 Monitoring Diversity

The use of species richness or diversity as an indicator of restoration success is a widespread practice in many cases; for example, in the late 1980s, Majer [49] suggested that local species richness of particular taxa may indicate ecosystem quality and, consequently, the success of restoration projects. Previous studies have used species diversity of some small mammals, birds, herpetofauna [50, 51], and invertebrates – especially ants, which are easily surveyed with minimal time and financial investment [45, 49, 52–55] – to evaluate restoration. However, several authors [56, 57] argue that using species alone as the indicator has several limitations since there is a strong likelihood of missing rare, cryptic, migratory, or seasonally active species. In particular, the presence of a vertebrate community in restored areas could indicate temporal habitat sufficiency or dispersion capacity for fauna but not site quality or habitat conditions to support resident fauna communities in the long term [58].

Recently, some authors have criticized the viewpoint of a passive role for fauna in restoration because animal–plant interactions frequently mediate life-history transitions that determine seedling recruitment and other plant community processes. So, some fauna groups can drive the developmental trajectories of floristic diversity and composition during revegetation [58]. For example, earthworm bioturbation activity can mediate the effect of tree species on soil development in post-mining sites [59, 60]. Less attention is placed on studies in which the only goal is to assess site biodiversity. Instead, there is an increasing trend to assess interactions of fauna with soil physicochemical properties.

In contrast to other faunal taxa mentioned, little is known about the colonization and community development of soil invertebrate species and the soil food web in rehabilitated mine sites. An initial characterization of a site using a diversity index could provide useful information about rehabilitation stages and succession. For restoration efforts to be useful for all soil fauna groups, assessments should move from using specific taxa as bioindicators to an extensive range of fauna and their functional roles in the ecosystem [16]. However, the use of too many fauna groups could be difficult to implement for restoration practitioners. Alternatively, using specific taxa that are widely studied in mine sites and restored habitats elsewhere, such as by ecosystem engineers, offers a promising opportunity to bring scientific knowledge into day-to-day rehabilitation practices for the mining industry.

4.4.2 Vegetation

Revegetation is one of the most common practices in mine site rehabilitation due to its effectiveness in stabilizing soils, pollution control, and visual improvement [46, 61, 62]. Measures of vegetation structure (e.g. ground cover, density, height, biomass, and litter) have been established as common methods for assessing restoration success [63]. Surveying vegetation cover is also widely used as an above-ground indicator for monitoring restoration success in rehabilitated mine sites [16, 45].

Some authors argue that if the vegetation structure is on a trajectory to be restored to a target, it usually means a restoration of soil fauna will follow [39, 64]. However, De Deyn et al. [65] have shown the role of below-ground diversity in above-ground plant development and succession;

therefore, regarding soil fauna only as passive respondents to vegetation restoration is not appropriate. Soil fauna strongly affect the composition of natural vegetation, and this knowledge might improve the restoration and conservation of plant species diversity [66]. In post-mining restoration, soil fauna can affect plants via trophic and non-trophic interactions or influence the microbial soil community and physical or chemical properties of the soil [59, 67]. In temperate grasslands, earthworms can also influence above-ground vegetation structures by increasing the spatial heterogeneity of grassland plant communities [68]. Earthworms' engineering activity and consequent development of topsoil layers may be an essential driver of plant community succession, enhances the bacterial food web channel (Frouz et al. [69]) while macrosaprophagous guilds of soil fauna play an important role in litter decomposition and soil mixing in post-mining sites [70, 71].

4.4.3 Ecological Processes

The sustainability of a restored system also depends on processes associated with carbon, nutrient, and water cycles; yet most assessments of forest restoration employ indicators for composition, richness, and abundance of plant species, which are insufficient predictors for long-term assessments. In an extensive literature review, Ruiz-Jaen and Mitchell Aide [72] found that measures of biological interactions were the most common ecological processes, followed by nutrient pools and soil organic matter (SOM). In some restoration projects, the plant community composition is restored, but ecological processes on which the persistence of the plant communities depend are not [73].

Soil indicators may be used for assessing the effectiveness of restoring ecosystem functioning and services. However, selecting the most suitable soil indicators follows a clear identification of the different and sometimes competing expectations for ecosystem services that the project aims to restore [63, 74]. For example, detecting initial signs of ecosystem recovery may require decades in gold and metalliferous tailings [40, 75]. The choice of a few easily measurable and unambiguous indicators will increase the likelihood that the monitoring will continue for the expected life of the project evaluation phase. There are many options of indicators (e.g. short-term and long-term), but making them more uniform would help mutual comparisons among restoration projects [76]. The relationship between ecological properties, patterns, and processes varies across ecosystems, making it difficult to select global short-term and long-term indicators. These need to be ecosystem- and site-specific indicators: for instance, in dryland ecosystems, soil fertility, water retention capacity, and water availability are often crucial soil indicators [63, 74]. According to Cohen-Shacham et al. [77], soil indices such as soil stability, infiltration, and nutrient cycling provide information on soil function that can be used to guide management decisions during restoration.

Costantini et al. [63] concluded that below-ground soil indicators might be used for assessing the effectiveness of restoring ecosystem functioning and services. The use of below-ground indicators for evaluating restoration success has been limited to physicochemical and microbiological indicators, with less attention paid to soil fauna. A small number of studies have also considered methods based on soil mesofauna composition (microarthropods) [78, 79] or have proposed macrofauna for the evaluation of soil ecosystem services [63]. For example, the presence or absence of terrestrial arthropods, mainly Coleoptera, can be used as a cost-effective restoration indicator for assessing short-term environmental changes in coal mine spoil heaps [80]. However, even if soil biodiversity is directly related to soil health, and healthy soils can offer a more extensive range of habitat niches to fauna than poor soils, monitoring restoration success also requires looking at the role that soil fauna play in soil functions [81].

4.5 Gaps in the Assessment of Restoration Success in Mine Sites

As mentioned previously, the three most common attributes in restoration assessment studies (diversity, vegetation, and ecological processes) are essential for the long-term persistence of an ecosystem [82]. Mining companies' previous efforts to speed up ecosystem recovery have failed when they focused on visual recovery rather than soil-related problems [80, 83]. Moreover, limited studies have measured the reproducing population or evaluated the self-sustainability of ecosystems because these are long-term attributes that can rarely be measured in the time frame of most restoration projects [13, 72, 84]. Few restoration projects predetermine criteria for success, and even fewer quantitatively monitor project outcomes [48]. This is also the case for most assessments of rehabilitation success in mine sites [16, 45, 46, 85]; literature relating to monitoring in mine site restoration is difficult to access [16], and there is no agreement or standardization of which indicators should be used to monitor development. Research has tended to focus on above-ground attribute assessment, typically plant communities and vegetation structures [72, 86, 87]. However, much of this research lacks elucidation of the above/below-ground relationship in ecosystems [86]; evaluating the long-term success and self-sustainability of rehabilitation programs in mine sites has also been neglected [41], and more studies are required across varying climatic regions [88]. Even when methods and criteria to evaluate and assess restoration are available in open-pit mines such as quarry rehabilitation, they are, in general, very specialized and time-consuming. A new perspective to assess mine site restoration success should move toward a simplified methodology that enables the non-scientific public to evaluate restored areas [5]. Long-term rehabilitation programs and assessments in metal mine tailing should develop a simplified methodology that could be applied in mine sites across biomes.

It is worth mentioning the important role of assessing ecosystem complexity. When evaluated separately from each other, individual attributes of the ecosystem, such as soil quality and vegetation performance, can be misleading for evaluating the sustainability of a rehabilitation strategy [25] because everything is part of the complex assemblage that contributes to ecosystem functioning and resilience. Even landform reconstruction of mine sites [89] shows that spoil heaps from brown coal mining mechanical terrain reshaping following overburden deposition affect the nutrient distribution in soils, influencing plant community composition and the presence of soil fauna and accelerating migration to sites. The concave-slope base sector, mimicking the landforms of the surrounding undisturbed landscape, promotes soil formation and the establishment of self-sustaining, functional ecosystems in restored quarries, contributing to habitat diversity and spontaneous arrival of vegetation [90].

Specific problems at mining sites include the lack of common biological indicators of ecosystem restoration success across biomes and mineral spoil types; in addition, the soil fauna response has been globally underestimated and disregarded [16]. An exception is efforts to assess macrofauna (Lumbricidae, Collembola, Diplopoda, and Chilopoda) responses in Lignite post-mining sites in Germany [91–93]. A considerable amount of literature has been published on Hymenoptera, especially Formicidae, as bioindicators [52, 75, 94, 95]. Ants have been widely used as indicators of restoration processes in post-mining coal sites [53] and bauxite mining sites [96, 97]. There have been few extensive studies of the role of ants in rehabilitated mine tailings, except Ribas et al. [75], who showed that ant species can be used to evaluate the stage of the rehabilitation process in gold tailings and that some species are indicative of later stages of succession. In Ireland, the findings of Courtney et al. [45] indicate that bauxite residues can be successfully restored to a functioning grassland-type ecosystem and that monitoring key arthropod species is recommended as part of the restoration evaluation process.

When considering the vegetation structure and diversity monitoring approach, soil fauna are viewed as passengers during restoration, responding passively to vegetation; it is assumed that vegetation attributes reflect habitat adequacy for fauna and that animals will be capable of dispersing to the restored habitat and establishing populations [58]. A substantial body of literature shows that fauna can also drive the developmental trajectories in post-mining site restoration [59, 96–102]. Studies such as those of Frouz et al. [103] revealed that in addition to plant establishment, successful colonization of post-mining sites by soil fauna is critical for soil development as a substantial part of newly formed soils consist of soil macrofauna feces.

Prach et al. [76] proposed that there are many options of indicators for mine rehabilitation, but making them more uniform would allow comparisons among projects, assisting the mining industry in assessing rehabilitation performance. There is more agreement about assessing soil physicochemical characteristics, and these remain the primary indicator of rehabilitation success, as in bauxite residues [25, 104].

Gatica-Saavedra et al. [105] found that the three most commonly used ecosystem function indicators were related to soils: chemical composition, physical characteristics, and assemblage of bioindicators (soil organisms that are indicators of ecosystem processes). Vasconcellos et al. [106] suggested that organic matter quality and physical soil attributes, in addition to management practices, can accelerate the forest recovery process during restoration. Given the functional importance of soil biodiversity and the role of soils in human wellbeing, soil biodiversity should be considered further in policy agendas and conservation actions by adapting management practices to sustain soil biodiversity [107]. The same general principle also applies to mine site restoration, where soil biodiversity should be considered when designing restoration strategies and monitoring and assessing mine site rehabilitation. To gain deeper insight into post-mining ecosystem evolution, the long-term study of soil and vegetation development and soil fauna assemblage in post-mining land rehabilitation in mine tailing from different ages is crucial. To increase the knowledge of soil multifunctionality and soil ecosystem services in these novel ecosystems, it is necessary to look into the role of soil fauna.

4.6 Increasing Restoration Success by Enhancing Soil Biodiversity and Soil Multifunctionality

Given that it is much easier to revive soil than it is to create it, salvaging and replacing topsoil is the simplest and most effective means of restoring functional soil following disturbance. Practices that ensure landform stability and enhance soil water-holding capacity, aeration, and accumulation of organic matter and nutrients also make soils more conducive for soil microbial and faunal communities [122]. Rehabilitation to restore degraded lands benefits both soil biodiversity and ecosystem functioning [107]. A growing body of literature suggests strongly that a functioning below-ground ecosystem including diverse microbial and soil faunal communities is critical for rehabilitation strategies to become sustainable [25, 85, 94, 99, 115, 123]. From all present knowledge of mine-site restoration, it is clear that the success of amelioration depends on the waste substratum characteristics [92]. However, modification of rehabilitation technologies in a way that would support site colonization by saprophagous soil macrofauna might enhance soil mixing and carbon storage in mineral sites [70].

Soil processes are controlled by a complex community of soil organisms and their environment, such as the linkages between C and N mineralization, soil stability and structure, and activity of soil fauna, microorganisms, and plant roots [124]. However, Scheu and Setälä [125] showed that macrofauna and mesofauna play a stronger role in energy flow through predation, grazing, and substrate processing than microbiota, despite their low metabolic contribution to soil processes [126]. Findings from Briones (2014) support that meso- and macrofauna activity as bioturbators exert control over other organisms, creating favorable environments for colonization.

Soil microarthropods have been used as soil quality indicators since they play a vital role in maintaining soil functions [78, 127–129]. Working in the context of mine site restoration, Majer et al. [96] concluded that Hemiptera, Coleoptera, Araneae, and Hemiptera reflected biophysical changes in the environment.

In Europe, previous studies on recovery, colonization, and succession of micro-, meso- and macrofauna, have focused on coal and lignite sites, examples of which can be found in Table 4.1. However, understanding soil fauna contributions to soil multifunctionality is still limited by many factors: difficulties in taxonomic identification and the lack of taxonomic specialists, as molecular information alone is insufficient for many ecological studies [81, 130].

One clear finding of research to date is that recovery cannot be adequately discerned by simple measurement of total abundance or species richness within a higher taxon because richness often recovers much faster than composition. To measure both the overall development of species composition in revegetated sites and their similarity to reference sites, a greater number of multitaxon studies would be desirable, or grouping species into functional response guilds [58]. Regarding the first option, this is not always feasible for logistic reasons in restoration assessment since there are time and financial limitations [15]. Looking at functional response guilds based on their typical habitat requirements or other ecological attributes and then how successful they seem to be is an encouraging prospect to assess restoration in mine sites. Speaking about terrestrial ecosystems in general, Brussaard et al. [131] highlighted that decomposition and production processes cannot be understood and/or manipulated without explicitly addressing the composition and activity of the soil food web. Similarly, it has been acknowledged that in terrestrial ecology, there is a need for a more complete assessment of the role of the entire soil biological community and its interaction with the above-ground components and mineral-organic matrix [132, 133]. These two fundamental insights also need to be applied to the restoration and rehabilitation of mine sites.

4.7 Using Keystone Species and Ecosystem Engineers in Restoration

Generally speaking, above-ground fauna can be passengers or drivers in revegetation; in the latter case, fauna can play different roles such as dispersing seeds and pollinating [58]. Considering below-ground fauna, while the density of some mesofauna is closely related to the presence of vegetation, there is no clear preference for vegetation type (e.g. grassy or woody vegetation) [120], whereas earthworms have been linked with driving vegetation regeneration and soil development [59, 70, 99]. Therefore, there is potential for developing novel restoration interventions that manipulate fauna as a technique of initiating or accelerating revegetation [58]. So, identifying key fauna groups that could potentially be used for field trial inoculation in rehabilitated mine sites seems promising for the restoration of metal mining sites. This has been proved successful in previous research studies for post-mining coal mine sites [116, 134–136], amended bauxite residues [85], and several landfills [83].

Herein, we follow the definition of Jones et al. [137] for *ecosystem engineers*, i.e. organisms that directly or indirectly modulate the availability of resources to other species by causing physical state changes in biotic or abiotic materials. Ecosystem engineers of any kind have the potential to enhance ecosystem functions in soil, probably more than in any other ecological medium [138]. The concept of *keystone species* was introduced by Paine [139] and refers to species that have a

Table 4.1 Target invertebrate taxa studied as indicators (biological soil quality indicator) in mining restoration in Europe during the past 20 years.

Country	Order/ Family/Species	Target	Mineral type	Ecological process	Soil physicochemical	Findings	References
Ireland	Coleoptera: Staphylinidae	Relative abundance of Coleoptera; soil metal concentrations	Non-ferrous metalliferous and Zn/Pb	Not assessed	Soil metal concentrations	Stable soil moisture conditions appear to be important for the recolonization of metalliferous soils by Staphylinidae.	Good [40]
Ireland	Isopoda Chilopoda Diplopoda Araneae Opiliones Collembola Dermaptera Hemiptera Coleoptera, Hymenoptera	Monitor and evaluate restoration success by analyzing the colonization and community development of soil macro-arthropods.	Bauxite residue	Not assessed	pH, EC, C : N, concentrations (P, K, Na, Ca, Mg)	Many species identified indicate later stages of succession. Bauxite residues can be successfully restored to a functioning grassland-type ecosystem.	Courtney et al. [45]
Ireland	Lumbricidae: *A. chlorotica*	Exposure tests to understand the potential for rehabilitated residues to support earthworm communities	Bauxite residue	Earthworm survival and population viability	Soil pH, EC, salinity, and sodicity	Capability of the rehabilitated residue to support earthworm populations indicates the development of a functioning soil system in rehabilitated residues.	Courtney et al. [85]
Ireland	Lumbricidae: *E. fetida* Collembola: *F. candida*	To investigate the ability of keystone meso/ macrofauna to colonize and establish populations	Bauxite residue	Growth and reproduction tests for Collembola	Soil pH, EC, ESP, available cations (Na, Ca, K, and Mg)	Amendment improves the residue's physicochemical properties, and data revealed that both gypsum and organic addition are required to promote conditions favorable to plant growth and mesofauna establishment. Bauxite residue can be transformed into a soil-like medium capable of supporting keystone species	Finngean et al. [104]

(Continued)

Table 4.1 (Continued)

Country	Order/ Family/Species	Target	Mineral type	Ecological process	Soil physicochemical	Findings	References
Ireland	Lumbricidae: *E. fetida*	To investigate the effects of bauxite residue properties on earthworm survival and viability for future inoculation to accelerate rehabilitation	Bauxite residue	Earthworm survival and population viability	Soil pH, EC, salinity, and sodicity	Earthworms exposed to soil containing ≥10% residue took up and accumulated elevated concentrations of Al, As, Cr, and V in comparison to the control earthworms.	Di Carlo et al. [108]
Germany	Lumbricidae: *D. octaedra* *D. rubidus* *L. castaneus* *L. rubellus* *L. terrestris* *A. chlorotica* *A. caliginosa* Diplopoda Isopoda	Spontaneous colonization of soil fauna and implications for microbial activity and soil quality	Lignite mining	Microbial activity and available macronutrients	NH_4, NO_3, PO_4, N, organic litter layer	In troughs, soil fauna directly enhanced microbial activity and indirectly increased the availability of macronutrients and improved soil quality by increasing pH values, increasing the ammonium–nitrogen content, and decreasing the content of aluminum ions. A multiple-species system can improve soil quality more effectively than a single-species system.	Topp et al. [109]
Germany	Lumbricidae: *D. dendrodrilus* *L. rubellus* Collembola Actinedida Oribatida Chilopoda Diplopoda Gamasida	Species abundance; earthworm biomass	Lignite mining	Decomposition and potential decomposition level of saprophagous, respiration, primary production	SOM	Soil faunal development allows assessment of biological soil quality based on species composition, biomass, metabolic activity of soil animal groups, and the kind of competition for food supply (e.g. litter decomposition) between saprophagous groups.	Dunger and Voigtlander [92]

Germany	Lumbricidae Aranea Opilion Collembola Diplopoda Chilopoda Amoebae Nematoda Oribatidae Formicidae Carabidae Staphylinidae	Species abundance, earthworm biomass, microarthropod density, nematodes, and amoebae density	Lignite mining	Decomposition and potential decomposition level of saprophagous, respiration, primary production	SOM	During succession of the mine site community, an interesting interrelationship between parts of the soil fauna can be demonstrated, especially between lumbricids and microarthropods (activity level of saprophagous fauna).	Dunger et al. [93]
UK	Lumbricidae: *L. terrestris* *A. longa* *A. caliginosa* *L. rubellus* *A. chlorotica*	To measure the effects on soil organic matter and associated soil aggregation after nine years of earthworm inoculation	Coal mining	Biomass-C, basal respiration, dehydrogenase activity, metabolic quotient, Cmic-to-Corg ratio, mineralization constant	SOM	The results emphasize the important influence of earthworm activity on aggregate and organic matter stabilization.	Scullion and Malik [110]
UK	Lumbricidae: *L. terrestris* *A. longa* *A. caliginosa* *L. rubellus* *A. chlorotica*	Compare soil properties in inoculated areas with those of controls to evaluate the contribution of casts to bulk soil aggregation, and SOM and root content responses to earthworm activity	Coal mining	Productivity and root content	WSA, SOM, soil aggregates	Results suggest that earthworms, rather than plant roots, initiate aggregation in severely degraded grassland soils.	Marashi and Scullion [111]

(*Continued*)

Table 4.1 (Continued)

Country	Order/ Family/Species	Target	Mineral type	Ecological process	Soil physicochemical	Findings	References
Czech Republic Germany	Lumbricidae: *D. octaedra* *L. rubellus* *A. caliginosa* *A. rosea and* *O. lacteum* Enchytraeidae Testacea Collembola Oribatida Isopoda Diplopoda dipteran larvae	Density and community composition of a wide spectrum of soil organisms, direct counts of bacteria, rate of cellulose decomposition, and microstructure of upper soil layers	Coal mining	Decomposition rate	Soil pH, thickness of litter, fermentation, and humus layers	Slower development of soil biota in initial stages of succession is caused by unfavorable physical and chemical conditions of heap substrates and by lower input and poor quality of litter in pine plantations.	Frouz et al. [112]
Czech Republic	Macrofauna	The effect of macrofauna on the removal of litter from the soil surface and accumulation of organic matter in the mineral soil layer	Coal mining	Litter decomposition	Accumulation of C in the mineral layer	The accumulation of C in the mineral layer resulted in the long-term enhancement of microbial biomass and respiration. Macrofauna enhance C removal from the litter layer through an effect on litter fragmentation and its accumulation on the mineral layer, but it does not affect litter mineralization.	Frouz [113]
Czech Republic Germany	*Enchytraeus crypticus*	Toxicity test of *E. crypticus* to evaluate the toxicity of brown coal mining areas	Coal mining	Survival and population growth	Soil pH, EC, Na, Ca, K, Al, Fe, loss of ignition, and polyphenol contents	Spoil with a pH below 3 did not support the survival of *E. crypticus*. However, some alkaline spoils with high conductivity and ion concentration were also toxic.	Frouz et al. [114]

Czech Republic	Hymenoptera	Abundance, species richness, and diversity index of ant communities	Coal mining	Not assessed	Fermentation layer, humus layer depth	A critical group of factors for ant community composition (plot age, depth of humus and fermentation layers, cover of bare soil, and soil substrate) corresponded to the successional age of the plot and the development of the soil.	Holec and Frouz [95]
Czech Republic	Hymenoptera: *Lasius* *Formica* *Myrmica*	Describing the spatial patterns of nest distribution and epigeic activity of ants in relation to the vegetation mosaic and species abundance	Coal mining	Not assessed	Not assessed	The results indicated that nesting and foraging may differ in their microclimatic requirements, and the formation of vegetation mosaics may be important to changes in the ant population during succession. Ant nest significantly enriched soil by available P.	Holec et al. [100]
Czech Republic	Lumbricidae Enchytraeidae Collembola	The effect of litter type and macrofauna community on litter decomposition and organic matter accumulation	Coal mining	Microbial respiration, decomposition rate	C : N ratio	In reclaimed sites where macrofauna was dominated by the sarcophagus, mainly earthworms, the macrofauna increased soil mixing. In sites where predators dominated, the macrofauna exclusion probably suppressed fragmentation and mixing activity of the mesofauna.	Frouz [70]
Czech Republic	Lumbricidae: *D. octaedra* *D. rubidus* *L. rubellus* Diplopoda Microorganism Collembola dipteran larvae	The effect of litter quality and soil faunal composition on organic matter dynamics	Coal mining	Microbial respiration, litter removal	SOM, C, and N concentrations in the mineral and litter layers	Removal of litter from the litter layer and accumulation of C and N in the mineral layer increased microbial respiration and biomass measured as total PLFA.	Frouz et al. [115] Frouz et al. [99]
Czech Republic	Lumbricidae: *D. octaedra* *D. rubidus* *L. rubellus*	To study soil macrofauna density, species composition, and the distribution of macrofauna feces in topsoil	Coal mining	Soil fauna density and activity	Formation of humus layer	Besides plant establishment, successful colonization of post-mining sites by soil fauna seems to be critical for soil development as the substantial part of newly formed soils consisted of soil macrofauna feces.	Frouz et al. [103]

(*Continued*)

Table 4.1 (Continued)

Country	Order/ Family/Species	Target	Mineral type	Ecological process	Soil physicochemical	Findings	References
Czech Republic	Lumbricidae Enchytraeidae Testacea Oribatida Collembola Tardigrades Nematodes	Guilds of soil fauna density in non-reclaimed sites and implications for soil development	Coal mining	Litter reserve on the soil surface, annual litter input herb biomass and root biomass, soil microstructure	Soil pH, total C, and N, contents of available Ca, K, and Na, WSP	Microstructure analysis indicated that earthworm activity, namely the mixing of organic and mineral layers, played a principal role in humus layer formation. Macrosaprophagous guilds of soil fauna, most important for litter decomposition and soil mixing, attained their highest density in the oldest sites.	Frouz et al. [39]
Czech Republic	Lumbricidae: *L. rubellus* *A. caliginosa*	To investigate if the introduction of earthworms into a post-mining soil enhances the growth of late-succession plant species	Coal mining	Vegetation biomass	Soil pH; total N, Ca, and C; exchangeable P, K	Massive changes in plant community composition occur around 23 years of succession when the heaps are colonized by the earthworms.	Roubickova et al. [116]
Czech Republic	Enchytraeidae Formicidae Chilopoda	To explore processes responsible for environmental heterogeneity	Coal mining	Vegetation biomass	Soil pH; EC; total C; total available P, K, and Ca; water-soluble P	The abundance of soil fauna was significantly related to microhabitat. Maintaining the spatial heterogeneity created by the heaping may enhance the spontaneous re-vegetation of post-mining sites.	Frouz et al. [117]
Czech Republic	Gastropoda Myriapoda Diplopoda Isopoda Heteroptera Coleoptera	To study the effects of pedological properties on species richness inhabiting soil to shrub	Coal mining	Not assessed	Soil pH; C; total N; exchangeable P, K, and Ca content	The species richness of various groups was affected by different soil properties either independently of other variables or in interaction with microclimatic conditions or management history.	Hendrychova et al. [118]

Czech Republic	Microflora Fungi Bacteria Micro-meso-macrofauna	To study the parameters characterizing the structure of the decomposer food web, biomass of soil microflora	Coal mining	Estimation of feeding rates in the soil food web	Soil pH, herb mass, fine root, litter input	Biomass of the soil microflora (bacteria and fungi) is the best predictor of the thickness of the fermentation layer, which corresponds with changes in topsoil microstructure driven by a combination of plant organic matter input and engineering effects of earthworms.	Frouz et al. [71]
Czech Republic	Lumbricidae: *L. rubellus* *A. caliginosa* Collembola	Test the effect of earthworms on plants and Collembola	Coal mining	Soil microbial respiration, microbial biomass	Soil pH, C, total N, and exchangeable P and K	Earthworms help drive succession of both plant and Collembola communities on post-mining sites.	Mudrak et al. [119]
Czech Republic	Lumbricidae	To study the effects of tree species on the chemical, micro-morphological, and biological properties of soil and the role of soil fauna	Coal mining	Microbial respiration, microbial biomass bioturbation activity	C : N ratio	The results indicate that the effect of tree species on soil development is substantially mediated by soil fauna activity and especially by earthworm bioturbation.	Frouz et al. [60]
Czech Republic	Enchytraeidae Lumbricidae Lepidoptera Araneae Coleoptera Aphids Hymenoptera Thysanoptera Orthoptera Dermaptera Chilopoda Diplopoda	To study variations in soil chemical properties, microbial biomass, and faunal populations as related to plant functional traits, patch types, and successional stage	Coal mining	Microbial biomass, PLFA analysis	Soil pH, total N, C, EC, and exchangeable P and K	Patch type was the only plant-related factor capable of explaining the mesofauna community variations. Epigeic macrofauna was found to correlate more strongly with plant functional traits than endogeic macrofauna. The abundance of most fauna correlated positively with an increase in plant cover and litter input and was significantly higher in vegetated patches and old sites.	Moradi et al. [120]

(*Continued*)

Table 4.1 (Continued)

Country	Order/ Family/Species	Target	Mineral type	Ecological process	Soil physicochemical	Findings	References
Czech Republic	Lepidoptera Araneae Coleoptera Hymenoptera Orthoptera Dermaptera Chilopoda Diplopoda Oniscidea	To study the effects of micro-topographic heterogeneity on biodiversity and conservation potential of arthropod communities	Coal mining	Not assessed	Not assessed	Surface dozing should be applied only where the technical approaches are expedient. Micro-topographic heterogeneity is responsible for the small-scale heterogeneity of various environmental conditions that support high biodiversity of groups with various needs.	Moradi et al. [121]
Czech Republic	Orabitidae Lumbricidae Diplopoda Diptera Coleoptera	Study the effect(s) of soil transplantation on survival and colonization	Coal mining	Microbial biomass, microbial respiration	C content, C : N ratio, pH and EC	Soil development of the spoil overburden might be even more critical than the migration barrier.	Moradi et al. [38]
Spain	Coleoptera	Analyze the potential use of terrestrial arthropods as indicators of successional restoration stages	Coal mining	Not assessed	Not assessed	Coleoptera may be useful as a cost-effective indicator for assessing short-term environmental changes.	Tizado and Núñez-Pérez [80]

SOM, Soil organic matter; EC, electrical conductivity; ESP, exchangeable sodium percentage; WSA, water-stable aggregation; PLFA, fatty acid profiles of the phospholipids; WSP, water-soluble phosphorus.

disproportionately large effect on ecosystem functioning in their environment relative to their biomass. This concept partly overlaps with ecosystem engineers – an ecosystem engineer can be a keystone species and vice versa. Earthworms, termites, and ants are considered the most essential ecosystem engineers in soil [67, 140, 141], followed by Collembola and mites [142, 143]. These groups also require, in general, less survey effort since they are relatively easy to sample, and there is extensive literature available that facilitates their (morphological) identification.

Ecosystem engineers may increase restoration success and guide soil. For instance, Filser et al. [144] reported that earthworms, Collembola, and other larger invertebrates strongly impact carbon turnover, sometimes even turning C sinks into C sources or vice versa. Tuma et al. [145] emphasized the importance, across all habitats, of termites for generating standing above-ground soil structures and earthworms for short-term soil turnover. Some studies have effectively used soil fauna or vegetation as ecosystem engineers by selecting adequate species and investigating the functioning of the whole system (applying ecosystem-based management) for ecosystem restoration [83, 146, 147]. Selected taxa will be discussed in more detail below. The use of selected soil invertebrate taxa as bioindicators by mineral spoil and their functional roles in restored soils are summarized in Figures 4.1 and 4.2, respectively.

4.7.1 Earthworms

Earthworms are one of the most studied groups of ecosystem engineers [148]. They are important drivers for the formation of soil structures and SOM dynamics [149], and they drive nutrient turnover and the functioning of detrital food webs and terrestrial ecosystems in general [81, 144, 150, 151].

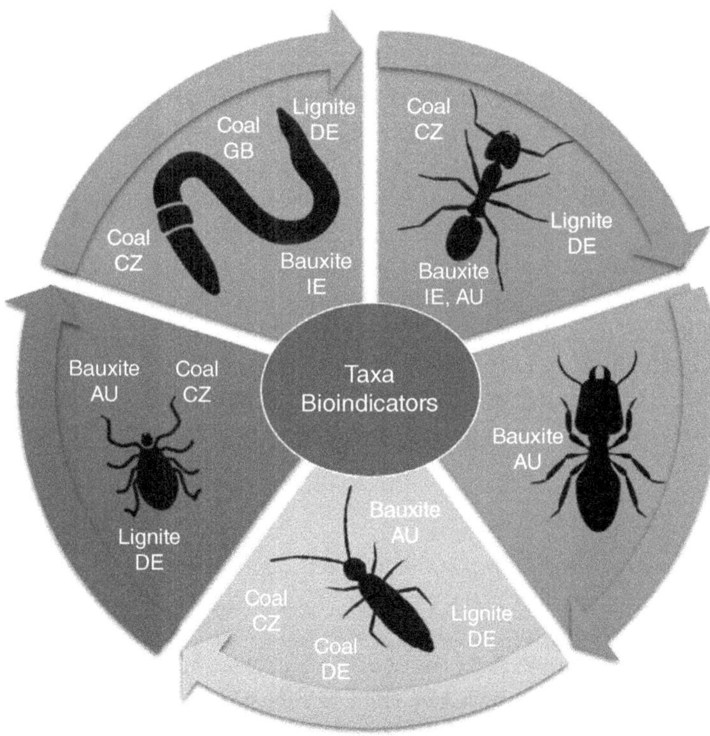

Figure 4.1 Primary soil ecosystem engineers used as bioindicators in different mineral spoil types in mine site rehabilitation worldwide. AU, Australia; CZ, Czech Republic; DE, Germany; GB, United Kingdom of Great Britain and Northern Ireland; IE, Ireland (according to ISO 3166 country codes).

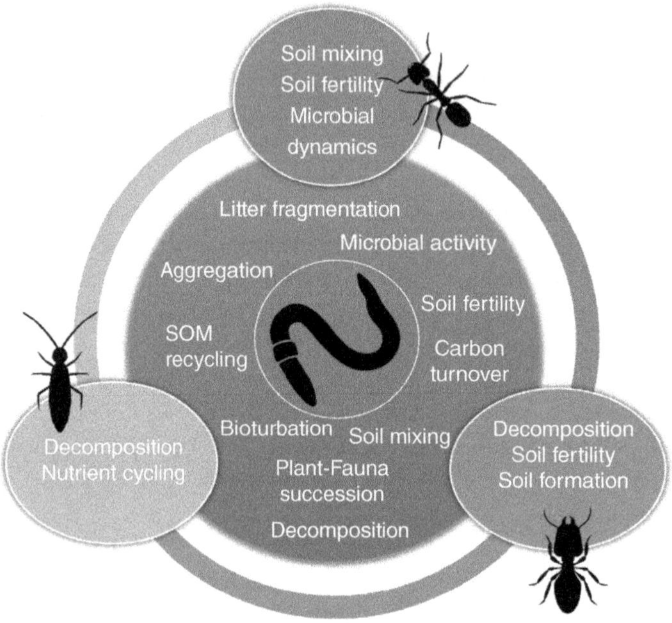

Figure 4.2 Primary ecosystem engineers used as indicators or drivers of restoration success and their roles in modifying soil physicochemical conditions in post-mining sites. Both graphs are based on the literature search conducted for this chapter, which includes assessments in rehabilitated coal mines, lignite, and bauxite site retrieved from Table 4.1 and literature in Section 4.7.

As detritivores, they are partially responsible for the breakdown and recycling of dead organic matter, while soil passing through earthworms' guts increases soil quality [83]. Marashi and Scullion [111] found that earthworms, rather than plant roots, initiate aggregation in severely degraded grasslands.

In post-mining coal sites, Frouz et al. [60] observed that the effect of tree species on soil development is substantially mediated by soil fauna activity and especially by earthworm bioturbation. Roubickova et al. [116] showed that dramatic changes in plant community composition occurred around year 23 of succession when the heaps were colonized by earthworms in post-mining coal sites. The findings from Scullion and Malik [110] highlight the important influence of earthworm activity on aggregate and organic matter stabilization in post-mining coal. Results from Mudrak et al. [119] indicate that earthworms can promote other species and, therefore, the restoration of post-mining sites, driving the succession of both Collembola and plants. In surfaces affected by coal mining, earthworm inoculation and subsequent activity and cast formation improved soil structure [111].

In metal mining tailings, the role of earthworms in soil development has been poorly investigated. However, approaches of this kind are promising and have various implications for practice: evaluation of soil inoculation necessity, an appropriate choice of species, timing, and inoculation method; coordination with labor; and monitoring of the earthworms (species, numbers, and dispersal), preferably linked with measurements of developing soil properties [83, 146].

4.7.2 Ants

Ants play many important roles in directly and indirectly controlling resource availability for other organisms. They are considered keystone species in many ecosystems, acting as important predators of insects and other invertebrates [152]. As Majer et al. [96] reported, Hemiptera are influenced by both vegetation and soil factors, which makes them good bioindicators of environmental conditions and biophysical changes. Ants have proved to be cost-effective to survey and potentially high in information for restored bauxite mines [97]. Ant density has been found to increase due to higher resource availability in restored lands [153]. Frouz et al. [154] confirmed that ants can also initiate soil modification; ant nests induce chemical changes, as mixing due to excavation of deeper soil layers, enrichment of substrate, and interactions of nests chemistry with other physical, chemical, and microbial parameters could guide the restoration outcomes in post-mining sites.

4.7.3 Termites

Termites are among the principal macro-invertebrate decomposers in the tropics. Due to highly evolved mutualism with microbes, termites play a significant role in decomposition processes and nutrient recycling as soil ecosystem engineers impacting soil formation and fertility [155, 156]. Termite-mediated processes are used to improve the soil–water balance [157]. In rehabilitated bauxite mines, forest development and the changing availability of food materials enabled colonization by termite groups. In restored bauxite mines, results showed that adding woody debris to restored areas might accelerate the return of termites and the associated ecosystem processes, decomposition, and nutrient recycling [96].

4.7.4 Collembola and Mites

Microarthropods, including mites (order Acari) and Collembolan (springtails, order Hexapoda), are the most abundant and diverse soil mesofauna. They play an essential role in processing organic matter via chemical, physical, and biological mechanisms, so they are considered necessary in shaping ecosystems [158].

Collembola are involved in nutrient cycling, especially in litter decomposition, by controlling microbial diversity and decomposition [159, 160] and play a role in the translocation of litter carbon to soil (Chamberlain et al. [161]). Meyer-Wolfarth et al. [162] demonstrated their potential as biological regulators of plant pathogens. Moreover, Collembola community composition, which is very sensitive to disturbances, was found to be a good environmental indicator in Anthrosols [78]. In coal waste deposits, Dunger and Voigtlaender [91] found that Collembola succession was dependent on rehabilitation age. In bauxite mines, collembolan populations mediate, or are at least associated with, ecosystem decomposition and nutrient cycling [96].

Mites contribute to the formation of soil structure and increase the activity and distribution of microorganisms, thereby enhancing decomposition processes [159]. Mites have essential impacts on community dynamics, ecosystem processes, and biodiversity within many insect-fungus systems [163]. In restored bauxite mines, Majer et al. [96] found a strong positive correlation between mite species richness and age of restoration.

Although some studies have evaluated the role of collembolans and mites as indicators of the successional stage, few studies have focused on how they modify soil physicochemical conditions and their effects on ecosystem function from rehabilitated metal mine sites.

4.8 Conclusions and Further Perspective for the Restoration of Metalliferous Tailings

We agree with Cross et al. (2019), who asserted that restoration assessment should move from traditional evaluation toward a re-evaluation of regulatory standards and include soil fauna monitoring in mine restoration projects and its role in soil ecosystem services. Investing in soil fauna assessments and the long-term evolution of rehabilitated mine sites will assist industry, promote further investment in restoration science, and accelerate recovery of soil functions.

More research is needed to investigate the importance of metalliferous tailing properties on soil fauna for future inoculation to accelerate rehabilitation. Mine restoration projects should implement common biological indicators to facilitate comparisons among projects [16] and simplified methodologies to monitor ecosystem restoration [5]. It may be reasonable to look at functional response guilds or the soil food web rather than species diversity to avoid difficulties in taxonomic identification. Identifying functional response guilds based on their typical habitat requirements or other ecological attributes and then looking at how successful they are appears to be an encouraging prospect for metal mine rehabilitation.

In this chapter, we presented evidence supporting the idea that assessments and monitoring of restoration success must target specific taxa, mainly those considered as ecosystem engineers and their effects on ecosystem function, rather than focus on ease of survey effort or the use of single taxa as biological indicators. Furthermore, identifying key fauna groups that could potentially be used for field inoculations in rehabilitated mine sites offers a promising opportunity for enhancing ecosystem restoration in metal mining sites. Even though inoculation with earthworms has been proven successful in some sites, it is necessary to conduct more long-term field trials to test rehabilitation techniques, preferably linked with measurements of developing soil properties across climatic regions, ecosystems, and mineral spoil types.

Acknowledgements

This work was supported by Science Foundation Ireland (SFI 17/CDA/4778).

References

1 Dominati, E., Patterson, M., and Mackay, A. (2010). A framework for classifying and quantifying the natural capital and ecosystem services of soils. *Ecological Economics* 69: 1858–1868.
2 Montanarella, L., Pennock, D.J., Mckenzie, N. et al. (2016). World's soils are under threat. *Soil* 2: 79–82.
3 Rodríguez, N., Mclaughlin, M., and Pennock, D. (2018). Soil Pollution: A Hidden Reality. Rome: Food and Agriculture Organization of the United Nations.
4 EUMICON & Franz Friesenbichler, A. H., Roman Stiftner, Sonja Bredl (2018). 25 Ideas For a Future Made in Europe. *European Mineral Resources Confederation*.
5 Carabassa, V., Ortiz, O., and Alcaniz, J.M. (2019). Restoquarry: indicators for self-evaluation of ecological restoration in open-pit mines. *Ecological Indicators* 102: 437–445.
6 Seleiman, M.F. and Kheir, A.M. (2018). Maize productivity, heavy metals uptake and their availability in contaminated clay and sandy alkaline soils as affected by inorganic and organic amendments. *Chemosphere* 204: 514–522.
7 Antoniadis, V., Golia, E.E., Shaheen, S.M., and Rinklebe, J. (2017). Bioavailability and health risk assessment of potentially toxic elements in Thriasio Plain, near Athens, Greece. *Environmental Geochemistry and Health* 39: 319–330.
8 Bradshaw, A. (1997). Restoration of mined lands – using natural processes. *Ecological Engineering* 8: 255–269.
9 Corlett, R.T. (2016). Restoration, reintroduction, and rewilding in a changing world. *Trends in Ecology & Evolution* 31: 453–462.
10 Higgs, E., Falk, D.A., Guerrini, A. et al. (2014). The changing role of history in restoration ecology. *Frontiers in Ecology and the Environment* 12: 499–506.

11 Moreno-Mateos, D., Barbier, E.B., Jones, P.C. et al. (2017). Anthropogenic ecosystem disturbance and the recovery debt. *Nature Communications* 8: 14163.

12 Anderson, R.M., Buitenwerf, R., Driessen, C. et al. (2019). Introducing rewilding to restoration to expand the conservation effort: a response to Hayward et al. *Biodiversity and Conservation* 28: 3691–3693.

13 Gann, G.D., Mcdonald, T., Walder, B. et al. (2019). International principles and standards for the practice of ecological restoration. Second edition. *Restoration Ecology* 27: S1–S46.

14 McDonald, T., Jonson, J., and Dixon, K.W. (2016). National standards for the practice of ecological restoration in Australia. *Restoration Ecology* 24: S6–S32.

15 Cross, A.T., Young, R., Nevill, P. et al. (2018). Appropriate aspirations for effective post-mining restoration and rehabilitation: a response to Kaźmierczak et al. *Environmental Earth Sciences* 77: 256.

16 Cross, S.L., Tomlinson, S., Craig, M.D. et al. (2019). Overlooked and undervalued: the neglected role of fauna and a global bias in ecological restoration assessments. *Pacific Conservation Biology* 25: 331–341.

17 Kaźmierczak, U., Lorenc, M.W., and Strzałkowski, P. (2017). The analysis of the existing terminology related to a post-mining land use: a proposal for new classification. *Environmental Earth Sciences* 76: 693.

18 Jones, H. and Boger, D.V. (2012). Sustainability and waste management in the resource industries. *Indonesia* 51: 10057–10065.

19 Tayebi-Khorami, M., Edraki, M., Corder, G., and Golev, A. (2019). Re-thinking mining waste through an integrative approach led by circular economy aspirations. *Minerals* 9: 286.

20 Courtney, R. (2013). Mine tailings composition in a historic site: implications for ecological restoration. *Environmental Geochemistry and Health* 35: 79–88.

21 Courtney, R. (2018). Irish mine site rehabilitation – a case study. In: Bio-Geotechnologies for Mine Site Rehabilitation (eds. M.N.V. Prasad, P.J. De Campos Favas and S.K. Maiti), 439–456. Amsterdam: Elsevier.

22 Council Directive (EC) 2006 2006/21/EC of the European Parliament and of the Council of 15 March 2006 on the management of waste from extractive industries and amending Directive 2004/35/EC. https://eur-lex.europa.eu/legal-content/EN/TXT/?uri=CELEX:02006L0021-20090807 [accessed 20 May 2020].

23 Desai, M., Haigh, M., and Walkington, H. (2019). Phytoremediation: metal decontamination of soils after the sequential forestation of former opencast coal land. *Science of the Total Environment* 656: 670–680.

24 Xie, L. and Van Zyl, D. (2020). Distinguishing reclamation, revegetation and phytoremediation, and the importance of geochemical processes in the reclamation of sulfidic mine tailings: a review. *Chemosphere* 252: 126446.

25 Di Carlo, E., Chen, C.R., Haynes, R.J. et al. (2019). Soil quality and vegetation performance indicators for sustainable rehabilitation of bauxite residue disposal areas: a review. *Soil Research* 57: 419–446.

26 Haigh, M. (2018). Building a cradle for nature: a paradigm for environmental reconstruction. In: Bio-Geotechnologies for Mine Site Rehabilitation (eds. M.N.V. Prasad, P.J.D.C. Favas and S.K. Maiti), 593–615. Amsterdam: Elsevier.

27 Makineci, E., Gungor, B., and Kumbasli, M. (2011). Natural plant revegetation on reclaimed coal mine landscapes in Agacli-Istanbul. *African Journal of Biotechnology* 10: 16. https://doi.org/10.5897/AJB10.2499.

28 Tordoff, G.M., Baker, A.J.M., and Willis, A.J. (2000). Current approaches to the revegetation and reclamation of metalliferous mine wastes. *Chemosphere* 41: 219–228.

29 Tripathi, P., Dwivedi, S., Mishra, A. et al. (2012). Arsenic accumulation in native plants of West Bengal, India: prospects for phytoremediation but concerns with the use of medicinal plants. *Environmental Monitoring and Assessment* 184: 2617–2631.

30 Macdonald, E. and King, E.G. (2018). Novel ecosystems: a bridging concept for the consilience of cultural landscape conservation and ecological restoration. *Landscape and Urban Planning* 177: 148–159.

31 Macdonald, S.E., Landhausser, S.M., Skousen, J. et al. (2015). Forest restoration following surface mining disturbance: challenges and solutions. *New Forests* 46: 703–732.

32 Belnap, J.Y.E., Ludwig, J.A., Wilcox, B.P. et al. (2012). Introduced and invasive species in novel rangeland ecosystems: friends or foes? *Rangeland Ecology & Management* 65: 569–578.

33 Collier, M.J. and Devitt, C. (2016). Novel ecosystems: challenges and opportunities for the Anthropocene. *The Anthropocene Review* 3: 231–242.

34 Murcia, C., Aronson, J., Kattan, G.H. et al. (2014). A critique of the 'novel ecosystem' concept. *Trends in Ecology & Evolution* 29: 548–553.

35 Curry, J.P. and Boyle, K.E. (1995). Restoring soil fauna in reclaimed land, with particular reference to earthworms in cutover peat. *Acta Zoologica Fennica* 196: 371–375.

36 Curry, J. and Good, J. (1992). Soil faunal degradation and restoration. In: Soil Restoration (eds. R. Lal and B.A. Stewart), 171–215. New York: Springer.

37 Brady, V.J., Cardinale, B.J., Gathman, J.P., and Burton, T.M. (2002). Does facilitation of faunal recruitment benefit ecosystem restoration? An experimental study of invertebrate assemblages in wetland mesocosms. *Restoration Ecology* 10: 617–626.

38 Moradi, J., Vicentini, F., Simackova, H. et al. (2018). An investigation into the long-term effect of soil transplant in bare spoil heaps on survival and migration of soil meso and macrofauna. *Ecological Engineering* 110: 158–164.

39 Frouz, J., Prach, K., Pizl, V. et al. (2008). Interactions between soil development, vegetation and soil fauna during spontaneous succession in post mining sites. *European Journal of Soil Biology* 44: 109–121.

40 Good, J.A. (1999). Recolonisation by Staphylinidae (Coleoptera) of old metalliferous tailings and mine soils in Ireland. *Biology and Environment: Proceedings of the Royal Irish Academy* 99B: 27–35.

41 Di Carlo, E., Boullemant, A., and Courtney, R. (2019). A field assessment of bauxite residue rehabilitation strategies. *Science of the Total Environment* 663: 915–926.

42 Carabassa, V., Domene, X., and Alcañiz, J.M. (2020). Soil restoration using compost-like-outputs and digestates from non-source-separated urban waste as organic amendments: limitations and opportunities. *Journal of Environmental Management* 255: 109909.

43 Leguédois, S., Séré, G., Auclerc, A. et al. (2016). Modelling pedogenesis of Technosols. *Geoderma* 262: 199–212.

44 You, F., Dalal, R., and Huang, L. (2018). Initiation of soil formation in weathered sulfidic Cu-Pb-Zn tailings under subtropical and semi-arid climatic conditions. *Chemosphere* 204: 318–326.

45 Courtney, R., O'neill, N., Harrington, T., and Breen, J. (2010). Macro-arthropod succession in grassland growing on bauxite residue. *Ecological Engineering* 36: 1666–1671.

46 Shu, W.S., Ye, Z.H., Zhang, Z.Q. et al. (2005). Natural colonization of plants on five lead/zinc mine tailings in Southern China. *Restoration Ecology* 13: 49–60.

47 Gardner, J. (2001). Rehabilitating mines to meet land use objectives: bauxite mining in the jarrah forest of Western Australia. *Unasylva (FAO).*

48 Suding, K.N. (2011). Toward an era of restoration in ecology: successes, failures, and opportunities ahead. *Annual Review of Ecology, Evolution, and Systematicsl* 42: 465–487.

49 Majer, J.D. (1989). Animals in Primary Succession: The Role of Fauna in Reclaimed Lands. Cambridge: Cambridge University Press.
50 Parmenter, R.R. and Macmahon, J.A. (1990). Faunal community development on disturbed lands: an indicator of reclamation success. In: Evaluating Reclamation Success: The Ecological Consideration (eds. J.C. Chambers and G.L. Wade), 73–89. USDA, Forest Service, Northeastern Forest Station.
51 Slabova, M., Broumova, H., and Pecharova, E. (2008). Communities of small mammals as indicators of biodiversity changes in reclaimed areas after coal mining. In: Rapantova, N. &Hrkal, Z., eds., *Mine Water and the Environment, Proceedings of the 10th International Mine Water Association Congress*, pp. 179–182. VSB – Technical University of Ostrava, Ostrava.
52 Andersen, A.N. and Sparling, G.P. (1997). Ants as indicators of restoration success: relationship with soil microbial biomass in the Australian seasonal tropics. *Restoration Ecology* 5: 109–114.
53 Andersen, A.N., Hoffmann, B.D., Muller, W.J., and Griffiths, A.D. (2002). Using ants as bioindicators in land management: simplifying assessment of ant community responses. *Journal of Applied Ecology* 39: 8–17.
54 Crouzeilles, R. and Curran, M. (2016). Which landscape size best predicts the influence of forest cover on restoration success? A global meta-analysis on the scale of effect. *Journal of Applied Ecology* 53: 440–448.
55 O'Grady, A., Breen, J., Harrington, T.J., and Courtney, R. (2013). The seed bank in soil from the nests of grassland ants in a unique limestone grassland community in Ireland. *Ecological Engineering* 61: 58–64.
56 Chiarucci, A., Bacaro, G., and Scheiner, S.M. (2011). Old and new challenges in using species diversity for assessing biodiversity. *Philosophical Transactions of the Royal Society B-Biological Sciences* 366: 2426–2437.
57 Hejda, M., Pyšek, P., and Jarošík, V. (2009). Impact of invasive plants on the species richness, diversity and composition of invaded communities. *Journal of Ecology* 97: 393–403.
58 Catterall, C.P. (2018). Fauna as passengers and drivers in vegetation restoration: a synthesis of processes and evidence. *Ecological Management & Restoration* 19: 54–62.
59 Frouz, J. (2014). Soil Biota and Ecosystem Development in Post Mining Sites. Boca Raton: CRC Press.
60 Frouz, J., Liveckova, M., Albrechtova, J. et al. (2013). Is the effect of trees on soil properties mediated by soil fauna? A case study from post-mining sites. *Forest Ecology and Management* 309: 87–95.
61 Wang, Y., Li, Y.-L., Shi, J.-H. et al. (2012). Effect of different vegetation restoration measures on the species diversity and soil properties of iron tailings. *Journal of Soil and Water Conservation* 26: 112–117,183.
62 Wong, M.H. (2003). Ecological restoration of mine degraded soils, with emphasis on metal contaminated soils. *Chemosphere* 50: 775–780.
63 Costantini, E.A.C., Branquinho, C., Nunes, A. et al. (2016). Soil indicators to assess the effectiveness of restoration strategies in dryland ecosystems. *Solid Earth* 7: 397–414.
64 Dalle Laste, K.C., Durigan, G., and Andersen, A.N. (2019). Biodiversity responses to land-use and restoration in a global biodiversity hotspot: ant communities in Brazilian Cerrado. *Austral Ecology* 44: 313–326.
65 De Deyn, G.B., Raaijmakers, C.E., and Van der Putten, W.H. (2004). Soil fauna enhances grassland restoration. *Levende Natuur* 105: 10–12.
66 De Deyn, G.B., Raaijmakers, C.E., Zoomer, H.R. et al. (2003). Soil invertebrate fauna enhances grassland succession and diversity. *Nature* 422: 711–713.
67 Anderson, J.M. (1988). Invertebrate-mediated transport processes in soils. *Agriculture, Ecosystems & Environment* 24: 5–19.
68 Orgiazzi, A., Dunbar, M.B., Panagos, P. et al. (2015). Soil biodiversity and DNA barcodes: opportunities and challenges. *Soil Biology and Biochemistry* 80: 244–250.

69 Frouz, J., Thebault, E., Pizl, V. et al. (2013). Soil food web changes during spontaneous succession at post mining sites: a possible ecosystem engineering effect on food web organization? *PLos One* 8: e79694. https://doi.org/10.1371/journal.pone.0079694.

70 Frouz, J. (2008). The effect of litter type and macrofauna community on litter decomposition and organic matter accumulation in post-mining sites. *Biologia* 63: 249–253.

71 Frouz, J., Pižl, V., Tajovský, K. et al. (2014). Soil macro- and mesofauna succession in post-mining sites and other disturbed areas. In: Soil Biota and Ecosystem Development in Post Mining Sites (ed. J. Frouz), 104–131. Boca Raton: CRC Press.

72 Ruiz-Jaen, M.C. and Mitchell Aide, T. (2005). Restoration success: how is it being measured? *Restoration Ecology* 13: 569–577.

73 Herrick, J.E., Schuman, G.E., and Rango, A. (2006). Monitoring ecological processes for restoration projects. *Journal for Nature Conservation (Jena)* 14: 161–171.

74 Nunes, A., Oliveira, G., Mexia, T. et al. (2016). Ecological restoration across the Mediterranean Basin as viewed by practitioners. *Science of the Total Environment* 566: 722–732.

75 Ribas, C.R., Schmidt, F.A., Solar, R.R.C. et al. (2012). Ants as indicators of the success of rehabilitation efforts in deposits of gold mining tailings. *Restoration Ecology* 20: 712–720.

76 Prach, K., Durigan, G., Fennessy, S. et al. (2019). A primer on choosing goals and indicators to evaluate ecological restoration success. *Restoration Ecology* 27: 917–923.

77 Cohen-Shacham, E., Andrade, A., Dalton, J. et al. (2019). Core principles for successfully implementing and upscaling nature-based solutions. *Environmental Science & Policy* 98: 20–29.

78 Magro, S., Gutiérrez-López, M., Casado, M.A. et al. (2013). Soil functionality at the roadside: zooming in on a microarthropod community in an anthropogenic soil. *Ecological Engineering* 60: 81–87.

79 Starý, J. (2002). Changes of oribatid mite communities (Acari: Oribatida) during primary cussession on colliery spoil heaps near Sokolov, North-West Bohemia, Czech Republic. In: Studies on Soil Fauna in Central Europe (eds. K. Tajovský, V. Balík and V. Pižl), 199–206. Ceské Budejovice: Institute of Soil Biology, Academy of Sciences of the Czech Republic.

80 Tizado, E.J. and Núñez-Pérez, E. (2016). Terrestrial arthropods in the initial restoration stages of anthracite coal mine spoil heaps in northwestern Spain: potential usefulness of higher taxa as restoration indicators. *Land Degradation & Development* 27: 1131–1140.

81 Briones, M.J.I. (2014). Soil fauna and soil functions: a jigsaw puzzle. *Frontiers in Environmental Science* 2: 7. https://doi.org/10.3389/fenvs.2014.00007.

82 Elmqvist, T., Setälä, H., Handel, S.N. et al. (2015). Benefits of restoring ecosystem services in urban areas. *Current Opinion in Environmental Sustainability* 14: 101–108.

83 Butt, K.R. (2008). Earthworms in soil restoration: lessons learned from United Kingdom case studies of land reclamation. *Restoration Ecology* 16: 637–641.

84 Cortina, J.W.B.G. (2018). SERE and the international standards for the practice of ecological restoration. SERE (Society for Ecological Restoration Europe).

85 Courtney, R., Di Carlo, E., and Schmidt, O. (2018). Soil properties and earthworm populations associated with bauxite residue rehabilitation strategies. *Environmental Science and Pollution Research* 27 (27): 33401–33409.

86 Hasegawa, M., Fujii, S., Kaneda, S. et al. (2017). Recent progress in ecological studies of soil fauna. *Japanese Journal of Ecology (Otsu)* 67: 95–118.

87 Koch, J.M., Grigg, A.H., Gordon, R.K., and Majer, J.D. (2010). Arthropods in coarse woody debris in jarrah forest and rehabilitated bauxite mines in Western Australia. *Annals of Forest Science* 67: 106–106.

88 Frouz, J., Jilkova, V., Cajthaml, T. et al. (2013). Soil biota in post-mining sites along a climatic gradient in the USA: simple communities in shortgrass prairie recover faster than complex communities in tallgrass prairie and forest. *Soil Biology & Biochemistry* 67: 212–225.

89 Walmsley Roubíčková, A., Vachová, P., and Vach, M. (2017). Topography of spoil heaps and its role in plant succession and soil fauna presence. *Scientia Agriculturae Bohemica* 48: 30–38.

90 Martín-Duque, J.F., Sanz, M.A., Bodoque, J.M. et al. (2010). Restoring earth surface processes through landform design. A 13-year monitoring of a geomorphic reclamation model for quarries on slopes. *Earth Surface Processes and Landforms* 35: 531–548.

91 Dunger, W. and Voigtlaender, K. (2009). Soil fauna (Lumbricidae, Collembola, Diplopoda and Chilopoda) as indicators of soil eco-subsystem development in post-mining sites of eastern Germany – a review. *Soil Organisms* 81: 1–51.

92 Dunger, W. and Voigtlander, K. (2005). Assessment of biological soil quality in wooded reclaimed mine sites. *Geoderma* 129: 32–44.

93 Dunger, W., Wanner, M., Hauser, H. et al. (2001). Development of soil fauna at mine sites during 46 years after afforestation. *Pedobiologia* 45: 243–271.

94 Courtney, R., Feeney, E., and O'Grady, A. (2014). An ecological assessment of rehabilitated bauxite residue. *Ecological Engineering* 73: 373–379.

95 Holec, M. and Frouz, J. (2005). Ant (Hymenoptera: Formicidae) communities in reclaimed and unreclaimed brown coal mining spoil dumps in the Czech Republic. *Pedobiologia* 49: 345–357.

96 Majer, J.D., Brennan, K.E.C., and Moir, M.L. (2007). Invertebrates and the restoration of a forest ecosystem: 30 years of research following bauxite mining in Western Australia. *Restoration Ecology* 15: S104–S115.

97 Orabi, G., Moir, M.L., and Majer, J.D. (2010). Assessing the success of mine restoration using Hemiptera as indicators. *Australian Journal of Zoology* 58: 243–249.

98 Frouz, J. (2017). Effects of soil development time and litter quality on soil carbon sequestration: assessing soil carbon saturation with a field transplant experiment along a post-mining chronosequence. *Land Degradation & Development* 28: 664–672.

99 Frouz, J., Elhottova, D., Kuraz, V., and Sourkova, M. (2006). Effects of soil macrofauna on other soil biota and soil formation in reclaimed and unreclaimed post mining sites: results of a field microcosm experiment. *Applied Soil Ecology* 33: 308–320.

100 Holec, M., Frouz, J., and Pokorny, R. (2006). The influence of different vegetation patches on the spatial distribution of nests and the epigeic activity of ants (Lasius niger) on a spoil dump after brown coal mining (Czech Republic). *European Journal of Soil Biology* 42: 158–165.

101 Majer, J.D. (2009). Animals in the restoration process – progressing the trends. *Restoration Ecology* 17: 315–319.

102 Moir, M.L., Brennan, K.E.C., Majer, J.D. et al. (2010). Plant species redundancy and the restoration of faunal habitat: lessons from plant-dwelling bugs. *Restoration Ecology* 18: 136–147.

103 Frouz, J., Pižl, V., and Tajovský, K. (2007). The effect of earthworms and other saprophagous macrofauna on soil microstructure in reclaimed and un-reclaimed post-mining sites in Central Europe. *European Journal of Soil Biology* 43: S184–S189.

104 Finngean, G., O'Grady, A., and Courtney, R. (2018). Plant assays and avoidance tests with collembola and earthworms demonstrate rehabilitation success in bauxite residue. *Environmental Science and Pollution Research* 25: 2157–2166.

105 Gatica-Saavedra, P., Echeverria, C., and Nelson, C.R. (2017). Ecological indicators for assessing ecological success of forest restoration: a world review. *Restoration Ecology* 25: 850–857.

106 Vasconcellos, R.L.F., Segat, J.C., Bonfim, J.A. et al. (2013). Soil macrofauna as an indicator of soil quality in an undisturbed riparian forest and recovering sites of different ages. *European Journal of Soil Biology* 58: 105–112.

107 Cameron, E., Martins, I., Lavelle, P. et al. (2019). Global mismatches in aboveground and belowground biodiversity. *Conservation Biology* 33: 1187–1192.

108 Di Carlo, E., Boullemant, A., Poynton, H., and Courtney, R. (2020). Exposure of earthworm (*Eisenia fetida*) to bauxite residue: implications for future rehabilitation programmes. *Science of The Total Environment* 716: 137126.

109 Topp, W., Simon, M., Kautz, G. et al. (2001). Soil fauna of a reclaimed lignite open-cast mine of the Rhineland: improvement of soil quality by surface pattern. *Ecological Engineering* 17: 307–322.

110 Scullion, J. and Malik, A. (2000). Earthworm activity affecting organic matter, aggregation and microbial activity in soils restored after opencast mining for coal. *Soil Biology and Biochemistry* 32: 119–126.

111 Marashi, A.R.A. and Scullion, J. (2003). Earthworm casts form stable aggregates in physically degraded soils. *Biology and Fertility of Soils* 37: 375–380.

112 Frouz, J., Keplin, B., Pizl, V. et al. (2001). Soil biota and upper soil layer development in two contrasting post-mining chronosequences. *Ecological Engineering* 17: 275–284.

113 Frouz, J. (2002). The effect of soil macrofauna on litter decomposition and soil organic matter accumulation during soil formation in spoil heaps after brown coal mining: a preliminary results. *Ekologia-Bratislava* 21: 363–369.

114 Frouz, J., Kristufek, V., Bastl, J. et al. (2005). Determination of toxicity of spoil substrates after brown coal mining using a laboratory reproduction test with Enchytraeus crypticus (Oligochaeta). *Water Air and Soil Pollution* 162: 37–47.

115 Frouz, J., Elhottova, D., Pizl, V. et al. (2007). The effect of litter quality and soil faunal composition on organic matter dynamics in post-mining soil: a laboratory study. *Applied Soil Ecology* 37: 72–80.

116 Roubickova, A., Mudrak, O., and Frouz, J. (2009). Effect of earthworm on growth of late succession plant species in postmining sites under laboratory and field conditions. *Biology and Fertility of Soils* 45: 769–774.

117 Frouz, J., Kalcik, J., and Velichova, V. (2011). Factors causing spatial heterogeneity in soil properties, plant cover, and soil fauna in a non-reclaimed post-mining site. *Ecological Engineering* 37: 1910–1913.

118 Hendrychova, M., Salek, M., Tajovsky, K., and Rehor, M. (2012). Soil properties and species richness of invertebrates on afforested sites after brown coal mining. *Restoration Ecology* 20: 561–567.

119 Mudrak, O., Uteseny, K., and Frouz, J. (2012). Earthworms drive succession of both plant and Collembola communities in post-mining sites. *Applied Soil Ecology* 62: 170–177.

120 Moradi, J., Mudrak, O., Kukla, J. et al. (2017). Variations in soil chemical properties, microbial biomass, and faunal populations as related to plant functional traits, patch types, and successional stages at Sokolov post-mining site – a case study. *European Journal of Soil Biology* 83: 58–64.

121 Moradi, J., Potocky, P., Kocarek, P. et al. (2018). Influence of surface flattening on biodiversity of terrestrial arthropods during early stages of brown coal spoil heap restoration. *Journal of Environmental Management* 220: 1–7.

122 Prescott, C.E., Frouz, J., Grayston, S.J. et al. (2019). Rehabilitating forest soils after disturbance. In: Developments in Soil Science, vol. 36 (eds. M. Busse, C.P. Giardina, D.M. Morris and D.S. Page-Dumroese), 309–343. Elsevier.

123 Biederman, L.A., Boutton, T.W., and Whisenant, S.G. (2008). Nematode community development early in ecological restoration: the role of organic amendments. *Soil Biology & Biochemistry* 40: 2366–2374.

124 Six, J., Bossuyt, H., Degryze, S., and Denef, K. (2004). A history of research on the link between (micro) aggregates, soil biota, and soil organic matter dynamics. *Soil and Tillage Research* 79: 7–31.

125 Scheu, S. and Setälä, H. (2002). Multitrophic interactions in decomposer food-webs. In: Multitrophic Level Interactions (eds. B.A. Hawkins and T. Tscharntke). Cambridge: Cambridge University Press.

126 Petersen, H. and Luxton, M. (1982). A comparative analysis of soil fauna populations and their role in decomposition processes. *Oikos* 39: 288–388.

127 Parisi, V., Menta, C., Gardi, C. et al. (2005). Microarthropod communities as a tool to assess soil quality and biodiversity: a new approach in Italy. *Agriculture, Ecosystems & Environment* 105: 323–333.

128 Tsiafouli, M.A., Kallimanis, A.S., Katana, E. et al. (2005). Responses of soil microarthropods to experimental short-term manipulations of soil moisture. *Applied Soil Ecology* 29: 17–26.

129 Van Straalen, N.M. (1998). Evaluation of bioindicator systems derived from soil arthropod communities. *Applied Soil Ecology* 9: 429–437.

130 Coleman, D. and Callaham, M.C. Jr. (2017). Fundamentals of Soil Ecology, 3e. London: Academic Press.

131 Brussaard, L., Pulleman, M.M., Ouédraogo, É. et al. (2007). Soil fauna and soil function in the fabric of the food web. *Pedobiologia* 50: 447–462.

132 Harris, J. (2009). Soil microbial communities and restoration ecology: facilitators or followers? *Science* 325: 573–574.

133 Harris, J.A., Hobbs, R.J., Higgs, E., and Aronson, J. (2006). Ecological restoration and global climate change. *Restoration Ecology* 14: 170–176.

134 Scullion, J. (1991). Re-establishing earthworm populations on former opencast coal mining land. In: Land Reclamation. An End to Dereliction? (ed. M.C.R. Davies), 377–386. London: Elsevier Applied Sciences.

135 Scullion, J. (1994). Earthworms and soil rehabilitation after opencast mining for coal. Proceedings of the 2nd International Symposium on Environmental Biotechnology, Brighton. Institution of Chemical Engineers Symposium Series, 49–51.

136 Scullion, J., Mohammed, A.R.A., and Ramshaw, G.A. (1988). Changes in earthworm populations following cultivation of undisturbed and former opencast coal-mining land. *Agriculture, Ecosystems & Environment* 20: 289–302.

137 Jones, C., Lawton, J., and Shachak, M. (1994). Organisms as ecosystem engineers. *Oikos* 69: 373–386.

138 Lavelle, P., Decaens, T., Aubert, M. et al. (2006). Soil invertebrates and ecosystem services. *European Journal of Soil Biology* 42: S3–S15.

139 Paine, R.T. (1969). A note on trophic complexity and community stability. *The American Naturalist* 103: 91–93.

140 Jouquet, P., Dauber, J., Lagerlof, J. et al. (2006). Soil invertebrates as ecosystem engineers: intended and accidental effects on soil and feedback loops. *Applied Soil Ecology* 32: 153–164.

141 Lavelle, P., Bignell, D., Lepage, M. et al. (1997). Soil function in a changing world: the role of invertebrate ecosystem engineers. *European Journal of Soil Science* 33: 159–193.

142 Fournier, V., Rosenheim, J.A., Brodeur, J. et al. (2003). Herbivorous mites as ecological engineers: indirect effects on arthropods inhabiting papaya foliage. *Oecologia* 135: 442–450.

143 Salmon, S., Geoffroy, J.-J., and Ponge, J.-F. (2005). Earthworms and collembola relationships: effects of predatory centipedes and humus forms. *Soil Biology and Biochemistry* 37: 487–495.

144 Filser, J., Faber, J.H., Tiunov, A.V. et al. (2016). Soil fauna: key to new carbon models. *Soil* 2: 565–582.

145 Tuma, J., Fleiss, S., Eggleton, P. et al. (2019). Logging of rainforest and conversion to oil palm reduces bioturbator diversity but not levels of bioturbation. *Applied Soil Ecology* 144: 123–133.

146 Butt, K.R., Frederickson, J., and Morris, R.M. (1995). An earthworm cultivation and soil inoculation technique for land restoration. *Ecological Engineering* 4: 1–9.

147 Byers, J.E., Cuddington, K., Jones, C.G. et al. (2006). Using ecosystem engineers to restore ecological systems. *Trends in Ecology & Evolution* 21: 493–500.

148 Angst, S., Mueller, C.W., Cajthaml, T. et al. (2017). Stabilization of soil organic matter by earthworms is connected with physical protection rather than with chemical changes of organic matter. *Geoderma* 289: 29–35.

149 Fonte, S.J., Quintero, D.C., Velásquez, E., and Lavelle, P. (2012). Interactive effects of plants and earthworms on the physical stabilization of soil organic matter in aggregates. *Plant and Soil* 359: 205–214.

150 Bohlen, P.J. and Edwards, C.A. (1995). Earthworm effects on N dynamics and soil respiration in microcosms receiving organic and inorganic nutrients. *Soil Biology & Biochemistry* 27: 341–348.

151 Uvarov, A.V., Ilieva-Makulec, K., Karaban, K. et al. (2019). Effects of intra- and interspecific interactions in earthworm assemblages: a comparative study. *Biology Bulletin* 46: 475–482.

152 O'Grady, A., Schmidt, O., and Breen, J. (2010). Trophic relationships of grassland ants based on stable isotopes. *Pedobiologia* 53: 221–225.

153 De Araújo, A.S.F., Eisenhauer, N., Nunes, L.A.P.L. et al. (2015). Soil surface-active fauna in degraded and restored lands of Northeast Brazil. *Land Degradation & Development* 26: 1–8.

154 Frouz, J., Holec, M., and Kalcik, J. (2003). The effect of *Lasius niger* (Hymenoptera, Formicidae) ant nest on selected soil chemical properties. *Pedobiologia* 47: 205–212.

155 König, H. (2006). Intestinal Microorganisms of Termites and Other Invertebrates. Springer Science & Business Media.

156 Spain, A., Hinz, D., and Tibbett, M. Colonisation of rehabilitated lands by termites (Dictyoptera), Rio Tinto Alcan Gove bauxite mine, Northern Territory, Australia. In: Fourie, A. B., Tibbett, M. & Wiertz, J., eds., *Proceedings Fifth International Conference on Mine Closure (Mine Closure* 2010*)*, pp. 23–26.Australian Centre for Geomechanics.

157 Pardeshi, M. and Prusty, B.K. (2010). Termites as ecosystem engineers and potentials for soil restoration. *Current Science* 99: 11.

158 Lee, K.E. and Foster, R.C. (1991). Soil fauna and soil structure. *Australian Journal of Soil Research* 29: 745–775.

159 Hättenschwiler, S., Tiunov, A.V., and Scheu, S. (2005). Biodiversity and litter decomposition in terrestrial ecosystems. *Annual Review of Ecology Evolution and Systematics* 36: 191–218.

160 Jeffery, S., Gardi, C., Jones, A. et al. (2010). European Atlas of Soil Biodiversity. Brussels: European Commission.

161 Chamberlain, P.M., McNamara, N.P., Chaplow, J. et al. (2006). Translocation of surface litter carbon into soil by Collembola. *Soil Biology and Biochemistry* 38 (9): 2655–2664.

162 Meyer-Wolfarth, F., Schrader, S., Oldenburg, E. et al. (2017). Collembolans and soil nematodes as biological regulators of the plant pathogen *Fusarium culmorum*. *Journal of Plant Diseases and Protection* 124: 493–498.

163 Hofstetter, R. and Moser, J. (2014). The role of mites in insect-fungus associations. *Annual Review of Entomology* 59: 537–557.

5

Ecological Engineering and Green Infrastructure in Mitigating Emerging Urban Environmental Threats

Florin-Constantin Mihai[1], *Petra Schneider*[2], *and Mihail Eva*[3]

[1] *CERNESIM Center, Department of Sciences, Interdisciplinary Research Institute, "Alexandru Ioan Cuza" University, Iasi, Romania*
[2] *Magdeburg-Stendal University of Applied Sciences, Magdeburg, Germany*
[3] *Department of Geography, Faculty of Geography and Geology, "Alexandru Ioan Cuza" University, Iasi, Romania*

5.1 Dimensions of Ecological Engineering in the Frame of Ecosystem Service Provision

Ecological engineering (EE) is concerned with the development of sustainable systems in accordance with ecological principles, whereby human activities are integrated into the natural environment for the benefit of both under the scope of sustainable use of the natural capital [60]. One main scope of EE is restoring ecosystem functions, and therefore the approach can make substantial contributions to the development of green infrastructure (GI). GI is defined and described by the European Commission Directorate-General for Environment as: "a strategically planned network of high quality natural and semi-natural areas with other environmental features, which is designed and managed to deliver a wide range of ecosystem services and protect biodiversity in both rural and urban settings. More specifically, green Infrastructure, being a spatial structure providing benefits from nature to people, aims to enhance nature's ability to deliver multiple valuable ecosystem goods and services, such as clean air or water" [11].

The core of this definition was also adopted by the International Union for Conservation of Nature (IUCN), which proposed the following definition: "Green Infrastructure is an innovative way of protecting biodiversity while simultaneously contributing to sustainable and smart growth. It is defined as a strategically planned network of natural and semi-natural areas, which deliver a wide range of ecosystem services in terrestrial and marine areas" [30]. An *ecosystem* was described by Moll and Petit [44] as a set of interacting species and their local abiotic environment functioning together to sustain life. *Ecosystem services* (ES) are nature's services to humankind, which are obtained through habitats and living things such as animals and plants [42]. ES are divided into four categories according to the nature of the benefits for humans [42]:

- *Provisioning services*: Products derived from ecosystems, such as food, water, raw materials, energy, etc.
- *Regulatory benefits*: Benefits deriving from the regulation of ecosystems, such as air and water purification, climate regulation, etc.

Handbook of Ecological and Ecosystem Engineering, First Edition. Edited by Majeti Narasimha Vara Prasad.

- *Cultural benefits*: Non-material benefits gained through ecosystems, such as the fulfillment of aesthetic, spiritual, and intellectual needs; recreation; and cultural heritage
- *Support services*: Services needed to produce all other ES, such as pollination, nutrient cycling, etc.

In recent decades, many ecosystems and their services have been substantially degraded in their ecosystem functions [70], mainly due to increasing urbanization. From 2000 to 2006, the annual land used for urbanization in Europe was around 116 000 ha [12]. Urban areas and industrial, commercial, and recreational areas accounted for around 70% of this total. Traffic areas accounted for around 10% of the changes. Compared to the previous period from 1990 to 2000, annual land use increased by around 19% [12]. More and more arable land is being rededicated and thus removed from the production of food or biomass. In cities, which are often favored for settlement, primarily fertile agricultural areas are lost. But forests and natural and semi-natural areas also have to give way to people's urge to expand. This goes hand in hand with the loss of valuable ecological functions.

The urbanized environment is a particular space with a modified water cycle and reduced soil functions. However, the urban realm also provides options for restoring ecosystem services through the establishment of urban green infrastructure (UGI) [18]. The question is how to combine urbanization processes with the creation of ES, and what kind of spaces and approaches to use to restore degraded ecosystems. In this regard, the alternative reuse of brownfields is a developing topic that offers multidimensional benefits.

Urban areas worldwide face serious challenges in dealing with environmental threats such as air, water, and soil pollution; sanitation and waste management activities; construction; and land use management in the context of rapid urbanization, industrial development, urban population increase, and demanding energy and natural resource consumption patterns. Furthermore, cities tend to be more vulnerable to natural hazards such as extreme weather events, regardless of their geographical location, due to climate changes coupled with environmental pollution and socioeconomic risks [21]. Crowded urban areas put significant pressures on infrastructure and public services to cope with daily operational tasks. Urban expansion requires more land than ever to construct new residential areas, offices, shops, industrial parks, and civil engineering projects (roads, bridges, etc.). Geographical barriers (e.g. islands, mountain areas) require some cities to reintegrate each piece of available land into new urban functional areas. In this context, old landfill sites, abandoned facilities, urban brownfields, and transport arteries passing through downtown areas are converted into green spaces, recreational areas, renewable energy farms, sports centers, etc. UGI development is becoming a necessity to respond to increasing air pollution levels associated with the transportation and construction sectors and mitigate the effects of urban heat islands. Smart solutions must rethink how to integrate old, undesired sites into new urban functional areas and use GI as a reliable alternative to deal with emerging environmental threats.

Against this background, EE solutions are further examined in this chapter in three main sections. The first section examines potential landfill afteruse options in urban landscapes as new green or recreational areas, highlighting best practices around the world. Almost every city has an old or an active landfill site that poses serious risks for the urban environment, since this waste management option still prevails across the globe [69]. These sites must be carefully closed, properly monitored, and reclaimed by urban land based on suitable EE solutions. The second section examines the role of EE in managing urban brownfields and old industrial contaminated sites as new land destined for urban consumption. This section highlights the role of brownfield recycling based on GI and presents a detailed case study of the transformation of Albert Railway Station in Dresden (Germany) into a green corridor (Weißeritz Greenbelt) with multiple positive effects: protection from flooding, biodiversity network function, urban climate regulation during hot summer days, and recreational areas. The last section deals with air pollution generated by the urban transportation sector and the role played by large GI in managing this environmental issue. This section points out that replacing old elevated expressways passing through downtown areas with blue/greenways delivers significant social, cultural, and environmental benefits. Therefore, this chapter points out reliable EE solutions based on GI options to combat critical environmental issues associated with landfill sites, urban brownfields, and air pollution related to the urban transportation sector.

5.2 Landfill Afteruse Practices Based on Ecological Engineering and Green Infrastructure

5.2.1 Old Landfill Closure and Rehabilitation Procedures

Landfilling of waste remains the primary option for waste disposal worldwide, particularly in low- and middle-income countries. Poor waste management infrastructure leads to uncontrolled waste disposal practices such as open dumps and conventional landfills without sound environmental protection measures [43]. Sanitary landfills – which are designed to protect the soil and groundwater infiltration, collect and treat the leachate, collect landfill gas, and monitor environmental factors – are replacing such old urban facilities. In some high-income countries, the shift from landfills to waste energy plants and recycling facilities is almost complete (e.g. Sweden, Germany, the Netherlands, Belgium, Japan); but in other countries, the transition toward a "zero waste to landfill" target is more difficult to achieve (UK, USA). Many old landfill sites occupy large areas in the context of increasing urban land consumption.

In some cases, geographical restrictions require cities to reclaim old landfill sites as new functional urban areas supported by innovative projects like those in Japan, Taiwan, and Hong Kong [69]. The transformation of old urban landfills into new recreational areas is being explored in EU countries like Estonia [50] and Croatia [48]. Similar opportunities are being considered in developing countries where public parks are well-received by local communities as a use for ex-urban landfills [63].

The afteruse of urban landfills is an important option for developing urban GI with environmental and socioeconomic benefits for citizens, including a wide range of activities and options [2, 72]. The closure of each landfill site must take into consideration several key actions: (i) using a pipeline network to collect gas that is used for thermal energy/electricity generation or flared; (ii) collecting leachate through pumping wells with a treatment site nearby, and using a capping system (e.g. synthetic geo-composite drainage and blocking layers coupled with soil and vegetation coverage); (iii) investigating slope stability to prevent the collapse of the landfill body and reduce rainwater infiltration; and (iv) monitoring wells, collecting probes, etc.

Following these post-closure measures, the landfill restoration process can begin. This stage may take several years to complete with proper financial resources. The aftercare period can last 30 years, a period in which the landfill can be used for recreational purposes and when landfill sites should be used a with a low density of buildings on top. Open dumps (old landfills) must be revegetated with plant species resistant to landfill gas, drought, and nutrient deficiency (e.g. Acacia spp.), while in the case of engineered landfills, the revegetation process relies on native species with adequate provision of organic amendment and soil organisms [8, 71]. Suitable exotic species such as Acacia spp. can maximize the performance of evapotranspiration to reduce water percolation and restore habitats in former urban landfills sites in humid regions [27].

This section points out some best practices worldwide and the primary challenges involved in transforming old landfill sites into new urban facilities based on EE solutions.

5.2.2 Landfill Restoration Examples Around the World

5.2.2.1 Conventional Landfill Closure (Campulung, Romania)

When in operation, Campulung landfill in Arges county (Romania) served about 44 500 rural and 38 200 urban inhabitants. Campulung landfill had a volume of 380 000 m^3 of municipal household waste on an area of 3.5 ha. The landfill was closed as a result of European and Romanian legislation and covered with an engineered clay capping including geomembrane and a gas ventilation layer. The slope inclination was designed as 1 : 3 with a runoff inclination of 5%.

The site was provided with leachate collection, a gas extraction and combustion system, and a wastewater pretreatment plant. As required in the legislation, an interim cover system including grass seed as erosion protection was built during the settlement period of the waste. Later, a final surface sealing was built, including settlement observation markers, along with a final surface water drainage system (Figure 5.1). The closure took place from 2008 to 2010. Beside the closed landfill, a waste transfer station was built with a total capacity of about 38 000 t/year. The area includes an administrative building, a composting station, and a garage for the machines and equipment used to operate the transfer station, as shown in Figure 5.1. Using former landfill locations for sustainable future waste management activities is quite common.

Figure 5.1 Closed Campulung landfill, Arges county, Romania. (a) Cows and horses grazing on the surface of the closed landfill; (b) biogas plant at the Campulung landfill; (c) the newly built waste transfer station beside the old landfill; view from the landfill body. *Source:* Photos by Petra Schneider. *See colour plate section for colour representation of this figure.*

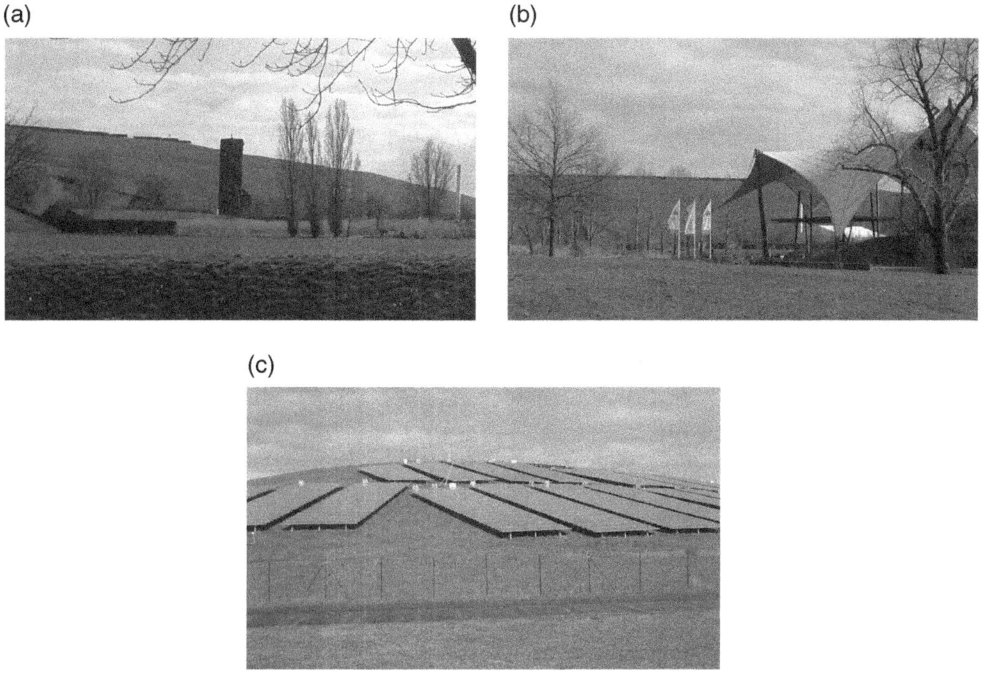

Figure 5.2 (a) Closed Am Cracauer Anger landfill in Magdeburg, Germany, with summer toboggan run and artificial climbing rock; (b) Magdeburg lake concert stage in the Elbauenpark, with the landfill in the background; (c) photovoltaic system on the landfill body. *Source:* Photos by Petra Schneider.

5.2.2.2 Elbauenpark Including Am Cracauer Anger Landfill (Magdeburg, Germany)

At the Am Cracauer Anger landfill in Magdeburg, Germany, a recycling center was built on the eastern part, while the western part of the rehabilitated landfill was included in the Elbauenpark recreation area. The Elbauenpark was established in 1990 for a Federal Garden Exhibition. It was created on a former barracks site that was buried under war ruins. The site covers over 90 ha. In 2000, the closed landfill was included in the park (Figure 5.2). The Cracauer Anger landfill is located in the eastern part of the city of Magdeburg (236 000 inhabitants) and was used from 1945 to 1998 for waste disposal. The landfill has a surface area of 40 ha, and 10.5 million m^3 of municipal and commercial waste was deposited in it. After the end of the storage operation, the landfill was decommissioned, and a surface seal was built in 1998. The surface sealing was completed in 2000. In 2011, a photovoltaic system with a capacity of around 8.5 MWp was put into operation at the landfill, as shown in Figure 5.2

The former landfill forms a unique viewpoint in the recreation park. The ecologically recultivated landfill measures 40 m high, making it the second-highest attraction in the Elbauenpark. Two stairs lead to the landfill mountain. From the north, there are 182 steps to a fallow deer enclosure. To the west, the climb is over 176 steps near a maze. Once at the top, there is a fantastic view to admire the garden art of the Elbauenpark from both east and west and see Magdeburg's skyline.

5.2.2.3 World Cup Park (Nanjido Landfill, Seoul, South Korea)

The Nanjido landfill served Seoul as a waste disposal site between 1978 and 1992, with a restoration period after closure between 1996 and 2002. The post-closure environmental protection

Figure 5.3 Korean world cup park view. *Source:* Vanessaliam, 2013. Licensed under CC-BY-SA 3.0.

measures are shown in Figure 5.3. The FIFA World Cup organized by South Korea and Japan in 2002 led to the transformation of this landfill site into a large eco-park called World Cup Park, which still receives 10 million visitors annually [61]. This eco-park is used for special events and festivals but has significant benefits for local ecosystems. Several years of monitoring found that the number of species (both plant and animal) jumped from 438 in 2000 to 1092 in 2013, and water quality of the Han river has improved [61].

Seoul received a Special Award from the United Nations Human Settlements Program (UN-HABITAT) for its landfill recovery project. This is international recognition of the importance of converting landfill sites into urban GI projects. Restoring landfill sites can play a key role in improving the ecological diversity of urbanized areas in South Korea [10].

5.2.2.4 Fudekeng Environmental Restoration Park (Taiwan)

Most landfill sites are restored as green spaces in the first stage to restore local ecosystems. Restored landfills can be used to construct ground-based photovoltaic parks since other buildings are restricted on these sites.

For example, the old landfill in Wenshan district (Taipei) with a total area of 37 ha was transformed into a green park in 2003 and updated as a renewable energy park in 2017 with 8000 solar panels covering 3 ha, as shown in Figure 5.4. The photovoltaic power station can generate 2 GWh of electricity annually. Such initiatives will be extended in other former landfill sites across Taiwan.

5.2.2.5 Hong Kong

The Hong Kong government initiated a campaign to close and restore 13 urban landfill sites between 1997 and 2006, transforming them into new sports centers, urban recreational areas, and green parks, as shown in Figure 5.5. Table 5.1 presents an overview of these measures.

Figure 5.4 Fudekeng Restoration Park (Taiwan). *Source:* Lai Liangming, 2013. Licensed under CC-BY-SA 3.0.

Figure 5.5 Jordan Valley Park restoration in (a) 2008; (b) 2013. *Sources:* (a) Shaded.b 27. October 2008 GNU Free Documentation. Licensed under CC-BY-SA 3.0; (b) Wing1990hk, 2013. Licensed under CC-BY-SA 3.0.

Besides the new urban facilities, the green areas of old landfills may favor the natural development of rare plant species, e.g. Bamboo Orchid in Pillar Point Valley, as pointed out by Environmental Protection Department of Hong Kong. Ecological restoration of closed landfill sites can be enhanced by mixing fast-growing exotic trees in the first phase (e.g. rapid erosion control) with a diverse range of native species, which ultimately increases the native ecosystem regeneration [8].

5.2.2.6 Hyria Landfill Site (Tel Aviv, Israel)

This "mountain garbage site" site located on Aylon Plain served Tel Aviv as a waste disposal site (118 ha), accumulating millions of tons of waste that polluted the local environment. The landfill site was closed in 1998, and a large project was initiated to transform it into a green park area. The landfill was stabilized using several terraces built with dry stones from recycled construction and demolition waste, covered by soil and fragrant, drought-resistant shrubs and herbs to protect the

Table 5.1 Landfill restoration process in Hong Kong. *Source:* Environment Protection Department (Government of Hong Kong), Landfill Restoration and Afteruse – Summary Information. © 2000, Environment Protection Department (Government of Hong Kong).

Landfill site restored	Area (ha)	Total waste received (Mt)	Recreational features and facilities
Jordan Valley	11	1.5	Jordan Valley Park (2010): Radio-controlled model car racing circuit, horticultural education center, community garden, children's play areas, elderly exercise corner, jogging track
Ma Yau Tong Central	11	1	Part of the site converted into a sitting area (since 2011)
Ma Yau Tong West	6	0.6	Part of the site converted into a sitting area opened to the public in September 2011.
Sai Tso Wan	9	1.6	Sports center (since 2004): baseball and soccer grass, baseball batting cages, environmental gallery, children's playground jogging track, changing rooms, car park
Ngau Chi Wan	8	0.7	Ngau Chi Wan Park (since 2010): archery field, jogging trail, elderly fitness corners, children's play areas, basketball court
Siu Lang Shui	12	1.2	Green area
Ma Tso Lung	2	0.2	For recreational use by the Tung Wah Group of Hospitals
Ngau Tam Me	2	0.15	Green area
Gin Drinkers Bay	29	3.5	Hong Kong Jockey Club International BMX Park (2009): A 350 m standard BMX track and a 60 m BMX cycling track, pro shop and management office, canteen facility, equipment and ancillary stores, changing rooms, parking facilities
Tseung Kwan O StageI	68	15.2	The waterfront of the former landfill was developed into a cycle track/footpath and opened to the public in June 2012.
Tseung Kwan O Stage II/III	42	12.6	The top platform of the former landfill was developed into a model airplane training field for the Hong Kong Air Cadet Crops to use during weekends and public holidays (2005).
Shuen Wan	55	15	The golf driving range (opened in 1999) provides 145 driving bays, one par 3 hole, sand bunkers, a putting green, a pro shop car park, and changing room.
Pillar Point Valley	65	11	Part of the site was developed into a shooting range in July 2016.

landfill from wind erosion and sun radiation. The recycling plant near the landfill site provided the recycled construction materials.

The top of the landfill is divided into several sections: a belvedere pointing toward the skyline of Tel Aviv; on the west, a plateau with several green branches acting as drainage systems; and on the east, a plateau that contains a lake (fed by a water retention system during the rainy season) surrounded by vegetation (herbs, shrubs, and tall trees) and lanes. Lush vegetation of palms and other trees, shrubs, and aquatic and flowering plants is planted in the center of the park as a symbol of nature and Mediterranean culture, replicating an oasis [36]. There is a visiting center building (café, information, wooden deck) on the most stable site of the former landfill.

5.2.2.7 Valdemingomez Forest Park (Madrid, Spain)

The closure of this large landfill, which served as Madrid's main waste disposal site, had three main stages: sealing the surface of the landfill, extracting biogas accumulated in the landfill for electricity generation, and transforming the surface of the landfill into a green park (110 ha) using indigenous species. This landfill accumulated more than 21 million tons of waste during 22 years of activity (1978–2000) with a maximum height of 652 m, around 40 m higher than the surrounding terrain [1]. On top of the landfill, the revegetation process focuses on plant species similar to surrounding areas such as Portuguese oaks, olive trees, pine trees, almond trees, holm oaks, cork oaks, and wetland areas created around two lakes (1500 m^2 each) with two bird-watching refuges [1]. These plants responded well to local conditions, and they expanded from initial small pockets of vegetation. On slopes, synthetic membranes and herbal vegetation were used to combat the erosion process and water percolation. In addition to vegetation coverage and landfill gas capture to generate electricity, the landfill plateau will act as a public park, which will be fully opened in 2030. Today, access to this landfill is restricted to guided tours in some areas. One key aim of this restoration process is to provide a landscape integration of this site with the surrounding Southeast Regional Park.

5.2.2.8 Freshkills Park – A Mega Restoration Project in the US

Transformation of what was formerly the world's largest landfill into the 2200-acre Freshkills Park (of which 1200 acres used to dispose of waste generated by New York City; see figure 5.6) is an example of the untapped resource potential of US landfills as urban open spaces [26]. Use of the landfill site for disposal of municipal waste stopped in 2001. The restoration process involves three central systems, according to the master plan draft [46]: (i) program – a variety of public spaces and facilities for social, cultural, and physical activity (sports fields; community events, mountain biking, canoeing, horseback riding, cycling, extreme sports, education, birding, nature trails, public, outdoor dining); (ii) habitats – wildlife, birds and plant communities, and natural recreational areas (salt marsh, native prairies, maritime oak forest, beech woodland, pine/oak barrens), and

Figure 5.6 Freshkills landfill flyover, Staten Island, New York. *Source:* US EPA, Environmental Protection Agency, 2012. Licensed under CC-BY-SA 3.0.

(iii) circulation – an expansive network of paths and recreational waterways, which will create an interconnected parkland (bikeways, pathways, running trails, 7 miles of park drives, recreational boating, horseback riding trails, etc.). This project is divided into several stages, and the full implementation will have a timeframe of around 20–30 years. Such a large project also involves acceptance from nearby communities. The current status can be consulted at https://freshkillspark.org/tour-information. The complexity of Freshkills Park also derives from political discourse of the local administration related to a landscape associated with waste, climate change, and ecological issues [28].

5.3 Role of Ecological Engineering in Transforming Brownfields into Greenfields

Each society that enters into the development process and starts urbanization activities forces industrialization to create economic benefits. Land is consumed for the extraction of materials that are used to construct housing and industrial facilities. Often this happens without responsible planning of the urbanization process and the protection of soil resources. As history has shown, during societal transformations, industrial sites may be de-industrialized, resulting in *brownfields*. The US Environmental Protection Agency [67], based on Public Law 107–118 [55], defined a brownfield as a property, the expansion, redevelopment, or reuse of which may be complicated by the presence or potential presence of a hazardous substance, pollutant, or contaminant.

In Germany, the Federal Soil Protection Act came into force in 1998 [3, 4]. This law aims to give soil the necessary protection against current and future claims and overexploitation. In the German law, soil is described as having natural functions, an archive function, and use functions. The profiles of the functions are characterized as follows [4].

Natural functions:

- Livelihood and habitat for humans, animals, plants, and soil organisms
- Part of natural cycles, particularly water and nutrient cycles
- Degradation, equalization, and regeneration medium for material effects due to soil's filter, buffer, and material conversion properties, in particular for the protection of groundwater

Archive function:

- Natural archive of cultural history and natural history

Use functions

- Raw material deposit
- Area for settlement and recreation
- Site for agricultural and forestry use
- Location for other economic and public uses, transport, supply, and disposal

Besides this, soil serves as an archive of past soil pollution. Often, waste has been buried in brownfields and endangers use of the land. Further, in brownfields with an industrial history, soil pollution is often encountered.

Polluted sites are complex issues that have arisen in the past, have not yet been mastered in the present, and threaten to pose further risks. Because of the material properties of contaminated sites, their persistence in the environmental media, and the resulting threats, experts are dealing with these areas in an interdisciplinary manner. There are an estimated 3.5 million industrial sites in the European Union, of which nearly 500 000 will need to be remediated [13]. France alone counts about 250 000 such fallow sites due to its industrial past. At least 15 000–20 000 ha of brownfield lands are located in urban areas and could be reused. Belgium also has around 20 000 sites in need of refurbishment. It is estimated that there are more than 450 000 brownfields in the US [67]. Whether old industrial companies, medium-sized factories, gas stations, or other contaminated areas, the benefits of restoration are obvious. Decontamination and transformation into residential and commercial spaces, or even into natural areas through ecosystem restoration, reduces the risks to public health and the environment. In addition, there is a structural shortage of building land in some cities, and some houses are very run down. Here new urban planning could be established.

Brownfields may contain hazardous pollutants from former industrial use. According to ISO/IEC Guide 51, the term *hazard* describes a state or situation in which there is a possibility that a property to be protected (e.g. a person or natural life form) may meet with a potential source of damage. Such danger can lead to damage: for example, to injury, illness, death, or loss of function. Depending on the previous use, brownfields may contain a spectrum of hazardous substances including metalloids, organic pollutants, or inorganic pollutants. Table 5.2 gives an overview of the potential pollutions by use; the main characteristics of the pollutants are given in Table 5.3.

Table 5.2 Overview of industry-typical pollutants (extract).

Type of former use	Operational materials	Pollutants
Petrol storages, petrol stations, depots, repair shops	Petrol, diesel, petroleum ether, cold cleaner	Mineral oils, BTEX, VOC, PAH, arsenic, lead
Gasworks	Coal, tar, ammonia	PAH, phenols, benzene, cyanides, mineral oils
Wood protection: manufacture and processing	Tar oils, solvents, herbicides, pesticides, fungicides	PAH, phenols / PCP, benzene, DDT, aldrin, lindane, chromium VI
Chemical cleaning	Solvents, wash petrol	VOC, BTEX
Metal processing, electroplating	Greases, solvents, hardeners	Mineral oils, VOC, cyanides, metalloids
Painting station	Paints, solvents	BTEX, VOC, mineral oils, metalloids
Railway line operations	Coal, pesticides	PAH, PCP, pesticides, metalloids
Barracks (military properties)	Petrol, diesel, kerosene, heating oil, cold cleaner	Mineral oils, BTEX, VOC
Blasting sites, dropping sites	Explosive compounds, explosive powder compounds	Trinitrotoluene (TNT), hexogen (RDX), pentaerythritol tetranitrate (PETN, nitropenta), picric acid, nitrocellulose (NC), dinitrotoluenes (DNT), nitroglycerin (Ngl), diethylene glycol dinitrate (DEGDN)
Transformer stations	Oils	Mineral oils, PCB
Scrapyards	Oils, hydraulic oils, brake oils, diesel, petrol, greases, solvents, cold cleaners	Mineral oils, BTEX, VOC, PCB, metalloids
Printing stations	Solvents, paints	BTEX, VOC, metalloids

Table 5.3 Main characteristics of the pollutants.

Pollutant	Habitus	Effect on...	Geoaccumulation?	Bioaccumulation?	Carcinogenic?	Toxic to humans?
Mineral oil	Liquid (petrol/diesel)	Groundwater, soil	Yes	Yes	No	No
Mineral oil	Solid (asphalt)	foundation	Yes	Yes	No	No
PAH	Liquid (oil)	Groundwater, soil	Yes	Yes	Yes	Yes
PAH	Solid (ashes, tar)	Groundwater, soil	Yes	Yes	Yes	Yes
BTEX	Liquid, gaseous	Groundwater, soil, gas	Yes	Yes	Yes	Yes
VOC	Liquid, gaseous	Groundwater, soil, gas	little	Yes	Yes	Yes
PCB	Liquid, solid	Groundwater, soil	Yes	Yes	Yes	Yes
Phenols	Liquid, solid	Groundwater, soil	Yes	Yes	Yes	Yes
Cyanides	Liquid, solid	Groundwater, soil	Yes	Yes	No	Yes
PCP	Liquid, solid	Groundwater, soil	Yes	Yes	No	Yes
Metalloids	Solid	Groundwater, soil	Yes	Cu, Ni, Zn: essential As, Cd, CrVI, Hg: not essential	As, Cd, CrVI, Ni, Hg: yes Pb, Zn: no	Yes

Some pollutants can be eliminated through biological processes of phytoremediation, using targeted plants for remediation purposes. Feasible plants are herbaceous species, grasses, trees, wetland species, and algae for the remediation of certain pollutants. The main processes are [19, 40, 51–54]:

- Phytostabilization, a process that contains the pollutants in plant material (typical pollutants: As, Cd, Cr, Cu, Pb, Zn)
- Rhizodegradation, a process that causes remediation by destroying organic compounds (PAHs, BTEX, pesticides, PCB)
- Phytoaccumulation, remediation by extraction and capture, applicable for metalloids and radionuclides
- Phytodegradation, a process that destroys organic compounds, chlorinated solvents, phenols, pesticides, and explosive compounds
- Phytovolatilization, a remediation process based on extraction from media and release into the air, applicable for chlorinated solvents and some inorganics (As, Hg)
- Evapotranspiration, which refers to containment and applies to water-soluble organics and inorganics

More than 400 plant species described in the literature have usable functions for phytoremediation. Preconditions for use are the bioavailability of the pollutants, reachability of the pollutants in the root zone, toxicity tolerance of the plants against the pollutants, and plant growth. Besides land farming as a remediation process for phytoremediation, GI can also be designed and applied to serve such a purpose. Some typical terrestrial plants are *Thlaspi arvense* and *Brassica juncea* (both accumulate metalloids) and *Salix caprea* (accumulates As, Cd, Pb, Zn), and an effective fluviatile plant is *Eichhornia crassipes* (accumulates Cd, Cr, Co, Ni, Pb, Hg, cyanides). A fluviatile plant that degrades organic compounds through rhizodegradation is *Typha specialis*.

5.3.1 UGI Options for Brownfield Recycling

The remediation of disused industrial areas is technically complex and expensive. In addition, numerous environmental regulations must be complied with. Generally, there are two approaches to brownfield recycling:

- Rehabilitation/revitalization, referring to industrial reuse
- Restoration, referring to returning the site to a near-nature green space or a nature protection area, including use as a biodiversity hub for a biotope network, reforestation, or urban farming

Brownfield recycling as part of the circular economy is naturally a dynamic process. Concerning the need to generate economic potential, there are often planning reservations about reuse as a GI, and revitalization is given preference. In this context, renaturation should also be understood as a potential interim solution until industrial infrastructure or residential development can give the area a new function. Nevertheless, there will always be areas that will be permanently removed from the revitalization process. However, this does not mean these areas do not contain any economic potential: for example, they can be developed as a land reserve pool for compensation measures in nature conservation. GI reuse for climate protection and CO_2 binding purposes are also conceivable, for example, by creating forests. Finally, the recreational function should be mentioned because nature conservation upgrades these areas.

In the United States, the concept of private equity funds working with technicians and financial experts has already proven its worth [13, 67]. For example, the Cherokee Fund, which has been in existence for over 20 years, has reused around 500 industrial wastelands. It now manages $2 billion [13]. On the other hand, in continental Europe, very few industrial wastelands are being reused with the support of private investors. Site owners, planners, and property developers work together only sporadically. While there are some examples in the UK, such as the London Olympic Stadium, the commercial and urban planning potential of such cooperation between public or private owners and industrial companies and planners/developers is far from exhausted [13].

The US Brownfields Program creates many benefits for local communities [67], as highlighted here:

- Leveraged $16.86 per EPA dollar expended
- Leveraged more than 144 800 jobs nationwide
- Can increase residential property values 5–15.2% near brownfields sites when cleanup is completed
- Can increase local tax revenues when cleanup is completed

5.3.2 Pilot Case: Restoration of a Brownfield to Provide ES – Albert Railway Station (Dresden, Germany) Transformation into the Weißeritz Greenbelt

Restoring a brownfield to provide ES is illustrated through the example case of the Albert railway station in Dresden, Germany. Dresden is the capital of the Freestate of Saxony with approx. 510 000 inhabitants in southeast Germany, close to the border with Poland and the Czech Republic. The

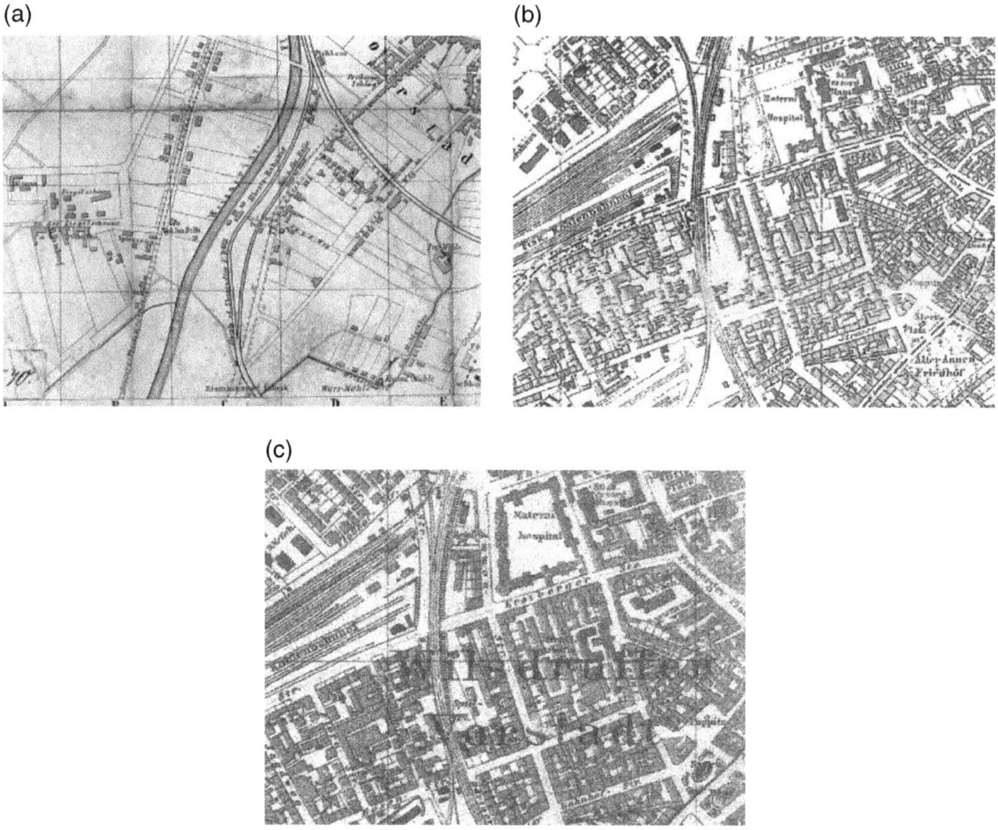

Figure 5.7 (a) Excerpt from a Dresden city map from 1856 showing the location of the Albert railway station in the center of the map; (b) the coal station on a map from 1898; (c) the coal station on a map from 1927. Sources: (a) *Source:* Hessler, Plan de Dresde, 1856 [25]. Licensed under CC-BY-SA 3.0; (b) Gerke survey director [17]; (c) *Source:* City Map historical version, 1927 [68]. Licensed under CC-BY-SA 3.0.

Albert station was in the Dresden Wilsdruff suburb. The railway station, which opened in 1855 as the terminus of the Albertsbahn railway, was used primarily to supply Dresden's households and industry with hard coal from surrounding areas [6]. In 1868, the railway line was integrated into the Bohemian railway station, and afterward the coal station was used only for goods provision [6]. After World War II, the station gradually deteriorated [22, 24]. Figure 5.7 illustrates the development of the railway line as part of the urbanization process in the period 1856–1927. Urban density increased substantially in that period. Figure 5.8 shows the coal station.

Almost all of Dresden's coal needs for industry and households were transferred from railway cars to horse-drawn carriages at the coal station in the nineteenth century. At the turn of the century, the station's handling capacity was around 500 000 tons of hard coal annually [22].

At the end of the 1960s, the new Dresden-Altstadt freight yard was built south of the coal station, connected to the Albert station via a double-track line that branched off at the Nossen bridge. When the construction of Dresden central station was underway three decades later, the freight yard was partially converted into a station for passenger trains [31]. In the station overview from 1922, it bears the name Dresden-Altstadt, Kohlenbahnhof. In the 1920s, most of the coal deposits in the Plauenschen Grund were mined, so the station became increasingly insignificant [22]. In 1945, it was badly damaged in the air raids on Dresden. After World War II, the Albertbahnhof was

Figure 5.8 (a) View of the former coal station; (b) workers breaking frozen coal from the railway in winter. Sources: (a) *Source:* Dresden city, Planning authority, photo service; (b) *Source:* Saxon State, State and University Library Dresden, photo library, Hähne, E. & Pohl, E. Public information panel photo by Petra Schneider.

poorly repaired and was only used as a storage station for scrapping decommissioned steam locomotives [22].

With the construction of the new thermal power station on the Nossener Bridge, the freight yard again served as a coal station for the storage of block trains with brown coal from the Cottbus district [31]. Until the mid-1960s, mainly 89.0 class tank locomotives were used at the coal station. Toward the end of the GDR era, there was only one track for local coal trade in the station. After 1990, the connecting track to the Dresden-Mitte power station was dismantled, and the track systems were used for the parking and maintenance of passenger coaches. In the first years after Germany's reunification, the area was fallow land with dismantled former industrial facilities.

The structural change since the early 1990s brought with it considerable functional losses [45]. Fallow areas, neglected green spaces, and public infrastructure in generally poor condition shaped the image of Dresden's southwest, where the Weißeritz River enters the urban area and flows west of the city center into the Elbe. The Weißeritz is Dresden's second-largest river. Its mouth was moved 3 km down the Elbe at the end of the nineteenth century to make room for the railway connection between Dresden central station and Dresden Neustadt. Subsequently, the areas on both sides of the river became industrial. Much of the course of the river was straightened and delimited by stone walls.

Against this background, the city of Dresden planned the "Weißeritz district development project" in 2000. It developed an integrated action plan for this urban area [45]. The 2002 flood caused the Weißeritz River to return to its old river bed and flood the neighboring districts. As a result, the urban renewal project received increased attention and the additional dimension of flood

Figure 5.9 Overview of the Weißeritz Greenbelt in Dresden. *Source:* Google Maps.

protection. In connection with the attractive landscape of the Dresden-Plauen district and industrial cultural heritage e a cross-urban green belt, the Weißeritzgrünzug, was developed: an attractive open green space that is connected to the city center by a footpath and bike path. Polluted soil in the zone of former industrial activities was replaced. The Weißeritz Greenbelt was officially opened in 2008 and extended in 2013. It is approximately 3 km long (Figure 5.9).

With the Weißeritz district development project, Dresden faced the simultaneous challenges of structural change and flood protection in an urban area [45]. Significant restrictions for water restoration in densely built-up urban areas have become clear. All the more remarkable is using flood protection measures and urban renewal for a local green space, footpath, and bike path connection, thus improving quality of life. This combination of measures to protect against flooding and environmentally related measures for urban renewal has a lasting impact [45]. Information boards were set up to explain the visible historical references along the green corridor (Figure 5.9). The construction was co-financed by the European Commission. The strategically planned network of the Weißeritz Greenbelt offers a migration path for fauna and flora, even in urban area (Figures 5.10–5.13).

The Weißeritz Greenbelt illustrates the role and the potential of EE in the brownfield/greenfield transformation process and the development potential for UGI in urban areas. It is a

(a) (b)

Figure 5.10 Information boards were set up to explain the visible historical references along the green corridor. *Source:* Photo by Petra Schneider.

(a) (b)

Figure 5.11 (a) Art installations with an industrial touch were set up to commemorate the history of the area; (b) gabions were built to create living space and increase inner-city biodiversity. *Source:* Photo by Petra Schneider.

(a) (b)

Figure 5.12 In addition to nature conservation elements (gabions, left), installations were built for human recreation: (a) gabions; (b) skater track. *Source:* Photos by Petra Schneider.

(a) (b)

Figure 5.13 (a) Piles of stones offer a retreat for insects and small reptiles (lizards). (b) Building material from the site's past industrial use was also used to build the gabions. *Source:* Photos by Petra Schneider.

multifunctional hub that provides all types of ES, including biodiversity network functions, protecting the city from flooding, and regulating the city climate on hot summer days, and it is an attractive place for the population to relax.

5.4 Green Infrastructures for Mitigating Urban Transport-Induced Threats

Transport is acknowledged as one of the primary sources of urban pollution and greenhouse gas emissions that cause global warming [7, 14, 16, 29, 65]. Urban transport-induced threats include human health concerns triggered by urban pollution and fatalities due to transport-related accidents. Furthermore, a car-centered way of life is associated with high levels of physical inactivity and, consequently, is causing unprecedented levels of health problems such as obesity and diabetes. Car-centered urban policies make it less safe for people to move freely within the urban space [9]. Today, around 1.35 million people die annually in car accidents, most of them in urban areas, and car crashes have recently become the primary cause of death for children aged 5–14 [66].

Against this backdrop, cities and governments worldwide are searching for solutions promoting safer and sustainable urban mobility. The first actions were taken following World War II and consisted of implementing regulations aimed at increasing traffic safety and reducing urban pollution. Initially, California (US) led the battle for a more sustainable transportation sector [47]. However, in recent decades, the European Union has emerged as the leading political entity in mitigating urban transportation-induced pollution. It has done so mainly by implementing various regulations [47] and promoting the adoption and implementation of sustainable urban mobility plans [58, 59].

On the practical side, EE and UGI play a significant role in mitigating urban transport-induced threats and promoting sustainable mobility. Solutions can take multiple forms. They range from the radical approach of completely replacing current major road arteries with linear parks/waterfronts to more limited interventions such as building green roofs for bus stops, greening tramway paths, building tree-lined streets, etc. Against this array of interventions, some countries and cities are taking a risk and trying cutting-edge EE solutions. One such approach replaced former urban expressways arteries with greenways by deploying extremely large GI.

This section discusses two such EE projects well known for their pioneering efforts and related controversies: the Rose Kennedy Greenway in Boston (Massachusetts, US) and the Cheonggyecheon river restoration in Seoul (South Korea). The primary purpose of these investments was to tackle traffic jams and related negative effects – such as noise and pollution – and therefore increase the value of the land and the quality of life for urban citizens. The Rose Kennedy Greenway is known for its approach going "against the backdrop of the planning system in the US" [49], while Cheonggyecheon river restoration is considered to be "the most known urban renewal showcase" that has taken space from roads and given it back to nature [9]. By implementing high-cost projects, the two cities recognized that an increasing number of cars jeopardizes the urban environment, thus affecting places where people live, work, and relax [9].

5.4.1 Transportation Heritage from the Industrial Period

During the first two decades following World War II, cities were designed around motor vehicles. Thriving metropolises built large urban transportation infrastructures to accommodate the increasing number of cars and facilitate vehicular flow. As a symbol of independence, freedom, and welfare, the car set out to conquer the world, including every urban street. In this context, streets have been mostly designed for cars, with increasingly larger spaces dedicated to them. In some cases, large elevated expressways were built even in the downtown areas of prosperous cities. Boston and Seoul were two such cases.

During the 1940s, Boston started to plan for the John F. Fitzgerald Expressway (the elevated Central Artery) passing through its downtown – a huge infrastructure project also known at the time as the "Highway in the Skies" [15]. The construction was carried out during the 1950s and built at the expense of a vibrant traditional neighborhood [49], with thousands of inhabitants displaced and hundreds of buildings demolished. Cars and road traffic clearly dominated urban planning. During the 1960s and 1970s, Seoul's municipality covered the Cheonggyecheon stream in concrete and asphalt and then elevated an expressway over the road "that had been built over the culverted river" [20]. This transformation of the Cheonggyecheon stream epitomizes the same urban development process in which cars play the central role and conquer almost every aspect of urban mobility.

The limitations of the two road arteries located in the heart of their respective cities soon became apparent. Traffic congestion, noise, pollution levels, and declining appeal of abutting areas called for dramatic changes in urban planning. Such extensive transport infrastructure became a threat to quality of life, pedestrian safety, and the global image of the host cities and their subsequent economic competitiveness. In this context, a shift from deploying large road transport infrastructure to deploying large GI projects favoring sustainable forms of mobility was about to happen. EE and GI emerged as leading approaches in promoting alternatives forms of mobility that could be part of solutions for tackling emerging urban environmental threats.

5.4.2 The Cases of the Rose Kennedy Greenway and Cheonggyecheon River Restoration

5.4.2.1 The Concept: Expressway-to-Greenway Conversion

Boston's solution consists of removing the elevated expressway that was passing through downtown and relocating it underground while also improving expressway connections between major interstates and the airport, including building long tunnels and new bridges. In the downtown area, the project created a substantial area of prime new urban land. About 27 acres of land were thus generated, and a land-use dilemma emerged concerning the use of new areas regained by the city [49]. It was finally resolved by deciding that at least 75% of the area would be dedicated to public spaces such as parks and urban plazas [5]. Indeed, 20 acres out of the 27 generated by removing the former expressway have been devoted to a linear park, widely known as *Boston's ribbon of contemporary parks* [49]. Although aerial images might suggest the existence of a linear park (Figure 5.14), the Boston Greenway is more a series of parks than a single linear one. It is composed of seven parks, each designed by a different architect and having different designs and sometimes different functionalities.

A similar solution was implemented in the case of the Cheonggyecheon area in Seoul. During 2003–2004, the elevated expressway was torn down to make room for the restored Cheonggyecheon stream and an urban linear park [32]. The initiative's primary purpose was to replace transport infrastructure that promoted a highly polluting form of mobility with an attractive urban amenity promoting quality of life and sustainable mobility [32].

Cheonggyecheon's revitalization is a policy intervention to promote sustainability in Seoul [34] and probably the most expensive urban river restoration project yet undertaken (Figure 5.15). An interesting fact about such large-scale GI is that it not only promotes healthier and safer urban mobility but also fosters social capital by transforming former urban barriers into public spaces. Although expressways were built during the 1950s and 1960s to ensure movement and accessibility, they paradoxically ended up fragmenting the urban landscape, disconnecting neighborhoods, and

Figure 5.14 Rose Kennedy Greenway in Boston after the Big Dig. *Source:* Greenway Conservancy, 2017. Licensed under CC-BY-SA 4.0.

impeding local movement. Their removal and replacement with a linear park/water stream led to fundamental changes in the vicinity: former urban barriers and unattractive areas become places of interaction and social life. For example, the Boston Greenway now connects some of its oldest and most vibrant neighborhoods. Furthermore, such large-scale GI incorporates a strong artistic and cultural dimension by promoting and integrating art into public space. Indeed, the Boston Greenway promotes innovative contemporary art and engages people in various cultural experiences [56]. Small and large festivals, concerts, free fitness classes, and food vendors animate the Greenway all year round. Art has even been recognized as fundamental, and a five-year public art strategy for the Greenway area was defined in 2012 for the period 2013–2017 [41]. This shows the increasing synergies between deploying urban GI and public art. GI mixed with public art facilities has the potential to rebrand cities and build new urban identities.

Looking at the impact of such large GI projects is probably the most efficient way of evaluating their current value and future perspective. The environmental, social, and economic impacts of these two projects have been widely discussed in the literature, although not every aspect was addressed.

5.4.2.2 Environmental Efficiency and Effectiveness

The two case studies show that a number of significant environmental benefits can be attributed to large GI investments. Researchers generally agree on the positive impact of mitigating urban heat islands, increasing biodiversity, and decreasing pollution and greenhouse gas emissions.

Figure 5.15 Cheonggyecheon after restoration. *Source:* Stari4ek, 2008. Licensed under CC BY-SA 2.0.

There is agreement that the Cheonggyecheon policy intervention *mitigated warming in the downtown area*. The policy intervention decreased daily mean temperatures in a buffer of about 300 m along the Cheonggyecheon stream and has also increased relative and specific humidity [35], further leading to decreased energy demand for cooling nearby buildings [34]. Some studies report that mean temperatures have fallen, such that the temperature is 3.3–5.9 °C cooler along the Cheonggyecheon stream compared to parallel roads located a few hundred meters away (Seoul Development Institute, as cited in [32]). We are not aware of similar studies on the effect of the Boston Greenway on air temperature and humidity. However, removing materials that accumulate solar heat (concrete and asphalt) and are responsible for urban heat islands has probably led to similar results there.

Studies also seem to agree on the *increased biodiversity generated* by the projects. Comparisons between the number of plant and animal species before and after the Cheonggyecheon project suggest a significant increase in biodiversity [23, 57]. However, other studies (e.g. [33]) found that some animal classes decreased in diversity in some periods after 2009. We are not aware of similar studies conducted for the case of Boston's Greenway. But what is interesting is that it is organically managed, which makes it a unique area in Boston and one of the few such parks in the United States. Organic management does not use toxic chemicals and thus favors a healthier environment for both the ecosystem and visitors interacting with it.

Both projects promote more sustainable means of transportation and thus contribute to *reducing greenhouse gas emissions*. In the case of Boston, the expressway was moved underground; simultaneously, expressway connectivity was improved, leading to fewer traffic jams, saved time, and less greenhouse gas emissions [64]. As the top of the underground expressway was covered by parks and urban plazas, increased walkability and cycling have been seen. Walkability, cycling, and public transportation have also been significantly enhanced in the case of Cheonggyecheon [37]. Furthermore, less vehicular traffic and more walking, cycling, and public transport usage lead to *less pollution*. Indeed, researchers found a significant positive impact in the case of Cheonggyecheon's revitalization. Levels of CO, NO_2, and O_3 significantly decreased after the inauguration of the park [34]. As of 2010, a 15% decrease in particulate matter has also been reported, compared to 10% in neighboring areas [37].

5.4.2.3 Social Impact

The impact of GI on the social life of the two cities varies depending on the issue. According to a review by Kim and Young [34], there was a significant positive impact on residents' mental health and improved quality of life for those residing near the Cheonggyecheon area. We are not aware of similar studies for the Greenway area in Boston. However, the results should be similar, as recent studies have found that happiness is significantly influenced by the quality of the landscape [62].

On the other hand, governance quality is highly debated. Kim and Young [34] found that public engagement of all stakeholders in decision-making is "one of the most contentious dimensions." Some argue that the Cheonggyecheon project fostered collaborative approaches, while others point to the fact that a few elites mainly drove the decision-making and some citizen groups and merchants were excluded from the entire process. On the other hand, in the case of the Boston Greenway, Mahmoud et al. [39] conducted a series of interviews and found that the governmental approach tried to conserve age, gender, social, and class diversity and strongly encouraged people to mix along the Greenway. All in all, the social impact of both projects on creating just cities and enhancing the right to the city was positive, although negative impacts such as gentrification do exist.

5.4.2.4 Economic Efficiency

Both projects are known for being extremely expensive. Boston's Greenway was part of a larger project commonly known as the *Big Dig*. The Big Dig project created underground expressways and bridges, as well as the Rose Kennedy Greenway. Hence, it involved huge, complex work over 17 years (1991–2008), and it experienced unexpected escalating costs. The whole Big Dig project was first estimated to cost less than $3 billion. However, the cost was subsequently revised to $14.6 billion in 2007 and further recalculated at $20 billion or $24.3 billion, depending on the year and source, and was sometimes compared to massive engineering projects like the Panama Canal [64]. With a cost of around $366 million, Cheonggyecheon is known for being the most expensive urban river revitalization project yet undertaken [20]. It should not be a surprise that the two projects' cost-benefit ratios were constantly under scrutiny. One study [38] employs an ex-post cost-benefit analysis using actual data and maintenance costs and finds a 0.75 benefit-cost ratio for the Cheonggyecheon revitalization project (compared to the 1.89 ratio estimated in the ex ante analysis). The same study argues that the benefits were lower because the project did not meet the expectations of citizens expecting it to restore the original/natural Cheonggyecheon stream. Instead, the intervention created an urban, artificial landscape. In Boston, Sorrin [64] takes a holistic approach by computing not only pure economic benefits (capital, labor, loan interests, property values, etc.) but also ES, air quality, value of time, and recreation. She finds that, according to the most conservative scenario, the *Big Dig's* benefits amount to 5% of the total costs. However, benefits could have been at least 20% if costs had been kept under control.

Despite significant positive impacts in both cases, the urban development models heralded by the two projects quickly receded. Such projects have not been replicated in the United States; in South Korea, the model has been replaced by a new focus on welfare [37], despite some significant initiatives aiming at replicating Cheonggyecheon at the national scale (e.g. Cheonggyecheon Plus 20). This suggests that large-scale green and blue infrastructure have their limits, and perhaps smaller-scale policy interventions should be implemented and replicated in a broader scale.

Did Cheonggyecheon's restoration and Boston's Greenway restore ecosystem functions? The answer is nuanced. The projects delivered blue and green infrastructure in the heart of two densely populated cities. However, the ecosystems put in place are more urban-like than natural. Did they put in place GI capable of delivering ES? In this case, the answer is clearly yes.

5.5 Conclusions

Urban land reclamation for GI projects based on EE solutions is expected to increase, to cope with emerging air, water, and soil pollution issues and provide new urban green spaces for crowded urban settings. Urban landfills and contaminated industrial sites need to be transformed from urban environmental pollution hotspots into new functional urban areas in an environmentally friendly way.

In the first phase, the proper closure of landfill sites is compulsory (e.g. leachate and biogas collection systems, capping system, monitoring wells), coupled with revegetation with indigenous species and/or drought- and erosion-resistant plants. This chapter has shown some best practices of global landfill restoration practices and afteruse options (green/forest parks, wetlands, recreational areas, sports centers, etc.) to point out opportunities for former urban landfills to become new urban GI projects with multiple benefits for social, economic, and environmental features that enhance the transition toward urban sustainability based on EE solutions.

Meanwhile, brownfield recycling is recognized as a social necessity since global land consumption is still at a high level. Despite the current strong demand for space, there are regions where there is no demand for land and revitalization of such spaces is unlikely. This situation can have various reasons; migration of the population from these typically structurally weak regions usually plays an important role. In this context, one has to think about which regional development policy perspectives exist for such regions. The probability of revitalizing all of these regions and equipping them with new industrial infrastructure is low. This allows us to think about alternatives, including new and/or different forms of land use. One of these alternatives is ecosystem restoration; this might involve natural processes (succession); purposeful greening of these areas, e.g. with green areas or parks; or an intermediate step between these two types of greening (steered succession). Both approaches are forms of renaturation. One way of adapting reuse in regions for which revitalization is unlikely in the near future is the concept of GI, which was developed in recent years. An essential feature of GI is multifunctionality, i.e. intersectoral multiple uses of areas with simultaneous advantages for nature and the environment.

The two example projects related to the road transportation sector show that leading cities worldwide started to recognize long ago that human communities depend very much on the benefits generated by Nature. Therefore, cities and governments are open to investing large amounts of money to correct errors made half a century ago and deploy blue and green infrastructure in places once dominated by large elevated expressways. The literature discussed in this chapter clearly suggests that the two urban GI projects are delivering at least two of the four types of ecosystem services as defined by the Millennium Ecosystem Assessment [42]: regulatory benefits (such as air and water purification and topo-climate regulation) and cultural benefits (recreation and fulfillment of aesthetic, spiritual, and intellectual needs).

References

1 Alba, I. (2018). A recovered landfill in the construction of a metropolis: the Valdemingómez project, over time. *Ri-Vista* 16 (1): 68–89. https://doi.org/10.13128/RV-22989.

2 Artuso, A., Cossu, E., and Stegmann, R. (2018). Afteruse of landfills. In: Solid Waste Landfilling. Concepts, Processes, Technologies (eds. R. Cossu and R. Stegmann), 915–936. Elsevier. Available at: https://arcoplan.it/wp-content/uploads/2019/04/SWL-Capitolo-16.3.pdf.

3 BGBl. (1998). Bundes-Bodenschutzgesetz as of 17.03 1998 (BGBl. I S. 502), Soil Protection Law of Germany, amended as of 27 September 2017 (BGBl. I S. 3465).

4 BGBl. (1998). Bundes-Bodenschutz und Altlastenverordnung as of 12. July 1999 (BGBl. I S. 1554), Soil Protection and Brownfield Legislation of Germany, amended as of 27 September 2017 (BGBl. I S. 3465).

5 Boston Redevelopment Authority (1991). Boston 2000 plan. https://archive.org/details/boston2000planfo00bost/page/8/mode/2up.

6 Canzler, A., Hauschild, A., and Neumann, L. (1878). Die Bauten, technischen und industriellen Anlagen von Dresden. Dresden: Meinhold & Söhne.

7 Chapman, L. (2007). Transport and climate change: a review. *Journal of Transport Geography* 15: 354–367.

8 Pang, C.-C., Lo, W.-F., Yan, R.W.-M., and Hau, B.C.-H. (2019). Plant community composition on landfill sites after multiple years of ecological restoration. *Landscape Research* 45: 458–469. https://doi.org/10.1080/01426397.2019.1674266.

9 Costa, C.S. (2014). Can we change processes in our cities? Reflections on the role of urban mobility in strengthening sustainable green infrastructures. *Journal of Traffic and Logistics Engineering* 2 (2): 146–155.

10 Do, Y., Kim, J.Y., Kim, G.-Y., and Joo, G.-J. (2014). Importance of closed landfills as green space in urbanized areas: ecological assessment using carabid beetles. *Landscape and Ecological Engineering* 10 (2): 277–284. https://doi.org/10.1007/s11355-013-0223-x.

11 European Comission (2013). Communication from the Commission to the European Parliament, the Council, the European Economic and Social Committee and the Committee of the Regions – Green INFRASTRUCTURE (GI) – Enhancing Europe's natural capital. Brussels: European Commission. https://eur-lex.europa.eu/resource.html?uri=cellar:d41348f2-01d5-4abe-b817-4c73e6f1b2df.0014.03/DOC_1&format=PDF.

12 European Environment Agency (2017). Land take assessment 2006–2012. https://www.eea.europa.eu/data-and-maps/indicators/land-take-2/assessment-1 (accessed 9 December 2019).

13 European Investment Bank. (2019). Neue Zukunft für alte Industriebrachen. https://www.eib.org/de/infocentre/stories/all/2015-february-01/cleaning-up-the-past-building-the-future.htm.

14 Eva, M., Mihai, F.C., and Munteanu, A.V. (2019). Sustainability of the transport sector during the last 20 years: evidences from a panel of 35 countries. *19th International Multidisciplinary Scientific GeoConference on Ecology, Economics, Education, and Legislation (SGEM 2019)*. doi: 10.5593/sgem2019/5.1/S20.085

15 Ferraiuolo, A. (2009). Religious Festive Practices in Boston's North End: Ephemeral Identities in an Italian American Community. New York: Sunny Press, State University of New York.

16 Gabbastiss, J. (2018). Transport becomes the most polluting UK sector as greenhouse gas emissions drop overall. *Independent*. www.independent.co.uk/environment/air-pollution-uk-transport-most-polluting-sector-greenhouse-gas-emissions-drop-carbon-dioxide-a8196866.html.

17 Gerke Survey Director (1898). Map of Dresden, edited by the surveying office of the city of Dresden. https://commons.wikimedia.org/w/index.php?curid=30681749.

18 German Federal Agency for Nature Conservation (2017). Urban green infrastructure. A foundation of attractive and sustainable cities. Pointers for municipal practice. https://www.bfn.de/fileadmin/BfN/planung/siedlung/Dokumente/UGI_Broschuere_Eng.pdf.

19 Glass, D.J. (1999). U.S. and International Markets for Phytoremediation, 1999–2000, 266. Needham, MA: D. Glass Associates.

20 Green, C. (2011). Input to deliverable 6.1.5–6 Comparative analysis of enabling factors for sustainable urban water management image: Christian.com Case study brief – The restoration of the river Cheonggyecheon, Seoul. http://www.switchurbanwater.eu/outputs/pdfs/W6-1_GEN_DEM_D6.1.6_Case_study_-_Seoul.pdf.

21 Grozavu, A. and Mihai, F.-C. (2018). Introductory chapter: environmental risks between conceptualization and action. In: Environmental Risks (eds. F.C. Mihai and A. Grozavu), 1–12. Intech https://doi.org/10.5772/intechopen.81072.

22 Hahn, A. and Neef, E. (1985). Werte Unserer Heimat – Dresden (Band 42). Akademie-Verlag Berlin.

23 Hanley, P. (2014). Eleven. Victoria: Frissen Press, 400 p.

24 Herrmann, L. (2019). Elbezweigbahn. www.dresdner-stadtteile.de/Zentrum/Friedrichstadt/Dresden_-_Berlin/Anschlussbahn_Kohlebahnhof/anschlussbahn_kohlebahnhof.html.

25 Hessler, O. (1856). Plan de Dresde, Ausschnitt Albertbahnhof, Lith. Rau: Anst. von J. H. C.

26 Hoefer, W., Gallagher, F., Hyslop, T. et al. (2016). Environmental reviews and case studies: unique landfill restoration designs increase opportunities to create urban open space. *Environmental Practice* 18 (2): 106–115. https://doi.org/10.1017/S1466046616000090.

27 Hui, L.C. and Chu, L.M. (2019). Identifying suitable tree species for evapotranspiration covers of landfills in humid regions using seedlings. *Urban Forestry & Urban Greening* 38: 157–164. https://doi.org/10.1016/j.ufug.2018.12.004.

28 Hutchinson, P. (2017). Exploring the connection between landscape and biopolitics: the story of Freshkills Park. *Landscape Review* 17 (1): 96–107.

29 IEA (2017). CO2 Emissions from Fuel Combustion. 2017 Statistics. Paris: International Environment Agency OECD.

30 International Union for Conservation of Nature (IUCN) (2014). Green infrastructure, a sustainable answer to Europe's challenges. https://www.iucn.org/content/green-infrastructure-sustainable-answer-europe%E2%80%99s-challenges.

31 Kaiß, K. and Matthias, H. (1994). Dresdens Eisenbahn: 1894–1994. Düsseldorf: Alba Publikation, 3-87094-350-5.

32 Kang, C.D. and Cervero, R. (2009). From elevated freeway to urban greenway: land value impacts of the CGC project in Seoul, Korea. *Urban Studies* 46 (13): 2771–2794.

33 Kang, T.-H., Yoo, S.-H., Kim, I.-K. et al. (2012). Change of avifauna following restoration and management in Cheonggye stream, Seoul, Korea. *Journal of Korean Nature* 5 (2): 107–114.

34 Kim, H. and Jung, Y. (2019). Is Cheonggyecheon sustainable? A systematic literature review of a stream restoration in Seoul, South Korea. *Sustainable Cities and Society* 45: 59–69.

35 Kim, K.R., Kwon, T.H., Kim, Y.-H. et al. (2009). Restoration of an inner-city stream and its impact an air temperature and humidity based on long-term monitoring data. *Advances in Atmospheric Science* 26 (2): 283–292.

36 Latz, T. (2018). Rehabilitation of the Hiriya landfill, Tel Aviv. *Ri-Vista* 16 (1): 54–67. https://doi.org/10.13128/RV-22991.

37 Lee, J.Y. and Anderson, C.D. (2013). The restored Cheonggyecheon and the quality of life in Seoul. *Journal of Urban Techonology* 20 (4): 3–22.
38 Lee, M. and Jung, I. (2015). Assessment of an urban stream restoration project by cost-benefit analysis: the case of Cheonggyecheon stream in Seoul, South Korea. *KSCE Journal of Engineering* 20 (1): 152–162.
39 Mahmoud, I.H., Appleyard, B., and Bevilacqua, C. (2019). From 'highway into greenway': how public spaces change zoning regulations. The case of Northern Park rose Fitzgerald Kennedy, Boston, USA. In: New Metropolitan Perspectives. Local Knowledge and Innovation Dynamics Towards Territory Attractiveness Through the Implementation of Horizon/E2020/Agenda2030, vol. 2 (eds. F. Calabrò, L. Della Spina and C. Bevilacqua), 200–210. Springer.
40 McCutcheon, S.C. and Schnoor, J.L. (2003). Phytoremediation, Transformation and Control of Contaminants. New York: Wiley.
41 McGregor, J. and Piechocki, R. (2012). Public art on the greenway. A five year strategy: 2013–2017. Boston.
42 Millennium Ecosystem Assessment (2005). Ecosystem and Human Well-Being: Bio-Diversity Synthesis. Washington, DC: World Resource Institute.
43 Mihai, F.-C. and Ichim, P. (2013). Landfills – territorial issues of cities from North-East region, Romania. *Forum Geografic* 12 (2): 201–210. https://doi.org/10.5775/fg.2067-4635.2013.244d.
44 Moll, G. and Petit, J. (1994). The urban ecosystem: putting nature back in the picture. *Urban Forests* (Oct/Nov): 8–15.
45 Nationale Stadtentwicklungspolitik. (2017). Dresden Stadtteilentwicklungsprojekt Weißeritz. https://www.nationale-stadtentwicklungspolitik.de/NSP/SharedDocs/Projekte/WSProjekte_DE/Dresden_Stadtteilentwicklungsprojekt_Wei%C3%9Feritz.html.
46 New York City Department of City Planning (2013). Fresh Kills Park project (draft master plan). https://freshkillspark.org/wp-content/uploads/2013/07/Fresh-Kills-Park-Draft-Master-Plan.pdf.
47 Nieuwenhuis, P. (2019). The foundations of sustainability and the implications for transport modes. In: Contemporary Operations and Logistics. Achieving Excellence in Turbulent Times (ed. P. Wells), 29–44. Palgrave Macmillan.
48 Opačak, M. and Wang, E. (2019). Estimating willingness to pay for a future recreational park atop the current jakuševec landfill in Zagreb, Croatia. *Sustainability* 11 (21): 6038. https://doi.org/10.3390/su11216038.
49 Palazzo, A.L. (2014). The Boston greenway: form and process issues at stake. *Journal of Urban Design* 19 (3): 352–367.
50 Pehme, K.-M. and Kriipsalu, M. (2018). Full-scale project – from landfill to recreational area. *Detritus* 1: 174–179. https://doi.org/10.26403/detritus/2018.17.
51 Prasad, M.N.V. and Strzalka, K. (2002). Physiology and Biochemistry of Heavy Metal Toxicity and Tolerance in Plants. Dordrecht: Kluwer Academic Publishers.
52 Prasad, M.N.V. (2004). Phytoremediation of metals in the environment for sustainable development. *Proceeding of the Indian National Science Acadamy* B70 (1): 71–98.
53 Prasad, M.N.V. (2007). Phytoremediation in India. In: Phytoremediation – Methods and Reviews (ed. N. Willey), 435–454. Totowa: Humana Press.
54 Prasad, M.N.V. (2012). Exploitation of weeds and ornamentals for bioremediation of metalliferous substrates in the era of climate change. In: Environmental Adaptation and Stress Tolerance of Plants in the Era of Climate Change (eds. P. Ahmad and M.N.V. Prasad), 487–508. Springer https://doi.org/10.1007/978-1-4614-0815-4.23.
55 Public Law 107-118 (H.R. 2869), the "Small Business Liability Relief and Brownfields Revitalization Act", signed into law January 11, 2002.

56 Rembeza, M. (2017). Integrating art into places in transition – Rose Kennedy Greenway in Boston as a case study. *IOP Conference Series: Materials Science and Engineering* 245: 052070.
57 Revkin, A.C. (2009). Peeling back pavement to expose watery havens. *New York Times*. https://www.nytimes.com/2009/07/17/world/asia/17daylight.html?_r=1.
58 Ruprecht Consult (2019). Guidelines for Developing and Implementing a Sustainable Urban Mobility Plan, 2e. Brussels: European Commission: Directorate-General for Mobility and Transport.
59 Ruprecht Consult (2013). Guidelines. Developing and Implementing a Sustainable Urban Mobility Plan. Brussels: European Commission: Directorate-General for Mobility and Transport.
60 Schneider, P. and Lüderitz, V. (2018). Integration of ecosystem services as part of the nexus approach into the applied teaching of ecological engineering. In: Handbook of Sustainability Science and Research. World Sustainability Series (ed. W. Leal), 369–387. Springer Cham https://doi.org/10.1007/978-3-319-63007-6_22, 978-3-319-63006-9).
61 Seoul Metropolitan Government. (2017) Landfill recovery project: Transformation of landfill to ecological park. https://seoulsolution.kr/en/content/landfill-recovery-project-transformation-landfill-ecological-park.
62 Seresinhe, C.I., Preis, T., MacKerron, G., and Moat, H.S. (2019). Happiness is greater in more scenic locations. *Nature Scientific Reports* 9: 4498.
63 Simis, M., Awang, A., and Arifin, K. (2016). From ex-landfill to Public Park: impact on local Community's quality of life and living environment. *Procedia-Social and Behavioral Sciences* 222: 763–771.
64 Sorrin, S. (2018). Environmental and economic implications of relocating urban highways underground. Master of Liberal Arts thesis, Harvard University Extension School.
65 Sultana, S., Salon, D., and Kuby, M. (2019). Transportation sustainability in the urban context: a comprehensive review. *Urban Geography* 40 (3): 279–308.
66 Wold Health Organization (2018). Global Status Report on Road Safety 2018. Geneva: World Health Organization, 404 p.
67 United States Environmental Protection Agency. (2019). Overview of EPA's brownfields program. https://www.epa.gov/brownfields/overview-epas-brownfields-program.
68 Verlag des Stadtplanes von Dresden. (1927). Dresden city map. https://commons.wikimedia.org/w/index.php?curid=30681807.
69 Weng, Y.-C., Fujiwara, T., Houng, H.J. et al. (2015). Management of landfill reclamation with regard to biodiversity preservation, global warming mitigation and landfill mining: experiences from the Asia–Pacific region. *Journal of Cleaner Production* 104: 364–373. https://doi.org/10.1016/j.jclepro.2015.05.014.
70 WHO – World Health Organization. (2012). Our planet, our health, our future. Human health and the Rio Conventions: biological diversity, climate change and desertification. https://www.who.int/globalchange/publications/reports/health_rioconventions.pdf.
71 Wong, M.-H. (2018). Ecological restoration of man-made habitats, with emphasis on metal-contaminated sites and domestic landfills. In: Twenty Years of Research and Development on Soil Pollution and Remediation in China (eds. Y. Luo and C. Tu), 15–37. Springer Singapore https://doi.org/10.1007/978-981-10-6029-8_2.
72 Wowrzeczka, B. (2019). Landscape forms in the architecture of waste management facilities in cities. *Acta Scientiarun Polonorum Architectura* 18 (3): 87–102. https://doi.org/10.22630/ASPA.2019.18.3.38.

6

Urban Environmental Issues and Mitigation by Applying Ecological and Ecosystem Engineering

Shailendra Yadav, Suvha Lama, and Atya Kapley

Director's Research Cell, National Environmental Engineering Research Institute, (CSIR: NEERI), Nagpur, Maharashtra, India

6.1 Urbanization

The concept of urban areas and cities first originated in 5500 B Cduring the Mesopotamian civilization. Earlier, cities were built to provide good governance and social security and enhance the quality of life. Urban areas act as hubs of knowledge, infrastructure, cultural and social diversity, jobs, education, and public services.

Urbanization can be defined as a multidimensional process that manifests itself by the rapid expansion of the human population and changing land cover. In the modern era, urbanization is an important feature of social development [1]. Urban land-cover change threatens biodiversity and affects ecosystem productivity through loss of habitat, biomass, and carbon storage. The global urban population is expected to reach 5 billion in 2030, and land-cover use will increase by 1.2 million km^2, three time the global urban land circa 2000. Very little information is available about the rate of expansion, magnitude, and locations of global urban sprawl [2]. According to the Intergovernmental Science-Policy Platform on Biodiversity and Ecosystem Services (IPBES), approximately 1 million plant and animal species face extinction due to urban activities. Approximately 77% of land and 66% of ocean area has been altered by anthropogenic activities [3]. Urbanization is adversely affecting the relationship between society and the environment. Unplanned urbanization combined with limited land has led to over-crowding, degraded natural environments, increased service delivery challenges, and increased the vulnerability of urban and rural communities. Urban areas also act as hotspots of resource use, pollution, and global climate change [4].

Climate and associated factors play an inevitable role in shaping the health of our mother planet, Earth, ecosystem services, and biodiversity. Climate change mediated by extensive urbanization has emerged as a potential threat to the global population. Climate change is not just a change in global temperature; it shifts weather toward extreme conditions [5]. It is a naturally evolving process and has been accelerated by anthropogenic activities including excessive urbanization, forest and land degradation, and pollution. Increasing temperatures also negatively affect multiple ecosystem services [6–9]. According to the Intergovernmental Panel on Climate Change (IPCC) report, the global mean temperature may increase between 1.4 and 5.8 °C by 2100 [1].

Handbook of Ecological and Ecosystem Engineering, First Edition. Edited by Majeti Narasimha Vara Prasad.

Land use contributes about 25% of total global anthropogenic greenhouse gas (GHG) emissions. The influx of greenhouse gases (methane) from agriculture fields, especially rice paddies, is expected to increase with rising temperatures in the near future [7, 9]. The increasing temperature is only effect; others include extreme weather conditions like frequent droughts and cyclones, changing patterns of lake freezing and melting, reduction in the size of glaciers and snow cover on mountains, sea-level rise, changes in heat stored in oceans, acidification of oceans, and changes in rainfall patterns. In 2011, 14 extreme weather events (tornadoes, temperature, and precipitation) happened in the US alone, resulting in losses totaling 1 billion dollars. Other countries such as Japan, Australia, China, Brazil, and Korea have also faced extreme weather events such as rainfall, drought, and wildfire in recent years, leading to substantial economic and biodiversity loss [10].

6.2 Global Trends of Urbanization and Its Consequences

The global urban population is expanding rapidly, and very little information is available regarding magnitudes and rates of urban expansion. During the eighteenth century, only 3% of the global population was living in urban areas. In 1950, two-thirds of the human population was living in rural areas. However, a rapid increase in urban population was observed between 1950 and 1965, when the urbanization rate was greater than 3%. In 2007, half of the world's population was living in urban areas. But in 2014, 54% of the global population lived in an urbanized area. This is expected to increase by 72% by 2050 – approximately 6.3 billion people will be urban in 2050 [11]. Geographically, urbanization is greatest in the developed world, located in the temperate zone. It has also been found that northern cities exhibit lower growth rates than southern ones. Also, urbanization in developing counties situated in tropics and subtropics is rapidly increasing due to the rapid migration of the population [12, 13].

Urbanization is an uneven process that affects environmental health both locally and globally. It alters land cover and usage patterns, hydrological cycles, ecosystem processes, phylogenic biodiversity and species richness, ecosystem services, and carbon storage [14]. Factors such as global climate change, agricultural expansion, temperate reforestation or afforestation, cropland intensification, and extensive urbanization act as drivers of land usage and land-cover change. The change in land-cover patterns affects the earth's overall energy balance and biogeochemical processes, which adversely affects the land's surface properties and the provisioning of ecosystem services. Song and co-workers analyzed the dynamics of global land change from 1982–2016 utilizing 35 years of satellite data. Despite the common assumption that forest areas have declined, tree cover increased to 2.24 million km^2 (7.1% more than 1982 level). Also, a 3.1% (1.16 million km^2) reduction in global bare ground was found, mostly in agricultural regions of Asia. The study predicted that 60% of land change resulted from anthropogenic activities, and 40% was the indirect outcome of natural phenomena such as climate change [13].

It has been predicted that rapid global urban expansion will occur in Asia, with India and China being the epicenters. In India, major urban expansion will be clustered around seven capital cities and small villages and towns of the Himalayan region, which have large areas and lower population density [4]. However, the pace of urban expansion will be slower in these areas. In China, urban expansion will create an 1800 km coastal urban corridor, connecting Hangzhou to Shenyang province. The impact of urban expansion will be more critical in Africa. During 2000, only 1% of all hotspots were urbanized, which will increase to 1.8% of total hotspots by 2030 [4]. Globally, five biodiversity hotspots – Japan (6%), the Caribbean Islands (4%), the Philippines (4%), the Western Ghats and Sri Lanka (4%), and the Guinean forests of West Africa (7%) – will suffer most from urban expansion. Loss of vegetation in areas with probable expansion will account for approximately 1.38 petagrams of carbon emissions equaling about 5% of carbon emissions from tropical deforestation and land cover change [2].

The modern era can be referred to as the *urban century* due to rapid urban expansion and transformation of rural areas into cities. The higher-quality infrastructure and connectivity in urban areas (water and energy supplies, sewage treatment plants, transportation [roads and railways], and telecommunications) lead to rapid urban migration, which may pose a risk to the functioning of the existing system. Currently, urban areas host half the world's population and are responsible for over 70% of global energy consumption and greenhouse gas (CH_4 and CO_2) emissions [15]. Ninety percent of the global population resides in coastal areas, making them more susceptible to climate change events: drought, flooding, heat stress, and extreme rainfall events (tornadoes, cyclones) [5, 16–18]. Urban areas also suffer higher fatalities and economic losses due to higher population density, compact buildings, and interconnected infrastructure.

Expansion in urban areas causes deterioration of the natural environment. Therefore, urbanization should be sustainable to ensure the survival of future generations. The concept of smart cities, sustainable cities, resilient cities, and low-carbon cities has been proposed to mitigate urbanization and climate change issues. However, it has led to confusion among local decision-makers and stakeholders [19]. In the future, urbanized areas may serve as a tool for the transition toward sustainability or as a potential threat to sustainability [15]. Hence, in the urban century, the emphasis should be on identifying emerging challenges associated with urbanization and planning and implementing government policy to ensure a sustainable urban future. Mitigation of urbanization consequences requires the coordinated participation of civic authorities, policymakers, and the public. Future research should address the knowledge gap between urbanization, socioeconomic development, and dynamics (feedback and interplay) of ecosystem functions [20, 21].

In its 2030 Sustainable Development agenda, the United Nations proposed sustainable development goals (SDGs) that included the economy, the environment, and society as three essential parameters of sustainability (https://www.un.org/sustainabledevelopment/cities) [22, 23]. The deadline for achieving the SDGs is 2030, and 60% of the global population will become urban during this period [4].

6.3 Urban Environmental Issues

Recently, urban ecology has gained the attention of scientists, society, and policymakers due to the impact of urbanization on the global ecosystem. The urban ecosystem differs from the natural environment in terms of human dominance and contains remnants of the natural ecosystem (microenvironments and biological assemblages) and human-built environments. The built environment contains both biotic and abiotic components. Biotic components mostly contain natural ecosystem analogs such as planted ornamental gardens and lawns with higher flora and fauna diversity. Abiotic components include materials constructed from concrete, asphalt, iron, and other building materials. Abiotic components negatively impact ecosystem services, which affects the urban ecosystem and human health [24].

Humans have always exploited and enjoyed ecosystem services provided by planet Earth. Ecosystem services are still necessary for our survival. However, with cities, human priorities have shifted from natural ecosystem services to artificial services (monetary). Although urbanization ensures a better quality of life (economic powers and opportunities), urban areas also act as hotspots of resource consumption and waste generation. Both expansion in urban areas and overexploitation of natural resources (environment) are causing biodiversity loss, global climate change, eutrophication, and degradation of ecosystems (terrestrial and aquatic) (Figure 6.1).

Extensive urbanization and increasing population density have adversely affected the supply and demand of ecosystem and non-ecosystem services. For example, agricultural crop biodiversity

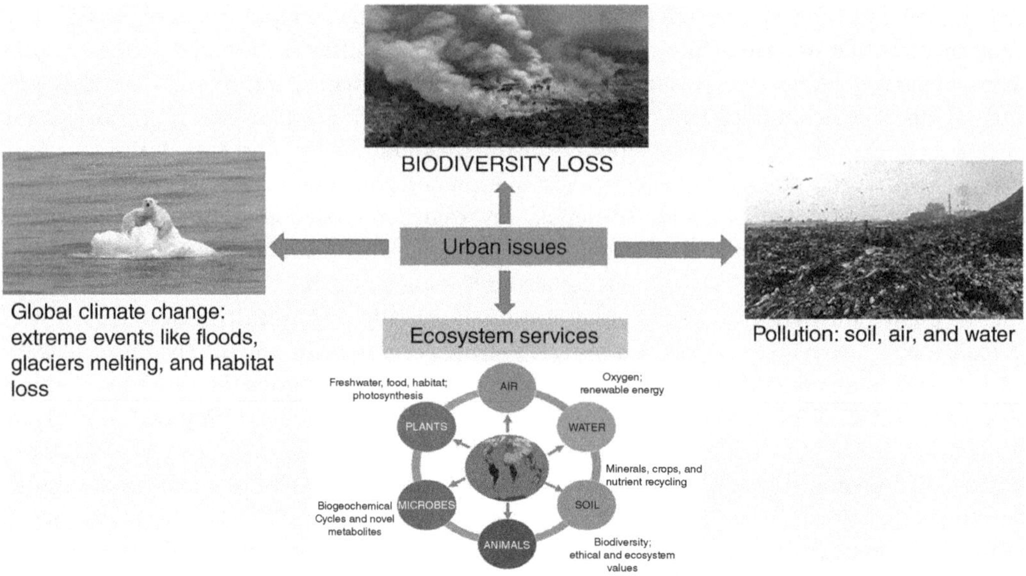

Figure 6.1 Urbanization and overexploitation, leading to biodiversity loss, climate change, and ecosystem degradation. *See colour plate section for colour representation of this figure.*

is declining, and excessive use of agrochemicals is causing the degradation of agricultural lands. The human population is expected to increase to 9.6 billion by 2050, and it will be complicated for the agriculture sector to meet increasing demand given global climate change [25, 26]. Due to higher population density, ecosystem pollution and waste management have emerged as potential threats to the global population. In other words, urbanization, climate change, ecosystem services, and human health are interlinked with each other as a complex nexus [27, 28].

Broadly, urban environmental issues can be classified in two categories: physical and biological. Urban environment physical issues include urban heat islands (UHIs), pollution (air, water, soil, noise) and waste management, shortages of drinking water, and degradation of the ecosystem. Biological and environmental issues include declining ecosystem services (quality and quantity), loss of biodiversity and species richness, and disease epidemiology [2].

6.3.1 Physical Urban Environmental Issues

6.3.1.1 Urban Heat Islands

Urbanization increases living standards but also raises social, economic, and environmental issues. UHIs are among the major urban environmental issues. The UHI effect can be described as the prevalence of higher temperatures in built-up urban areas compared to the surrounding natural environment and suburban and rural areas [27]. It is mainly caused by higher absorption and solar energy storage in construction materials (concrete, asphalt, etc.) and low vegetation cover. The intensity of the UHI effect depends on the size of the city and population density. The impact of UHI can be significant during both day and night. The UHI effect has biological implications, too, causing leaf emergence and flowering to occur early in temperate zone cities. The higher temperature in urban areas leads to increased electricity consumption mediated by air conditioning loads, leading to higher morbidity and mortality [29]. Increased electrical consumption in the form of air conditioning also increases ground-level ozone production.

Table 6.1 UHI intensity of the 10 cities with the greatest UHI effect worldwide (2018).

S.no.	City	Surface area	Daytime UHI	Measurement	Reference
1	Delhi	1484 km^2	8	Ground and satellite	[33]
2	Mumbai	603.4 km^2	8–10	Satellite	[33]
3	Nagpur	227.4 km^2	8–10	Satellite	[33]
4	Mexico	1973 km^2	7.19	Satellite	[33]
5	Sao Paulo	1521 km^2	6.94	Satellite	[33]
6	Los Angeles	1302 km^2	3.51	Satellite	[33]
7	Singapore	721.5 km^2	2.18	Satellite	[33]
8	Berlin	891.8 km^2	2.04	Satellite	[33]
9	Paris	105.4 km^2	1.98	Satellite	[33]
10	London	1572 km^2	1.14	Satellite	[33]

Heatwaves are successive hot days. It was found that heatwaves contributed approximately 50% of total heat-related mortality in the West Midlands of the UK [29]. Similarly, in Ahmedabad, India, a heatwave was associated with a 43.1% increase in mortality in 2010 [30]. Mazdiyasini et al. in 2017 analyzed the variation of summer temperature, the severity of heatwaves, and associated mortality in India between 1960 and 2009. It was found that the Indian mean temperature increased by 0.5 °C annually during this period and is expected to reach 2.2 and 5.5 °C by the end of the twenty-first century. They proposed that even a slight increase in mean temperature will increase heatwave frequency and associated mortality in India [31]. Appropriate measures should be taken by society and governments to improve the resilience of heatwave-vulnerable regions.

The UHI effect has emerged as a primary environmental concern due to rapid urban expansion and migration. It has been estimated that 65% of the global population (6 billion people) will migrate to urban areas by 2050. These people will be prone to the high-intensity UHI effect [32]. The intensity of this effect can be measured by (i) measuring the temperature of pavements and buildings, i.e. urban skin temperature; or (ii) measuring air temperature, i.e. the meteorological UHI. The magnitude of the UHI effect varies depending on the geographic location and local climatic conditions [121][122]. The daytime UHI of global cities having a greater UHI effect is given in Table 6.1.

6.3.1.2 Urban Flooding

Urban flooding occurs in cities in the form of flash floods and waterlogging on streets due to intense rainfall. Precipitation is enhanced in the downwind of cities due to the higher level of particulate nuclei in the urban environment. Precipitation is 5–10% greater in cities with higher fog and clouds. The extent of damage caused by urban flooding has increased over time due to higher rainfall runoff volume and broad exposure of people and assets in urban areas. On comparing urban areas to rural areas, urban runoff increases from 10 to 30%, and groundwater decreases from 50 to 32%.

The risk of urban flooding varies from city to city. Due to less canopy in cities, the flowrate of urban runoff increases and reduces the lag period between the onset of storms and the start of runoff. Urban flooding poses a threat to the sustainability of the economy, ecosystems, and human safety.

Ecological degradation of streams draining urban areas is referred to as *urban stream syndrome*. It is characterized by a higher level of nutrients (nitrogen, phosphorus, and salts), organic and inorganic compounds, and altered biotic communities. Urban runoff may negatively impact aquatic biodiversity due to elevated levels of nutrients, temperature, and flashy hydrology [34].

Most major cities worldwide evolved around coastal areas, which makes them vulnerable to climate change and flood events. The earth's temperature is rising, causing the melting of the cryosphere, and the amount of permafrost has decreased by 7%. Due to this, the sea level is rising approximately 1.8 mm per year, leading to coastal/urban flooding [35]. Also, urban flooding occurs due to heavy rainfall (caused by cloudbursts, cyclones, or thunderstorms) and overflow from dams, rivers, and drainage systems. The UHI effect also plays a vital role in urban flooding by inducing convective rainfall around cities [18].

The steep increase in urbanization, economy, and climate variability has made the United Kingdom and China more vulnerable to urban flooding [36]. India has also witnessed several incidences of extreme urban flooding. India comprises 28 states, 9 union territories, 7935 towns, and 468 urban agglomerations (UAs). According to the 2011 census, 377 million people live in urban areas, and this number is expected to reach 600 million by 2031. Extensive urbanization and urban migration reduce bodies of water, groundwater recharge, and the capacity of urban drainage systems and increase paved areas. Rapid urban population growth has stressed existing sewage treatment facilities, resulting in the discharge of excess wastewater into rivers and lakes. During monsoon season, sewage mixes with river water. Due to heavy rainfall caused by cloudbursts or thunderstorms, water overflows, resulting in cities flooding and the outbreak of epidemics [37].

6.3.1.3 Urban Pollution (Air, Water, Noise) and Waste Management

Urbanization enhances quality of life but also negatively affects environmental and human health. About half of the world's population resides in urban areas. The rapid expansion of urban dwellers (i.e. population density) in megacities exerts pressure on existing infrastructure such as waste management (solid waste management, water/sewage treatment facility), leads to shortages of drinking water, and causes pollution (air, water, and noise). Urbanization act as a driver of climate change and pollution. Due to increasing population density, urban areas act as centers for waste generation. Poor waste management and inadequate disposal of waste cause soil, air, and water pollution. Open landfill sites play a significant role in water and air pollution and transmission of diseases. Solid waste disposal sites release approximately 5–20% of global methane emissions and contribute around 700 Mt of CH_4, which contributes to the global climate change process [38].

Waste management has emerged as a significant challenge of the urban century [39]. Land use and land-cover change (LULC) also act as drivers of global climate change. The impact of LULC is severe in areas with high population density, urbanization, industrialization, and deforestation. Variations in LULC alter regional microclimates, cause heat and cold waves, and alter rainfall patterns.

A *microclimate* is the difference between the temperature of urban areas (local scale) and their natural counterparts. It is formed due to dense construction, environmental degradation, and anthropogenic activities in urban areas. In 2019, Gogoi et al. evaluated the impact of LULC on the surface temperature of Odisha between 1981 and 2010. It was found that the mean temperature of the state increased by 0.3 °C in the last three decades. However, the increase in surface temperature was greater in the last decade (2001–2010) [40].

Plastic wastes are now ubiquitous in both terrestrial and aquatic ecosystems. Plastics are nonbiodegradable and often accumulate as microplastics in the marine environment, posing a threat to aquatic organisms. The accumulation of pharmaceuticals and personal care products that are nonbiodegradable as been observed in bodies of water. The following contaminants also play a crucial role in the emission of microplastics [41, 42].

6.3.1.3.1 *Air Pollution* Air pollution has emerged as a significant threat to urban areas. After the Industrial Revolution, the release of hazardous pollutants caused rapid deterioration of environmental quality globally [43]. In the nineteenth century, sulfur compounds generated by the burning of coal were potent air pollutants. Coal is still a significant pollutant in developing countries. According to the World Health Organization (WHO), 37 Indian cities, including Delhi, Raipur, and Lucknow, are among the top 10 world cities with the worst PM_{10} pollution [44, 45]. Urban society, mainly in developing countries, depends on fossil fuels to meet energy requirements for domestic and industrial purposes. Photochemical smog due to heavy traffic, excessive industrialization, power generation, and agriculture waste burning has also gained the attention of the scientific community and policymakers. Air pollution contributes significantly to climate change and causes urban and regional haze and public health hazards [46].

Exposure to air pollution can be occur indoors or outdoors. Indoor air pollution poses a risk to the health of the urbanized world as people spend most of their time indoors. The time spent indoors depends upon age, gender, season, profession, and health. Indoor air quality is directly affected by outdoor air quality. Outdoor air pollutants include nitrogen oxide (NOx), sulfur dioxide (SO_2), ozone (O_3), carbon monoxide (CO), particulate matter (PM), and microplastics of various sizes. Indoor air pollutants contain volatile organic compounds (VOCs), radon, microbes, and outdoor pollutants. The impact of indoor air pollutants on human health depends upon their concentration, exposure time, and toxicity [47]. Outdoor air pollution, particularly exposure to PM and ozone, has been attributed to over 695 000 premature deaths in India in 2010. However, in India, the national ambient CO concentration is better than the WHO guideline. The concentration of PM_{10} and $PM_{2.5}$ is higher than the WHO guideline values [48].

6.3.1.3.2 *Water Pollution* Rivers act as the interface between watersheds and seas. They also carry various pollutants of natural and anthropogenic origin. Urbanized areas have a significant impact on surface water quality. Untreated discharge of industrial effluents and urban runoff cause water pollution and loss of biodiversity. In India, approximately 61 754 million liters per day (MLD) of sewage is generated, and treatment facilities exist for only 22 963 MLD. About 38 791 MLD of sewage remains untreated and is discharged into bodies of water [49]. Thus, it is evident that a significant gap exists between wastewater generation and treatment [49]. Also, the migration rate of urban dwellers exceeds the building rate of sewage systems and wastewater treatment plants, which puts stress on existing system.

Due to the strain on existing facilities, sewage and wastewater are released into surrounding surface bodies of water like rivers and lakes, which leads to organic enrichment and water pollution. Organic enrichment due to wastewater discharge contaminates river water and poses a threat to the nearby population. Untreated sewage contains many pathogenic bacteria that cause wider dissemination of infectious diseases and morbidity and mortality [50]. During monsoons, increased urban water runoff and cloudbursts also cause urban flooding.

Further, rapid urban expansion is putting stress on existing freshwater reserves. It has been proposed that with an increase in the urban population, the need for municipal water supplies will also increase [51]. A study was conducted on Poland's Lyna river to explore the impact of urban areas on the physicochemical parameters of water quality and bacterial communities. The results divided the Lyna river into two zones: (i) upper, the unpolluted section from the headwater; and (ii) lower, which had anthropogenic contamination from urban areas. It was found that the water quality of the upper region was a benchmark for monitoring the physicochemical and microbiological quality of river water [52].

6.3.1.3.3 Noise Pollution The urban soundscape consists of noise from diverse sources such as the built environment, street traffic, construction, industries and rail, aircraft, loudspeakers, and honking vehicles. Noise pollution has also caused severe problems in urban areas and may impose adverse health effects [53, 54]. In urban settings, compact structures, high density, and high-rise buildings are significant contributors to noise pollution [55]. Exposure to noise pollution may cause health issues like cardiovascular disease and autonomic homeostasis. It alters the function of multiple organs, increases blood pressure and heart rate, and affects sleep patterns. In 1972, the WHO recognized noise as a significant pollutant. In India, with the increasing number of vehicles on the street, it has become hard to ignore and escape street noise. As per the Central Pollution Control Board (CPCB), noise pollution is highest in Mumbai, followed by Hyderabad, New Delhi, Chennai, and Kolkata [56].

6.3.2 Biological Urban Environmental Issues

6.3.2.1 Declining Urban Ecosystem Services Due to Loss of Biodiversity

The term *ecosystem services* was coined in 1970 to establish the link between ecosystem functions and human society. Ecosystem services can be defined as the benefits humans derive directly or indirectly from the ecosystem [57]. They can be classified into four types: (i) provisioning services (food, freshwater, raw materials, etc.), (ii) regulating services (climate and air quality, carbon sequestrations and storage), (iii) habitat and life support services, and (iv) cultural services (mental and physical health, aesthetic and spiritual services).

Ecosystem services can be used as an indicator for measuring the impact of urbanization on ecology [1]. Urban ecology is essential in maintaining the sustainability of the ecosystem. Urban ecosystem services are mediated by blue and green areas. *Blue areas* consist of lakes, ponds, and water-retention ponds. *Green areas* include parks, urban forests, gardens, and campus areas. Blue and green urban areas provide essential ecosystem services such as regulating microclimate, water regulation, pollution reduction, greater public health, and cultural services. Urban green space also acts as a mosaic where the diversity of exotic flora and fauna becomes rich in comparison with their natural counterparts [58].

Urbanization, biodiversity loss, and declining ecosystem services are interlinked in multiple ways. Urbanization impacts biodiversity and ecosystem services directly and indirectly. The direct impact of urbanization causes ecosystem degradation, habitat loss (physiological transformations), and loss of biodiversity. Indirect impacts include changes in water and nutrient availability, increases in abiotic stressors such as air pollution, increases in competition from non-native species, and changes in herbivory and predation rates. Urban expansion occurs rapidly in low-lying coastal areas with higher biodiversity. It has been proposed that approximately 25% of endangered or critically endangered species will be affected by urban expansion [59–61]. Further, 10% urban growth in ecoregions would result in 80% loss of species during 2000–2030 [60, 62, 63].

Urbanization also affects regulatory ecosystem services such as hydrological services. Urban expansion often occurs at the cost of deforestation, which increases the volume of surface runoff, exposing downstream communities to flooding. Urban landscapes with less vegetation cover lose 40–83% of urban rainfall compared to forest landscapes. Also, ecosystem services (recreational values and cultural services) provided by urban green spaces and forests are vital for human (physical and mental) health but are often neglected during urban planning. Recreational ecosystem services provided by the urban ecosystem can be considered the most valued ecosystem services in cities [60, 63]. Demand for urban ecosystem services is mainly due to the large number of immediate beneficiaries.

Urban areas/cities constitute a small fraction of the earth's surface but have a huge environmental impact on both the local and global scales. Urban areas act as hotspots for resource consumption and waste generation, driving climate change and affecting ecosystem services [25]. According to the Millennium Ecosystem Assessment (MA), more than 60% of ecosystem services either are deteriorating or were overexploited in urban areas [64]. The majority of countries failed to achieve the goals of the Convention of Biological Diversity (CBD) 2010; therefore, at the 10th conference held in Nagoya, Japan CBD 2020 proposed an action plan to achieve five strategic goals comprising 20 targets, including foods, fuels, and fibers (targets 6, 7, and 13); and carbon sequestration and resilience (target 15). The five goals were to (i) address causes of biodiversity loss; (ii) reduce pressure on biodiversity and promote sustainability; (iii) improve biodiversity by conserving ecosystem, species, and genetic diversity; (iv) enhance the benefits of ecosystem services to all; and (v) enhance the adaption and implementation of a national biodiversity strategy and action plan and knowledge dissemination [65].

The pace of urbanization will be the greatest in the coming 30 years. In the urbanization era, it is a huge challenge to design resilient cities that can adapt to climate change and improve ecosystem services [34].

6.3.2.2 Increasing Disease Epidemiology

Urbanization characterized by intensification of agriculture and ecological fragmentation (LULC) may impose a considerable impact on the epidemiology of infectious diseases. In the developed world, improved living standards, sanitation, and a one-health approach (a transdisciplinary approach working at global and local scale with a goal of achieving optimal health conditions considering interconnection between animals, plants, and the environment) have significantly reduced the morbidity and mortality rates of infectious diseases. But in developing nations with lower per-capita income, even in urban settings, economic inequality acts as a major driver of infections [66]. Urbanization creates an interface between wildlife, livestock, and humans, which plays a critical role in transmitting pathogens and in the evolution of infectious diseases [67]. Also, urban areas often have high population density and are located near coastal areas, making them vulnerable to infectious/zoonotic disease. In addition, the impact of climate change in these areas is often high, which further exacerbates the epidemic of climate-sensitive infectious diseases, potentially changing their spatial distributions, affecting annual/seasonal cycles, or altering disease incidence and severity [68]. Infectious diseases account for about 11 million global deaths per year, which is 51% of life lost worldwide annually.

Pathogen prevalence, richness, and spatial distribution vary among regions in response to climate conditions. The sensitivity of pathogens to climate change determines whether an infectious disease will respond to climate change [68]. Diseases respond differently to climate change phenomenon. The severity of infectious diseases is higher in poor countries. Per capita, mortality from disease is 300 times higher in developing countries than in the developed world. According to the WHO, the global distribution of vector-borne disease is accelerating due to rapid unplanned urbanization, deforestation, and urban migration [69]. Urbanized areas may serve as incubators of new epidemics and zoonotic diseases. Chronic diseases like chronic obstructive lung disease, diarrheal diseases, ischemic heart disease, stroke, and lower respiratory infections are leading causes of global deaths. The higher density of urbanized areas and frequency of contact between people turn these sites into hotspots of emerging infectious diseases like avian flu, Ebola virus, and severe respiratory acute syndrome (SARS) [70]. As evidenced in 2020, with the outbreak of the coronavirus (Covid-19) pandemic, regions with higher density and poor sanitation are more prone to the infection. Also, increasing urbanization leads to encroachment on natural habitats, which increases the chances of humans encountering wild animals and zoonoses and the opportunity for zoonotic infections [71]. The only way to control the spread is to take precautions and social distance.

The following associations between changing climate conditions and vector-borne diseases (malaria, dengue, and cholera) have been reported [72–74].

- *Malaria*: Malaria is a mosquito-borne infectious disease caused by a parasitic protozoan of the genus *Plasmodium vivax and Plasmodium falciparum*. Female mosquitos belonging to the Anopheles species act as vectors for the transmission of malaria. The distribution and transmission of malaria are linked to seasonal patterns and are sensitive to global climate change. The epidemic of malaria is prevalent in both developing and developed economies, but the severity is notably higher in tropical and low-income countries [73, 75].
- *Dengue*: Dengue is a viral infectious disease transmitted by *Aedes* mosquitoes. It poses a threat to global health [76] and is the most significant arboviral disease in the world. Incidences of dengue have increased 30-fold in the past 50 years. Annually, about 500 000 people suffer from severe dengue and require hospitalization [77]. Annual precipitation and temperature play a significant role in the spread of dengue [78]. A study carried out in Guangzhou province of southern China during 2005–2015 found that rainfall and the climate combined with urbanization directly impact dengue transmission and dynamics in the human population. People living substandard lives below the poverty line were more vulnerable to dengue [79]. In the last 10 years, the incidence of dengue has increased in India. A study conducted in 2017 found that average increase in temperature, incubation period, and rainfall drive the outbreak and transmission of dengue virus [80].
- *Cholera*: Cholera is an intestinal infection caused by the bacterium *Vibrio cholerae*, which spreads through the consumption of water contaminated with feces. Cholera poses a threat to human health worldwide, especially in South Asia, Africa, and Latin America. The problem is severe in areas with inadequate sanitation and water quality [81]. The WHO aims to reduce deaths from cholera by 30% by 2030, requiring the development of new vaccines and improved access to clean drinking water [82]. Discharge of municipal wastewater and urban runoff causes organic enrichment, salinity, and greater pathogenic loads in adjacent bodies of water (rivers, lakes), which modulates both the frequency and intensity of infections.
- *Antibiotic resistance*: Antibiotic resistance is a threat to global health. Climate change also accelerates the dissemination of antimicrobial resistance in urban areas. MacFadden and co-workers studied the impact of climate change (temperature) on the emergence and distribution of antibiotic resistance in the United States. It was found that increasing local temperature and population density are associated with higher incidences of antibiotic resistance. An average 10 °C increase in temperature led to a 4.2, 2.2, and 2.7% increase in antibiotic resistivity among the common pathogens *Escherichia coli, Klebsiella Pneumoniae*, and *Staphylococcus aureus*, respectively [83]. Similarly, the resistome profile of urbanized watersheds was monitored for five years in Houxi River, China. The study's outcome indicated that the abundance of antibiotic resistance genes (ARGs) was greater downstream compared to midstream and upstream and increased with urbanization level. Also, land use and the percentage of urbanized areas indirectly impact the ARG profile distribution [84].
- *Urban associated disease (UAD)*: Urban associated diseases can be defined as any disease whose prevalence and severity increases with an urban lifestyle and urban growth. UADs

include non-communicable diseases like cardiovascular diseases and obesity, diabetes, chronic pulmonary diseases, cancer, and mental illness. Cardiovascular diseases have accounted for about one-third of global deaths in 2015. The association between urban lifestyle, physiological conditions such as smoking, abnormal lipids, hypertension, diabetes mellitus, abdominal obesity, and mortality caused due to CVD in urban areas was high, particularly in developing nations. Cancer is the second leading cause of global deaths. Urban habits such as smoking, alcohol consumption, a diet low in fruits and vegetables, and physical inactivity has been attributed to higher mortality from cancer, obesity, and diabetes in urbanized areas. The uncontrolled rapid growth of urban areas has been linked to the prevalence of type 2 diabetes. Mental disorders such as depression, anxiety, and schizophrenia have also increased in urban areas during the last decade [85].

Urbanized areas, climate change, and loss of biodiversity are interlinked and often act as a nexus. Urban environmental issues such as the UHI effect, urban flooding, and biodiversity loss can be attributed to climate change, which is a direct consequence of LULC, deforestation, and insufficient green space. Policymakers and society often neglect the ecosystem services provided by green urban spaces. These urban environmental issues can be mitigated by restoring the ecosystem and adopting an ecosystem engineering approach, which provides an eco-friendly, economical, sustainable option for the mitigation of urban environmental issues.

6.4 Ecosystem Engineering

The concept of ecological engineering emerged to provide goods and services for human welfare by applying engineering principles and techniques and reducing ecosystem degradation. It considers human survival dependent on ecosystem services drawn from the natural environment, and overexploitation of ecosystem services due to the growing population has damaged and will put stress on the functioning (biogeochemical processes) of the ecosystem. The term *ecological engineering* was coined by H.T. Odum in 1960. According to Odum, ecological engineering is the "management of nature." According to Mitsch and Jørgensen, ecological engineering may be defined as "the framework of sustainable ecosystems that integrate human society with its natural environment for mutual benefit" [86]. Ecological engineering combines both basic and applied sciences to restore ecosystems. Its objectives are:

- Restoration of anthropogenically degraded ecosystems (environmental pollution and LULC)
- Designing sustainable ecosystems by integrating both human and ecological importance

The concept of ecosystem engineering can be categorized into five points:

1) *Self-designing (organizational) capacity of the ecosystem*: Self-organization is the ability of a system to reorganize itself in a given environment that is unstable and non-homogeneous. It is suitable for ecosystems in which species interactions are high (i.e. species are continually introduced and deleted) and the environment itself changes. It often involves developing new ecosystems and pilot-scale models to test the self-organization behavior.
2) *Acid test of ecological theories*: Ecological restoration is the process of recovery of a degraded ecosystem [87]. Restoration ecologists have always stressed the association between basic research and ecosystem restoration. The best way to understand an ecosystem is to reassemble it, repair it, or readjust it so that it functions properly. Restoring a degraded ecosystem can be viewed as the acid test of our understanding of that system. Ecological principles proposed

earlier must act as the baseline of ecological engineering practices. Ecological engineering offers ecological research of fundamental importance and ensures advancement in the field of ecology.

3) *System approaches*: Ecological engineering emphasizes following a holistic approach, i.e. considering entire ecosystems, not only associated factors. In restoration ecology, researchers study the entire ecosystem rather than components of an ecosystem in isolation with each other.
4) *Non-renewable energy resource conservation*: Ecological engineering focuses on constructing ecosystems that have self-sustainable potential and conserve fossil fuels. Most ecosystems constructed by ecotechnology are solar-based and consume very little non-renewable energy resources.
5) *Ecosystem conservation*: Ecological engineering advocates working in symbiotic association with nature. Humankind should use ecosystem services for survival but not at the cost of nature. Often, solving problems for humanity create issues for nature. Therefore, ecological engineering also considers ecosystem conservation as a goal. It attempts to cope with environmental problems by utilizing the self-rejuvenating potential of natural ecosystems.

6.5 Approaches for Mitigation of Urban Environmental Issues

6.5.1 Nature-Based Solutions

Conserving nature and ecosystem services in urban areas has become a significant challenge of the urban century. Due to higher population density, cities act as the prime sites of resource and ecosystem service consumption. Acquisition of land for human settlements and subsequent soil sealing (covering of soil by impermeable material) also affect ecosystem functioning. The last three decades have witnessed the most rapid transformation of rural areas into urban areas, both in terms of human population and surface area. Urbanization also causes a reduction in human-nature interaction, which has a major impact on human mental and physical health [88, 89]. Necessary actions should be taken to ensure the conservation of nature and ecosystem services. Maintaining quality of life, reducing the ecological footprint, and adaptation to climate change are three significant challenges of the urbanized world that need to be answered.

A nature-based solution (NBS) is the application of nature in solving global challenges such as climate change, food security, and disaster management [90]. NBS relies on the mitigation of urban environmental issues (UHI effect, flooding, etc.) by biodiversity conservation and enhancement of ecosystem services. It ensures better adaptability in response to climate change, improved ecosystem services, and increasing green infrastructure to conserve the ecosystem. Blue and green spaces of urban areas play a crucial role in the provisioning of ecosystem services. NBS promotes cities' resilience to climate change by promoting blue and green spaces in urban areas [90]. For example, NBS provides urban green infrastructure such as lawns, parks, and street trees that may reduce temperatures in cities and control air and water flows. Adopting NBS and ecorestoration (integration of nature and important biogeochemical processes) in an urban environment can provide solutions for urban environmental issues such as climate change mitigation and human health and well-being.

6.5.1.1 Green Infrastructure (GI)

Green infrastructure can be defined as an interlinked network of various types of green spaces, including green roofs, green walls, and urban forests, which help deliver ecosystem services and ensure the sustainability of nature. The concept of GI was first used in the United States and Europe. GI plays an important role in the sustainability of modern communities and is an essential component of urban planning. It plays multifunctional roles in urban settings such as mitigating the UHI effect, carbon sequestration, stormwater, and urban runoff; prevents urban flooding; allows nature and biodiversity conservation; and promotes human physical and mental health.

Urban trees act as environmental engineers and help increase the resilience of cities by maintaining hydrological cycles and controlling air, water, soil, and noise pollution [91]. Therefore, GI has now become an essential component of urban policy, planning, and management. Mostly urban cities are located within a 50 km range from a forest area, and GI in the form of an urban corridor acts as a buffer zone and also helps reduce the transmission of zoonotic diseases. GI can be of various forms and sizes depending upon the location and available space. Some of them are discussed here:

- *Urban forests*: An urban forest is a collection of trees, gardens, parks, and pervious soils in a complex-built environment with human dominance. In urban areas, the distribution of plant (quality and quantity) vegetation is controlled by humans who are the main drivers of their types, amounts, and distribution. Urban trees and forests perform several essential functions in an urban ecosystem: reducing heatwaves, pollution control, biofiltration of stormwater and urban runoff [92], nutrient (phosphate and nitrogen) removal, and soil-carbon sequestration, in addition to provisioning cultural and recreational ecosystem services [93–96]. Also, urban trees reduce socioeconomic losses by limiting urban runoff [93]. The ozone level is rising in urban areas due to anthropogenic activities such as higher power consumption in response to heatwaves. A higher ozone concentration is responsible for several diseases including lung inflammation lungs, respiratory disorders, and cardiovascular disease. Previous studies have shown that urban forests and trees help remove ozone from urbanized areas [97].
- *Green roofs*: Green roofs are an urban greening system that involves growing green plants on rooftops. The concept of green roofs has existed since ancient times, but its application has become more relevant in the urban century [98]. These plants compensate for the loss of vegetation caused due to the construction of buildings and restore urban ecosystem services by creating a pleasant environment in which to live and work. They also provide multiple benefits that may help mitigate urban environmental issues like the UHI effects, stormwater management, urban runoff, noise and air pollution, and increasing urban biodiversity [99]. Urban roofs can be categorized into two types:
  - *Intensive green roofs* are commonly designed for public places and often include exotic trees, ornamental plants, shrubs, and hardscapes. They require a surface depth of approximately 15 cm and intense maintenance. Intensive roofs are also a cost-intensive approach that requires a strong infrastructure to support the weight.
  - *Extensive green roofs* are rarely used and require a substrate depth of less than 15 cm. Due to this shallow depth, only limited plants such as grasses and drought-resistant plants are used. Extensive green roofs require less maintenance.
- *Green walls*: Green walls have green vegetation on the vertical surfaces of buildings. They can be classified into two types:
  - *Green facades* place green plants directly or indirectly on facades with the help of supporting structures.
  - *Living wall systems* are advanced greening systems that use modular panels containing soil for plant growth. Such a system supports multiple plant species such as ferns and shrubs in addition to climbers.

The addition of green walls on the surfaces of buildings provides multiple benefits, including thermal insulation, abatement of air pollution by absorbing carbon monoxide, and reducing the concentration of particulate matter and noise pollution [100].

6.5.1.1.1 Health Benefits of Urban Green Infrastructure Human populations living in cities are often exposed to chronic stress due to limited interaction with nature, lack of physical activity, and environmental pollution. Urban GI like urban gardens, urban forests, etc. boosts mental and physical health and reduces morbidity and mortality caused by chronic diseases such as cardiovascular disease and respiratory disorders [90, 101, 102]. Contact with nature helps restore mental health by stimulating a positive emotional state in people suffering from mental illness [88, 103]. Increasing contact with nature also promotes the immune system. It has been found that urban forests boost the immune system through the production of anticancerous proteins, thus reducing the health risk of cancer. Citizens living in urban areas with more GI are least prone to asthma and allergic reactions. Infants with higher exposure to allergens, bacteria, and pathogens in early life have strong immunity against infectious and autoimmune diseases and allergic reactions [58, 104]. It has been found that increased physiological activity in nature promotes physical and mental health better than physical activity in other environments [103]. Recreational walking in nature also promotes physical activity and performance of individuals at work [103].

6.5.1.2 Urban Wetlands and Riparian Forests

Wetlands and riparian forests are among the most productive and vulnerable ecosystems in urbanized areas due to their complex watershed systems. They often have very high biodiversity. In the urban century, these ecosystems are under intense pressure due to LULC and surface sealing. Wetlands and riparian forests are very efficient for water and matter regulation, pollutant fixation, and floodwater retention. These endangered ecosystems also act as buffers between surface runoff and rivers and reduce urban flooding events.

A 1 acre wetland has the potential to store around 1 million gallons of water and may reduce urban flooding by 60%, which can prevent morbidity, mortality, and socioeconomic losses in urban areas. It can also act as a water purification system and remove pollutants, organic nutrients, and phosphate and nitrogen from urban runoff. By the time urban runoff reaches rivers or lakes, the water is mostly clean. Thus, wetlands also act as natural wastewater treatment plants and represent perfect nature-based solutions for risk mitigation and adaptation of extreme events such as floods and drought. Wetlands and riparian forests also provide a cooling effect and reduce the intensity of heatwaves in cities [105].

Wetlands and floodplain forests serve as recreational areas and provide multiple benefits such as fresh air, moisture, oxygen, and biogenic essentials and cultural services to urban dwellers. Wetlands also provide a habitat for a vast number of plants, animals, and endangered species. However, natural ecosystems of prime significance are on the verge of extinction due to anthropogenic activities, and wetlands are declining in response to global climate change [90, 95, 101].

6.5.1.3 Solar Energy

Adaptation to climate change is the biggest challenge of the urban century. Harnessing nature by applying ecosystem-based engineering may help combat several urban environmental issues. Adopting solar energy as a renewable energy source will reduce our dependence on fossil fuels for energy, reducing greenhouse gases and helping combat global warming and climate change [106]. Solar energy includes photovoltaic cells system and solar thermal systems. Photovoltaic cell systems directly convert solar energy into electrical energy, whereas solar thermal systems convert solar energy into heat and are often used for heating water and cooking food using solar cookers [107]. They can also be used in microgrids as an alternative source of electricity at a local scale, such as for small towns, buildings, and hospitals [108].

6.5.2 Artificial Engineering Approaches

6.5.3 Landfill Gas as an Alternative Source of Energy: Waste to Wealth

Waste generation increased 10-fold in the last century due to rapid urbanization. Annually, approximately 1.3 billion tons of waste are generated globally, and this is expected to increase to 2.2 billion tons by 2025. Waste management is essential for protecting the environment and human health [109]. The amount of waste generated is one aspect of significant issues with the segregation and proper disposal of wastes.

Often, in developing countries, municipal waste, biomedical waste, and construction materials from cities are collected and disposed of in unplanned landfills. These landfill sites play a major role in global warming and climate change because landfill gases contain methane. Methane is 25 times more powerful than carbon dioxide in terms of global warming potential. Methane emissions from landfills represent the largest source of greenhouse gas (GHG) emissions from the waste sector, contributing around 700 Mt of CH_4 [38]. Organic waste makes up 40–70% of municipal solid waste in developing countries, which is more than in the developed world, further exacerbating the potential for methane emissions. Also, dumping waste in open landfills causes human exposure, leading to the spread of infectious diseases and aquifer contamination by the formation of leachate.

Abatement of methane emissions from landfills by using methane-rich landfill gas as a renewable energy source will help mitigate urban environmental issues such as greenhouse gas emissions, global warming, climate change, and environmental pollution [110]. It will also improve the quality of life for people living in the vicinity of landfill sites [111]. Engineered landfill sites can be used to generate electricity using landfill gas. According to the US Environmental Protection Agency (EPA), about 11 billion Kilowatt-hours (kWh) of electricity have been generated from 270 billion cubic feet of landfill gas collected from 352 landfill sites.

6.5.3.1 Wastewater/Sewage Treatment Plants as Sources of Energy

Wastewater/sewage treatment plants can also be used as sources of energy recovery. Obtained energy can be utilized for the plants' operation. An activated sludge process is commonly used for waste/sewage treatment. Bacteria are essential for the efficiency of the treatment process, and biogas is often produced as a byproduct. Biogas is also rich in methane and is used for the production of energy. As mentioned, energy recovered from treatment plants is often used for plant operation, thus enhancing the energy efficiency of existing facilities [112, 113]. The process also reduces the abundance of pathogenic bacteria entering the environment and mitigates the spread of infectious diseases [114, 115].

6.5.3.2 Rainwater Harvesting

Due to rapid climate change, freshwater sources and the groundwater table are declining. Rapid urbanization and industrialization put additional pressure on government and civic systems for adequate drinking water in terms of quality and quantity [116, 117].

In the last few decades, India has witnessed rapid urban expansion. According to the Ministry of Housing and Urban Affairs, urban populations increased from 29 crore to 37.70 crore during the period from 2002 to 2011. As a result, water demand has rapidly increased in urban areas. Cities in India receive heavy rainfall during monsoons. Despite higher precipitation during monsoons, cities such as Delhi, Mumbai, Kolkata, Chennai, and Bengaluru are experiencing a drinking water shortage. Therefore, Niti Aayog has considered rainwater harvesting as a sustainable water conservation option and to recharge groundwater.

Rainwater is considered an alternative drinking water source after freshwater reserves such as rivers, lakes, and glaciers. Rainwater is suitable for drinking because it is often free from geogenic contaminants (fluoride, arsenic, and uranium) and pathogens. Therefore, rain harvesting may provide a solution to water scarcity in urbanized areas [118, 119].

6.5.3.3 Constructed Floating Islands for Water Treatment

Constructed floating islands (CFIs) are composed of three layers: the upper leaves and stems of plants, the middle floating body (floating beds/buoyant mat, sponge), and the bottom root system and associated microbial communities. The mechanism of action of CFIs depends on the plant nutrient assimilation rate and microbial metabolism to remove pollutants. Because of its low cost and convenient maintenance, CFI technology has been widely applied to treatments of variant eutrophic waters worldwide.

The concept of the floating island system originated in Europe to treat the riparian ecosystem. In the 1980s, artificial floating islands of water hyacinth (*Eichhornia crassipes*) and alligator weeds were used to clean water polluted with cobalt, silver, and strontium metals in Germany. Recently, CFI systems have been used to remove nitrogen, phosphorus, and cyanobacterial biomass from diverse aquatic systems. For example, in China, a CFI system was used to treat rural sewage: it reduced the chemical oxygen demand (COD), total nitrogen (TN), and total phosphorus (TP) concentrations by 79.59, 61.06, and 77.14%, respectively. Enhanced floating island/CFI systems are also being used for nutrient removal and improving the water quality of eutrophic lake water.

6.5.3.4 Microgrids

A *microgrid* is a small power system independent from the bulk electric network (macrogrid). It can operate using renewable energy sources like biogas and fuel cells. Microgrids can function as primary power systems at a small scale, such as in hospitals, universities, and small communities. Using microgrids can reduce the consumption of fossil fuels as energy sources in urban areas, leading to a reduction in greenhouse gas emissions and the mitigation of climate change phenomena. Therefore, they can increase the resilience of urbanized areas [108, 120].

6.6 Future Perspective

Urbanization is beneficial for improving quality of life but should not be at the cost of nature. Ecosystem health and services are essential for human survival. The NBS/ecological engineering approach provides an economical and sustainable option to mitigate urban environmental issues. Previous studies have shown the positive impact of contact with nature on human physical and mental health. Green infrastructure also improves the provisioning of ecosystem services and biodiversity in urban areas. Also, most urban environmental issues can be attributed to a lack of vegetation and green space, which exacerbates climate change. Thus, climate change, urban areas, and ecosystem services act as a nexus.

Ecological engineering and restoration can be used as tools for finding solutions to emerging urban environmental issues. Further, instead of developing resilient/smart cities, the emphasis should be at the micro level, i.e. on developing resilient towns with proper green infrastructure. This will ensure better adaptability and resilience in mitigating environmental issues.

Acknowledgments

The authors wish to acknowledge the Director, CSIR-NEERI for providing the necessary facilities for this work.

References

1 Peng, J., Tian, L., Liu, Y. et al. (2017). Ecosystem services response to urbanization in metropolitan areas: thresholds identification. *Science of the Total Environment* 607–608: 706–714. https://doi.org/10.1016/j.scitotenv.2017.06.218.

2 Seto, K.C., Güneralp, B., and Hutyra, L.R. (2012). Global forecasts of urban expansion to 2030 and direct impacts on biodiversity and carbon pools. *Proceedings of the National Academy of Sciences* 109 (40): 16083. https://doi.org/10.1073/pnas.1211658109.

3 Tollefson, J. (2019). One million species face extinction. *Nature* 9.

4 Elmqvist, T., Andersson, E., Frantzeskaki, N. et al. (2019). Sustainability and resilience for transformation in the urban century. *Nature Sustainability* 2 (4): 267–273. https://doi.org/10.1038/s41893-019-0250-1.

5 McPhillips, L.E., Chang, H., Chester, M.V. et al. (2018). Defining extreme events: a cross-disciplinary review. *Earth's Future* 6 (3): 441–455. https://doi.org/10.1002/2017ef000686.

6 Núñez-Montero, K. and Barrientos, L. (2018). Advances in Antarctic research for antimicrobial discovery: a comprehensive narrative review of bacteria from Antarctic environments as potential sources of novel antibiotic compounds against human pathogens and microorganisms of industrial importance. *Antibiotics (Basel, Switzerland)* 7 (4): 90. https://doi.org/10.3390/antibiotics7040090.

7 Pecl, G.T., Araújo, M.B., Bell, J.D. et al. (2017). Biodiversity redistribution under climate change: impacts on ecosystems and human well-being. *Science* 355 (6332): eaai9214.

8 Perrings, C. (2010). Biodiversity, ecosystem services, and climate change. The Economic Problem/Environment Department Papers (120), 1–39. World Bank Environment Department.

9 Rhind, S.M. (2009). Anthropogenic pollutants: a threat to ecosystem sustainability? *Philosophical Transactions of the Royal Society, B: Biological Sciences* 364 (1534): 3391–3401. https://doi.org/10.1098/rstb.2009.0122.

10 Scheffer, M., Carpenter, S.R., Lenton, T.M. et al. (2012). Anticipating critical transitions. *Science* 338 (6105): 344–348. https://doi.org/10.1126/science.1225244.

11 Zhang, X.Q. (2016). The trends, promises and challenges of urbanisation in the world. *Habitat International* 54: 241–252. https://doi.org/10.1016/j.habitatint.2015.11.018.

12 Bren d'Amour, C., Reitsma, F., Baiocchi, G. et al. (2017). Future urban land expansion and implications for global croplands. *Proceedings of the National Academy of Sciences* 114 (34): 8939. https://doi.org/10.1073/pnas.1606036114.

13 Song, X.-P., Hansen, M.C., Stehman, S.V. et al. (2018). Global land change from 1982 to 2016. *Nature* 560 (7720): 639–643. https://doi.org/10.1038/s41586-018-0411-9.

14 Bai, X., McAllister, R.R.J., Beaty, R.M., and Taylor, B. (2010). Urban policy and governance in a global environment: complex systems, scale mismatches and public participation. *Current Opinion in Environmental Sustainability* 2 (3): 129–135. https://doi.org/10.1016/j.cosust.2010.05.008.

15 Seto, K.C., Golden, J.S., Alberti, M., and Turner, B.L. (2017). Sustainability in an urbanizing planet. *Proceedings of the National Academy of Sciences* 114 (34): 8935. https://doi.org/10.1073/pnas.1606037114.

16 Green, T.R., Taniguchi, M., Kooi, H. et al. (2011). Beneath the surface of global change: impacts of climate change on groundwater. *Journal of Hydrology* 405 (3–4): 532–560.

17 Grimmond, S.U.E. (2007). Urbanization and global environmental change: local effects of urban warming. *The Geographical Journal* 173 (1): 83–88. https://doi.org/10.1111/j.1475-4959.2007.232_3.x.

18 Miller, J.D. and Hutchins, M. (2017). The impacts of urbanisation and climate change on urban flooding and urban water quality: a review of the evidence concerning the United Kingdom. *Journal of Hydrology: Regional Studies* 12: 345–362. https://doi.org/10.1016/j.ejrh.2017.06.006.

19 Sanchez Rodriguez, R., Ürge-Vorsatz, D., and Barau, A.S. (2018). Sustainable development goals and climate change adaptation in cities. *Nature Climate Change* 8 (3): 181–183. https://doi.org/10.1038/s41558-018-0098-9.

20 Das, M. and Das, A. (2019). Dynamics of urbanization and its impact on urban ecosystem services (UESs): a study of a medium size town of West Bengal, eastern India. *Journal of Urban Management* https://doi.org/10.1016/j.jum.2019.03.002.

21 Wu, J. (2014). Urban ecology and sustainability: the state-of-the-science and future directions. *Landscape and Urban Planning* 125: 209–221. https://doi.org/10.1016/j.landurbplan.2014.01.018.

22 Srinivasan, V., Seto, K.C., Emerson, R., and Gorelick, S.M. (2013). The impact of urbanization on water vulnerability: a coupled human–environment system approach for Chennai, India. *Global Environmental Change* 23 (1): 229–239. https://doi.org/10.1016/j.gloenvcha.2012.10.002.

23 Vardoulakis, S., Dear, K., and Wilkinson, P. (2016). Challenges and opportunities for urban environmental health and sustainability: the HEALTHY-POLIS initiative. *Environmental Health* 15 (1): S30. https://doi.org/10.1186/s12940-016-0096-1.

24 Pataki, D.E. (2015). Grand challenges in urban ecology. *Frontiers in Ecology and Evolution* 3: 57.

25 Cumming, G.S., Buerkert, A., Hoffmann, E.M. et al. (2014). Implications of agricultural transitions and urbanization for ecosystem services. *Nature* 515 (7525): 50–57.

26 Kang, Y., Khan, S., and Ma, X. (2009). Climate change impacts on crop yield, crop water productivity and food security–a review. *Progress in Natural Science* 19 (12): 1665–1674.

27 Chiabai, A., Quiroga, S., Martinez-Juarez, P. et al. (2018). The nexus between climate change, ecosystem services and human health: towards a conceptual framework. *Science of the Total Environment* 635: 1191–1204. https://doi.org/10.1016/j.scitotenv.2018.03.323.

28 Romero-Lankao, P., McPhearson, T., and Davidson, D.J. (2017). The food-energy-water nexus and urban complexity. *Nature Climate Change* 7: 233. https://doi.org/10.1038/nclimate3260.

29 Heaviside, C., Vardoulakis, S., and Cai, X.-M. (2016). Attribution of mortality to the urban heat island during heatwaves in the West Midlands, UK. *Environmental Health* 15 (1): S27. https://doi.org/10.1186/s12940-016-0100-9.

30 Azhar, G. S., Mavalankar, D., Nori-Sarma, A., Rajiva, A., Dutta, P., Jaiswal, A., Group, A.H.S. (2014). Heat-related mortality in India: excess all-cause mortality associated with the 2010 Ahmedabad heat wave. *PLoS One*, 9(3): e91831.

31 Mazdiyasni, O., AghaKouchak, A., Davis, S.J. et al. (2017). Increasing probability of mortality during Indian heat waves. *Science Advances* 3 (6): e1700066.

32 Phelan, P.E., Kaloush, K., Miner, M. et al. (2015). Urban heat island: mechanisms, implications, and possible remedies. *Annual Review of Environment and Resources* 40: 285–307.

33 Stocker, T. (2014). *Climate Change 2013: The Physical Science Basis: Working Group I Contribution to the Fifth Assessment Report of the Intergovernmental Panel on Climate Change*. Cambridge University Press.

34 Pickett, S.T.A., Cadenasso, M.L., Grove, J.M. et al. (2011). Urban ecological systems: scientific foundations and a decade of progress. *Journal of Environmental Management* 92 (3): 331–362. https://doi.org/10.1016/j.jenvman.2010.08.022.

35 Vaughan, D.G., Comiso, J.C., Allison, I. et al. (2013). Observations: cryosphere. *Climate Change* 2103: 317–382.

36 Rubinato, M., Nichols, A., Peng, Y. et al. (2019). Urban and river flooding: comparison of flood risk management approaches in the UK and China and an assessment of future knowledge needs. *Water Science and Engineering* 12 (4): 274–283. https://doi.org/10.1016/j.wse.2019.12.004.

37 Gupta, K. (2020). Challenges in developing urban flood resilience in India. *Philosophical Transactions of the Royal Society A* 378 (2168): 20190211.

38 Bogner, J., Pipatti, R., Hashimoto, S. et al. (2008). Mitigation of global greenhouse gas emissions from waste: conclusions and strategies from the intergovernmental panel on climate change (IPCC) fourth Assessment report. Working Group III (mitigation). *Waste Management & Research* 26 (1): 11–32.

39 Koop, S.H.A. and van Leeuwen, C.J. (2017). The challenges of water, waste and climate change in cities. *Environment, Development and Sustainability* 19 (2): 385–418. https://doi.org/10.1007/s10668-016-9760-4.

40 Gogoi, P.P., Vinoj, V., Swain, D. et al. (2019). Land use and land cover change effect on surface temperature over eastern India. *Scientific Reports* 9 (1): 8859. https://doi.org/10.1038/s41598-019-45213-z.

41 van Wezel, A., Caris, I., and Kools, S.A. (2016). Release of primary microplastics from consumer products to wastewater in the Netherlands. *Environmental Toxicology and Chemistry* 35 (7): 1627–1631.

42 Yadav, S., Jadeja, N.B., Dafale, N.A. et al. (2019). Pharmaceuticals and personal care products mediated antimicrobial resistance: future challenges. In: *Pharmaceuticals and Personal Care Products: Waste Management and Treatment Technology* (eds. M.N.V. Prasad, M. Vithanage and A. Kapley), 409–428. Butterworth-Heinemann.

43 Power, A.L., Tennant, R.K., Jones, R.T. et al. (2018). Monitoring impacts of urbanisation and industrialisation on air quality in the Anthropocene using urban pond sediments. *Frontiers in Earth Science* 6 (131) https://doi.org/10.3389/feart.2018.00131.

44 WHO. (2014a). Public Health, Environmental and Social Determinants of Health (PHE). Ambient (Outdoor) Air Pollution in Cities Database 2014.

45 WHO (2014b). *WHO Guidelines for Indoor Air Quality: Household Fuel Combustion*. World Health Organization.

46 Chow, J.C., Watson, J.G., Shah, J.J. et al. (2004). Megacities and atmospheric pollution. *Journal of the Air & Waste Management Association* 54 (10): 1226–1235.

47 Leung, D.Y.C. (2015). Outdoor-indoor air pollution in urban environment: challenges and opportunity. *Frontiers in Environmental Science* 2 (69) https://doi.org/10.3389/fenvs.2014.00069.

48 Guttikunda, S.K., Goel, R., and Pant, P. (2014). Nature of air pollution, emission sources, and management in the Indian cities. *Atmospheric Environment* 95: 501–510. https://doi.org/10.1016/j.atmosenv.2014.07.006.

49 Shrivastava, M., Ghosh, A., Bhattacharyya, R., & Singh, S. (2018). Urban Pollution in India. In (pp. 341).

50 Wen, Y., Schoups, G., and van de Giesen, N. (2017). Organic pollution of rivers: combined threats of urbanization, livestock farming and global climate change. *Scientific Reports* 7 (1): 43289. https://doi.org/10.1038/srep43289.

51 McDonald, R.I., Weber, K., Padowski, J. et al. (2014b). Water on an urban planet: urbanization and the reach of urban water infrastructure. *Global Environmental Change* 27: 96–105. https://doi.org/10.1016/j.gloenvcha.2014.04.022.

52 Glińska-Lewczuk, K., Gołaś, I., Koc, J. et al. (2016). The impact of urban areas on the water quality gradient along a lowland river. *Environmental Monitoring and Assessment* 188 (11): 624.

53 McAlexander, T.P., Gershon, R.R.M., and Neitzel, R.L. (2015). Street-level noise in an urban setting: assessment and contribution to personal exposure. *Environmental Health* 14 (1): 18. https://doi.org/10.1186/s12940-015-0006-y.

54 de Paiva Vianna, K.M., Alves Cardoso, M.R., and Rodrigues, R.M.C. (2015). Noise pollution and annoyance: an urban soundscapes study. *Noise & Health* 17 (76): 125–133. https://doi.org/10.4103/1463-1741.155833.

55 Yuan, M., Yin, C., Sun, Y., and Chen, W. (2019). Examining the associations between urban built environment and noise pollution in high-density high-rise urban areas: a case study in Wuhan, China. *Sustainable Cities and Society* 50: 101678. https://doi.org/10.1016/j.scs.2019.101678.

56 Münzel, T., Gori, T., Babisch, W., and Basner, M. (2014). Cardiovascular effects of environmental noise exposure. *European Heart Journal* 35 (13): 829–836. https://doi.org/10.1093/eurheartj/ehu030.

57 Barot, S., Yé, L., Abbadie, L. et al. (2017). Ecosystem services must tackle anthropized ecosystems and ecological engineering. *Ecological Engineering* 99: 486–495.

58 Elmqvist, T., Setälä, H., Handel, S.N. et al. (2015). Benefits of restoring ecosystem services in urban areas. *Current Opinion in Environmental Sustainability* 14: 101–108. https://doi.org/10.1016/j.cosust.2015.05.001.

59 Elmqvist, T., Fragkias, M., Goodness, J. et al. (2013). *Urbanization, Biodiversity and Ecosystem Services: Challenges and Opportunities: A Global Assessment*. Springer.

60 McDonald, R., Marcotullio, P., and Güneralp, B. (2013). Urbanization and global trends in biodiversity and ecosystem services. In: *Urbanization, Biodiversity and Ecosystem Services: Challenges and Opportunities*. Springer.

61 McPhearson, T., Pickett, S.T.A., Grimm, N.B. et al. (2016). Advancing urban ecology toward a science of cities. *BioScience* 66 (3): 198–212. https://doi.org/10.1093/biosci/biw002.

62 Bodrossy, L., Kovács, K.L., McDonald, I.R., and Murrell, J.C. (1999). A novel thermophilic methane-oxidising γ-Proteobacterium. *FEMS Microbiology Letters* 170 (2): 335–341.

63 McDonald, R., Guneralp, B., Zipperer, W., and Marcotullio, P. (2014a). The future of global urbanization and the environment. *Solutions* 2014 (November–December): 60–69.

64 Assessment, M.E. (2005). *Ecosystems and Human Well-Being*, vol. 5. Washington, DC: Island press.

65 Perrings, C., Naeem, S., Ahrestani, F. et al. (2010). Ecosystem services for 2020. *Science* 330 (6002): 323–324.

66 Alirol, E., Getaz, L., Stoll, B. et al. (2011). Urbanisation and infectious diseases in a globalised world. *The Lancet Infectious Diseases* 11 (2): 131–141.

67 Hassell, J.M., Begon, M., Ward, M.J., and Fèvre, E.M. (2017). Urbanization and disease emergence: dynamics at the wildlife–livestock–human Interface. *Trends in Ecology & Evolution* 32 (1): 55–67. https://doi.org/10.1016/j.tree.2016.09.012.

68 McIntyre, K.M., Setzkorn, C., Hepworth, P.J. et al. (2017). Systematic Assessment of the climate sensitivity of important human and domestic animals pathogens in Europe. *Scientific Reports* 7 (1): 7134. https://doi.org/10.1038/s41598-017-06948-9.

69 Myers, S.S., Gaffikin, L., Golden, C.D. et al. (2013). Human health impacts of ecosystem alteration. *Proceedings of the National Academy of Sciences of the United States of America* 110 (47): 18753–18760. https://doi.org/10.1073/pnas.1218656110.

70 Neiderud, C.-J. (2015). How urbanization affects the epidemiology of emerging infectious diseases. *Infection Ecology & Epidemiology* 5: 10.3402/iee.v3405.27060. https://doi.org/10.3402/iee.v5.27060.

71 Lee, V.J., Ho, M., Kai, C.W. et al. (2020). Epidemic preparedness in urban settings: new challenges and opportunities. *The Lancet Infectious Diseases* 20 (5): 527–529.

72 Amutha, D. and Juliet, M. (2017). Impact of climate changes on human health in India. *SSRN 3071055.*

73 Ikeda, T., Behera, S.K., Morioka, Y. et al. (2017). Seasonally lagged effects of climatic factors on malaria incidence in South Africa. *Scientific Reports* 7 (1): 2458. https://doi.org/10.1038/s41598-017-02680-6.

74 Morand, S., Owers, K.A., Waret-Szkuta, A. et al. (2013). Climate variability and outbreaks of infectious diseases in Europe. *Scientific Reports* 3: 1774. https://doi.org/10.1038/srep01774.

75 Caminade, C., Kovats, S., Rocklov, J. et al. (2014). Impact of climate change on global malaria distribution. *Proceedings of the National Academy of Sciences* 111 (9): 3286–3291. https://doi.org/10.1073/pnas.1302089111.

76 Bhatt, S., Gething, P.W., Brady, O.J. et al. (2013). The global distribution and burden of dengue. *Nature* 496: 504. https://doi.org/10.1038/nature12060.

77 Campbell-Lendrum, D., Manga, L., Bagayoko, M., and Sommerfeld, J. (2015). Climate change and vector-borne diseases: what are the implications for public health research and policy? *Philosophical Transactions of The Royal Society B Biological Sciences* 370 (1665): 20130552.

78 Ebi, K.L. and Nealon, J. (2016). Dengue in a changing climate. *Environmental Research* 151: 115–123. https://doi.org/10.1016/j.envres.2016.07.026.

79 Xu, L., Stige, L.C., Chan, K.-S. et al. (2017). Climate variation drives dengue dynamics. *Proceedings of the National Academy of Sciences* 114 (1): 113–118.

80 Mutheneni, S.R., Morse, A.P., Caminade, C., and Upadhyayula, S.M. (2017). Dengue burden in India: recent trends and importance of climatic parameters. *Emerging Microbes & Infections* 6: e70. https://doi.org/10.1038/emi.2017.57.

81 Constantin de Magny, G., Murtugudde, R., Sapiano, M.R.P. et al. (2008). Environmental signatures associated with cholera epidemics. *Proceedings of the National Academy of Sciences* 105 (46): 17676–17681. https://doi.org/10.1073/pnas.0809654105.

82 Weill, F.-X., Domman, D., Njamkepo, E. et al. (2017). Genomic history of the seventh pandemic of cholera in Africa. *Science* 358 (6364): 785–789. https://doi.org/10.1126/science.aad5901.

83 MacFadden, D.R., McGough, S.F., Fisman, D. et al. (2018). Antibiotic resistance increases with local temperature. *Nature Climate Change* 8 (6): 510–514. https://doi.org/10.1038/s41558-018-0161-6.

84 Peng, F., Guo, Y., Isabwe, A. et al. (2020). Urbanization drives riverine bacterial antibiotic resistome more than taxonomic community at watershed scale. *Environment International* 137: 105524. https://doi.org/10.1016/j.envint.2020.105524.

85 Flies, E.J., Mavoa, S., Zosky, G.R. et al. (2019). Urban-associated diseases: candidate diseases, environmental risk factors, and a path forward. *Environment International* 133: 105187. https://doi.org/10.1016/j.envint.2019.105187.

86 Mitsch, W.J. and Jørgensen, S.E. (2003). Ecological engineering: a field whose time has come. *Ecological Engineering* 20 (5): 363–377.

87 Aronson, J., Clewell, A., and Moreno-Mateos, D. (2016). Ecological restoration and ecological engineering: Complementary or indivisible? *Engineering* 91: 392–395.

88 Bratman, G.N., Anderson, C.B., Berman, M.G. et al. (2019). Nature and mental health: an ecosystem service perspective. *Science Advances* 5 (7): eaax0903. https://doi.org/10.1126/sciadv.aax0903.

89 Keniger, L.E., Gaston, K.J., Irvine, K.N., and Fuller, R.A. (2013). What are the benefits of interacting with nature? *International Journal of Environmental Research and Public Health* 10 (3): 913–935. https://doi.org/10.3390/ijerph10030913.

90 Kabisch, N., Frantzeskaki, N., Pauleit, S. et al. (2016a). Nature-based solutions to climate change mitigation and adaptation in urban areas: perspectives on indicators, knowledge gaps, barriers, and opportunities for action. *Ecology and Society* 21 (2).

91 Ziter, C.D., Pedersen, E.J., Kucharik, C.J., and Turner, M.G. (2019). Scale-dependent interactions between tree canopy cover and impervious surfaces reduce daytime urban heat during summer. *Proceedings of the National Academy of Sciences* 116 (15): 7575–7580.

92 Berland, A., Shiflett, S.A., Shuster, W.D. et al. (2017). The role of trees in urban stormwater management. *Landscape and Urban Planning* 162: 167–177. https://doi.org/10.1016/j.landurbplan.2017.02.017.

93 Denman, E., May, P., and Moore, G. (2016). The potential role of urban forests in removing nutrients from stormwater. *Journal of Environmental Quality* 45 (1): 207–214.

94 Escobedo, F.J., Kroeger, T., and Wagner, J.E. (2011). Urban forests and pollution mitigation: Analyzing ecosystem services and disservices. *Environmental Pollution* 159 (8): 2078–2087. https://doi.org/10.1016/j.envpol.2011.01.010.

95 Escobedo, F.J., Giannico, V., Jim, C.Y. et al. (2019). Urban forests, ecosystem services, green infrastructure and nature-based solutions: nexus or evolving metaphors? *Urban Forestry & Urban Greening* 37: 3–12. https://doi.org/10.1016/j.ufug.2018.02.011.

96 Livesley, S.J., McPherson, E.G., and Calfapietra, C. (2016). The urban Forest and ecosystem services: impacts on urban water, heat, and pollution cycles at the tree, street, and city scale. *Journal of Environmental Quality* 45 (1): 119–124. https://doi.org/10.2134/jeq2015.11.0567.

97 Calfapietra, C., Morani, A., Sgrigna, G. et al. (2016). Removal of ozone by urban and peri-urban forests: evidence from laboratory, field, and modeling approaches. *Journal of Environmental Quality* 45 (1): 224–233.

98 Herrera-Gomez, S.S., Quevedo-Nolasco, A., and Pérez-Urrestarazu, L. (2017). The role of green roofs in climate change mitigation. A case study in Seville (Spain). *Building and Environment* 123: 575–584. https://doi.org/10.1016/j.buildenv.2017.07.036.

99 Rowe, D.B. (2011). Green roofs as a means of pollution abatement. *Environmental Pollution* 159 (8): 2100–2110. https://doi.org/10.1016/j.envpol.2010.10.029.

100 Palermo, S. and Turco, M. (2020). Green wall systems: where do we stand? IOP Conference Series: Earth and Environmental Science.

101 Eggermont, H., Balian, E., Azevedo, J.M.N. et al. (2015). Nature-based solutions: new influence for environmental management and research in Europe. *GAIA-Ecological Perspectives for Science and Society* 24 (4): 243–248.

102 Pereira, G., Foster, S., Martin, K. et al. (2012). The association between neighborhood greenness and cardiovascular disease: an observational study. *BMC Public Health* 12 (1): 466.

103 Mitchell, R. (2013). Is physical activity in natural environments better for mental health than physical activity in other environments? *Social Science & Medicine* 91: 130–134. https://doi.org/10.1016/j.socscimed.2012.04.012.

104 Douglas, I. (2015). Ecosystems and human well-being. In: *Reference Module in Earth Systems and Environmental Sciences* (ed. S. Elias). Elsevier http://dx.doi.org/10.1016/B978-0-12-409548-9.09206-X.

105 Ma, L., Jha, R. K., Mondal, P. K., & Li, W. (2019). Wetlands restoration engineering in the metropolitan area. In *Ecological Wisdom Inspired Restoration Engineering* (ed. V. Achal and A. Mukherjee), 197–221. Springer.

106 Jones, H.P., Hole, D.G., and Zavaleta, E.S. (2012). Harnessing nature to help people adapt to climate change. *Nature Climate Change* 2 (7): 504–509. https://doi.org/10.1038/nclimate1463.

107 Drennen, T.E., Erickson, J.D., and Chapman, D. (1996). Solar power and climate change policy in developing countries. *Energy Policy* 24 (1): 9–16. https://doi.org/10.1016/0301-4215(95)00117-4.

108 Vine, D. and Morsch, A. (2017). Microgrids: What Every City Should Know. *Center for Climate and Energy Solutions.*

109 Hoornweg, D., Bhada-Tata, P., and Kennedy, C. (2013). Environment: waste production must peak this century. *Nature News* 502 (7473): 615.

110 Yadav, S., Kundu, S., Ghosh, S. K., & Maitra, S. (2015). Molecular analysis of methanogen richness in landfill and marshland targeting 16S rDNA sequences. *Archaea*, 2015, Article ID 563414, 9.

111 Narayana, T. (2009). Municipal solid waste management in India: from waste disposal to recovery of resources? *Waste Management* 29 (3): 1163–1166. https://doi.org/10.1016/j.wasman.2008.06.038.

112 Gandiglio, M., Lanzini, A., Soto, A. et al. (2017). Enhancing the energy efficiency of wastewater treatment plants through co-digestion and fuel cell systems. *Frontiers in Environmental Science* 5: 70.

113 Liu, F., Ouedraogo, A., Manghee, S., and Danilenko, A. (2012). A primer on energy efficiency for municipal water and wastewater utilities. http://hdl.handle.net/10986/18060.

114 Al-Gheethi, A.A., Mohamed, R.M., Efaq, A.N. et al. (2017). Reduction of bacteria in storage system of sewage effluents. *Sustainable Water Resources Management* 3 (2): 193–203. https://doi.org/10.1007/s40899-017-0102-y.

115 Al-Gheethi, A.A., Efaq, A.N., Bala, J.D. et al. (2018). Removal of pathogenic bacteria from sewage-treated effluent and biosolids for agricultural purposes. *Applied Water Science* 8 (2): 74. https://doi.org/10.1007/s13201-018-0698-6.

116 Bates, B. (2009). *Climate Change and Water: IPCC Technical Paper VI.* World Health Organization.

117 Lal, R. (2012). Urban ecosystems and climate change. In: *Carbon Sequestration in Urban Ecosystems* (eds. R. Lal and B. Augustin), 3–19. Dordrecht: Springer Netherlands.

118 Bhawan, P. and Nager, A. (2001). *Concepts and Practices for Rainwater Harvesting.* Ministry of Environment & Forests, Govt. of India, Delhi.

119 Ndeketeya, A. and Dundu, M. (2019). Maximising the benefits of rainwater harvesting technology towards sustainability in urban areas of South Africa: a case study. *Urban Water Journal* 16 (2): 163–169. https://doi.org/10.1080/1573062x.2019.1637907.

120 Lasseter, R.H. and Paigi, P. (2004). Microgrid: A conceptual solution. IEEE 35th Annual Power Electronics Specialists Conference (IEEE Cat. No. 04CH37551).

121 Sultana, S. and A. Satyanarayana, Urban heat island intensity during winter over metropolitan cities ofIndia using remote-sensing techniques: Impact of urbanization. *International journal of remote sensing*, 2018.

122 Chakraborty, T., Hsu, A.,Manya, D., & Sheriff, G. (2019). Disproportionately higher exposure tourban heat in lower-income neighborhoods: a multi-city perspective. *Environmental Research Letters*, 14(10):105003.

7

Soil Fertility Restoration, Theory and Practice

V. Matichenkov[1] *and E. Bocharnikova*[2]

[1] *Department of Ecology and Physiology of Autotrophic Organisms, Institute Basic Biological Problems, Pushchino, Russia*
[2] *Department of Functional Ecology, Institute Basic Biological Problems, Pushchino, Russia*

7.1 Introduction

Soil degradation and reduction of soil fertility are twenty-first-century global problems. *Soil fertility* is defined as the ability of a soil to provide the conditions required for plant growth. It is determined by complex physical, chemical, and biological processes that act together to deliver nutrients, water, aeration, and stability to plants and freedom from any substances that may inhibit growth [1–4]. The fertility of agricultural soil can be defined as its capacity to produce the crops desired [5]. The application of mineral fertilizers is the most common tool to provide plant nutrition and enhance soil productivity [6]. In the mid and late twentieth century, the Green Revolution, based on mineral fertilizers and synthetic pesticides and herbicides, provided a remarkable increase in agriculture productivity and food supply in many countries.

Intensive agriculture using large inputs of chemicals has increased crop yields but led to severe environmental problems, including pollution and soil degradation [7]. The fertilizer industry is considered a source of natural radionuclides and heavy metals. Many types of mineral fertilizers contain heavy metals, including Hg, Cd, As, Pb, Cu, Ni, and others, along with the radionuclides ^{238}U, ^{232}Th, and ^{210}Po [8]. For example, in the EU countries, 54–58% of Cd entered the environment from phosphorous fertilizers [9].

Despite the positive effects of inorganic fertilizers on crop yields, there can be indirect adverse impacts on soil quality. Soil fertilization with nitrogen may lower soil pH. Among the consequences for the soil, acidification leads to leaching, nutrient deficiency, and soil degradation [10, 11]. Leon and Osorio [12] report that degradation decreased soil ecosystem services by 60% between 1950 and 2010. Globally, 33% of the earth's land surface is affected by soil degradation [13]. Soil quality deterioration is associated with reductions in soil organic matter, soil adsorption capacity, microorganism populations, and other harmful processes [14]. Degradation of soil fertility leads to increased application of agrochemicals, thus accelerating the negative consequences [15]. In addition to negatively impacting agronomic productivity, soil degradation also dampens economic growth, especially in countries where agriculture is the engine for economic development [16].

Handbook of Ecological and Ecosystem Engineering, First Edition. Edited by Majeti Narasimha Vara Prasad.

About 1.5 billion hectares are used by modern world agriculture, and reserves of undeveloped land that could be involved in agriculture total about 1.3 billion ha [17]. In the entire history of civilization (approximately 10 000 years), we have lost at least 2 billion hectares [18]. Over the past 50 years, the rate of annual losses of soil resources was 30 times higher than the average historical rates and reached almost 20 million ha [17].

It should be noted that the negative impact of degradation of agricultural soils is almost invisible at the local level. Farmers cultivate soil and generally do not report soil problems but observe reductions in yield quality and quantity. If we look at global soil fertility, we reveal a critical situation posing a real threat to the environment and food security. As a result, attention to the need to improve soil fertility is increasing, with a focus on soil management technologies.

The most popular technologies for soil fertility restoration apply organic materials [19, 20]. The addition of fresh or treated manure is a very old technology. In recent years, biochar has become increasingly popular as a soil fertility enhancer [21, 22]. Fertilization with organic amendments provides micro- and macronutrients required for plant growth and enhances microbial biomass, diversity, and activity [10]. Numerous biochemical processes that occur in the soil, such as organic matter decomposition, humus formation, nutrient transformation, and cycling, involve the participation of soil microorganisms; they play a vital role in maintaining soil productivity [23]. Soil biota improves physical properties such as structure, porosity, aeration, and water infiltration by forming and stabilizing soil aggregates. Microbial activity and diversity directly determine the level of soil fertility. One promising approach to improving soil quality is based on microbial preparations [14, 24].

There is data that Si-rich soil amendments or Si fertilizers can be exploited for increasing soil fertility and productivity [25, 26]. Si substances make up the soil skeleton and affect many soil properties such as pH, the stability of soil organic matter, microbial population, nutrient mobility, and bioavailability [27]. Soluble forms of Si and Si-containing minerals reduce the bioavailability of organic and inorganic pollutants [28]. Additional Si nutrition increases plant resistance to biotic and abiotic stresses and affects the transport of macro-and microelements as well as pollutants [29]. All this makes Si and its compounds indispensable in soil fertility management and for agricultural productivity.

Soil is a complex natural system in which inorganic and organic substances and living organisms have symbiotic interactions [30]. Therefore, soil fertility restoration strategies should be based on an integrated approach, not just adding organic compounds (treated manure), active Si, or microbial fertilizers. This study aimed to compare conventional agricultural manure-related technology and complex technology – including organic matter, Si-rich substances, and microorganisms – on the yield of tomatoes, peppers, and cucumbers; soil fertility; and farm economy.

7.2 Materials and Methods

In the greenhouse experiment, tomatoes (*Solanum lycopersicum*, Aurora variety), cucumbers (*Cucumis sativus*, Herman variety), and sweet peppers (*Capsicum annuum*, Agapovsky variety) were grown in medium loam gray forest soil.

The following treatments were conducted:

1) Control (no substances were applied)
2) NPK
3) Composted chicken manure (compost)
4) Diatomite (DE)

5) Activated diatomite (A-DE)
6) Microbial preparation (M)
7) Compost + A-DE + M

The area of each experimental plot was 4 m^2. Each treatment had four replications.

The composted chicken manure contained nitrogen (N) – 4.0%, phosphorus (P) – 2.7%, and potassium (K) – 4.01% on dry weight and was applied at 1 t of dry weight/ha. Application rates for conventional mineral fertilizers were calculated according to their content in chicken compost. Nitrogen was added as ammonium nitrate (NH_4NO_3) at 114 kg/ha, P as double superphosphate ($Ca(H_2PO_4)_2 \cdot H_2O$) at 110 kg/ha, and K as potassium chloride (KCl) at 71.7 kg/ha.

Diatomite from the Inza Deposit (Ulyanovsk region, Russia) was used as a Si-rich substance and applied at 500 kg/ha. Diatomite had the following chemical composition: SiO_2 88.1–88.3%, Fe_2O_3 1.1–1.5%; CaO 2.2–2.5%; MgO 1.0–1.3%; Na_2O 0.8–1.1%; pH 6.1–6.3; particle size 0.1–0.2 mm.

Because natural diatomite has an insufficient content of plant-available Si – 45–44 mg/kg of water-soluble Si and 640–650 mg/kg of acid-soluble (0.1 n HCl) Si – [31], it was activated by treatment with concentrated monosilicic acid (Ekkor, Orekhovo-Zuyevo, Russia). Diatomite was mixed with concentrated monosilicic acid (15% Si) at a proportion of 50 : 1 (DE:monosilicic acid) and then dried at +65 °C. This treatment increased the plant-available Si to 125–128 mg/kg for water-soluble Si and 1200–1230 mg/kg for 0.1 n HCl-soluble Si. This Si levelcan be classified as high [31]. Activated diatomite was applied at 250 kg/ha.

The microbial preparation contained *Bacillus amyloliquefasciens* (Ecomicrobial Co., USA). Before applying, 1 g of dry microbial preparation was mixed with 1 l of distilled water and kept for 48 hours at room temperature. The bacteria were added at a rate of 6 mg/m^2 three times for a month.

Fertilizers and other additives were applied to the soil before planting. Seeds were pre-soaked in 0.1% hydrogen peroxide solution for 20 minutes and then put into distilled water for 24 hours at a temperature of +24° ± 3 °C. Germinated seeds were planted 25 cm apart at 2–3 cm depth. The soil was irrigated with distilled water daily to maintain moisture of 20–25%. The air temperature in the greenhouse was kept at 24 ± 2 °C during the day and 20 ± 2 °C at night. The light period was 12 hours; the light intensity was 130 μmol photons/m^2/s^1. The relative air humidity was 65 ± 5% during the day and 70 ± 5% at night.

Harvesting vegetables began 1.5 months after planting and continued for two months. In addition to yield, vegetable quality was assessed. The last harvested vegetables were analyzed for Brix (sugars) using a refractometer (KH-512ATC, Kehang, China) and nitrates using an ionometer (Twin NO_3^-, B-343, Horiba, Japan).

After harvesting, the soil was sampled and analyzed for pH, total carbon (C), and cation exchange capacity (CEC) using standard methods [32]. Soil tests were performed in triplicate. The least significant difference was calculated at a 5% level with Excel 2010.

7.3 Results

The application of all additives increased the yields of tomatoes, cucumbers, and peppers (Figure 7.1). The yield increases provided by conventional fertilizers were 62.9, 89.6, and 31.0% for tomatoes, cucumbers, and peppers, respectively. The effects of compost were similar to those of NPK. The DE- and bacteria-mediated yield increases were lower compared to other additives. Higher yields were observed with combined compost + A-DE + M treatment, with increases of 119.6, 201.5, and 89.8% for tomatoes, cucumbers, and peppers, respectively.

The nitrate and Brix content are important parameters for the vegetable quality assessment. Conventional mineral fertilizers generally reduce the quality of vegetables [33]. In our experiment, the NPK application significantly increased nitrates (Figure 7.2) and reduced Brix values in all vegetables (Figure 7.3). Nitrates increased by 96.7, 68.5, and 92.3% in tomatoes, cucumbers, and peppers,

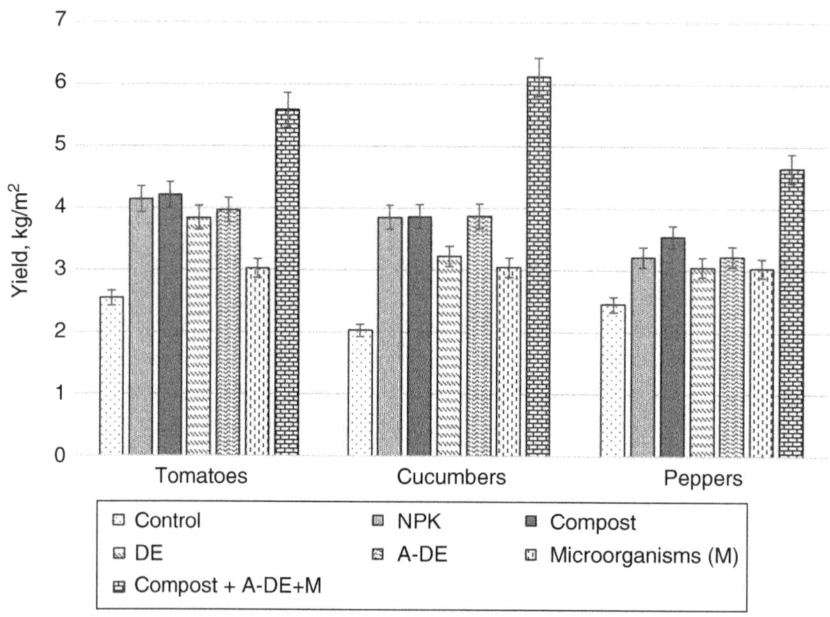

Figure 7.1 Yield of tomatoes, cucumbers, and peppers in the greenhouse test, kg/m^2.

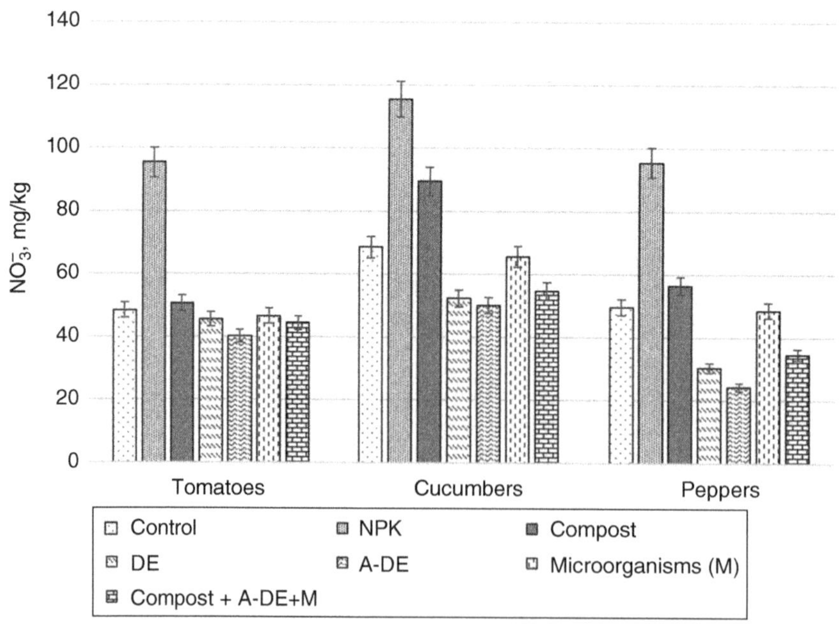

Figure 7.2 Nitrates in tomatoes, cucumbers, and peppers, NO^{-3} mg/kg.

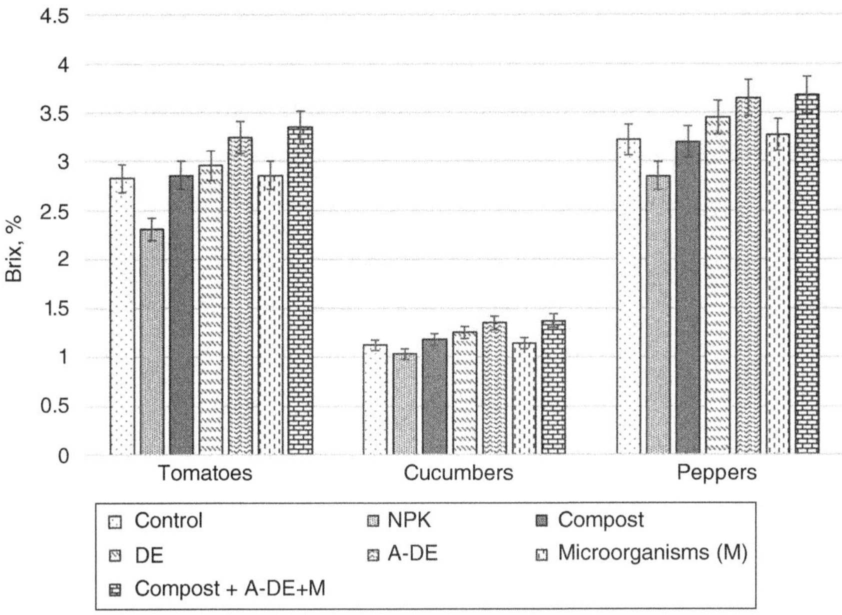

Figure 7.3 Brix in tomatoes, cucumbers, and peppers, %.

respectively, while Brix decreased by 18.4, 8.1, and 11.5%. Compost significantly increased nitrates only in cucumbers (by 30.6%). DE, A-DE, and bacteria (separately or together with compost) reduced nitrates. A greater reducing effect (by 51.2%) was observed in peppers with bacteria application.

The application of DE and A-DE separately or together with compost significantly increased the Brix values in tomatoes (by 4.6–18.4%), cucumbers (by 11.6–22.3%), and peppers (by 7.1–14.3%). For all crops, maximum increases in carbohydrates were obtained with combined treatment.

The results of soil analysis are shown in Figures 7.4–7.6. Conventional fertilizers reduced pH from 6.73 to 6.10. All other additives insignificantly increased soil pH, with a greater effect from A-DE.

Organic carbon (C_{org}) is considered a primary soil fertility-related parameter [34]. As evident from many reports, long-term application of mineral fertilizers commonly leads to decreased soil organic matter [35, 36]. The current investigation also demonstrates that NPK fertilizers reduced total C_{org} in soil (Figure 7.5). However, because of the short experiment time, the reduction was not significant. A significant increase in C_{org} was observed only in the soils where compost was applied separately or with A-DE + M (from 0.85 to 0.97 and 0.94%).

CEC is an important characteristic of soil fertility and plant nutrition [37]. Soil CEC after the vegetable experiment is shown in Figure 7.6. Considering that CEC is a very stable parameter and the experiment was conducted during a single season, CEC was significantly altered only in the soil where A-DE was applied alone or together with compost and bacteria. The application of NPK slightly reduced the CEC value.

7.4 Discussion and Conclusions

The data obtained showed that conventional NPK fertilization and compost application had approximately the same effect on the yield of tested crops, while crop quality and soil fertility were better when compost was used. Application of DE, A-DE, and bacteria separately also benefitted

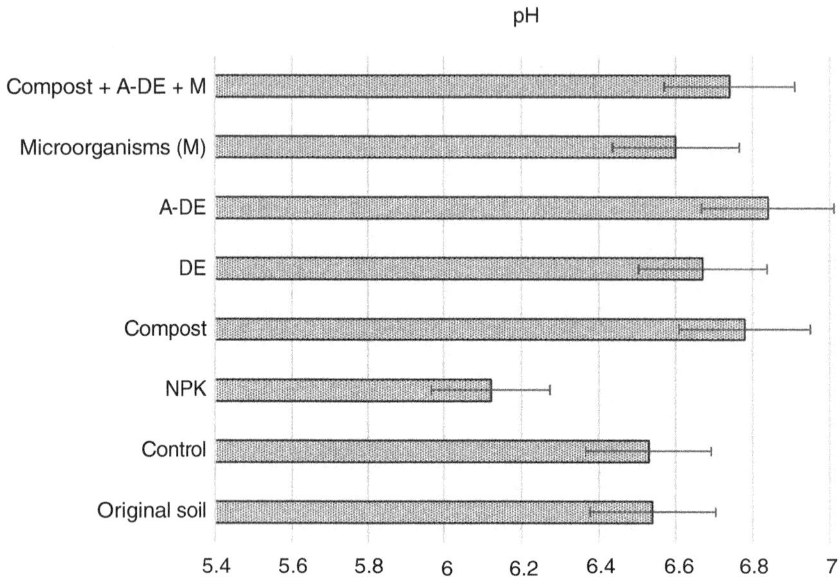

Figure 7.4 Effect of applied materials on soil pH.

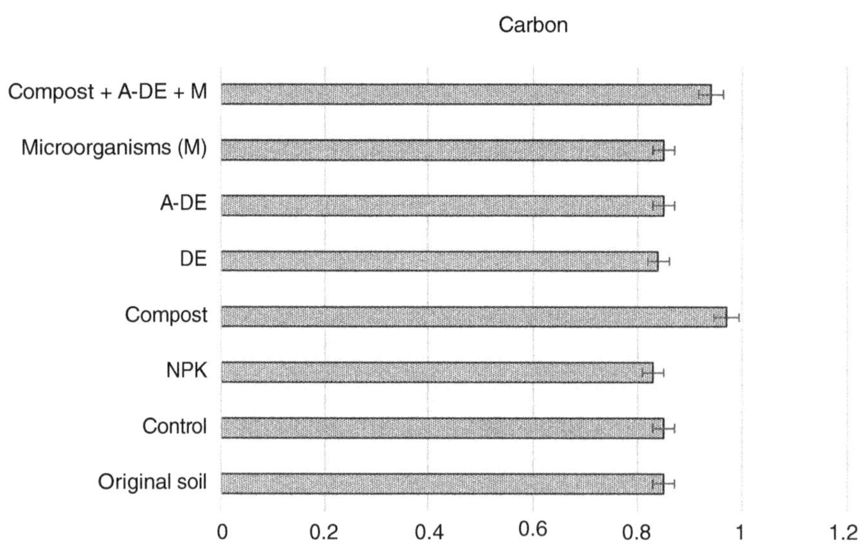

Figure 7.5 Effect of applied materials on C_{org} in soil, %.

crop quality and soil fertility, whereas combined application of compost, A-DE, and microorganisms had the best effect on crop quality and quantity. It is important to note that soil fertility was enhanced as well, thus contributing to sustainable and highly productive agriculture in the future.

There is intense discussion about the cost-effectiveness of conventional and organic or natural agriculture [38–40]. To assess the economic efficiency of growing selected vegetables, we used the average warehouse price for vegetables in Russia in 2019 (https://agrobazar.ru/vegetables/wholesale). The cost of conventional fertilizers was obtained from a Russian website selling fertilizers for large farms (Flagma.ru). The price for ammonium nitrate was 13 000 rubles/ton, double superphosphate

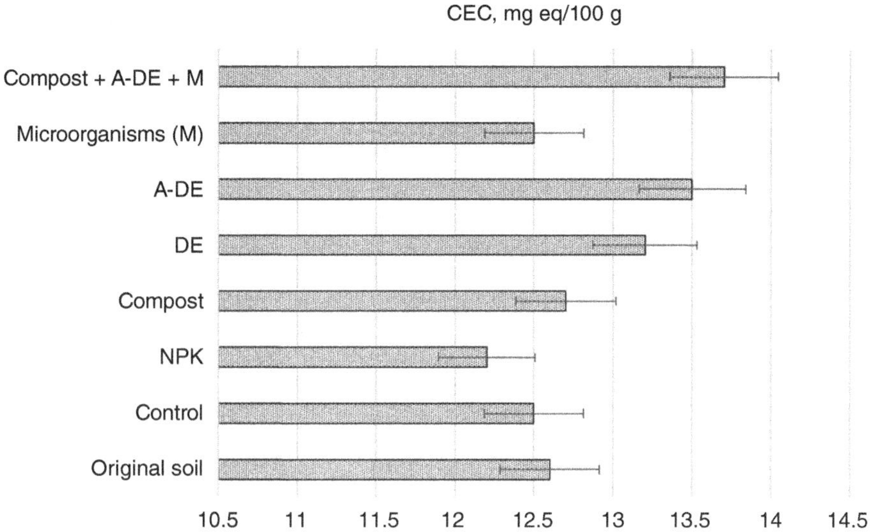

Figure 7.6 Effect of applied materials on soil CEC, mg eq/100 g.

20 500 rubles/ton, and potassium chloride 20 700 rubles/ton. In 2019, 1 US$ = 65 rubles. The cost of chicken compost was 7000 rubles/ton, diatomite 2600 rubles/ton, activated diatomite 14 800 rubles/ton, and microbial preparation 8000 rubles/kg. Based on this information, the cost of the vegetables grown (fertilizer cost, rub/m^2, divided by yield, kg/m^2) was calculated (Table 7.1).

The lowest cost for growing tomatoes was obtained with the microbial application (0.05 rubles/m^2) and A-DE and DE (0.09 rubles/m^2). The highest cost for tomatoes was with compost + A-DE + M. Microbial application provided the lowest cost of growing cucumbers and peppers, whereas the highest cost was with compost + A-DE + M. However, if only microorganisms and DE are applied, essential nutrients like N, P, and K will not be replenished entirely, thereby resulting in decreasing productivity over time.

To correctly assess fertilization's economic efficiency, it is essential to know the cost of additional yield calculated as the difference between yields from fertilizer-added and control plots [41, 42]. The cost of additional yield was assessed according to the following equation: additional cost = cost of applied materials, rub/m^2/yield from treated plot minus yield from control plot, kg/m^2. The minimum cost of growing additional tomatoes was with DE application (0.26–0.27 rub/m^2). The maximum cost was calculated with compost (0.42 rubles/m^2), while the cost of growing vegetables with compost + A-DE + M was 0.40 rubles/m^2.

The lowest cost (0.14 rubles/m^2) for cucumbers was obtained with bacteria application, whereas the cost was twice as high (0.29 rubles/m^2) with traditional NPK fertilizers. The complex of environmentally friendly preparations allowed the crop to have a lower cost (by 0.08 rubles/m^2) than compost alone.

The highest cost for additional peppers was observed with conventional mineral fertilizers (0.69 rub/m2). The minimum cost came with using bacteria (0.25 rubles/m^2). Compost or compost + A-DE + M made it possible to obtain the crop significantly cheaper than with NPK.

The calculated costs do not include the water, electricity, transportation, and wages but allow us to compare the effectiveness of tested additives. According to the results, one of the main factors that inhibits the widespread use of compost is the high cost. In addition, commercial compost commonly lacks active forms of Si and microorganisms – components essential for soil fertility and plant growth.

Table 7.1 Cost of vegetable production, rub/m^2.

	Tomatoes			Cucumbers			Peppers		
	Additional yield, kg/m^2	Total cost of yield, rub/m^2	Cost of additional yield, rub/m^2	Additional yield, kg/m^2	Total cost of yield, rub/m^2	Cost of additional yield, rub/m^2	Additional yield, kg/m^2	Total cost of yield, rub/m^2	Cost of additional yield, rub/m^2
NPK	1.6	0.13	0.33	1.82	0.14	0.29	0.76	0.16	0.69
Compost	1.67	0.17	0.42	1.83	0.18	0.38	1.09	0.20	0.64
DE	1.3	0.09	0.27	1.19	0.11	0.29	0.6	0.11	0.58
A-DE	1.42	0.09	0.26	1.84	0.10	0.20	0.77	0.11	0.48
Microorganisms (M)	0.48	0.05	0.30	1.01	0.05	0.14	0.58	0.05	0.25
Compost + A-DE + M	3.04	0.22	0.40	4.09	0.20	0.30	2.2	0.26	0.55

Interestingly, compost was applied in many ancient civilizations, including the Roman Empire, China, and the Middle East [43–45]. Using manure or compost to make highly fertile soil on thousands of hectares is familiar in a number of countries, such as Belorussia. However, restoring soil fertility with this technology takes a long time (about 20–30 years), negatively affecting farm profitability and requiring governmental subsidies [46]. These technologies are based on the addition of huge doses (up to 20 t/ha and higher) of organic matter, reducing economic efficiency and environmental safety [47].

Intensive agriculture needs new advanced technologies based on understanding soil processes and restoring soil fertility or formation of highly fertile soil in a short time – two to five years. The current study demonstrates that components such as Si fertilizers and special microbial preparations should be included in soil management strategies to maintain the high productivity of cultivated plants, high quality of crops, and soil fertility.

Organic wastes are common by-products in any world region. Si fertilizers and Si-based soil amendments can be produced from many industrial by-products and natural minerals, like diatomite, zeolites, and clays with high levels of Si. Microbe-growing technologies are common today and can be used worldwide. Therefore, the tested technology could be applicable in any region and successfully used to increase soil fertility level and farm economic efficiency.

Acknowledgment

This study was supported by the Ministry of Science and Higher Education of the Russian Federation, theme #AAAA-A17-117030110137-5 and #AAAA-A17-117030110139-9.

References

1 De Datta, S.K., Gomez, K.A., and Descalsota, J.P. (1988). Changes in yield response to major nutrients and in soil fertility under intensive rice cropping. *Soil Science* 146 (5): 350–358.
2 Klimashevskiĭ, Ė.L. (1990). Theory of plant agrochemical efficiency. *Agrokhimiya* 1: 131–148.
3 Ottman, M.J. and Welch, L.F. (1989). Planting patterns and radiation interception, plant nutrient concentration, and yield in corn. *Agronomy Journal* 81 (2): 167–174.
4 Stockdale, E.A., Shepherd, M.A., Fortune, S., and Cuttle, S.P. (2002). Soil fertility in organic farming systems–fundamentally different? *Soil Use and Management* 18: 301–308.
5 Cooke, G.W. (1967). The Control of Soil Fertility, 526. London: Crosby Lockwood & Son Ltd.
6 Bhattacharyya, R., Ghosh, B.N., Mishra, P.K. et al. (2015). Soil degradation in India: challenges and potential solutions. *Sustainability* 7 (4): 3528–3570.
7 Tilman, D. (1999). Global environmental impacts of agricultural expansion: the need for sustainable and efficient practices. *Proceedings of the National Academy of Sciences* 96 (11): 5995–6000.
8 FAO (2009). Resource STAT-Fertilizer. Food and Agriculture Organization of the United Nations. http://faostat.fao.org/site/575/Desktop Default.aspx? PageID=575#ancor, 12.03.2009.
9 Alloway, B.J. (1995). Chapter 6: Cadmium. In: Heavy Metals in Soils, 122–151. Glasgow: *Blackie*.
10 Franciolini, D., Schulz, E., Lentendu, G. et al. (2016). Mineral vs organic amendmnets: activity and abundance of agricultural microbes are driven by long-term fertilization strategies. *Frontier in Microbiology* 7: 1446.

11 Ulén, B., Pietrzak, S., Ramnerö, B., and Strand, L. (2016). Agricultural soil acidity and phosphorus leaching risk at farm level in two focus areas. *Acta Agriculturae Scandinavica, Section B – Soil & Plant Science* 66 (4): 359–368.

12 León, J.D. and Osorio, N.W. (2014). Role of litter turnover in soil quality in tropical degraded lands of Colombia. *The Scientific World Journal.*

13 Bini, C. (2011). Soil: a precious natural resource. In: Conservation of Natural Resources (ed. N.J. Kudrow), 1–48. Hauppauge, NY: Nova Science Publishers.

14 Rashid, M.I., Mujawar, L.H., Shahzad, T. et al. (2016). Bacteria and fungi can contribute to nutrients bioavailability and aggregate formation in degraded soils. *Microbiological Research* 183: 26–41.

15 Black, C.A. (2013). Soil Fertility Evaluation and Control. CRC Press.

16 Lal, R. (2015). Restoring soil quality to mitigate soil degradation. *Sustainability* 7 (5): 5875–5895.

17 Kerzhentsev, A.C. and Kuzmenchuk, U.A. (2009). We have no other land. *Bulletin of the Russian Academy of Sciences* 79 (4): 312–319.

18 Dobrovolsky, G.V. (1999). Global Character of Modern Degradation of the Soil Cover: The Structural Function of Soil in the Biosphere, 209–216. Moscow: GEOS.

19 Chirinda, N., Olesen, J., and Porter, J. (2008). Effects of organic matter input on soil microbial properties and crop yields in conventional and organic cropping systems. 16th IFOAM Organic World Congress, Modena, Italy, June 16–20.

20 Kay, B.D. (2018). Soil structure and organic carbon: a review. In: Soil Processes and the Carbon Cycle, 169–197. CRC Press.

21 Ameloot, N., Sleutel, S., Das, K.C. et al. (2015). Biochar amendment to soils with contrasting organic matter level: effects on N mineralization and biological soil properties. *GCB Bioenergy* 7 (1): 135–144.

22 Ding, Y., Liu, Y., Liu, S. et al. (2016). Biochar to improve soil fertility. A review. *Agronomy for Sustainable Development* 36 (2): 36.

23 Liu, Z., Rong, Q., Zhou, W., and Liang, G. (2017). Effects of inorganic and organic amendment on soil chemical properties, enzyme activities, microbial community and soil quality in yellow clayey soil. *PLoS one* 12 (3): e0172767.

24 An, T., Schaeffer, S., Zhuang, J. et al. (2015). Dynamics and distribution of 13 C-labeled straw carbon by microorganisms as affected by soil fertility levels in the Black soil region of Northeast China. *Biology and Fertility of Soils* 51 (5): 605–613.

25 Ma, J.F. and Takahashi, E. (2002). Soil, Fertilizer, and Plant Silicon Research in Japan. The Netherlands: Elsevier.

26 Xing, Y., Jiang, Z., Tan, Y. et al. (2016). Effects of silicon-calcium-phosphorus fertilizer on sugarcane growth and soil fertility of acidified filed. *Journal of Southern Agriculture* 47 (9): 1495–1499.

27 Kulikova, A.C. (2013). Silicon and Highly Siliceous Rocks in the System of Fertilization of Agricultural Crops. Ulyanovsk: Ulianovskagrichalture Academy Press.

28 Meharg, C. and Meharg, A.A. (2015). Silicon, the silver bullet for mitigating biotic and abiotic stress, and improving grain quality, in rice? *Environmental Experimental Botany* 120: 8–17.

29 Frew, A., Weston, L.A., Reynolds, O.L., and Gurr, G.M. (2018). The role of silicon in plant biology: a paradigm shift in research approach. *Annals of Botany* 121 (7): 1265–1273.

30 Volkova, I. (2017). Ecological Pedology. Litres.

31 Bocharnikova, E.A., Matichenkov, V.V., and Pogorelov, A.G. (2011). Comparative characteristics of some silicon fertilizers. *Agrochemistry* 11: 25–30.

32 Sparks D.L., Page A.L., Helmke P.A., Loeppert R.H., Soltanpour P.N., Tabatabai M.A., . . . Sumner M.E. (1996) Methods of Soil Analysis Part 3 – Chemical Methods. SSSA Book Ser. 5.3.

33 Bhugaloo, R.A., Lalouette, J.A., Bachraz, D.Y., and Sukerdeep, N. (1999). Effect of different levels of nitrogen on yield and quality of pineapple variety Queen Victoria. Annual Meeting of Agricultural Scientists.

34 Lehmann, J. and Kleber, M. (2015). The contentious nature of soil organic matter. *Nature* 528 (7580): 60–68.

35 Cheng, W., Padre, A.T., Shiono, H. et al. (2017). Changes in the pH, EC, available P, SOC and TN stocks in a single rice paddy after long-term application of inorganic fertilizers and organic matters in a cold temperate region of Japan. *Journal of Soils and Sediments* 17 (7): 1834–1842.

36 Singh Brar, B., Singh, J., Singh, G., and Kaur, G. (2015). Effects of long-term application of inorganic and organic fertilizers on soil organic carbon and physical properties in maize–wheat rotation. *Agronomy* 5 (2): 220–238.

37 Moral, F.J. and Rebollo, F.J. (2017). Characterization of soil fertility using the Rasch model. *Journal of Soil Science and Plant Nutrition* 17 (2): 486–498.

38 Kamali, F.P., Meuwissen, M.P., de Boer, I.J. et al. (2017). Evaluation of the environmental, economic, and social performance of soybean farming systems in southern Brazil. *Journal of Cleaner Production* 142: 385–394.

39 Pimentel, D., Hepperly, P., Hanson, J. et al. (2005). Environmental, energetic, and economic comparisons of organic and conventional farming systems. *BioScience* 55 (7): 573–582.

40 Wachter, J.M. (2017). Economic and environmental performance of mixed crop-livestock farming systems on the Palouse. Thesis, Washington State University.

41 Belous, I.N. and Prishep, D.N. (2009). Agro-economic efficiency of potato cultivation technologies// Agroeconomic efficiency of potato cultivation technologies. *Bulletin of the Bryansk State Agricultural Academy* 6: 1–6.

42 Vetrova, L.N. and Buhvostov, Y.V. (2007). Calculating the cost of protected ground vegetables. *Everything for an Accountant* 5: 16–20.

43 Conner, C.D. (2009). A People's History of Science: Miners, Midwives, and Low Mechanics. UK: Hachette.

44 Cui, Z., Dou, Z., Chen, X. et al. (2014). Managing agricultural nutrients for food security in China: past, present, and future. *Agronomy Journal* 106 (1): 191–198.

45 Scheidel, W., Morris, I., and Saller, R.P. (eds.) (2007). The Cambridge Economic History of the Greco-Roman World, 38. Cambridge: Cambridge University Press.

46 Gorbacheva, E.V., Azarenok, T.N., and Shulgina, S.V. (2010). The influence of humus state of cultured agrozem on its structural-aggregate composition. *Bulletin of the Belarussian State Agricultural Academy: The Guidance Journal* 3: 71–74.

47 Lednev, A.B. and Lednev, N.A. (2008). The effect of fertilizers on the reproduction of fertility of disturbed sod-podzolic soils. *Agrarian Science of the Euro-North-East* 11: 152–155.

8

Extracellular Soil Enzymes Act as Moderators to Restore Carbon in Soil Habitats

Rupinder Kaur and Anand Narain Singh

Soil Ecosystems and Restoration Ecology Lab, Department of Botany, Panjab University, Chandigarh, India

8.1 Introduction

Soil is the foundation of all processes required for the sustenance of life on earth. All living beings are dependent on nutrient cycles operating in the biosphere regulated by the soil system. Soil is a living entity with a complex metabolic system operated and regulated by various physical, chemical, and biological components. Any disturbance in this system can cause a potential threat to inhabiting flora and fauna and cause a significant disturbance in the biogeochemical cycling of nutrients in the biosphere. Due to its high endurance and resilience, soil has long tolerated devastating environmental and anthropogenic alterations. But recently, soil quality is deteriorating at unprecedented rates in many ecosystems due to the never-ending pressure of anthropogenic activities.

Changes in land use have altered a significant part of the natural world; for example, forest cover and grasslands have been converted into agricultural lands, urban areas, industrial areas, massive mining sites, etc. [1, 2]. Soil habitats are also degrading because of various environmental and geographical factors [3–6]. However, every soil type has different properties and faces different kinds of challenges and needs appropriate management practices [7–9]. To achieve stable soil health and support soil biota and nutrient cycling, proper information regarding soil development, factors responsible for soil fertility, etc. are required to understand soil's mechanistic role while considering its soil fertility in a changing environment. Various edaphic, environmental, and biological factors control the formation, functioning, fertility, and deterioration of soil systems:

- *Geographical location*: Geographical location, i.e. the latitude and longitude of a soil habitat, is the primary factor that affects the type and formation of soil. Soils in the northern hemisphere are different from the southern hemisphere. Their geographical location also decides the climate of the habitat and whether the soil belongs to an arid, subtropical, or tropical zone [10–12].
- *Topography*: This factor explains the structure and shape of the land surface, e.g. slope, undulating, flat, patchy, plane, etc. Sloping or undulating surfaces may experience more soil erosion, which is worse in the absence of soil-binding vegetation. Therefore, topography significantly affects the types of vegetation and soil quality. Soil erosion has been a challenge since the start of human civilization, even given our advanced knowledge of soil biology [11, 13, 14].

Handbook of Ecological and Ecosystem Engineering, First Edition. Edited by Majeti Narasimha Vara Prasad.

- *Climate*: This is one of the most important factors controlling the formation and deterioration of soil worldwide. Soil habitats with more periodic climatic conditions support more diversity than soil habitats that face extreme cold or hot temperatures. Therefore, tropical and subtropical environments have more vegetation diversity than arid zones and temperate habitats [4, 15–18].
- *Precipitation*: Precipitation in the form of rainfall affects the abundance and diversity of vegetation in the soil. Excessive rainfall may cause erosion in sloping areas. It also controls the microbial populations in soil; waterlogged and dry soil habitats show low microbial activities compared to humid soil habitats [17, 19–21].
- *Wind*: Wind velocity is an environmental factor that accelerates the pace of soil erosion. The flow and direction of winds affect the soil vegetation and structure, as severe winds may erode the top layers of soil and affect both vegetation and microbes in the soil [22].
- *Time of irradiance* (day: night ratio): The length of the day is a periodic phenomenon that alters the activity of soil biota and also controls the type of vegetation. Soil habitats where days are shorter have different vegetation than habitats with longer days [23, 24].
- *Altitude*: Altitude is measured as the height from mean sea level (msl). Climate varies not only with latitude and longitude but also with altitude. For example, soil habitats at high altitudes have more extreme climatic conditions, such as glaciers, and soils nearest to sea level are comparatively saline [10, 11].
- *Soil formation history*: The geological history of soil formation is the primary regulator of soil type and texture. Whether the soil is formed by sedimentation or rocks disintegration determines the texture, flora and fauna, and nutrient cycling in soil habitats [25].
- *Soil type*: The type of soil is an essential determinant of all other factors such as porosity, cation exchange capacity (CEC), bulk density, water and nutrient retention capacity, etc. It thus influences all the biotic components of the soil habitat [21, 26, 27].
- *Porosity*: Porosity refers to the air spaces in the soil particles. Soil with high porosity (sand) has poor nutrients and water-holding capacity, while soil with medium porosity (loam) tend to contains more biota in comparison to sandy or tightly packed clayish soil (low porosity) [28, 29].
- *CEC*: CEC is a key factor (Ca^{++}, Na^{+}, K^{+}, and Mg^{++}) that alters the chemical structure of the soil. It is described as the capacity of soil particles to chemically bind molecules on their surface: i.e. how many cations can bind soil particles with macro and micro elements such as C, N, P, iron, sulfur, etc. It is also used as an important parameter for soil fertility, as well as maintaining a buffer state in the soil environment. Therefore, CEC involvement in soil reflects the amount of nutrients available to soil biota for their activities [30, 31].
- *Water and nutrient retention capacity*: All the physical properties jointly regulate the water and nutrient holding capacity of the soil. It determines the diversity, population, and activity of soil biota, influencing soil fertility [21, 32–34].
- *Types and diversity of microbes and vegetation*: Diversity and types of vegetation control the soil microbial diversity and population in the rhizosphere of contrasting habitats [35, 36].
- *Microbial diversity and biomass in soil*: Microorganisms in the soil are key regulators of nutrient cycling and soil quality. They respond to changes in environmental conditions [37–40].
- *Soil organic matter (SOM)*: SOM refers to organic matter added by the detritivore community (saprotrophs) by decomposing plant litter and animal residues. It is the most essential component of all soil habitats and regulates soil's physical and chemical properties [41, 42].
- *pH*: The soil pH determines the hydrogen (H^{+}) and hydroxyl ions (OH^{-}) in the soil medium. Therefore, pH regulates the types of ions and nutrients in the soil. Furthermore, it helps maintain acidity or alkalinity and regulates the type of biota in soils [31, 43–45].

- *Available and total nutrients*: Available nutrients and total nutrients (mineral nutrients like NO_3-N, NH_4-N and available Pi and total nutrients like soil organic carbon [SOC], Total Kjeldahl nitrogen (TKN), and total phosphorus (TP) in the soil habitat) determine plant and microbial colonization on the soil surface. All nutrients in the soil layer (0–20 cm in depth) are known as the soil's *mineral zone* and form the top layer of soil. These nutrients are the true indicator of soil fertility when abundant soil flora and fauna with rich nutrients are available [7, 9, 21, 46].
- *Extracellular enzymes*: All the previous parameters are the determinants of soil type, quality, fertility, nutritional composition, soil biota, and vegetation, but they do not explain the exact cycling of nutrients without the involvement of soil enzymes. Extracellular soil enzymes are the key parameter that regulates the functioning fertility of soil habitats by performing vital processes such as carbon sequestration and N-mineralization [1, 17, 47–50].

Due to all these various factors, in different soil habitats, extensive study of contrasting habitats is required to determine the role of soil enzymatic activities [9, 17, 49, 51, 52].

The literature contains a large number of studies regarding enhancement of soil fertility by incorporating various kinds of amendments [53–56], management practices like tillage and crop rotation, and revegetation and reforestation [36, 57, 58]. The goal of all the studies is to enhance the soil nutrient concentration and cycling to support and enhance microbial activities in the soil [59, 60]. Activities of microbes are determined by analyzing biochemical and biological processes in the soil such as CO_2–efflux, rate of N–mineralization and nitrification, carbon sequestration, and activities of soil enzymes [61–63]. All these processes require substrates and soil enzymes to function properly [58, 64].

The most widely available substrate in most terrestrial ecosystems is plant litter. Decomposition in soil releases nutrients (C, N, and P) that were mobilized/immobilized in soil biota and plant components through various processes [65]. For example, C is photosynthetically sequestered from the atmosphere, fixed in plant biomass, and then sequestered from plant litter through decomposition by extracellular enzymatic activities into the soil sink and microbial biomass [66–72].

8.2 Soil Organic Matter (SOM)

SOM consists of plant litter and animal waste in various stages of decomposition. The quantity and quality of SOM in soil determines the biological activity and nutrient cycling of the habitat and varies significantly in contrasting soils [42, 73]. The total SOM consists of three layers: the first layer is fresh decomposing plant litter and has high biological activity; this layer is the main zone of nutrient cycling and is home to most soil biota [25, 74]. The second is the mineral layer, also known as humus, composed of physically protected forms of nutrients; it may take many years to decompose. The third layer is chemically stable and also known as the *soil sink*, as nutrients bound in this layer do not return to cycling for hundreds or thousands of years [71, 73, 75].

The quality of SOM determines the physicochemical, biochemical, biological, and microbiological properties of soil such as porosity, bulk density, water-holding capacity, nutrient retention, cycling, microbial biomass carbon (MBC), respiration, rates of N–mineralization and carbon sequestration, etc.[29, 63, 65, 76]. Thus, SOM helps with soil functioning by maintaining soil porosity, water- and nutrient-holding capacity [73, 74], soil aggregates and stabilization, etc. [41, 77]. Since plant biomass is the primary source of SOM, the microbes in soil decompose the root and leaf litter of plants, convert nutrients for plant and microbial use, and release CO_2 as a by-product into the atmosphere [78–80].

SOM is the most important component of soil habitats and the home of all the microbial and biogeochemical processes [29, 81]. Hence it needs to be maintained to achieve good soil quality and productivity. All the soil's available nutrients are part of the SOM layer; however, almost half of the total SOM is composed of carbon [82]. Generally, SOM contains two to three times more carbon than collectively found in the earth's vegetation. Therefore, it plays a vital role in the biogeochemical cycling of carbon on the earth, as it's the principal regulator of the carbon sequestration process [73–75, 77, 82].

8.3 Soil Organic Carbon (SOC)

SOC is the carbon content in the SOM layer of soil habitats. It is the largest carbon store on the earth, as it holds about three times as much carbon as the atmosphere and vegetation collectively [25, 83]. However, the distribution of carbon in different soils is not the same: some soils are very rich in carbon content, while some are very poor [84, 85]. This difference among soils is one of the major determinants of the type of soil biota and vegetation and, therefore, acts as the key regulator of soil fertility of various soil habitats [29, 59, 71].

Vegetation or plant litter is the major contributor of SOM in all soils, and this plant biomass is the total amount of organic matter derived via the photosynthetic process. It is decomposed by microbes and enzymes in the soil and releases CO_2, which goes back to the atmosphere, and organic carbon in the form of monosaccharides. This carbon is either consumed by soil microbes for their growth or adsorbed on the surface of soil particles to form humus, which later goes to the soil sink, where it remains for thousands of years [86–89]. However, SOC, which is the central part of the SOM released to the soil through the complex procedure of soil biota and available in a microbial-consumable form (glucose), is the most important step in determining nutrient cycling because the other two forms are resistant to decomposition and are not readily available for soil microorganisms [59, 70].

Various global studies have focused on enhancing the SOC content of soil for two reasons: (i) to increase organic carbon content via the addition of a substrate supply (plant litter, residue, etc.) in soil habitats to attract more soil microbes to enhance nutrient cycling and improve soil fertility; and (ii) to sequester more carbon from the atmosphere via vegetation and increase its mean residence time in the soil, to mitigate the increasing concentration of CO_2 in the atmosphere due to anthropogenic activities [4, 36, 49, 90].

In contrasting habitats, the amount of organic carbon varies significantly and indicates the carbon sequestration capacity of different soils [59]. Various strategies are being designed by environmentalists and researchers worldwide to replenish declining SOC from soil habitats. These include increasing SOC content by adding organic matter such as manure, compost, animal waste, etc. [8, 9]. In addition to supplying organic matter inputs, experiments were also designed to increase the mean residence time of carbon in the soil [3, 29, 91, 92].

8.4 Soil Carbon Sequestration

Soil carbon sequestration is the process of sequestering carbon from one source to another and finally depositing it in the soil carbon sink. It is chemically stable and resistant to decomposition and hence stays there for hundreds or thousands of years [29, 84, 93, 94]. This process includes sequestration of carbon from the atmosphere into the plant biomass via the process of photosynthesis, then sequestration from plant components (biomass) into the soil and microbial biomass by decomposition, after which it is stabilized on soil aggregates such as mineral and clay and forms physically stabilized complexes [29, 58, 95] (Figure 8.1). These complexes condense to form recalcitrant forms of carbon such as charcoal; these forms are biochemically stable and are known as a carbon sink. They store carbon for a very long time and do not allow it to return to the atmosphere easily [88, 96].

The process of carbon sequestration helps with the following:

- Controlling increasing levels of CO_2 in the atmosphere due to anthropogenic activities such as the burning of plant biomass and fossil fuel combustion, hence improving environmental quality
- Increasing plant biomass and soil productivity
- Enhancing microbial diversity in the soil
- Stabilizing soil aggregates and preventing further degradation and soil erosion by maintaining proper aeration and porosity and enhancing nutrients and water-holding capacity
- Trapping carbon in the soil sink, where it forms a reservoir as organically bonded carbon stock in other formats

Sequestration of carbon in the soil can be accomplished via management practices that increase plant biomass and add plant litter to the soil. An increase in organic matter in the soil can increase microbial activity and improve soil quality [25, 36, 97]. Studies of carbon sequestration mostly focus on the increasing residence time of carbon in soil habitats because, during the process of decomposition, some CO_2 is also released back to the atmosphere (Figure 8.2). By increasing carbon turnover time, soil carbon does not reach the point of saturation easily and, therefore, can sequester more carbon from the atmosphere to fulfill microbial demand, which is achieved by incorporating organic matter into the soil [98].

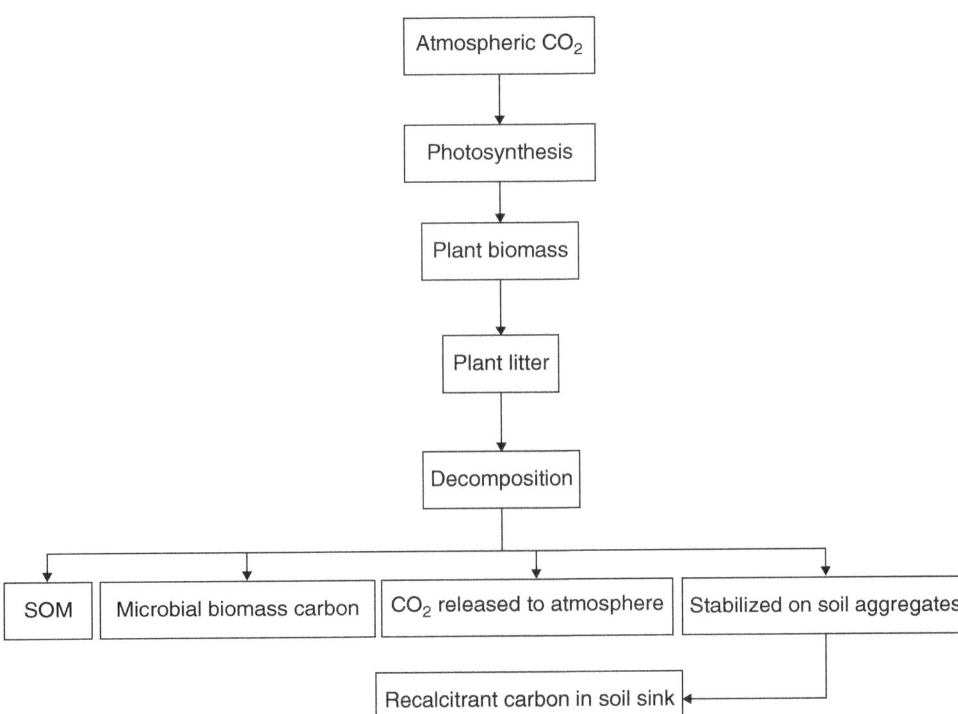

Figure 8.1 A generalized outline of the carbon sequestration mechanism from the atmosphere into the soil sink.

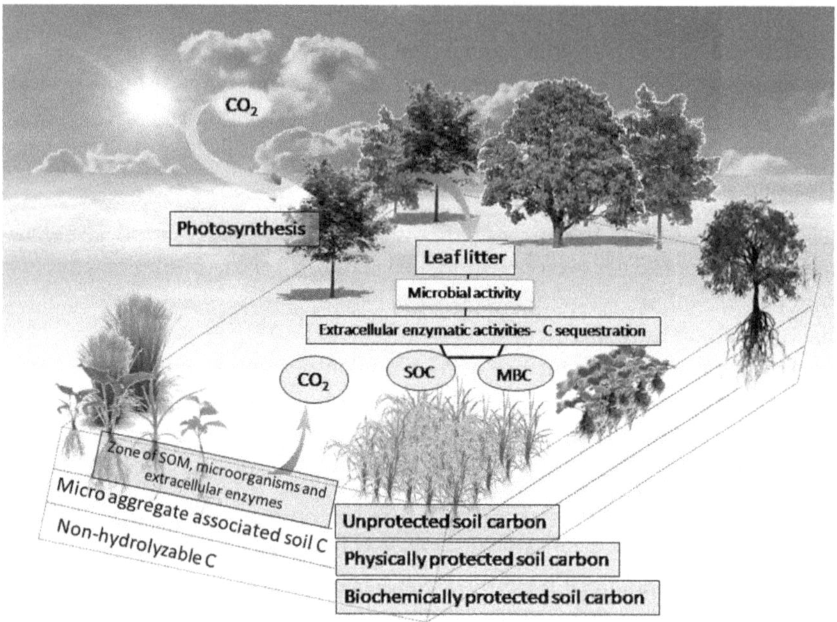

Figure 8.2 A generalized diagram showing carbon dynamics in a typical terrestrial habitat. *See colour plate section for colour representation of this figure.*

Due to its significance, soil carbon sequestration has been widely studied [25, 71]. Management practices have been proposed to enhance carbon sequestration rates in contrasting soil habitats [92, 99–102].

8.5 Extracellular Soil Enzymes

In general, soil's extracellular enzymes are released by plants, animals, and microorganisms; however, microbes are considered major contributors in the soil extracellular enzymatic pool [48, 50, 103]. Once these enzymes are released in the soil, their origin is difficult to determine; however, all the enzymes are either involved in the process of carbon sequestration and mineralization as soon as they are released or are stabilized by making complexes with clay or mineral particles of soil (Figure 8.3). These clay-enzyme complexes are resistant to degeneration from the activity of proteases [60, 104].

Based on the reactions they mediate and bonds they break, enzymes in the soil are of two types: oxidoreductases and hydrolases. Enzymes that are involved directly in carbon sequestration are mainly hydrolases as they break hydrogen bonds that exist in the chains of polysaccharide complexes [50, 105]. The function of these enzymes is to degrade large chains of complex polysaccharides such as cellulose, hemicelluloses, and starch. They act on plant litter, dead microbial cells, animal waste, and dead and decaying bodies and release nutrients into simple available forms [106]. These available forms of nutrients are further utilized by soil biota and plants to grow and develop [107–109]. However, one enzyme cannot mediate the total breakdown of those multifaceted forms from complex to simple forms. A group of enzymes is required to break different type of bonds: for example, cellulase is a group of enzymes including endocellulases that break down

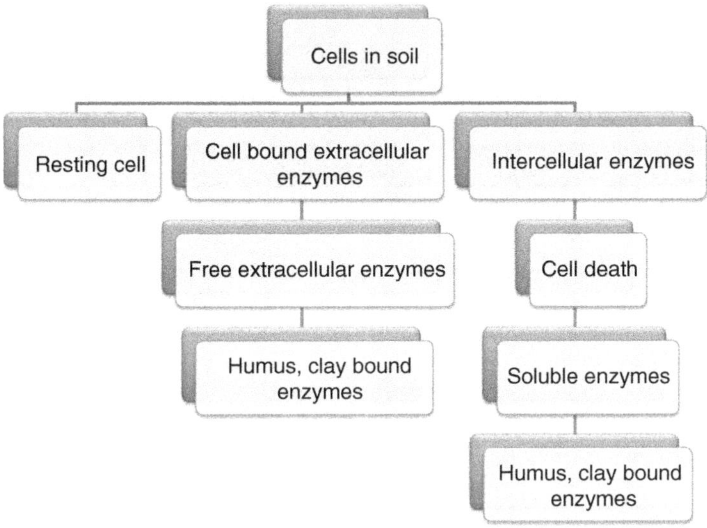

Figure 8.3 Origin and involvement of extracellular enzymes in soil. Source: Based on Gianfreda and Rao [1].

glucose units from (random) middle positions of a chain, while exocellulases release glucose from the ends of cellulose chains. The activity of a single enzyme-mediated complex cannot describe the functioning of a whole process of sequestration or mineralization [110, 111].

Many researchers have studied the activities of these enzymes involved in various reactions. Although the first report regarding the existence of extracellular enzymes in the soil was given by Woods in 1899 at an annual meeting of the American Association for the Advancement of Science in Columbus, Ohio (USA), the major work on the activities of these enzymes and their estimation methods was developed after 1950 [112]. Activities of extracellular enzymes in soil have now been studied for more than a century, and their importance in the metabolism of the living system of soil has been explored in various soil ecosystems since then. In the mid and late 1950s, a substantial number of investigators devoted considerable time to developing accurate methods to measure enzymatic activities and used theoretical approaches to understand the origin and state of enzymes in the soil (Figure 8.4).

A number of reports in the literature state that plant roots are a source of extracellular enzymes in the soil [113]. In some reports, researchers have also correlated the presence of enzymes in the rhizosphere and its regulation by the diversity of vegetation. But they have not distinguished the contribution of plant roots and microflora in the rhizosphere. According to the reports of Vlasyuk et al. [114], enzymes in the rhizosphere depend on the type of vegetation: e.g. urease activity was higher in the legume rhizosphere as compared to other crops. With the progress in soil enzymology, much work was reported on microflora as a significant source of most free enzymes in the soil [115]. The progress of medical, general, and industrial microbiology has supported the notion of the release of extracellular enzymes by microbes. Therefore, all the studies from past decades are based on the microbial release of extracellular enzymes determined by various methodologies.

Kuprevisch and Shcherbakova [116] and Galstyan in the Soviet Union carried out extensive research measuring several enzyme activities with different vegetative covers, soils, and climates. Skujins [112] stated that "it is evident enzymes are substrate specific and individual enzymatic

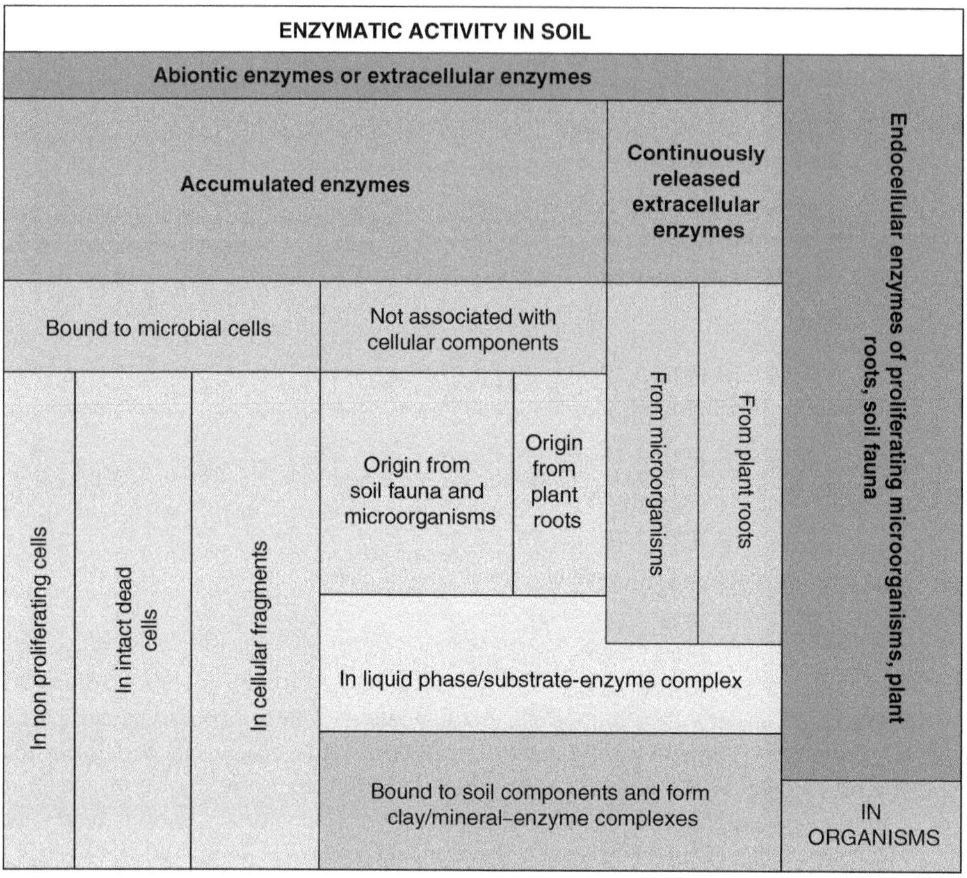

Figure 8.4 Origin of enzymatic activities in soil. Source: Skujins [112] © 1978, Elsevier.

measurements can not reflect the total nutrient status of the soil"; this should be considered prior to any study involving measurements of enzyme activity in the soil.

Hofmann and Seegerer [117] have suggested that determining soil enzymatic activity is more important to describe soil fertility than soil microbiological measurement and carbon dioxide production. Although enzymatic activities are substrate-specific and alone may not be sufficient to describe the total nutrient status of soil, they can yield valuable information regarding the availability of specific organic compounds and their degradation over time. The activities and significance of enzymes secreted or released into the soil were investigated and debated for decades [118][48, 50, 103, 118–122].

8.6 Interactive Role of Extracellular Soil Enzymes in Soil Carbon Transformation

Extracellular enzymes play a pivotal role in the process of carbon sequestration. Plant biomass is a significant reservoir of organic matter and carbon in the biosphere, but sequestration of carbon from plant biomass into the soil is mainly mediated by extracellular enzymes [123, 124]. Therefore, the activities of these enzymes are vital in the cycling of carbon. The activity of extracellular enzymes increases, in the shortage of consumable forms of carbon (glucose) and availability of complex substrates (cellulose/starch) [105, 106].

When a complex substrate is added to the soil surface, first the soil biota breaks down organic matter and makes it available to microbial action. Then, extracellular enzymes are secreted, which hydrolyze the bonds of long and complex chains. The resulting monosaccharides are used by microbes for growth and reproduction. However, during this process, polysaccharide chains are stabilized by binding to clay and mineral particles of the soil [76, 124]. The entire decomposition process leads to the overall sequestration of carbon in the soil and stabilizes its physical structure by binding organic matter to soil particles, which improves soil porosity. Improved soil porosity helps in proper soil aeration, root penetration, and improved water-holding capacity, which can efficiently reduce soil erosion [29]. Along with this, enzymatic activities also influence atmospheric CO_2 by efficiently trapping it in the soil sink. CO_2 is also released in the atmosphere during microbial respiration, but enhanced soil productivity due to the availability of nutrients helps trap atmospheric carbon in plant biomass. Therefore, the soil system plays a vital role in the cycling of carbon in the biosphere, wherein extracellular soil enzymes are the functional entities that mediate its sequestration [36, 63, 65, 124].

Carbon is the primary element in most plant components because it forms the backbone of all the chains that are joined together by hydrogen bonds [125]. Cellulose is the most abundant organic compound (polysaccharide) on earth, and starch is also a polysaccharide used for energy storage in green plants. Xylan is the third most abundant polysaccharide in nature and is a structural unit of cell walls [126]. All these compounds are in plant biomass. They are degenerated into their monosaccharide units by the activities of extracellular enzymes in the soil habitats, which further determine the presence of SOC in the soil. Therefore, extracellular enzymes play the most critical role in carbon sequestration from plant biomass to soil organisms and finally to the soil sink [64].

8.6.1 Cellulase

Cellulose is a structural polysaccharide and the most abundant organic compound on the earth, composed of glucose units joined together by β 1,4 glycosidic bonds (Figure 8.5). Cellulose is the major component of plant biomass, formed by fixation of atmospheric CO_2 by photosynthesis [126, 127]. Therefore, plant litter is the primary source of cellulose on the soil surface, and because of its abundance, microbes largely depend on this polysaccharide to fulfill their energy demands. Thus, to break down the β 1,4 glycosidic bonds and release the glucose monomers from cellulose chains, microorganisms release a cellulase enzyme complex into the soil [128, 129]. This enzyme is a complex of three forms: endocellulase breakdown the bonds that join cellulose chains together, exocellulase breaks long cellulose chains into smaller saccharides by acting on random positions and cellobiase releases glucose monomers for the consumption of soil biota [130]. This enzyme in the soil is released by the microbial community in plant residues and released by fungal interactions.

Cellulase activity in soil habitats is significantly affected by biotic and abiotic factors such as temperature, rainfall, vegetation, and microbial diversity [131, 132]. In addition, the availability of substrates and microbial demand for glucose are the main factors that regulate cellulase activity in soil habitats. Due to these variables, this enzyme's activity may vary significantly in different soil habitats [99, 132–134].

Pancholy and Rice [133] investigated cellulase activity in three contrasting ecosystems: tall grass prairie, post Oak-blackjack oak forest, and oak pine forest in Oklahoma, Kansas (USA) with respect to different successional stages and climax community composition. A seasonal effect on cellulase activity was also observed by Zafar et al. [134] in five soil habitats in Pakistan; they found the lowest cellulase activity in July and the maximum in September. Higher cellulase activity in reforested *Pinus* forest soil was observed by Bini et al. [99] in Brazil. In one study from China by Yin et al. [132],

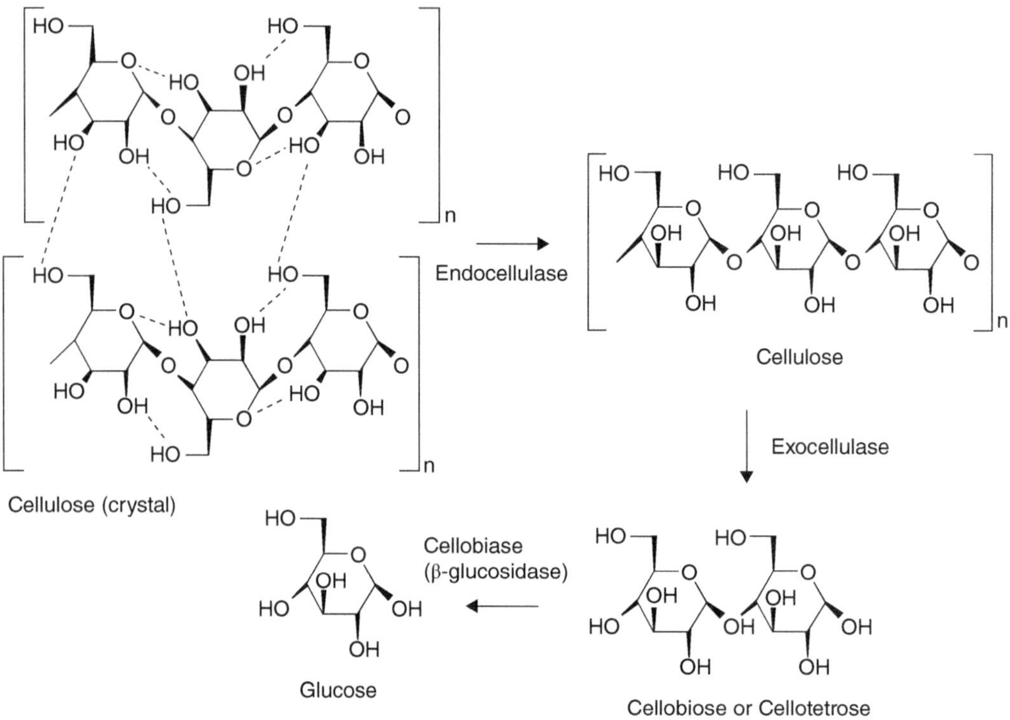

Figure 8.5 Mediated reactions by the cellulase enzyme complex in the breakdown of cellulose polysaccharide.

they found cellulase activity in *Pinus massoniana* was less than in soils of *Lespedeza bicolor* and indicated a specific effect of enzymatic activities at a particular level.

Carreiro et al. [135] observed the direct and indirect effects of nitrogen deposition on plant litter and organic matter decomposition in different tree species from New York (USA). According to their observations, adding ammonium and nitrate to fresh litter stimulates the decomposition of cellulose, whereas adding them to the humus stage of decomposition suppresses the activities of cellulose-degrading enzymes.

Cellulase activity was observed by Bini et al. [99] at six different sites in Parana, in southern Brazil. They found the highest cellulase levels in soils in *Araucaria angustifolia* and *Pinus elliottii* plantations, followed by native mixed forest, secondary mixed forest, and clear-cut *P. elliottii* stands, respectively. They found the lowest cellulase activity among the ecosystems in agricultural soils with a maize cropping system.

Cellulase activities on a mass basis showed the highest rates in the soil under *Fagus sylvatica* (beech). Cellulase activity showed somewhat comparable mean values about threefold higher in soils in beech and Holm oak (*Quercus ilex*) forests than those analyzed in the soils of turkey oak (*Quercus cerris*) stands of the Matese area in Apennines (southern Italy). According to Grosso et al. [131], in this study, cellulase activity is affected by the quantity as well as quality of organic matter, because the production of cellulase also depends on the C and N demand of microorganisms. The effect of temperature and climate change on the activity of endocellulase was observed by Baldrian et al. [136] in the *Quercus* forest of Prague (Czech Republic).

Thus, cellulase activity been studied worldwide in response to different soil habitats and varying environmental conditions [97, 99, 100, 131, 135, 137] and management practices [138–143].

8.6.2 β-Glucosidase

β–glucosidase is the most important enzyme for carbon metabolism in the soil as it mediates the breakdown of cellobiose, a disaccharide, by hydrolyzing β 1,4 glycosidic linkages (Figure 8.6) [144, 145]. It plays a crucial role in the complete decomposition of plant residue into utilizable glucose forms. The activity of β–glucosidase thus directly influences microbial growth in the soil. Therefore, it acts as the functional indicator of soil carbon sequestration and biological fertility [146]. β–glucosidase is further considered a sensitive indicator of climatic variation, vegetation type, nutrient depletion and augmentation, soil amendments, and land-use changes [17, 61, 147–149]. Thus is widely studied enzyme for determining soil quality and functions.

Contrasting soil habitats were found to have significant differences in this enzyme's activity: for example, forest and pasture lands have higher activity of the corresponding enzymes compared to agricultural lands [150]. Moreover, different types of forests also have significantly different enzymatic activity [151]. β–glucosidase is also very sensitive to soil amendments such as complex and simple nutrient amendments [105, 147, 152–156], and environmental and climatic variations alter the activity of this enzyme significantly [136, 157–159].

Trasar-Cepeda et al. [160] have investigated β–glucosidase activity in intense agriculture, i.e. maize fields and vineyards, oak forest soils, pine and eucalyptus reforested soils, and grassland soils. β–glucosidase activity is higher in reforested pine and eucalyptus soils than oak forests and soils under intense vegetation. According to Acosta-Martinez et al. [161], β–glucosidase activity was affected by land-use patterns and soil orders in soils of the Rio Grande de Arecibo watershed in north-central Puerto Rico. They have observed higher β–glucosidase activity in oxisols when compared to inceptisols soil. The β–glucosidase activity was observed to be higher in pasture soils than soils in forest and agriculture systems. The agriculture sites they studied were primarily used for long-term intensive coffee production, and pasture sites had grasses like *Eriochloa polytachya*, *Urochloa mutica*, *Cynodon nlemfuensis*, and *Digitaria eriantha*. Grandy et al. [162] observed significantly higher β–glucosidase activity in alpine soil than mixed forest soils from two distinct basins of the San Juan Mountains in southwestern Colorado, with dominating trees of Engelmann spruce, subalpine fir, and occasionally cork bark fir. The alpine sites studied were dominated by shrubs and lush herbaceous plants, including alpine avens, cinquefoils, Indian paintbrush, and patches of bistort. They have explained that higher enzymatic activity in the alpine ecosystem supports a more rapid turnover of C compared to the subalpine forest.

In light of the literature survey, it can be considered that β–glucosidase activities are the functional indicator of soil fertility and carbon sequestration of contrasting soil habitats [99, 145].

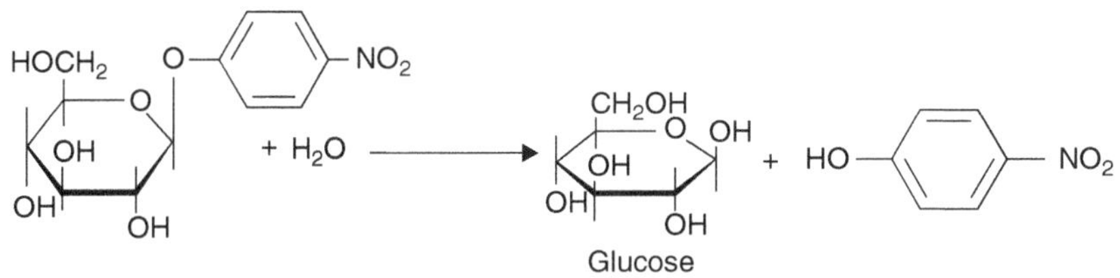

Figure 8.6 Mediated reaction by the β-glucosidase enzyme in the breakdown of β-D-glucoside.

8.6.3 Invertase

One of the most abundant sugars in plants is sucrose; it is a disaccharide of fructose and glucose linked by glycosidic bonds (α, β 1→2 glucose- fructose linkage). Unlike most disaccharides and oligosaccharides, sucrose has a free anomeric carbon atom and behaves like a non-reducing sugar (Figure 8.7). It is much more easily hydrolyzed into monomers of glucose and fructose than other disaccharides. In the soil ecosystem, the invertase or saccharase enzyme catalyzes the hydrolysis of sucrose and is widely distributed in microorganisms, animals, and plants. Invertase was the first enzyme to be studied among carbohydrases, which may be due to its ease of detection. In the 1950s, much effort was directed toward examining the relationship of invertase activity to soil fertility.

Invertase activity was shown to be related to the microbial biomass in different soil habitats. The corresponding enzyme is also a good indicator of environmental changes, soil nutrient concentration, vegetation type, and diversity. Invertase is reported to show significantly varying activity under contrasting habitats, such as greater activity in forest and grassland soils than agricultural lands [133, 134, 163]. Considerable variation was also observed in the activity of the same enzyme in response to changes in environmental conditions [152, 164–166] as well as changes in soil nutrient composition via soil amendments [163, 167–169] and various management strategies [138, 170]. Therefore, invertase activity is considered an important parameter to determine the carbon sequestration capacity of contrasting soils during the present study.

8.6.4 Amylase

Amylase is responsible for the breakdown of starch, the storage sugar of plants. It hydrolyses the α– and β– glycosidic linkages to release glucose monomers or dimmers (Figure 8.8). This enzyme consists of two components: α–amylase and β–amylase [171, 172]. Studies in the literature have indicated that the origin of α-amylases can be plants, animals, and microorganisms, while β-amylase is mainly synthesized by plants [173, 174].

This enzyme was first reported by Hofmann and Seegerer [117] in soil, and its extracellular activity in soil was first reported by Galstyan [175]. Presumably, several other enzymes are involved in the hydrolysis of starch in soil, but of significant importance are α–amylase, which converts starch to glucose or oligosaccharides, and β–amylase, which hydrolyzes starch into maltose [174]. However, some studies have indicated that amylase activity in soil may be influenced by factors including land-use changes, type of vegetation, environmental factors, and soil type [99, 133, 134, 176]. For example, plants may influence soil amylase activity by directly excreting enzymes from

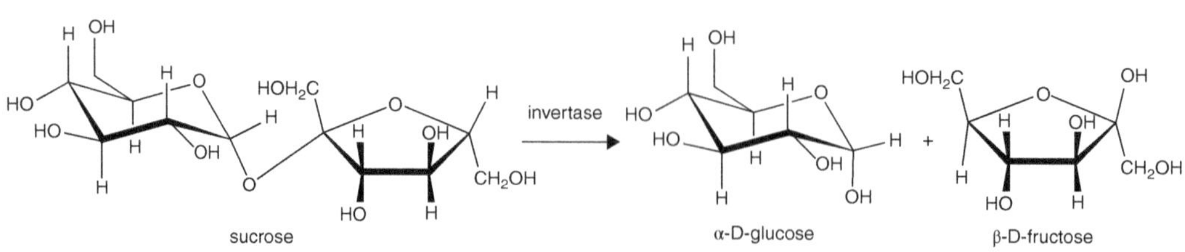

Figure 8.7 Mediated reaction by the invertase enzyme in the breakdown of sucrose.

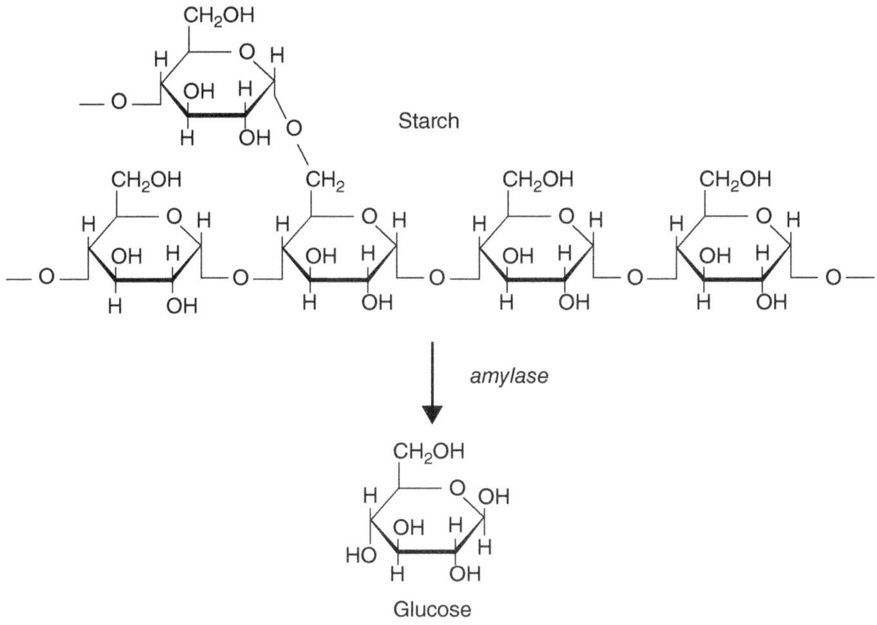

Figure 8.8 Mediated reaction by amylase enzyme in the breakdown of starch.

their residues or providing substrates for enhancing the microbial release of extracellular enzymes. Therefore, understanding the role of the amylase enzyme and its relationship with chemical, biological, physical, and climatic factors in the soil will further enable proper management techniques to be devised to maximize the benefits derived from such enzymes. Scientists worldwide have studied amylase activity in various ecosystems such as agriculture [134], grasslands [133], forests, and even anthropogenically degraded ecosystems [99].

8.6.5 Xylanase

The xylanase enzyme mediates the breakdown of xylan into xylose monosaccharide. Xylan is one of the most abundant polysaccharides and hemicellulose in plant cell walls and is composed of xylose units joined together by linear β–1,4 linkages, while side branches of xylans consist of α–glucuronic acids and α–arabinofuranose (Figure 8.9). Xylanase activity was observed to be significantly altered by environmental conditions such as dry weather [177], vegetation type [164], and management practices [168, 178].

Schinner and VonMersi [179] have studied xylanase activity in seven soils with different properties and SOM content and found that along with SOM, pH and temperature indicated a positive response. However, considerable work has been done on soil extracellular enzymes and factors influencing their activities in various ecosystems [110]. A relationship between soil carbon sequestration, SOM, and soil enzymatic activities has also been observed with patterns of litter decomposition [18, 180–182]. Agroecosystems and forest ecosystems are most discussed regarding soil biochemical properties and their relationships with enzymatic activities [109, 123, 124]. The literature survey has indicated that soil extracellular enzymatic activities are not fully understood in their soil carbon sequestration abilities in contrasting ecosystems.

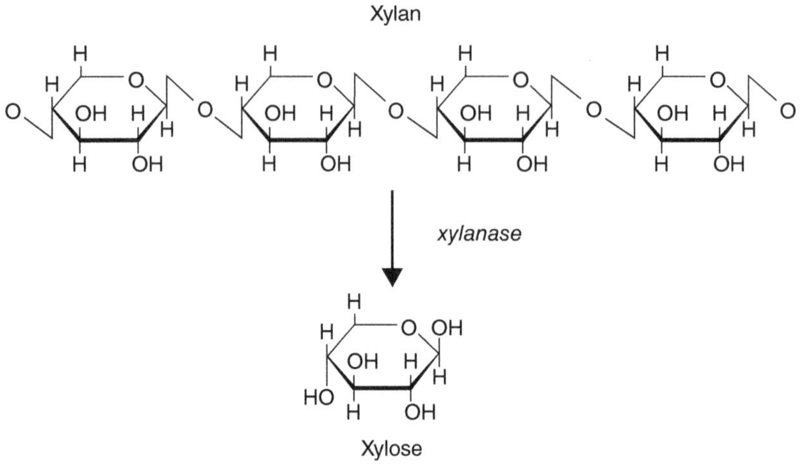

Figure 8.9 Mediated reaction by the xylanase enzyme in the breakdown of xylan.

8.7 Conclusion

Soil carbon sequestration has been studied for decades, but an explanation of the process is still unclear due to the large variety of regulators and moderators involved. Understanding the complexity of the various factors is necessary today because it is a key process that controls the cycling of carbon in the biosphere and can cause irreversible damage to ecosystems. A literature survey has indicated that globally, activities of extracellular enzymes in soils are key mediators. If understood and managed sustainably, they can presumably play an important role in carbon channelization into the soil.

References

1 Gianfreda, L. and Rao, M.A. (2014). *Enzymes in Agricultural Sciences*. Foster: OMICS Group International.
2 Harris, J.A. (2003). Measurements of the soil microbial community for estimating the success of restoration. *European Journal of Soil Science* 54: 801–808.
3 Carvalhais, N., Forkel, M., Khomik, M. et al. (2014). Global covariation of carbon turnover times with climate in terrestrial ecosystems. *Nature* 514 (7521): 213.
4 Davidson, E.A. and Janssens, I.A. (2006). Temperature sensitivity of soil carbon decomposition and feedbacks to climate change. *Nature* 440: 165–173.
5 Doran, J.W. and Zeiss, M.R. (2000). Soil health and sustainability: managing the biotic component of soil quality. *Applied Soil Ecology* 15: 3–11.
6 Holt, J.A. (1997). Grazing pressure and soil carbon, microbial biomass and enzyme activities in semi-arid northeastern Australia. *Applied Soil Ecology* 5 (2): 143–149.
7 Gil-Sotres, F., Trasar-Cepeda, C., Leiros, M.C., and Seoane, S. (2005). Different approaches to evaluating soil quality using biochemical properties. *Soil Biology and Biochemistry* 37: 877–887.
8 Marinari, S., Masciandaro, G., Ceccanti, B., and Grego, S. (2000). Influence of organic and mineral fertilisers on soil biological and physical properties. *Bioresource Technology* 72 (1): 9–17.

9 Wiesmeier, M., Urbanski, L., Hobley, E. et al. (2019). Soil organic carbon storage as a key function of soils- a review of drivers and indicators at various scales. *Geoderma* 333: 149–162.

10 De Deyn, G.B., Cornelissen, J.H., and Bardgett, R.D. (2008). Plant functional traits and soil carbon sequestration in contrasting biomes. *Ecology Letters* 11 (5): 516–531.

11 Herold, N., Schoning, I., Berner, D. et al. (2014). Vertical gradients of potential enzyme activities in soil profiles of European beech, Norway spruce and Scots pine dominated forest sites. *Pedobiologia* 57 (3): 181–189.

12 Hu, Y., Veresoglou, S.D., Tedersoo, L. et al. (2019). Contrasting latitudinal diversity and co-occurrence patterns of soil fungi and plants in forest ecosystems. *Soil Biology and Biochemistry* 131: 100–110.

13 Feller, C., Albrecht, A., Blanchart, E. et al. (2001). Soil organic carbon sequestration in tropical areas. General considerations and analysis of some edaphic determinants for lesser Antilles soils. *Nutrient Cycling in Agroecosystems* 61: 19–31.

14 Suriyavirun, N., Krichels, A.H., Kent, A.D., and Yang, W.H. (2019). Microtopographic differences in soil properties and microbial community composition at the field scale. *Soil Biology and Biochemistry* 131: 71–80.

15 Gao, W. and Yan, D. (2019). Warming suppresses microbial biomass but enhances N recycling. *Soil Biology and Biochemistry* 131: 111–118.

16 Hendriksen, N.B., Creamer, R.E., Stone, D., and Winding, A. (2016). Soil exo-enzyme activities across Europe- the influence of climate, land-use and soil properties. *Applied Soil Ecology* 97: 44–48.

17 Henry, H.A.L. (2012). Soil extracellular enzyme dynamics in a changing climate. *Soil Biology and Biochemistry* 47: 53–59.

18 Chen, J., Elsgaard, L., van Groenigen, K.J. et al. (2020). Soil carbon loss with warming: new evidence from carbon-degrading enzymes. *Global Change Biology* 26: 1944–1952.

19 Buttler, A., Mariotte, P., Meisser, M. et al. (2019). Drought-induced decline of productivity in the dominant grassland species *Lolium perenne* L. depends on soil type and prevailing climatic conditions. *Soil Biology and Biochemistry* 132: 47–57.

20 Ren, C., Zhao, F., Shi, Z. et al. (2017). Differential responses of soil microbial biomass and carbon-degrading enzyme activities to altered precipitation. *Soil Biology and Biochemistry* 115: 1–10.

21 Page, K.L., Dang, Y.P., Dalal, R.C. et al. (2020). The impact, identification and management of dispersive soils in rainfed cropping systems. *European Journal of Soil Science*: 1–20. https://doi.org/10.1111/ejss.13070.

22 Kumar, A., Mishra, S., Rawat, K.S., and Singh, V. (2011). Assessment of Aeolian sand affected wasteland area in Sirsa District using remote sensing and GIS. *Journal of Agricultural Physics* 11: 84–87.

23 Veenendaal, E.M., Swaine, M.D., Lecha, R.T. et al. (1996). Responses of west African forest tree seedlings to irradiance and soil fertility. *Functional Ecology* 10: 501–511.

24 Vilhar, U., Rozenbergar, D., Simoncic, P., and Diaci, J. (2015). Variation in irradiance, soil features and regeneration patterns in experimental forest canopy gaps. *Annals of Forest Science* 72: 253–266.

25 Stockmann, U., Adams, M.A., Crawford, J.W. et al. (2013). The knowns, known unknows and unknowns of sequestration of soil organic carbon. *Agriculture, Ecosystems and Environment* 164: 80–99.

26 Birkhofer, K., Schoning, I., Alt, F. et al. (2012). General relationships between abiotic soil properties and soil biota across spatial scales and different land-use types. *PLoS One* 7 (8): e43292.

27 Sakin, E. (2012). Relationships between of carbon, nitrogen stocks and texture of the Harran Plain soils in southeastern Turkey. *Bulgarian Journal of Agricultural Science* 18 (4): 626–634.

28 Saeedifar, Z. and Asgari, H.R. (2014). Effects of soil compaction on soil carbon and nitrogen sequestration and some physico-chemical features. *Ecopersia* 2 (4): 743–755.

29 Page, K.L., Dang, Y.P., and Dalal, R.C. (2020). The ability of conservation agriculture to conserve soil organic carbon and the subsequent impact on soil physical, chemical, and biological properties and yield. *Frontiers in Sustainable Food Systems* 4 (31) https://doi.org/10.3389/fsufs.2020.00031.

30 Liang, B., Lehmann, J., Solomon, D. et al. (2006). Black carbon increases cation exchange capacity in soils. *Soil Science Society of America Journal* 70 (5): 1719–1730.

31 Pratt, P.P. and Bair, F.L. (1961). A comparison of three reagents for the extraction of aluminum from soils. *Soil Science* 91 (6): 357–359.

32 Bachmann, J., Guggenberger, G., Baumgartl, T. et al. (2008). Physical carbon sequestration mechanisms under special consideration of soil wettability. *Journal of Plant Nutrition and Soil Science* 171 (1): 14–26.

33 Das, S., Brian, R.K., Kelly, H.L. et al. (2019). Lower mineralizability of soil carbon with higher legacy soil moisture. *Soil Biology and Biochemistry* 130: 94–104.

34 Preece, C., Verbruggen, E., Liu, L. et al. (2019). Effects of past and current drought on the composition and diversity of soil microbial communities. *Soil Biology and Biochemistry* 131: 28–39.

35 Hattenschwiler, S., Tiunov, A.V., and Scheu, S. (2005). Biodiversity and litter decomposition in terrestrial ecosystems. *Annual Review of Ecology, Evolution, and Systematics* 36: 191–218.

36 Singh, J. and Kumar, S. (2021). Responses of soil microbial community structure and greenhouse gas fluxes to crop rotations that include winter cover crops. *Geoderma* 385: 114843.

37 Gougoulias, C., Clark, J.M., and Shaw, L.J. (2014). The role of soil microbes in the global carbon cycle: tracking the below-ground microbial processing of plant-derived carbon for manipulating carbon dynamics in agricultural systems. *Journal of the Science of Food and Agriculture* 94 (12): 2362–2371.

38 Upton, R.N., Bach, E.M., and Hofmockel, K.S. (2019). Spatio-temporal microbial community dynamics within soil aggregates. *Soil Biology and Biochemistry* 132: 58–68.

39 Wieland, G., Neumann, R., and Backhaus, H. (2001). Variation of microbial communities in soil, rhizosphere and rhizoplane in response to crop species, soil type, and crop development. *Applied and Environmental Microbiology* 67 (12): 5849–5854.

40 Anderson, T.H. (2003). Microbial eco-physiological indicators to asses soil quality. *Agriculture, Ecosystems and Environment* 98: 285–293.

41 Gregorich, E.G., Carter, M.R., Angers, D.A. et al. (1994). Towards a minimum data set to assess soil organic matter quality in agricultural soils. *Canadian Journal of Soil Science* 74: 367–385.

42 Kallenbach, C.M., Frey, S.D., and Grandy, A.S. (2016). Direct evidence for microbial-derived soil organic matter formation and its ecophysiological controls. *Nature Communications* 7: 13630.

43 Fekadu, E., Kibret, K., Bedadi, B. et al. (2018). Organic and inorganic amendments on soil chemical properties at different period of incubation of acidic soil. *Eurasian Journal of Soil Science* 7 (3): 273–283.

44 Malik, A.A., Puissant, J., Buckeridge, K.M. et al. (2018). Land use driven change in soil pH affects microbial carbon cycling processes. *Nature Communications* 9 (1): 3591.

45 Rousk, J., Brookes, P.C., and Baath, E. (2009). Contrasting soil pH effects on fungal and bacterial growth suggest functional redundancy in carbon mineralization. *Applied and Environmental Microbiology* 75 (6): 1589–1596.

46 Musinguzi, P., Tenywa, J.S., Ebanyat, P. et al. (2013). Soil organic carbon thresholds and nitrogen management in tropical agroecosystems: concepts and prospects. *Journal of Sustainable Development* 6 (12): 31–43.

47 Burns, R.G. (1983). Extracellular enzyme–substrate interactions in soil. In: *Microbes in Their Natural Environment* (eds. J.H. Slater, R. Wittenbury and J.W.T. Wimpenny), 249–298. London: Cambridge University Press.

48 Burns, R.G., DeForest, J.L., Marxsen, J. et al. (2013). Soil enzymes in a changing environment: current knowledge and future directions. *Soil Biology and Biochemistry* 58: 216–234.

49 Meng, C., Tian, D., Zeng, H. et al. (2020). Global meta-analysis on the responses of soil extracellular enzyme activities to warming. *Science of the Total Environment* 705: 135992.

50 Dick, R.P. (ed.) (2020). *Methods of Soil Enzymology*. Wiley.

51 Delgado-Baquerizo, M., Maestre, F.T., Gallardo, A. et al. (2013). Decoupling of soil nutrient cycles as a function of aridity in global drylands. *Nature* 502 (7473): 672.

52 Delgado-Baquerizo, M., Grinyer, J., Reich, P.B., and Singh, B.K. (2016). Relative importance of soil properties and microbial community for soil functionality: insights from a microbial swap experiment. *Functional Ecology* 30 (11): 1862–1873.

53 Agbenin, J.O. and Goladi, J.T. (1997). Carbon, nitrogen and phosphorus dynamics under continuous cultivation as influenced by farmyard manure and inorganic fertilizers in the savanna of northern Nigeria. *Agriculture, Ecosystems & Environment* 63 (1): 17–24.

54 Bhattacharyya, R., Prakash, V., Kundu, S. et al. (2009). Soil properties and their relationships with crop productivity after 30 years of different fertilization in the Indian Himalayas. *Archives of Agronomy and Soil Science* 55 (6): 641–661.

55 Bowles, T.M., Acosta-Martínez, V., Calderon, F., and Jackson, L.E. (2014). Soil enzyme activities, microbial communities, and carbon and nitrogen availability in organic agroecosystems across an intensively managed agricultural landscape. *Soil Biology and Biochemistry* 68: 252–262.

56 Singh, A. and Ghoshal, N. (2013). Impact of herbicide and various soil amendments on soil enzymes activities in a tropical rainfed agroecosystem. *European Journal of Soil Biology* 54: 56–62.

57 Hombegowda, H.C., Straaten, O.V., Kohler, M., and Holscher, D. (2016). On the rebound: soil organic carbon stocks can bounce back to near forest levels when agroforests replace agriculture in southern India. *Soil* 2 (1): 13–23.

58 Jandl, R., Vesterdal, L., Olsson, M. et al. (2007). Carbon sequestration and forest management. *CAB Reviews: Perspectives in Agriculture, Veterinary Science, Nutrition and Natural Resources* 2: 1–16.

59 Dawson, J.J.C. and Smith, P. (2007). Carbon losses from soil and its consequences for land-use management. *Science of the Total Environment* 382: 165–190.

60 Ma, W., Li, J., Gao, Y. et al. (2020). Responses of soil extracellular enzyme activities and microbial community properties to interaction between nitrogen addition and increased precipitation in a semi-arid grassland ecosystem. *Science of the Total Environment* 703: 134691.

61 Cenini, V.L., Fornara, D.A., Mcmullan, G. et al. (2016). Linkages between extracellular enzyme activities and the carbon and nitrogen content of grassland soils. *Soil Biology and Biochemistry* 96: 198–206.

62 Urakawa, R., Ohte, N., Shibata, H. et al. (2016). Factors contributing to soil nitrogen mineralization and nitrification rates of forest soils in the Japanese archipelago. *Forest Ecology and Management* 361: 382–396.

63 Piotrowska-Długosz, A. (2020). Significance of the enzymes associated with soil C and N transformation. In: *Carbon and Nitrogen Cycling in Soil* (eds. R. Datta, R. Meena, S. Pathan and M. Ceccherini), 399–437. Singapore: Springer.

64 Xiao, W., Chen, X., Jing, X., and Zhu, B. (2018). A meta-analysis of soil extracellular enzyme activities in response to global change. *Soil Biology and Biochemistry* 123: 21–32.

65 Ghosh, A., Singh, A.B., Kumar, R.V. et al. (2020). Soil enzymes and microbial elemental stoichiometry as bio-indicators of soil quality in diverse cropping systems and nutrient management practices of Indian Vertisols. *Applied Soil Ecology* 145: 103304.

66 Cai, A., Xu, H., Shao, X. et al. (2016). Carbon and nitrogen mineralization in relation to soil particle-size fractions after 32 years of chemical and manure application in a continuous maize cropping system. *PLoS One* 11 (3): e0152521.

67 Chen, Y., Day, S.D., Wick, A.F. et al. (2013). Changes in soil carbon pools and microbial biomass from urban land development and subsequent post-development soil rehabilitation. *Soil Biology and Biochemistry* 66: 38–44.

68 Cleveland, C.C. and Liptzin, D. (2007). C:N:P stoichiometry in soil: is there a "Redfield ratio" for the microbial biomass? *Biogeochemistry* 85: 235–252.

69 Drake, J.E., Gallet-Budynek, A., Hofmockel, K.S. et al. (2011). Increases in the flux of carbon belowground stimulate nitrogen uptake and sustain the long-term enhancement of forest productivity under elevated CO_2. *Ecology Letters* 14: 349–357.

70 Fontaine, S., Bardoux, G., Abbadie, L., and Mariotti, A. (2004). Carbon input to soil may decrease soil carbon content. *Ecology Letters* 7: 314–320.

71 Scharlemann, J.P.W., Tanner, E.V., Hiederer, R., and Kapos, V. (2014). Global soil carbon: understanding and managing the largest terrestrial carbon pool. *Carbon Management* 5 (1): 81–91.

72 Visser, S. and Parkinson, D. (1992). Soil biological criteria as indicators of soil quality: soil microorganisms. *American Journal of Alternative Agriculture* 7 (1–2): 33–37.

73 Schmidt, M.W.I., Torn, M.S., Abiven, S. et al. (2011). Persistence of soil organic matter as an ecosystem property. *Nature* 478: 49–56.

74 Paul, E.A. (2016). The nature and dynamics of soil organic matter: plant inputs, microbial transformations, and organic matter stabilization. *Soil Biology and Biochemistry* 98: 109–126.

75 Campbell, E.E. and Paustian, K. (2015). Current developments in soil organic matter modeling and the expansion of model applications: a review. *Environmental Research Letters* 10 (12): 123004.

76 Olagoke, F.K., Kalbitz, K., and Vogel, C. (2019). Control of soil extracellular enzyme activities by clay minerals – perspectives on microbial responses. *Soil Systems* 3 (4): 64.

77 Kleber, M. (2010). What is recalcitrant soil organic matter? *Environmental Chemistry* 7 (4): 320–332.

78 German, D.P., Chacon, S.S., and Allison, S.D. (2011). Substrate concentration and enzyme allocation can affect rates of microbial decomposition. *Ecology* 92 (7): 1471–1480.

79 Tian, L., Dell, E., and Shi, W. (2010). Chemical composition of dissolved organic matter in agroecosystems: correlations with soil enzyme activity and carbon and nitrogen mineralization. *Applied Soil Ecology* 46: 426–435.

80 Xu, G., Long, Z., Ren, P. et al. (2020). Differential responses of soil hydrolytic and oxidative enzyme activities to the natural forest conversion. *Science of the Total Environment* 716: 136414.

81 Fang, C., Smith, P., Moncrieff, J.B., and Smith, J.U. (2005). Similar response of labile and resistant soil organic matter pools to changes in temperature. *Nature* 433 (7021): 57.

82 Sollins, P., Homann, P., and Caldwell, B.A. (1996). Stabilization and destabilization of soil organic matter: mechanisms and controls. *Geoderma* 74: 65–105.

83 Schlesinger, W.H. (1997). *Biogeochemistry: An Analysis of Global Change*. California: Academic Press.

84 Amundson, R. (2001). The carbon budget in soils. *Annual Review of Earth and Planetary Sciences* 29 (1): 535–562.

85 Batjes, N.H. (1996). Total carbon and nitrogen in the soils of the world. *European Journal of Soil Science* 47: 151–163.

86 Dick, R.P. (1994). Soil enzyme activities as indicators of soil quality. In: *Defining Soil Quality for a Sustainable Environment* (eds. J.V. Doran, D.C. Coleman, D.F. Bezdicek and B.A. Stewart), 107–124. Madison: Soil Science Society of America. American Society of Agriculture.

87 Ekschmitt, K., Kandeler, E., Poll, C. et al. (2008). Soil carbon preservation through habitat constraints and biological limitations on decomposer activity. *Journal of Plant Nutrition and Soil Science* 171: 27–35.

88 Mrabet, R., Ibno-Namr, K., Bessam, F., and Saber, N. (2001). Soil chemical quality changes and implications for fertilizer management after 11 years of no-tillage wheat production systems in semiarid Morocco. *Land Degradation & Development* 12 (6): 505–517.

89 Schlesinger, W.H. (2000). Carbon sequestration in soils: some cautions amidst optimism. *Agriculture, Ecosystems and Environment* 82: 121–127.

90 Kauppi, P. and Sedjo, R (2018). Technological and economic potential of options to enhance, maintain, and manage biological carbon reservoirs and geo-engineering. *Economics of Forestry*: 373.

91 Bloom, A.A., Exbrayat, J.F., van der Velde, I.R. et al. (2016). The decadal state of the terrestrial carbon cycle: global retrievals of terrestrial carbon allocation, pools, and residence times. *Proceedings of the National Academy of Sciences* 113 (5): 1285–1290.

92 Jastrow, J.D., Amonette, J.E., and Bailey, V.L. (2007). Mechanisms controlling soil carbon turnover and their potential application for enhancing carbon sequestration. *Climatic Change* 80 (1–2): 5–23.

93 Lal, R. (2004). Soil carbon sequestration in India. *Climatic Change* 65 (3): 277–296.

94 Srinivasarao, C., Lal, R., Kundu, S. et al. (2014). Soil carbon sequestration in rainfed production systems in the semiarid tropics of India. *Science of the Total Environment* 487: 587–603.

95 Lorenz, K., Lal, R., Preston, C.M., and Nierop, K.G. (2007). Strengthening the soil organic carbon pool by increasing contributions from recalcitrant aliphatic bio (macro) molecules. *Geoderma* 142 (1–2): 1–10.

96 Liu, L. and Greaver, T.L. (2010). A global perspective on belowground carbon dynamics under nitrogen enrichment. *Ecology Letters* 13 (7): 819–828.

97 Debosz, K., Rasmussen, P.H., and Pedersen, A.R. (1999). Temporal variations in microbial biomass C and cellulolytic enzyme activity in arable soils: effects of organic matter input. *Applied Soil Ecology* 13 (3): 209–218.

98 Six, J., Conant, R.T., Paul, E.A., and Paustian, K. (2002). Stabilization mechanisms of soil organic matter: implications for C-saturation of soils. *Plant and Soil* 241 (2): 155–176.

99 Bini, D., dos Santos, C.A., do Carmo, K.B. et al. (2013). Effects of land use on soil organic carbon and microbial processes associated with soil health in southern Brazil. *European Journal of Soil Biology* 55: 117–123.

100 Brackin, R., Robinson, N., Lakshmanan, P., and Schmidt, S. (2013). Microbial function in adjacent subtropical forest and agricultural soil. *Soil Biology and Biochemistry* 57: 68–77.

101 Chen, L., Gong, J., Fu, B. et al. (2007). Effect of land use conversion on soil organic carbon sequestration in the loess hilly area, loess plateau of China. *Ecological Research* 22: 641–648.

102 Li, S., Su, J., Liu, W. et al. (2015). Changes in biomass carbon and soil organic carbon stocks following the conversion from a secondary coniferous forest to a pine plantation. *PLoS One* 10 (9): e0135946.

103 Caldwell, B.A. (2005). Enzyme activities as a component of soil biodiversity: a review. *Pedobiologia* 49 (6): 637–644.

104 Allison, S.D., Gartner, T.B., Holland, K. et al. (2007). Soil enzymes: linking proteomics and ecological processes. In: *Manual of Environmental Microbiology*, 3e, 704–711. American Society of Microbiology.

105 Allison, S.D. and Vitousek, P.M. (2005). Responses of extracellular enzymes to simple and complex nutrient inputs. *Soil Biology and Biochemistry* 37: 937–944.

106 Sinsabaugh, R.L., Lauber, C.L., Weintraub, M.N. et al. (2008). Stoichiometry of soil enzyme activity at global scale. *Ecology Letters* 11: 1252–1264.

107 Li, L. and Wang, G. (2019). Enzymatic origin and variouscurvatures of metabolic scaling inmicrobes. *Scientific Reports* 9 (1): 4082.

108 Sinsabaugh, R.L., Moorhead, D.L., and Linkins, A.E. (1994). The enzymic basis of plant litter decomposition: emergence of an ecological process. *Applied Soil Ecology* 1 (2): 97–111.

109 Chuan, X., Carlyle, C.N., Bork, E.W. et al. (2020). Extracellular enzyme activity in grass litter varies with grazing history, environment and plant species in temperate grasslands. *Science of the Total Environment* 702: 134562.

110 Rosinger, C., Rousk, J., and Sanden, H. (2019). Can enzymatic stoichiometry be used to determine growth-limiting nutrients for microorganisms? A critical assessment in two subtropical soils. *Soil Biology and Biochemistry* 128: 115–126.

111 Sinsabaugh, R.L., Belnap, J., Findlay, S.G. et al. (2014). Extracellular enzyme kinetics scale with resource availability. *Biogeochemistry* 121 (2): 287–304.

112 Skujins, J. (1978). Soil enzymology and fertility index- a fallacy? History of abiotic soil enzyme research. In: *Soil Enzymes* (ed. R.G. Burns). London, UK: Academic Press.

113 Collet, G.F. (1975). Exsudationsracinairesd'enzymes. *Bulletin de la SocieteBotanique de France* 122: 61–75.

114 Vlasyuk, P.A., Dobrotvorskaya, K.M., and Gordienko, S.A. (1959). Activity of enzymes in the rhizosphere of agricultural plants. *Nauch. Trudy Ukrain. Inst. Fiziol. Rast.* 20: 12–12.

115 Ladd, J.N. (1978). Origin and range of enzymes in soil. In: *Soil Enzymes* (ed. R.G. Burns), 51–96. New York: Academic Press.

116 Kuprevisch, V.F. and Shcherbakova, T.A. (1971). Comparative enzymatic activity in diverse types of soil. In: *Soil Biochemistry*, vol. 2 (eds. A.D. McLaren and J. Skujins), 167–201. New York: Marcel Dekker.

117 Hofmann, E. and Seegerer, A. (1950). Soil enzyme as measure of biological activity. *Biochemische Zeitschrift* 321: 97.

118 Burns, R.G. (1982). Enzyme activity in soil: location and a possible role in microbial ecology. *Soil Biology and Biochemistry* 14: 423–427.

119 Burns, R.G. and Dick, R.P. (eds.) (2002). *Enzymes in the Environment: Activity, Ecology, and Applications*. CRC Press.

120 Howard, P.J.A. (1972). Problems in the estimation of biological activity in soil. *Oikos* 23 (2): 235–240.

121 Nannipieri, P., Kandeler, E., and Ruggiero, P. (2002). *Enzyme Activities and Microbiological and Biochemical Processes in Soil. Enzymes in the Environment*, 1–33. New York: Marcel Dekker.

122 Tabatabai, M.A. (1994). Soil enzymes. In: *Methods of Soil Analysis, Part 2. Microbiological and Biochemical Properties* (ed. S.H. Mickelson), 775–833. Madison, Wisconsin: Soil Science Society of America.

123 Tischer, A., Sehl, L., Meyer, U.N. et al. (2019). Land-use intensity shapes kinetics of extracellular enzymes in rhizosphere soil of agricultural grassland plant species. *Plant and Soil* 437 (1–2): 215–239.

124 Miao, F., Li, Y., Cui, S. et al. (2019). Soil extracellular enzyme activities under long-term fertilization management in the croplands of China: a meta-analysis. *Nutrient Cycling in Agroecosystems* 114 (2): 125–138.

125 Knicker, H. (2011). Soil organic N- an under-rated player for C sequestration in soils? *Soil Biology and Biochemistry* 43: 1118–1129.

126 Hilden, L. and Johansson, G. (2004). Recent developments on cellulases and carbohydrate binding modules with cellulose affinity. *Biotechnology Letters* 26 (22): 1683–1693.

127 Eriksson, K.E.L., Blancbette, R.A., and Ander, P. (1990). Biodegradation of cellulose. In: *Microbial and Enzymatic Degradation of Wood and Wood Components* (eds. K.E.L. Eriksson, R.A. Blancbette and P. Ander), 89–180. New York: Springer.

128 Alef, K. and Nannipieri, P. (1995). Methods in applied soil microbiology and biochemistry. *Academic Press* 631 (46): M592ma.

129 Bayer, E.A., Chanzy, H., Lamed, R., and Shoham, Y. (1998). Cellulose, cellulases and cellulosomes. *Current Opinion in Structural Biology* 8 (5): 548–557.

130 Hope, C.F.A. and Burns, R.G. (1987). Activity, origins and location of cellulases in a silt loam soil. *Biology and Fertility of Soils* 5 (2): 164–170.

131 Grosso, F., Temussi, F., and De Nicola, F. (2014). Water-extractable organic matter and enzyme activity in three forest soils of the Mediterranean area. *European Journal of Soil Biology* 64: 15–22.

132 Yin, R., Deng, H., Wang, H., and Zhang, B. (2014). Vegetation type affects soil enzyme activities and microbial functional diversity following revegetation of a severely eroded red soil in sub-tropical China. *Catena* 115: 96–103.

133 Pancholy, S.K. and Rice, E.L. (1973). Soil enzymes in relation to old field succession: amylase, cellulase, invertase, dehydrogenase, and urease. *Soil Science Society of America Journal* 37 (1): 47–50.

134 Zafar, Y., Malik, K.A., and Chaudhary, A.H. (1981). Activities of some carbohydrases in agriculture soils. *Pakistan Journal of Agricultural Research* 2 (1): 41–45.

135 Carreiro, M.M., Sinsabaugh, R.L., Repert, D.A., and Parkhurst, D.F. (2000). Microbial enzyme shifts explain litter decay responses to simulated nitrogen deposition. *Ecology* 81 (9): 2359–2365.

136 Baldrian, P., Snajdr, J., Merhautova, V. et al. (2013). Responses of the extracellular enzyme activities in hardwood forest to soil temperature and seasonality and the potential effects of climate change. *Soil Biology and Biochemistry* 56: 60–68.

137 Hou, S., Xin, M., Wang, L. et al. (2014). The effects of erosion on the microbial populations and enzyme activity in black soil of northeastern China. *Acta EcologicaSinica* 34 (6): 295–301.

138 Chen, Y., Mu, S., Sun, Z. et al. (2016). Grassland carbon sequestration ability in China: a new perspective from terrestrial aridity zones. *Rangeland Ecology & Management* 69 (1): 84–94.

139 De Varennes, A. and Torres, M.O. (2011). Post-fallow tillage and crop effects on soil enzymes and other indicators. *Soil Use and Management* 27 (1): 18–27.

140 Geng, Y., Dighton, J., and Gray, D. (2012). The effects of thinning and soil disturbance on enzyme activities under pitch pine soil in New Jersey pinelands. *Applied Soil Ecology* 62: 1–7.

141 Iyyemperumal, K. and Shi, W. (2008). Soil enzyme activities in two forage systems following application of different rates of swine lagoon effluent or ammonium nitrate. *Applied Soil Ecology* 38 (2): 128–136.

142 Lagomarsino, A., Grego, S., Marhan, S. et al. (2009). Soil management modifies micro-scale abundance and function of soil microorganisms in a Mediterranean ecosystem. *European Journal of Soil Science* 60: 2–12.

143 Li, S., Zhang, S., Pu, Y. et al. (2016). Dynamics of soil labile organic carbon fractions and C-cycle enzyme activities under straw mulch in Chengdu plain. *Soil and Tillage Research* 155: 289–297.

144 Eivazi, F. and Tabatabai, M.A. (1988). Glucosidases and galactosidases in soils. *Soil Biology and Biochemistry* 20 (5): 601–606.

145 Stott, D.E., Andrews, S.S., Liebig, M.A. et al. (2010). Evaluation of β-glucosidase activity as a soil quality indicator for the soil management assessment framework. *Soil Science Society of America Journal* 74 (1): 107–119.

146 Makoi, J.H.J.R. and Ndakidemi, P.A. (2008). Selected soil enzymes: examples of their potential roles in the ecosystem. *African Journal of Biotechnology* 7 (3): 181–191.

147 Bastida, F., Torres, I.F., Hernandez, T. et al. (2012). Can the labile carbon contribute to carbon immobilization in semiarid soils? Priming effects and microbial community dynamics. *Soil Biology and Biochemistry* 57: 892–902.

148 Fenner, N., Freeman, C., and Reynolds, B. (2005). Observations of a seasonally shifting thermal optimum in peatland carbon-cycling processes; implications for the global carbon cycle and soil enzyme methodologies. *Soil Biology and Biochemistry* 37 (10): 1814–1821.

149 Galvez, A., Sinicco, T., Cayuela, M.L. et al. (2012). Short term effects of bioenergy by products on soil C and N dynamics, nutrient availability and biochemical properties. *Agriculture, Ecosystems & Environment* 160: 3–14.

150 Acosta-Martínez, V., Acosta-Mercado, D., Sotomayor-Ramírez, D., and Cruz-Rodríguez, L. (2008). Microbial communities and enzymatic activities under different management in semiarid soils. *Applied Soil Ecology* 38 (3): 249–260.

151 Adamczyk, B., Kilpelainen, P., Kitunen, V., and Smolander, A. (2014). Potential activities of enzymes involved in N, C, P and S cycling in boreal forest soil under different tree species. *Pedobiologia* 57 (2): 97–102.

152 Bandick, A.K. and Dick, R.P. (1999). Field management effects on soil enzyme activities. *Soil Biology and Biochemistry* 31: 1471–1479.

153 Bhattacharyya, P., Roy, K.S., Neogi, S. et al. (2012). Effects of rice straw and nitrogen fertilization on greenhouse gas emissions and carbon storage in tropical flooded soil planted with rice. *Soil and Tillage Research* 124: 119–130.

154 Garcia-Gil, J.C., Plaza, C., Soler-Rovira, P., and Polo, A. (2000). Long-term effects of municipal solid waste compost application on soil enzyme activities and microbial biomass. *Soil Biology and Biochemistry* 32 (13): 1907–1913.

155 Geisseler, D., Joergensen, R.G., and Ludwig, B. (2012). Potential soil enzyme activities are decoupled from microbial activity in dry residue-amended soil. *Pedobiologia- International Journal of Soil Biology* 55: 253–261.

156 Zhao, S., Li, K., Zhou, W. et al. (2016). Changes in soil microbial community, enzyme activities and organic matter fractions under long-term straw return in north-central China. *Agriculture, Ecosystems & Environment* 216: 82–88.

157 Bell, T.H. and Henry, H.A.L. (2011). Fine scale variability in soil extracellular enzyme activity is insensitive to rain events and temperature in a mesic system. *Pedobiologia- International Journal of Soil Biology* 54: 141–146.

158 Dominguez, M.T., Holthof, E., Smith, A.R. et al. (2017). Contrasting response of summer soil respiration and enzyme activities to long-term warming and drought in a wet shrubland (NE Wales, UK). *Applied Soil Ecology* 110: 151–155.

159 Jain, R. and Pandey, A. (2016). Soil enzymes and microbial endophytes as indicators of climate variation along an altitudinal gradient with respect to wheat rhizosphere under mountain ecosystem. *Rhizosphere* 2: 75–84.

160 Trasar-Cepeda, C., Leiros, M.C., and Gil-Sotres, F. (2008). Hydrolytic enzyme activities in agricultural and forest soils. Some implications for their use as indicators of soil quality. *Soil Biology & Biochemistry* 40: 2146–2155.

161 Acosta-Martínez, V., Cruz, L., Sotomayor-Ramírez, D., and Pérez-Alegría, L. (2007). Enzyme activities as affected by soil properties and land use in a tropical watershed. *Applied Soil Ecology* 35 (1): 35–45.

162 Grandy, A.S., Neff, J.C., and Weintraub, M.N. (2007). Carbon structure and enzyme activities in alpine and forest ecosystems. *Soil Biology and Biochemistry* 39 (11): 2701–2711.

163 Jiao, X., Gao, C., Lu, G., and Sui, Y. (2011). Effect of long-term fertilization on soil enzyme activities under different hydrothermal conditions in Northeast China. *Agricultural Sciences in China* 10 (3): 412–422.

164 Drissner, D., Blum, H., Tscherko, D., and Kandeler, E. (2007). Nine years of enriched CO_2 changes the function and structural diversity of soil microorganisms in a grassland. *European Journal of Soil Science* 58: 260–269.

165 Liu, X., Li, Q., Liang, W., and Jiang, Y. (2008). Distribution of soil enzyme activities and microbial biomass along a latitudinal gradient in farmlands of Songliao plain, Northeast China. *Pedosphere* 18 (4): 431–440.

166 Li, J., Tong, X., Awasthi, M.K. et al. (2018). Dynamics of soil microbial biomass and enzyme activities along a chronosequence of desertified land revegetation. *Ecological Engineering* 111: 22–30.

167 Balasubramanian, A., Siddaramappa, R., and Rangaswami, G. (1972). Effect of organic manuring on the activities of the enzymes hydrolysing sucrose and urea and on soil aggregation. *Plant and Soil* 37 (2): 319–328.

168 Stemmer, M., Gerzabek, M.H., and Kandeler, E. (1999). Invertase and xylanase activity of bulk soil and particle-size fractions during maize straw decomposition. *Soil Biology and Biochemistry* 31: 9–18.

169 Zhang, P., Chen, X., Wei, T. et al. (2016). Effects of straw incorporation on the soil nutrient contents, enzyme activities, and crop yield in a semiarid region of China. *Soil & Tillage Research* 160: 65–72.

170 Gianfreda, L., Rao, M.A., Piotrowska, A. et al. (2005). Soil enzyme activities as affected by anthropogenic alterations: intensive agricultural practices and organic pollution. *Science of the Total Environment* 341: 265–279.

171 Galstyan, A.S.H. (1970). A method of determining the activity of hydrolytic enzymes in soil. *Soviet Soil Science* 2: 170–175.

172 Ross, D.J. 1968. Activities of enzymes hydrolysing sucrose and starch in some grassland soils. Trans 9th International Congress Soil Science **3**: 299-308.

173 Pazur, J.H. (1965). Enzymes in the synthesis and hydrolysis of starch. In: *Starch: Chemistry and Technology, Fundamental Aspects* (eds. R. Whistler and E.F. Paschall), 133–175. New York: Academic Press.

174 Thoma, J.A., Spradlin, J.E., and Dygert, S. (1971). Plant and animal amylases. In: *The Enzymes*, vol. 5 (ed. P.D. Boyer), 115–189. Netherlands: International Society of Soil Science.

175 Galstyan, A.S.H. (1965). Methods of determining activities of hydrolytic enzymes of soil. *Pochvovedenie* 2: 68–74.

176 Rose, D.J. (1975). Studies on a climosequence of soils in tussock grasslands-5. Invertase and amylase activities of topsoils and their relationships with other properties. *New Zealand Journal of Science* 18: 511–518.

177 Kandeler, E., Mosier, A.R., Morgan, J.A. et al. (2006). Response of soil microbial biomass and enzyme activities to the transient elevation of carbon dioxide in a semi-arid grassland. *Soil Biology and Biochemistry* 38 (8): 2448–2460.

178 Leinweber, P., Jandl, G., Baum, C. et al. (2008). Stability and composition of soil organic matter control respiration and soil enzyme activities. *Soil Biology and Biochemistry* 40: 1496–1505.

179 Schinner, F. and VonMersi, W. (1990). Xylanase-, CM-cellulase-and invertase activity in soil: an improved method. *Soil Biology and Biochemistry* 22 (4): 511–515.

180 Dilly, O. and Munch, J.C. (1996). Microbial biomass content, basal respiration and enzyme activities during the course of decomposition of leaf litter in a black alder (Alnusglutinosa (L.) Gaertn.) forest. *Soil Biology and Biochemistry* 28 (8): 1073–1081.

181 Moorhead, D.L. and Sinsabaugh, R.L. (2000). Simulated patterns of litter decay predict patterns of extracellular enzyme activities. *Applied Soil Ecology* 14: 71–79.

182 Sajjad, M.H., Lodhi, A., and Azam, F. (2002). Changes in enzyme activity during the decomposition of plant residues in soil. *Pakistan Journal of Biological Sciences* 5 (9): 952–955.

9

Ecological Engineering for Solid Waste Segregation, Reduction, and Resource Recovery – A Contextual Analysis in Brazil

Luís P. Azevedo[1], *Fernando G. da Silva Araújo*[1], *Carlos A.F. Lagarinhos*[2], *Jorge A.S. Tenório*[3], *Denise C.R. Espinosa*[3], *and Majeti Narasimha Vara Prasad*[4]

[1] Materials Engineering Department, Federal University of Ouro Preto, Ouro Preto, Minas Gerais, Brazil
[2] Polytechnic School of University of São Paulo, Metallurgical and Materials Department, São Paulo, Brazil
[3] Polytechnic School of University of São Paulo, Chemical Engineering Department, São Paulo, Brazil
[4] School of Life Sciences, University of Hydebarad, Hyderabad, Telangana, India

9.1 Introduction

Globally, consumption of natural resources is a significant problem. Responsible consumption of natural resources and minimizing industrial and urban wastes is a priority agenda item in line with UN sustainable development goals (SDGs). One successfully implemented initiative is the use of scrap automobile tires in clinker kilns for cement production; the tires replace coal and have a calorific value equivalent to most fossil fuels: 1 kg of tire material generates 8000 kcal/kg of energy. Recent advancements in combustion technology and air pollution control systems (APCS) make incineration an attractive and sustainable solution for municipal solid waste (MSW) management [21, 24, 31, 36, 37, 42, 57, 62, 67].

As per a World Bank survey report, the composition of MSW plays an important role in its management. For example, half of the waste in South Asia is organic, whereas the organic fraction is less than two-thirds in OECD (European Community Development Organization Countries) (Table 9.1). Cities must adopt segregation of waste at the source and minimize dumping in landfills.

It is a general practice to send MSW to landfills, which are subjected to an array of biological, chemical, and physical processes that lead to gaseous and liquid emissions. MSW management is a herculean task in all developing countries, and a logistical system is required to collect and dispose of MSW. In developing countries, the collected MSW is sent to landfills. Landfill facilities do not have enough capacity or complete engineering facilities; efforts to segregate, reduce, and recover resources from MSW are being considered in developing and developed countries.

Several countries in Europe already burn solid urban waste to generate electricity and steam, such as Sweden, the European Union (EU), and Japan. In Europe, 23% of MSW was landfilled in 2017. Despite targets that define waste and landfill requirements, there is still high variability in the waste management performance between the EU member states. Urban expansion and erosion processes are also serious environmental concerns, that impacts waste management in Nova Lima, Minas Gerais (Brazil) [20].

Handbook of Ecological and Ecosystem Engineering, First Edition. Edited by Majeti Narasimha Vara Prasad.

Table 9.1 Comparison of MSW in South Asia and OECD nations.

Component	% in South Asia	% in OECD countries
Organic	50	28
Paper	4	34
Plastics	6	12
Metals	6	1
Others	33	18
Glass	1	7

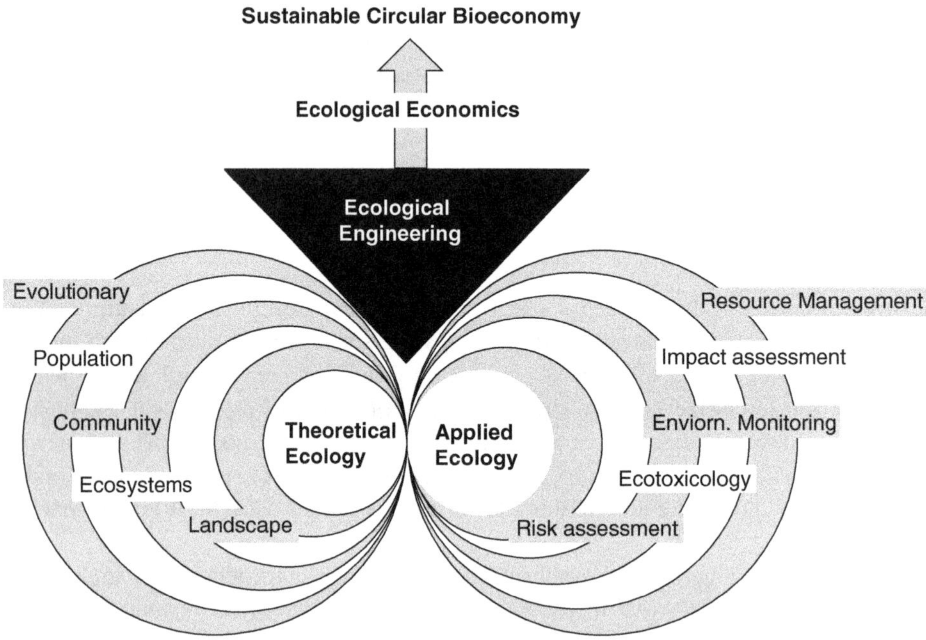

Figure 9.1 Theoretical and applied aspects of ecological and ecosystem engineering for boosting a sustainable circular bioeconomy.

Municipal waste segregation and resource recovery would eliminate landfills, saving many land resources and providing significant social, environmental, and economic benefits. MSW automatic sorting systems are suitable for a variety of domestic uses [3, 17, 44]. Figure 9.1 shows aspects of a sustainable circular bioeconomy.

The states of São Paulo and Rio de Janeiro have enough MSW to generate electric energy (waste to energy, WTE). The economic feasibility of installing heat treatment plants (energy recuperation units – ERUs) is presented in this chapter. The implications of harmful gases, such as carbon dioxide (CO_2), hydrogen (H_2), oxygen (O_2), nitrogen (N_2), and sulfur dioxide (SO_2), and the cost of energy generated are investigated. ERUs are an alternative for the generating energy using MSW and avoiding landfills. A recent literature search on approaches for ecological engineering of MSW aiming at waste reduction and resource recovery is shown in Table 9.2.

Table 9.2 Recent literature search results for approaches for ecological engineering of MSW, aiming at waste reduction and resource recovery.

Zero-waste approach in municipal solid waste incineration	[2]
Potential utilization of municipal solid waste as a renewable energy source in Lahore, Pakistan	[3]
Valorization of extracted organic matter from municipal solid waste leachate: application to soils from France and Togo	[4]
Integrated solid waste management in municipalities of west Bengal	[7]
Municipal solid waste management, household and local government participation: a cross-country analysis	[6]
A mixed-integer linear programming model for long-term planning of municipal solid waste management systems	[8]
Waste management practice in Malaysia and future challenges	[10]
Economic analysis of a shared municipal solid waste management facility in a metropolitan region	[11]
Recycling municipal solid waste incineration slag and fly ash as precursors in low-range alkaline cement	[12]
Decision model for selecting advanced technologies for municipal solid waste management	[58]
Implementation of municipal solid waste regulations in Israel	[13]
Municipal waste management in Turkey	[14]
A municipal solid waste indicator for environmental impact: assessment and identification of best management practices	[15]
Analytical framework and data for a municipal solid waste environmental performance assessment	[16]
Composting and recycling municipal solid waste	[17]
Incineration of municipal solid waste in Brazil: an analysis of the economically viable energy potential	[18]
Comparative study of municipal solid waste disposal in three representative Chinese cities	[19]
Urban Expansion and Erosion Processes in an Area of Environmental Protection in Nova Lima, Minas Gerais State, Brazil.	[20]
Comparative LCA of treatment options for US scrap tires: material recycling and tire-derived fuel combustion	[21]
Assessment of municipal solid waste selective collection scenarios with geographic information systems in Bolivia	[22]
Using blockchain to improve solid waste management in small municipalities	[23]
Vitrification of municipal solid waste incineration fly ash with B2O3 as a fluxing agent	[25]
Sustainable municipal solid waste disposal	[26]
Bioleaching for resource recovery from low-grade wastes like fly and bottom ashes from municipal incinerators	[27]
Bioreactor landfills for leachate decontamination and municipal solid waste stabilization	[28]
China's urbanization on the quantity of municipal solid waste	[29]
Sustainability evaluation of a municipal solid waste management system for Hanoi (Vietnam)	[30]
Complete determination of the material composition of municipal solid waste incineration bottom ash	[32]
Proposal of social indicators to assess the social performance of waste management systems in developing countries: a Brazilian case study	[33]

(Continued)

Table 9.2 (Continued)

A review of municipal solid waste management in the (Brazil, Russia, India and South Africa) BRIC and high-income countries	[34]
Assessing the sustainable municipal solid waste (MSW) to electricity generation in Pacific Small Island Developing States (PSIDS)	[35]
Environmental life cycle assessment of different biorefinery platforms valorizing municipal solid waste to bioenergy, microbial protein, lactic, and succinic acid	[38]
Heavy metal removal from municipal solid waste fly ash through chloride volatilization using poly (vinyl chloride) as a chlorinating agent	[40]
Utilization of pretreated municipal solid waste incineration fly ash for cement-stabilized soil	[41]
Environmental and economic assessment of sustainable municipal solid waste management strategies in Iran	[43]
Municipal solid waste prevention: a review of market-based instruments in six European Union countries	[44]
Municipal solid waste management with cost minimization and emission control objectives: a case study of Ankara	[45]
Optimal process design for integrated municipal waste management with energy recovery in Argentina	[46]
Energy and resource recovery through integrated sustainable waste management	[47]
Multi-objective optimization of life cycle assessment of integrated waste management based on genetic algorithms: a case study in Tehran	[48]
Composition analysis of major organic fractions of municipal solid waste generated from Chennai	[49]
Existing realities and sustainable pathways for solid waste management in Ghana	[50]
An efficient waste management system with a municipal solid waste incineration plant	[51]
Municipal solid waste management: integrated analysis of environmental and economic indicators based on life cycle assessment	[52]
Municipal solid wastes – a promising sustainable source of energy	[53]
Review of environmental implications of incineration of municipal solid waste and ash disposal	[54]
Solid waste management system in Dindigul Corporation of Tamilnadu and feasibility analysis of waste-to-energy projects	[55]
Circular economy of municipal solid waste in Korea	[56]
The environmental impacts of municipal solid waste landfills in Europe	[59]
Mapping the incinerability of municipal solid waste in the Indian subcontinent	[60]
Influence of recycling on the incinerability of municipal solid waste	[61]
Use of tire rubber particles as addition to cement paste. Cement and Concrete Research	[62]
Reducing methane emissions from municipal solid waste landfills by using mechanical, biological treatment	[63]
Used tires recycle management and processing: a review	[64]
Solid waste management and its impact on the environment	[65]
Analysis and advanced characterization of municipal solid waste vermicompost maturity for a green environment	[66]
Insights into pyrolysis of waste tire in fixed bed reactor: thermal behavior.	[67]
Evaluation of existing solid waste management system in Una town, India	[68]
A robust green location-allocation-inventory problem to design an urban waste management system under uncertainty.	[69]

Table 9.2 (Continued)

Robust information for effective municipal solid waste policies:	[70]
A causal municipal solid waste management model for sustainable cities in Vietnam under uncertainty: a comparison	[71]
Implementing a circular economy in a municipal solid waste treatment systems	[72]
Greenhouse gas emissions from municipal solid waste	[73]
Combination of pyrolysis and catalytic cracking of municipal solid waste for the production of syngas and solid refuse-derived fuels	[74]
Substitution modeling in life cycle assessment of municipal solid waste management	[75]
Solid waste management in the Republic of Serbia	[76]
Life cycle assessment of municipal solid waste management in Nottingham, England	[77]
Future improvements in the performance of EU landfill directive-driven municipal solid waste management for a city in England	[78]
Waste tire recycling assessment: road application potential and carbon emissions reduction analysis of crumb rubber modified asphalt in China	[79]
Characterization of the volatile compounds emitted from municipal solid waste and identification of the key volatile pollutants	[80]
Municipal solid waste recovered fuel by thermogravimetric and kinetic analysis	[81]
A genetic-algorithm-aided fuzzy chance-constrained programming model for municipal solid waste management	[82]
Sustainable collection and transportation of municipal solid waste in urban centers	[83]
Municipal solid waste management and treatment in China	[84]
Designing an integrated municipal solid waste management network: a case study	[86]

Table 9.3 Parameters used for electrical generation capacity and value.

MSW processing capacity	1200	tons/day
Calorific power of MSW	1900	kcal/kg
Installed power	23	MW
Average power available for sale	18.3	MW
Total area of plant	30 000	m^2
Energy	500	KWh/ton
Operation/maintenance costs	1.20	US$/ton
Average tariff paid to suppliers in São Paulo	0.1	US$/KWh
Average freight cost / km driven	0.33	US$/km

Source: Adapted from CETESB. Companhia de Tecnologia de Saneamento Ambiental. (2005). Inventário estadual de resíduos sólidos domiciliares, São Paulo, Brazil.

In this chapter, we evaluate the thermal valorization of solid waste in São Paulo, Brazil (Table 9.3). A cost-benefit analysis is included for environmental gain in relation to reduction in fossil fuel consumption, the costs of reverse logistics, and the total environmental and economic gains from the process of thermal treatment of MSW with regard to the quantity of MSW transported to operate an ERU in São Paulo from different locations.

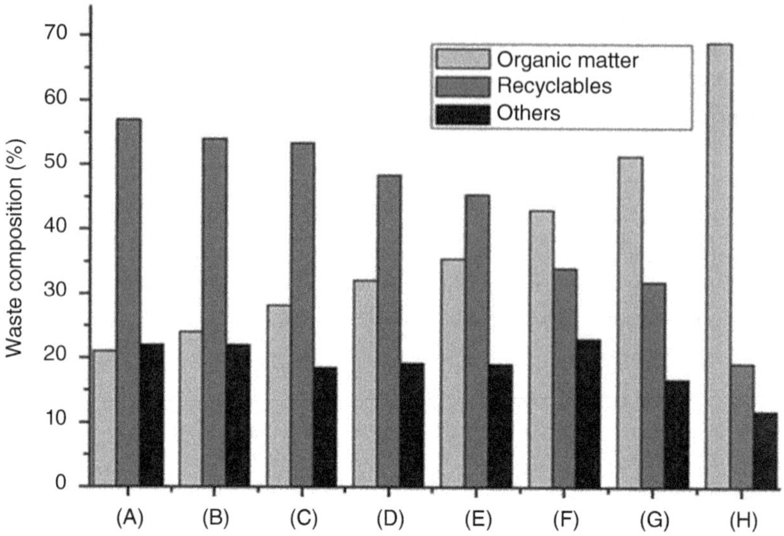

Figure 9.2 Composition of MSW from different countries: (a) Denmark; (b) Varna, Bulgaria; (c) USA; (d) Germany; (e) Portugal – urban region; (f) Istanbul, Turkey; (g) Brazil; (h) Ho Chi Minh, Vietnam. *Source:* R.G. de Souza Marotta Alfaia, A.M. Costa, and J. C. Campos, "Municipal solid waste in Brazil: A review," *Waste Management & Research* 2017, 35 (12): 1195–1209. © SAGE publications.

The amount of plastic consumed by today's society has been responsible for reducing marine fauna, silting of rivers, and other environmental damage. Recycling has been encouraged, but it is relatively limited because the physical characteristics of the recycled material have lower quality than new material (Figure 9.2). Such low-grade material is usually sent to landfills.

In Brazil, the reverse logistics chain is followed by the social inclusion of collectors, aiming at recycling. The savings in terms of reduced consumption of fossil fuels has been assessed. This concept has been integrated into steel scrap conversion furnaces that use waste tires and other examples like using aerobic digestion in rural areas to produce methane and fertilizers.

9.2 Municipal Solid Waste in Brazil

The management of MSW is marked by its complexity, diversity, and composition [18]. Its treatment and valorization become relevant for public agencies to implement an integrated system for handling MSW. Characterizing the negative environmental and energy impacts associated with the system's various components is important for a better scientific understanding of these issues.

To understand energy recovery, it is necessary to first understand the entire treatment chain of MSW (of which waste electrical and electronic equipment [WEEE] is not a part) in addition to solid industrial waste, medicines, oils, and pesticides. The selective collection of organic, biodegradable material is carried out by large-scale producers of this type of waste, such as restaurants and homes, hotels, supply markets, and retailers, in cleaning gardens and other green spaces. This collection aims to generate a high-quality organic compound, without chemical additives, which

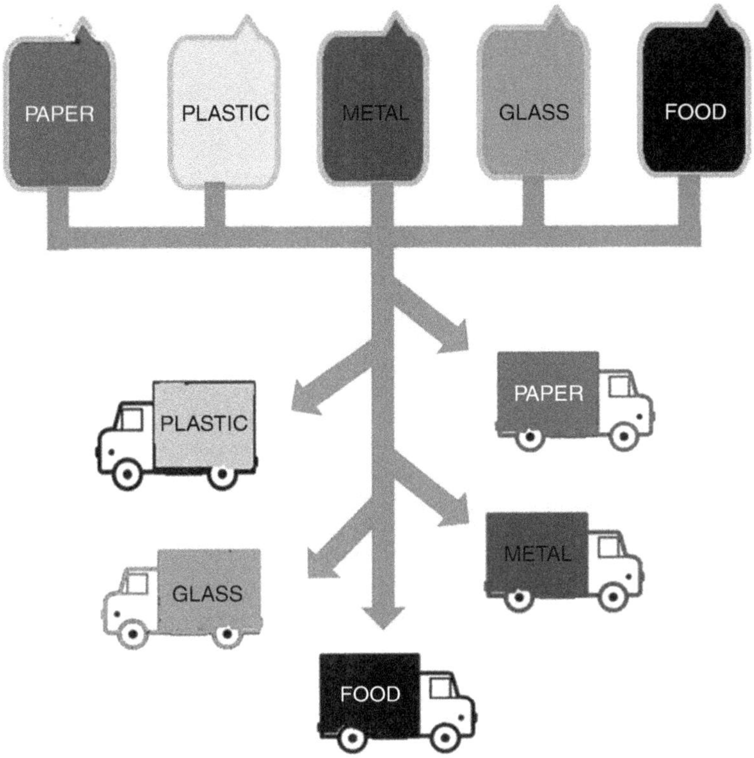

Figure 9.3 Segregation of MSW and transport to a resource recovery site. *See colour plate section for colour representation of this figure.*

can later be used in agriculture as a fertilizer. Non-selective collections of MSW must be sent to energy or organic recovery units and undergo an efficient sorting process.

Implementing selective collection of recyclable materials requires recyclable collection points: paper and cardboard, glass, packaging (plastic and metal), and batteries (not classified as MSW). To dispose of other physically large waste, such as used oil, wood, WEEE, batteries, and fluorescent lamps, after these materials are separated, each group is stored until it is dispatched to the respective recycling industry. Figure 9.3 depicts the ecological cycle of reducing pollution in the environment:

Plastic packaging, due to its diversity, must be separated into HDPE (low-density polyethylene), PET (polyethylene terephthalate), PVC (polyvinyl chloride), and plastic film (traditional plastic bags).

9.3 Compostable Waste

Composting is still incipient in Brazil. In the composting process, after mechanical and manual presorting of the MSW, organic matter is degraded by aerobic digestion [17, 66]. In Brazil, however, this type of recycling occurs for only 4% of the fraction generated, with more than 60% of the total mass of waste generated by Brazilians in 2011. In addition, selective collection in Brazil does not emphasize the previous separation of the organic fraction of waste.

Prescreening separates collected MSW into three constituents:

- Biodegradable organic matter destined for the composting chamber
- Recyclable materials (ferrous, non-ferrous, and others) that are sent to recyclers
- Rejected waste that goes to a landfill

Organic waste is generated primarily in rural areas. In the composting chamber, in the presence of oxygen and a hot, humid environment, microorganisms degrade the organic matter of the waste in a closed compartment where the O_2 content, temperature, humidity, pH, organic content, and stability of the final product are systematically monitored. The aerobic process is generically described by the following chemical equation: $C_x\ H_y\ O_z + O_2 \rightarrow \alpha CO_2 + \beta H_2\ O + \text{biogas}$, or hydrocarbon + oxygen = carbon dioxide + water + biogas.

9.4 Anaerobic Digestion

In an anaerobic digestion process, prescreening upstream from the anaerobic digesters has the same function as in composting, and the same constituents are separated. In the digesters, the degradation of organic matter is carried out by anaerobic microorganisms: that is, in a humid, oxygen-free medium, with the parameters of temperature and percentage of humidity controlled appropriately. Biogas is cogenerated during this process, and a turbine produces electrical energy for self-consumption by the installation or to export to the national electricity system. In addition, as in composting, the process generates a compound suitable for use in agriculture as a fertilizer (Kumar and Samadder 2020).

9.5 Recycling

Recycling involves several steps or processes and is not a low-cost activity. Therefore, it is essential that, along with its implementation, the formation of a market for recycled material is encouraged in order to make the process more efficient and profitable [17]. Transforming waste into new inputs and raw materials is an economic activity that is part of an industrialized system carried out by private companies that must rely on a physical, technical, and economic-fiscal infrastructure to contribute effectively to the reuse of materials and conserve natural resources, as shown in Figure 9.4.

Recycling changes the physical, physical-chemical, or biological properties of solid waste to transform it into inputs or new products. In other words, it consists of the processing and reuse of materials. Recycling allows the substitution of inputs whose production usually requires a large consumption of energy. In principle, recycling is an environmentally efficient way to use MSW [1].

9.6 Burning Waste Tires

Waste tires are an excellent alternative fuel source [9, 24]. This can be verified by comparing a tire's calorific value with other materials. This option is much sought after by cement kilns in Brazil. It is an excellent alternative source of energy, complementary with burning MSW at plants. Small-scale studies of outdoor tire combustion between 100 and 2000 °C revealed that the empty spaces in the middle of waste tires transports oxygen, controlling the process of combustion.

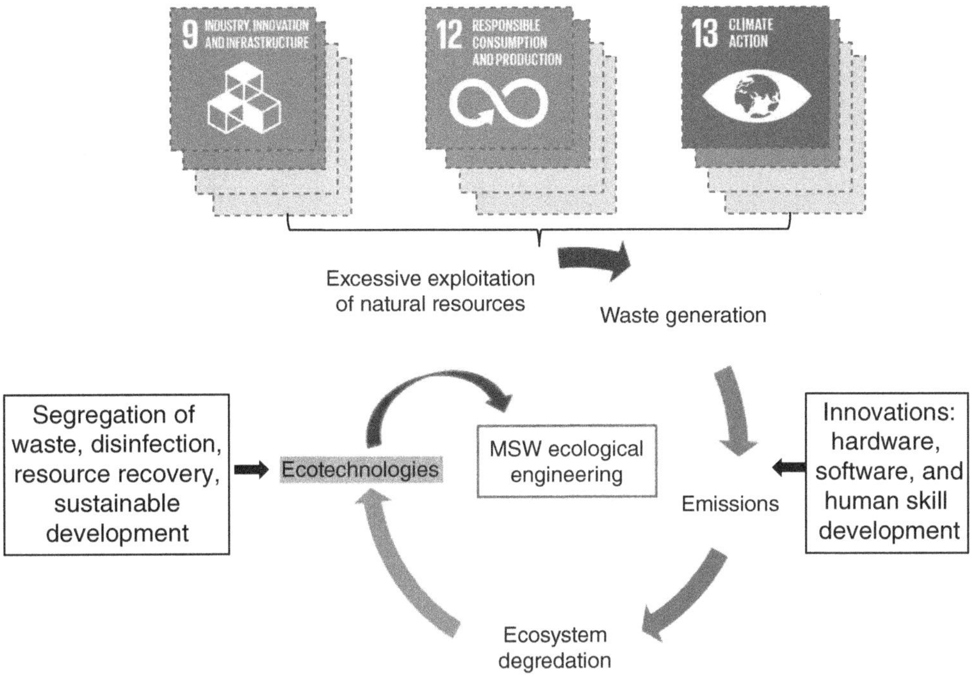

Figure 9.4 Role of ecological and ecosystem engineering in solid waste reduction, resource recovery, and compliance with UN SDGs 9, 12, and 13. *See colour plate section for colour representation of this figure.*

CONAMA (National Council for the Environment) Resolution No. 416/09 (30 September 2009 changed the calculation method for recycling waste tires from the number of tires produced to the number of tires sold in the aftermarket. Distributors, resellers, recipients, consumers, and public authorities must participate together with manufacturers and importers to implement procedures to collect tires.

9.7 Energy Recovery

An MSW incineration plant burns undifferentiated waste that is not likely to be recycled. This waste is used as fuel, burned with oxygen from the air, at temperatures in the range of 900–1200 °C. It is necessary to exhaustively control the temperature and the cooling time of the gases resulting from combustion, which has to be extremely fast (about two to five seconds) to minimize the formation of dioxins and furans resulting from the presence of chlorine in polymeric residues [9, 31, 64, 77].

The incineration of MSW produces combustion gases, which are a source of thermal energy thanks to the generation of superheated steam in heat-recovery boilers. After exchanging heat inside the boiler, these gases are treated to reduce pollutants (NOx, SOx, HCl, etc.) according to the limits required by current legislation. Monitoring and control of pollutant emissions are carried out using continuous analysis systems installed in the chimney.

In heat-treatment plants, MSW is discharged into the silo of the plant and thrown into hoppers by mechanical grabbers. From the hoppers, the garbage is gradually pushed into an incinerator that operates at temperatures between 750 and 1000 °C. The heat produced by burning the waste is

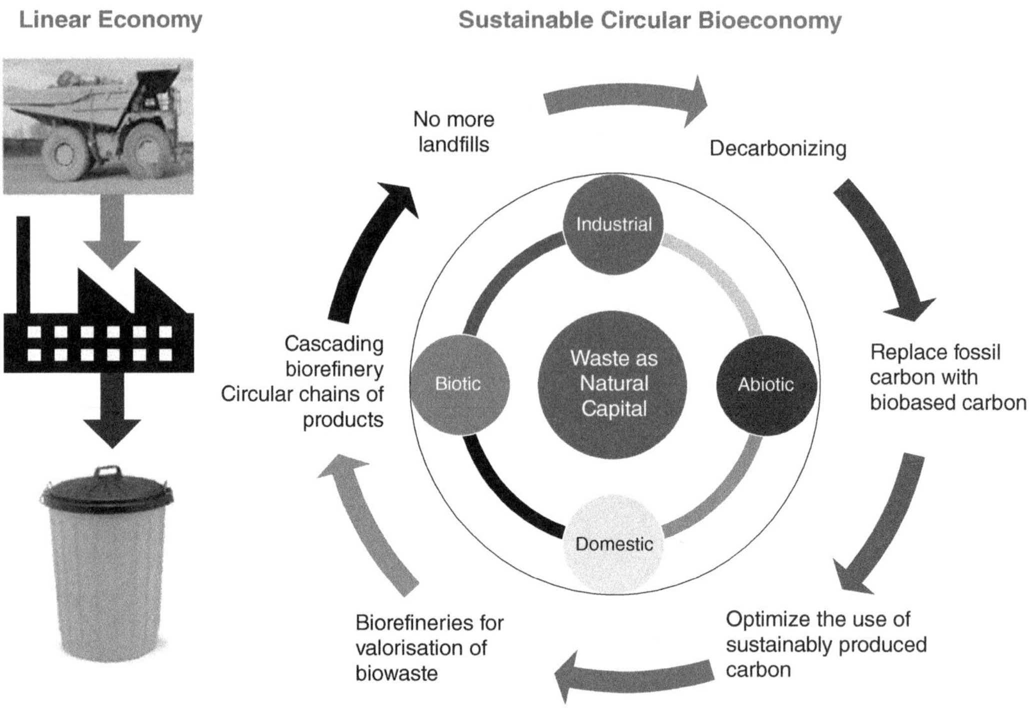

Figure 9.5 Simple model of MSW recycling.

used in boilers, and the steam generated is conducted by pipes to a system of turbines and generators for the production of electric energy. The process is part of the sustainable circular bioeconomy in Figure 9.5.

9.8 Coprocessing Industrial Waste in Cement Kilns

In São Paulo, re-collection centers (RCs) and screening centers (SCs) are installed. In the specific case of the city of São Paulo, the existing MSW collection network could be used, and waste could be transported directly to the heat-treatment plant. The RCs would be dispensed with due to the need to transport MSW quickly to avoid generating bad smells, leachate, and so on from organic material. Logistical costs have yet to be calculated, but travel will be only between neighboring municipalities and the capital. Figure 9.6 shows a model for storing and sorting 12 000 tons/month.

Extending the scope of work to the whole country, there is the amount of solid urban waste generated in tons/day by region and municipality size. The southeast region is the largest generator of MSW; it is also the most developed and has the most extensive road network. Table 9.4 lists 2008 estimates from the IBGE (Brazilian Institute of Geography and Statistics) for the quantity of MSW sent for final disposal and the difference between the amount collected and quantities destined for final disposal.

From Table 9.4, it can be seen that power worth millions of dollars per day is generated in the 15 chosen municipalities. Waste tires could also go to clinker furnaces for energy recovery (not included in this work). Batteries and fluorescent lamps must receive other treatment; they are costly and dangerous items.

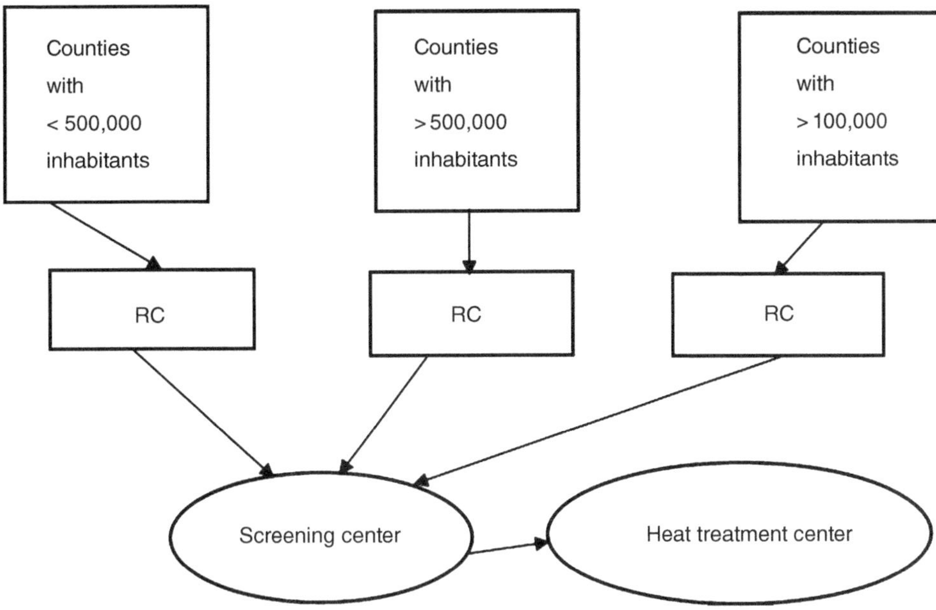

Figure 9.6 Model for reverse logistics proposed in São Paulo.

WEEE must be treated separately, preferably by recyclers that separate usable polymers, metals, and integrated circuit boards. The best valorization for these would send them to large enterprises in Europe, for example, to recycle noble metals and generate foreign currency that would return to Brazil.

The calculated values show the feasibility of WTE in Brazil's southeast region; they are probably overestimated but are sufficient to confirm that the conversion of WTE is an excellent alternative to other methods of waste management. This approach reduces the volume of waste, generates energy without pollution, and, by recycling, keeps waste pickers at work. The amount of energy is on the order of dozens of megawatts.

Energy demand in Brazil today is 1.2 gigawatts, and WTE can be part of the energy matrix. However, the factor that we consider vital for the success of a project of this type is a constant and uninterrupted flow of MSW (MSW and waste from rural areas), because the industrial lines cannot stop under penalty of generating losses due to energy needed to relight furnaces.

9.9 Conclusions

A simulation was carried out for the 15 municipalities with the largest populations in São Paulo, and their viability was verified. It is necessary to plan the management of MSW in other regions of the country that are less developed and with greater distances to developed regions, such as the northeast region, the north region (including the Amazon rainforest), and river transport (the only viable means of transport in the Amazon region). When analyzing Brazil's global data, the numbers point to the possibility of the amount of MSW being used for energy; however, logistics may make the process unfeasible for some regions.

Table 9.4 Electric energy produced from MSW in 15 cities/towns in Brazil.

N	City/Town	Population 2010	Estimated population in 2016	MSW (tons/day)	Distance to São Paulo city(Km)	Transport costs (US$)	Sales of energy (US$/day)
More than 1 000 000 inhabitants							
1	São Paulo	11 253 503	12 038 175	120 381.8	30.0	216687.1	7136 230.1
2	Guarulhos	1 221 979	1 330 878	1337.0	21.5	172 484.2	799 444.3
3	Campinas	1 080 113	1 173 330	1773.0	92.4	65 049.4	6516 205.4
More than 500 000 inhabitants							
4	S. Bernardo do Campo	765 463	822 242	822 732	24	11 840.2	490 385.1
5	Santo André	676 407	712 749	71 275	25	10 691.2	424 655.8
6	Osasco	666 740	696 382	696 380	48	20 055.8	405 294.3
7	S. José dos Campos	629 921	695 992	695	134	55 957.75	369 154.15
8	Ribeirão Preto	60 468	67 440	67 441	314	127 057.9	284 868.6
9	Sorocaba	586 625	652 481	652 480	99.5	38 953.1	359 582.2
More than 200 000 inhabitants							
10	Mauá	422 064	457 696	45 770	27	74 146.7	272 146.0
11	S. José do Rio Preto	40 828	446 649	44 665	439	117 647.3	155 165.8
12	Santos	419 400	434 353	43 435	80	20 848.9	244 453.8
13	Mogi das Cruzes	387 779	429 321	42 932	64	16 485.9	245 743.3
14	Diadema	386 089	415 180	41 518	21.3	5306	248 285.9
15	Jundiaí	370 126	405 740	40 574	58	14 119.7	233 706.2
						Total US$	12320736.7

Source: IBGE, Instituto Brasileiro de Geografia e Estatística. (2011). Indicadores de Desenvolvimento Sustentável Brasil 2011. www.ibge.gov.br/home/geociencias/Recursosnaturais/ids/default_2010.shtm.

SABESP. (2015). COMPANHIA, DE SANEAMENTO BÁSICO DO ESTADO DE SÃO PAULO (SABESP):Complexo metropolitano. Available at: http://site.sabesp.com.br/site/interna/Default.aspxBrazil.

Note: Only the cost of operation/ maintenance (OPEX) was taken into account because the Capex cost would be a long-term investment with a rate of return on investment depending on financial factors.

MSW is a potentially huge source of foreign exchange, which in addition to reducing imports of fossil fuels would be another alternative energy source for the country and would reduce pollution generated by the large volumes of waste generated daily and often disposed of inappropriately. The survey showed interesting results from economic and financial points of view, and we found the need for a good waste management plan operated by municipalities, private initiatives, and citizens.

References

1 Ahmad Bukhari, N.A., Fauzi, M.F., Muin, A. et al. (2020). Towards zero waste: technologies and minimisation strategies. *MAEH Journal of Environmental Health (Short Communication)* 1 (1): 4–7.

2 Assi, A., Bilo, F., Zanoletti, A. et al. (2020). Zero-waste approach in municipal solid waste incineration: reuse of bottom ash to stabilize fly ash. *Journal of Cleaner Production* 245: 118779.

3 Azam, M., Jahromy, S.S., Raza, W. et al. (2020). Status, characterization, and potential utilization of municipal solid waste as renewable energy source: Lahore case study in Pakistan. *Environment International* 134: 105291.

4 Baccot, C., Pallier, V., Thom, M.T. et al. (2020). Valorization of extracted organic matter from municipal solid waste leachate: application to soils from France and Togo. *Waste Management* 102: 161–169.

5 Banerjee, S. and Sarkhel, P. (2020). Municipal solid waste management, household and local government participation: a cross country analysis. *Journal of Environmental Planning and Management* 63 (2): 210–235.

6 Banerjee, K., Ray, S.K., and Ghosh, S.K. (2020). Integrated solid waste management under Japan international co-operation agency project in municipalities of West Bengal. In: *Sustainable Waste Management: Policies and Case Studies*, vol. 1, 187–201. Singapore: Springer.

7 Batur, M.E., Cihan, A., Korucu, M.K. et al. (2020). A mixed integer linear programming model for long-term planning of municipal solid waste management systems: against restricted mass balances. *Waste Management* 105: 211–222.

8 Bittencourt, E.S., Fontes, C.H., Rodriguez, J.L., and Ferreira, A.M. (2020). Modeling the socioeconomic metabolism of end-of-life tires using structural equations: a Brazilian case study. *Sustainability* 12 (5): 2106.

9 Chua, H.S. and Bashir, M.J. (2020). Waste management practice in Malaysia and future challenges. In: *Hand Book of Research on Resource Management for Pollution and Waste Treatment*, 531–549. IGI Global.

10 Colvero, D.A., Ramalho, J., Gomes, A.P. et al. (2020). Economic analysis of a shared municipal solid waste management facility in a metropolitan region. *Waste Management* 102: 823–837.

11 Cristelo, N., Segadães, L., Coelho, J. et al. (2020). Recycling municipal solid waste incineration slag and fly ash as precursors in low-range alkaline cements. *Waste Management* 104: 60–73.

12 Daskal, S., Ayalon, O., and Shechter, M. (2020). Implementation of municipal solid waste regulations in Israel. In: *Sustainable Waste Management: Policies and Case Studies*, 279–290. Singapore: Springer.

13 Demir, A., Basturk, A., Ozkaya, B., and Bilgili, M.S. (2020). Municipal waste Management in Turkey. In: *Sustainable Waste Management Challenges in Developing Countries*, 441–465. IGI Global.

14 Deus, R.M., Mele, F.D., Bezerra, B.S., and Battistelle, R.A.G. (2020a). A municipal solid waste indicator for environmental impact: assessment and identification of best management practices. *Journal of Cleaner Production* 242: 118433.

15 Deus, R.M., Mele, F.D., Bezerra, B.S., and Battistelle, R.A.G. (2020b). Analytical framework and data for a municipal solid waste environmental performance assessment. *Data in Brief* 3: 105085.

16 Diaz, L.F., Golueke, C.G., Savage, G.M., and Eggerth, L.L. (2020). *Composting and Recycling Municipal Solid Waste.* CRC Press.

17 dos Santos, I.F., Mensah, J.H., Gonçalves, A.T., and Barros, R.M. (2020). Incineration of municipal solid waste in Brazil: an analysis of the economically viable energy potential. *Renewable Energy* 149: 1386–1394.

18 Duan, N., Li, D., Wang, P. et al. (2020). Comparative study of municipal solid waste disposal in three Chinese representative cities. *Journal of Cleaner Production* 120134.

19 Ercoli, R.F., Matias, V.R., and Zago, V.C. (2020). *Urban Expansion and Erosion Processes in an Area of Environmental Protection in Nova Lima*, Frontiers in Environmental Science. Brazil: Minas Gerais State.

20 Feraldi, R., Cashman, S., Huff, M., and Raahauge, L. (2013). Comparative LCA of treatment options for US scrap tires: material recycling and tire-derived fuel combustion. *The International Journal of Life Cycle Assessment* 18 (3): 613–625.

21 Ferronato, N., Preziosi, G., Portillo, M.A. et al. (2020). Assessment of municipal solid waste selective collection scenarios with geographic information systems in Bolivia. *Waste Management* 102: 919–931.

22 França, A.S., Neto, J.A., Gonçalves, R.F., and Almeida, C.M. (2020). Proposing the use of blockchain to improve the solid waste management in small municipalities. *Journal of Cleaner Production* 244: 118529.

23 Gamboa, A.R., Rocha, A.M., dos Santos, L.R., and de Carvalho Jr, J.A. (2020). Tire pyrolysis oil in Brazil: potential production and quality of fuel. *Renewable and Sustainable Energy Reviews* 120: 109614.

24 Gao, J., Dong, C., Zhao, Y. et al. (2020). Vitrification of municipal solid waste incineration fly ash with B2O3 as a fluxing agent. *Waste Management* 102: 932–938.

25 Ghalehkhondabi, I. and Maihami, R. (2020). Sustainable municipal solid waste disposal supply chain analysis under price-sensitive demand: a game theory approach. *Waste Management & Research* 38 (3): 300–311.

26 Gomes, H.I., Funari, V., and Ferrari, R. (2020). Bioleaching for resource recovery from low-grade wastes like fly and bottom ashes from municipal incinerators: a SWOT analysis. *Science of The Total Environment* 715: 136945.

27 Gu, Z., Chen, W., Wang, F., and Li, Q. (2020). A pilot-scale comparative study of bioreactor landfills for leachate decontamination and municipal solid waste stabilization. *Waste Management* 103: 113–121.

28 Han, X., Hu, C., and Lin, L. (2020). A study on the impact of China's urbanization on the quantity of municipal solid waste produced. *Waste Management & Research* 38 (2): 184–192.

29 Hoang, N.H. and Fogarassy, C. (2020). Sustainability evaluation of municipal solid waste management system for Hanoi (Vietnam) – why to choose the "waste-to-Energy"Concept. *Sustainability* 12 (3): 1085.

30 Hoang, A.T., Nguyen, T.H., and Nguyen, H.P. (2020). Scrap tire pyrolysis as a potential strategy for waste management pathway: a review. *Energy Sources, Part A: Recovery, Utilization, and Environmental Effects* 28: 1–8.

31 Huber, F., Blasenbauer, D., Aschenbrenner, P., and Fellner, J. (2020). Complete determination of the material composition of municipal solid waste incineration bottom ash. *Waste Management* 102: 677–685.

32 Ibañez-Forés, V., Bovea, M.D., and Coutinho-Nóbrega, C. (2020). Proposal of social indicators to assess the social performance of waste management systems in developing countries: a Brazilian case study. In: *Perspectives on Social LCA*, 95–103. Cham: Springer.

33 Iyamu, H.O., Anda, M., and Ho, G. (2020). A review of municipal solid waste management in the BRIC and high-income countries: a thematic framework for low-income countries. *Habitat International* 95: 102097.

34 Joseph, L.P. and Prasad, R. (2020). Assessing the sustainable municipal solid waste (MSW) to electricity generation potentials in selected Pacific Small Island developing states (PSIDS). *Journal of Cleaner Production* 248: 119222.

35 Juma, M., Koreňová, Z., Markoš, J. et al. (2006). Pyrolysis and combustion of scrap tire. *Petroleum and Coal* 48 (1): 15–26.

36 Kandasamy, J. and Gökalp, I. (2015). Pyrolysis, combustion, and steam gasification of various types of scrap tires for energy recovery. *Energy & Fuels* 29 (1): 346–354.

37 Khoshnevisan, B., Tabatabaei, M., Tsapekos, P. et al. (2020). Environmental life cycle assessment of different biorefinery platforms valorizing municipal solid waste to bioenergy, microbial protein, lactic and succinic acid. *Renewable and Sustainable Energy Reviews* 117: 109493.

38 Kurashima, K., Kumagai, S., Kameda, T. et al. (2020). Heavy metal removal from municipal solid waste fly ash through chloride volatilization using poly (vinyl chloride) as chlorinating agent. *Journal of Material Cycles and Waste Management* 30: 1–4.

39 Liang, S., Chen, J., Guo, M. et al. (2020). Utilization of pretreated municipal solid waste incineration fly ash for cement-stabilized soil. *Waste Management* 105: 425–432.

40 Machin, E.B., Pedroso, D.T., and de Carvalho Jr, J.A. (2017). Energetic valorization of waste tires. *Renewable and Sustainable Energy Reviews* 68: 306–315.

41 Maghmoumi, A., Marashi, F., and Houshfar, E. (2020). Environmental and economic assessment of sustainable municipal solid waste management strategies in Iran. *Sustainable Cities and Society* 102161.

42 Magrini, C., D'Addato, F., and Bonoli, A. (2020). Municipal solid waste prevention: a review of market-based instruments in six European Union countries. *Waste Management & Research* 38: 0734242X19894622.

43 Mohsenizadeh, M., Tural, M.K., and Kentel, E. (2020). Municipal solid waste management with cost minimization and emission control objectives: a case study of Ankara. *Sustainable Cities and Society* 52: 101807.

44 Morero, B., Montagna, A.F., Campanella, E.A., and Cafaro, D.C. (2020). Optimal process design for integrated municipal waste management with energy recovery in Argentina. *Renewable Energy* 146: 2626–2636.

45 Moustakas, K., Rehan, M., Loizidou, M. et al. (2020). Energy and resource recovery through integrated sustainable waste management. *Applied Energy* 261: 114372.

46 Movahed, Z.P., Kabiri, M., Ranjbar, S., and Joda, F. (2020). Multi-objective optimization of life cycle assessment of integrated waste management based on genetic algorithms: a case study of Tehran. *Journal of Cleaner Production* 247: 119153.

47 Mozhiarasi, V., Raghul, R., Speier, C.J. et al. (2020). Composition analysis of major organic fractions of municipal solid waste generated from Chennai. In: *Sustainable Waste Management: Policies and Case Studies*, 143–152. Singapore: Springer.

48 Oduro-Appiah, K., Scheinberg, A., Miezah, K. et al. (2020). Existing realities and sustainable pathways for solid waste Management in Ghana. In: *Sustainable Waste Management Challenges in Developing Countries*, 115–143. IGI Global.

49 Ozturk, M. and Dincer, I. (2020). An efficient waste management system with municipal solid waste incineration plant. *Greenhouse Gases: Science and Technology* 10 (4): 855–864.

50 Paes, M.X., de Medeiros, G.A., Mancini, S.D. et al. (2020). Municipal solid waste management: integrated analysis of environmental and economic indicators based on life cycle assessment. *Journal of Cleaner Production* 254: 119848.

51 Parashar, C.K., Das, P., Samanta, S. et al. (2020). Municipal solid wastes—a promising sustainable source of energy: a review on different waste-to-energy conversion technologies. In: *Energy Recovery Processes from Wastes*, 151–163. Singapore: Springer.

52 Raghav, S. and Kumar, D. (2020). Review environmental implications of incineration of municipal solid waste and ash disposal. In: *Handbook of Research on Environmental and Human Health Impacts of Plastic Pollution*, 59–79. IGI Global.

53 Raj, J.A., Asif, A., Vimalesh, S.S. et al. (2020). Study on solid waste management system in Dindigul Corporation of Tamilnadu and Feasibility Analysis on waste-to-energy projects. In: *Sustainable Waste Management: Policies and Case Studies*, 135–141. Singapore: Springer.

54 Rhee, S.W. (2020). Circular economy of municipal solid waste (MSW) in Korea. In: *Circular Economy: Global Perspective*, 317–332. Singapore: Springer.

55 Rodovalho, E., Quaglio, O., Junior, W.S. et al. (2020). Reducing GHG emissions through efficient tire consumption in open pit mines. *Journal of Cleaner Production* 255: 120185.

56 da Roza, D.A., Aguiar, G.T., de Lima, E.P. et al. (2020). Decision model for selecting advanced technologies for municipal solid waste management. In: *International Business, Trade and Institutional Sustainability*, 201–220. Cham: Springer.

57 Sauve, G. and Van Acker, K. (2020). The environmental impacts of municipal solid waste landfills in Europe: a life cycle assessment of proper reference cases to support decision making. *Journal of Environmental Management* 261: 110216.

58 Sebastian, R.M., Kumar, D., and Alappat, B. (2020a). Mapping incinerability of municipal solid waste in Indian sub-continent. *Clean Technologies and Environmental Policy* 22 (1): 91–104.

59 Sebastian, R.M., Kumar, D., and Alappat, B.J. (2020b). Influence of recycling on the Incinerability of municipal solid waste. In: *Recent Developments in Waste Management*, 461–472. Singapore: Springer.

60 Segre, N. and Joekes, I. (2000). Use of tire rubber particles as addition to cement paste. *Cement and Concrete Research* 30 (9): 1421–1425.

61 Shahin, H. (2020). Reducing methane emissions from municipal solid waste landfills by using mechanical, biological treatment (case study Wady Alhaddeh (MBT) plant, in Tartous). In: *Waste Management in MENA Regions*, 241–258. Cham: Springer.

62 Sitepu, M.H., Matondang, A.R., and Sembiring, M.T. (2020). Used tires recycle management and processing: a review. *IOP Conference Series: Materials Science and Engineering* 801 (1): 012116.

63 Srivastava, R. (2020). Solid waste management and its impact on the environment. In: *Handbook of Research on Environmental and Human Health Impacts of Plastic Pollution*, 389–400. IGI Global.

64 Srivastava, V., Goel, G., Thakur, V.K. et al. (2020). Analysis and advanced characterization of municipal solid waste vermicompost maturity for a green environment. *Journal of Environmental Management* 255: 109914.

65 Taleb, D.A., Abd Hamid, H., Deris, R.R. et al. (2020). Insights into pyrolysis of waste tire in fixed bed reactor: thermal behavior. *Materials Today: Proceedings* 31 (1): 178–186.

66 Thakur, D., Ganguly, R., Gupta, A.K., and Ghali, V. (2020). Evaluation of existing solid waste management system in Una town, India. In: *Sustainable Waste Management: Policies and Case Studies*, 367–381. Singapore: Springer.

67 Tirkolaee, E.B., Mahdavi, I., Esfahani, M.M., and Weber, G.W. (2020). A robust green location-allocation-inventory problem to design an urban waste management system under uncertainty. *Waste Management* 102: 340–350.

68 Torrente-Velásquez, J.M., Chifari, R., Ripa, M., and Giampietro, M. (2020). Robust information for effective municipal solid waste policies: identifying behaviour of waste generation across spatial levels of organization. *Waste Management* 103: 208–217.

69 Tsai, F.M., Bui, T.D., Tseng, M.L., and Wu, K.J. (2020). A causal municipal solid waste management model for sustainable cities in Vietnam under uncertainty: a comparison. *Resources, Conservation and Recycling* 154: 104599.

70 Van Fan, Y., Klemeš, J.J., Walmsley, T.G., and Bertók, B. (2020). Implementing circular economy in municipal solid waste treatment system using P-graph. *Science of The Total Environment* 701: 134652.

71 Verma, N., Kaur, M., and Tripathi, A.K. (2020). Greenhouse gas emissions from municipal solid waste management practice. In: *Environmental Concerns and Sustainable Development*, 399–408. Singapore: Springer.

72 Veses, A., Sanahuja-Parejo, O., Callén, M.S. et al. (2020). A combined two-stage process of pyrolysis and catalytic cracking of municipal solid waste for the production of syngas and solid refuse-derived fuels. *Waste Management* 101: 171–179.

73 Viau, S., Majeau-Bettez, G., Spreutels, L. et al. (2020). Substitution modelling in life cycle assessment of municipal solid waste management. *Waste Management* 102: 795–803.

74 Vujić, G. and Tot, B. (2020). Solid waste management in the Republic of Serbia. In: *Sustainable Waste Management Challenges in Developing Countries*, 306–326. IGI Global.

75 Wang, D., He, J., Tang, Y.T. et al. (2020). Life cycle assessment of municipal solid waste management in Nottingham, England: past and future perspectives. *Journal of Cleaner Production* 251: 119636.

76 Wang, D., Tang, Y.T., Long, G. et al. (2020). Future improvements on performance of an EU landfill directive driven municipal solid waste management for a city in England. *Waste Management* 102: 452–463.

77 Wang, Q.Z., Wang, N.N., Tseng, M.L. et al. (2020). Waste tire recycling assessment: road application potential and carbon emissions reduction analysis of crumb rubber modified asphalt in China. *Journal of Cleaner Production* 249: 119411.

78 Wu, C., Shu, M., Liu, X. et al. (2020). Characterization of the volatile compounds emitted from municipal solid waste and identification of the key volatile pollutants. *Waste Management* 103: 314–322.

79 Wu, L., Jiang, X., Lv, G. et al. (2020). Interactive effect of the sorted components of solid recovered fuel manufactured from municipal solid waste by thermogravimetric and kinetic analysis. *Waste Management* 102: 270–280.

79 Xu, Y., Liu, X., Hu, X. et al. (2020). A genetic-algorithm-aided fuzzy chance-constrained programming model for municipal solid waste management. *Engineering Optimization* 52 (4): 652–668.

80 Yadav, V. and Karmakar, S. (2020). Sustainable collection and transportation of municipal solid waste in urban centers. *Sustainable Cities and Society* 53: 101.

81 Yan, M., Wibowo, H., Liu, Q. et al. (2020). Municipal solid waste management and treatment in China. In: *Sustainable Waste Management Challenges in Developing Countries*, 86–114. IGI Global.

82 Yousefloo, A. and Babazadeh, R. (2020). Designing an integrated municipal solid waste management network: a case study. *Journal of Cleaner Production* 244: 118824. https://doi.org/10.1016/j.jclepro.2019.118824.

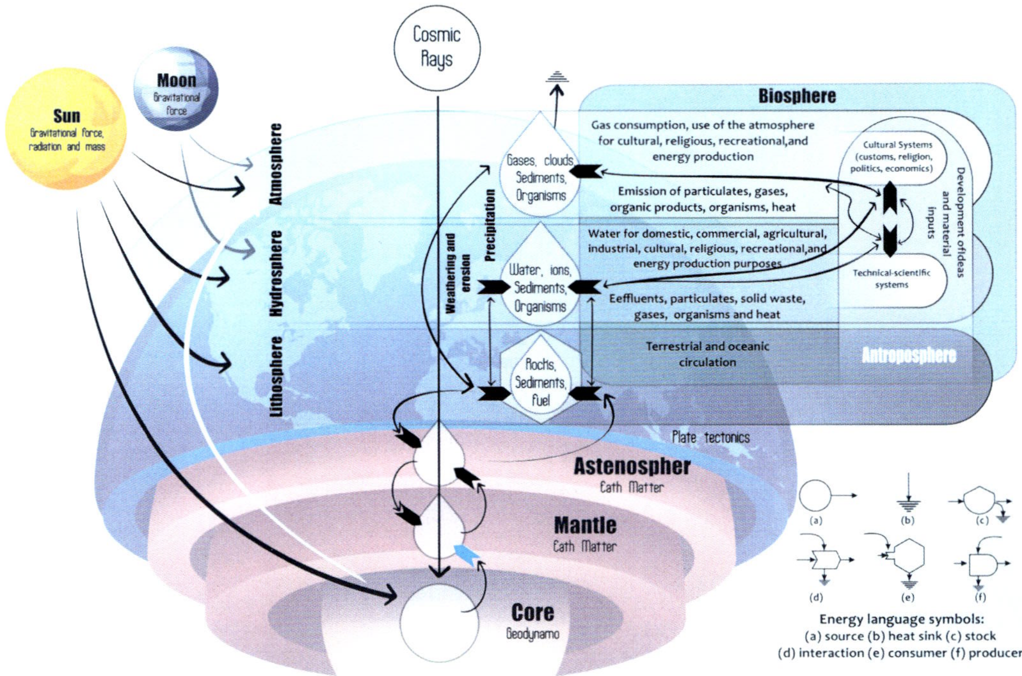

Figure 1.1 Global systems and their interactions.

Handbook of Ecological and Ecosystem Engineering, First Edition. Edited by Majeti Narasimha Vara Prasad.

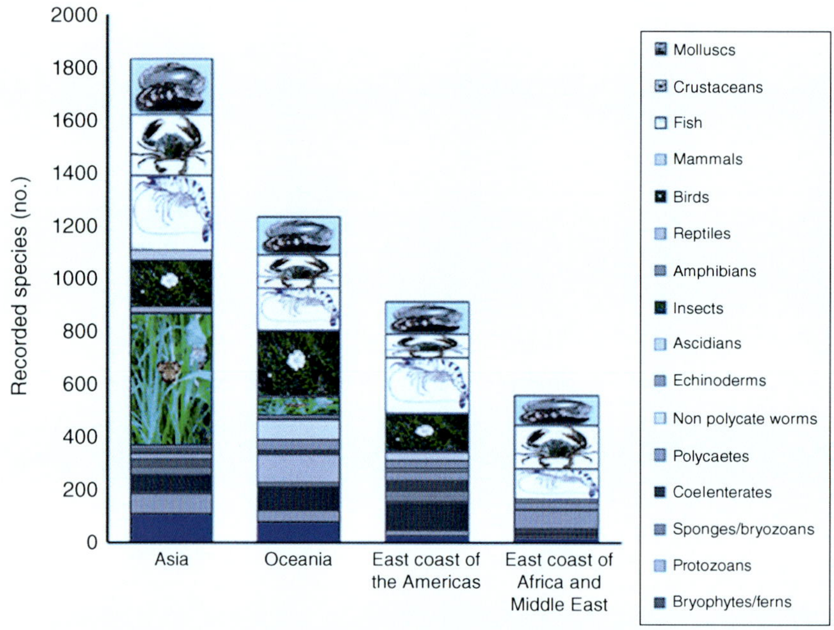

Figure 3.2 Schematic representation of the geographical distribution of mangrove biodiversity. *Source:* Adapted from Biswas et al. [3]. © 2009, Springer Nature.

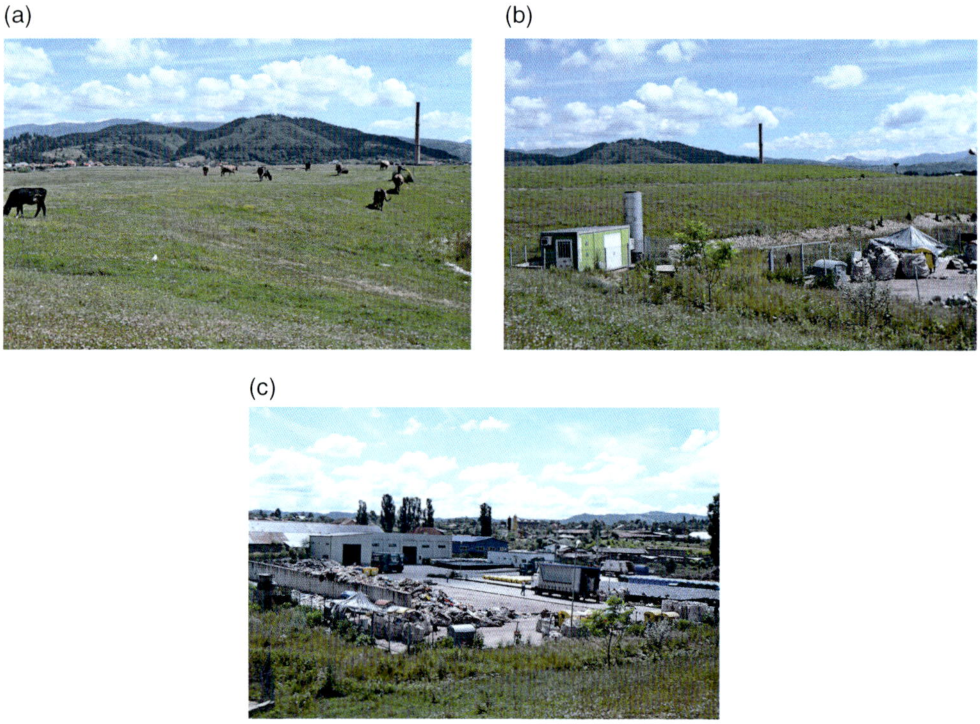

Figure 5.1 Closed Campulung landfill, Arges county, Romania. (a) Cows and horses grazing on the surface of the closed landfill; (b) biogas plant at the Campulung landfill; (c) the newly built waste transfer station beside the old landfill; view from the landfill body. *Source:* Photos by Petra Schneider.

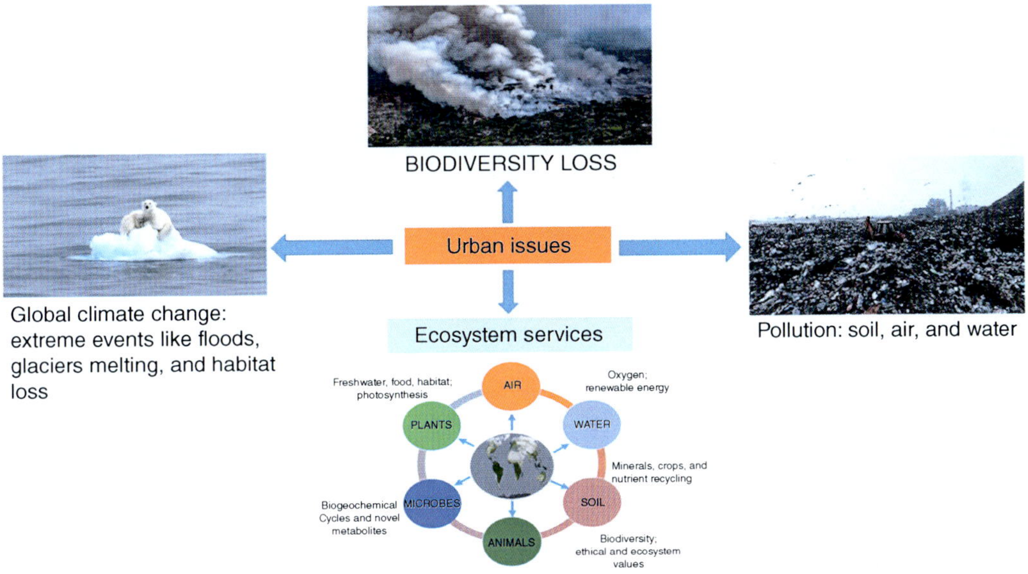

Figure 6.1 Urbanization and overexploitation, leading to biodiversity loss, climate change, and ecosystem degradation.

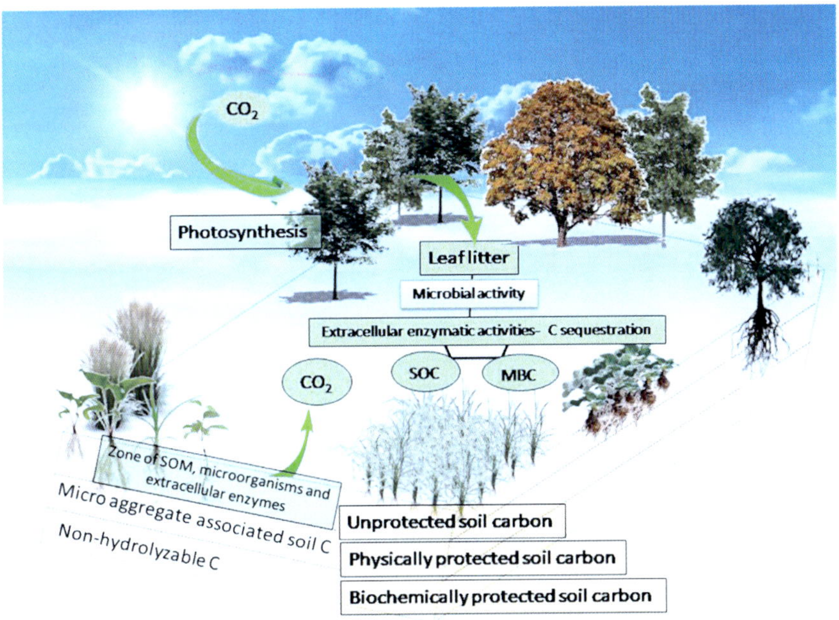

Figure 8.2 A generalized diagram showing carbon dynamics in a typical terrestrial habitat.

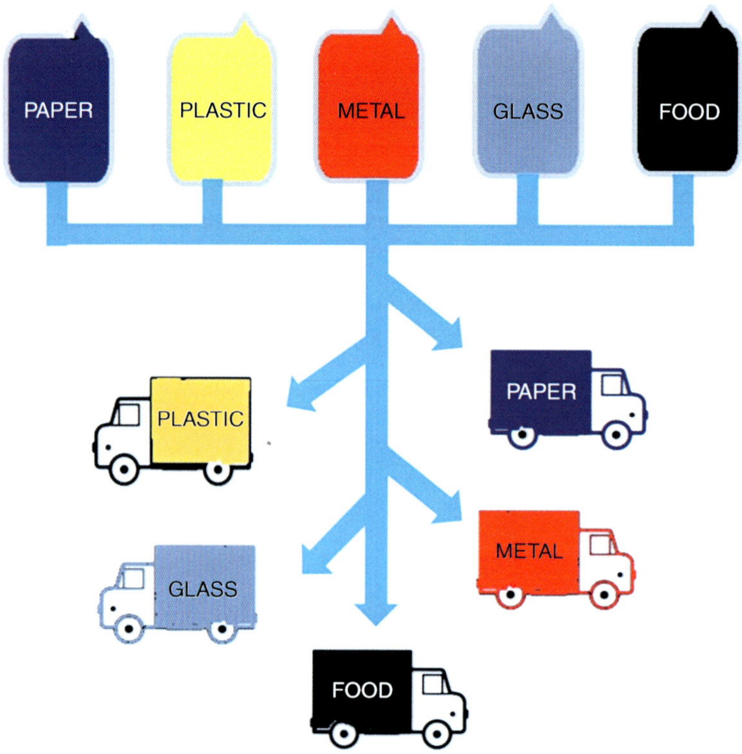

Figure 9.3 Segregation of MSW and transport to a resource recovery site.

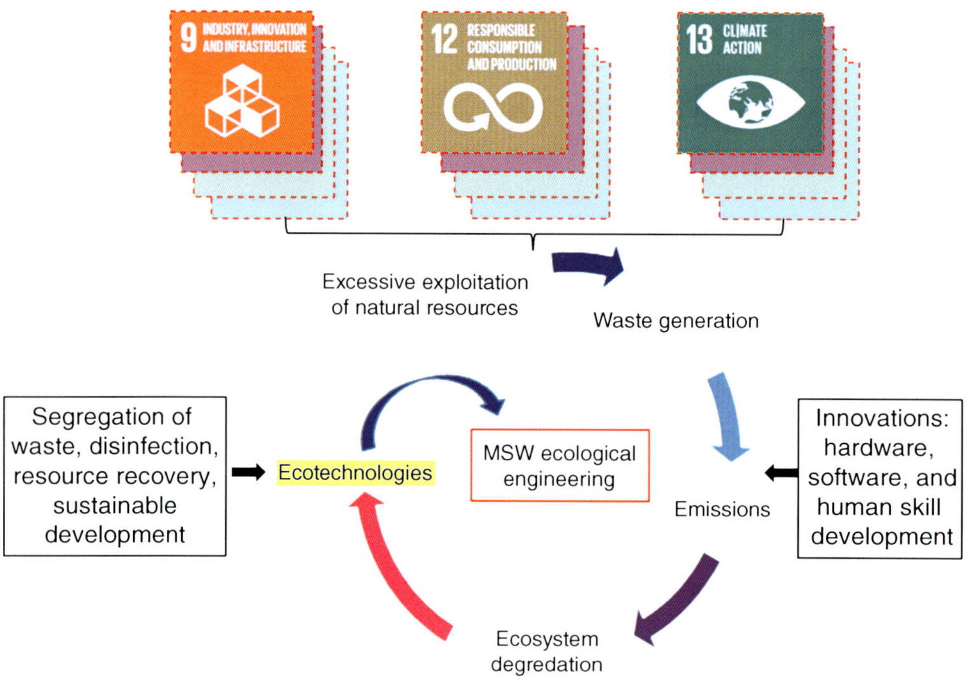

Figure 9.4 Role of ecological and ecosystem engineering in solid waste reduction, resource recovery, and compliance with UN SDGs 9, 12, and 13.

Figure 10.2 A typical flood in an urban area (Assam, India).

Figure 11.1 An example of a region with intensive mining and a changing landscape caused by ongoing operations: open pit Lignite Mine in Bełchatów (central Poland). *Source:* Photo by Marcin Pietrzykowski.

Figure 11.4 Varied mine soil substrate and differentiation in redeveloped habitat conditions: (a) loamy sands; (b) Miocene acidic sands with lignite inclusions; (c and d) carboniferous shales with a high content of fossil coal; (e) quaternary sands; (f) loamy sands; (g and h) sulphurized mix of Neogene clays and sands. *Source:* Photos by Marcin Pietrzykowski.

Figure 12.1 Steps in surface mining and ecological restoration of mine degraded sites: (1) natural forest before mining, (2) deforested land, (3) removing topsoil and overburden materials, (4) excavation by surface mining methods (5) backfilling of overburden materials, (6) regenerating backfilled sites with a grass-legume mixture, (7) growth of tree species during the revegetation phase, and (8) developing site characteristics along the age of reclamation.

Figure 12.4 Photos of the study site showing (a) unreclaimed dump with no vegetation; and development of vegetation cover on (b) 3-year-old afforested dump; (c) 7-year-old afforested site; (d) 10-year-old afforested site; (e) 15-year-old afforested site; (f) natural forest in the area.

Figure 14.8 Baby dolphin jumping.

Figure 14.9 Intentional killing of dolphins.

Figure 14.10 Sand mining in a river bed.

Figure 15.2 (a) An aerial view of Loktak with the floating *athaphoom* or traditional fishing ponds; and (b) manual removal of *phoomdi* for composting.

Figure 15.3 Degraded ecological conditions of the Nambul river (a main tributary of Loktak): (a) Khwairamband Bazaar; (b, c) Keishamthong stretch, the greater Imphal region.

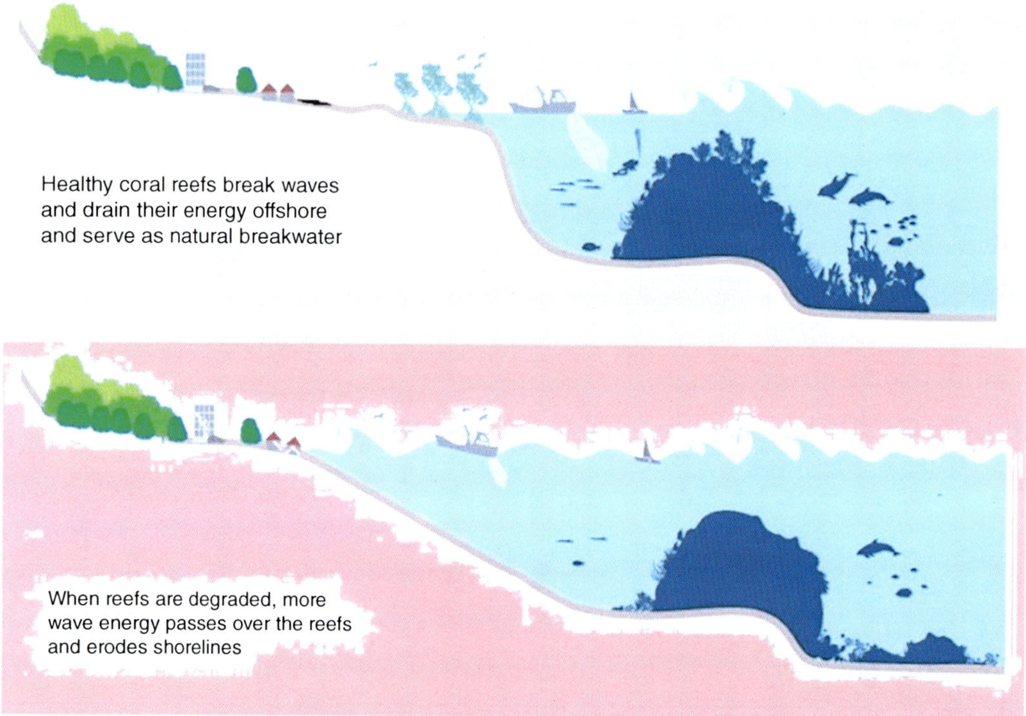

Figure 16.5 Comparison between healthy and degraded coral reefs in shoreline protection.

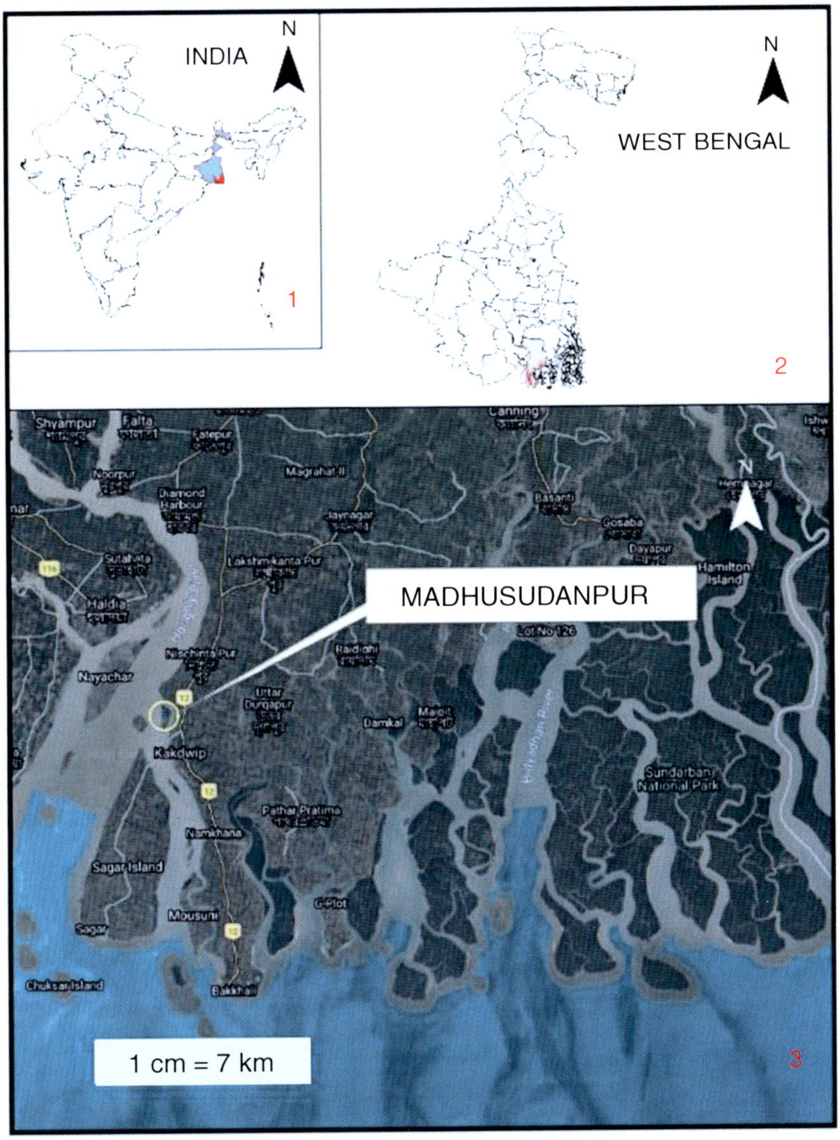

Figure 17.1 Geographical position of the island Madhusudanpurchor. Latitude: 21° 56′ 30.82″ N; longitude: 88° 10′ 26.37″ E.

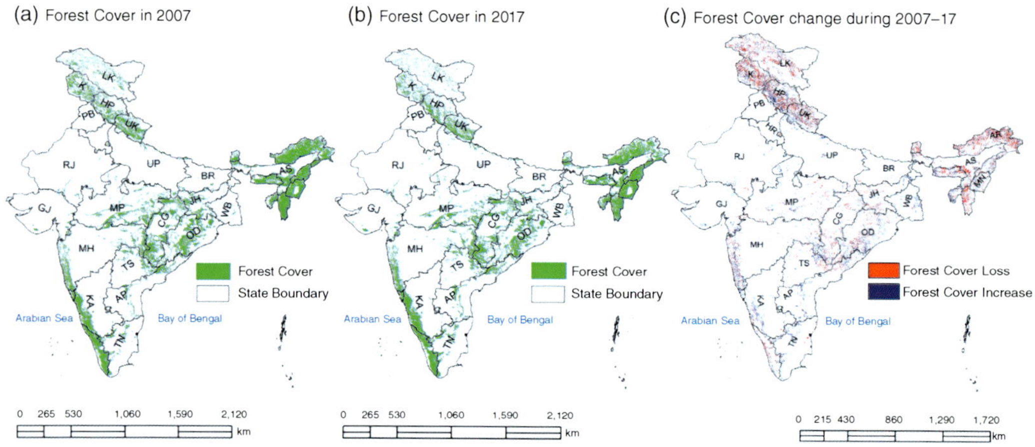

Figure 18.1 ALOS PALSAR satellite image. (a) Forest cover in 2007; (b) forest cover in 2017; (c) forest cover change during 2007–2017.

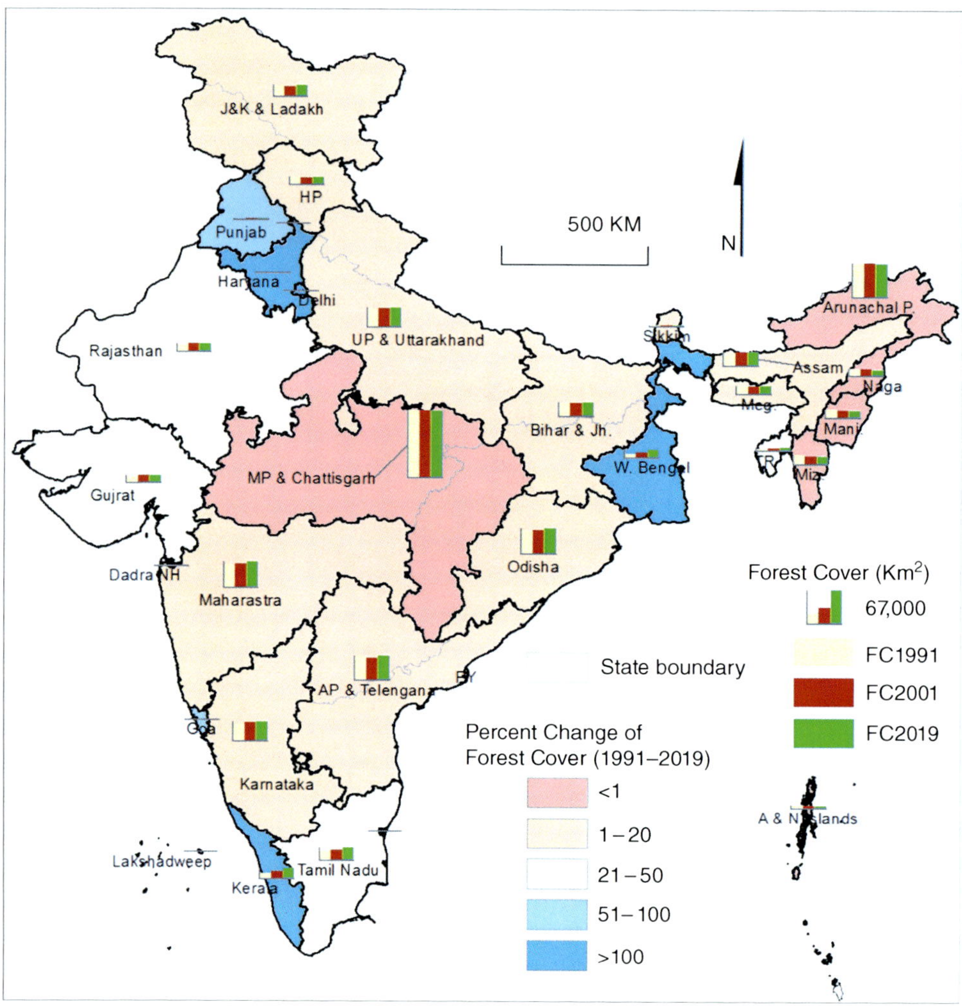

Figure 18.4 Forest cover (1991, 2001, 2019) and forest cover change during 1991–2019 in the states of India.

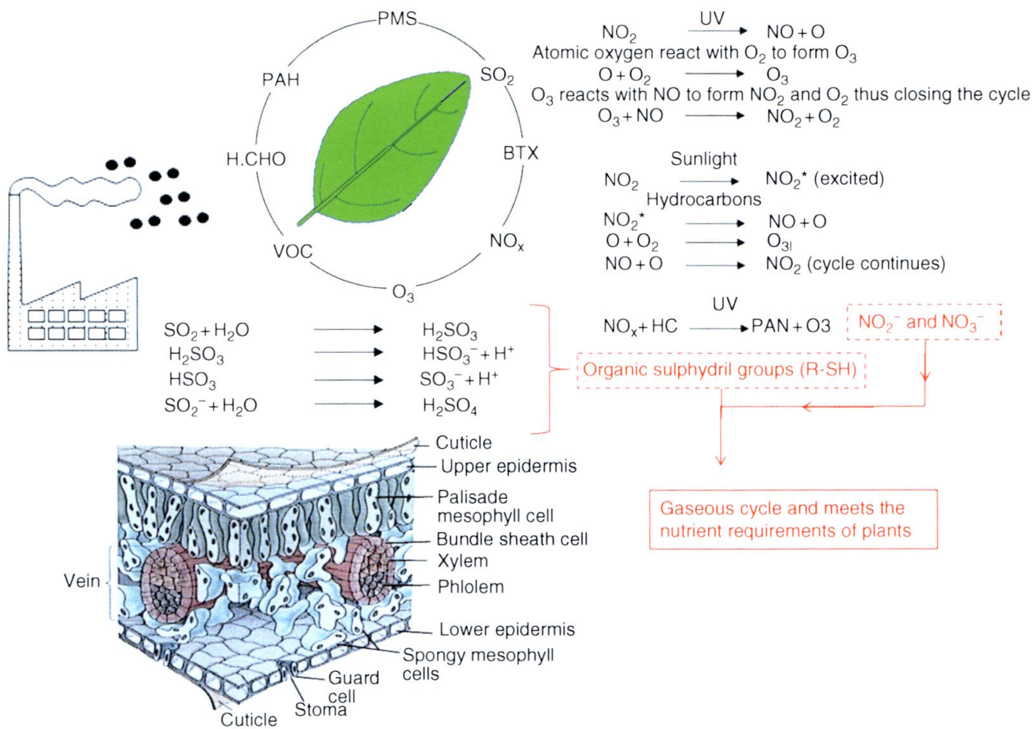

Figure 22.4 Leaf tissue involvement and chemistry in mitigating gaseous pollutants and their conversion to nutrients required by plants.

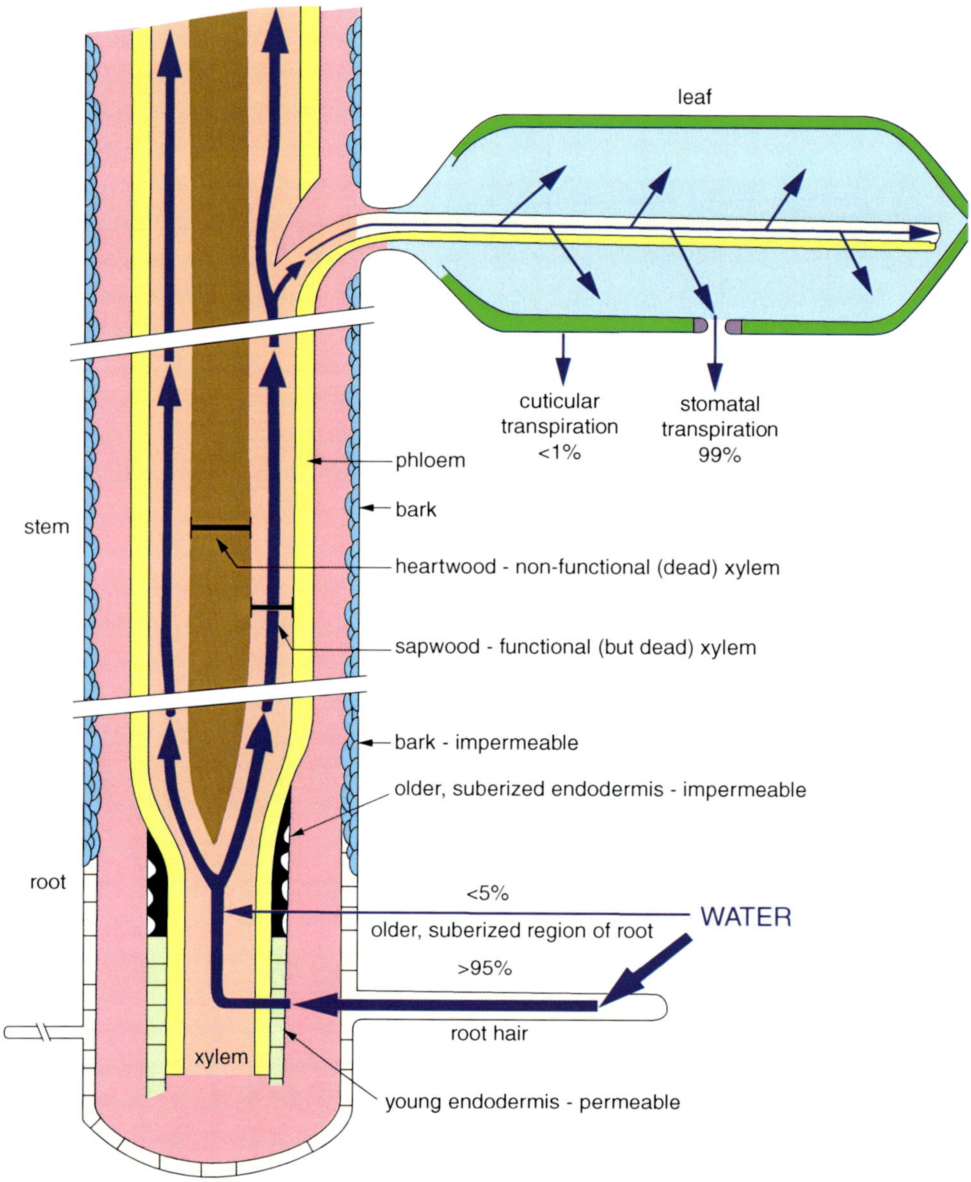

Figure 22.6 Schematic representation of moisture condition surrounding the phyllosphere. *Source:* Shrivastava et al. [10], Central Pollution Control Board, MOEF&CC, Govt. of India, New Delhi.

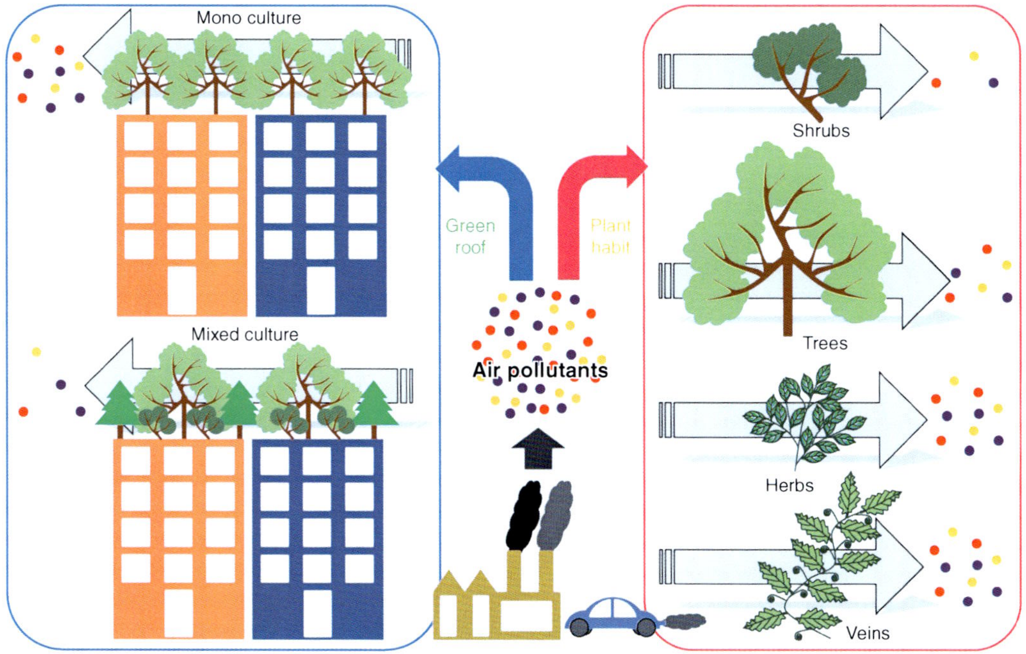

Figure 24.1 A schematic illustration of the role of leaf architecture in removing air pollutants.

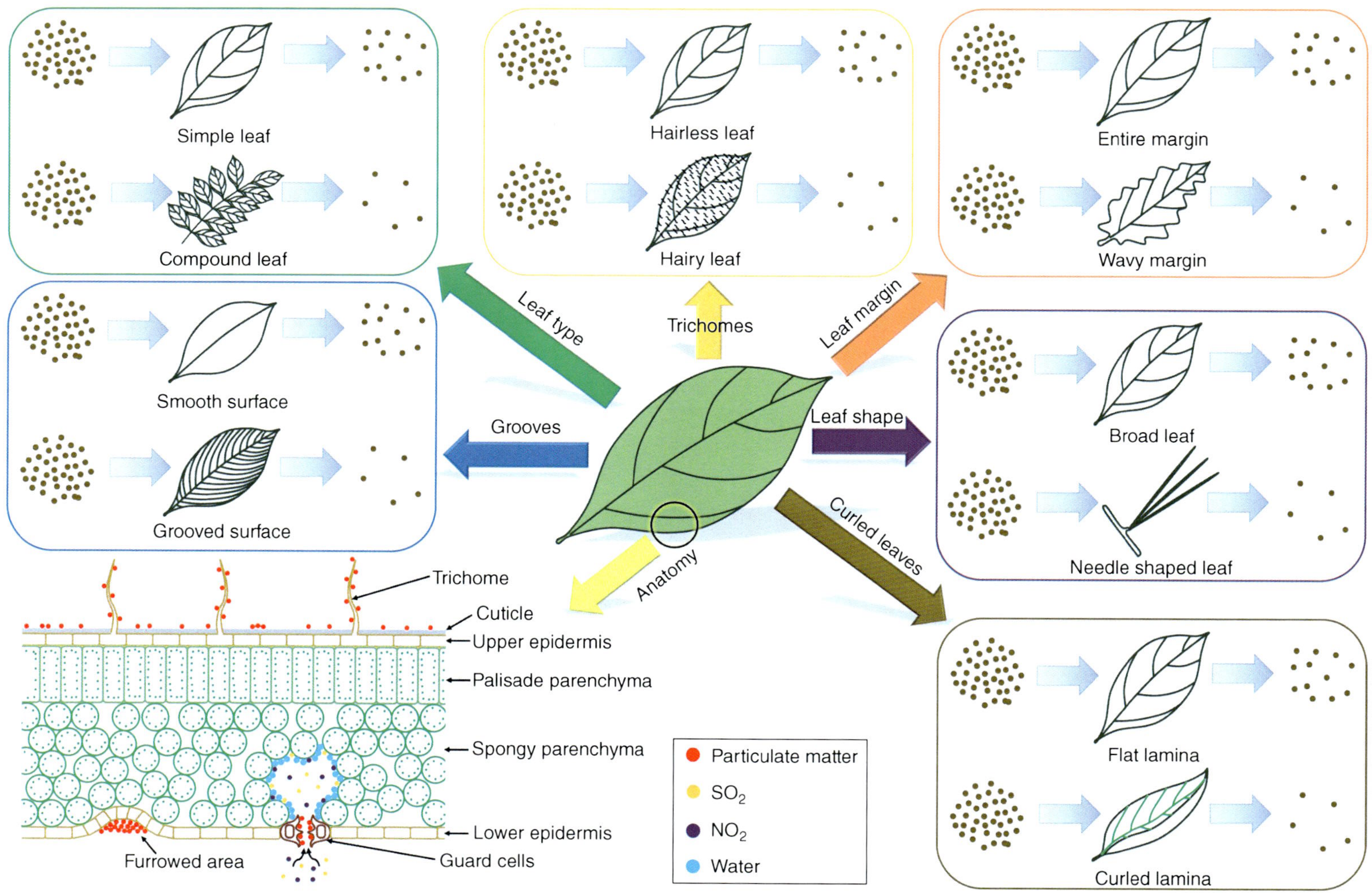

Figure 24.2 Air quality improvement in cities with green infrastructure.

10

Urban Floods and Mitigation by Applying Ecological and Ecosystem Engineering

Jyotirmoy Sarma[1] and Sanchayita Rajkhowa[2]

[1] *Department of Chemistry, Kaziranga University, Jorhat, Assam, India*
[2] *Department of Chemistry, Jorhat Institute of Science & Technology (JIST), Jorhat, Assam, India*

10.1 Sustainable Ecosystems through Engineering Approaches

"Ecological engineering is the design of supportable ecosystems that integrate human society with its natural environment for the advantage of both" [1]. Ecological engineering and ecosystem restoration are two vital aspects that we need to take care of in this century. Ecological engineering helps restore the natural environment by developing sustainable ecosystems. Developing sustainable natural ecosystems depends on the proper application of basic and applied sciences. Ecological engineering and ecotechnology focus on restoring ecosystems, which are frequently disturbed by anthropogenic activities like environmental pollution, and creating new sustainable ecosystems.

According to the World Commission on Environment and Development, sustainable development maintains a balance between present and future needs. Ecological engineering works on the construction of the natural environment via the application of basic and applied sciences. The environment is a self-designing ecosystem made up of the biological species of the world. The design of self-designing sustainable ecosystems as part of ecological engineering helps achieve goals for discrete projects such as urban flood mitigation. Ecological engineering focuses on the creation, restoration, and management of ecosystems [1]. Stream restoration in this context of ecological engineering is an example of ecological services.

The innovative field known as *ecological engineering* was introduced in the 1960s by ecologist Howard T. Odum [2]. Odum stated concisely that management of nature and, more specifically, a partnership with nature is the fundamental meaning of ecological engineering. Also in the 1960s, the concept of ecological engineering was introduced by Ma Shijun. As per Shijun, ecological engineering is based on the principles of species symbiosis and the cycling and regeneration of substances in an ecological system [1]. The process of ecological research is from systems analysis to system design and construction that mitigates global environmental problems. For example, ecological engineering was used in China for integrated fish culture management because of its attention to using and recycling waste [3].

Ecological engineering is applicable to various natural resource and environmental issues like fisheries, shallow ponds, wetlands, and agriculture, and wastewater treatment with coastline protection (Figure 10.1). In the last case, the applications for wastewater treatment are described as

Handbook of Ecological and Ecosystem Engineering, First Edition. Edited by Majeti Narasimha Vara Prasad.

Figure 10.1 Example of a Maryland stream restoration project that removed trees and tree roots from a coastal stream bank. *Source:* Palmer et al. [4].

"environmentally responsible technology generating little or no sludge, producing valuable by-products, avoiding hazardous chemicals during the process and removing synthetic chemicals from the wastewater" [1].

More specifically, ecological engineering follows these concepts [2, 5]:

1) Most important, it uses the self-designing capacity of ecosystems.
2) It is dependent on ecological theories.
3) It depends on system approaches.
4) Conservation of non-renewable energy sources is essential.
5) Ecological and biological conservation need to be established.

Ecological engineering is known by names such as *ecosystem restoration*, *ecosystem rehabilitation*, *ecotechnology*, *artificial ecology*, *biomanipulation*, *nature engineering* (in Holland), *hydroecology* (in Eastern Europe), and *bioengineering* (originated in Germany) [3]. Ecological engineers construct wetlands, lakes, and rivers, keeping in mind the conservation of the ecosystem. In association with ecologists, engineers work on rivers, lakes, ponds, dams, etc. to manage flooding. In doing so, they should have a proper understanding of the components of the ecosystem and the link or connection between them. Ecological engineering approaches should focus on preserving natural ecosystems and conserving non-renewable energy resources to reduce the adverse effects of pollution [5]. In this context, the evolution of solar-based ecosystems is significant.

10.2 Flooding and, Specifically, Urban Flooding as a Problem of Interest

Conservation of ecosystems and their species depends on the acknowledgment of ecosystem values. Wetlands are directly related to flood control and enhancing water quality in the context of ecosystem values (Table 10.1). Hence wetland protection is of utmost importance [6]. The ability of ecosystems

Table 10.1 Elements of all-inclusive urban flood management.

Pre-flood preparedness	Operational flood management	Post-flood response
Flood risk management, considering all possible causes of flooding	Determining the likelihood of a flood	Relief for the immediate needs of those affected by the disaster
Construction of physical flood defense infrastructure	Forecasting future river flow conditions from hydrometeorological observations	Reconstructing damaged buildings, infrastructure, and flood defenses
Legislation	Issuing warnings to appropriate authorities and the public regarding the extent, severity, and timing of the flood	Recovery and regeneration of the environment and economic activities in the flooded area
Investment in research and development on floods	Emergency rescue of people and property from flooded areas	Review of flood-management activities to improve the process and planning for future events
Development of controls in the flood plains	Emergency protection of levees from breaches and overtopping	
Increasing source control, infiltration, and storage/retardation facilities in urban basins	Strengthening defenses	
Land-use planning and management	Deciding to operate reservoirs and retardation ponds; issuing prior warnings about emergency spills to the people who will be affected	
Implementing building codes and floodproofing		
Implementing flood forecasting and warning arrangements		
Public communication and education about the extent of flood risks and actions to take in a flood emergency		
Disaster contingency planning		
Maintenance of preparedness of community self-protection activities		
Initiating insurance plans		

Source: Re-created based on Kundzewicz and Takeuchi [29].

to self-purify is vital for pollution control and hence ecosystem conservation. Developing environmental and ecological models through the ecological engineering approach helps enhance the self-purification capacity of ecosystems. Clean technology is an essential part of ecological engineering and handles recycling by-products or final waste products during production processes, which is an environmentally friendly approach [3].

Flooding is a type of natural disaster. Every year, flooding causes significant problems in most of the developing countries across the world. The fatality rate in developed countries is significantly lower than in developing ones [7]. The harmful effects of flooding, especially to transportation, agriculture, and health, cannot be ignored. Government and non-governmental organizations have adopted preventive measures for flood control. Engineering approaches to flood prevention are noteworthy. However, the prime concern of engineering should be the conservation and sustainability of natural ecosystems [8].

Climate change plays a significant role in harmful flooding across the world [9]. Heavy rainfall is also a primary cause of flooding, and regions situated in the monsoon climate zone are the most vulnerable. Hence, flooding is believed to be associated with hydrometeorological events [7]. Flooding causes significant hardships primarily in low-income economies in developing countries. Asia is comparatively more vulnerable to flooding as per the global flood damage record [8]. Waterlogging in urban areas creates serious problems for inhabitants related to their social, work, environment, and economic activities [10]. Additionally, flooding imposes enormous economic damage and health hazards in catchment areas. Waterlogging causes significant urban infrastructure issues, especially for roads, railways, communications, housing, and institutions.

During high floods, lakes, canals, and low-lying areas serve as retention basins for water [11]. All of these are part of drainage systems that create hazardous flooding. Adequate flood management is of utmost importance around the world. Flood policies should transition from simple flood control to effective flood management in integrated water management [8]. Integrated water management comprises four basic elements: sustainability, integration, river basins, and interaction. As part of integrated water management, there must be consideration of social and environmental consequences of water management interventions. Because of the significant risk of flooding hazards, vulnerability, and damage in urban areas, sustainable research and development through an ecological approach in the context of urban flood mitigation are very important.

10.3 Causes and Impacts of Urban Flooding

Numerous metropolises and urban areas are located in flood plains. This is attributed to the fertile and flat land, suitable for agriculture and urban development. Cities have a considerable percentage of impervious areas that avert effective infiltration of rainfall into soil. This is the root cause of runoff and high flood levels, as shown in Figure 10.2. This, in turn, leads to inundation and significant flood damage [12]. Urban areas, due to their high population, possess potential economic value and hence, during flooding, may experience unrecoverable disaster.

Urban floods are categorized as floods caused by locally heavy rainfall, overflowing river banks, and high tides or storm surges [8]. Rivers are the primary source of water for domestic, industrial, and irrigation purposes. They also provide convenient means for navigation, transportation, and communication. Insufficient or poor drainage systems trigger floods due to locally heavy rainfall. Floods due to overflow arise when the river level rises above river banks. Extreme river levels are usually caused by runoff from upstream and the backwater effect of high tides at the river mouth. Similarly, the level of rivers is increased by discharge from upstream rivers and the backwater effect instigated by the sea from downstream rivers [13]. However, according to Chowdhury [14], these high water levels are the result of four mechanisms:

- Reduction in infiltration and enhancement surface runoff due to increased impervious areas
- Deviations in drainage flow patterns due to the construction of roads and other infrastructures
- Filling of low-lying areas, normal drainage, and retention systems (canals and lakes)
- Insufficient drainage paths and improper operation and maintenance of the prevailing stormwater drainage system.

Figure 10.2 A typical flood in an urban area (Assam, India). *See colour plate section for colour representation of this figure.*

Flood damage is worse due to the construction of cities in flood plains, which reduces water storage and blocks floodways. Flood dikes in cities can rupture as a result of high flood levels and cause severe damage [11]. In cities in coastal areas, low-lying areas can be drained by pumping. Flood drainage to the sea is hindered by high tides or storm surges, initiating prolonged flooding with polluted floodwater and health complications in cities [15].

Floods are triggered by natural factors or by an amalgamation of natural and human factors. Especially in urban areas, changes in land use, vegetation removal, and increasing impervious rates lead to considerably greater surface runoff volumes [16]. Urbanization significantly disturbs the urban environment or ecosystem because the production and living activities of human society can lead to improper land use. Urbanization transforms wet, vegetation-covered lands into concrete as the result of construction in urban areas [15].

Flood risk is the possibility of loss, which can be expressed as
Risk = Hazard × Vulnerability (Figure 10.3)

Flood hazard relies on flood magnitudes like flood depth, velocity, and duration [17]. *Vulnerability* may be defined as the proneness of a community to the impact of hazards, depending on physical, social, economic, and environmental factors. The vulnerability of people and infrastructure in urban areas is vital for predicting the degree of harm and damage during flooding.

The warm, humid nature of urban areas in Asia has been attributed to factors such as rice paddy agriculture, habitation in flood plains, large populations, rapid urbanization, industrialization, environmental concerns, etc. [14]. Asia is comparatively more vulnerable to flooding than Europe. In developing countries such as Dhaka (Bangladesh), almost 28% of the population lives below the poverty line [18]. Moreover, high urbanization rates have resulted in the development of Dhaka city, which was to a large degree unplanned. The increased rate of urbanization is generally attributed to high rural-to-urban migration, loss of green spaces, changes in land use, and unplanned

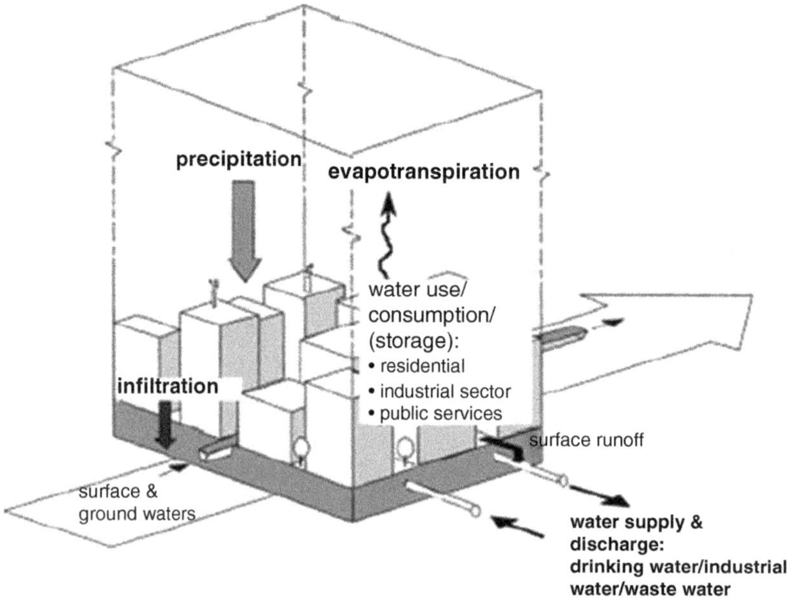

Figure 10.3 Urban water cycle for probable flood risk. *Source:* Tingsanchali [8] © 2012, Elsevier.

urban expansion. Anthropological intrusions such as encroachment on marginal lands, filling bodies of water, waste disposal, etc., have influenced the natural flow and function of rivers and thus caused environmental degradation and a noteworthy increase in flood risk [8]. The conversion of paddy fields (which served as flood-retardation ponds) into residential areas accelerates the incidence of floods.

The natural recharge of groundwater has been reduced significantly because of demands for water and an increase in impermeable surfaces [14], leading to greater runoff. Thus damage in an urban context is multifaceted, as it damage is directly proportional to vulnerability and flood type, especially in terms of depth, flow velocity, and duration. The large population and significant assets account for the increase in urban flood damage. Cities in many developing countries are growing rapidly. Large populations in cities due to rural-to-urban migration are related to increasing human settlements, industrial growth, and infrastructure development. Due to limited space, urban growth in flood plains expands over floodways, reducing flood drainage and channel capacity.

The risk of flooding is dependent on climate change and variability. As mentioned earlier, heavy rainfall is a significant cause of floods. As per an; intergovernmental panel on climate change, rainfall intensity increases with greater greenhouse gas concentrations [19]. Intense rainfall may increase runoff, erosion, and sediment yield, blocking drainage systems in urban areas. This, in turn, contributes to the elevated risk of floods. The magnitude and timing of runoff and the intensity of floods and droughts are influenced by the nonlinear nature of evaporation, vegetation, and soil-moisture variations. Greater runoff, vigorous erosion, and heavy transport of water, sediments, and other materials are also caused by an increase in the proportion of impervious area, deforestation, and channel enhancement, e.g. straightening of water passages [8]. All of the reasons for urban flooding are inter-linked.

In mountainous areas, development of residential areas, deforestation, road construction, etc. extends to hilly slopes subjected to landslide and debris flows. Some floods are also attributed to failures of manmade infrastructure (e.g. bursting dams and river dikes).

Flooding has the most extensive effect when its extent and duration are great. For example, the 1998 flood in Bangladesh was the most dangerous in terms of extent and duration. It was the result of both river water and stormwater. The city was affected by flooding for more than two months [20]. Extreme runoff, leakage of the incomplete flood wall, failure of sluice gates, spilling from rivers, and floodwater invasion through open underground pipes and drains were the primary causes of the Dhaka flood [21]. Moreover, the entry of floodwater through unblocked culverts and unclosed regulators contributed to flooding. Intensive rainfall in this context is a significant source of stormwater flooding. River floods are either rainfall-induced or snowmelt induced or a mixture of both. For instance, floods in deltas are jeopardized by storm surges, ice jam, and dam breaks [15].

In urban areas, especially, flooding creates significant health hazards and leads to economic losses and property damage. During flooding, water and waterlogging create problems for urban transport and communication infrastructure including roads, railways, communication lines, formal and informal housing, and institutions. Flash floods are a source of suspended sediments, heavy metals, hydrocarbons, nutrients, and water-borne pathogens, resulting in water pollution [22]. Hence it is clear that flood protection and prevention must be included in the political agenda of every country. Direct and indirect economic losses result from urban floods.

Developing cities have had to manage floods and drainage systems since ancient times. Drainage systems are always of prime concern in the context of city infrastructure and urban life. The failure of drainage systems in urban areas leads to flooding, possible environmental degradation, sanitation and health problems, and disruption of city services [16]. Hence the impacts of urban floods can be categorized as physical, economic, social, and environmental effects. Better land use can prevent both direct and indirect primary potential losses, which will impact potential secondary losses [15]. But better flood emergency response mechanisms are effective against potential secondary losses. On the other hand, in rural areas, damage due to floods is mostly related to the loss of agricultural production. Urban floods are deadly: they account for high death tolls in addition to significant property damage. The number of fatalities and the strength of typhoons causing floods are directly related. Landslides (including steep slope collapses) and debris flow can also cause major destruction in urban areas [8].

Rising anthropopressure impacts in terms of flood risk and losses are noteworthy. Pressure on the environment from anthropogenic activities like massive deforestation, urbanization, and river regulation decreases the available water storage capacity and accelerates flood flows [23]. The risk of flooding is increasing because ecosystem conservation is ignored during the engineering of flood protection and management systems. Based on this discussion of the causes and impacts of floods, it can be seen that flooding is a significant and emergent threat to societies. Hydrologists, engineers, and water management workers have significant responsibilities with respect to flood preparedness, flood control, and mitigation. As a whole, there is a necessity for holistic flood management approaches during the entire flood cycle.

10.4 Protection Against and Mitigation of Urban Flooding in the Context of Sustainability

Sustainability in water resource development is the long-term protection of the most important value that the water system delivers to society, including social, economic, and environmental value. The focus of sustainable water governance is on long-term water resource policy, flood management, and spatial planning [24].

10.4.1 Living with Floods as a Sustainable Approach

The phrase "living with floods" means adjusting to and living with flooding, as the complete elimination of flooding is impossible. There have been several times in history when people have changed their lifestyle and societal structure to cope with the natural disaster of floods. For example, farmers relying on regular seasonal floods have to plan their farming processes. Accordingly, villagers used to arrange boats in their houses to be used during flooding. Homemade constructions such as circular dikes also help people during flooding [25]. In this way, individuals learn to remain in place during floods or protect themselves during floods.

As it is not possible for a flood protection system to provide total safety, it is essential to live with an awareness of the possibility of floods. In this context, developing flood consciousness and communicating risks to the public are essential. Moreover, as the magnitude of flooding is random, it is also difficult to have protection devices accordingly. In this context, people can try to determine the probable maximum flood (PMF) in a particular region [24].

10.4.2 Urban Flood Risk Management

Flood risk cannot be avoided entirely, so it needs to be managed. Flood management is related to flood mitigation rather than the elimination of flood hazards. Urban flood risk management focuses on minimizing human loss and economic damage. Tasks must be assessed before and after their implementation to properly evaluation their effectiveness. So in urban flood risk management, assessing risk before and after implementing flood-mitigation measures is essential [17]. Risk-assessment plans must consider all the possible water-related hazards in an integrated fashion (Figure 10.4).

Flood management is based on a few steps as part of a management cycle at both the individual and societal levels. The necessary steps are awareness, readiness, reaction/response, and recovery/rehabilitation [24]. Preparedness measures work to prevent potential risks that, in turn, can become disasters. Awareness works to mitigate flood risks to a satisfactory and reasonable level and develop activities for dealing with residual risks. Flood fighting is a type of reactive or response measure adopted by people during flooding to decrease flood damage [26]. Response measures are generally implemented during or immediately after a flooding incident. To respond to an emergency, one should ensure advance planning and preparedness. Recovery and rehabilitation are critical steps after the flood impact.

In developing countries, flood disaster management consists mostly of reactive responding: emergency responses and recovery to deal with disaster situations. Independent efforts are implemented without proper coordination or integration. There needs to be a transition of reactive responses to proactive responses to manage and control lives and property. Proactive disaster management involves integrated flood management [27] and the interaction between governmental,

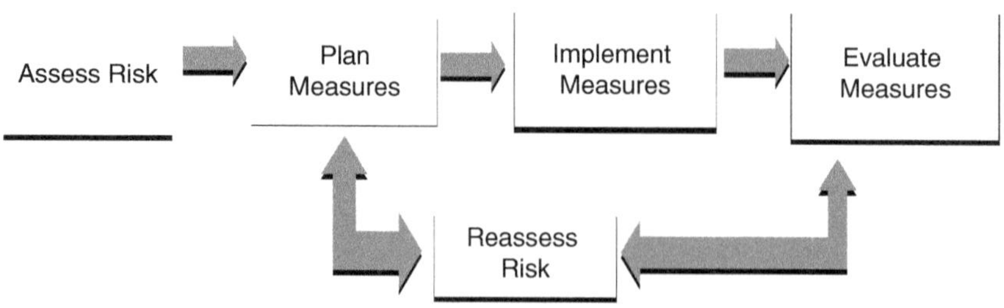

Figure 10.4 Steps in the risk management process. *Source:* Redrawn based on WMO/GWP [17].

non-governmental, and private agencies and the public sector. Obviously, it requires a higher budget, equipment, facilities, human resources, and effort and time. Hence, with respect to the ecological concerns of society, integrated flood disaster management is based on a strategic framework: (i) awareness before flooding, such as flood forecasting and warnings; (ii) readiness when a flood occurs; (iii) emergency measures during flood impacts; and (iv) recovery and rehabilitation after flooding [27, 28]. Urban flood disaster and risk management must implement related concepts, policies, plans, and tasks.

Proper urban planning and design are very important in urban flood mitigation as they provide more effective economic solutions than outdated approaches like developing exposed areas and then protecting them somehow [30]. During urban planning, there should be a combination of flood models with urbanization scenarios. Moreover, reservoir operation development planning is an essential preparedness step for floods. By increasing awareness of flood devastation, online distributed flood forecasting models can provide valuable information and necessary education [31]. Assessment and improvement of flood hazard preparedness can be achieved by the regionalization of flood estimation methods. *Regionalization* estimates regional floods during different return periods.

Hydraulic and engineering factors and socioeconomic and environmental features are fundamental for successful flood control and risk management in ecological and ecosystem engineering approaches. Inclusive and integrated water and land management are essential for flood and drought management and ecological and environmental quality control [15]. Cooperation between stakeholders and concerned authorities such as urban planners, civil and water resources engineers, civil disaster defense authorities, health and social services, etc., is vital for effective flood management [26].

10.4.3 Integrated and Interactive Flood Management

An integrated approach to mitigating flood damage under a comprehensive flood control management strategy is necessary. As per integrated flood management, flood control should focus on the river basin as a whole, not just the river, in enhancing its retardation capacity [27, 28]. In this context, in urban river basins, flood retardation and infiltration services are introduced in addition to drainage functions. The amount of flood discharge should not increase as the result of new development in urban areas. This is also known as the *law of discharge conservation* [17]. The storage and retardation functions of vanishing paddy fields can be replaced by district disaster-prevention ponds and rainfall storage. Additionally, the infiltration capacity of natural forests, grass, and farmlands can be achieved by infiltration facilities at individual buildings. Park and school playgrounds can be used on an emergency basis for flood retardation.

The modern, interactive approach to managing floods has been adopted by countries worldwide. This framework's primary factors are integration, river basin approaches, sustainability, and interaction [33]. The institutional arrangements for flood management are dependent on the above factors of interactive water management. Organizations associated with water management processes should not independently decide on plans, projects, and policies in a democratic society. Instead, they should follow an interactive water management approach by involving the stakeholders in public participation processes [8] (Figures 10.5 and 10.6). Replacing experienced water authorities with inexperienced people is also is a matter of concern as it can weaken societal efforts against floods.

To mitigate urban floods, stormwater retention measures are necessary. The construction of basins or ponds can preserve stormwater by temporarily storing surface runoff and releasing it at a measured rate [34]. In addition, preserving green spaces in urban areas by increased infiltration and evapo-transpiration from catchment areas can successfully reduce surface runoff.

10.4.4 Structural and Nonstructural Measures for Flood Control

Structural measures for flood control generally involve construction. However, such construction is often opposed by environmental groups. This type of criticism is due to a lack of ecological and ecosystem protection concerns in engineering. Engineering projects related to flood control include construction of retardation reservoirs, flow-diversion canals, dams, levees, embankments, river dikes and flood diversions, floodways, expansion of river channels, upstream sediment control, and even removing vegetation from rivers [8, 24]. Reservoirs store excess water or runoff, which results in uniform distribution of streamflow over time. Thus the construction of a reservoir is one remedy to relieve the flood problem [23]. Even unregulated flood-control reservoirs, which generally retain a portion of floodwater, can conquer the destructive flood peak. The role of checking dams is to control the discharge of rocks, sand, and sediment through storage. However, structural flood-mitigation measures alone can't guarantee absolute protection.

Structural approaches to flood control may conflict with ecological factors. However, nowadays, water authorities related to government and non-government organizations pay attention to preserving the environment's ecological value and, most important, the ecosystem when implementing structural flood-mitigation projects [33]. Alternative means include nonstructural measures. Some flood-protection measures are criticized during sustainable development because flood-control solutions should not cause unacceptable disturbances in ecosystems, to protect future generations [35].

Prevention and mitigation of flood losses include structural flood-control measures such as dams or river dikes and nonstructural measures such as flood forecasting and warnings, flood hazard and risk management, public participation, institutional arrangements, and evacuation by community self-protection teams [26, 33]. Mitigation of urban floods due to river bank overflow can be achieved by implementing structural measures to detain or divert runoff. In this context, structural measures like polders and basins, bypass channels, dredging and widening streams and channels, levees, and embankments work satisfactorily [8, 24]. Tidal barriers are employed as preventive measures against floods due to high tides. However, natural tsunamis and storm surges cannot be eliminated. In such cases, flood disaster management plays a significant role in mitigation [36].

There exist numerous retardation facilities as flood-mitigation strategies for small urban basins. All of these have been developed while keeping in mind the protection of the ecosystem. They include district ponds, house and building storage, emergency use of school playgrounds and parks, and underground retardation and drainage pipes [19]. Infiltration facilities such as ponds, conduits with infiltration holes, permeable pavement, etc. are also effective. Better drainage facilities, including drainage pumps in lowlands, floodproofing by elevating house foundations, and district walls protecting subways from flood influxes, also serve as lifelines in the context of urban flood mitigation [15, 20].

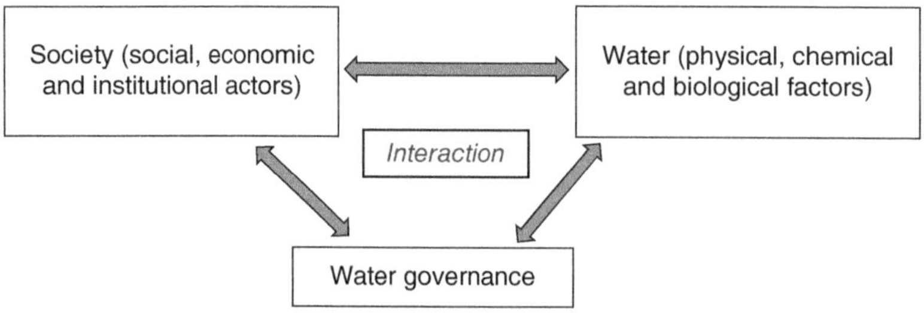

Figure 10.5 Cooperative water management. *Source:* Redrawn based on Van Ast and Boot [32].

Nonstructural measures for flood mitigation teach people to live with and accommodate flooding rather than try to eliminate it. These measures also decrease vulnerability to flood damage. A few of the important nonstructural measures are zoning, building codes, floodproofing, flood forecasting and warnings, evacuation, institutional arrangements, and flood hazard areas with flood risk maps and models [8, 35]. As a precaution, only low-value infrastructure should be built on flood plains. An effective part of nonstructural measures is interactive or integrated flood management, as discussed in detail earlier in this chapter. Almost all of these nonstructural measures help conserve the ecosystem.

Generally, surplus water is stored by polders or flood-retardation ponds. These act as off-stream reservoirs consisting of a flood plain and a dike. Resources of flood plains contribute to the restoration and enhancement of wetlands [25]. In some risk areas, permanent evacuation of the existing infrastructure from flood-prone areas is an effective nonstructural measure. However, an evacuation plan should include evacuation shelters and supplies of fresh water, food, medicine, etc., for affected people [35]. Evacuation plans should be developed in close cooperation with the affected people to reduce their physical vulnerability. Restoration of flood plains is possible without unnecessary development. Also, proper care should be taken, such as impeding floodwater, floodproofing, and relocating hazardous locations to minimize damage in case of ongoing development [7, 24]. People in flood-prone areas should be aware of preventive measures like floodproofing, land-use zoning, dike protection, and evacuation plans. In earlier days, local radio networks and community patrols were the only the available media for warnings and preparedness or flood forecasting [8].

An important nonstructural measure is building codes, which play a significant role in decreasing the physical vulnerability of houses and infrastructure. Building codes provide guidelines concerning construction in flood-prone areas, such as the type of construction materials used and/or the structural features of the construction, so that efficient mitigation plans may be implemented [26]. Building codes also provide information about the occupancy and use of houses.

Flood warning systems can be made effective for communities at risk by transforming forecast information into practice and disseminating warnings to people. However, the performance of flood warning and forecasting systems depends on people's understanding of flood risk and their awareness of emergency response to incoming floods [23]. Moreover, quantitative precipitation forecasts are a potential subject for study and development.

In the context of ecosystem preservation, source controls like watershed management that include land use and soil conservation minimize surface runoff, erosion, and sediment transport as a flood-mitigation measure [37]. River protection and re-naturalization are also important nonstructural tasks implemented in developed countries for flood mitigation [15].

Another more interesting nonstructural measure is flood insurance. Its basic concept is to distribute flood risks and losses among a wide range of individuals over a long period of time [8]. An effective flood-mitigation system consisting of forecasting, warnings, dissemination of public information, evacuation, relief, and post-flood recovery can substantially reduce losses.

10.4.5 River and Wetland Restoration

Artificial restoration of a comprehensive hydrological cycle is a useful step for the benefit of the environment. This can be made possible by improving the retardation and infiltration of rainfall in urban areas and river basins [38]. This approach provides an efficient solution for both ecological and water resource requirements. Management of a sound hydrological cycle also supports the implementation of nature-oriented river improvement works.

River and wetland restoration plays a significant role in environment and ecosystem conservation. In Japan, the River Law was passed by the Parliament in 1997 and declared for the first time that the river environment should be improved and conserved and that the green belts around reservoirs and rivers are under the regulation of the law [39].

Restoration of rivers and wetlands is frightening in regions worldwide with high human population density, urbanization, and harsh environmental conditions [39]. To protect human life and support economic growth, river regulation and flood control have long been recognized as serious issues. The comparable and integrated importance of environmental, flood control, and water resource management issues needs to be recognized [25]. River regulation and flood control can be achievable by developing river basin flood programs, building higher continuous levees, and constructing dikes and floodproof buildings [40]. Engineering activities during urban development and anthropogenic impacts such as river engineering, agricultural development, and incursion by non-native species contribute significantly to the degradation of aquatic ecosystems [41].

Habitats of aquatic living organisms are in danger because of agricultural modernization that transforms wet paddy fields into dry paddy fields [42]. So, restoring wet paddy fields is an important task. Traditional paddy fields support many aquatic species, increase nutrient retention, control floods, and recharge groundwater. In the context of modern agricultural development, engineering approaches such as reconnecting rice paddies with channels and restoring their hydrological conditions preserve ecological and ecosystem value [43]. For effective flood management, most downstream river sections in Japan are leveed (Figure 10.7), and most riparian forests have been transformed into paddy fields [44]. Moreover, several riparian forests have been cut to improve the channel's flood-routing capacity as part of stream restoration.

High, wide levees along major rivers protect urban areas. Urban buildings and traffic facilities are developed on the levees, resulting in the redevelopment of urban areas with increasing value [8, 45].

The amount of river engineering work and risk management required by ecologists, civil engineers, and river managers is significantly reduced by nonstructural actions such as developing flood hazard maps, local flood-fighting corps, and flood warning systems [8, 35]. The concept of river management by ecological restoration has been implemented and accepted by countries worldwide. Urban restoration can increase the quality of human life (e.g. recreational value). However, the framework for many urban restoration projects depends on the successful integration of ecological, economic, and social factors [15, 25].

Numerous valuable ecosystem services, such as water clarification, climate regulation, and flood mitigation, are dependent on ecosystems of freshwater wetlands, including shallow lakes, river flood plains, marshes, and rice paddies. These types of ecosystems also serve as habitats for many animals and plants [6, 46]. Densely populated watersheds like freshwater wetlands are significantly affected by anthropogenic activities. So, implementing and planning for restoration of ecological wetlands such as lakes has been undertaken by ecologists and engineers, as mentioned previously. Conservation and recovery of wetland biodiversity by restoring degraded wetland ecosystems are critical global issues. In this context, a novel approach of revegetation is worth mentioning: with this technique, soil seed banks are employed as the source of plant recolonization [47] (Figure 10.8). Restoration of lakeshore vegetation using such a technique at Lake Kasumigaura, Japan, in 2002 furnishes an excellent example in this context.

It is anticipated that soil seed banks from past and present wetlands contain species and genetically diverse entities that can help conserve and restore the biodiversity of wetlands. Based on this, soil seed banks have been used as plant materials as a common revegetation technique for wetland restoration.

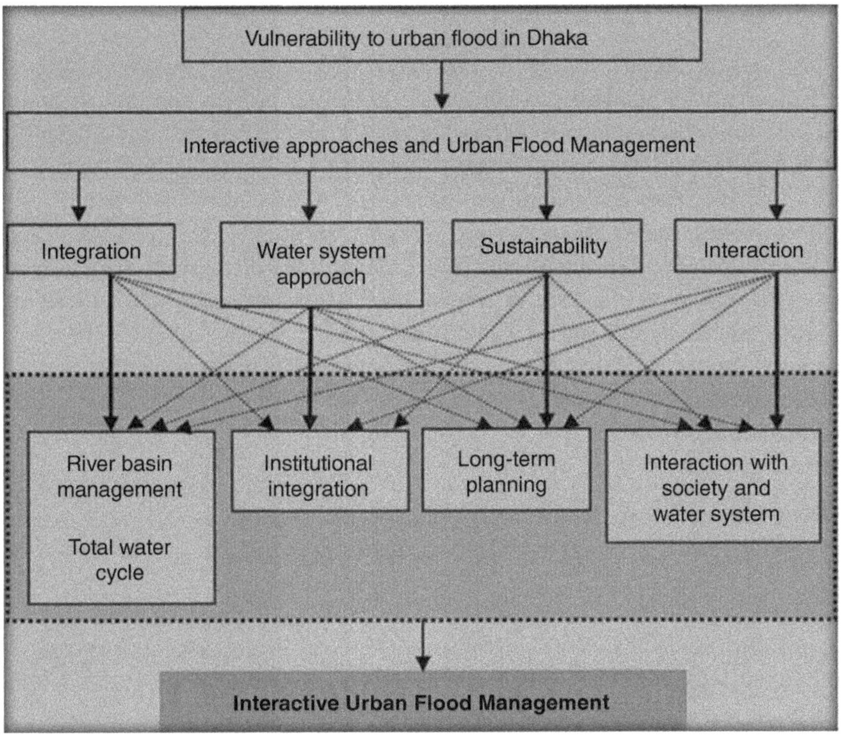

Figure 10.6 Cooperative urban flood management. *Source:* Redrawn based on Barua and Van Ast [33]).

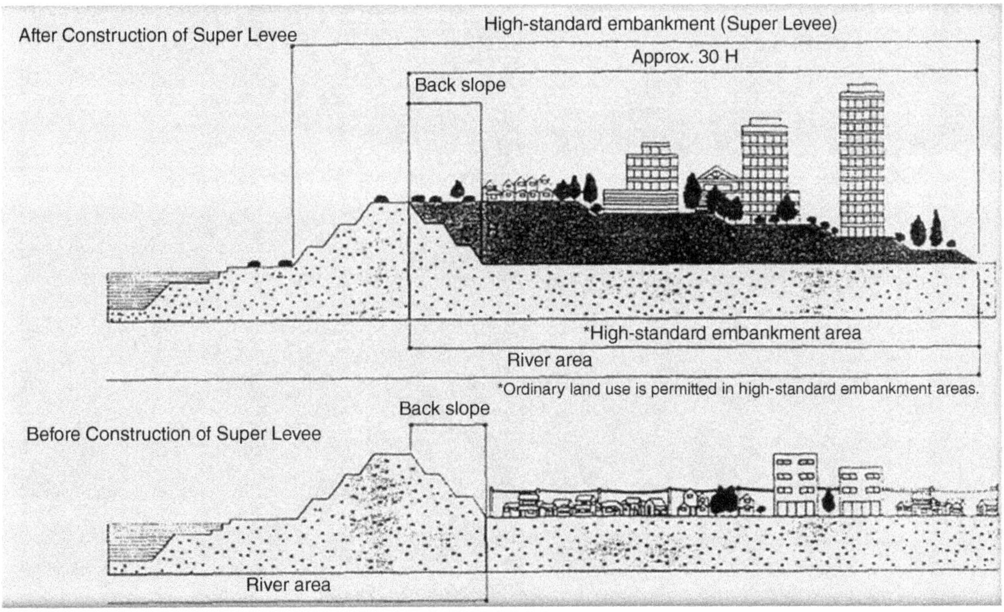

Figure 10.7 Example plan of a levee. *Source:* Luo, P., He, B., Takara, K., et al. (2015). "Historical assessment of Chinese and Japanese flood management policies and implications for managing future floods." *Environ. Sci. Pol.* 48: 265–277. © 2015, Elsevier.

10.4.6 Low Impact Development (LID) and Best Management Practices (BMPs)

The structural and nonstructural measures adopted to weaken the hydrological response to urbanization and decrease the jeopardy of flood disasters are collectively referred to as best management practices (BMPs) and low-impact development (LID) approaches. LID techniques can efficiently reduce the risk of flood disasters, pollution, and urban runoff [48]. LID practices increase urban greenbelt areas, water areas, and water permeability, thereby protecting a city's ecological environment [49]. Several LID practices, such as bioretention, pervious pavements, and grassed swales, can expand the area of green land and water surface [50]. This, in turn, increases the amount of pervious surface area. The foundation of LID is, of course, BMPs. LID approaches, in connection with BMPs, can efficiently restore hydrological mechanisms to their original state when they are damaged by, for example, stormwater-related disasters [48]. Restoration is made possible by reducing impervious surface area via LID design strategies such as bio-retention, penetration laid, green roofs, rain gardens, and vegetated swales [50].

Briefly, BMP-influenced LID practices are based on the following four principles:

1) Reduce surface runoff, minimize impervious rates, and maintain green areas.
2) Increase flow paths and surface roughness to preserve concentration times of pre-development discharge volumes.

Figure 10.8 Example plan of a lakeshore restoration site. *Source:* Prosser, D.J., Jordan, T.E., Nagel, J.L. et al. (2018). Impacts of coastal land use and shoreline armoring on estuarine ecosystems: an introduction to a special issue. *Estuaries and Coasts* 41: 2–18. Licensed under CC BY 4.0.

3) Use retention reservoirs to control peak discharge and increase water quality.
4) Use additional detention reservoirs to prevent flooding, if needed.

These postulates are also applicable for essential structural flood-mitigation measures and sustainable drainage systems in urban areas [15]. Sustainable drainage systems also belong to the category of LID practices. To achieve a sustainable drainage system in the context of integrated city development, sustainability should focus on related fields such as hydraulic engineering, architecture and urbanism, city planning and management, social disciplines, and economic affairs. The traditional method of using drainage systems as a standard sanitation measure to carry stormwater and wastewater is unsustainable.

A sustainable drainage system should be able to [51, 52]:

- Control runoff amounts and rates, thereby reducing the effect of urbanization on flooding
- Boost natural groundwater recharge
- Ensure water quality
- Improve the facility and cleanliness of developed areas
- Create an environment for wildlife in urban areas, thereby developing biodiversity
- Address environmental and local community requirements

Enhancing flow generation and end-of-pipe solutions are essential parameters of a sustainable drainage system. Technical measures (hydraulics aspects) are not sufficient for sustainable drainage systems; rather, integrated urban hydrological cycle management is essential [15]. Solutions for drainage systems are sustainable only if they account for urban renewal, river restoration, and improved quality of life. Acceptance, support, and continuity of these solutions need to be obtained via institutional arrangements.

10.5 Conclusions and Future Scope

Worldwide urbanization is unstoppable. This unavoidable urbanization causes deterioration of the urban ecosystem as a result of anthropogenic activities. A natural disaster like an urban flood is a direct result of the urbanization process. Cities commonly face the challenges of urban flood mitigation, and mitigation measures must achieve ecological balance or, more specifically, concentrate on protecting the ecosystem. Internationally, water authorities have adopted sustainable structural engineering and nonstructural flood-control measures in this context. Structural measures cannot guarantee the end of flooding. However, nonstructural measures at least teach people to become accustomed to flooding. Impacts on natural systems can be limited with the help of principles from restoration science and ecological engineering.

The problem of urban flooding is gaining attention because of its many disadvantages for city growth. Urban flood risk is a combination of flooding hazards and flood vulnerability. The impacts of floods in urban areas are significant and include health hazards, economic loss, and infrastructure damage. It is essential to implement urban flood risk management. Interactive water management and integrated urban flood management strategies work effectively in the context of sustainable urban flood mitigation. Institutionalized connections between flood monitoring authorities facilitate cooperative planning. Community participation in flood risk management plans should be given priority. To improve interactive flood management, one can identify relevant stakeholders, work to achieve more public participation, and provide education about flood-related issues. Flood-mitigation measures such as green roofs, rainwater harvesting or underground storage, rain gardens, and permeable pavements can reduce runoff and are environment friendly. Increasing green areas can contribute to quality of life in urban spaces.

The development of infiltration devices minimizes annual runoff and surface water issues. Rehabilitation and re-excavation of encroaching rivers, canals, and wetlands are valuable techniques for flood mitigation. Conserving and recovering wetland biodiversity can be achieved by restoring degraded wetland ecosystems. Low-impact development (LID) strategies based on best management practices (BMPs) increase urban greenbelt area, water area, storage capacity, and water permeability, thereby protecting cities' ecological environment. LID in connection with BMPs can efficiently restore damaged hydrological mechanisms to their original state. Sustainable urban drainage systems (SUDs) can significantly contribute to making cities and their populations more resilient.

References

1 Mitsch, W.J. (2012). What is ecological engineering? *Ecol. Eng.* 45: 5–12.

2 Odum, H.T. and Odum, B. (2003). Concepts and methods of ecological engineering. *Ecol. Eng.* 20: 339–361.

3 Mitsch, W.J. and Jørgensen, S.E. (2003). Ecological engineering: a field whose time has come. *Ecol. Eng.* 20: 363–377.

4 Palmer, M.A., Filoso, S., and Fanelli, R.M. (2014). From ecosystems to ecosystem services: stream restoration as ecological engineering. *Ecol. Eng.* 65: 62–70.

5 Bergen, S.D., Bolton, S.M., and Fridley, J.L. (2001). Design principles for ecological engineering. *Ecol. Eng.* 18: 201–210.

6 Mitsch, W.J. and Gosselink, J.G. (2000). *Wetlands*, 3e. New York: Wiley.

7 Biswas, A.K. (2006). Water management for major urban centers. In: *Water Management for Large Cities* (eds. C. Tortajada, O. Varis and A.K. Lundqvist), 3–17. London/New York: Routledge.

8 Tingsanchali, T. (2012). Urban flood disaster management. *Procedia Eng.* 32: 25–37.

9 Mirza, M.M.Q. (2002). Global warming and changes in the probability of occurrence of floods in Bangladesh and implications. *Glob. Environ. Chang.* 12: 127–138.

10 Alam, M. and Rabbani, M.G. (2007). Vulnerabilities and responses to climate change for Dhaka. *Environ. Urban.* 19: 81–97.

11 Khan, M.S.A. (2006). Storm water flooding in Dhaka City: causes and management. *J. Hydro. Meteor.* 3: 77–85.

12 Pauleit, S. and Duhme, F. (2000). Assessing the environmental performance of land cover types for urban planning. *Landsc. Urban Plan.* 52: 1–20.

13 Islam, I. (2008). Wetlands of Dhaka metro area: A study from social, economic and institutional perspectives. PhD thesis, J.S.P. S. Japan Coll. Pol. Sci. Ritsumeikan University.

14 Chowdhury, N.T. (2010). Water management in Bangladesh: an analytical review. *Water Policy* 12: 32–51.

15 Miguez, M. G., Veról, A. P., and Carneiro, P. R. F. (2012). Sustainable Drainage Systems: An Integrated Approach, Combining Hydraulic Engineering Design, Urban Land Control and River Revitalisation Aspects. In: Prof. Muhammad Salik Javaid (Ed.), Drainage Systems. InTech publishers, ISBN: 978-953-51-0243-4.

16 Lamond, J. E., Rose, C. B., and Booth, C. A. (2014). Evidence for improved urban flood resilience by sustainable drainage retrofit. *Proc. Inst. Civil Eng. Paper 1300022*, 1–11.

17 WMO/GWP (2008). *Urban Flood Risk Management, a Tool for Integrated Flood Management*. Associated Programme on Flood Management, World Meteorological Organization.
18 Haq, K.A. (2006). Water management in Dhaka. In: *Water Management for Large Cities* (eds. C. Tortajada, O. Varis, J. Lundqvist and A.K. Biswas), 107–127. London/NewYork: Routledge.
19 Kundzewicz, Z.W. (2011). Global change and flood risk management. Large-scale Flood Report.
20 Huq, S. and Alam, M. (2003). Flood management and vulnerability of Dhaka City. In: *Building Safer Cities: The Future of Disaster Risk, Disaster Risk Management Series* (eds. A. Kreimer, M. Arnold and A. Carlin), 121–135. Washington, D.C.: The World Bank.
21 Jahan, S. (2000). Coping with flood: the experience of the people of Dhaka during the 1998 flood disaster. *Aust. J. Emerg. Manag.* 15: 16–20.
22 Wright, D.B., Smith, J.A., Villarini, G., and Baeck, M.L. (2012). Hydroclimatology of flash flooding in Atlanta. *Wat. Resour. Res.* 48: W04524.
23 Merz, B., Kundzewicz, Z.W., Delgado, J. et al. (2012). Detection and attribution of changes in flood Hazard and risk. In: *Changes in Flood Risk in Europe*, Special Publication No. 10 (ed. Z.W. Kundzewicz), 25. Wallingford, Oxfordshire, UK, Ch: IAHS Press.
24 Tingsanchali, T. Flood disaster and risk management. *Invited Paper, Proc. 1st EIT Inter. Confer. Wat. Resour. Eng.* Petchaburi, Thailand, August, 2011, 15–24.
25 Chowdhury, J.U. (2008). Integrated water resources management: Concept and actions needed for Bangladesh. Inst. Water & Flood Manag. BUET, Dhaka.
26 Menzel, L. and Kundzewicz, Z.W. (2003). Non-structural flood protection – a challenge. ECO-FLOOD Conf.: Towards Natural Flood Reduction Strategies. Warszawa, 6–13 September.
27 Associated Programme on Flood Management, (APFM). Integrated flood management – Concept paper. *WMO Series No. 1047, Geneva, Switzerland*, 2009, 27.
28 Associated Programme on Flood Management, (APFM). Integrated Flood Management, APFM Technical Document No. 1, Concept Paper. *APFM-WMO, Geneva, Switzerland*, 2004.
29 Kundzewicz, Z.W. and Takeuchi, K. (1999). Flood protection and management: quo vadimus? *Hydrol. Sci. Jour.* 44: 417–432.
30 Moll, R. (41-48, 2005). Co-operation within Europe on flood management and spatial planning. In: *Urban Flood Management* (eds. A. Szöllösi-Nagy and C. Zevenbergen). A.A. Balkema Publishers.
31 Herath, S. Geographical information systems in disaster reduction. *Inst. Indus. Sci. The University of Tokyo, Japan,* 2001.
32 Van Ast, J.A. and Boot, S. (2003). Participation in European water policy. *Phys. Chem. Earth* 28 (12–13): 555–562.
33 Barua, S. and Van Ast, J.A. (2011). Towards interactive flood management in Dhaka, Bangladesh. *Water Policy* 13: 693–716.
34 Chanan, A. and Woods, P. (2005). Managing the water cycle in Sydney metropolitan: local governments do matter! In: *Integrated Concepts in Water Recycling* (eds. S.J. Khan, A.I. Schäfer and M.H. Muston).
35 Kundzewicz, Z.W. (2002b). Nonstructural flood protection and sustainability. *Water Inter.* 27: 3–13.
36 Tingsanchali, T.; Keokhumcheng, Y. Flood damage functions for surrounding area of Second Bangkok International Airport. *Proc. Intern. Sym. Urban Safe. Mega Cities in Asia.* Phuket, Thailand, November, 2006, 291–300.
37 Mirchi, A., Watkins, D. Jr., and Madani, K. (2009). Modeling for watershed planning, management, and decision making. In: *Watersheds: Management, Restoration and Environmental* (ed. J.C. Vaughn), 1–25. Nova Science Publishers, Inc.
38 Nakamura, K., Tockner, K., and Amano, K. (2006). River and wetland restoration: lessons from Japan. *Bio-Sci.* 56: 419–429.

39 Yoshikawa, K. (2004). On the progress of river restoration and the future view in Japan and Asia. Third European Conference on River Restoration, Zagreb, Croatia, 17–21 May.

40 Takeuchi, K. (2002). Flood management in Japan—from rivers to basins. *Water Int.* 27: 20–26.

41 Yoshimura, C., Omura, T., Furumai, H., and Tockner, K. (2005). Present state of rivers and streams in Japan. *Riv. Res. Appl.* 21: 93–112.

42 Sato, H. (2001). The current state of paddy agriculture in Japan. *Irrig. Drain.* 50: 91–99.

43 Hata, K. (2002). Perspectives for fish protection in Japanese paddy field irrigation systems. *Jpn. Agric. Res. Q.* 36: 211–218.

44 Nakamura, F., Jitsu, M., Kameyama, S., and Mizugaki, S. (2002). Changes in riparian forests in the Kushiro mire, Japan, associated with stream channelization. *River Res. Appl.* 18: 65–79.

45 Kundzewicz, Z.W. (2002a). Flood protection in the context of sustainable development. In: *The Extremes of the Extremes: Extraordinary Floods* (eds. A. Snorrason, H.P. Finnsdottir and M.E. Moss), 351–366. Wallingford, Oxfordshire: IAHS Publ.

46 Van der Valk, A.G. (2006). *The Biology of Freshwater Wetlands*. Oxford, UK: Oxford University Press.

47 Nishihiro, J., Nishihiro, M.A., and Washitani, I. (2006). Restoration of wetland vegetation using soil seed banks: lessons from a project in Lake Kasumigaura, Japan. *Landsc. Ecol. Eng.* 2: 171–176.

48 Clary, J., Quigley, M., and Poresky, A. (2011). Integration of low-impact development into the international Stormwater BMP database. *J. Irrig. Drain. Eng. ASCE* 137: 190–198.

49 Gomez-Ullate, E. (2011). Analysis and contrast of different pervious pavements for Management of Storm-Water in a parking area in northern Spain. *Water Resour. Manag.* 25: 1525–1535.

50 Gao, C., Liu, J., Zhu, J., and Wang, Z. (2013). Review of current research on urban low-impact development practices. *Res. J. Chem. Environ.* 17: 209–214.

51 Langenbach, H.; Eckart, J.; Schröder, G. Water Sensitive Urban Design – Results and Principles. *Proc. 3rd SWITCH Sci. Meet.* Belo Horizonte, Brazil, 2008.

52 Wong, T.H.F. (2006). Water sensitive Urban Design – the journey thus far. *Aust. J. Water Resour.* 10: 213–222.

11

Ecological Engineering and Restoration of Mine Ecosystems

Marcin Pietrzykowski

Department of Forest Ecology and Silviculture, Faculty of Forestry, University of Agriculture in Krakow, Krakow, Poland

11.1 Background and Definitions

Mining for coal and lignite, oil shale, mineral sands and rock, sulfur, and metallic mineral deposits affect all ecosystem components through diverse geomorphological, hydrological, and chemical impacts. Globally, coal and lignite in particular play an essential role in the energy mix, and post-mining landscapes are large-scale transformations and disturbances of the ecosystem and are of worldwide interest and concern [1]. In regions with intensive mining, the landscape is changed by ongoing operations (Figure 11.1) and also by closed-down mines with waste rock dumps, contaminated sites, brownfields, and underutilized fallow land (Figure 11.2). The pre-mining landscape is transformed into a new technogenic environment: the post-mining landscape, which is often far from a state of equilibrium and ecologically fragile [2, 3].

In addition to the site-related impacts of mining, numerous indirect effects put pressure on the surrounding landscape. Among these *off-site impacts,* disturbance of watersheds and pollutant emissions are most important and have ecological consequences far beyond the end of the mining operation.

All these aspects are discussed in numerous books and scientific journals, but the most important issue is the possibility of ecosystem reconstruction, restoration, or rehabilitation. Activities aimed at improving damaged ecosystems are variously referred to as *reclamation*, *restoration*, *rehabilitation*, *remediation*, *mitigation*, etc. [1]. In practice, these terms are often used synonymously. Nevertheless, their definitions vary by author (e.g. scientific literature, R&D projects) or by regulatory and implementing agencies (e.g. legislation, regulations, technical reports). According to the SER [4], *restoration* is defined as the process of assisting the recovery of an ecosystem that has been degraded, damaged, or destroyed. For example, according to the Surface Mining Control and Reclamation Act of 1977 (United States Congress), *reclamation* includes restoration of the local topography and soils using technical and biological methods. It is recommended, if possible from a technological point of view, that the top layer of natural soil removed from the overburden should be used to cover the tipped rock (this is called *topsoiling*). A number of agrotechnical treatments follow, and finally revegetation, i.e. the restoration of vegetation cover. The type of vegetation depends on the local conditions and natural plant formations originally found in the mined territory.

Handbook of Ecological and Ecosystem Engineering, First Edition. Edited by Majeti Narasimha Vara Prasad.

Figure 11.1 An example of a region with intensive mining and a changing landscape caused by ongoing operations: open pit Lignite Mine in Bełchatów (central Poland). *Source:* Photo by Marcin Pietrzykowski. *See colour plate section for colour representation of this figure.*

Figure 11.2 A hard coal mine spoil heap and hydrological and geomorphological transformation caused by mining operations in upper Silesia (Poland). *Source:* Photo by Marcin Pietrzykowski.

From an ecological point of view, the goal of reclaiming post-mine sites is to develop a long-term sustainable ecosystem that is native to the area and includes all functions and ecosystem structures [5] (Figure 11.3).

(a)

(b)

Figure 11.3 From sandpit mining operation to manmade ecosystem via ecological engineering (Southern Poland): (a) first phase of the technical and biological reclamation (2007); (b) dynamically developing ecosystem (2015). *Source:* Photo by Marcin Pietrzykowski.

In this chapter, *restoration* is used with a broad meaning, and *reclamation* is defined as a restoration tool and a general process for repairing disturbed, damaged, degraded, or destroyed land to its former or future productive use [1]. In this context, terrestrial ecosystem reclamation is conducted by accelerating and directing the ecosystem succession process in which introduced plant and animal communities interact and transform the biotope (i.e. post-mine environment with overburden and waste rock) [5]. Reclamation is considered successful when the result is a self-sustaining bio-ecological system (ecosystem) in which an exchange of matter and energy occurs [6].

11.2 Ecological Criteria for Successful Mine Site Restoration

Constant monitoring of restored ecosystems, particularly soil and plant interactions, is an essential part of the reclamation process and successful restoration [7]. For example, afforestation introduces tree stands that radically modify the developing environment (microclimate, lighting, and biochemical conditions) for other organisms, which appear by way of natural succession. The health and growth of tree stands introduced on reclaimed sites are directly dependent on the substrate's potential (parent rock) to meet the nutritional needs of developing soils as they gradually increase along with an increase in biomass. An impaired nutrient circulation cycle means an inadequate proportion of organic matter is formed in situ, which in natural conditions provides a significant reservoir of nutrients [5, 8].

Developing mine soils to meet the nutritional requirements of plants (fertility) depends on the share of nutrients in forms available to plants (N, K, Ca, Mg, S, and micronutrients) in the total stock of a given element. In organic horizons, these values depend on the rate of decomposition and mineralization of organic matter forming in situ; mineral horizons depend on the rate of weathering of minerals in mine soil substrates. As already mentioned, in natural ecosystems, soils are characterized by high biological efficiency, and nutrients are stored primarily in organic-mineral and organic horizons and gradually released during the process of mineralization. However, soil organic matter (SOM) in the initial phase of accumulation in a forest ecosystem also plays an essential role in the nutrition balance of tree stands on reclaimed sites, so mineral fertilization and weathering of parent rock are the primary sources of vegetation nutrition [5].

A full assessment of the restoration process should consider many other ecological factors [9, 10]. The basic criteria for assessing the effectiveness of restoration include SOM, soil organic carbon (SOC) and nitrogen (N) accumulation rates, soil biological activity, rate of plant and soil biota succession, biodiversity of communities, aboveground and fine root biomass, and others ecological factors [11–14]. To assess the stability of emerging ecosystems, an essential criterion is the efficiency of element availability measured as the ratio of accumulated nutrients in vegetation biomass to the pool of nutrients available in soil [15] or nutrient leaching rate and tree stand nutrient supply [14, 16, 17]. However, the best solution for a comprehensive assessment of restoration effectiveness is to consider both environmental and economic factors [18]. This assessment includes the energy efficiency balance (EEB), calculated as the energy input and energy accumulation in the ecosystem and its components [6]. In this case, the EEB is calculated based on the carbon content of ecosystem components such as vegetation biomass and SOM. This approach makes it possible to characterize ecosystem structures from an individual to populations because it is then possible to reduce its basic differentiation to a single unit, i.e. joules (J) or grams (g) [19].

A case study of the EEB and comparison of afforestation and unreclaimed sites with natural succession was conducted on a sand mine pit in southern Poland [6]. The study demonstrated that forest ecosystems restored in a complete reclamation, including technical development of biotope, biological reclamation, and afforestation, significantly exceeded the EEB in the emerging ecosystem: about twofold during the first 25 years from the commencement of reclamation [6]. On the other hand, including the process of natural succession in developed ecosystems and using diverse geological overburden mostly deposited unselected reduces costs and increases biodiversity by a range of geological formations, but also usually results in less tree growth and biomass. Therefore, the measures taken in specific conditions should follow from an analysis of environmental and economic conditions and goals; the optimal solution is between the two extremes [5]. This means natural succession should be taken into account (included in the developed system mosaic), introducing monocultures should be avoided, and production of biomass should be optimized (through a diversity of suitable species, sufficient dose neutralizers in acidic soil, and mineral fertilization) [20].

11.3 Examples of Reclamation Technology and Afforestation in Mining Areas

In temperate zones in the northern hemisphere, forests are the best organized and dominant ecosystems, called a *climax biome*. For this reason, in Europe, North America, South America, and Asia, most post-mining landscapes are reclaimed to forest [10, 21–24]. Thus afforestation is potentially the best sustainable strategy for reclaiming mine lands to their former state [10, 22]. The development of the ecological afforestation strategy depends on the primary goal and approaches: the greater the investment, the more significant the effects that may be achieved in the form of tree stands and biomass [5, 25]. Afforestation can be achieved by forming habitats consisting of homogeneous substrates that have the best potential for productivity and introducing monocultures. However, this is not beneficial due to low biodiversity and stability and potential susceptibility to fungus and insects.

Thus the design and detailed plans for reclamation treatments are determined by the aim of reclamation (how it is to be used later) and natural conditions: i.e. the climate, availability of soil substrates, water balance, and technical (mining) management, including management of rock overburden and relief of the site for reclamation. Thus reclamation is an integral part of mine investment planning, which is of paramount importance to its implementation and outcomes. This is because mining technologies, including removing overburden, and depositing on spoil heaps, and forming opencast excavations, directly impacts the specific habitat conditions in areas subjected to reclamation [26]. In addition to these conditions, which can be categorized as technical (mining) factors dependent on humans, the degree of difficulty of biological reclamation and efficiency of afforestation also depend on ecological factors, independent of people, such as the climate, type of water balance, and quality of overburden rocks [5]. Thus, much depends on the work during the conceptual and design phase of a project and the shape of the landscape produced by mining.

In some cases, the soil substrate created by mining operations may create new biotopes not found in the natural environment (Figure 11.4). For example, these could be mixtures of Neogene loam with sands; layers of sulfurous Miocene sands that require neutralization, which occur in the spoil heap top horizon; carboniferous shales and sandstones; and other rocks together with fossils [20]. As a result, the technical phase of reclamation may increase the biodiversity of the developed ecosystems.

In areas with an appropriately formed landscape, favorable hydrological conditions, and potentially productive soils, reclamation may consist of targeting and accelerating soil formation

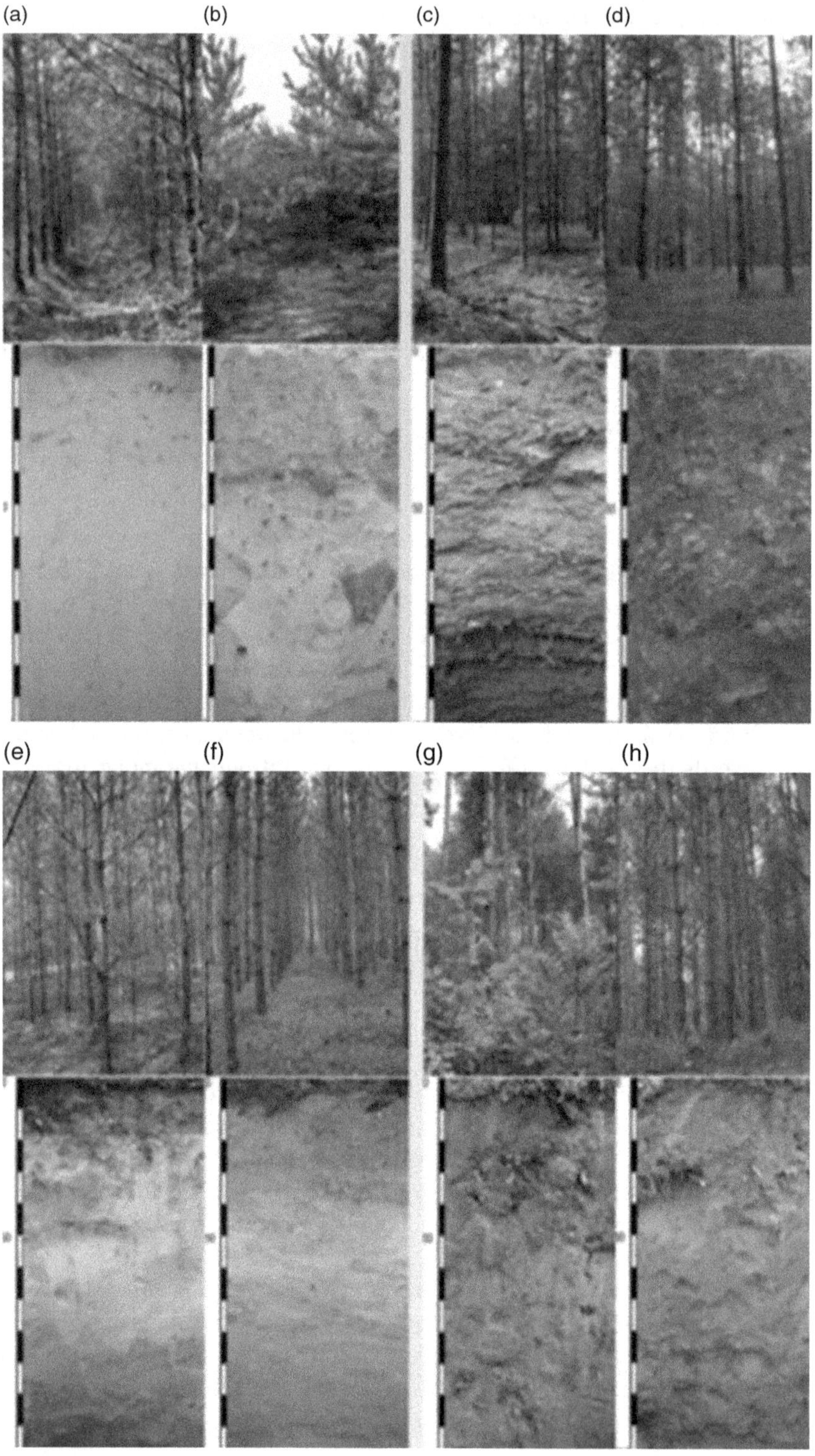

Figure 11.4 Varied mine soil substrate and differentiation in redeveloped habitat conditions: (a) loamy sands; (b) Miocene acidic sands with lignite inclusions; (c and d) carboniferous shales with a high content of fossil coal; (e) quaternary sands; (f) loamy sands; (g and h) sulphurized mix of Neogene clays and sands. *Source:* Photos by Marcin Pietrzykowski. *See colour plate section for colour representation of this figure.*

processes using biological methods such as agricultural practices and mineral fertilization, and introducing appropriate SOM-forming vegetation. By contrast, in areas consisting of barren or phytotoxic soils (such as the already mentioned acidic and sulfurous Miocene sands), the basic reclamation procedure, in addition to suitable landscape formation and regulation of hydrological conditions, consists of sealing and neutralization followed by biological reclamation.

Mining (production, transport, dumping, and mixing of material) causes changes in the original soils of the overburden, and therefore the quality and spatial variability assessment may only be described in detail in a formed site. Frequently, significant spatial variability of microhabitat conditions hinders reclamation. Diagnosing habitat conditions (soil fertility) before the biological reclamation phase has a significant impact on the choice of methods and the effectiveness of introducing vegetation, particularly afforestation [26].

Correct identification of habitat conditions at mine sites and planning species composition for afforestation both have a fundamental effect on the stability of reclaimed stands and developed forest ecosystems [16, 20]. Diagnosing and classifying habitats on mine soils reclaimed for forestry is paramount for selecting species composition at afforested sites [20, 26]. Proper selection of tree species means their ecological requirements are appropriately adjusted to habitat conditions (climate, soil fertility, and humidity).

The fertility of mining soil should be considered at two levels: baseline fertility and potential fertility. Fertility is influenced by soil particle size, the ability to retain water available for plants, soil-air conditions, pH, sorption capacity, and nutrient storage and availability to plants [20]. However, assessing the reclamation suitability of deposits in the first phase should exclude phytotoxicity (over-acidification, alkalization, salinity, sulphation, concentration of heavy metals, etc.) and indicate possible methods of detoxification, neutralization, or sealing of deposits that are unwanted on mine land.

An objective comparison of mine soil quality and spatial and temporal variability is possible based on quantitative indices (e.g. soil quality index or SQI). This is a key issue for sustainable reclamation management and monitoring of developed ecosystem services [20]. In practice, soil quality can be evaluated based on its chemical, physical, and biological properties [27]. The Mine Soil Quality Index (MSQI) was proposed by Pietrzykowski [20] to assess initial mining soils and classify habitats on post-mining sites afforested with a monoculture of pine in Europe. The MSQI index is the sum of sub-indices weighted with a relevant validity coefficient factor (*wt*) for a given feature, where the maximum value of 1.0 corresponds to 100%. A general equation was applied to calculate the MSQI:

$$\text{MSQI} = \left(\text{Ist}\,wt(0.3)\right) + \left(\text{Ina}\,wt(0.3)\right) + \left(\text{IP}\,wt(0.1)\right) + \left(\text{Iac}\,wt(0.2)\right) + \left(\text{Iba}\,wt(0.1)\right) \tag{11.1}$$

wt is the validity coefficient factor for individual sub-index. Ist is a sub-index of soil texture, which is the sum of Isi (the silt-sized fraction stock (SFS, 0.05–0.002-mm fraction) and Icl, the clay SFS (CFr, <0.002 mm fraction) calculated according to the total stock of these fractions in soil up to 110 cm deep (Mg/ha 110/cm). Ina, the nutrient availability index, refers to the nutrient content in a volume of soil (exchangeable cation status [ESC] for Ca^{2+}, Mg^{2+}, K^{+}, and Na^{+} up to 110 cm deep), $wt = 0.3$. IP is the available phosphorus sub-index based on available phosphorus stock (APS) in a volume of soil (Mg/ha 110/cm), $wt = 0.1$. Iac is the acidity sub-index based on the hydrogen stock (H^{+} S) in a volume of soil (kg/ha 110/cm); however, H^{+} S is calculated based on pH_{KCl}, which is a measure of hydrogen ions in a soil-water solution with ratio 1 : 2.5. This manner of determining acidity allowed the researchers to avoid taking the arithmetic mean of the pH from soil horizons (pH is a logarithmic value); $wt = 0.2$. Iba is the soil biological activity index based on the C_{org}-to-N_t ratio in the OLf horizon, $wt = 0.1$.

To automate the calculation of MSQI based on a valuation scale, the *IF* simple function was used to perform logic tests on values and formulas in a Microsoft Excel spreadsheet. The MSQI is useful as a tool for planning species composition for both afforestation and pine monoculture conversion in the next generation of tree stands.

11.4 Selected Reclamation Practices Versus Mining Extraction and Environmental Conditions

There are numerous examples of well-performed reclamation treatments and appropriate monitoring of restored ecosystems all over the world. It is difficult to cite examples of all the best practices from Europe, North America, Asia, and Australia.

As mentioned, mining extraction (coal, sand, rocks, metal ores, oil) affects reclamation conditions and strategies. The practice of reclamation and various factors considered during reclamation design can be exemplified by lignite, sand, and metalliferous ore mining, as contrast post-mine objects and environmental conditions. In formed lignite spoil heaps, there is a significant diversity of habitats further compounded by the steepness of slopes and exposure, which determines their susceptibility to erosion. The protection of natural ecosystems demands that spoil heaps take up as little land as possible. Such sites are usually shaped like a truncated cone with slopes that have an average steepness of about 1:4. In lignite mining, the highest degree of reclamation difficulty is on the spoil heaps with infertile, phytotoxic Miocene sands and high carbon and sulfur content (pH <3.0), susceptible to water and wind erosion. Such spoil heaps cover a vast area in lower Lusatian in Germany and in Poland close to the western border. These areas, on both the Polish and German sides of the border, were reclaimed to forest in the 1970s [28]. The basic reclamation procedure was primarily liming and intensive fertilization. The over 20-year-old pine monocultures are currently a significant problem, with conifer root rot (Heterobasidion annosum) observed at these sites, particularly with pH above 6.5 [29].

On slopes of acidic sands, the following series of treatments can be carried out: neutralization of excessively acidic sediments, sowing grass and legume seed mixtures using the traditional method and hydroseeding, multi-species forestation, and crop tending. To protect the slopes from erosion and rapidly initiate soil-forming processes, the hydroseeding method is recommended. It consists of dropping biosolids (an appropriate quantity and quality of fermented hydrated sludge) in a dose of approximately 30 t/ha mixed with a selected amount of herbaceous plant seeds (grasses and legumes).

An example of well-performed restoration monitoring is the lignite extraction areas of the Powder River Basin (PRB) in Wyoming (western United States), which has dry, cold climate conditions. The region produces a significant percentage (approximately 40%) of the total sub-bituminous coal in the US [30]. Such massive opencast coal mining involves a large-scale transformation of the land and landscape. These are mainly geomechanical and hydrological transformations, which have a significant impact on local ecosystems. Thus, in the PRB, restorative treatment implemented by way of reclamation is regarded on an equal basis with mining operations.

The opencast excavation floors and slopes after sand extraction are mainly made up of sandy deposits, including loose, coarse sands. The opencast excavations have different depths, but they average 5–25 m, and the slope varies from about 1:3 to 1:2. The difficulties of biological reclamation with afforestation are conditioned by the depth at which groundwater occurs, fertility, and substrate abundance in nutrients available for plants. Biological reclamation methods appropriate for afforestation include simultaneous improvement of infertile soils by mineral fertilization and legume crops (especially lupine) with green manure [31]. In the afforestation of sandpits, the main species in poorer habitats in Central European conditions can be used: Scots pine (*Pinus sylvestris*), silver birch (*Betula pendula*) with an admixture of black alder (*Alnus glutinosa*). In the past some species as gray alder (*Alnus incana*), red oak (*Quercus rubra*), wild cherry (*Padus avium*), and mountain ash (*Sorbus aucuparia*) were used, as well. In better-quality habitats with soils with a higher proportion of clays, European larch (*Larix decidua*), English oak (*Quercus robur*), and sycamore (*Acer pseudoplatanus*) are introduced in groups or clumps in addition to the black alder [31]. As part of technical reclamation, it is recommended to introduce some altitude differences and small bodies of water to differentiate and enrich the emerging ecosystem.

The non-ferrous metal industry, i.e. copper, zinc, and lead, produces a large amount of waste. Low metal content in the mined ore means that waste from the ore enrichment process constitutes 80–90% of the total amount of processed material. The mining of these ores results in geomechanical, hydrological, and chemical transformations. Post-mining, geomechanically transformed sites may be divided into post-mining areas (spoil heaps), post-processing areas (sedimentation ponds, flotation ponds), and metallurgical waste heaps. The spoil heaps are made of barren rock: dolomite, marl limestone, and various material from preparatory mining work. Approximately 98% of the rock mined in such facilities finds its way to mine and metallurgical spoil heaps and sedimentation ponds at various stages of mining and preparation processes.

While the biological reclamation of barren rock heaps is not a major problem, the situation is different for waste in flotation ponds. Factors that hinder introducing vegetation are poor physical properties of the waste; shortage of air and water available for plants; high pH; excess amounts of Pb, Zn, Cu, and other heavy metal compounds; high susceptibility to wind erosion; and almost complete lack of available phosphorus, potassium, and nitrogen. As a result of years of experience [32], biological restoration methods have been developed for such sites. Methods for introducing vegetation directly on sediments following the flotation of copper ore included using extensive, appropriately selected mineral fertilization (especially with phosphorus). In sedimentation ponds, introducing turf vegetation following the flotation of lead and zinc ores did not produce the same positive effects as in the case of waste from processing copper ore [32]. Using high doses of phosphoric fertilizer was based on the assumption that the solubility of heavy metal salts in water should be reduced to minimize the negative effects of heavy metal contamination [32]. Layering horizons proved effective and can be used for biological reclamation in these conditions: a sediment sealing horizon in the form of hydrosilicates and about a dozen centimeters of mineral soil that had been appropriately fertilized and sown with a special blend of turf vegetation [32].

11.5 Final Comments and Remarks

Criteria for reclamation assessment of post-mine sites depend on how the needs and objectives are defined during the process of ecosystem reconstruction, as well as on environmental conditions (mainly the climate and geological structure, which affect the properties of the substrates deposited in the overburden from mines) and deposition technology and construction of post-industrial facilities (selection of geological sediments, optimization of tip and opencast construction using the best possible sediments in the deposits, insulation of phytotoxic sediments). Obviously, numerous ecological criteria for the assessment of restoration success may be used, including the diversity and structure of communities; productivity expressed as an increase in biomass and wood raw material; dynamics of the soil formation process, particularly the quantity and properties of the accumulated organic matter and formation of humus (organic soil horizon subsequence); carbon sequestration and its dynamics in various elements of the ecosystem; dynamics of biogeochemical element cycles and the flow of matter and energy in the ecosystem; and indicators of soil microbial activity. There are also economical-ecological criteria based on relationships of energy input and gain in reclamation and accumulation in the ecosystem. The issue of the quality of the restored ecosystem compared to the pre-mining ecosystem remains controversial; however, given the demand for energy and minerals, mine sites will continue to require reclamation. Therefore, it is of prime importance to improve reclamation technology and monitoring systems for newly developed ecosystems to mitigate the effects of mining.

References

1 Favas, P.J.C., Martino, L.E., and Prasad, M.N.F. (2018). Abandoned mine land reclamation – challenges and opportunities (holistic approach). In: Bio-Geotechnologies for Mine Site Rehabilitation (eds. M.N.V. Prasad, P.J.C. Favas and S.K. Maiti), 3–32. Amsterdam-Oxford-Cambridge: Elsevier.

2 Martín-Duque, J.F., Sanz, M.A., Bodoque, J.M. et al. (2010). Restoring earth surface processes through landform design. A 13-year monitoring of a geomorphic reclamation model for quarries on slopes. *Earth Surface Processes and Landforms* 35: 531–548.

3 Mudroch, A., Stottmeister, U., Kennedy, C., and Klapper, H. (eds.) (2002). Remediation of Abandoned Surface Coal Mining Sites. Berlin, Heidelberg: Springer Verlag.

4 Society for Ecological Restoration Science and Policy Working Group (2002) The SER primer on ecological restoration. Society for Ecological Restora-tion, www.ser.org/.

5 Pietrzykowski, M. (2015). Reclamation and reconstruction of terrestrial ecosystems on mine sites -ecological effectiveness assessment. (Chapter 5). In: Series: Energy Science and Technology, Coal Energy, vol. 2 (eds. J.N. Govil, B. Khorramdel, B. Khorramdel, et al.), 121–151. New Delhi, Houston: Studium Press LLC.

6 Pietrzykowski, M. and Krzaklewski, W. (2007). An assessment of energy efficiency in reclamation to forest. *Ecological Engineering* 30: 341–348.

7 Ludwig, J.A., Hindley, N., and Barnett, G. (2003). Indicators for monitoring minesite rehabilitation: trends on waste rock dumps, northern Australia. *Ecological Indicators* 3: 143–153.

8 Pietrzykowski, M., Woś, B., and Haus, N. (2013). Scots pine needles macronutrient (N, P, K, CA, MG, and S) supply at different reclaimed mine soil substrates – as an indicator of the stability of developed forest ecosystems. *Environmental Monitoring Assessment* 185: 7445–7457.

9 Hüttl, R.F. and Bradshaw, A. (2000). Aspect of reclamation ecology. *Landscape Urban Plan* 51: 73–74.

10 Zipper, C., Burger, J., Skousen, J. et al. (2011). Restoring forests and associated ecosystem services on Appalachian coal surface mines. *Environmental Management* 47: 751–765.

11 Frouz, J., Prach, K., Pižl, V. et al. (2008). Interactions between soil development, vegetation and soil fauna during spontaneous succession in post mining sites. *European Journal of Soil Biology* 44 (1): 109–121.

12 Pietrzykowski, M. and Krzaklewski, W. (2007). Soil organic matter, C and N accumulation during natural succession and reclamation in an opencast sand quarry (southern Poland). *Archives of Agronomy and Soil Science* 53 (5): 473–483.

13 Šourková, M., Frouz, J., Fettweis, U. et al. (2005). Soil development and properties of microbial biomass succession in post mining sites near Sokolov (Czech Republic) and near Cottbus (Germany). *Geoderma* 129: 73–80.

14 Świątek, B., Pietrzykowski, M., Józefowska, A. et al. (2019). Fine root biomass and the associated C and nutrient pool under the alder (Alnus spp.) plantings on reclaimed technosols. *Geoderma* 337: 1021–1037.

15 Pietrzykowski, M. (2008). Soil and plant communities development and ecological effectiveness of reclamation on a sand mine cast. *Journal of Forest Science* 12 (54): 554–565.

16 Knoche, D., Embacher, A., and Katzur, J. (2002). Water and element fluxes of red oak ecosystems during stand development on post-mining sites (Lusatian Lignite District). *Water Air Soil Pollution* 141: 219–231.

17 Woś, B., Pietrzykowski, M., and Józefowska, A. (2018). Reclaimed mine soil substrates and tree stands vs. successional forest floor vegetation: a case study of developing ecosystems on afforested mine sites. *Ecological Engineering* 120: 504–512.

18 Rodrigue, J.A., Burger, J.A., and Oderwald, R.G. (2002). Forest productivity and commercial value of pre-law reclaimed mined land in the eastern United States. *Northern Journal Applied Forestry* 19 (3): 106–114.

19 Krebs, C.J. (2001). Ecology: The Experimental Analysis of Distribution and Abundance, 5e. San Francisco: Benjamin Cummings.

20 Pietrzykowski, M. (2014). Soil quality index as a tool for scots pine (*Pinus sylvestris*) monoculture conversion planning on afforested, reclaimed mine land. *Journal of Forestry Research* 25 (1): 63–74.

21 Torbert, J.L. and Burger, J. (2000). Forest land reclamation. In: Reclamation of Drastically Disturbed Lands, 371–398. American Society of Agronomy, Publishers.

22 Dutta, R. and Agrawal, M. (2003). Restoration of opencast coal mine spoil by planting exotic tree species: a case study in dry tropical region. *Ecological Engineering* 21: 143–151.

23 Kaar, E. (2002). Coniferous trees on exhausted oil shale opencast mines. *MetsanduslikudUurimused* 36: 120–125.

24 Knoche, D. (2005). Effects of stand conversion by thinning and underplanting on water and element fluxes of a pine ecosystem (*P. sylvestris* L.) on lignite mine spoil. *Forest Ecology and Management* 212: 214–220.

25 Pietrzykowski, M. and Socha, J. (2011). An estimation of Scots pine (Pinus sylvestris L.) ecosystem productivity on reclaimed post-mining sites in Poland (Central Europe) using of allometric equations. *Ecological Engineering* 37: 381–386.

26 Krzaklewski, W. and Pietrzykowski, M. (2007). Diagnosis of sites on post-mine sites reclaimed to forest, with special attention to phytosociological-soil method. (Diagnoza siedlisk na terenach pogórniczych rekultywowanych dla leśnictwa, ze szczególnym uwzględnieniem metody fitosocjologiczno-glebowej, in Polish, English summary). *Sylwan* 151 (1): 51–57.

27 Schoenholtz, S., Miegroet, H., and Burger, J. (2000). A review of chemical and physical properties as indicators of forest soil quality: challenges and opportunities. *Forest Ecology and Management* 138: 335–356.

28 Krzaklewski, W. and Pietrzykowski, M. (2001). Evaluation of selected components of the forest ecosystem played in the course of reclamation on the internal dumping ground pit "D" of the former mine "PrzyjaźńNarodów" in Łęknica. *Environmental Enginiering* 125: 187–197.

29 Knoche, D. and Ertle, C. (2010). Infection of scots pine afforestations (Pinus sylvestris L.) by annosum root rot (Heterobasidion annosum (fr.) Bref.) in the eastern german lignite district. *Civil and Environmental Engineering Reports* 4: 35–45.

30 Foulke, R. (2002). The Sea Voyage Narrative. New York: Routledge.

31 Pietrzykowski, M., Gruba, P., and Sproull, G. (2017). The effectiveness of yellow lupine (Lupinusluteus L.) green manure cropping in sand mine cast reclamation. *Ecological Engineering* 102: 72–79.

32 Krzaklewski, W. and Pietrzykowski, M. (2002). Selected physico-chemical properties of zinc and lead ore tailings and their biological stabilisation. *Water, Air, Soil Pollution* 141: 125–141.

12

Ecological Restoration of Abandoned Mine Land

Theory to Practice

Jitendra Ahirwal and Subodh Kumar Maiti

Restoration Ecology Laboratory, Department of Environmental Science and Engineering, Indian Institute of Technology (Indian School of Mines), Dhanbad, Jharkhand, India

12.1 Introduction

Natural resources are vital to fulfilling ever-growing global energy demands. Most energy generation has shifted from conventional sources to unconventional sources like solar and nuclear energy. However, some energy resources, including metals or fossil fuels, are still extracted by mining. Surface mining inevitably results in deforestation, land degradation, and increased CO_2 emissions into the atmosphere. Mining-induced land-use changes are common during surface mining activities in forest areas and result in wastelands, agricultural land, and bodies of water, ultimately altering the ecosystem. To offset surface mining's adverse effects on the environment, degraded mine lands have to be restored by developing vegetation cover such as forest, grassland, and agriculture. Revegetation of post-mining land has proven to be a viable end use for land; however, the magnitude of technical and biological reclamation practices plays an important role in restoration success.

In the 1980s, as curiosity about ecological functions and understanding of biodiversity conservation and ecosystem management increased, restoration ecology became a distinct field of applied science. *Restoration ecology* is the scientific process of developing a theory to guide restoration practices and primarily deals with the study of disturbed ecosystems. On the other hand, *ecological restoration* is the process of assisting the recovery of an ecosystem that has been degraded, damaged, or destroyed [1]. Even 40 years after the emergence of ecological restoration as a scientific discipline, people often argue the practicality of reference site selection and the potential of restoration efforts in this rapidly changing environment. There is a pressing need to develop standard restoration policies and implement proper environmental management plans to restore degraded ecosystems.

Rapid industrialization and the growing consumption of energy resources have increased greenhouse gases (GHGs). As part of the goal of reducing the global emission of GHGs under the Kyoto protocol, the involved parties decided to reduce GHGs emissions by at least 18% below the 1990 level between 2013 and 2020 [2]. Surface mining increases the global CO_2 concentration by clearing forests and degrading soil. Other undesirable impacts of surface mining on the environment include the clearing of large areas of forest; altered geological settings, topography, and land cover; ecosystem fragmentation, resulting in loss of biodiversity, water recharge potential and water balance; and release of hazardous pollutants. Although surface mining can severely damage the ecosystem, it also provides an opportunity to study the reestablishment of the post-mining

Handbook of Ecological and Ecosystem Engineering, First Edition. Edited by Majeti Narasimha Vara Prasad.

ecosystem. Planting forests at post-mining sites can enhance soil quality and land productivity and accumulate atmospheric CO_2 in the soil and biomass, thus reducing global atmospheric CO_2[3–5].

Forest growth on post-mining land is always a challenging for restoration ecologists. Many factors affect the restoration process at post-mining sites, but selection of tree species, geo-mining conditions, and the nature of mine soil are particular concerns.

Recently, rather than just planting a forest, the emphasis has shifted to quantifying the role of biomass and soils in ecosystem development. The three components of restoration practices that require attention are (i) remodeling physical aspects of the habitat like drainage, water recharge potential, and stability of constructed land; (ii) remodeling chemical aspects such assoil chemical properties, nutrient content related to soil fertility, and toxicity caused by metal and acid mine drainage; and (iii) replacing missing species such as native plants and common wildlife. The steps in surface mining and ecological restoration of degraded mine sites should include referencing natural forests and deforested lands, removing topsoil and overburden materials, excavation with surface mining methods, dumping overburden material and backfilling of mine voids, revegetation backfilled sites with a grass-legume mixture, planting selected tree species, developing vegetation cover, and developing the site characteristics along the age of reclamation(Figure 12.1). Working on all such ecosystem components (physical, chemical, and biological) and functions during restoration results in the creation of a self-sustaining ecosystem. Thus, ecological restoration of post-mine land is vital because it stores the attributes of an ecosystem, recuperates soil quality and biodiversity, and offsets increased greenhouse gas emissions.

Figure 12.1 Steps in surface mining and ecological restoration of mine degraded sites: (1) natural forest before mining, (2) deforested land, (3) removing topsoil and overburden materials, (4) excavation by surface mining methods (5) backfilling of overburden materials, (6) regenerating backfilled sites with a grass-legume mixture, (7) growth of tree species during the revegetation phase, and (8) developing site characteristics along the age of reclamation. *See colour plate section for colour representation of this figure.*

12.2 Integration of Ecology Theory, Restoration Ecology, and Ecological Restoration

Restoration ecology draws on several ecological concepts, discussed in this section.

12.2.1 Disturbance

A *disturbance* is a change in environmental conditions that interferes with the functioning of an ecological system. The degree of disturbance may be spatial or temporal.

12.2.2 Succession

Ecological succession is the process by which the composition of species changes over time. It is the overall status of an ecological community and how the structure and functions change. After a disturbance, an ecosystem usually progresses from a simple level of organization (i.e. few dominant species) to a more complex community.

The three stages of ecological succession are (1) primary, when an ecological community first enters a new form of habitat that it has not been present in before; (ii) secondary, when a habitat has been established but is then disturbed or changed in some fashion, after which a new community moves in; and (iii) climax, the last stage, when the ecosystem has become balanced and there is little risk of an interfering event or change to mutate the environment.

12.2.3 Fragmentation

Fragmentation is the appearance of spatial discontinuities in an ecosystem. Land-use changes break the continuity of an ecosystem and divert it into smaller fragments that are more fragile and support fewer species.

12.2.4 Ecosystem Functions

Functions are the potential of an ecosystem to provide supporting services like pedogenesis, photosynthesis, and nutrient cycling; provisioning services like timber production; regulating services like climate, water, and soil quality; and cultural services such as recreational and spiritual functions.

The link between ecological theory, restoration ecology, and ecological restoration (Figure 12.2) and the principles of restoration ecology are often based on the 5R terminology–restoration, reclamation, rehabilitation, regeneration, and recovery – an understanding of which is essential for researchers working in this field.

12.2.5 Restoration

Restoration establishes a desired ecosystem that matches the species composition, productivity, and overall structure of the original ecosystem present at the site before disturbance. This process assists in the recovery of an ecosystem that has been degraded, damaged, or destroyed [1]. Ecological restoration refers to the rehabilitation, reclamation, recreation, and recovery of degraded lands.

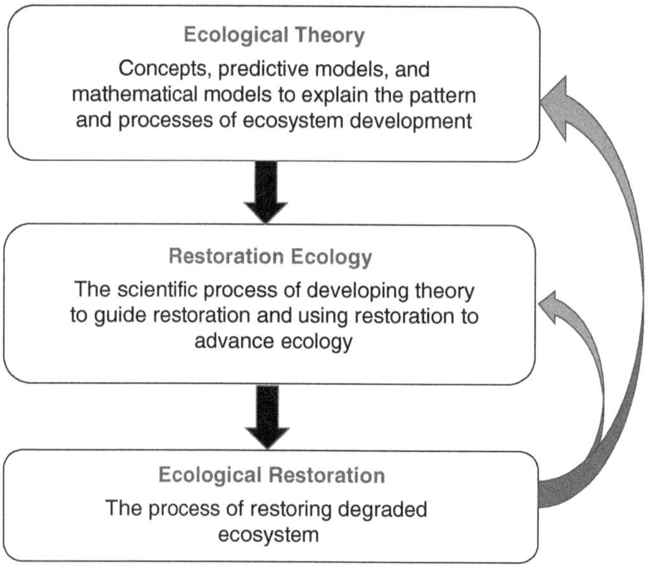

Figure 12.2 The linkage between ecological theory, restoration ecology, and ecological restoration.

12.2.6 Reclamation

In British terminology, reclamation is the process of gaining or recovering land, bringing it into a condition for cultivation or other use. Reclamation operations usually get underway as soon as minerals are removed from a mine site. The term *reclamation* describes the general process whereby the land surface is returned to some form of beneficial use. The process includes restoring the land so it is close to its original appearance. This can be done by using exotic species, and as time passes, productivity may reach the level of the original ecosystem. Reclamation encompasses the stabilization of the terrain, assurance of public safety, and overall aesthetics.

12.2.7 Rehabilitation

Rehabilitation is the re-creation of a stable and self-sustaining ecosystem, but not necessarily the one that existed before mining. Rehabilitation emphasizes the reparation of ecosystem processes, productivity, and services, re-establishing some, but not necessarily all, of the plant and animal species originally present at a site.

12.2.8 Regeneration

Regeneration is the growth of native species after a site has been degraded, as a result of protecting the area from biotic interference. Regeneration may come about naturally or result from human intervention.

12.2.9 Recovery

During recovery, the site returns naturally to the state it was in before degradation, without any intervention from humans. Recovery of habitat is highly dependent on the ecological succession that occurs at the site.

The link between ecological theory, restoration ecology, and ecological restoration offers benefits including opportunities to study ecosystem elements in a context that can be manipulated and test and expand theories based on ecology principles. It provides a logical background for restoration, explains multiple interactions between ecosystem components, and improves the quality and efficiency of restoration efforts.

12.3 Restoration Planning

Mining not only disturbs the environment but also has a socioeconomic effect on society. Thus, restoration ecologists and land managers should consider all the aspects of mining while they plan to restore the degraded ecosystem. Before revegetation planning, land planners should know about the native ecosystem, the vision for the desired ecosystem, and restoration methodology. Technical aspects like dump stability, topsoil protection, and erosion control measures should be considered while preparing the restoration plan. All technical and biological measures should follow sustainability principles because the ultimate aim of any restoration project is to create a self-sustainable ecosystem. Many mining projects have not been approved due to flaws in restoration planning; today, government agencies and regulatory authorities are very strict about approving mining projects due to their impact on communities.

Ecological restoration planning should include the following steps (Figure 12.3):

1) Identify a reference system, including the benchmark ecosystem or native landscape that was altered due to the mining project.
2) Identify restoration needs. Major components that need human attention should be tackled on a priority basis. This step involves proper planning to restore ecosystem functions, biodiversity, and the area's overall aesthetics
3) Prioritize and set goals. Time required to achieve restoration goals, the cost of restoration, a cost-benefit analysis, and the engagement of scientific expertise should take priority.

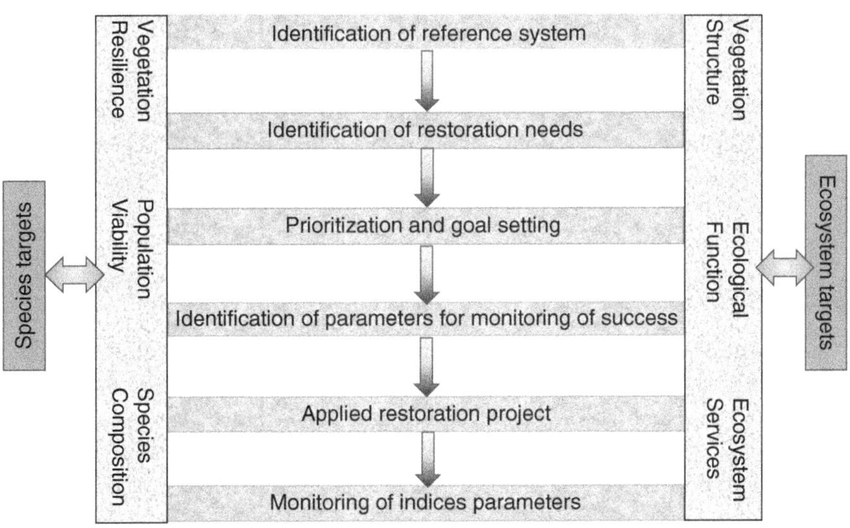

Figure 12.3 Steps of restoration planning to fulfill species and ecosystem targets by developing vegetation structures, ecological functions, and ecosystem services.

4) Identify parameters for monitoring success. Relevant, simple, reliable parameters that can indicate the status of ecosystem recovery are important to determine the state of the ecosystem.
5) Implement the restoration project in a scientific manner and observe ground restoration practices. This is vital for any restoration project.
6) Monitor index parameters. Continuous measurement of indicator parameters, follow-up case studies, and experiments that fulfill the ecosystem targets are valuable.

Revegetation planning should be based on ecological principles and include a clear definition of the type of species used for afforestation; descriptions of native species; a detailed analysis of mine spoil; atmospheric conditions like precipitation and temperature; and the reference site. The objectives of the project should mention the final land use of the restored ecosystem. Restoration planning also includes community involvement, such as the participation of local and regional authorities and local citizens, and scientific advisors should be appointed at the initial stage of the project. The restoration plan also contains detailed planning of physical aspects like drainage, erosion control, slop formation, and inputs to increase land stability. Issues from past mining operations such as underground mine fires and other possible human disturbances should be addressed in the restoration plan.

12.4 Components of Restoration

The ecosystem is destroyed when surface mining activities take place in a forest or river valley. Factors affecting the functioning of the ecosystem should be considered when the restoration process begins. Several factors govern the recovery of a degraded ecosystem, some natural and others anthropogenic. It is vital to point out the major controlling factors that are not recovered naturally.

12.4.1 Natural Processes

Restoration research has shown that natural processes that help to recover degraded ecosystems should be preferred because they are self-sustainable, cost nothing, and can be applied on a larger scale. Although it is understood that natural processes can potentially restore the ecosystem, they may take longer and sometimes need human assistance. Natural succession processes demonstrate that nature can attain restoration unaided over time and develop fully functioning soils [6]. In practical projects, we cannot completely depend on natural succession to sow desirable species, eliminate undesirable species, and manage practices. The natural process of plant succession takes time. And most of the soil characteristics of mine land require a long time to develop through natural processes; thus, it is indispensable to incorporate soil management tools that can accelerate the recovery of mine soil.

12.4.2 Physical and Nutritional Constraints

Mined overburden waste materials contain large rock fragments and have higher bulk density and low moisture, which inhibits the establishment of plant species. Therefore, mechanical actions are needed to reduce soil compactness and increase the water-retention capacity of mine soils. Mine soils do not have soil horizons, soil physical and chemical properties, and microbiological activities [7].They are young soil systems [8] with a higher abundance of relatively unaltered spoil that lacks structured development.

Soil structure is a key factor in the functioning of soil and moderating environmental quality. It influences water infiltration, soil erosion, nutrient recycling, root growth, and crop yields.

Moreover, adverse physicochemical properties, lack of certain nutrients, and toxicity of mine spoil inhibit soil-forming processes and plant growth [6, 9].Salinity, acidity, poor water-holding capacity [10] and lack of organic matter retard the process of natural succession. The adverse properties of coal mine spoil cause environmental pollution and perturbation of nutrient cycles and land-use patterns. The formation of good soil structure in degraded landscapes can be accelerated by adding organic matter [11].

The physicochemical properties of mine soils are unique and differ from natural soil [12, 13]. The infiltration rate (IR) affects water content and the amount of surface runoff from mine soils. Mine soils usually have low IR due to their high bulk density and low porosity [14].

12.4.3 Species Diversity

Species composition is a significant factor that determines the trajectory of a restored ecosystem. It is often reported that exotics species are used to restore degraded land. Although they are fast-growing and quickly adapt to hostile environmental conditions, they may not perform ecosystem functions as well as native species. During the reclamation phase, some native species are naturally colonized and increase the biodiversity and functions of a restored ecosystem. Trees can establish a sustainable bionetwork through processes such as accumulation of organic matter, rhizosphere growth, and improved soil biological activity. Planting trees significantly accelerates soil redevelopment on mine spoil [15].

The soil microbial community also has a significant impact on soil functioning and is vital for instigating restoration processes such as litter decomposition. However, soil organisms have received relatively little attention as agents of restoration. Invertebrates are often included because they represent many different functional groups [16] and play a critical role in nutrient cycling. Vertebrates are included when wildlife recovery is the primary motivation of the restoration project [17].

12.5 Afforestation of Mine-Degraded Land

Afforestation is the process of planting trees on barren land to create a forest. In comparison, *reforestation* is the process of planting native trees in a forest that has decreasing numbers of trees. Afforestation creates a "new" forest. Planting trees is beneficial whether it takes place in a barren land or is used to regenerate a depleted forest. Large-scale afforestation can control problems caused by burning fossil fuels, industrialization, and climate change. Afforesting is a positive effort in curbing the over-use and destruction of natural forests. If done with proper planning and at appropriate sites, it can become a commercially viable solution for many human needs without harming the balance of nature.

12.5.1 Miyawaki Planting Methods

Akira Miyawaki is a Japanese botanist known as a specialist in the restoration of natural vegetation on degraded land (1928-present). He developed, tested, and refined a method of ecological engineering known as the *Miyawaki method*. Most experts believed that the rapid restoration of a forest was impossible or very difficult on laterite and decertified soil following destruction. Miyawaki, on the other hand, showed that rapid restoration of forest cover and soil is possible by judicious choice of pioneer and secondary indigenous species in dense planting and mycorrhizal growth. Using local species that play key and complementary roles in the normal tree community is essential during restoration. These species are accompanied by various accompanying species (40–60 or more types of plants in the tropics) for support, along with and potential natural vegetation (PNV – vegetation that would occur naturally in the absence of human intervention).

The Miyawaki method demonstrated the development in only 10 years of a forest equivalent to a 100-year-old forest. According to Miyawaki, a forest is a place so dense with trees that a person – no matter how big or small –can't just walk into it.

Miyawaki's method is based on a theory called PNV [18]. It employs six steps:

1) *Survey the native species.* For instance, if a forest is to be created, identifying and understanding native plant species is essential. Plants that have adapted to the local climate are also taken into account.
2) *Identify the soil's nutrition and fertility.* Here, three parameters are considered: the soil's water retention capacity, porosity, and available nutrition content. The process starts with a soil survey to determine what the soil is missing? The minimum land size should be 100 m^2. The first step is the analysis of the soil. The soil is touched, felt, and even tasted to identify what properties it lacks. If the soil is made up of small particles (i.e. higher clay content or clay soil), the soil becomes so compact that water cannot seep through or the surface may be so compacted and hard that it restricts water infiltration.
3) *Collect or buy saplings.* The initial survey studies the native plant species and biomass. After the survey, saplings are prepared in a nursery. Using native tree species is one of the key concepts in the Miyawaki system of forestry. Once the trees are selected, they are divided into four layers: shrub, sub-tree, tree, and emergent layer. The ratio of each layer is fixed, and then the percentage of each tree species in the mixture is determined. For example, for a fruit forest, the percentage of fruit-bearing trees is increased.
4) *Prepare the soil.* For a Miyawaki forest to be successful, a minimum of 100 m^2 is needed, on which 300 plants can thrive. Once the area is identified, the soil is mixed with biomass to make it more fertile. To restore nutrients in the soil, microorganisms should be added because they decompose the biomass, release the nutrients available to plants, and increase soil fertility.
5) *Plant and mulch.* Plant 50 to 100 varieties of native species at a density of 3–5 m^2. Then, collect seeds and germinate saplings. Ensure that trees belonging to the same layer are not planted next to each other, or vertical space will constrain their growth. Plant saplings close to each other. A thick layer of mulch is added on the surface so the soil stays moist during hot, dry seasons.
6) *Perform maintenance.* For the next two to three years, there is a continuing need to water the plants and remove weeds. At this stage, not a single drop of water that falls onto the forest floor will evaporate back into the atmosphere. The dense forest condenses the moist air and retains its moisture. Gradually, watering of the trees is stopped. The forest floor remains moist and sometimes even dark. Now, when a leaf falls on this forest floor, it immediately starts decaying. This decayed biomass forms humus, which is food for the forest. Once established, such a forest will regenerate, and exponential growth can be observed. The forest needs no maintenance and becomes self-sustainable.

In a natural forest like this, no management is the best management. The forest grows collectively as a single unit. If the same trees were planted independently, they wouldn't grow as fast – and thus, a 100-year-old forest grew in just 10 years. The tangible benefits are innumerable. These forests are rich in biodiversity and so dense that, as mentioned, one can't directly walk into them. Naturally, they support many bird species, small animals, and microorganisms. During the rainy season, a negligible amount of water evaporates because of the density; and since the forest is naturally conserving rainwater, the ground table rises and has a significant impact on fruits and vegetation.

12.6 Methods of Evaluating Ecological Restoration Success

Restoration of any disturbed ecosystem is performed by combining technical and biological restoration practices and later focusing on forestry and agriculture as end land use. Three basic goals that any restoration plan should include are [19]:

- Stabilizing newly reclaimed lands against accelerated wind and/or water erosion
- Developing target-specific revegetation programs
- Implementing achievable, sustainable land use by enforcing specific minimum performance standards

The assessment of restoration success is always difficult because it depends on numerous functions and ecological processes that cannot be restored quickly. According to Pecharová et al. [20], the process should allow spontaneous succession or use technical restoration by sowing or planting target species and restoring or improving the site conditions.

12.6.1 Criteria for Restoration Success

What are the criteria for restoration success? It is vital to answer this question in order for restoration ecologists to determine the efficacy of restoration efforts. The Society for Ecological Restoration International (SERI) addresses this issue by taking under consideration nine ecosystem-related parameters to measure restoration success [21]:

- Diversity and community structure that are similar to reference sites
- Presence of local species
- Functional groups necessary for long-term stability
- Capacity of the physical environment to sustain viable populations
- Regular functioning
- Integration with the landscape
- Removing potential threats
- Resilience to natural disturbances
- Self-sustainability

The evaluation of key criteria to assess restoration success was reviewed by Ahirwal et al.[22] and Bandyopadhyay and Maiti [23]. They concluded that development of ecosystem structure and soil functions can be easily monitored at restored sites of different ages, as discussed here:

- *Ecosystem structure:* Plant growth, species composition, and diversity are the most critical factors on which the structure of an ecosystem rely. What should the species composition of the newly developed site, and what is the similarity among ecosystems? It depends on the types of ecosystem and can vary among ecosystems. The United States Surface Mining Control and Reclamation Act (SMCRA) expects 90% species similarity between old and newly developed ecosystems.
- *Soil function*: Soil is a complex, dynamic system that incorporates the interaction between abiotic and biotic factors. Abiotic factors such as weathering, leaching, freezing, thawing, wetting, and drying and biotic processes such as the capture of energy, accumulation of organic material by plants, and its decomposition by microorganisms accelerate the soil development process. Soil structure and functions govern in the development of a self-sustaining ecosystem by supporting biomass production and nutrient cycling and regulating environmental interactions. Adverse soil conditions in post-mining land harm the soil's physical and biochemical properties.

To offset the impact of mining on soil, soil amendments such as biochar, fly ash, and compost have been used. The speedy recovery of decomposers can be achieved by adding organic mulch and topsoil and implementing afforestation, all of which are practiced in India. Juwarkar and Jambhulkar [24] studied the effect of effluent treatment plant (ETP) sludge as an organic amendment, biofertilizer, and mycorrhizal fungi along with suitable plant species and observed an improvement in the physicochemical properties of coal mine spoil and reduction in heavy metal toxicity. Decomposers are heterotroph and detritus feeders; thus, accumulating a litter layer is essential to developing a self-sustaining ecosystem.

- *Time*: When restoration targets are set, the time required to achieve those targets is not easily fixed. Five years is taken as the period for bond release under SMCRA. But for the mine restoration process, five years is not long enough to judge whether interventions are adequate. It has been observed that even after 16 years of reclamation and afforestation on mine-degraded lands, soil carbon stocks are still not the same amount as natural forest [25]. The recovery of ecosystem attributes depends on geoclimatic conditions, types of plant species, nature of the mine spoil, aftercare and maintenance, and magnitude of disturbance by anthropogenic activity. Ecorestoration of mine-degraded site seeks to stimulate natural succession processes that lead to forest. All vegetation types are established at the initial stage of reclamation (grasses and legumes seed are sown between tree rows). As time passes, grasses and legume cover enhance the yield of organic matter and nitrogen for fast-growing trees and multipurpose trees (MPTs), which gradually mature and develop forest.

12.6.2 Indicator Parameters of a Restored Ecosystem

Soil characteristics, particularly physicochemical characteristics, are often used to evaluate mine soil quality, primarily as a measure of contamination. Since soil biological properties respond quickly to contaminants, they may be considered a useful indicator of mine soil quality and restoration success [26, 27]. Biological properties include microbial activity and diversity and their role in soil development, decomposition, and nutrient mineralization processes ([7]). The presence and progression of these easily recognizable indicators can reflect the early success of the restoration.

It is important to set an index or parameters forth objectives and adjust management practices accordingly [28]. Moreover, the indicators that determine the early success of restoration should be simple and easy to recognize. Researchers use physicochemical parameters extensively to measure mine soil quality, but these parameters change very slowly, and changes take several years to measure. However, biological properties change rapidly and respond quickly to minor changes in mine soil quality. Biological indicators include microbial biomass content, diversity, and function and their role in soil development, decomposition, and nutrient mineralization [7]. The overall characteristics of the reclaimed site, such as dump physicochemical status of plant growth; nutrient content such as soil organic matter (SOM), Soil organic carbon (SOC), nitrogen, phosphorous, and potassium; ecological factors like standing biomass, litterfall, and litter decomposition; microbial status, including density of bacteria, nitrifiers, actinomycetes, fungi, and mycorrhiza; and density of microfauna.

Monitoring and assessing the potential impact of mining that perturbs soil quality is a high priority. Restoring mine-degraded lands may contribute to mitigating the negative impacts of mining activities. Any strategy to evaluate the disturbance and restoration of land requires criteria for judging restoration progress. Many studies have attempted to develop soil quality indicators and assess mine restoration success by measuring soil characteristics. As there is no unanimously used indicator to determine soil quality, it is important to standardize the frequently used indicators.

They should also specify why soil is not functioning as desired. An index of a biological soil property or group of properties may reflect the structure and functions of ecosystem processes sensitive to environmental changes and variations in management practices.

As enzymes are strongly correlated with living cells and play an essential role in nutrient cycling, their activity depends on environmental factors that influence microbial functions. Dehydrogenase activity is most widely used to determine soil health because it oxidizes organic compounds and transports electrons to generate energy for microbial growth and microbial respiration, which represent the oxidative capacity of soil microorganisms. A rapid, reliable method for analyzing mine soil quality is needed to help with developing soil-protection policies and assessing the actual condition of the ecosystem.

12.6.3 Soil Quality Index

Soil is a dynamic biomaterial that can be characterized by numerous properties. Soil *quality* can be defined as the soil's ability to perform a given function in the system. Karlen et al. [29] defined soil quality as "the capacity of a specific kind of soil to function, within natural or manages ecosystem boundaries, to sustain plant and animal productivity, and support human health and habitation." Mine soils are heterogeneous and may reflect soil characteristics that are different from the original soils. Therefore, assessing mine soil quality is vital for predicting the path of restoration and recuperation of the soil.

The development of soil quality indexes to assess soil quality is widely practiced, and numerous quality indexes have been developed in recent years. Soil quality indexes (SQIs) integrate mathematical tools and goals of soil productivity; therefore, it is a tedious task to put both objectives together. Formulation of a SQI involves the following steps:

1) Screen the soil properties (minimum data set, MDS), which can reflect changes in soil using principal component analysis (PCA).
2) Transform the MDS into a numeric score that can be provided in various units as numbers called *indicators*. To calculate the score of an individual soil parameter, we used an equation based on a sigmoidal curve with an asymptote ranging from 0 to 1:

$$S = \frac{a}{\left[1 + \left(\frac{x}{xo}\right)^{b}\right]} \tag{12.1}$$

where x is the value of an individual soil property, a is the maximum score (1) of the soil property, x_0 is the mean value of each soil property, and b is the slope of the equation.
The slope used was −2.5 for "more is better" and +2.5 for "less is better" to obtain a sigmoidal curve [30, 31].
3) Calculate the using scoring values and a weighting factor for each PC with the following weighted addition equation [32]:

$$SQI = \sum_{i=1}^{n} W_i S_i \tag{12.2}$$

where Wi is the weighting factor of the SQIs and Si is the nonlinear score.

The SQI was developed to evaluate reclaimed mine soil quality with different tree covers using PCA analysis and found a high SQI for *Dalbergia sissoo* stands [33]. The authors also reported that coarse fractions, soil pH, SOC, DHA, Ca, EC, P, and S are significant parameters that indicate changes during reclamation and soil quality. A reclaimed mine soil index (RMSI) was developed to screen out plant species for reclamation of coal mine degraded land and found that *Cassia seamea* and *D. sissoo* have higher rhizosphere soil quality and are recommended for reclamation [34].

Sinha et al. [35] used PCA analysis to derive a rhizosphere soil microbial index (RSMI) and found dehydrogenase, BSR/MBC (basal soil respiration / microbial biomass carbon), MBC/SOC (microbial biomass carbon / soil organic carbon), electrical conductivity (EC), phenol oxidase, and AMBC (active microbial biomass carbon) to be the most critical properties. The tree species grown in mine soil show different RSMI values and are as follows in decreasing order: *Aegle marmelos* (0.718) > *Azadirachta indica* (0.715) > *Bauhinia veriegeta* (0.693) > *Butea monosperma* (0.611) > *Eugenia jambolana* (0.601) > *Moringa oleifera* (0.565) > *D. sissoo* (0.498) > *Tamarindus indica* (0.488) > *Morus alba* (0.415) > *Ficus religiosa* (0.291) > *Eucalyptus sp.* (0.232) > *Tectona grandis* (0.181). Mukherjee and Lal [36] assessed the comparative performance of three SQIs: (i) simple additive SQI (SQI-1), (ii) weighted additive SQI (SQI-2), and (iii) statistically modeled SQI (SQI-3) based on PCA and reported that SQI-3 was most strongly correlated ($r = 0.74$–0.78) with crop yield.

12.7 Development of a Post-Mining Ecosystem: A Case Study in India

A field study was carried out to restore a post-mining ecosystem by afforestation [37]. The study is based on the hypothesis that mine soil quality in the post-mining ecosystem depends on tree growth and time since afforestation. This study's objective was to analyze the change in plant biodiversity and mine soil quality along with the chronosequence age of afforestation and estimate the carbon sequestration potential of the post-mining ecosystem. The authors also developed the Technosol quality index (TQI) to assess the mine soil quality and compared the restored ecosystem's attributes with the natural forest. This case study was conducted on reclaimed and afforested mined land in Jharkhand, India. The authors selected one unreclaimed site (0 years) that contained fresh overburden material and four chronosequence afforested mined areas of age 3, 7, 10, and 15 years (Figure 12.4). A field survey was conducted to study plant species composition and biomass carbon stock, and species diversity was calculated using different biodiversity indices. Soil samples were collected from all the study sites and analyzed for physicochemical and biological characteristics.

A total of 15 tree species (10 families) were identified at the afforested site; the natural forest had 4 species, each of which belonged to a unique family (Table 12.1). Plant diversity indices showed that species diversity was highest at the 15-year-old afforested site and lowest in the forest. Tree biomass carbon was lowest in the 3-year-old afforested site (1.43 Mg C/ha), increased significantly with the age of afforestation, and was greatest in the 15-year-old afforested site (67.6 Mg C/ha). Differences in unreclaimed, afforested, and natural forest soils were significant in terms of pH, sand, silt and clay, bulk density, SOC, total nitrogen, NPK (nitrogen, phosphorus, potassium), microbial biomass carbon, and dehydrogenase enzyme activity (Table 12.2).Soil organic carbon stock showed a fourfold increase after 3 to 15 years of afforestation. However, the reference forest site still exhibits an ecosystem carbon pool 58% greater than the 15-year-old afforested site. Soil CO_2 flux showed a linear increase with the age of afforestation and was highest in the 15-year-old afforested site, which was 20% lesser than the natural forest (3.99 μmol $CO_2/m^2/s$).

Figure 12.4 Photos of the study site showing (a) unreclaimed dump with no vegetation; and development of vegetation cover on (b) 3-year-old afforested dump; (c) 7-year-old afforested site; (d) 10-year-old afforested site; (e) 15-year-old afforested site; (f) natural forest in the area. *See colour plate section for colour representation of this figure.*

Table 12.1 Plant species growing at the afforested mine land and natural forest sites.

Afforested mine land	Natural forest
Acacia catechu (L.f.) Willd.	*Acacia catechu* (L.f.) Willd.
Azadirachta indica A. Juss	*Butea monosperma* (Lam.)
Acacia auriculiformis A.Cunn. ex Benth.	*Ficus benghalensis* L.
Butea monosperma (Lam.)	*Shorea robusta* Roth
Bombax ceiba L.	
Dalbergia sissoo Roxb.	
Eucalyptus spp.	
Ficus racemose L.	
Heterophragma adenophyllum Seem.	
Holoptelea integrifolia (Roxb.) Planch	
Leucaena leucocephala (Lam.) de Wit	
Melia azedarach L.	
Senna siamea (Lam.) Irwin et Barneby	
Tremma orientalis (L.) Br.	
Ziziphus jujuba Mill	

The authors developed the TQI using PCA and identified soil indicator parameters that define the afforested post-mining ecosystem's soil quality. PCA showed higher loading of dehydrogenase activity, available nitrogen, and silt content, and thus these parameter loadings were used to develop the minimum dataset. Then the score of each individual parameter and weight of the respective parameter were used to calculate the TQI. The results showed that DHA had a

Table 12.2 Changes in selected soil properties in chronosequence afforested soil and natural forest soil.

		Chronosequence afforested mine land				
Soil properties	**Unreclaimed dump**	**3 year**	**7 year**	**10 year**	**15 year**	**Natural forest**
pH	8.34	6.38	5.61	5.48	5.38	5.18
BD (Mg/m^3)	2.14	1.85	1.71	1.68	1.53	1.64
SOC (%)	0.15	0.20	0.33	0.48	0.58	0.90
Total N (%)	0.016	0.018	0.10	0.17	0.24	0.31
Av-N (mg/kg)	18.7	26.7	71.3	76.3	116	88.8
Av-P (mg/kg)	1.68	2.79	4.43	4.65	6.81	5.07
Ex-K (mg/kg)	17.8	65.2	72.7	63.4	127	141
MBC (mg/kg)	18.2	58.3	122	168	254	224
DHA (μg TPF/g/h)	7.21	14.6	36.6	49.9	65.7	61.7

BD, bulk density; SOC, soil organic carbon; Av-N, available nitrogen; Av-P, available phosphorus; Ex-K, exchangeable potassium; MBC, microbial biomass carbon; DHA, dehydrogenase activity.

maximum contribution in the TQI, followed by available nitrogen and silt content. After that, linear regression analysis was performed between the TQI values and vegetation characteristics to determine the relationship between plant and soil characteristics. The greater R^2 values for the TQI and Shannon diversity index showed that soil quality is highly dependent on plant biodiversity in an afforested post-mining ecosystem. Overall, the study concluded that afforestation of the post-mining ecosystem increases soil fertility, the ecosystem carbon pool, and technosols quality where DHA, available nitrogen, and silt content act as indicator parameters.

12.8 Conclusions and Future Research

Surface mining methods are widely used to excavate minerals of interest, which results in ecosystem disturbance and environmental pollution. To offset pollution, mine-degraded lands should be reclaimed and ecologically restored by integrating ecological theories and restoration ecology. Ecological restoration practices must restore every aspect of the ecosystem; however, some ecosystem services simply cannot be engineered.

Restoration of degraded ecosystems has often lacked scientific recommendations. The scope of ecological restoration must shift from establishing a naturalistic biological community in the degraded ecosystem to restoring ecological functions, particularly those that impart ecosystem services. Such systems should have functioning food chains, nutrient cycling, energy flow, prey-predator relationships, and feedback loops. In all cases, ecological restoration will improve the biological diversity of degraded landscapes and enhance landscape connectivity, reduce pollution, and contribute to human wellbeing. Ecological evaluation should integrate all of the necessary management options to achieve economies of time, energy, and cost.

Research related to the role of a restored ecosystem in climate change and greenhouse gas emissions needs much attention in this continuously changing environment. Again, shifting from invasive species to native species to restore the biodiversity and functions of an ecosystem is more beneficial to a productive ecosystem. Advanced satellite technology and real-time monitoring of ecosystems need to be implemented at restored sites. Setting realistic, socio-environmental, and scientific goals for restoration is of prime importance under the dynamic restoration approach. Successful restoration needs to include a participatory approach involving the local community, scientific personnel, economists, and land managers.

References

1 Society for Ecological Restoration (SER). (2004). SER international primer on ecological restoration. Society for Ecological Restoration International Science & Policy Working Group, Version 2.

2 UNFCC (2012). *United Nations Framework Convention on Climate Change*. Doha: Amendment.

3 Ahirwal, J., Maiti, S.K., and Singh, A.K. (2017). Changes in ecosystem carbon pool and soil CO2 flux following post-mine reclamation in dry tropical environment, India. *Science of the Total Environment* 583: 153–162.

4 IPCC, 2010. Mastrandrea, M.D., C.B.Field, T.F.Stocker, O.Edenhofer, K.L.Ebi, D.J.Frame, H.Held, E.Kriegler, K.J.Mach, P.R.Matschoss, G.-K.Plattner, G.W.Yohe, and F.W.Zwiers, Guidance note for lead authors of the IPCC Fifth Assessment Report on Consistent Treatment of Uncertainties. Intergovernmental Panel on Climate Change (IPCC).

5 Ussiri, D.A. and Lal, R. (2005). Carbon sequestration in reclaimed minesoils. *Critical Reviews in Plant Sciences* 24 (3): 151–165.

6 Bradshaw, A. (1997). Restoration of mined lands—using natural processes. *Ecological Engineering* 8 (4): 255–269.

7 Maiti, S.K. (2013). *Ecorestoration of the Coalmine Degraded Lands*. Springer Science & Business Media, Springer India.

8 Roberts, J.A., Daniels, W.L., Burger, J.A., and Bell, J.C. (1988). Early stages of mine soil genesis in a Southwest Virginia spoil lithosequence. *Soil Science Society of America Journal* 52 (3): 716–723.

9 Sencindiver, J.C. and Ammons, J.T. (2000). Minesoil genesis and classification. *Reclamation of Drastically Disturbed Lands* 41: 595–613.

10 Jha, A.K. and Singh, J.S. (1991). Spoil characteristics and vegetation development of an age series of mine spoils in a dry tropical environment. *Vegetatio* 97 (1): 63–76.

11 Brady, N.C. and Weil, R.R. (1996). Soils and chemical pollution. In: *The Nature and Properties of Soils* (eds. N.C. Brady and R.R. Weil), 897–951. USA: Prentiss Hall Intnal.

12 Johnson, C.D. and Skousen, J.G. (1995). Minesoil properties of 15 abandoned mine land sites in West Virginia. *Journal of Environmental Quality* 24 (4): 635–643.

13 Shrestha, R.K. and Lal, R. (2006). Ecosystem carbon budgeting and soil carbon sequestration in reclaimed mine soil. *Environment International* 32 (6): 781–796.

14 Ahirwal, J. and Maiti, S.K. (2016). Assessment of soil properties of different land uses generated due to surface coal mining activities in tropical Sal (*Shorearobusta*) forest, India. *Catena* 140: 155–163.

15 Singh, A.N. and Singh, J.S. (2006). Experiments on ecological restoration of coal mine spoil using native trees in a dry tropical environment, India: a synthesis. *New Forests* 31 (1): 25–39.

16 Jouquet, P., Dauber, J., Lagerlöf, J. et al. (2006). Soil invertebrates as ecosystem engineers: intended and accidental effects on soil and feedback loops. *Applied Soil Ecology* 32 (2): 153–164.

17 Nichols, O.G. and Nichols, F.M. (2003). Long-term trends in faunal recolonization after bauxite mining in the jarrah forest of southwestern Australia. *Restoration Ecology* 11 (3): 261–272.

18 Miyawaki, A. (1992). Restoration of evergreen broad-leaved forests in the Pacific region. In: *Ecosystem Rehabilitation. 2. Ecosystem Analysis and Synthesis* (ed. M.K. Wali), 233–245. The Hague: SPB Academic Publishing.

19 Powell, J.A. (1988). Biological objectives and the management of living marine resources: a compilation of international agreements and US laws. In: *Fishery Science and Management* (ed. W.S. Wooster), 295–330. New York, NY: Springer.

20 Pecharová, E., Broumová-Dušáková, H., Novotná, K., and Svoboda, I. (2011). Function of vegetation in new landscape units after brown coal mining. *International Journal of Mining, Reclamation and Environment* 25 (4): 367–376.

21 Hendrychová, M. (2008). Reclamation success in post-mining landscapes in the Czech Republic: a review of pedological and biological studies. *JournalofLandscape Studies* 1: 63–68.

22 Ahirwal, J., Maiti, S.K., and Singh, A.K. (2016). Ecological restoration of coal mine-degraded lands in dry tropical climate: what has been done and what needs to be done? *Environmental Quality Management* 26 (1): 25–36.

23 Bandyopadhyay, S. and Maiti, S.K. (2019). Evaluation of ecological restoration success in mining-degraded lands. *Environmental Quality Management* 29 (1): 89–100.

24 Juwarkar, A.A. and Jambhulkar, H.P. (2008). Phytoremediation of coal mine spoil dump through integrated biotechnological approach. *Bioresource Technology* 99 (11): 4732–4741.

25 Ahirwal, J. and Maiti, S.K. (2018). Development of Technosol properties and recovery of carbon stock after 16 years of revegetation on coal mine degraded lands, India. *Catena* 166: 114–123.

26 Masto, R.E., Sheik, S., Nehru, G. et al. (2015). Environmental soil quality index and indicators for a coal mining soil. *Solid Earth Discussions* 7 (1).

27 Mummey, D.L., Stahl, P.D., and Buyer, J.S. (2002). Microbial biomarkers as an indicator of ecosystem recovery following surface mine reclamation. *Applied Soil Ecology* 21 (3): 251–259.

28 Jaunatre, R., Buisson, E., Muller, I. et al. (2013). New synthetic indicators to assess community resilience and restoration success. *Ecological Indicators* 29: 468–477.

29 Karlen, D.L., Andrews, S.S., Wienhold, B.J., and Zobeck, T.M. (2008). Soil quality assessment: past, present and future. *Journal of Integrative Biosciences* 6: 1.

30 Bastida, F., Moreno, J.L., Hernandez, T., and García, C. (2006). Microbiological degradation index of soils in a semiarid climate. *Soil Biology and Biochemistry* 38 (12): 3463–3473.

31 Mukhopadhyay, S., Maiti, S.K., and Masto, R.E. (2014). Development of mine soil quality index (MSQI) for evaluation of reclamation success: a chronosequence study. *Ecological Engineering* 71: 10–20.

32 Andrews, S.S., Karlen, D.L., and Mitchell, J.P. (2002). A comparison of soil quality indexing methods for vegetable production systems in Northern California. *Agriculture, Ecosystems & Environment* 90 (1): 25–45.

33 Mukhopadhyay, S., Masto, R.E., Yadav, A. et al. (2016). Soil quality index for evaluation of reclaimed coal mine spoil. *Science of the Total Environment* 542: 540–550.

34 Mukhopadhyay, S., Maiti, S.K., and Masto, R.E. (2013). Use of Reclaimed Mine Soil Index (RMSI) for screening of tree species for reclamation of coal mine degraded land. *Ecological Engineering* 57: 133–142.

35 Sinha, S., Masto, R.E., Ram, L.C. et al. (2009). Rhizosphere soil microbial index of tree species in a coal mining ecosystem. *Soil Biology and Biochemistry* 41 (9): 1824–1832.

36 Mukherjee, A. and Lal, R. (2014). Comparison of soil quality index using three methods. *PLoS One* 9 (8).

37 Ahirwal, J., Kumar, A., Pietrzykowski, M., and Maiti, S.K. (2018). Reclamation of coal mine spoil and its effect on Technosol quality and carbon sequestration: a case study from India. *Environmental Science and Pollution Research* 25 (28): 27992–28003.

13

Wetland, Watershed, and Lake Restoration

Bhupinder Dhir

School of Sciences, Indira Gandhi National Open University, New Delhi, India

13.1 Introduction

Aquatic systems and water resources have been tremendously altered and exploited by humans to fulfill needs of water, food, energy, and recreation. Anthropogenic activities have modified the structure and functioning of water bodies and caused deterioration in water quality (through the addition of contaminants such as chemicals, organic compounds, and metals). Changes in land-use practices and overexploitation of water resources have led to a drastic degradation in the status of freshwater resources [1]. Over the years, watercourses have faced excessive deterioration because of the input of pollutants such as chemicals, fertilizers, pesticides, and organic compounds and the inflow of wastewater, sewage, and other wastes [2]. In recent years, bodies of water have undergone changes in morphology, such as channelization; thus, revival and restoration of freshwater resources are essential to ensure ecosystem integrity and sustainability [3]. This alarming situation has prompted a search for methodologies that can renovate wastewater or restore existing water resources.

Restoration is a multifaceted process and involves knowledge of various disciplines such as landscape ecology, geomorphology, hydrology, soil, water physics, chemistry, and biology [4, 5]. Restoring of water bodies can be planned only if we know what modifications affected the water's natural processes. These include changes in physical properties such as water flow velocity and reduced water retention [6].Restoring bodies of water is important because it helps to mitigate water loss, increases the functioning of water bodies, improves groundwater recharge, maintains hydrology, and improves the capacity for water retention [7, 8]. One of the primary aims of restoring water resources is to reestablish habitat and native species, eliminate undesirable species, and remove contamination [4, 9, 10]. The holistic approach involves maintaining the physical and chemical conditions of soil and water and manipulating biological attributes (mainly via reintroduction of native flora and fauna) [11]. The recovery of a degraded, damaged, or destroyed water ecosystem is the aim of ecological restoration.

13.2 Renovation of Wastewater

Treating wastewater facilitates the recycling and recovery of valuable components. Wastewaters from different sources such as municipalities, industrial operations, domestic and agricultural activities are contaminated with solids such as sewage and additives such as detergents, grease,

Handbook of Ecological and Ecosystem Engineering, First Edition. Edited by Majeti Narasimha Vara Prasad.

oils, chemicals, heavy metals, etc. Renovation of wastewater improves its quality so that it can be safely disposed of and hazardous environmental effects can be avoided. Various physical, chemical, and biological techniques have been used to renovate or treat wastewater for decades [12–14].

13.2.1 Physical Methods

Physical methods such as screening, sedimentation, and skimming remove solids from wastewater. The sedimentation process removes insoluble/heavy particles. Insoluble/heavy material settles at the bottom and later is removed. Filtration removes contaminants or insoluble particles from the surface of the wastewater. Adsorption removes a wide range of organic materials, including detergents and toxic compounds. Adsorbents such as activated alumina, clay colloids, hydroxides, resins, and activated carbon remove soluble molecules (adsorbate) by attaching them to the surface of the solid substrate (adsorbent).

13.2.2 Chemical Methods

Chemical treatments follow the primary (physical) treatment of wastewater [15]. These methods remove large amounts of dissolved, colloidal materials and toxic compounds. The chemical treatment processes include precipitation (coagulation, flocculation), ion exchange, neutralization, and disinfection (chlorination/dechlorination, ozone, UV light). Dissolved inorganic components can be removed by adding an acid or alkali. Neutralization regulates the pH of the wastewater. Lime (CaO), calcium hydroxide ($Ca(OH)_2$), sodium hydroxide (NaOH), and sodium carbonate (Na_2CO_3) are commonly used for neutralizing or adjusting pH. The precipitate formed after treatment can be removed by sedimentation or flotation.

In the disinfection method, water is treated with chemical disinfectants such as chlorine, hypochlorite, ozone, ultraviolet radiation, and bromine that remove or inactivate pathogens. Ozone (O_3) is a powerful oxidant and oxidizes inorganic and organic compounds in wastewater. Ultraviolet (UV) radiation possesses good bactericidal properties and kills microorganisms.

In the ion exchange process, a charged ion in a solution is exchanged with a similarly charged ion attached to an immobile solid particle. Wastewater is introduced into a bed of resin, and sodium is exchanged with cations in the solution. The bed becomes saturated with the exchanged ions. The process results in the softening of the water. The high cost of chemicals and environmental issues such as sludge generation limit the applicability of this process for wastewater treatment.

13.2.3 Biological Methods

Secondary treatment, i.e. biological treatment of wastewater, involves removing suspended solids using microorganisms such as algae, fungi, or bacteria under aerobic or anaerobic conditions. Microorganisms transform dissolved organic matter into a dense biomass. The microorganisms use dissolved organic matter as food. Aerobic treatment includes oxidation ponds, aeration lagoons, aerobic bioreactors, activated sludge, percolating or trickling filters, and biological filters. During the aerobic process, bacteria decompose organic matter in the presence of oxygen and convert it into carbon dioxide that can be used by plants. In a bioreactor, organic matter is utilized by microorganisms such as bacteria (aerobically or anaerobically), algae, and fungi (aerobically). In the activated sludge process, a thick mixture of bacterial populations is suspended in the wastewater under aerobic conditions [16].

Membrane bioreactors (MBRs) show high treatment efficiency for removing a wide range of organic and inorganic contaminants, including pesticides, pharmaceutical compounds, biodegradable organic compounds, industrial chemicals, and nutrients.

Anaerobic/aerobic systems have been widely used at both the municipal and industrial levels to treat wastewater for many years. These processes use less energy and chemicals, produce less sludge, and are simpler to operate (require less equipment).

Bioremediation processes such as phytoremediation, bioaugmentation, and biostimulation convert hazardous materials into less toxic or non-toxic materials. Phytoremediation involves using of plants for the abatement of pollutants in contaminated soils or water. The plants possess the capacity to accumulate, degrade, or eliminate contaminants such as heavy metals, pesticides, solvents, explosives, organic compounds, and many others. Root zone technology involves the engineered use of green plants such as grasses and woody species to remove environmental contaminants from soil/water or render them harmless [17]. Plant-influenced biological, chemical, and physical processes help in the uptake, sequestration, and metabolism of contaminants by plants or free-living organisms present in the rhizosphere. The plant root system's unique, selective uptake capabilities assist in the translocation, bioaccumulation, and degradation of contaminants. During transpiration, nutrients are taken up from the soil to leaves and stems. Along with transpired water, contaminants in the soil are also taken up, sequestered, metabolized, or vaporized out of the leaves. Domestic and industrial effluents are treated effectively and naturally with this technology.

Phytoremediation has been used in constructed wetlands to treat wastewater and remove contaminants via processes such as rhizofiltration, rhizodegradation, phytodegradation, phytoaccumulation, and phytotransformation. The leaves and stems of wetland plants absorb oxygen from the atmosphere and transfer it down to the stems and roots. The plants conduct oxygen through their stems into their root systems and create favorable conditions for bacteria growth. The roots transfer this oxygen to the soil through their weak membranes. Wastewater flowing through the root zone horizontally or vertically degrades the organic pollutants by means of bacteria present in the rhizosphere of plants. Organic pollutants are removed in significant amounts from wastewater, and reduced elemental forms are found abundantly. The application of water hyacinth as a low-cost, efficient water purifier is increasing worldwide because of its potential to remove toxic organic wastes, heavy metals, and radioactive wastes. It is considered a solution for fulfilling the need for water supply, environment protection, energy, fertilizer, and raw material for various industries.

Biological treatments have some limitations, such as sensitivity to variations in hydraulic loads and alteration in functioning because of effects on microorganism growth due to temperature changes. A decline in water temperature by 10 °C reduces the rate of biological reactions by half.

13.2.4 Other Methods

Reverse osmosis has been widely used for the desalination of water and wastewater. Chemical oxidation (O_3, Fenton reagents), photochemical oxidation (ultraviolet UV/O_3, UV/H_2O_2), heterogeneous photocatalysis (UV/TiO_2), and electrolysis are other modern techniques used for improving water quality. The photocatalytic treatment of wastewater using titanium oxide (TiO2) has been very effective in purifying water. Other semiconductors used for water treatment include ZnO, Fe_2O_3, CdS, and ZnS. The crystalline structure, surface area, hydroxyl groups, and adsorption/desorption characteristics of these semiconductors play a vital role in the photocatalytic degradation process. Bacteria, fungi, and viruses are killed very quickly.

The photodegradation technology for treating wastewater uses sunlight. Green plants trap solar energy from sunlight, and a series of enzyme-catalyzed redox processes convert the CO_2 into water and carbohydrates by releasing oxygen into the atmosphere. Treatment with UV-A and UV-C radiation assists in the degradation of many environmental contaminants.

Membranes are used as a thin layer barrier for removing contaminants. These are used along with chemical and biological treatment systems. In this technique, a semipermeable membrane acts as a barrier, controls the movement of some components by penetration, and assists in the rejection of others through pores. Microfiltration (MF), ultrafiltration (UF), and nanofiltration (NF) techniques are used today to treat water effectively. Microfiltration is a type of sieve mechanism that retains macromolecules or particles larger than 0.1 μm. It helps remove suspended solids, bacteria, some viruses, protozoan cysts, and turbidity. Nanofiltration generally follows chemical precipitation and is removes organic compounds, phosphorus, hardness, polyvalent cations, metals, reduction in color, tannins, and turbidity. Membrane technology has low energy requirement but high wastewater treatment efficiency. Technology applications include microbial fuel cells, removing organic and inorganic components, disinfection, pathogen removal, and desalination. Although advantages include no sludge production, quick reaction rate, and low operating and maintenance costs, there are also limitations such as the consumption of large amounts of oxidant.

All kinds of water treatment procedures help reduce the level of contamination and restore water quality. Combined water treatment using primary, secondary, and tertiary techniques is effective in removing suspended solids, biochemical oxygen demand (BOD), nutrients such as nitrates, and phosphates from municipal wastewater. An amalgamation of conventional wastewater treatment methods (physical, chemical, biological) and modern-day technologies is useful in treating wastewater to the level that it is non-hazardous while improving its quality.

13.3 Restoration of Bodies of Water

The restoration process includes reducing/eliminating degradation sources, recovery, and maintenance of the natural potential, i.e. regeneration. Restoration helps in maintenance of ecological integrity of the system [18]. *Ecological integrity* is the maintenance of native communities and the establishment of ecosystems based on natural processes. Restoration can be of two types: (i) reestablishment, in which the water body returns to its original state, and (ii) rehabilitation, in which a degraded/altered body of water shows improvement. The restoration of water resources should include a range from single river corridors to entire watersheds. Both passive and active processes improve bodies of water and overcome degradation by incorporating specific changes that help mitigate or compensate for loss of water, increase the functioning of water bodies, maintain hydrology, and improve water retention [19–23]. The key processes of restoration include maintaining water levels, flow patterns, and physical features such channels; changing dynamics of sediment erosion; and establishing native biological communities [24].

Aquatic resources that require restoration show alterations in physical characteristics such as channels and flow regimes as well as habitat degradation. A restoration project needs to focus on structural alterations such as stream channelization, shoreline modifications, and detachment from adjacent ecosystems; maintaining structures; composition; and natural processes of biotic communities and their physical environment. An ecosystem with integrity is a resilient and self-sustaining natural system that can accommodate stress and change. To restore any water resource, we need to know the following:

- Structure of the degraded site
- Conditions that existed before degradation
- Changes in land use

- Changes in the hydrological regime and nutrient fluxes
- Topography of site
- Structural (state) and functional (process) attributes of the process

Restoration, conservation, and mitigation need to be carried out on a watershed scale. Reestablishing the natural structure of river corridors, lakes, wetlands, estuaries, and other aquatic resources can bring back beneficial functions. This is because the structure and function of water resources are closely linked. Mitigation design should incorporate hydrologic functioning, while biological dynamics should be evaluated in terms of the population present in the region and the ecological requirements.

Preventing degradation should be the first step of any restoration project. Other methods that need to be used in a restoration project should include:

- Removing encroachments near or around the body of water.
- Cleaning the body of water using in situ measures such as de-silting, de-weeding, bioremediation, etc.
- Regular monitoring of inlets and outlets of the body of water.
- Treatments such as afforestation, water drainage management, and silt treatment in the catchment area.
- Preventing sewage inflow into the body of water.
- Preventing dumping of solid waste near watershed areas.
- Bioengineering techniques that use the potential of live and dead plants or inorganic materials to create systems that prevent erosion, control sediment, stabilize banks, mitigate floods, and treat water. The restoration of vegetation on river banks enhances the natural decontamination of runoff.
- Using invasive, non-native species to overcome problems associated with the vulnerability to invasion.

13.3.1 Watersheds

An area from where water drains to a common waterway such as a lake, river, or ocean is called a *watershed* [25, 26]. The flow of water through the watershed carries along with it materials like sediment and nutrients present in the drainage basin. Anthropogenic activities have added pollutants such as fertilizers, pesticides, and sewage to watersheds, thereby degrading them along the drainage basin [27–29]. Watersheds have been altered by changes in water input, sediments, nutrients, and the entry of contaminants into the main channel.

Restoring a watershed requires a design-based approach. Activities throughout the watershed need to be rectified to prevent adverse effects on the aquatic resource. Reference sites can be used as models for restoration projects. *Reference sites* are areas that are comparable in structure and function to the restoration site before its degradation [30].Various practices proposed for the management of watershed include:

- Knowledge of watershed components and processes (data gathering)
- Knowledge of the problems associated with the site (requirements for protection or restoration)
- Knowledge of methods of management and their implementation
- Assessment of environmental and economic impacts of management
- Feasibility of techniques
- Development of good management strategies and comprehensive implementation procedures as approved by decision-makers [31, 32]

13.3.2 Wetlands

Wetlands occupy about 6.4% of the earth's surface. They perform various functions beneficial to the environment and life on earth, including improving water quality, providing habitat for plants and animals, and providing breeding grounds and feeding sites (nursery) for invertebrates, amphibians, birds, and mammals [33]. Constructed wetlands provide many services to the ecosystem [34, 35] (Figure 13.1). They have been found effective in treating municipal wastewaters. The primary functions of wetlands include:

- Improving water quality by removing suspended solids and inorganic and organic contaminants from surface waters. Pollutants are removed via sedimentation and chemical detoxification processes.
- Effective storage of rainwater and a primary source for recharging groundwater supplies.
- Control of erosion via vegetation.
- Contributing to shoreline stabilization and storm protection via dissipation of force and protection against wind and wave action.
- Retention of excess floodwaters.
- Retention of nutrients like nitrogen and phosphorus that accumulate in the sub-soil, thereby decreasing eutrophication.
- Socioeconomic services such as water supply, fisheries, fuel wood, livestock grazing, energy resource, recreation, and tourism.

The functional properties of wetlands play a role in maintaining ecological balance. Wetland efficiency for wastewater treatment depends on the substrate, hydrology, surface loading rate, composition of vegetation, and climatic factors such as temperature.

About 50% of the world's wetlands have been lost in the past century because of urbanization, increase in development (excavation, filling, draining), and agricultural activities [36]. The loss of wetland area, changes in water regime, and decreased water quality have been noted as the major problems associated with wetlands [37]. It is essential to manage and restore wetlands and prevent wetland degradation to protect biodiversity and regional climates and maintain ecological security.

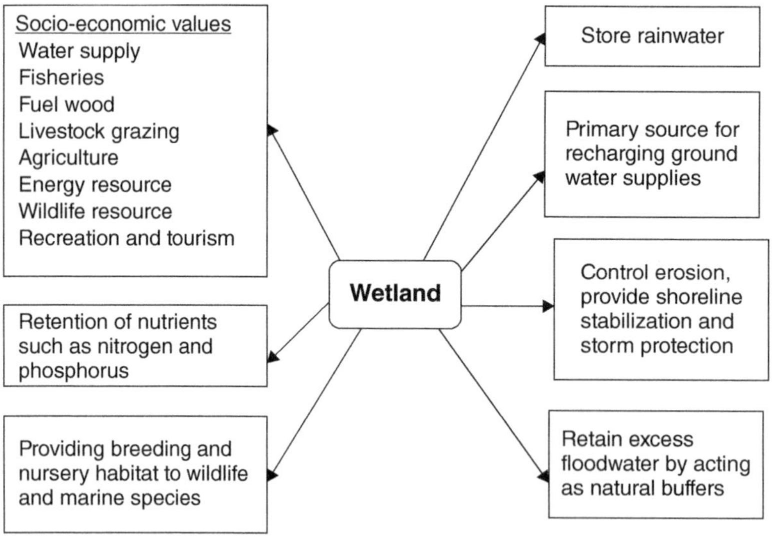

Figure 13.1 Major functions of a wetland. *Sources:* Stewart et al. [34]; Watson et al. [35].

The restoration of wetlands is required to curtail the following:

- Hydrologic alterations such as changes in wetland structure and functioning
- Changes in surface drainage that reduce water due to alteration in surface water inflows and channelization [36]
- Increased pollutant loading from non-point sources and point sources (wastewater discharges)
- Increase in sedimentation, nutrient loading, and organic matter deposition
- Changes in characteristics of wetland flora

The process of restoring a wetland involves reviving it where it formerly existed (reestablishment) or reestablishing it in nearby (rehabilitation). Restoration includes protection and preservation to improve water quantity and quality [38]. The hydrologic and nutrient cycles and nitrogen dynamics (nitrogen fixation and denitrification) must be retained to restore wetlands.

13.3.2.1 Methods of Restoring Wetlands

Restoring wetlands mainly includes methodologies such as (i) reestablishment or management of hydrology, (ii) elimination or control of contaminants, and (iii) reestablishment or management of native biota. Hence, at a simple level, restoration involves reestablishing native biota, while at a complex level, it involves maintaining hydrology and reintroducing native flora and fauna [38–42]. Restoration techniques can be applied once an analysis of wetland functions has been done and their replacements have been designed. The restoration program should include checking changes in water hydrology, maintaining hydrologic regimes including soil, topography, reduction in loss of surface waters (due to dewatering of the landscape through ditching and groundwater pumping), and establishing vegetation in proper proportions. Removing contaminants to reestablish bodies of water should also be included in the restoration project. Restoring the bottom part of the wetland can play an important role in reestablishing the hydrological regime and nutrient fluxes. The requirements for restoration of a wetland differ depending on whether it is coastal-estuarine, riverine, depressional, or freshwater [39]. After restoration, the wetland performs the same functions and possesses the same biological conditions as existed before the damage.

13.3.3 Rivers

Rivers serve various roles in people's lives. These include generating electricity, water supply, food production, and sediment transport [43, 44]. Rivers have experienced tremendous pressure and degradation from human activities in the last few decades. Construction of dams hampers hydrological connectivity and alters flow regimes and ecological operations. The main problems associated with rivers include changes in hydrological processes and deterioration of water quality resulting from pollution.

Serious consideration must be given to river dynamics, hydro-morphological processes (determine the living conditions in rivers), and realistic conditions before initiating a restoration project. Restoring rivers includes management ranging from structural modifications such as stabilizing banks to changes in ecosystem processes and biota [45, 46]. Projects related to the restoration of rivers involve modifying channels, adjacent riparian zones, floodplains, water status, sediment, and solute inputs [47–49].Measures such as removing built-up structures (embankments), fixing beds, and constructing structures that re-divert water or maintain flow back into river channels are the initial steps [50].Removing contaminant inflow and dams is also effective [51].Restoration techniques for rivers involve physical methods such as sewage interception, dredging, water diversion, etc.; chemical methods such as flocculation and adding chemicals like lime; and biological-ecological methods including aquatic plant restoration, microbial enhancement, biomembranes, and activated sludge. Restoration actions also include removing sediment deposits. Dredging removes pollutants from the bottom of bodies of water, thereby improving water quality [52]. All the treatment steps improve hydrologic, geomorphic, and/or ecological processes of the degraded water system and return it to its natural state [53–55].

Other steps in river restoration include straightening river channels to return them to a natural watercourse, removing bank fixations, shaping the structure of a natural watercourse, widening the corridor to increase structural variability, reviving the natural hydro-morphological character (from static to dynamic), and planting trees and shrubs on adjacent land (forestation in buffer zones) [55, 56]. Restoration also emphasizes promoting channel-floodplain connectivity, longitudinal connectivity, and partial restoration of water and sediment fluxes [23, 57, 58].

Restoration returns rivers and streams to their natural state and improves their ecological status. Restoration activities and measures involve municipalities, water management authorities, water maintenance associations, fishing associations, and citizen initiatives. Ecological research and restoration are essential for river restoration [59].

River restoration activities have increased in countries like the US, Australia, and many European countries in the past few decades [27, 57, 60–62].The Kissimmee River restoration project in Florida is considered a good example of reestablishing hydrology [63]. Restoration has resulted in significant changes in the water quality and ecosystem functions of these bodies of water [64].

13.3.4 Lakes

Lakes provide innumerable services to humanity, including water for irrigation, drinking, industry, hydroelectric power generation, recreation, and aesthetic pleasure. Changes in agriculture, riparian land use, forestry, and increasing demand for ecosystem services have affected these systems on a large scale. Acceleration in the flow of elements such as phosphorus from sewage and agricultural activities, overfishing, and the growth of exotic species have resulted in the degradation of lakes. Since increasing human impacts affect the ecosystem services provided by lakes, measures need to be taken to restore them [65].

Lake restoration mainly includes removing stress loadings such as contaminants, preventing runoff from agricultural and urban areas, and improving aquatic ecosystems and water quality. The quality of water in lakes can be maintained by reducing nutrient retention time. The reduction in retention time reduces the input of nutrients and decreases eutrophication. Diversion of water and canalization also help prevent eutrophication of lakes. Other processes involved in restoration include regulating hydrology to prevent floods.

13.3.5 Streams

The physical and biological degradation of streams has been caused by alterations in corridors and channels [58, 66, 67].Channelization affects energy dissipation, erosion, sediment transport, and deposition [52, 68]. The design of restoration for stream habitats and ecosystems includes steps such as assessment of problems and conditions leading to changes in structure and functioning. The main restoration programs include the reversal of watershed changes (such as channel widening) and maintaining physical habitat and biological communities [69, 70].

Effective restoration depends on stream reestablishment to the extent that it maintains a dynamic equilibrium [71, 72].Passive restoration reestablishes stable channels and floodplains, helps with regrowth of riparian vegetation, and improves river and stream habitats.

13.3.6 Case Studies

Few examples of lake restoration are found in the literature. One of the most studied cases is of Lake Washington located in King County. It is the second largest natural lake in the State of Washington. The lake showed deterioration in water quality after receiving discharges of municipal wastewater. The recovery of the lake occurred after the rapid flushing rate, low enrichment capacity, and low phosphorus content were altered. Another example is freshwater Lake Sammamish (Washington, US). The lake showed degradation because of eutrophication when waste from the municipal treatment plant was put into it. The lake showed slow recovery when steps such as a shift in nutrient dynamics, algal populations, and water clarity were followed.

Springfield Lake (Illinois, US) is an example of a restored reservoir. Depositions of agricultural soil over time (51 years) led to loss of storage capacity. Algal blooms, weeds, and turbidity affected the water quality and increased the cost of water treatment. The loading of silt and nutrients was found, and these were removed by hydraulic dredging and pumping water to adjacent farmland.

Lake Baldwin, located on the Orlando Ridge in Orlando, Florida (US), is another example of lake restoration. The lake became eutrophic and was stocked with 34 grass carp and *Hydrilla*. The plant was eradicated, and the growth of *Hydrilla* was controlled using grass carp biomass, thus restoring its natural potential.

Lake Trummen (Sweden) showed degradation after it received domestic wastewater and flax mill discharge for many years. Algal blooms and fish kills were noted in the lake. The loading was diverted from the lake, but no improvement was noted as the nutrient loading continued. A sediment-skimming treatment removed the enriched sediment materials. Carp, which became abundant, were also removed.

In Waseca, Minnesota (US), Clear Lake became eutrophic from sanitary sewage and urban and agricultural runoff. About 50% of the water load and 55% of the phosphorus load were diverted into a marsh. About 40% of the lake's annual phosphorus load was retained in the marsh. Excessive phosphorus loading led to high hypolimnetic oxygen demand and high internal phosphorus recycling in the lake.

Small bodies of water along the watershed of the Miyun reservoir in Beijing and Hebei (China) were restored. The restoration projects were undertaken to improve flood control, the ecology of the river with riparian zones and floodplains, and enhance the beauty of the area. Control measures such as enlarging river space were undertaken to control floods.

13.4 Problems Encountered in Restoration Projects

Restoration projects encounter many problems including site selection, topographic contouring, transplantation, inadequate nutrient supplies for plant reestablishment, improper drainage, inappropriate transplantation, human disturbances, insufficient monitoring, inappropriate removal of exotic species, and poor management of restored sites. Some of the primary limitations include an incomplete understanding of the physical and ecological processes in bodies of water.

Restoration projects do not address changes in water, sediment, nutrients, and contaminant inputs. River restoration remains constrained because of a lack of fundamental understanding of sediment dynamics. The most significant concern in restoration is its long-term success and their maintenance over time. Another subject of concern is the resilience of transformed water bodies such as watersheds and lakes. State agencies and university researchers should be included in planning, implementing, and evaluating restoration projects.

13.5 Conclusion

The success of restoration depends upon knowledge of hydrology, ecology, and engineering and cooperation with stakeholders such as water management authorities, nonprofit organizations, and other groups such as non-governmental organizations (NGOs). Effective restoration of aquatic habitats depends on reestablishing watershed processes to the extent that a dynamic equilibrium is maintained. It is necessary to analyze the site's recovery needs and determine the time and natural processes to assist in restoration. Analysis of a method's effectiveness for restoration, proper planning, and project management can ensure successful restoration.

Restoration projects need to be assessed considering ecological criteria (aiming to improve ecosystems). Restoration requires removing resilience mechanisms to maintain valuable systems such as wetlands and riparian waters. Restoring watershed scan be achieved using techniques that favor natural processes and sustenance of communities with native ecosystems over time. Resilience stabilizes valuable ecosystems. Restoration and renovation can ensure the supply of water and maintain water resources. Hence restoration, when combined with protection and preservation, can help achieve improvement in water resources.

References

1 U.S.EPA. (2011) Exposure Factors Handbook 2011 Edition (Final Report). U.S. Environmental Protection Agency, Washington, DC, EPA/600/R-09/052F.

2 Malmqvist, B. and Rundle, S. (2002). Threats to running water ecosystems of the world. *Environ. Conserv.* 29: 134–153.

3 Strayer, D.L. and Dudgeon, D. (2010). Freshwater biodiversity conservation: recent progress and future challenges. *J. N. Am. Benthol. Soc.* 29: 344–358.

4 van Andel, J., Grootjans, A.P., and Aronson, J. (2012). Unifying concepts. In: *Restoration Ecology: The New Frontier*, 2e (eds. J. van Andel and J. Aronson), 9–22. Oxford: WileyBlackwell.

5 Roni, P., Hall, J.E., Drenner, S.M., and Arterburn, D. (2019). Monitoring the effectiveness of floodplain habitat restoration: a review of methods and recommendations for future monitoring. *Wiley Interdiscip. Rev. Water* 6: 4.

6 Lewis, R.R. (2005). Ecological engineering for successful management and restoration. *Ecol. Eng.* 24: 403–418.

7 Beechie, T.J., Pess, G., Roni, P., and Giannico, G. (2008). Setting river restoration priorities: a review of approaches and a general protocol for identifying and prioritizing actions. *N. Am. J. Fish. Manag.* 28: 891–905.

8 Van Diggelen, R., Grootjans, A.P., and Harris, J.A. (2001). Ecological restoration: state of the art or state of the science? *Restor. Ecol.* 9: 115–118.

9 Palmer, M.A., Filoso, S., and Fanelli, R.M. (2014a). From ecosystems to ecosystem services: stream restoration as ecological engineering. *Ecol. Eng.* 65: 62–70.

10 Palmer, M.A., Hondula, K.L., and Koch, B.J. (2014b). Ecological restoration of streams and rivers: shifting strategies and shifting goals. *Annu. Rev. Ecol. Evol. Syst.* 45: 247–269.

11 Hwang, S., Lee, S., and Yoo, B. (2014). Ecological conservation and the restoration of freshwater environments in Korea. *Paddy Water Environ.* 12: 1–5.

12 Shukla, S.P., Kumar, S., Gita, S. et al. (2018). Recent technologies for wastewater treatment: a brief review. In: *Wastewater Management Through Aquaculture* (eds. B. Jana, R. Mandal and P. Jayasankar), 225–234. Singapore: Springer.

13 Tetteh, E.K., Rathilal, S., Chetty, M. et al. (2019). Treatment of water and wastewater for reuse and energy generation-emerging technologies. In: *Water and Wastewater Treatment*, Chapter 4 (ed. M. Eyvaz), 53–80. IntechOpen https://doi.org/10.5772/intechopen.84474.
14 Topare, N.S., Attar, S.J., and Manfe, M.M. (2011). Sewage/wastewater treatment technologies: a review. *Sci. Revs. Chem. Commun.* 1: 18–24.
15 Wang, Y.H., Yuan, B., Lü, S., and Zhu, Y.C. (2004). Preliminary study on river water treatment by chemically enhanced primary treatment technology. *Environ. Sci. Technol.* 27: 95–96.
16 Samer, M. (2015). Biological and chemical wastewater treatment processes. In: *Wastewater Treatment Engineering*, Chapter 1 (ed. M. Samer), 1–50. IntechOpen https://doi.org/10.5772/61250.
17 Trivedy, R.K. (2007). Low cost and energy saving technologies for water and wastewater treatment. *J. Ind. Pollut. Control.* 23: 403–411.
18 Mitsch, W.J., Lefeuvre, J., and Bouchard, V. (2002). Ecological engineering applied to river and wetland restoration. *Ecol. Eng.* 18: 529–541.
19 Cosgrove, W.J. and Loucks, D.P. (2015). Water management: current and future challenges and research directions. *Water Resour. Res.* 15: 4823–4839.
20 Davis, A.P., Hunt, W.F., Traver, R.G., and Clar, M. (2009). Bioretention technology: An overview of current practice and future needs. *J. Environ. Eng.* 135: 109–117.
21 Jansson, R., Nilsson, C., and Malmqvist, B. (2007). Restoring freshwater ecosystems in riverine landscapes: the roles of connectivity and recovery processes. *Freshw. Biol.* 52: 589–596.
22 Richter, B.D. and Thomas, G.A. (2007). Restoring environmental flows by modifying dam operations. *Ecol. Soc.* 12: 12.
23 Shafroth, P.B., Wilcox, A.C., Lytle, D.A. et al. (2010). Ecosystem effects of environmental flows: modelling and experimental floods in a dryland river. *Freshw. Biol.* 55: 68–85.
24 Hughes, F.M.R., Colston, A., and Mountford, J.O. (2005). Restoring riparian ecosystems: the challenge of accommodating variability and designing restoration trajectories. *Ecol. Soc.* 10: 12.
25 DeBarry, P.A. (2004). *Watersheds: Processes, Assessment, and Management.* Hoboken, NJ: Wiley.
26 Folliott, P.F., Gregersen, H.M., and DeBano, L.F. (2003). *Hydrology and the Management of Watersheds*, 3e. Ames, IA: Iowa State University Press.
27 Cooke, G.D., Welch, E.B., Peterson, S.A., and Newroth, P.R. (2005). *Restoration and Management of Lakes and Reservoirs*, 3e. Boca Raton, FL: Taylor and Francis.
28 Bohn, B.A. and Kershner, J.L. (2002). Establishing aquatic restoration priorities using a watershed approach. *J. Environ. Manag.* 64: 355–363.
29 Mitsch, W.J. and Gosselink, J.G. (2007). *Wetlands*, 4e. Hoboken, NJ: Wiley.
30 Weller, D.E. and Baker, M.E. (2014). Cropland riparian buffers throughout Chesapeake Bay watershed: spatial patterns and effects on nitrate loads delivered to streams. *J. Am. Water Resour. Assoc.* 50: 696–712.
31 Heathcote, I.W. (2009). *Integrated Watershed Management: Principles and Practice*, 2e, 453. New Jersey: Wiley.
32 Zhang, C. and Li, Y. (2016). Verification of watershed vegetation restoration policies, arid China. *Sci. Reports* 6: 30740.
33 Suhad, A.A.A.N.A., Abed, S.N., and Scholz, M. (2018). Wetlands for wastewater treatment and subsequent recycling of treated effluent: a review. *Environ. Sci. Pollut. Res. Int.* 25: 23595–23623.
34 Stewart, F.M., Mulholland, T., Cunningham, A.B. et al. (2008). Floating islands as an alternative to constructed wetlands for treatment of excess nutrients from agricultural and municipal wastes–results of laboratory-scale tests. *Land Contam. Reclamat.* 16: 25–33.
35 Watson, J.E.M., Jones, K.R., Fuller, R.A. et al. (2016). Persistent disparities between recent rates of habitat conversion and protection and implications for future global conservation targets. *Conserv. Lett.* 0: 1–9.

36 Moreno-Mateos, D., Power, M.E., Comín, F.A., and Yockteng, R. (2012). Structural and functional loss in restored wetland ecosystems. *PLoS Biol.* 10: e1001247.

37 Bai, J., Cui, B., Cao, H. et al. (2013). Wetland degradation and ecological restoration. *Sci. World J.* 2013: 523632.

38 An, S., Zhang, X., Song, S. et al. (2019). Wetland restoration in China: principles, techniques, and practices. In: *Wetlands: Ecosystem Services, Restoration and Wise Use. Ecological Studies (Analysis and Synthesis)*, vol. 238 (eds. S. An and J. Verhoeven). Cham: Springer.

39 Chou, Y., Shang-Lien, L., Jeff, K. et al. (2018). Design and actual performance of constructed wetlands for river renovation in Taiwan. *Desalin. Water Treat.* 131: 114–122.

40 Mitsch, W.J., Gosselink, J.G., Anderson, C.J., and Zhang, L. (2009). *Wetland Ecosystems*. Hoboken, NJ: Wiley.

41 Morimoto, J., Shibata, M., Shida, Y., and Nakamura, F. (2017). Wetland restoration by natural succession in abandoned pastures with a degraded soil seed bank: natural wetland restoration. *Restor. Ecol.* 25: 1005–1014.

42 Singh, N.K., Gourevitch, J.D., Wemple, B.C. et al. (2019). Optimizing wetland restoration to improve water quality at a regional scale. *Environ. Res. Lett.* 14: 064006.

43 Pan, B.Z., Wang, Z.Y., and Xu, M.Z. (2012a). Macroinvertebrates in abandoned channels: assemblage characteristics and their indications for channel management. *River Res. Appl.* 28: 1149–1160.

44 Pan, B.Z., Wang, Z.Y., Xu, M.Z., and Xing, L.H. (2012b). Relation between streamhabitat conditions and macroinvertebrate assemblages in three Chinese rivers. *Quat. Int.* 282: 178–118.

45 Schneider, P., Vogt, T., Schirmer, M. et al. (2011). Towards improved instrumentation for assessing river-groundwater interactions in a restored river corridor. *Hydrol. Earth Syst. Sci. Discuss.* 8: 2503–2553.

46 Ward, J.V., Tockner, K., Uehlinger, U., and Malard, F. (2001). Understanding natural patterns and processes in river corridors as the basis for effective river restoration. *Regul. Rivers Res. Manag.* 17: 311–323.

47 Bennett, V.A., Doerr, V.A.J., Doerr, E.D. et al. (2013). Habitat selection and behaviour of a reintroduced passerine: linking experimental restoration, behaviour and habitat ecology. *PLoS One* 8 (1): e54539.

48 Wohl, E., Bledsoe, B.P., Jacobson, R.B. et al. (2015a). The natural sediment regime in rivers: broadening the foundation for ecosystem management. *Bioscience* 65: 358–371.

49 Wohl, E., Lane, S.N., and Wilcox, A.C. (2015b). The science and practice of river restoration. *Water Resour. Res.* 51: 5974–5997.

50 Pan, B., Yuan, J., Zhang, X. et al. (2016). A review of ecological restoration techniques influvial rivers. *Int. J. Sediment Res.* 31: 110–119.

51 Palmer, M., Allan, J.D., Meyer, J., and Bernhardt, E.S. (2007). River restoration in the twenty-first century: data and experimental knowledge to inform future efforts. *Restor. Ecol.* 15: 472–481.

52 Zhong, J., Fan, C., and Zhang, L. (2010). Significance of dredging on sediment denitrification in Meiliang Bay, China: a yearlong simulation study. *J. Environ. Sci.* 22: 68–75.

53 Hart, D.D., Johnson, T.E., Bushaw-Newton, K.L. et al. (2002). Dam removal: challenges and opportu-nities for ecological research and river restoration. *Bioscience* 8: 669–682.

54 Palmer, M.A., Bernhardt, E., Allan, J.D. et al. (2004). Standards for ecologically successful river restoration. *J. Appl. Ecol.* 42: 208–217.

55 Wohl, E., Angermeier, P.L., Bledsoe, B. et al. (2005). River restoration. *Water Resour. Res.* 41: W10301.

56 Kondolf, G.M., Boulton, A.J., O'Daniel, S. et al. (2006). Process-based ecological river restoration: visualizing three-dimensional connectivity and dynamic vectors to recover lost linkages. *Ecol. Soc.* 11: 5.

57 Gumiero, B., Mant, J., Hein, T. et al. (2013). Linking the restoration of rivers and riparian zones/ wetlands in Europe: sharing knowledge through case studies. *Ecol. Eng.* 56: 36–50.

58 Konrad, C.P., Olden, J.D., Lytle, D.A. et al. (2011). Large-scale flow experiments for managing river systems. *Bioscience* 61: 948–959.

59 Lake, P.S. (2001). On the maturing of restoration: linking ecological research and restoration. *Ecol. Manag. Restor.* 2: 110–115.

60 Muhar, S., Januschke, K., Kail, J. et al. (2016). Evaluating good-practice cases for river restoration across Europe: context, methodological framework, selected results and recommendation. *Hydrobiologia* 769: 3–19.

61 Nilsson, C., Jansson, R., Malmqvist, B., and Naiman, R.J. (2007). Restoring riverine landscapes: the challenge of identifying priorities, reference states, and techniques. *Ecol. Soc.* 12: 16.

62 Wyżga, B., Zawiejska, J., Radecki-Pawlik, A., and Hajdukiewicz, H. (2012). Environmental change, hydromorphological reference conditions and the restoration of Polish Carpathian rivers. *Earth Surf. Process. Landf.* 37: 1213–1226.

63 Whalen, P.J., Toth, L.A., Koebel, J.W., and Strayer, P., K. (2002). Kissimmee River restoration: a case-study. *Water Sci. Technol.* 45: 55–62.

64 Theiling, C.H., Janvrin, J.A., and Hendrickson, J. (2015). Upper Mississippi River restoration: implementation, monitoring, and learning since 1986. *Restor. Ecol.* 23: 157–166.

65 Lee, Y.K., Kim, J.K., Jung, S.M. et al. (2014). Development of a water quality index model for lakes and reservoirs. *Paddy Water Environ.* 12: 19–28.

66 Saldi-Carolmile, K., Bates, K., Skidmore, P. et al. (2004). *Stream Handbook Restoration Guidelines.* Olympia, WA: Prepared for Washington State Aquatic Habitat Guidelines Program.

67 Yocom, K. (2014). Building watershed narratives: An approach for broadening the scope of success in urban stream restoration. *Lands. Res.* 39: 698–714.

68 Roni, P., Hanson, K., and Beechie, T. (2008). Global review of the physical and biological effectiveness of stream habitat rehabilitation techniques. *N. Am. J. Fish. Manag.* 28: 856–890.

69 Jeong, K.S., Hong, D.G., Byeon, M.S. et al. (2010). Stream modification patterns in a river basin: field survey and self-organizing map (SOM) application. *Ecol. Inform.* 5: 293–303.

70 Tuttle, A.K., McMillan, S.K., Gardner, A., and Jennings, G.D. (2014). Channel complexity and nitrate concentrations drive denitrification rates in urban restored and unrestored streams. *Ecol. Eng.* 73: 770–777.

71 Kaushal, S.S., Groffman, P.M., Mayer, P.M. et al. (2008). Effects of stream restoration on denitrification in an urbanizing watershed. *Ecol. Appl.* 18: 789–804.

72 Violin, C.R., Cada, P., Sudduth, E.B. et al. (2011). Effects of urbanization and urban stream restoration on the physical and biological structure of stream ecosystems. *Ecol. Appl.* 21: 1932–1949.

14

Restoration of Riverine Health

An Ecohydrological Approach – Flow Regimes and Aquatic Biodiversity

S.P. Biswas

Department of Life Sciences, Dibrugarh University, Dibrugarh, Assam, India

14.1 Introduction

Rivers are the most essential freshwater source for ecosystem health, which underpins economies, cultures [1], and human sustenance [2]. Rivers are intrinsically related to human civilizations, providing innumerable ecological services [3] and supporting rich biodiversity, yet freshwater systems (including rivers) are among the most degraded habitats on earth. Burgeoning population growth and consequent fishing pressure to feed additional people coupled with alterations and shrinkage of fish habitats have changed the riverine habitat. Hydrological changes in a river system manifest in bank erosion, sediment deposition, channel evulsions, modification of floodplain morphology, and shifting of flooding areas – all of which are common in the third world.

The ecohydrological health of a river is determined by water quality and quantity, including natural flow patterns, physicochemical properties, sediment transport, and drainage basin runoff. Increased use of surface water and pesticides for crop production and the regulation of rivers in the form of embankments for checking flood waters have harmed riverine fauna. Any structural intervention, like the construction of an embankment or damming of a river, negatively impacts the riverine ecosystem, aquatic biota, and riparian life [4]. Despite all these issues, it is necessary to emphasize the underlying processes to understand where there are trade-offs and synergies [70] and how these vary with global climate change [5].

14.2 Habitat Ecology

A river is a four-dimensional freshwater ecosystem characterized by the continuous transport of organic matter and nutrients from a catchment area, including linked ecosystems. The river ecosystem's structure and function are broadly regulated by climatic conditions and the geomorphology of the watershed. Free flow of water and sediment discharges are two characteristic features of any natural lotic system. Together, they control riverine fish diversity and species richness in different geographical locations [6].

Handbook of Ecological and Ecosystem Engineering, First Edition. Edited by Majeti Narasimha Vara Prasad.

It is estimated that 60–80% of freshwater fish are obtained from open waters in Asia, Africa, and Latin America. The integrity and stability of rivers and their linked ecosystems depend on the spatial and temporal variability of the natural flow regime and riverine fisheries [7]. However, the primary driving force is the river discharge, which determines the degree of connectivity and the exchange of nutrients and organisms across river floodplain gradients.

14.2.1 Riverine Habitats

Riverine habitats have been studied from hydrological, climatic, and ecobiological perspectives [8]. The quantum of discharge, the gradient of the river, and water quality determine the assemblage and density of the riverine biota. For example, based on hydro-geomorphological conditions, fish habitats in Himalayan rivers may be categorized as follows:

- *Bedrock reaches* (fast-flowing rivers): Rapid streams with a relatively steep gradient and rocky bed that harbor small specialized genera that have adhesive organs.
- *Step pools* (upstream pools and lakes): Sluggish, deeper parts of upland streams and high-altitude lakes. Coldwater fishes are dominant here.
- *Regime reaches* (river meanderings and confluences): The eddy countercurrent system at the junction of two rivers is an ideal place for fish to assemble. The confluences are also passageways for upstream fish migration. Similarly, channel meanderings sustain many fish species.
- *Braided reaches* (adjoining floodplain lakes): These are weed-infested shallow bodies of water that are temporarily or permanently connected with the main river.
- *Plane-bed reaches* (open river): Large fish species – mostly carp, giant catfish, and other megafauna like *Platanista* and *Nilssonia* (*Trionix*) – are mainly encountered in the open river.
- *Ephemeral streams*: These streams are "alive" only during rainy months. Varieties of small and medium-sized fish are found here.
- *Seasonal water bodies*: Low-lying paddy fields, derelict ponds, swamps, and roadside canals are also temporary homes for a variety of small fish species.

14.2.2 Linked Ecosystems

Floodplains are typically inundated during the rainy season, primarily by overspill from adjacent rivers, thereby helping with flood mitigation. Floodplain lakes (FPLs) are feeding and breeding grounds (Figure 14.1) for many riverine species [9] and among the most productive fisheries [66]; they are sources of livelihoods for millions of rural people across the world. For example, fisheries are the most critical source of income for floodplain dwellers in the Amazon [10, 11] and rural households in Cambodia and Laos [12]. The shallow, eutrophic floodplain lakes of the Ganga-Brahmaputra-Meghna (GBM) basin in South Asia are rich in fishery resources [13] and livelihood options for an estimated 10 million fisher-folk in the Indian subcontinent. Further, alterations of land-use patterns [14] in the watersheds are the most important early warning indicators of eutrophication [15].

14.3 Riverine Issues

Freshwater (rivers and associated links) covers less than 1% of the earth's surface, but about 15% of world fisheries come from the freshwater system [16]; thus the riverine system is arguably the most critical [17] habitat on the planet. Population booms and consequent pressure on land and water to feed the additional people coupled with uncertainty due to climate change have been common

Figure 14.1 A typical flood-plain lake in the GBM basin.

Figure 14.2 River bank erosion is common in the Himalayan region.

in the third world. In addition to biodiversity and ecosystem services, rivers may also cause political [18], economic, and societal issues. Migration of poverty-driven populations as the result of civil wars and pandemic diseases will likely add more pressure on land and water resources in third-world countries.

14.3.1 Bank Erosion, Siltation, and Aggradations of Rivers

Rapid siltation of rivers and *beels* affects aquatic megafauna, particularly the river dolphin, which is undoubtedly the keystone species of the GBM system. In ancient times, rivers were the only route for communication and ferrying goods long distances in many countries. But a high rate of siltation and consequently raised riverbed made many rivers unsuitable for navigation year round.

Bank erosion is a significant source of siltation in rivers. Usually, erosion-prone banks in the GBM basin are dominated by silt and fine sand and are almost perpendicular (Figure 14.2). Erosion

Figure 14.3 Aggradation of the river due to significant siltation of the river bed.

occurs during rainy months but more intensely during receding floods. Eroded soil is deposited downstream, forming sand bars; consequently, the river becomes aggraded (Figure 14.3) and wider and shallower.

Asian and Latin American rivers are characterized by innumerable floodplain lakes and high-intensity floods, causing immense loss of life and property (along the Yellow River, for example) for centuries. In the Hindu Kush Himalayas (HKH) region, glaciers at high elevations (>3500 m) are the water source for highland people. High-intensity rainfall in the HKH region is common and often causes flash floods. Further, glacial lakes have caused some of the world's most devastating floods; and in recent decades, flooding patterns in the mountains [19] have changed and now pose a threat to all concerned.

Erosion of river banks is natural, but it may be hastened by human-induced processes like the sudden release of a large amount of water from a reservoir, leading to a flash flood and bank erosion downstream. The calamity caused by riverbank erosion in the GBM basin is symptomatic of global problem. Mitigating bank erosion is currently restricted to bank strengthening through reinforced concrete or bamboo porcupines (Figure 14.4) and other conventional structural methods. It is important to develop a viable solution that can respond to varied situations and is environmentally friendly.

14.3.2 Deforestation in Catchment Areas

Developing nations with ever-increasing populations are under heavy pressure in the face of climate change, deforestation, and, more recently, the Covid-19 pandemic. An age-old farming practice is the slash-and-burn technique (Figure 14.5) used in forested upland areas of many countries across continents; it is considered a sustainable livelihood practice when the population in a given area is small. Using slash and burn, a particular hillock was once left out for 15–20 years to grow vegetation. But now the gap is only for 3–5 years, which is not adequate for re-vegetation. This

Figure 14.4 Structural devices used for bank protection.

Figure 14.5 Slash-and-burn farming in upland areas of the eastern Himalayan region.

degraded land is subjected to topsoil weathering and responsible for siltation in adjacent river basins and linked ecosystems.

The HKH region's rivers are geologically unstable and highly prone to bank erosion due to their fragile rock structure and seismic activity. The floodplains in the region consist primarily of alluvial soil. Therefore, during monsoons, particles and aggregates are washed away by fluvial action, and riverbanks collapse. Large-scale bank erosion leads to widening and aggregation of the channel and raised river beds downstream due to the deposition of coarse soil particles. The phenomenon is natural, but its impacts are disastrous on the riparian community's life and livelihoods as well as fauna [20]. A collective approach involving stakeholders in joint management of the land (the watershed) is probably the best option to sustain livelihoods and ensure the ecorestoration of catchment areas.

14.3.3 River Pollution and Invasive Species

Rivers are naturally resilient and can tolerate a certain level of toxicity. Unfortunately, many rivers are so polluted that they are unfit for domestic use and no longer support commercial fisheries. Similarly, clandestinely introduced fish species pose serious threats to native fish biodiversity in many Asian countries. Alien species include Thailand magur (*Clarias gariepinus*), crocodile fish, alligator gar (*Atractosteus spatula*), African pangus (*Pangasius sutchi*), bighead carp (*Aristichthys nobilis*), and Nile tilapia (*Oreochromis nilotica*), Amazon piranha (*Piaractus brachypomus*) are often reported sneaking into natural water from inundated culture ponds during high floods. The risk of an alien species establishing itself in rivers and wetlands outside its distributional range depends on *propagule pressure* [21] and the number of pathways by which a species can be spread to the wild [22]. In addition to potential ecological impacts on endemic habitats and biodiversity, many exotic fish also pose pathogenic and genetic risks for the native fish fauna [65].

14.3.4 Fishing Pressure

Riverine biodiversity regulates ecosystem processes that underpin ecosystem services, and greater biodiversity is associated with increased ecosystem functions [23]. Biodiversity loss is one of the biggest challenges threatening humanity's future and may be more severe than climate change [24]. Riverine ecosystems and fisheries are subject to intense pressure from a wide range of anthropogenic disturbances [25]: altered land use and flow regimes, shrinkage of habitat, water pollution, exotic species invasions, and intensive exploitation of fish stocks. For instance, Africa is rich in fish resources, yet the continent's contribution is only about 5% of the global fish trade. Emerging challenges such as sustainability of water resources [26] and overexploitation of fishery resources are straining the fisheries sector. Another dimension of the problem is indiscriminate exploitation of endemic and rare species, many of which are illegally exported as ornamental fish [20]. Unfortunately, sustainability and conservation aspects of riverine fisheries across the world are largely overlooked [27].

Any substantial change in the natural flow may lead to degradation of riverine habitats, which will have an immense impact on the biota, including human beings dependent on riverine ecosystem services [28].In the GBM basin, increased use of surface water and pesticides for crop production and regulation of rivers by building dikes to check flood waters harms the basin's overall aquatic diversity. Fishing using explosives in streams is one of the most common causes of mass mortality of upland fishes in the Himalayan rivers. Rampant use of pesticides and insecticides in agriculture and dumping of solid wastes causes fish deaths (Figure 14.6) in many wetland and slow-moving water systems. As a consequence, riverine and other freshwater systems are much more threatened than other ecosystems. The global need to invest in research and production of food cannot be underestimated, but its application will require different approaches at a regional level.

Figure 14.6 Mass mortality of fish due to toxic substances and/or pollutants.

14.3.5 Status of Wetlands (FPLs)

The degradation of floodplains resulting from human interferences due to urbanization [31], construction of roads and embankments, dumping of waste products, encroachment for agricultural production, and natural causes (siltation, drought, cyclone, and intrusion of saline water) have had negative impacts on fish diversity and aquatic fauna in wetlands worldwide [32]. As a result, the current rate of wetland loss is three times that of forest loss [33].

14.3.6 Regulated Rivers and Their Impacts

Construction of dams across rivers for flood control and agricultural purposes is a very old practice. Although embankments provide temporary respite from floods on a particular stretch of a river, these structures block longitudinal hydrological connectivity across the floodplain, which hinders movement between rivers and wetlands, thus reducing fish stocks and species diversity. In the HKH region, the area where most dams are proposed is characterized by fragile bedrock, steep slopes, and a high rate of surface erosion. There is always a danger of flash floods downstream and devastation.

The Brahmaputra, a transboundary river flowing through China, India, and Bangladesh, carries 30% of India's water resources and has the potential to generate over 40% of its hydroelectric production. India's energy requirements are increasingly rapidly and projected to increase several fold in the coming decades. Thus India will have to find alternative energy sources. Mountainous rivers have an immense potential to generate hydropower, known as *green energy*, at a relatively low cost. This is why economically weak and land-locked countries (for instance, Bhutan and Laos) utilize their water resources to generate electricity and sell it to neighboring countries.

River dams hold back large amounts of water in reservoirs created for this purpose. When a series of dams operate simultaneously, they alter the natural flow of water. Usually, the flow of the downstream part of the river is greatly reduced when water is stored in the reservoir and a flood-like situation arises during peak hours when power generation is operative in storage hydroelectric projects. This diurnal variation of "very high" and "very low" discharge is likely to have a devastating effect on downstream biota. Similarly, a large volume of water released during flooding often has a catastrophic effect on life and crops downstream. Alterations of the flow regime can also adversely affect ecosystems [34]. Therefore, the sustainability of hydropower projects should be investigated thoroughly so that it does not cause flash floods downstream due to the forceful release of reservoir water or dam breaks as a result of tremors etc. [35].

14.4 Ecorestoration of River Basins

14.4.1 Environmental Flow

Environmental flow is a term that is used frequently in recent times to define agreed-on water availability for development projects (including agricultural and fisheries use) and maintenance of minimum flow required for existing riverine biota. This environmental flow is necessary to provide an adequate flow of water in rivers and groundwater systems to maintain a healthy ecosystem and its biota. Riverine fisheries have been adversely impacted in most large rivers as a result of unpredictable flow in recent years due to the changing pattern of monsoonal rain [9], permafrost melting [36], and damming of rivers. But the species probably hit hardest by siltation and habitat alteration has been the Gangetic dolphin; this endangered cetacean population has decreased alarmingly in the last 30 years or so.

Failure to maintain environmental flows may lead to a decline in the health of a water-dependent ecosystem. Although wetland fish can adjust their feeding and spawning behavior to a certain extent, delay or a shortfall of monsoonal rain may result in spawning failure of riverine fish.

14.4.2 Success Story of a Conservation Effort for Aquatic Fauna

14.4.2.1 River Dolphins

In a riverine system, tertiary consumers (fourth trophic level) like large carnivorous fish, turtles, crocodiles, otters, porpoises, and dolphins are the apex of the food chain. However, the most fascinating is the river dolphin. Like a tiger in a forest, the presence of dolphins indicates the health of the river.

The South Asian river dolphin has two subspecies: the Gangetic dolphin (*Platanista gangetica gangetica*) and the Indus dolphin (*Platanista gangetica minor*). The pink dolphin (*Inia geoffrensis* or boto) of the Amazon basin is the other freshwater dolphin that now exists in nature. The Yangtze River dolphin (Baiji-*Lipotes vexillifer*), often confused with the finless porpoise, is probably extinct in its natural habitat.

The Gangetic dolphin, the national aquatic animal of India, is found in the GBM river basin. It has a robust fish-like body and may attain 2 m and weigh 100 kg or more. Calves and juveniles are dark in color, while the adults are slate gray. They prefer sandy river beds and a moderate current and are often encountered in small groups (Figure 14.7)of three or four in deep pools (>3 m), usually at river confluence [37] and also at river meanders during lean months. The long rostrum bears a series of homodont teeth for holding prey fish. Dolphins are a secondary aquatic mammal and have to surface at regular intervals for breathing. They roll over quietly, exposing the melon (forehead) and part of the back for a second or so. The mean time between two surfacing has been recorded as 38–45 seconds for adults. The infants are very playful, surface more frequently (15–20 seconds between intervals), and often jump out of the water (Figure 14.8).Gangetic dolphins are virtually blind, but they are champion swimmers even in a highly turbid river. They possess a sonar (echolocation) system located on the melon.

The presence of dolphins is a clear indication of the good health of the river. They were once abundant in the Brahmaputra system, but the population fell drastically in the 1980s primarily due to intentional killing (Figure 14.9) of dolphins. Dolphin oil is believed by some to be a drug that

Figure 14.7 Dolphin habitat: R. Brahmaputra.

Figure 14.8 Baby dolphin jumping. *See colour plate section for colour representation of this figure.*

Figure 14.9 Intentional killing of dolphins. *See colour plate section for colour representation of this figure.*

can treat various ailments. In one instance, more than 50 dolphins were killed, and their body oil was extracted by professional killers at Narain Dahr upstream on the Barak (Meghna river system) in 1985–1986. This has led to the virtual extinction of dolphins from the Barak River in southern Assam [38]. The oil is also used for fish bait. Only a fragmented population of dolphins can be sighted in certain protected areas like Dibru-Saikhowa National Park, R. Subansiri, Kaziranga National Park, and R. Kulsi in the Upper Brahmaputra basin.

Siltation of river beds (especially pools), loss of a prey base, and intentional and accidental killing have taken a heavy toll on dolphins. Fortunately, the number of Ganges dolphins over their entire distributional range is increasing due to conservative measures in the last two decades.

14.4.2.2 Hilsa Fishery

Another commendable effort is the restoration of the hilsa fishery in Bangladesh. Hilsa (*Tenualosa ilisha*), the iconic fish of the GBM basin, is a highly prized anadromous fish mostly found in Bangladesh, India, and Myanmar (R. Irrawaddy). Although hilsa is widely distributed from the Persian Gulf to the China Sea, 95% of the landings are in Bangladesh, Eastern India, and Myanmar. The costliest fish in eastern India and Bangladesh, hilsa is distributed mostly in the upper region of the Bay of Bengal, stretching from estuaries in Myanmar to Mahanadi and the Godavari estuary on the eastern coast up to Narmada and the Tapti estuary on the western coast of India.

Most catches (over 80%) in India are recorded during monsoon season when gravid fish enter rivers for spawning, which is reported to occur throughout the year [39]; but in Bangladesh, the peak (about 60–70%) hilsa fishery is in October–November with other small phases in February–March and July–August in various rivers and estuaries in the region [40]. More than 3 million people are directly or indirectly involved in hilsa fishing in the subcontinent. Fortunately, hilsa landing has significantly increased in Bangladesh in recent years due to effective measures taken by authorities[41], including a ban on hilsa fishing for a specific period of the year. In India, too, strict implementations of rules are needed to curb overfishing, as has been done in Bangladesh.

14.4.3 Biomonitoring of Riverine Health and Ecosystem Engineering

Climate, topography, and gradients are critical to regulate the variety and composition of fish biota. Again, riverine health is determined by water quality and quantity, including natural flow characteristics, physicochemical properties, sediment transport, and drainage basin runoff [20]. Water quality is subjected to frequent changes due to economic activities like oil drilling and stone and sand mining (Figure 14.10), and regular monitoring of river health is necessary.

Biomonitoring is an option for assessing riverine health [42]. *Ecosystem engineering* is the process by which living biota interact with and modify the habitat. Ecosystem engineers significantly alter an ecosystem and can have a large impact [43] on an area's landscape patterns. An ecosystem engineer (plant, animal, or microbe) interacts with the trophic structure and functions and forms an *engineering web* that regulates the ecosystem and the associated biota. Assessing aquatic health through pollution-tolerant organisms, notably macroinvertebrate- and fish-based stream health indices [44], is a widely used technique. Segment-based ecoregions [45] using fish to restore the riverine environment and floating agriculture [46] have also been suggested as climate change adaptations.

Weed-infested (Figure 14.11) tropical FPLs are eutrophic due to an excessive load of nutrients and decomposed biota. The most common weed is *Eichhornia crassipes* (water hyacinth), a highly invasive free-floating perennial hydrophyte native to South America. However, under controlled conditions, *E. crassipes* and other hydrophytes can provide spawning substrate [47], act as a food source, maintain clean water [48], reduce nitrogen and phosphorus [49] and total suspended solids

Figure 14.10 Sand mining in a river bed. *See colour plate section for colour representation of this figure.*

Figure 14.11 Weed infested flood-plain lake.

and heavy metals, and reduce biological and chemical oxygen demand [50]. Further, hydrophytes in FPLs can determine the food chain length [69] and complexity of food webs [51, 52].

14.4.4 Integrated River Basin Management

Integrated river basin management (IRBM) has emerged as a holistic approach to managing river basins. It involves coordinated development and management of water, land, and related resources within a river basin to maximize economic and social benefits equitably without compromising the sustainability of riverine ecosystems. In most Asian and African countries, fishermen are economically and socially less-privileged sections of society, and relatively little emphasis are given to riverine fisheries. Of course, inadequate information about the population dynamics of most endemic species hinders the development of appropriate conservation action plans [53]. However, since the 1980s, fishery issues have been targeted in several aid-funded programs to ameliorate the fishermen's complex livelihoods.

Riverine floods caused extensive damage to landscapes, economies, and societies. Various structural and non-structural measures have been taken to contain riverine floods (Figure 14.12) worldwide. New concepts for ecological designs of river channels have also gained ground in recent decades. Restoration of riverine habitat by ecological maneuvering and manipulating of urban landscape design has yielded desired outcomes [54, 55]. Restoration plans for mitigating the impact of global warming [56] and restoring and creating floodplain grazing marsh has been advocated for biodiversity conservation and carbon storage [57].

A common feature of IRBM and decision-making is a coordinated approach by authorities. Integration of the most relevant management approach usually focuses on the local, regional, or national level. To achieve this goal, the cooperation of all stakeholders [64] and a good understanding of the river basin are essential for developing water management institutions [58]. This requires adequate and updated scientific data about the river in question. A package of suggestions may be considered for mitigating flash floods in the HKH region by strengthening the Disaster Management System due to climate change [59] by (i) raising and strengthening bank embankments downstream; (ii) coordinating hydrological monitoring with the meteorological departments and installing warning systems; (iii) monitoring and controlling bank erosion and establishing a green belt in the riparian zone (thick vegetation cover to mitigate the impact of flash floods); (iv) establishing flood shelters for downstream people; and (v) offering vocational training for local youth for alternative livelihoods like aquaculture, (vi) providing for the passage of migratory fish, and (vii) establishing live gene banks for endemic/endangered fish species.

Figure 14.12 River bank protection: (a) & (b) structural protection; (c) & (d) non-structural protection.

14.5 Summary and Conclusion

Rivers are the most important freshwater source on earth and are essential for ecosystem health and underpin the economies and lifeline of human sustenance. A recent UN report [67] focuses on adaptation, climate change mitigation, and improving resilience for better water management. The report suggests specific water management interventions in agriculture, wetland protection, and other nature-based solutions to reduce greenhouse gas emissions.

River water has both consumptive and ecological value. Unfortunately, we often forget the ecological issues of rivers, and the problem starts there. Habitat alteration due to the erratic nature of water discharge and high rates of erosion and siltation as the result of dam construction [60] have negative impacts on riverine fisheries. But the species that has perhaps been hit hardest by siltation and habitat alteration has been the Ganges dolphin; this endangered cetacean's population has decreased alarmingly throughout its distributional range. Delay or shortfall of monsoonal rain has also led to the spawning failure of many riverine species, resulting in reduced fish landings in the next season.

Large-scale felling of trees in catchment areas and construction of embankments and other developmental projects coupled with siltation of drainage systems have impacted riverine biota. Aquatic species have evolved strategies [25] primarily in direct response to natural flow. Although certain fish species, especially the air-breathing species of tropical and subtropical FPLs, can adjust their spawning behavior in response to delayed or reduced seasonal floods, maintaining natural flow and lateral connectivity is essential to the viability of edible and ornamental fish species.

Minimum flow is the flow required to maintain a river ecosystem's features and limit water loss during the dry season, and it is practiced in many Asian and African countries. The concept is a useful tool to balance the needs of the environment and other users in a river basin. Environmental flow is the water left in a river ecosystem or released into it for the specific purpose of managing the condition of that ecosystem. There is growing resentment of large river dams in the Amazon, the Congo, and the Brahmaputra basin, because a dam break [61] would have a catastrophic effect on the lives and livelihoods of people downstream. Therefore, maintaining the basic pattern of hydrology and retaining lateral and longitudinal connectivity are key factors of managing a river basin. Dredging the Himalayan rivers is essential – it will facilitate much-needed cover for aquatic megafauna, and natural connectivity with FPLs should be retained for autostocking by providing sluice gates.

A bioengineering approach for protecting the environment should be economically viable and acceptable to all the stakeholders. It is not always easy to convince an entire society to accept a new technology that is not proven. Phytoremediation is an emerging technology that has yet to prove its effectiveness on a field scale. However, risk-based pilot projects [62] on bank stabilization in a small, vulnerable river stretch may be demonstrated by planting soil-binding, ecologically suitable [68], fast-growing herbs and shrubs on a properly sloped river bank.

Transboundary collaboration for the management of flash floods, including information sharing and the establishment of early warning systems in the event of dam breaks or release of excess water from reservoirs, is necessary. In recent decades, frequent storms and the intrusion of salty water into agricultural lands and fishery ponds have also caused concern. The policy of land-use patterns in developing countries requires a paradigm shift to maintain the ecological integrity aquatic ecosystems [63]. Waterlogging in rainy months is a serious problem in almost all Indian cities and townships, primarily due to the non-implementation of the Environment, Forest, and Fishery Acts in letter or spirit [20].

The river connects life, which is why riparian communities have unique and innovative ways to live with floods and erosion. Therefore, comprehensive and critical scrutiny by scientists, engineers, bureaucrats, sociologists, and other stakeholders is necessary before giving the nod to any new hydro or mining exploration project in a river valley. Humans should consider ourselves part of the ecosystem, equal to other biota (deep ecology, deep understanding, and deep commitment), and must utilize natural resources, especially freshwater biota, very judiciously.

Acknowledgments

The author gratefully acknowledges the assistance rendered and personal collections shared by Dr. Sanchita, Dr. Anand, Dr. Shrutidhara, Dr. N. Dutta, Dr. L.P. Hazarika, Dr. Kaushal, Dr. Akash, Jyotirmoy, Anurag, Rimle, Abhijit, Shyamal, and Nipen during the preparation of the manuscript.

The author also declares that there is no conflict of interest.

References

1 Lokgariwar, C., Chopra, R., Smakhtin, V. et al. (2014). Including cultural water requirements in environmental flow assessment: an example from the upper Ganga River, India. *Water International* 39 (1): 81–96. https://doi.org/10.1080/02508060.2013.863684.

2 Green, P.A., Vörösmarty, C.J., Harrison, I. et al. (2015). Freshwater ecosystem services supporting humans: pivoting from water crisis to water solutions. *Global Environmental Change* 34: 108–118. https://doi.org/10.1016/j.gloenvcha.2015.06.007.

3 Liquete, C., Cid, N., Lanzanova, D. et al. (2016). Perspectives on the link between ecosystem services and biodiversity: the assessment of the nursery function. *Ecological Indicators* 63: 249–257. https://doi.org/10.1016/j.ecolind.2015.11.058.

4 Krause, F. and Strang, V. (2016). Thinking relationships through water. *Society and Natural Resources* 29 (6): 633–638. https://doi.org/10.1080/08941920.2016.1151714.

5 Arlidge, W.N., Bull, J.W., Addison, P.F.E. et al. (2018). A global mitigation hierarchy for nature conservation. *Bioscience* 68 (5): 336–347. https://doi.org/10.1093/biosci/biy029.

6 Arthington, A.H. (2012). Environmental Flows: Saving Rivers in the Third Millennium. Berkeley, CA: University of California Press.

7 Dudgeon, D. (2010). Prospects for sustaining freshwater biodiversity in the 21st century: linking ecosystem structure and function. *Current Opinion in Environmental Sustainability* 2: 422–430. https://doi.org/10.1016/j.cosust.2010.09.001.

8 Acreman, M., Arthington, A.H., Colloff, M.J. et al. (2014). Environmental flows for natural, hybrid, and novel riverine ecosystems in a changing world. *Frontiers in Ecology and the Environment* 12: 466–473. https://doi.org/10.1890/130134.

9 Biswas, S.P. (2014). Conservation plan for aquatic fauna in the Brahmaputra Basin. In: Rivers for Life – Proceedings of the International Symposium on River Biodiversity: Ganges-Brahmaputra-Meghna River System, Ecosystems for Life, a Bangladesh-India Initiative (eds. R.K. Sinha and B. Ahmed), 45–53. IUCN, International Union for Conservation of Nature.

10 Castello, L. and Macedo, M.N. (2016). Large-scale degradation of Amazonian freshwater ecosystems. *Global Change Biology* 22 (3): 990–1007. https://doi.org/10.1111/gcb.13173.

11 Van-der-Sleen, P. and Albert, J. (eds.) (2017). Field Guide to the Fishes of the Amazon, Orinoco, and Guianas (Princeton Field Guides). Princeton, NJ: Princeton Univerisity Press.

12 De Silva, S.S., Ingram, B.A., and Wilkinson, S. (eds.) (2015). Perspectives on culture-basedfisheries developments in Asia. *NACA Monograph Series No. 3*, NACA, Bangkok.
13 Dudgeon, D. (2011). Asian river fishes in the Anthropocene: threats and conservation challenges in an era of rapid environmental change. *Journal of Fish Biology* 79 (6): 1487–1524. https://doi.org/10.1111/j.1095-8649.2011.03086.x.
14 Choden, K., Keenan, R.J., and Craig, R.N. (2020). An approach for assessing adaptive capacity to climate change in resource dependent communities in the Nikachu watershed, Bhutan. *Ecological Indicators* 114: 249–257. https://doi.org/10.1016/j.ecolind.2020.106293.
15 Ma, C., Huo, S., Sun, W. et al. (2016). Establishment of physico-chemical variables and Chl *a* criteria based on land-use patterns and terrestrial ecosystem health. *Ecological Engineering* 97: 355–362. https://doi.org/10.1016/j.ecoleng.2016.10.009.
16 van Zwieten, P.A.M., Béné, C., Kolding, J. et al. (2011). Review of tropical reservoirs and their fisheries – The cases of Lake Nasser, Lake Volta and Indo-Gangetic Basin reservoirs. FAO Fisheries and Aquaculture Technical Paper No. 557.
17 Anderson, E.P., Jackson, S., Rebecca, T.T. et al. (2019). Understanding rivers and their social relations a critical step to advance environmental water management. *WIREs Water* 6e: 1381. https://doi.org/10.1002/wat2.1381.
18 Naz, F. (2014). Water: a cause of power politics in South Asia. *Transactions on Ecology and the Environment* l178: 101–109. https://doi.org/10.2495/WS13009.
19 Huggel, C., Cochachin, A., Drenkhan, F. et al. (2020). Glacier Lake 513, Peru: lessons for early warning service development. *WMO Bulletin* 69 (1): 45–52.
20 Biswas, S.P., Boruah, S., and Sharma, A. (2018). Environmental protection of the Brahmaputra River from environment, ecological and legal perspectives. *Soochow Law Journal, Taipei* 15 (2): 135–157.
21 Blackburn, T.M., Prowse, T.A., Lockwood, J.L. et al. (2013). Propagule pressure as a driver of establishment success in deliberately introduced exotic species: fact or artefact? *Biological Invasions* 15 (7): 1459–1469. https://doi.org/10.1007/s10530-013-0451-x.
22 Gallardo, B., Clavero, M., Sanchez, M.I. et al. (2015). Global ecological impacts of invasive species in aquatic ecosystems. *Global Change Biology* 22 (1): 151–163. https://doi.org/10.1111/gcb.13004.
23 Moss, B. (2018). Ecology of Freshwaters: Earth's Bloodstream. Wiley, Science.
24 Rockström, J., Schellnhuber, H.J., Hoskins, G. et al. (2016). The world's biggest gamble. *Earth's Future* 4: 465–470. https://doi.org/10.1002/2016EF000392.
25 Garcia-Moreno, J., Harrison, I.J., Dudgeon, D. et al. (2014). Sustaining freshwater biodiversity in the anthropocene. In: The Global Water System in the Anthropocene (eds. A. Bhaduri et al.), 247–270. Springer Water. https://doi.org/10.1007/978-3-319-07548-8_17.
26 Konar, M., Evans, T.P., Morgan, L. et al. (2016). Water resources sustainability in a globalizing world: who uses the water? *Hydrological Processes* 30: 3330–3336. https://doi.org/10.1002/hyp.10843.
27 Cooke, S.J., Allison, E.H., Beard, T.D. et al. (2016). On the sustainability of inland fisheries: finding a future for the forgotten. *Ambio* 45: 753–764. https://doi.org/10.1007/s13280-016-0787-4.
28 Gopal, B. (2016). A conceptual framework for environmental flows assessment based on ecosystem services and their economic valuation. *Ecosystem Services* 21: 53–58. https://doi.org/10.1016/j.ecoser.2016.07.013.
29 Hermelingmeier, V. and Nicholas, K.A. (2017). Identifying five different perspectives on the ecosystem services concept using Q methodology. *Ecological Economics* 136: 255–265. https://doi.org/10.1016/j.ecolecon.2017.01.006.

30 Lillford, P. and Hermansson, A.M. (2020). Global missions and the critical needs of food science and technology. *Trends in Food Science and Technology* 99 https://doi.org/10.1016/j.tifs.2020.04.009.

31 Santoro, A., Chambers, J.M., Robson, B.J. et al. (2020). Land use surrounding wetlands influences urban populations of a freshwater turtle. *Aquatic Conservation: Marine and Freshwater Ecosystems* 30 (5) https://doi.org/10.1002/aqc.3324.

32 Shimabukuro, E.M. and Henry, R. (2018). Benthic distribution in small tropical lakes: the presence of macrophytes matters. *Tropical Ecology* 59 (1): 69–81.

33 Gardner, R. and Finlayson, C. (2018). Global Wetland Outlook: State of the World's Wetlands and Their Services to People. Gland, Switzerland: Ramsar Convention Secretariat.

34 Scherer, L. and Pfister, S. (2016). Global water footprint assessment of hydropower. *Renewable Energy* 99: 711–720. https://doi.org/10.1016/j.renene.2016.07.021.

35 Nautiyal, H. and Goel, V. (2020). Sustainability assessment of hydropower projects. *Journal of Cleaner Production* 265 (in press). doi:https://doi.org/10.1016/j.jclepro.2020.121661.

36 Peng, H., Ma, W., Mu, Y. et al. (2015). Impact of permafrost degradation on embankment deformation of Qinghai-Tibet highway in permafrost regions. *Journal of Central South University* 22: 1079–1086. https://doi.org/10.1007/s11771-015-2619-2.

37 Baruah, D., Hazarika, L.P., Bakalial, B. et al. (2012). A grave danger for the Ganges dolphin (*Platanista gangetica* Roxburgh) in the Subansiri River due to a large hydroelectric project. *The Environmentalist* 32: 85–90. https://doi.org/10.1007/s10669-011-9375-0.

38 Singha, T.P., Dutta, B.K., and Biswas, S.P. (2013). The status of the Ganges River Dolphin *Platanista gangetica gangetica* in the River Barak, Assam, India. *Journal of the Bombay Natural History Society* 110 (2): 129–134.

39 Das, B.C., Hossain, M.A.R., Haq, M.S. et al. (2016). Reproductive biology of riverine populations of Tenualosa ilisha in Bangladesh. *Proceedings of 3rd International Conference on Natural Science and Techechnology*. Chittagong, Bangladesh.

40 Das, I., Hazra, S., Das, S. et al. (2019). Present status of the sustainable fishing limits for Hilsa Shad in the northern Bay of Bengal, India. *Proceedings of National Academy of Science, India, Section B Biological Sciences* 89: 525–532. https://doi.org/10.1007/s40011-018-0963-3.

41 Islam, M.M., Mahammed, E.Y., and Ali, L. (2016). Economic incentives for sustainable Hilsa fishing Bangladesh: an analysis of the legal and institutional framework. *Marine Policy* 68 (8–22) https://doi.org/10.1016/j.marpol.2016.02.005.

42 Mangadze, T., Bere, T., and Mwedzi, T. (2016). Choice of biota in stream assessment and monitoring programs in tropical streams: a comparison of diatoms, macroinvertebrates and fish. *Ecological Indicators* 63: 128–143. https://doi.org/10.1016/j.ecolind.2015.11.029.

43 Bytyçi, P.S., Zhushi Etemi, F.N., Ismaili, M.A. et al. (2018). Biomonitoring of water quality of river nerodime based on physicochemical parameters and macroinvertebrates. *Rasayan Journal of Chemistry* 11 (2): 554–568. https://doi.org/10.31788/RJC.2018.1122087.

44 Herman, M.R. and Nejadhashemi, A.P. (2015). A review of macroinvertebrate- and fish-based stream health indices. *Ecohydrology and Hydrobiology* 15 (2): 53–67. https://doi.org/10.1016/j.ecohyd.2015.04.001.

45 Itsukushima, R. and Shimatani, Y. (2016). Study of segment-based ecoregions using fish fauna for conserving river environments in Kyushu, Japan. *Ecological Engineering* 97: 465–475. https://doi.org/10.1016/j.ecoleng.2016.10.024.

46 Chowdhury, R.B. and Moore, G.A. (2017). Floating agriculture: a potential cleaner production technique for climate change adaptation and sustainable community development in Bangladesh. *Journal of Cleaner Production* 150: 371–389. https://doi.org/10.1016/j.jclepro.2015.10.060.

47 Taylor, M.D., Chick, R.C., Lorenzen, K. et al. (2017). Fisheries enhancement and restoration in a changing world. *Fisheries Research* 186: 407–412. https://doi.org/10.1016/j.fishres.2016.10.004.

48 Wang, Z. and Yan, S.H. (2016). Direct and strong influence of water hyacinth on aquatic communities in natural waters. In: Water Hyacinth: Environmental Challenges, Management and Utilization (eds. Y. Shaohua and J. Guo), 44–65. CRC press, Taylor & Francis.

49 Wang, B., Liu, S., Li, F. et al. (2017). Removal of nitrate from constructed wetland in winter in high-latitude areas with modified hydrophyte biochars. *Korean Journal of Chemical Engineering* 34 (3): 717–722. https://doi.org/10.1007/s11814-016-0308-y.

50 Prajapati, M., van Bruggen, J.J.A., Dalu, T. et al. (2017). Assessing the effectiveness of pollutant removal by macrophytes in a floating wetland for wastewater treatment. *Applied Water Science* 7: 4801–4809. https://doi.org/10.1007/s13201-017-0625-2.

51 da Silva, C.V. and Henry, R. (2020). Aquatic macroinvertebrate assemblages associated with two floating macrophyte species of contrasting root systems in a tropical wetland. *Limnology* 21 (1): 107–118. https://doi.org/10.1007/s10201-019-00588-w.

52 Pace, R.L., Chambers, J.M., and Robson, B.J. (2016). Potential of submerged macrophytes to support food webs in lowland agricultural streams. *Marine and Freshwater Research* 68: 549–562. https://doi.org/10.1071/MF15391.

53 Luiz, O.J., Woods, R.M., Madin, E.M.P. et al. (2016). Predicting IUCN extinction risk categories for the world's data deficient groupers (Teleostei: Epinephelidae). *Conservation Letters* 9 (5): 342–350. https://doi.org/10.1111/conl.12230.

54 Man, L. and Wang, C. (2013). An explorative study on ecological designs of the landscape of river channels. *International Conference on Remote Sensing, Environment and Transportation Engineering (RSETE 2013).* https://doi.org/10.2991/rsete.2013.21.

55 Shang, J.Y. (2018). Riverfront landscape based on the idea of river restoration: a case study of the riverfront landscape of Duoba New District in Xining city. *Journal of Building Construction and Planning Research* 6: 151–161. https://doi.org/10.4236/jbcpr.2018.63011.

56 Gonzalez, E., Sher, A.A., Tabacchi, E. et al. (2015). Restoration of riparian vegetation: a global review of implementation and evaluation approaches in the international, peer-reviewed literature. *Journal of Environmental Management* 158: 85–94. https://doi.org/10.1016/j.jenvman.2015.04.033.

57 Sozanska-Stantona, M., Carey, P.D., Griffiths, G.H. et al. (2016). Balancing conservation and climate change – a methodology using existing data demonstrated for twelve UK priority habitats. *Journal for Nature Conservation* 30: 76–89. https://doi.org/10.1016/j.jnc.2016.01.005.

58 Giakoumis, T. and Voulvoulis, N. (2018). The transition of EU water policy towards the water framework directive's integrated river basin management paradigm. *Environmntal Management* 62: 819–831. https://doi.org/10.1007/s00267-018-1080-z.

59 Pont, D., Logez, M., Carrel, G. et al. (2015). Historical change in fish species distribution: shifting reference conditions and global warming effects. *Aquatic Sciences* 77: 441–453. https://doi.org/10.1007/s00027-014-0386-z.

60 Rubin, Z.K., Kondolf, G.M., and Carling, P.A. (2015). Anticipated geomorphic impacts from Mekong basin dam construction. *International Journal of River Basin Management* 13 (1): 105–121. https://doi.org/10.1080/15715124.2014.981193.

61 Huber, A. (2019). Hydropower in the Himalayan hazardscape: strategic ignorance and the production of unequal risk. *Water* 11: 414–436. https://doi.org/10.3390/w11030414.

62 McGregor, G.B., Marshall, J.C., Lobegeiger, J.S. et al. (2018). A risk-based ecohydrological approach to assessing environmental flow regimes. *Environmental Management* 61: 358–374. https://doi.org/10.1007/s00267-017-0850-3.

63 Negus, P., Blessing, J., Clifford, S. et al. (2020). Adaptive monitoring using causative conceptual models: assessment of ecological integrity of aquatic ecosystems. *Australasian Journal of Environmental Management*: 1–17. https://doi.org/10.1080/14486563.2020.1750494.

64 Wehn, U., Collins, K., Anema, K. et al. (2017). Stakeholder engagement in water governance as social learning: lessons from practice. *Water International* 43: 1–26. https://doi.org/10.1080/02508060.2018.1403083.

65 Wei, H.Y., Hu, S., Li, F. et al. (2019). A review of freshwater fish introductions to the Guangdong province, China. *Aquatic Ecosystem Health and Management* 22 (4): 396–407. https://doi.org/10.1080/14634988.2019.1685850.

66 Welcomme, R.L., Baird, I.G., Dudgeon, D. et al. (2016). Fisheries of the rivers of Southeast Asia. In: Freshwater Fisheries Ecology (ed. J.F. Craig), 363–376. Wiley https://doi.org/10.1002/9781118394380.ch29.

67 WWAP. (2020). Water and Climate Change: The United Nations World Water Development Report 2020. World Water Assessment Programme, UNESCO. http://www.unesco.org/water/wwap/wwdr.

68 Wollny, J.T., Otte, A., and Harvolk-Schöning, S. (2019). Dominance of competitors in riparian plant species composition along constructed banks of the German rivers Main and Danube. *Ecological Engineering* 127: 324–337. https://doi.org/10.1016/j.ecoleng.2018.11.013.

69 Ziegler, J.P., Solomon, C.T., Finney, B.P. et al. (2015). Macrophyte biomass predicts food chain length in shallow lakes. *Ecosphere* 6: 1–16. https://doi.org/10.1890/ES14-00158.1.

70 Ziv, G., Baran, E., Nam, S. et al. (2012). Trading off fish biodiversity, food security, and hydropower in the Mekong River Basin. *Proceedings of the National Academy of Sciences of the United States of America* 109: 5609–5614. https://doi.org/10.1073/pnas.1201423109.

15

Ecosystem Services of the Phoomdi Islands of Loktak, a Dying Ramsar Site in Northeast India

Sijagurumayum Geetanjali Devi[1], Niteshwori Thongam[1], Maibam Dhanaraj Meitei[1], and Majeti Narasimha Vara Prasad[2]

[1] *Department of Forestry and Environmental Science, Manipur University, Imphal, Manipur, India*
[2] *School of Life Sciences, University of Hyderabad, Hyderabad, Telangana, India*

15.1 What Are Ecosystem Services?

The Millennium Ecosystem Assessment defined ecosystem services as the benefits people obtain from ecosystems [1]. Daily defined ecosystem services as conditions and processes through which natural ecosystems and the species that form them sustain and fulfill human life. They help maintain biodiversity and produce ecosystem goods [2]. Likewise, Costanza described ecosystem goods and services as the benefits humans derive, directly or indirectly, from ecosystem functions [3].

The Millennium Ecosystem Assessment categorized ecosystem services in four groups: provisioning, regulating, cultural, and supporting services [1]. Provisioning services represent the products (such as food, fodder, medicinal usage, fuel, fibers, and building materials) obtained from ecosystems. Regulating services include benefits from the regulation of ecosystem processes, such as the recharge of aquifers, purification of water, climate regulation, disease control, soil and sediment retention, etc. Cultural services are the non-material benefits that people obtain from ecosystems through recreation, aesthetic experiences, spiritual enrichment, and reflection. The supporting services are the support systems that are necessary for the production of all other ecosystem services. Such examples can be soil formation, primary production, and nutrient cycling in the ecosystems.

The Ramsar Convention has classified the ecosystem services provided by the wetlands under flood control, groundwater replenishment, shoreline stabilization, sediment and nutrient retention and export, water purification, reservoirs of biodiversity, wetland products, cultural values, recreation and tourism, and climate change mitigation [4].

15.2 *Phoomdi* Islands of Loktak

Wetlands are scattered in all climatic regions and cover about 6% of the earth, close to 2 billion acres (800 million ha) [5]. India supports and sustains diverse and unique wetland habitats [6]. The Ramsar Convention on Wetlands designated 42 wetlands in India as internationally

Handbook of Ecological and Ecosystem Engineering, First Edition. Edited by Majeti Narasimha Vara Prasad.

important [7]. Manipur, a landlocked state in the eastern Himalayas, has extremely rich and diverse wetlands that spread across 524.51 km^2, sharing 2.35% of the total geographical area within the Barack and Chindwin-Irrawaddy-Manipur river systems. Loktak, a Ramsar site (no. 463), is the largest wetland in the Manipur river basin and covers around 61% of the total identified wetlands of Manipur (Figure 15.1). A unique character of the wetland is the floating plant biomass known as *phoomdi* [8].

Phoomdi are heterogeneous masses of soil, vegetation, and organic matter in different stages of decay (Figure 15.2). This habitat has a large diversity of aquatic, semi-aquatic, and terrestrial plants, representing 128 species from 46 families [9]. Among the dominant plants, *Phragmites karka* alone constitutes about half the vegetation of Keibul Lamjao National Park (approx. 40 km^2). *Phoomdi* has different sizes, with thickness varying from a few centimeters to 2.5 m, and occupy nearly two - thirds of the wetland. Several plants help in the formation of *phoomdi*, with grasses, sedges, and other emergents giving out long floating rhizomes and runners with upright shoots representing the primary initiation stages. *Phoomdi* formation is initiated with a small mass of undecomposed organic matter or dense growth of floating plants accumulating in suspended silts that are progressively colonized by grasses and other herbaceous plants. In the later stages, as more organic matter is added from the death and decay of vegetation on and around the *phoomdi* and soil accumulated from the suspended silt load, the *phoomdi* grows in both shape and size and ultimately supports increasing amounts of diverse vegetation.

Structurally, *phoomdi* can be grouped into three vertical zones: upper root zone, middle mat zone, and lower peat zone. The upper zone is 0–15 cm thick across the wetland. The middle zone's thickness varies from 25 to 40 to 60 cm. It is a zone of compactly interwoven living, dead, and decaying roots with plant litter accumulated on top. The lower zone ranges from 10 to 25 cm. Below the lower zone is water; the distance between the *phoomdi* and the bottom of the wetland fluctuates with the water level [6].

The mat floats, with about one-fifth above and four-fifths below water level. During monsoons, when the water level rises, the *phoomdi* float on the wetland and eventually sink in the dry season. The process helps plants draw nutrients from the bottom soil and grow throughout the year. Even in the driest months of February and March, *phoomdi* retain their swampy top layer, thereby contributing to a high level of nutrient cycling.

15.3 Ecosystem Degradation of Loktak

Today, wetlands worldwide are endangered by nutrient enrichment and heavy metal pollution from numerous anthropogenic origins [10–13]. Annually, around 0.28 million people along the Nambul river catchment generate an estimated 72.23 million tons of solid waste and 31 027 m^3 of sewage. Likewise, 4.9 million tons of solid waste and 2121 m^3 of sewage are generated by inhabitants along the Nambul river and drain annually into Loktak (Figure 15.3) [6]. Adding to the problem, deforestation and shifting cultivation in catchment areas; unrestrained use of fertilizers, pesticides, and insecticides on agricultural lands in the valley; and domestic waste discharge and sewage runoff from human habitations (including people living in *phoomdi* huts) contribute to the extensive addition of silt, nutrients, and other pollutants into the wetland. These factors accelerate the aging of Loktak by rapid siltation and excessive *phoomdi* biomass production.

The wetland's ecological regime underwent a major shift when it was converted into a reservoir by building a barrage at Ithai for the generation of 105 MW of hydel power in 1983. The wetland

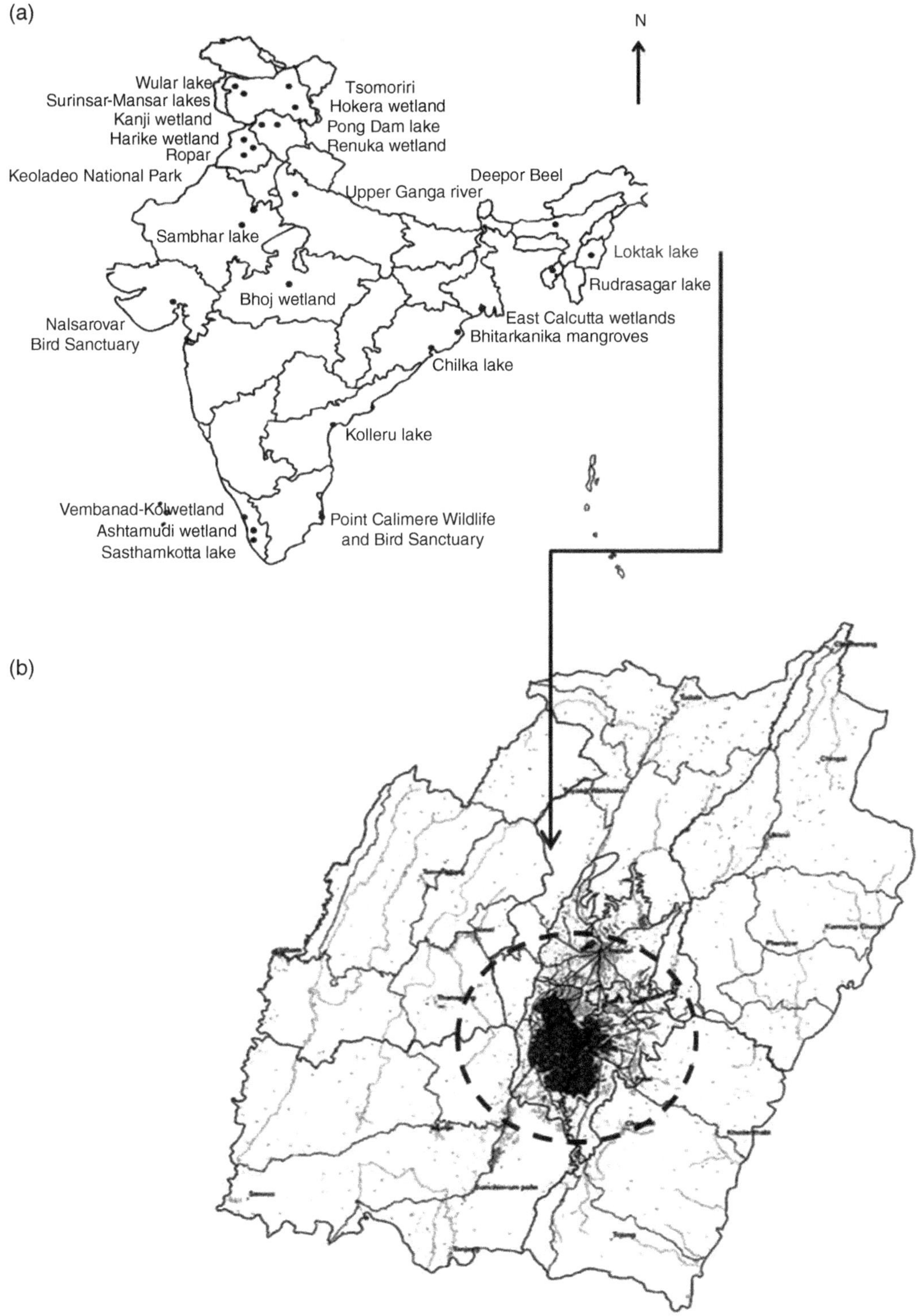

Figure 15.1 (a) Ramsar sites in India; (b) map of Manipur with Loktak.

(a)

(b)

Figure 15.2 (a) An aerial view of Loktak with the floating *athaphoom* or traditional fishing ponds; and (b) manual removal of *phoomdi* for composting. *See colour plate section for colour representation of this figure.*

now has no way for excess silt, nutrients, and metal ions to flow downstream. Further, there is a continuous loss of vegetation in the catchment areas of the 36 rivers draining into Loktak due to *jhoom* or shifting cultivation.

The Loktak basin receives 1152.14 tons of N/km^2/year and 137.74 tons of P from its catchment region [6]. Today, the rate of fertilizer application on surrounding agricultural fields is estimated at between 100 and 200 kg/ha, making the wetland more eutrophic. Kangabam et al. reported a higher NO_2^- concentration than the permissible limit set by the WHO in all four seasons, with a maximum of 4.12 mg/l in Thanga [14]. Water Quality Index (WQI) values ranged from 64 to 77, reflecting the poor status of the wetland water quality. The water of Loktak today exceeds the threshold limit for Fe in drinking water (0.37–0.57 mg/l) set by the WHO [15].

The Nambul river, the main tributary, is the primary source of pollution in Loktak [16]. The Nambul river water's physicochemical characteristics showed low pH, high turbidity, high electrical conductivity, low dissolved oxygen (DO), and high nitrate and phosphate concentrations (Table 15.1) [17]. Electrical conductivity was higher than the permissible limit of 300 μs/cm set by the WHO [18]. The high conductivity may be mainly due to high nutrient loads and sewage

Figure 15.3 Degraded ecological conditions of the Nambul river (a main tributary of Loktak): (a) Khwairamband Bazaar; (b, c) Keishamthong stretch, the greater Imphal region. *See colour plate section for colour representation of this figure.*

Table 15.1 Water characteristics of the Nambul River, a main tributary of Loktak. Source: Thongam and Meitei [17].

	Sampling locations during 2018-2019						
Parameters	**NT**	**SB**	**TN**	**KP**	**PT**	**LT**	**HT**
pH	6.41	6.44	6.58	6.65	7.06	6.74	6.74
EC (μS/m)	441	521	866	972	898	895	844
TDS (mg/l)	229	283	466	485	481	457	462
DO (mg/l)	2.21	0.99	0.55	0.61	1.07	1.02	1.18
Turbidity(NTU)	21	20.9	22.5	20.1	23.1	18.9	19.7
Total hardness (mg/l)	40.67	37.99	43.33	45.99	45.99	48.33	45.66
PO_4^{3-}(mg/l)	1.08	1.23	1.68	1.83	1.74	1.77	1.61
NO_3^-(mg/l)	4.24	6.61	9.14	8.74	7.02	5.52	5.98

NT – Naoremthong, SB – Sagolband, TN – Thong Nambonbi, KP – Keishampat, PT – Pishumthong, LT – Langthabal, and HT – Hiyangthang.

discharge, especially via Naga Nullah and from the surrounding area. Naga Nullah, which flows into the Nambul in the Khwairamband stretch, is one of the heavily polluted streams that run through the densely populated Imphal valley. High turbidity was recorded in the Nambul River, ranging from 18.9 to 23.1 nephelometric turbidity units (NTU). Also, high nitrate and phosphate levels in the range of 4.24–9.14 mg/l and 1.08–1.83 mg/ l are reported, indicating the eutrophic nature of the river that drains into Loktak.

Today, the overgrowth of *phoomdi* creates a harmful impact on the ecosystem processes and functions. The floating biomass displaces water due to its buoyancy, thereby affecting the wetland's overall elevation-area-storage capacity relationship [6]. A 1 m^3 *phoomdi* displaces 0.75 m^3 of water, and today, the floating biomass occupies approximately 107 km^2 of the wetland. Thereby, increasing *phoomdi* will ultimately reduce the wetland's storage capacity and affect water resources and the wetland environment.

In addition, the wetland area covered by *phoomdi* is responsible for a significant loss of water through evapotranspiration. It is estimated that 73.5 km^2 of open water area can cause a water loss of 75.3 million cubic meters (Mm^3) due to evaporation, and the loss due to evapotranspiration from 107 km^2 is 136.4 Mm^3 [8]. The *phoomdi* in the channels choke and interfere with the flow of the current, leading to stagnation upstream. *Phoomdi* are a primary cause of flooding in low-lying wetland regions. Also, increasing *phoomdi* turnover adds decaying and sinking plants in the wetland bottom and makes Loktak more eutrophic.

Kangabam et al. reported the changes in land use and land cover between 1977 and 2015 in Loktak and showed overall increases in open bodies of water, agricultural area, and settlement regions by 10.94, 10.26, and 2.23%, respectively, whereas *phoomdi* with thick and thin vegetation decreased by 3.48 and 28.89%, respectively [19]. Land encroachment is rampant throughout the wetland, converting it to human habitations and agricultural fields. The consequences of human actions are the loss of biodiversity, infestation by invasive species, decreased fisheries, reduction of wetland products, and pollution that threaten the survival of the wetland. These factors have led to Loktak being listed by the Ramsar Convention as a dying or degrading ecosystem in the Montreux records since 1993.

15.4 Ecosystem Services Provided by the *Phoomdi* Islands of Loktak

Loktak is referred to as the lifeline of the Manipuri valley communities because of the ecosystem services it provides to the local population (Figure 15.4). Loktak water is used for drinking, domestic purposes, irrigation, navigation, and power generation. Wetland plants are harvested for food, medicine, fodder, fiber, fuel, handicrafts, and making house [20]. Loktak supports rich biodiversity: around 233 plant species and 425 animal species [6, 21]. The wetland acts as a refuge habitat for thousands of birds that belong to 116 species (21 species are migratory) [21]. Among the fish species, *Monopterus albus* and *Osteobroma belangeri* are restricted to Manipur, Myanmar, and the Yunan state of China. The wetland is also a breeding habitat for migratory fish species such as *Labeo dero*, *Cirrhinus reba*, and *O. belangeri* that journey from the Chindwin-Irrawaddy river system of Myanmar to the upstream regions of the Manipur river and select breeding grounds in the shallow wetlands of the Manipur valley [8].

Adding to the uniqueness of the *phoomdi* is the Keibul Lamjao National Park in the southern zone of the wetland. It is the only floating national park in the world and the lone natural habitat of the endangered Eld's deer species, *Rucervus eldi eldi*.

15.5 *Phoomdi* and Provisioning Services

A significant population that lives along the wetland shore harvests aquatic plants from Loktak for a wide range of uses. In total, 27 *phoomdi* species from 15 families are extracted as food (Table 15.2) [9]. The edible species are consumed for their fruits, leaves, whole plant, roots/

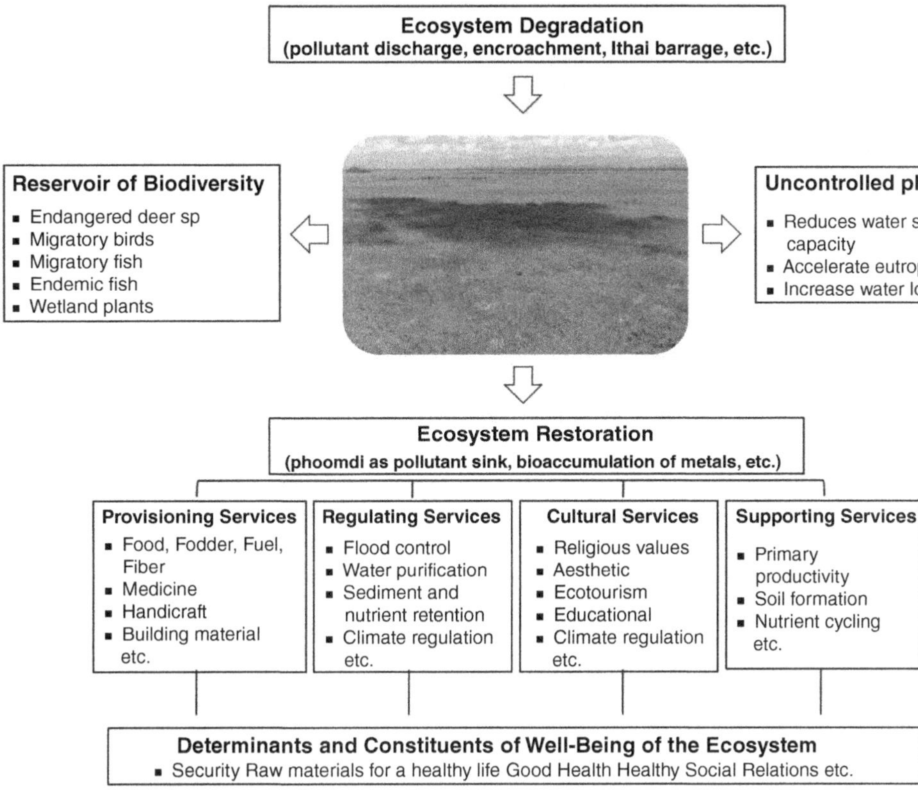

Figure 15.4 *Phoomdi* of Loktak: the four ecosystem services for the restoration of the wetland ecosystem.

rhizomes/tubers, and shoots. For example, *Alocasia cucullata, Centella asiatica, Hedychium coronarium, Neptunia oleracea, Oenanthe javanica,* and *Polygonum barbatum* are most commonly consumed to prepare traditional Manipuri delicacies such as iromba, kangsoi, kangsu, shak, singju, and utti.

In total, 25 species were recorded as being used by the local traditional healers known as *maiba* to treat ailments [9]. *Phoomdi* medicinal plants can cure 18 diseases and ailments using the whole plant or inflorescence, shoots, rhizomes/roots/tubers, and seeds, respectively. For generations, local medicinal practitioners have treated cut and wounds, liver troubles, stomach disorders, blood troubles, nervous disorders, rheumatism, skin diseases, fever, boils, cough, dropsy, piles, menstrual problems, epilepsy, asthma, and leprosy using *phoomdi* plants. The luxuriant growth of the *phoomdi* biomass provides fodder for a wide range of domestic animals, such as cows, pigs, buffalo, and goats, with commonly preferred harvested fodders including *O. javanica, Enhydra fluctuans, Echinochloa stagnina, Zizania latifolia,* and *Imperata cylindrica*.

Most households along the wetland shores today depend on the harvest from *phoomdi* for fuel. Local canoes harvest *N. porphyrocoma, Arundo donax, Scirpus lacustris,* and *P. karka*. Among the many hand-woven handicraft items of the indigenous Meitei community, mats or *phak*, which are woven using *S. lacustris*, known as *kouna*, are sold in local markets. The material is also used to make cushions, pillows, baskets, and floor mats and helps shape the economy of many households.

Table 15.2 Provisioning services of *phoomdi* from Loktak. Source: Meitei and Prasad [9].

Scientific name	Family	Vernacular name (Manipuri)	Use
Ageratum conyzoides L.	*Asteraceae*	*Khongjainapi*	MD: Shoot paste on cuts and wounds. MC: Used in *chenghi*.
Alpinia galanga Willd.	*Zingiberaceae*	*Kanghoo*	MD: Rhizome paste for fever and rheumatism. WE: Rhizomes in *iromba*. Flowers eaten fresh.
Alternanthera philoxeroides Griseb.	*Amaranthaceae*	*Kabonapi*	MD: Leaf paste on cuts and injuries. WE: Used in *utti*. FD: Whole plant.
Argyreia nervosa (Burm.f.) Bojer	*Convolvulaceae*	*Uri tujombi*	MD: Root extracts for rheumatism and nervous disorders. Leaf extract on cuts, wounds, and skin diseases.
Arundo donax L.	*Araceae*	*Yenthou*	MD: Rhizome extracts to stimulate menstrual discharge. HM: Culms. FD: Leaves. FL: Culms.
Centella asiatica (L.) Urb.	*Apiaceae*	*Peruk*	MD: Whole plant foreskin diseases and leprosy. WE: Whole plant for *kangsu*. MC: Used for *chenghi*.
Colocasia esculenta (L.) Schott	*Araceae*	*Lampal*	MD: Petiole juice on cuts and wounds. WE: Rhizome for *iromba* and leaves for *utti*.
Crotalaria juncea L.	*Papilionaceae*	*U hawai maton*	MD: Seed powder as a blood purifier. WE: Leaves for *kangsoi*.
Cynodon dactylon (L.) Pers.	*Poaceae*	*Tingthou*	MD: Shoot paste on cuts and wounds and root extract for dropsy and epilepsy.
Desmodium motorium (Houtt.) Merr.	*Papilionaceae*	*Lam hawai*	MD: Root extract for asthma and fever.
Enhydra fluctuans Lour.	*Asteraceae*	*Komprek tujombi*	MD: Shoot extract for skin and nervous diseases. WE: Tender shoots for *singju*.
Gynura cusimbua S. Moore	*Asteraceae*	*Terapaibi*	MD: Shoot paste on cuts and wounds. WE: Shoots. MC: Used for *chenghi*.
Hedychium coronarium J. Koenig.	*Zingiberaceae*	*Loklei*	MD: Rhizome paste for stomach and liver trouble. WE: Tender shoots and rhizomes for *iromba*.
Hedychium spicatum Lodd.	*Zingiberaceae*	*Takhellei*	MD: Rhizome extracts for stomach disorders and liver troubles. WE: Tender shoots and rhizomes for *iromba*.
Ipomoea aquatica Forssk.	*Convolvulaceae*	*Kolamani*	MD: Shoot extract as a purgative and emetic. WE: Tender shoots for *shak*.

Table 15.2 (Continued)

Scientific name	Family	Vernacular name (Manipuri)	Use
Neptunia oleracea Lour.	*Mimosaceae*	*Ishing-ikaithabi*	WE: Shoots for *iromba* or *singju*.
Oenanthe javanica DC.	*Apiaceae*	*Komprek*	WE: Tender shoots for *singju*.
Phragmites karka (Retz.) Trin. ex Steud.	*Poaceae*	*Tou*	HM: Culms. FL: Culms.
Polygonum barbatum L.	*Polygonaceae*	*Yellang*	MD: Shoot paste for stomach disorders. WE: Aerial parts for *singju*.
Saccharum munja Roxb.	*Poaceae*	*Khoimom*	MD: Shoot extract for treating blood troubles.
Scirpus lacustris L.	*Cyperaceae*	*Kouna*	HM: Aerial parts. FL: Shoots. HC: *Kouna*, a handicraft material.
Zizania latifolia Turcz. Ex. Stapf.	*Poaceae*	*Ishing kambong*	WE: Tender shoots. MD: Inflorescence extracts to treat indigestion. HM: Aerial parts.

MD –medicinal, WE –wild edible, FD –fodder, FL –fuel, HM –house making, HC –handicrafts, MC –miscellaneous.

Households surrounding the wetland have for time immemorial also depended on *phoomdi* plants as a source of raw material for building their homes. Eight plants are commonly used to build floating huts on the *phoomdi* that serve as temporary camps for fishermen during the harvest season [8, 9]. These huts are built on the floating biomass, which moves from one location to another as the water level and current change daily. The huts are typically built out of light raw materials such as bamboo, wooden planks, and reeds found floating in the *phoomdi*. The floors are mostly wooden planks or bamboo, made thicker (2–3 m) by adding several layers of *phoomdi* biomass. The frames, made of bamboo or light wooden pillars, are then pushed into the *phoomdi*. Finally, the four walls are covered with straw or reeds such as *P. karka, N. porphyrocoma, A. donax*, or *S. lacustris*, making the hut light weight, and *I. cylindrica, Cymbopogon nardus, Erianthus arundinaceous,* and *Z. latifolia* are used to cover the top of the hut. After the hut is built, it is anchored in a convenient fishing spot with bamboo poles on all sides so it cannot be displaced by currents or wind. On the floating *phoomdi*, there are approximately 700 floating huts; more than 80% of locals do not have a home in the mainland community [22].

15.6 *Phoomdi* as Reservoirs of Biodiversity

The *phoomdi* of Keibul Lamjao National Park (KLNP), which is 40 km^2, serve as the natural habitat of brow-antlered-deer (*Rucervus eldi eldi*). The floating park is situated in the southern part of Loktak (93°53′-93°60′ E, 24°14′-24°26′ N) at 764–769 m above mean sea level and has an average depth of 2.7 m [8]. Most of the park (around 90%) is a vast stretch of *phoomdi*, with three hills rising out of it as islands and many elevated strips of land that serve as refuges for animals during monsoons [8]. Subspecies of Eld's deer have evolved unique morphological traits such as divided hooves and long pasterns that allow the animals to walk over the floating *phoomdi* without being submerged into the wetland. The habitat provides luxurious growth of the deer's food plants, such as *Z. latifolia*, *H. coronarium*, *A. galanga*, *O. javanica*, *E. ferox*, *A. philoxeroides*, *Nymphaea* sp., *I. aquatica*, and *T. natans*.

The park today supports around 260 deer. The population has increased gradually from 14 in 1974 to 76 in 1986, 106 in 1991, and 162 in 2000 [23]. In addition, the park shelters 81 species of birds, 22 species of reptiles, and 22 species of mammals [21].

15.7 *Phoomdi* and Fisheries

The wetland's *phoomdi* habitat provides shelter to 54 fish species from 17 families [21]. The traditional method of floating pond or *athaphoom* fishing generally involves preparing circular enclosures and is carried out in two stages. In the first stage, called *phoom thaba*, large circular *phoom* (2 m wide and 1 m thick) are prepared and are left to float. Many of the floating *phoomdi* are then cut into long strips (2 m wide and 1 m thick), which are transplanted along the circular structure with a circumference of around 300–350 m. The traditional fisher folk then place into the circular *phoom* pieces of cut *phoomdi* or aquatic macrophytes such as *Eichhornia crassipes* or *S. cucullata,* and the ponds are left undisturbed for months.

During the second stage, called *phoom namba*, harvesting is done by a team of 30–40 fisher folk in hand-carved canoes. The circular *phoom* is encircled with dip nets touching the wetland bottom, leaving no room for fish to escape. The fisher folk then remove the floating plants from the enclosed circle for the easy operation of the fishing equipment. At the same time, the wetland bottom in the encircled *phoom* is disturbed with bamboo poles or canoes to deoxygenate the water column. Fish in the *athaphoom* go to the surface in search of oxygen, thereby making them easy to catch. This traditional process of *athaphoom* fishing accounted for 39% of the total harvest of 19 000 tons of fish from Loktak and adjoining wetlands of Manipur in 2010–2011 [24].

15.8 *Phoomdi* and Cultural Services

The islands' scenic beauty, the circular patterns of *athaphoom* fishing ponds in the wetland, the migratory waterfowl's etc. make the place a hotspot for ecotourism. The habitat is unique due to the floating *phoomdi* scattered over the wetland, and the ecosystem is an example for ecological studies, research, and education.

In addition, a number of *phoomdi* species are closely associated with the customs and traditional ethos of Manipuri society [9]. For example, the shoots of *E. stagnina* are offered to the Hindu deity during *Govardhan puja*; the twigs of *C. dactylon* are necessary for a number of ritual practices by the local spiritualist or *maiba*; and tender shoots of *O. javanica* are part of the offering to the deities during the *Cheirauba* and *Cheirauba* and *Imoinu puja* Manipuri festivals [9].

Given the natural significance and cultural importance of the area for the indigenous population, the Keibul Lamjao Conservation Area (KLCA) – consisting of a core area of KLNP (40 km^2) and a buffer of Loktak Lake (140 km^2) and Pumlen Pat (43 km^2) – was nominated by the government of India in 2016 to be included as a World Heritage Site by UNESCO.

15.9 *Phoomdi* and Regulating Services

Loktak is a highly eutrophic ecosystem because of the indiscriminate, untreated sewage dumps, agricultural runoff, and industrial discharge drained by its 36 tributaries and non-point sources [15]. The Loktak Development Authority (LDA) manages Loktak, focusing on enhancing the economic utility of the wetland and its beautification, but has ignored the fact that the nutrient regime of the wetland should be managed so that removing *phoomdis* does not induce phytoplankton blooms and cause water quality deterioration.

The *phoomdi* biomass absorbs nutrients and other forms of pollutants poured into the wetland [15]. More than 50% of the mineral nutrients present in the wetland system are locked within *phoomdi* tissues, reducing the concentration of nutrients in the water and suppressing algal growth. A nitrogen concentration of 16511.4 mg/kg by *Spirodela polyrhiza* and 19 308.7 mg/kg by *O. javanica* is reported in Loktak. A high bioaccumulation factor (BF) of 9.7 × 104 and 7.9 × 104 for N and P was observed in *Spirodela polyrhiza* and *Z. latifolia*, which accumulate significant amount of nutrients in their biomass. The *phoomdi* species *S. polyrhiza, P. barbatum, E. crassipes, O. javanica,* and *A. philoxeroides,* showed average BFs of 9.7×10^2, 9.4×10^2, 8.7×10^2, 8.6×10^2, and 8.1×10^2 during summer and 4.2×10^2, 5.2×10^2, 2.1×10^2,1.9×10^2, and 4.9×10^2 in winter, revealing a good ability to phytoextract nitrogen from the wetland. The *phoomdi* plants *T. natans, O. javanica, P. karka, A. philoxeroides,* and *Z. latifolia* showed average BFs of 3.6×10^4, 3.4×10^4, 3.3×10^4, 3.3×10^4, and 3.3×10^4 during summer and 4.5×10^4, 3.0×10^4, 3.2×10^4, 2.4×10^4, and 7.9×10^4 in winter, showing promise for removing phosphorus from the wetland water column [15].

For the conservation of Loktak, maintaining low nutrient levels in the river will help protect the wetland at the end. Macrophytes species such as *B. mutica* > *A. philoxeroides* > *A. sessilis* > *E. crassipes* > *P. stratiotes* ˃*H. verticillata* in the Nambul showed average BFs of $1.1 \times 10^4 > 6.8 \times 10^3 > 4.2 \times 10^3 > 4.1 \times 10^3 > 2.4 \times 10^{3>}$ 1.4×10^3 for phosphorus, revealing good phytoextraction ability. Maximum translocation factors of 0.82, 0.59, and 0.58 were observed for *A. sessilis, E. crassipes,* and *P. stratiotes*, respectively. The macrophytes strip nutrients from the water and can then be harvested to remove those nutrients. As *E. crassipes* and *P. stratiotes* are free-floating macrophytes with high biomass yield, they provide an added advantage for stripping nutrients through regular harvest from the river [25].

Nutrient accumulation by *phoomdi* creates a situation where regular harvesting of aboveground *phoomdi* parts could serve as a useful, practical approach for the control and management of eutrophication associated with Loktak. Another advantage of *phoomdi* plants is the suppression of algal blooms. Dominant *phoomdi* species such as *E. crassipes, P. karka, P. barbatum, S. cucullata,* and *Z. latifolia* act as phytosanitants and help to improve the water quality of the wetland as aquatic macrophytes are good phytofilter of nutrients (Table 15.3). If the huge amount of pollutants brought in by Nambul and 35 other rivers were not accumulated and filtered by the *phoomdi*, they would have led to the degradation of water quality, making it unfit for drinking, fisheries, and aquatic biodiversity. Thus, the nutrient-stripping mechanism of the *phoomdi* slows down Loktak's aging.

15.10 *Phoomdi* and Supporting Services

Phoomdi play a significant role in the ecological processes and functions of the wetland. They provide a biological sink for nutrients, thereby governing water quality and nutrient dynamics [6]. The *phoomdi* biomass's rich nutrients make it a perfect spot for fisheries and agriculture on the

Table 15.3 Regulating services of macrophytes as biological sinks for nutrients.

Species	Functions	References
Alocasia cucullata Schott	Removal of TN and TP in micro scale wetlands.	[26]
Alternanthera sp.	Reduction of 84, 76, and 86% NO_3-N, PO_4-P and NH_4-N from Ganga river water in constructed wetlands.	[27]
Arundo donax L.	Nitrogen input of 225–329 kg/ha/yr.	[28]
Canna indica L.	Removal of TN and TP up to 26.2, 30.8, and 27.0 g/kg and 7.1, 4.9, and 4.5 g/kg in stems, leaves, and roots from constructed wetlands.	[29]
	Uptake of TN and TP up to 36.84–44.13 and 1.98–3.61 g/kg biomass in vertical flow constructed wetlands.	[30]
Colocasia esculenta (L.) Schott	Removal efficiency of 64.6–97.6% and 65.38–98.16% for NO_3-N and PO_4-P from subsurface flow constructed wetlands.	[31]
Eichhornia crassipes (Mart.) Solms	Phosphorous uptake of 5.76–9.97 mg/g dw.	[32]
Hedychium coronarium J. Koenig	NH_4-N, NO_3-N, and PO_4-P removal efficiency of 92, 89, and 92% from wastewater in the Parismina river.	[33]
Imperata cylindrica Raeusch.	Nutrient content 100 and 12 kg/ha at 23 months for TN and TP.	[34]
Oenanthe javanica (Blume) DC.	Accumulation of TN and TP up to 23.15 and 1.67 g/kg dw from artificial floating beds.	[35]
	Nitrogen and phosphorous uptake of 2.09–4.08 and 0.17–0.28 g/kg from wastewater in Hanoi.	[36]
Phragmites karka (Retz.) Trin. ex Steud.	Removal efficiency of 49 and 8.5% TN and TP from shallow lake water in Baiyangdian.	[37]
Pistia stratiotes L.	Removal efficiency of 99.2, 64.2, 87.6, and 48.5% of NH_4-N, TN, and TP and soluble reactive P from dairy manure wastewater.	[38]
	Total P accumulation of 0.047–0.059 g/m^2 in tissues from Parana river water.	[39]
Polygonum sp.	Reduction of 84, 76, and 86% NO_3-N, PO_4-P, and NH_4-N from Ganga river water in constructed wetlands.	[27]
Zizania latifolia (Griseb.) Turcz. ex Stapf	Removal efficiency of 46.4, 48.0, and 47.9 for total N, total dissolved N, and NH_4-N from eutrophic water of lake Taihu.	[40]

surrounding plains [41]. The high nutrient accumulation and cycling in the *phoomdi* enriches the fertility of the surrounding lands and increases the potential output of fisheries in the wetland. The luxuriant growth of *phoomdi* favors the support of biodiverse life forms in Loktak, accommodating 233 species of macrophytes, 55 species of zooplankton, 425 species of animals (249 vertebrates and 176 invertebrates), 116 species of birds (21 migratory), 64 species of fish (many migratory), and a sizable population of two endangered mammals, *Axis porcinus* and *Rucervus eldi eldi* [6, 8, 9, 21].

15.11 Conclusion

The Ramsar site plays a significant part in providing ecological and economic security in the central region of the Manipur valley. The site's unique characteristic feature, the *phoomdi*, sustains a significant population living in and around the wetland. Several indigenous *phoom* - based fishing techniques have been developed over the generations by *phoom* - dwelling locals. A wide variety of *phoom* species are utilized by locals as sources of food, fuel, and fodder and have become part of the indigenous Manipuri culture and heritage. Also, *phoomdi* play an essential role in the wetland ecosystem's ecological processes and functions by providing key nutrients and governing the wetland's water quality and nutrient regime.

With its numerous floating islands, the wetland supports rich biological diversity, and the Keibul Lamjao National Park serves as the lone natural habitat of the remaining population of the endangered *Rucervus eldi eldi*. However, despite the tremendous economic utility of the *phoomdi*, their rapid proliferation in recent years has been a cause of serious concern. Today, there is a need to find a middle path where the *phoomdi* of Loktak are maintained in the wetland without disturbing the natural balance. *Phoomdi* management techniques that integrate the traditional knowledge of communities with technical know-how to provide secure livelihoods for locals and ecological integrity for the wetland ecosystem are badly needed before the dying wetland degrades further and threatens the livelihood and ecosystem security of the Manipuri society.

Acknowledgments

The authors thank the Department of Forestry and Environmental Science, Manipur University for providing a working environment.

References

1 Millennium Ecosystem Assessment. (2003). Ecosystems and human well-being: a framework for assessment. Island Press, 1718 Connecticut Avenue, N.W., Suite 300, Washington, DC 20009.

2 Daily, G.C. (1997). Introduction: what are ecosystem services? In: Nature's Services: Societal Dependence on Natural Ecosystems (ed. G.C. Daily), 1–10. Washington, DC: Island Press.

3 Costanza, R., D'Arge, R., de Groot, R.S. et al. (1997). The value of the world's ecosystem services and natural capital. *Nature* 387 (6630): 253–260.

4 Ramsar Convention, R. (2011). Wetland: Ecosystem Services. Ramsar Convention Secretariat, Gland.

5 Reddy, K.R. and De Laune, R.D. (2008). Biogeochemistry of Wetlands: Science and Applications, 1–26. New York: CRC press, Taylor and Francis group.

6 Trishal, C.I., and Manihar, T. (2002). Management of *phoomdis* in Loktak lake. Manipur and proceedings of a workshop held at Imphal. Loktak Development Authority.

7 Ramsar Convention (2020). Ramsar Sites Around the World. Ramsar Convention Secretariat, Gland.

8 Trishal, C.I. and Manihar, T. (2004). Loktak –the Atlas of Loktak Lake. New-Delhi, India: WISA-LDA.

9 Meitei, M.D. and Prasad, M.N.V. (2015). Phoomdi – a unique plant biosystem of Loktak lake, Manipur, North-East India: traditional and ecological knowledge. *Plant Biosyst.* 149 (4): 777–787.

10 Xu, X., Zhao, Y., Zhao, X. et al. (2014). Sources of heavy metal pollution in agricultural soils of a rapidly industrializing area in the Yangtze delta of China. *Ecotoxicol. Environ. Saf.* 108: 161–167.

11 Manolaki, P. and Papastergiadou, E. (2012). Responses of aquatic macrophyte assemblages to nutrient enrichment in a lowland river basin of Western Greece. *Plant Biosyst.* 146 (4): 1065–1077.

12 Meitei, M.D. and Prasad, M.N.V. (2013a). Lead (II) and cadmium (II) biosorption on *Spirodela polyrhiza* (L.) Schleiden biomass. *J. Environ. Chem. Eng.* 1 (3): 200–207.

13 Meitei, M.D. and Prasad, M.N.V. (2014). Adsorption of Cu (II), Mn (II) and Zn (II) by *Spirodela polyrhiza* (L.) Schleiden: equilibrium, kinetic and thermodynamic studies. *Ecol. Eng.* 71: 308–317.

14 Kangabam, R., Devi, B.S., Suganthi, K., and Govindaraju, M. (2017). Development of a water quality index (WQI) for the Loktak Lake in India. *Appl. Water Sci.* 8 (27). https://doi.org/10.1007/s13201-017-0579-4

15 Meitei, M.D. and Prasad, M.N.V. (2016a). Bioaccumulation of nutrients and metals in sediments, water and phoomdi from Loktak lake (Ramsar site), Northeast India; phytoremediation options and risk assessment. *Environ. Monit. Assess.* 188: 329.

16 Singh, N.K.S., Devi, C., Sudarshan, M. et al. (2013). Influence of Nambul river on the water quality of freshwater in Loktak lake. *Inter. J. Water Res. Environ. Eng.* 5 (6): 321–327.

17 Thongam, N. and Meitei, M.D. (2019). Potential of macrophytes to control the water quality of Nambul river. Masters project, Manipur University.

18 World Health Organization (WHO). (2017). Guidelines for drinking-water quality: fourth edition incorporating the first addendum. Geneva: World Health Organization; 2017. Licence: CC BY-NC-SA 3.0 IGO.

19 Kangabam, R.D., Selvaraj, M., and Govindaraju, M. (2019). Assessment of land use land cover changes in Loktak Lake in Indo-Burma biodiversity hotspot using geospatial techniques, Egypt. *J. Remote Sensing Space Sci.* 22: 137–143.

20 Meitei, M.D. and Prasad, M.N.V. (2013b). Phytotechnological applications of *phoomdi*, Loktak lake, Manipur, Northeast India. *Curr. Sci.* 105 (5): 569–570.

21 Singh, T.H. and Singh, R.K.S. (1994). Ramsar Sites of India, Loktak Lake. New Delhi, India: WWF-India.

22 Loktak Development Authority. (2011). Annual report. Loktak Development Authority, Government of Manipur, India.

23 Forest Department. (2011). Annual report. Department of Forest, Government of Manipur, India.

24 Fisheries Department (2011) Annual administrative report. Department of Fisheries, Government of Manipur, India.

25 Singh, S.H. and Meitei, M.D. (2018). Phosphorus amount assessment in water, sediment & macrophytes from Nambul river, Manipur. Masters project, Manipur University.

26 Lai, W.L., Zhang, Y., and Chen, Z.H. (2012). Radial oxygen loss, photosynthesis and nutrient removal of 35 wetland plants. *Ecol. Eng.* 39: 24–30.

27 Rai, U.N., Tripathi, R.D., Singh, N.K. et al. (2013). Constructed wetland as an ecotechnological tool for pollution abatement for conservation of Ganga river. *Bioresour. Technol.* 148: 535–554.

28 Borin, M., Barbera, A.C., Milani, M. et al. (2013). Biomass production and N balance of giant reed (*Arundo donax* L.) under high water and N input in Mediterranean environments. *Euro. J. Agro.* 51: 117–119.

29 Zhang, Z., Rengel, Z., and Meney, K. (2008). Interactive effects of N and P growth but not on resource allocation of *Canna indica* in wetland microcosms. *Aquat. Bot.* 89: 317–323.

30 Cui, L., Ouyang, Y., Lou, Q. et al. (2010). Removal of nutrients from wastewater with *Canna indica* L. under different vertical flow constructed wetland conditions. *Ecol. Eng.* 36: 1083–1088.

31 Bindu, T., Sylas, V.P., Mahesh, M. et al. (2008). Pollutant removal from domestic wastewater with taro (*Colocasia esculenta*) planted in a subsurface flow system. *Ecol. Eng.* 33: 68–82.

32 Xie, Y. and Yu, D. (2003). The significance of lateral roots in phosphorous acquisition of water hyacinth (*Eichhornia crassipes*). *Aquat. Bot.* 75: 311–321.

33 Nahlik, A.M. and Mitsch, W. (2006). Tropical treatment wetlands dominated by free floating macrophytes for water quality improvement in Costa Rica. *Ecol. Eng.* 28: 246–257.

34 Hartemink, A.E. (2001). Biomass and nutrient accumulation of *Piper aduncus* and *Imperata cylindrica* fallows in the humid lowlands of Papua New Guinea. *Forest Ecol. Management* 44: 19–32.

35 Zhu, L., Li, Z., and Ketola, T. (2011). Biomass accumulations and nutrient uptake of plants cultivated on artificial floating beds in China's rural area. *Ecol. Eng.* 37: 1460–1466.

36 Khai, N.M., Ha, P.Q., and Öborn, I. (2007). Nutrient flows in small-scale peri-urban vegetable farming systems in Southeast Asia-a case study in Hanoi. *Agric. Ecosyst. Environ.* 122 (2): 192–202.

37 Zhao, F., Yang, W., Zeng, Z. et al. (2012). Nutrient removal efficiency and biomass production of different bioenergy plants in hyper eutrophic water. *Biom. Bioen.* 42: 212–218.

38 Sooknah, R.D. and Wikie, A.C. (2004). Nutrient removal by floating aquatic macrophytes cultured in anaerobically digested flushed dairy manure wastewater. *Ecol. Eng.* 22: 27–42.

39 Hadad, H.R. and Maine, M.A. (2007). Phosphorous amount in floating and rooted macrophytes growing in wetlands from the Middle Parana River floodplain (Argentina). *Ecol. Eng.* 31: 251–258.

40 Hu, L., Hu, W., Deng, J. et al. (2010). Nutrient removal in wetlands with different macrophyte structures in eastern Lake Taihu, China. *Ecol. Eng.* 36 (12): 1725–1732.

41 Meitei, M.D. and Prasad, M.N.V. (2016b). Metal bioavailability in phoomdi-compost from Loktak lake (Ramsar site), Manipur, North-East India. In: Trace Elements in Temporary Waterlogged Soils and Sediments (eds. J. Rinklebe, A.S. Knox and M. Paller), 341–351. CRC press – Taylor & Francis Group.

16

The Application of Reefs in Shoreline Protection

Anu Joy[1] *and Anu Gopinath*[2]

[1] *Department of Chemistry and Research Centre, St Albert's College, Ernakulam, Kerala, India*
[2] *Department of Aquatic Environment Management, Kerala University of Fisheries and Ocean Studies, Kochi, Kerala, India*

16.1 General Introduction

Oceans cover approximately 71% of the Earth's surface, yet less than 0.1% of the world's ocean surface is covered by coral reefs [1]. Although corals first appeared over 400 million years ago, the corals that form reefs in tropical waters today appeared about 240 million years ago. Coral reefs generally exist in tropical and subtropical seas around the Equator where the water temperature is stable, around 21–29 °C, and are restricted to regions with water temperatures between 18 and 36 °C [2]. Azooxanthellate cold-water corals are found in cooler waters [3]. All corals need plenty of light to grow and grow best in waters less than 60 m in depth.

Coral reefs are living, colorful, multifaceted, vibrant underwater ecosystems that provide a home to 25% of marine species including fish, cnidarians, crustaceans, mollusks, sponges, tunicates, and echinoderms [4]. They have the highest biodiversity of any ecosystem globally in terms of the numbers of phyla and classes per hectare [5] and are often called the rainforests of the sea. It has been estimated that at least 100 000 species from almost every phylum live on tropical coral reefs; 4000 species from 100 families of fish have been documented at this time, and more species are yet to be discovered.

Coral reefs are colonies of groups of individual living animals called *polyps*. They are soft-bodied and radially symmetrical marine invertebrates of the phylum coelenterate. Reef-building corals host endosymbiotic dinoflagellates (zooxanthellae) of the genus Symbiodinium. Coral hosts profit from this mutualistic relationship by obtaining high-energy photosynthetic products in the form of sugars, amino acids, carbohydrates, and small peptides from the algae, while the symbiotic algae receive inorganic plant nutrients, carbon dioxide for photosynthesis by respiration, refuge, and protection within the polyp tissues [6, 7]. The symbiosis serves the primary purpose of restricting nutrient outflow into the surrounding oligotrophic water column. As a result, the host is endowed with substantially more energy than would otherwise be available to heterotrophs, enabling corals to extract calcium carbonate from surrounding waters and secrete it as a calcareous skeleton that forms the reef structure upon which they live. Millions of coral skeletons cemented together over a period ranging from a few thousand to millions of years give rise to reefs.

Handbook of Ecological and Ecosystem Engineering, First Edition. Edited by Majeti Narasimha Vara Prasad.

16.2 Types of Coral Reefs

Coral reefs are generally classified into four types: fringing reefs, barrier reefs, atolls, and platform reefs. Fringing reefs, also referred to as shore reefs, are formed from a calcium carbonate platform's vertical growth along a shelving coastline. These are the most common type of reefs and grow near coastlines around continents and islands. Shallow, narrow lagoons separate these reefs from the shore. The basic structure of fringing reefs includes a shallow platform that extends out to a defined edge, called the *reef crest*, where it begins to slant down to the seafloor [8, 9]. This is the most common type of reef in the Caribbean and the Red Sea.

Barrier reefs are like fringing reefs situated close to a non-reef landmass but several kilometers offshore, separated from the landmass by a lagoon or channel, often about 50 m deep [9]. These reefs can breathe water's surface by forming a barrier to navigation at their shallowest point. Some barrier reefs are more or less circular, surrounding an island. Larger barrier reefs, such as those along the Red Sea coast and Australia's Great Barrier Reef, are complex linear features consisting of chains of reef patches; some of them are elongated into ribbon reefs. Barrier reefs can be formed from fringing reefs that continue to grow around a subsiding landmass. As the fringing reef grows upward, a lagoon forms between the reef and the coastline and creates a barrier reef. In other cases, a lagoon can form between a fringing reef and the coastline as the result of sea-level rise, creating a barrier reef [9].

Atolls, also referred to as coral islands or lagoon islands, are circular reef structures enclosing a small or large lagoon. They develop from the upward growth of a barrier reef system that forms around a subsiding landmass [9]. Atolls are only found in the tropics and subtropics. Most of the world's atolls are in the Pacific Ocean (Tuamotu Islands, Caroline Islands, Marshall Islands, Coral Sea Islands, and the island groups of Kiribati, Tuvalu, and Tokelau) and Indian Ocean (the Atolls of the Maldives, the Lakshadweep Islands, the Chagos Archipelago, and the Outer Islands of Seychelles).

Platform (patch) reefs are isolated reefs that emerge from the open bottom of the continental shelf or island platform. They have irregular table-like or pinnacle features. Smaller patches occur inside atoll lagoons, while larger patches occur as isolated parts of larger developments of any other three reef categories [10]. They vary widely in size and seldom reach the water surface. These are common in the Caribbean islands, the Bahamas, Bermuda, and the Pacific Islands.

16.3 Global Distribution of Coral Reefs

Coral reefs distributed among 100 countries worldwide cover an estimated area of 284 300 km^2 of ocean. This is only 0.089% of the world's oceans and less than 1.2% of the continental shelf area. Thus, on a global scale, coral reefs are a rare habitat. The majority of coral reefs are found in the Indo-Pacific region, including the Red Sea, Arabian Sea, Indian Ocean, Southeast Asia, and Pacific Ocean (91.9%). The global distribution of coral reefs [9] is shown in Figure 16.1.

16.4 Benefits of Coral Reefs

Globally, half a billion people live within 100 km of a coral reef and benefit from its growth and protection [11]. Coral reefs are one of the most productive ecosystems and produce a wide range of ecosystem services (Figure 16.2). Some of the most important benefits are as follows:

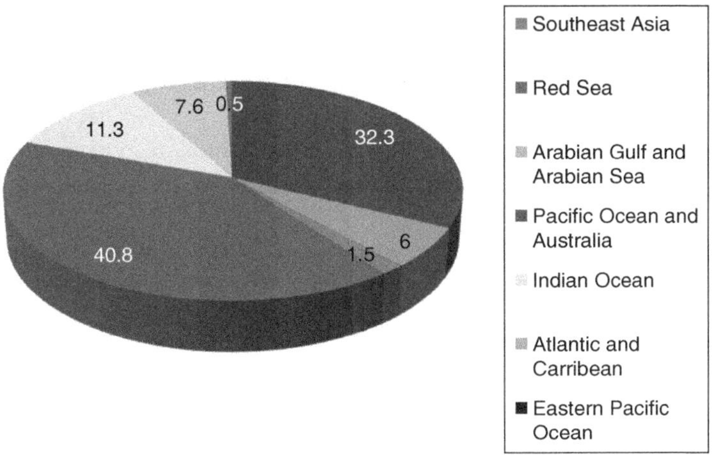

Figure 16.1 Global distribution of coral reefs. *Source:* Modified from Spalding, Ravilious, and Green [9].

Figure 16.2 Coral reef ecosystem services.

- *Vital habitat*: Coral reefs provide a vital habitat for more than 1 million diverse marine species, including 4000 fish species. Scientists estimate that coral reefs host more than a quarter of all marine life. Hence they are storehouses of immense biological wealth, making them one of the most diverse ecosystems on Earth [12].
- *Important source of food*: Coral reefs are an essential food source for many human communities since over 275 million people live within 20 miles of a coral reef. About 500 million people rely directly and indirectly on reefs for their food, livelihoods, and other resources [13]. The abundance of edible sea life in healthy coral reefs makes them essential to human nutrition. Hence, fisheries depend on coral reefs either directly or indirectly as a source of livelihood.
- *Coastal protection*: Coral reefs run along more than 90 000 miles of coastline, and due to their location in shallow, coastal waters, they are crucial to the protection of the coasts of more than 100 countries. Their construction makes them very effective for absorbing wave energy and reducing coastal erosion, property damage, and loss of life. They provide critical protection against storms, hurricanes, typhoons, and even tsunamis. Thus the reefs save billions of dollars each year in terms of reduced insurance and reconstruction costs and reduced need to build costly coastal defenses [12]. They also protect coastal wetlands, ports, harbors, and the economies they support.
- *Economic significance*: Coral reef ecosystems have an estimated value of 2.2% of all global ecosystem services like tourism and fisheries [14]. Tourism contributes more than 30% of export earnings in more than 20 countries, and in many small islands, 90% of new economic development is dependent on coastal tourism. The income generated from coral reef fisheries amounts to $29.8 billion globally, which is important to small and developing economies. Coral reefs form the habitat, spawning, and nursery grounds for about a quarter of the ocean's fish, thus providing revenue for local communities and national and international fishing fleets.
- *Medical research*: Coral reefs consist of multiple species containing pharmaceutical compounds that can potentially treat dangerous illnesses and diseases. Unique compounds obtained from coral reefs have helped treat cardiovascular diseases, ulcers, arthritis, human bacterial infections, Alzheimer's disease, leukemia, skin cancer, and HIV [12]. Corals have also been used for bone grafts and anti-cancer agents due to the similarity between coral skeletons and human bones.

16.5 Threats to Coral Reefs

Coral reefs are not stable, static communities living in benign environments, but rather are very dynamic ecosystems subject to frequent natural physical disturbances on time scales that vary from minutes to years [15, 16]. They are susceptible to stress from both global and local sources. The level of risk in major coral reef regions is given in Figure 16.3. Many scientists now believe the very existence of coral reefs may be in jeopardy unless we intensify our efforts to protect this ecosystem [17].

16.5.1 Global Threats

Global warming and resultant climate change pose an emerging and severe threat to coral reefs [13]. The greatest global threats to coral reef ecosystems are increasing temperatures and changing ocean chemistry due to warmer atmospheric temperatures and increasing levels of carbon dioxide in seawater:

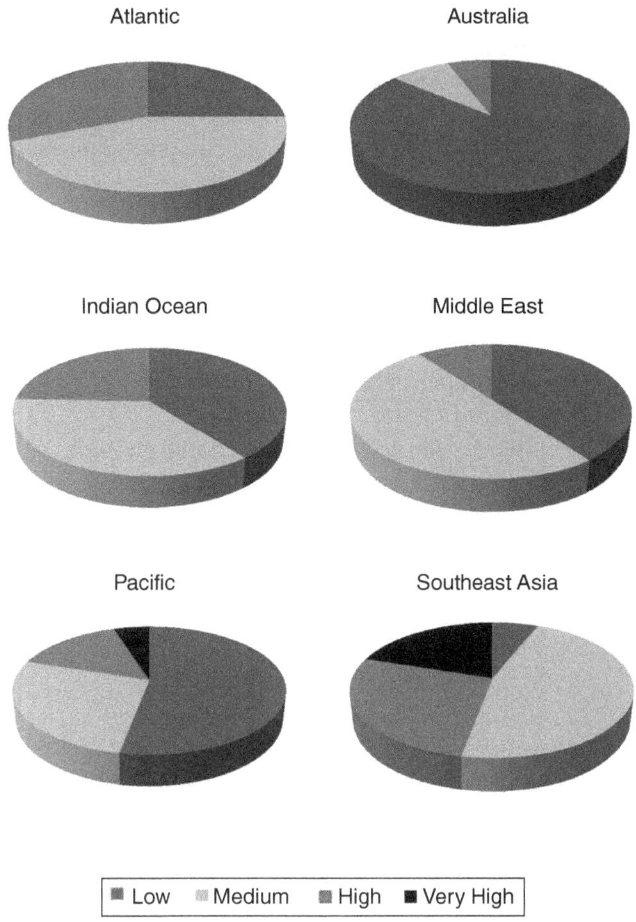

Figure 16.3 Risk analysis in the major coral reef regions.

16.5.1.1 Ocean Acidification

The amount of carbon dioxide in the atmosphere is in equilibrium with that in seawater. Hence, when atmospheric concentrations of CO_2 increase, oceanic concentrations also increase and lead to a decrease in ocean ph. Ocean acidity has increased by about 30% and is expected to increase by an additional 40% above present levels by the end of this century. Ocean acidification reduces the ability of corals to form a calcium carbonate structure; and under severe circumstances, coral skeletons dissolve.

16.5.1.2 Coral Bleaching

Coral bleaching is now a global phenomenon, and events are increasing in frequency and intensity. During mass bleaching events in the 1980s, coral reefs in tropical regions of the world were bleached white. Bleaching can be triggered by high sea surface temperatures associated with climate change, global warming, and El Nino events [18]. High water temperatures cause corals to lose the symbiotic zooxanthellae that produce food and nutrients and put stress on the corals. Thus, the coral loses its color, and coral bleaching occurs. Severe bleaching conditions can kill coral colonies or leave them more susceptible to infectious diseases.

16.5.1.3 Cyclones

Climate impacts such as sea-level rise, increased frequency and intensity of tropical storms, and altered ocean circulation patterns can also affect coral reefs. Powerful waves and winds generated during cyclones affect ocean currents and increase inshore ocean turbidity through the suspension of sediments and can seriously damage coral reefs and shorelines.

16.5.2 Local Threats

Coral reefs are particularly vulnerable to the effects of human activities [19], both through direct exploitation of reef resources and through indirect impacts from adjacent human activities on land and in coastal zones. Many human activities that degrade coral reefs are closely related to social, cultural, and economic impacts on regional coastal communities [11, 20].

16.5.2.1 Over-Fishing and Destructive Fishing Methods

Overfishing can alter the food-web structure and cause cascading effects, such as reducing the numbers of grazing fish that keep corals clean of algal overgrowth. Over-fishing of herbivorous fish on or adjacent to coral reefs can lead to high levels of algal growth and thereby affect the reef's ecological balance and biodiversity. Fishing practices such as blast or dynamite fishing, bottom trawling, cyanide fishing, and banging on a reef with sticks are destructive and unsustainable and can destroy entire reefs.

16.5.2.2 Coastal Development

Coastal populations have some of the fastest growth rates in tropical countries, and the associated development has a range of negative impacts on coral reefs. Airports and buildings are often built on land reclaimed from the sea. Sensitive habitats like coral reefs can be destroyed or disturbed by the dredging of deep-water channels or marinas and through the dumping of waste materials [21].

16.5.2.3 Recreational Activities

Physical damage to coral reefs can occur as the result of snorkeling, diving, fishing with people touching reefs, careless boating, collecting coral, stirring up sediment, and dropping anchors on reefs. Tourist resorts and infrastructure constructed directly on top of reefs and sewage or other wastes from resorts reaching directly into the water around coral reefs can also damage them.

16.5.2.4 Sedimentation

Sedimentation has been identified as a primary stressor for the existence and recovery of coral species and their habitats. Erosion caused by construction on coasts and inland, mining, urban storm water runoff, logging, forestry, and farming contributes to increasing sedimentation in rivers. This ends in the ocean and can strangle corals by depriving them of light and interfering with their ability to feed, grow, and reproduce.

16.5.2.5 Coral Mining and Harvesting

Live coral is extracted from reefs and used as bricks, road-fill, and cement to construct new buildings. Corals sold to exporters and tourists as souvenirs and harvested for the live rock trade, aquarium trade, and jewelry can lead to over-harvesting of specific species, destruction of reef habitat, and reduced biodiversity. Although corals have been listed as endangered species since 1985, illegal trade is increasing.

16.5.2.6 Pollution

Pollution from human activities inland can damage coral reefs when transported by rivers into coastal waters. Reefs are contaminating by sewage, urban and industrial waste, agrochemicals, and oil pollution. There are many types and sources of pollution from land-based activities. Some of them are as follows:

- *Excess nutrients* from agricultural and residential fertilizer use, sewage discharge, and animal waste can lead to the growth of algae that blocks sunlight and consumes oxygen needed for respiration. This often results in an imbalance affecting the entire ecosystem. Excess nutrients can also support the growth of microorganisms like bacteria and fungi that can be pathogenic to corals.
- *Pathogens* from inadequately treated sewage, storm-water and runoff from livestock pens can cause disease in corals, especially if they are stressed by other environmental conditions like bleaching. Coral disease occurs in healthy ecosystems, but the input of pathogen-containing pollution can exacerbate the frequency and intensity of disease outbreaks.
- *Toxic substances* including trace metals, organic chemicals, and pesticides found in industrial discharges, sunscreens, urban and agricultural runoff, antifouling paints, mining activities, and runoff from landfills can affect coral reproduction, growth, defensive responses, and other physiological processes.
- *Trash and micro-plastics* from improper disposal and storm-water runoff that make their way into the ocean can snag on corals and block the sunlight needed for photosynthesis, or entangle and kill reef organisms and break or damage corals. Eight million tons of plastic rubbish enters the world's oceans every year. Much discarded plastic is broken down into tiny pieces known as *micro-plastics*. Degraded plastics and micro-plastics can be consumed by coral and other reef animals, blocking their digestive tracts and potentially introducing toxics.

16.5.2.7 Invasive Species

The aggregate effects of the above stressors can decrease the reef's resilience overall and increase susceptibility to invasive species. For example, Crown-of-thorns starfish (COTS) are marine invertebrates that feed on coral polyps. They occur naturally on reefs throughout the Indo-Pacific region, and when conditions are right, they can reach plague proportions and devastate hard coral communities. Nutrient enrichment may lead to COTS outbreaks because elevated nutrient levels cause phytoplankton blooms, which provide a vital food source for COTS larvae [22]. Removal of COTS predators by overfishing can also lead to COTS outbreaks, and they can reach densities such that the starfish consume coral tissue faster than the corals can grow. The damage caused by invasive species can be devastating and includes alteration of ecosystem dynamics, biodiversity loss, reduction of ecosystem resilience, and loss of resources, with environmental, economic, and sociocultural impacts.

Risk classifications based on significant threats to coral reefs are shown in Figure 16.4.

16.6 Important Coral Reefs of the World

The world's oceans are filled with thousands of dazzling coral reefs, but a few are well known and have become popular diving spots due to their location and beauty. Some of the world's most famous coral reefs are presented in Table 16.1.

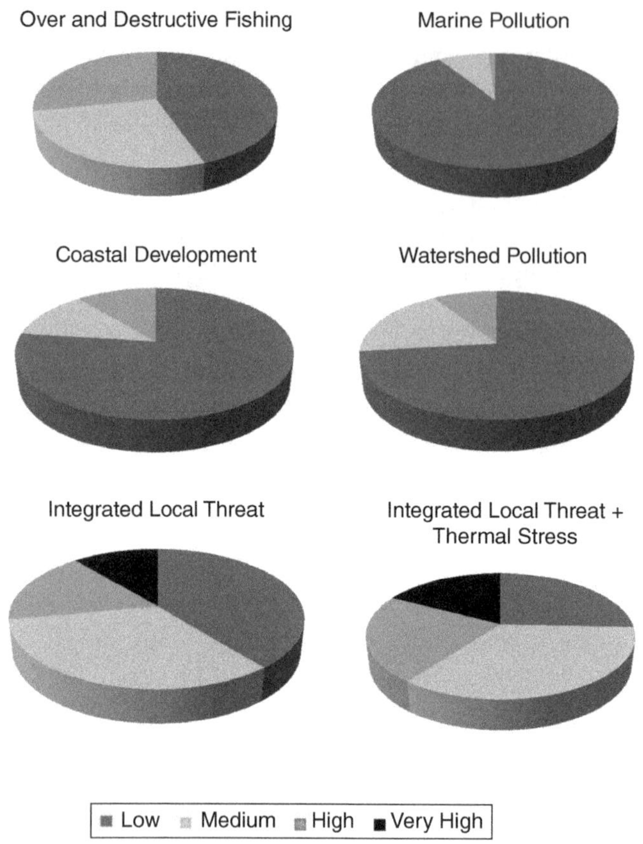

Figure 16.4 Major threats to coral reefs.

Table 16.1 The world's most popular coral reefs.

Name of coral reef	Location	Notes
Great Barrier Reef	Australia	World's largest coral reef; 300 species of hard corals
Belize Barrier Reef	Caribbean	World's second-largest coral reef; 106 species of corals
Palancar Reef	Mexico	"The underwater Garden of Eden"
Raja Ampat	Indonesia	1500 small islands; 600 species of hard corals
Grand Central Station	Fiji	"Soft coral capital of the world"; 400 species of corals
Tubbataha Reefs Natural Park	Philippines	A top diving site; 360 coral species
Bonaire Reef	Caribbean Netherlands	"The Diver's Paradise"; 60 species of corals
Red Sea Coral Reef	The northernmost part of the Indian ocean	300 species of hard corals; reef can withstand extreme temperature changes
New Caledonia Barrier Reef	Australia	Second-largest double barrier reef in the world

Table 16.1 (Continued)

Name of coral reef	Location	Notes
Maldives	Chagos-Laccadive Ridge, Indian Ocean	1200 islands, 26 atolls, and 187 species of corals
Rainbow Reef	Fiji	230 hard and soft corals; one of the top diving destinations in the world
Wakatobi Islands	Indonesia	750 coral reef species
The great Chagos Archipelago	Indian Ocean	64 islands; 371 species of corals
Lord Howe Island	Australia	90 coral species
Apo Reef	Philippines	Largest coral reef in the Philippines; 197 species of corals

16.7 The Application of Reefs in Shoreline Protection

The shoreline is a highly dynamic place, constantly affected by the vagaries of winds, waves, tides, and now global climate change [23]. Hence both natural and human-induced factors change shorelines on a wide range of temporal and spatial scales [24]. Although coastal areas with altitude less than 10 m (low-elevation coastal zone, LECZ) cover only 2% of the world's landmass, they contain 10% of the world's population [25]. But in tropical environments, coastal communities are rapidly at risk from both climate change and environmental degradation. The current rate of sea-level rise, about 2–3 mm per year, will increase dramatically in future decades as ice melting accelerates, perhaps with catastrophic surges.

Consequently, almost all of the world's beaches are vanishing – not just being inundated but retreating, washing away into the sea [23]. Since climate change links up with coastal developments and natural geomorphic shifts, major global problems like coastal erosion and flooding are becoming more acute [26–29]. About 40% of the world's population lives within 100 km of the coast. Hence, increasing natural hazards from flooding, coastal storms, and rising sea levels pose social, economic, and ecological threats of global significance [30]. According to the United Nations global disaster risk reduction report, concerns about economic loss associated with tropical cyclones and floods are increasing globally [31].

During the past century, over 85% of the barrier beaches providing natural physical defenses against flooding have experienced erosion along the east coast of the United States [32]. The most serious effects are in the world's lowest-lying islands like the Maldives in the Indian Ocean, and Tuvalu, Kiribati, Tokelau, and the Marshall Islands in the Pacific, along with thousands of other low-lying islands around the world, which could vanish entirely in the coming generation [23]. The adverse effects of flooding from sea-level rise and inundation, tropical cyclones, and storms are expected to increase substantially, and hence, huge investments are being made in coastal hazard mitigation. But the concerns are significant for many of the world's poorest communities, where direct dependence on coastal ecosystems for basic needs such as food and livelihoods is great. The alternatives and capacity to cope with climate change, natural disasters, or other disturbances are far less than in wealthier communities [33–35]. Hence, governments are intensely interested in assessing whether nature-based solutions can be used cost-effectively as part of a coastal defense strategy and an option to human-made defenses like seawalls and breakwaters [30].

16.7.1 Coral Reefs

Coral reefs are complex habitats that support millions of people worldwide, providing shoreline protection from waves and acting as an effective natural barrier [36]. They constitute the primary line of defense against flooding and erosion through the production and retention of sand and wave attenuation [30, 37–39]. Nearly 197 million people living below 10 m elevation and within 50 km of a reef receive the benefits of coastline protection from coral reefs. Around 100 million people living below 10 m elevation and within 10 km of a reef also receive risk-reduction benefits from reefs. Indonesia, India, and the Philippines are the three countries with the greatest number of people at risk and gain risk-reduction benefits from reefs. Nearly 50% of the global population living in low, exposed areas near reefs are in these three countries [30].

Evidence indicates the vital role of reefs in wave attenuation, but few direct links have been found between this and impacts on shorelines. The connections between factors like reef degradation, coastal development, and artificial defenses are evident from studies in the Maldives, Red Sea, Cancun (Mexico), and Bali (Indonesia) [30]. Only a few direct studies have shown a cause-and-effect relationship between coral reefs and shoreline stability [40, 41]. The causality is difficult to establish since variations in coral reefs modify wave energy propagation, sediment transport, and currents in complex spatial ways beyond wave attenuation [28, 42]. Moreover, other factors like coral mined for construction and sand further disrupt the shoreline and destroy reefs by sedimentation.

The physical structure of coral reefs is the primary focus of nearshore hydrodynamic studies [43]. The three-dimensional structure of live coral communities on coral reefs plays a crucial role in dissipating wave energy and thereby reducing the impact of large waves before they reach the shore [44–46]. Due to the presence of reef organisms, coral reefs often build a steep transition from particularly deep to shallow water and generate a very rough bottom surface [43]. Compared to other coastal settings, the high structural complexity of coral reefs results in high hydraulic roughness and greater frictional dissipation of waves [44]. The geophysics of wave energy attenuation and wave dynamics are influenced by geometry and reef parameters [28]. The high frictional dissipation on coral reefs, in conjunction with waves breaking on the reef rim, results in high rates of wave energy dissipation over relatively short distances; therefore, the coastal protection service provided by coral reefs is greater than many other marine ecosystems [43]. Live, healthy coral provides shallower geometrical complexity and more surface roughness to the reef, dispersing wave energy through friction and breaking waves [47]. Reef roughness plays an essential role in the mass transfer of energy from overlying waters into the reef structure, significantly dampening wave action [42, 48], and has an important influence on reducing coastal erosion [35, 49]. But coral reefs are a highly fragile marine habitat, and more than half of global reefs are under medium or high potential risk [11]. Reefs' ability to protect adjacent coastlines is threatened by sea-level rise and slowing reef accretion rates. Reefs are also affected by local threats like unsustainable coastal development, logging and mining activities leading to sedimentation, agricultural runoff, overexploitation of marine resources, pollution, nutrient enrichment, destructive fishing methods, careless anchoring of boats, and recreational damage. Coral mortality accelerates the wave energy reaching coasts, as the reef provides less friction to waves, and the removal of coral skeletons increases the water depth over the reef [49].

Ferrario et al. [30] reviewed 255 studies on the protective nature of coral reefs and analyzed the role, importance, and cost-effectiveness of reefs for wave attenuation and risk reduction. A meta-analysis report from 27 independent publications examining reefs across the Pacific, Atlantic, and Indian Oceans revealed that coral reefs reduce wave energy by 97% on average, causing the waves that reach the shoreline to be significantly calmer than they would have been without the reefs. The shallowest part of the reef (reef crest) where the waves break first dissipates 86% of wave energy on its own, and the whole reef reduces wave height by 84%. Beck and Lange [50] also reported that natural fringing reef crests perform just like low-crested breakwaters by dissipating wave energy. Coral reefs produce fine coral sand for shores as the result of physical forces and biota [28].

Reguero et al. [28] examined a comprehensive approach to determine the protective function of coral reefs and explore approaches focused on the reef's effect on wave propagation patterns on the shoreline of Grenville Bay, Grenada. The assessment of shoreline changes in the bay over the past six decades from bathymetry and benthic surveys, historical wave and sea level data, historical imagery, and wave energy modeling showed that the live and well-developed network of coral reefs in the southern bay keeps the shoreline stable and in balance. But the degraded reefs in the northern bay are related to severe coastal erosion. The comparison of wave dynamics modeling for past bathymetry showed that reef degradation thoroughly explained erosion from changes in climate and historical sea-level rise.

The reaction of shoreline stability to the effects of strong typhoons is also evidence that supports the role of coral reefs in shoreline protection. Cuttler et al. [44] studied the reaction of a fringing reef shoreline to the direct influence of a severe category 3 tropical cyclone, Olwyn, which affected Ningaloo Reef (Australia's largest fringing reef system and a UN World Heritage site) in 2015. It was observed that the shoreline remained largely unscathed, and large waves were effectively dissipated because of the protection provided by the offshore reef. A comparison of similar cyclone impacts on coastlines without reefs indicated that the beaches were eroded up to 10 times more than the beach at Ningaloo. The findings suggested that the erosion was due to locally produced wind waves within the lagoon instead of offshore waves dissipated on the reef crest. Hence, healthy and well-developed reefs can provide shoreline protection even under strong wind and wave conditions during cyclones [40, 51, 52]. Harris et al. [36] investigated the future ability of coral reefs at Moorea and Tahiti, French Polynesia, to protect coastal zones from average wave conditions under scenarios of changing coral reef structural complexity and sea-level rise by using a one-dimensional wave dissipation model. Their study revealed that preserving the structural complexity of coral reefs is essential to ensure shoreline stability on tropical coastlines in the future.

Zhao et al. [39] studied the potential role of coral reefs in maintaining the nearby coastline stability of north-eastern Hainan Island (Puqian Bay and Hainan Bay) in the South China Sea. It was reported that coral communities in the Hainan Bay with greater coverage and higher diversity had a relatively stable coastline. But coral reefs at the southern Puqian Bay with lower cover and diversity were in worse condition and related to severe coastline erosion. Hence this study gives some direct evidence that degraded reefs lead to severe coastal erosion and cannot maintain shoreline stability. The efficacy of coral reefs in providing natural shoreline protection reveals the correlation between severe coral reef degradation and adverse coastal erosion. In terms of shore profile stability and cross-shore sediment exchange, Martins et al. [53] evaluated the effect of a fringing reef on changes in beach morphodynamic behavior along 39 cross-shore profiles at Pontal do Cupe beach (Northeastern Brazil). The sheltering effect of the reef at Pontal do Cupe is evident from the fact that the wave energy that reaches the unprotected coast is approximately 50% greater than the energy reaching the beach protected by the nearshore reef. A comparison between healthy and degraded coral reefs for shoreline protection is shown in Figure 16.5.

It is estimated that reefs are responsible for preventing more than $4 billion in damages from storms per year [54]. Such damage is avoided through wave energy attenuation, which creates a sheltering effect facilitating sediment accumulation and beach stability. However, reefs are rarely

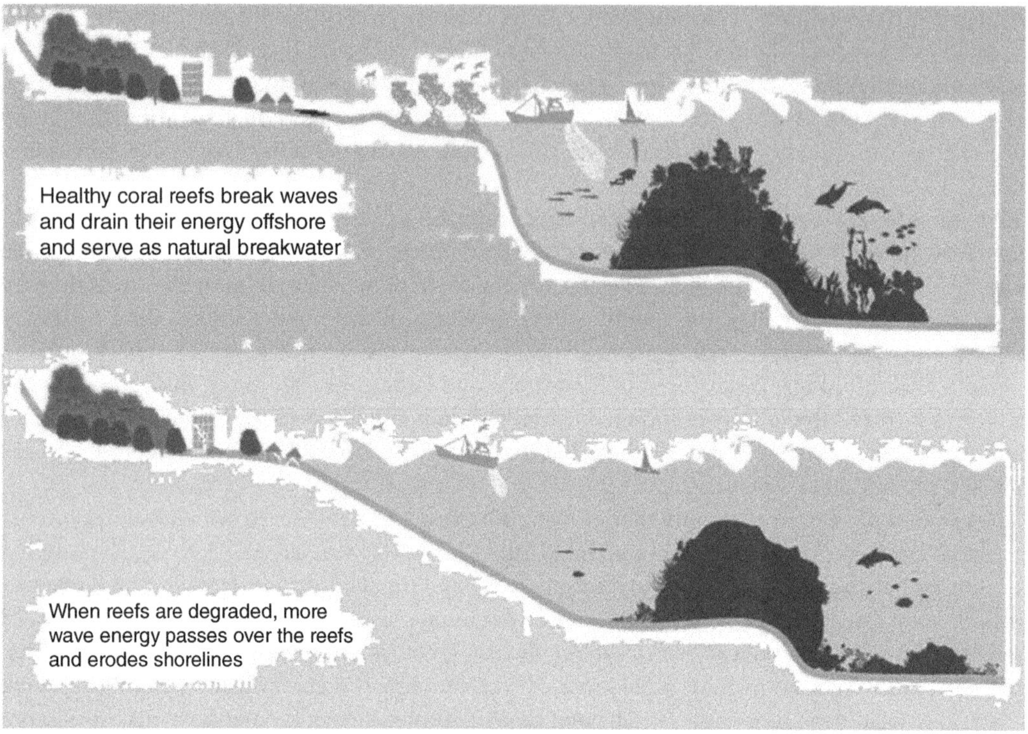

Figure 16.5 Comparison between healthy and degraded coral reefs in shoreline protection. *See colour plate section for colour representation of this figure.*

examined for their coastal defense capabilities and thus are not maintained for their infrastructure benefits while extensive degradation and damage continue [28]. Coral reefs in US waters, from Florida and the Caribbean to Hawaii and Guam, provide more than $1.8 billion in flood protection benefits every year. They reduce direct flood damages to public and private property worth more than $800 million annually and help avert other costs to lives and livelihoods worth an additional $1 billion. The flood protection benefits that reefs provide across the US are similar to those in more than 60 other nations, and the global cost of storm damage to the world's coastlines would double without reefs. Florida receives more than $675 million in annual flood-protection from reefs, and Puerto Rico gets $183 million in protection yearly. In Honolulu alone, reefs provided more than $435 million in protection from a catastrophic storm large enough that it would be expected to occur only once in 50 years [55]. These reports show the significant value of reef benefits and relevance of protecting coral reefs as climate change alters coral reefs' status globally.

In recent decades, coral reefs have experienced a significant decline worldwide, which has subsequently decreased their capacity to provide coastal defense. Degraded coral environments lead to a decrease in carbonate production and the biological geomorphological complexity of the reef framework [56, 57]. This endangers the growth and bottom roughness of the reefs, which are crucial for protecting shorelines by dissipating energy and breaking waves [39, 46]. Researchers estimate that millions of people living in coastal areas benefit from the protection afforded by coral reefs and will bear the cost if reefs are lost or degraded. To protect coral reefs and maintain their shoreline protection services, it is necessary to reduce local threats and establish marine reserves [39].

16.7.2 Oyster Reefs

Oyster reefs are considered coastal wetlands like coral reefs. They protect against coastal erosion and wave erosion [58]. Due to their complex structure, these natural barriers reduce water velocity, increase sedimentation rates, and provide improved conditions for settlement and retention of propagules, thereby improving the chances of species survival [59]. The ability of these reefs to deposit sediment and buffer wave energy can enhance and maintain adjacent habitats such as salt marshes and seagrasses, further increasing shoreline stabilization [35, 60]. Oyster reefs are adaptive to environmental changes, in addition to erosion control [61]. They can recover quickly from major storm events and accrete at a rate equal to or greater than sea-level rise [62] or local subsidence [63, 64]. But oyster reefs are also threatened by rapid environmental change, making it very important to protect and restore them. It is estimated that up to 80% of the world's oyster reefs have been lost, a rate unprecedented for any other marine habitat [58], which represents an enormous reduction in the various ecosystem services provided by these reefs.

16.7.3 Artificial Reefs

Increasing natural disasters due to flooding, coastal storms, and rising sea levels are leading to major investments worldwide in coastal defense structures such as seawalls and breakwaters. Wide-crested submerged breakwaters assist with shoreline stabilization by wave attenuation and wave refraction mechanisms, reducing the erosion of beaches [65]. They are constructed of rock, concrete, steel, and other traditional hard materials. Human-made defenses like sea walls, artificial breakwaters, and groynes are short-lived and expensive and can damage adjoining habitats and harm species that rely on them [55]. Seawalls and breakwaters are shore-parallel structures that reflect wave energy and constrain the shoreline's natural inland migration in response to erosion, ultimately causing beaches to become narrower and the beach seaward of the structure to drown [66, 67]. This can further lead to habitat loss and disrupt normal sediment dynamics [66, 68]. Hence these cannot satisfy coastal protection benefits, particularly at a level equivalent to natural coastal ecosystems and in a complex energetic environment similar to natural circumstances [28].

Reef restoration is cost-effective as opposed to the construction of artificial submerged breakwaters [38]. Artificial reefs and physical and biological techniques are used for structural and functional restoration of degraded coral reefs [28, 39, 69]. Multipurpose artificial surfing reefs (MPASR) are increasingly being adopted for coastal protection because of their associated advantages such as coastal protection, beach recreation, habitat for marine organisms, and increased socioeconomic prosperity. Although they are soft barriers, their success depends primarily on the quality of planning and management before their implementation [70].

Artificial reefs are aggregations of natural reefs and submerged structures, and they mimic the protection and ecological benefits of a natural reef like submerged breakwaters [71]. Reefs function like low-crested breakwaters, with hydrodynamic behavior well-characterized by coastal engineering models [47, 72, 73]. They can act as buffers like natural reefs in diffracting incident wave energy, offering friction, and leading to attenuation. The degree of attenuation of incident wave energy depends on the shape, size, and material of the artificial reefs.

The geometry and design of artificial reefs vary depending on the primary functions of the structure and materials used. These reefs favor conditions for recreational activities related to tourism and act as shelter and habitat for fish and algae, thereby improving marine biodiversity and ecosystem resilience. All kinds of materials such as steel, reinforced or pre-stressed concrete, fiberglass, and composite materials have been used to construct artificial reefs. Old wrecked cars, airplanes, military tanks, used truck and car tires, junked appliances, docks, old boats, ballistic missiles, decommissioned ships, and obsolete oil rigs have been sunk and designated as artificial reefs. Non-eco-friendly conventional methods used in coastal erosion mitigation and protection, such as seawalls, breakwaters, groynes, and gabions, have been shifted toward these soft and novel eco-friendly solutions. Geo-systems such as geo-bags, geo-tubes, geo-containers, geo-curtains, geo-grids, etc., and other systems such as Reef Balls, Aqua Reef, prefabricated units, and beach drainage have gained popularity in recent years because of their simplicity in placement, cost-effectiveness, and less-adverse environmental effects [70, 74].

Ranasinghe et al. [75] conducted numerical modeling to find the shoreline response to submerged breakwaters and reported that depending on the offshore distance to the artificial reef, the mode of shoreline response can vary between erosion and deposition of sand on beaches. Shin and Oh [76] conducted 2D limit equilibrium analysis and hydraulic model study tests of geo-tubes for Young-Jin beach along the east coast of Korea. It was reported that compared to extending the shoreline, the magnitude of re-erosion is lower. It was also concluded that the use of geo-tubes with zero-water depth above crest was more stable and effective for wave absorption than any other design plans. Alvarez et al. [77] designed low-crested structures using geo-tubes to reduce incident wave energy on the beaches of the northern coast of Yucatan in Mexico. These geotextile tubes were found to be an effective alternative for maintaining the dynamic balance on the shoreline by controlling wave breaking.

Blenkinsopp and Chaplin [78] conducted experiments and suggested that a submerged reef with a relatively low seaward gradient more significantly dissipates wave energy through wave breaking than sloping breakwaters. Strusinska and Oumeraci [79] examined the feasibility of an artificial reef as coastal protection against tsunami impacts and compared the evidence in the presence and absence of reefs through numerical studies. Harris [65] introduced Reef Ball units to construct submerged breakwaters along the southern Caribbean shore of the Dominican Republic, which helped with beach equilibrium, environment growth, and eco-tourism. This approach was very effective for stabilizing the beach, with a significant increase in beach width and elevation along the project shoreline during a direct hit by Hurricane Georges (Category 3) and large waves from Hurricane Mitch (Category 5). Reef Ball submerged breakwaters have also been constructed in other parts of the Caribbean, including the Cayman Islands; they have remained stable during waves from major hurricanes, including Category 5 Hurricane Ivan's direct hit in 2004 and large waves from Category 5 Hurricane Wilma in 2005. Boc and Burg [80] stated that offshore reef technology utilizing shelf material was effective in reducing erosion at the Sacred Falls on the island of Oahu, Hawaii. It had applicability in emergency and short-term situations in shallow-water island environments to protect infrastructure.

16.7.4 Coral Reef Restoration

In addition to natural marine growth, hard and soft corals can be transplanted and propagated on artificial reef units. Coral transplantation methods are often used to increase the rate of coral colonization at damaged sites and eliminate or mitigate the source of reef impacts [81]. Coral reef transplantation ideally starts with collecting fragments of living coral rock as soon as possible and storing them in a suitable location until they can be moved to the restoration site [81, 82] or attached to a suitable substrate. In difficult locations, artificial structures can be installed to provide a stable foundation for coral transplants. Transplantation success depends on the species, transplant shape and type, status of substrate attachment, and environmental conditions [82]. Coral may also be transplanted from nearby reef locations or from coral nurseries prepared in advance for restoration needs.

Corals can be moved from areas that have been or will be damaged by ship groundings, storms, dredging, or other natural or human-made activities, and natural coral growth will occur on artificial reefs [65]. Case studies show that creating large-scale manmade coral reefs is possible, and they can replicate many of the functions provided by naturally occurring coral reefs [58]. A synthesis study of the restoration of coral reefs worldwide exhibited the greatest success, with an annual survival rate of 64.5% [83].

16.7.5 Oyster Reef Restoration

Similar to coral reefs, oyster reefs also can be constructed artificially to replicate their natural functions. The physical structure and natural growth of oyster reefs have led to the consideration of oyster reef breakwaters as a cost-effective alternative to limestone rock breakwaters [35]. Depending on where restoration occurs and the range of services achieved, Grabowski et al. [84] analyzed the cost-benefit ratio of oyster reefs and found that restoration costs are typically recovered in 2–14 years. Oyster reef living shorelines have been proposed as an effective alternative to traditional coastal defense structures, along with other co-benefits.

Oysters can grow and form a reef that stops coastal erosion and enhances biodiversity by placing artificial substrates along the beach. Oyster reefs can be constructed using sustainable native materials, have the potential to increase in size over time, and are less likely to require long-term replenishment to remain effective [85]. The establishment and effectiveness of these reefs for shoreline protection are influenced by local hydrodynamic regimes, with the most protective benefits documented in low-energy environments [86, 87].

Compared to conventional structures like seawalls, oyster reefs can be rebuilt, upgraded, and maintained in response to a changing climate at significant expense [88]. The development of oyster habitat suitability models can help predict locations for successful oyster growth and oyster reef living shorelines [89, 90]. A structure that remains intact and can provide coastal defense through energy attenuation and shoreline stabilization is needed for oyster reef living shorelines. These structures may be built on the footprint of dead natural reefs, or the reefs can be deployed onto soft sediments if no previous hard substrate is present. There are also reef substrates made of steel, riprap, limestone, and crushed or pre-cast concrete. Oyster reefs' adaptive ability depends on successful oyster colonization and growth on the reef substrate [64].

The variable performance of oyster reef living shorelines regarding both oyster colonization and shoreline stabilization is evident from studies. A case study of oyster reef restoration projects in the northern Gulf of Mexico assessed a $150 million investment over 10 years in restoring 160 km of oyster reefs: the initial investment was returned twofold in the period via gains in fisheries, avoidance of damage to property and public infrastructure, new jobs, and goods and services delivered to local communities [58]. La Peyre et al. [91] showed that reefs constructed of ReefBlk^{SM} in Louisiana promoted shoreline accretion at one site and reduced shoreline erosion at a second site, where oysters were recruited. But there was no effect on shoreline stabilization at a third site, where oysters were absent. The effect of wave- and current-induced turbulence on spat settlement on the reef was reported from a study of living shoreline hydrodynamics. Flow-structure interaction over newly deployed reefs created with bagged oyster shells increased shoreline velocities by over an order of magnitude compared to two nearby control shoreline sites [92]. It was evident from the study of Salvador de Paiva et al. [93] that sediment accumulation by Pacific oyster (*Crassostreagigas*) reefs is dependent on oyster density as well as the length-to-width ratio of the reefs, where longer and narrower reefs with higher oyster density tend to trap more sediment and can provide sustained coastal defense over time.

Evidence from the studies of Morris et al. [64] along the east and gulf coasts of the United States suggested that research and implementation of restored oyster reefs were applicable for shoreline protection. Under similar flow conditions and production and dissipation of turbulent energy, in situ hydrodynamic measurements were an order of magnitude greater on existing healthy oyster reefs than on degraded reefs with no live oysters [94]. According to Chowdhury [95], developing the oyster population on the concrete rings off the coast of the Bengal island of Kutubdia to protect the coastline led to a self-sustaining oyster reef that functioned effectively as breakwaters. The force of waves between 0.5 and 1 m high is reduced, while waves lower than 0.5 m are blocked.

Oyster reef living shorelines can be successful for protecting coasts and restoring ecosystem services only if engineering and ecological principles are combined to construct persistent and efficacious oyster reefs [64].

16.8 Conclusion

Shore maintenance and protection are probably the largest single cost of global warming, but international and government agencies do not take much responsibility for preventing it. According to the report by Ferrario et al. [30], the median cost of building artificial wave defenses is $19 791 per meter, while the median cost of reef restoration projects is $1290 per meter; hence, maintaining the health of coral reefs is far less expensive than installing artificial defenses. Natural reefs are dynamic and in many ways resilient to threats from sea level rise and storms, because they can self-repair with minimal maintenance [96]. Compared to conventional human-made defenses, healthy and restored reefs enhance their surroundings by protecting shorelines and supporting fisheries and recreation [55]. Instead of granting one-time emergency aid to desperate governments, recognizing the importance of reefs as natural infrastructure with an emphasis on coastal protection could create further funding such as climate adaptation funds and technical opportunities to restore coral reefs. Living shorelines have been proposed as a solution to both ecological (loss of habitats) and engineering (non-adaptive traditional structures) challenges along increasingly human-impacted coasts [97]. The accelerated rate of global climate change requires particular consideration of the long-term fate of restored reefs. Knowledge of the appropriate benthic topography to create optimum recruitment conditions, like hydrodynamics, settlement surface, protection from predators, sedimentation, etc., and how this might need to be altered due to changes in climate [98] will increase the chances of creating self-sustaining reefs and maintaining coastal defenses [99]. However, many studies have shown that in most cases, investments in coral and oyster reef protection yield manifold benefits, including socioeconomic co-benefits.

In conclusion, these scientific studies and reports stated that existing and restored reefs could be cost-effective shoreline protection alternatives to conventional manmade structures under future climate change and coastal development scenarios. They also furnish direct proof of the importance of reef status in coastal protection, management, and shoreline stability. Hence, there is an urgent need for preservation and robust local action to reduce stressors and risks of climate change and strengthen these guardian structures' resiliency to protect our shorelines.

References

1 Pauly, D., Christensen, V., Guénette, S. et al. (2002). Towards sustainability in world fisheries. *Nature* 418 (6898): 689–695.

2 Hubbard, D.K. (1997). Reefs as dynamic systems. In: *Life and Death of Coral Reefs* (ed. C. Birkeland), 43–67. Chapman & Hall.

3 Jörgensen, T.L. (2016). Coral reef habitats and fish connectivity: Implications for coastal management and fishery. Doctoral dissertation, Stockholm University.
4 Hoegh-Guldberg, O. (2011). Coral reef ecosystems and anthropogenic climate change. *Regional Environmental Change* 11: 215–227.
5 Knowlton, N. and Jackson, J.B. (2008). Shifting baselines, local impacts, and global change on coral reefs. *PLoS Biology* 6 (2): e54.
6 Lesser, M.P. (2004). Experimental biology of coral reef ecosystems. *Journal of Experimental Marine Biology and Ecology* 300 (1–2): 217–252.
7 Muscatine, L. (1990). The role of symbiotic algae in carbon and energy flux in reef corals. *Coral Reefs* 25: 1–29.
8 Lalli, C.M. and Parsons, T.R. (1997). *Biological Oceanography: An Introduction*, 2e. Vancouver, Canada: University of British Columbia.
9 Spalding, M.D., Ravilious, C., and Green, E.P. (2001). *World Atlas of Coral Reefs*. Berkeley, USA: University of California Press.
10 Britannica, E. (2008). *Britannica Concise Encyclopedia*. Encyclopaedia Britannica, Inc.
11 Burke, L., Reytar, K., Spalding, M., and Perry, A. (2011). Reefs at risk revisited. World Resources Institute, Washington, DC.
12 Baswapoor, S. and Irfan, Z.B. (2018). *Current Status of Coral Reefs in India: Importance, Rising Threats and Policies for its Conservation and Management*. Chennai, India: Madras School of Economics.
13 Wilkinson, C. (2004). *Status of Coral Reefs of the World, 2004*, vol. 1. Townsville, Australia: Australian Institute of Marine Science.
14 Spalding, M., Burke, L., Wood, S.A. et al. (2017). Mapping the global value and distribution of coral reef tourism. *Marine Policy* 82: 104–113.
15 Furnas, M., Mitchell, A., Skuza, M., and Brodie, J. (2005). In the other 90%: phytoplankton responses to enhanced nutrient availability in the great barrier reef lagoon. *Marine Pollution Bulletin* 51 (1–4): 253–265.
16 Joy, A., Anoop, P.P., Rajesh, R. et al. (2019). Spatial variation of phosphorus fractionation in the coral reef sediments of Lakshadweep archipelago, Indian Ocean. *Chemistry and Ecology* 35 (7): 592–612.
17 Frieler, K., Meinshausen, M., Golly, A. et al. (2013). Limiting global warming to 2 C is unlikely to save most coral reefs. *Nature Climate Change* 3 (2): 165–170.
18 Douglas, A.E. (2003). Coral bleaching – how and why? *Marine Pollution Bulletin* 46: 385–392.
19 Joy, A., Anoop, P.P., Rajesh, R. et al. (2020). Spatial distribution and contamination assessment of trace metals in the coral reef sediments of Kavaratti Island in Lakshadweep archipelago, Indian Ocean. *Soil and Sediment Contamination* 29 (2): 209–231.
20 US EPA. (n.d.). Threats to coral reefs. United States Environmental Protection Agency. https://www.epa.gov/coral-reefs/threats-coral-reefs.
21 ICRI, International Coral Reef Initiative. (n.d.). Status of and threat to coral reefs. https://www.icriforum.org/about-coral-reefs/status-of-and-threat-to-coral-reefs.
22 Brodie, J., Fabricius, K., De'ath, G., and Okaji, K. (2005). Are increased nutrient inputs responsible for more outbreaks of crown-of-thorns starfish? An appraisal of the evidence. *Marine Pollution Bulletin* 51 (1–4): 266–278.
23 Goreau, T.J., Hilbertz, W., Azeez, A., and Hakeem, A. (2004). Maldives shorelines: Growing a beach. Global Coral Reef Alliance. http://www.globalcoral.org/maldives-shorelines-growing-a-beach.
24 Stive, M.J.F., Aarninkhof, S.G.J., Hamm, L. et al. (2002). Variability of shore and shoreline evolution. *Coastal Engineering* 47 (2): 211–235.

25 McGranahan, G., Balk, D., and Anderson, B. (2007). The rising tide: assessing the risks of climate change and human settlements in low elevation coastal zones. *Environment and Urbanization* 19 (1): 17–37.

26 Hallegatte, S., Green, C., Nicholls, R.J., and Corfee-Morlot, J. (2013). Future flood losses in major coastal cities. *Nature Climate Change* 3 (9): 802–806.

27 Kron, W. (2013). Coasts: the high-risk areas of the world. *Natural Hazards* 66 (3): 1363–1382.

28 Reguero, B.G., Beck, M.W., Agostini, V.N. et al. (2018). Coral reefs for coastal protection: a new methodological approach and engineering case study in Grenada. *Journal of Environmental Management* 210: 146–161.

29 Reguero, B.G., Losada, I.J., Díaz-Simal, P. et al. (2015). Effects of climate change on exposure to coastal flooding in Latin America and the Caribbean. *PLoS One* 10 (7): e0133409.

30 Ferrario, F., Beck, M.W., Storlazzi, C.D. et al. (2014). The effectiveness of coral reefs for coastal hazard risk reduction and adaptation. *Nature Communications* 5 (1): 1–9.

31 UNISDR. (2011). Global assessment report on disaster risk reduction, vol. 178. United Nations International Strategy for Disaster Reduction, Geneva, Switzerland.

32 Zhang, K., Douglas, B.C., and Leatherman, S.P. (2004). Global warming and coastal erosion. *Climatic Change* 64: 41–58.

33 Hale, L.Z., Meliane, I., Davidson, S. et al. (2009). Ecosystem-based adaptation in marine and coastal ecosystems. *Renewable Resources Journal* 25 (4): 21–28.

34 UNISDR. (2009). Global assessment report on disaster risk reduction. United Nations International Strategy for Disaster Reduction, Geneva, Switzerland.

35 Spalding, M.D., Ruffo, S., Lacambra, C. et al. (2014). The role of ecosystems in coastal protection: adapting to climate change and coastal hazards. *Ocean and Coastal Management* 90: 50–57.

36 Harris, D.L., Rovere, A., Casella, E. et al. (2018). Coral reef structural complexity provides important coastal protection from waves under rising sea levels. *Science Advances* 4 (2): eaao4350.

37 Elliff, C.I. and Silva, I.R. (2017). Coral reefs as the first line of defense: shoreline protection in face of climate change. *Marine Environmental Research* 127: 148–154.

38 Pascal, N., Allenbach, M., Brathwaite, A. et al. (2016). Economic valuation of coral reef ecosystem service of coastal protection: a pragmatic approach. *Ecosystem Services* 21: 72–80.

39 Zhao, M., Zhang, H., Zhong, Y. et al. (2019). The status of coral reefs and its importance for coastal protection: a case study of Northeastern Hainan Island, South China Sea. *Sustainability* 11 (16): 4354.

40 Alegria-Arzaburu, A.R.D., Mariño-Tapia, I., Enriquez, C. et al. (2013). The role of fringing coral reefs on beach morphodynamics. *Geomorphology* 198: 69–83.

41 Frihy, O.E., El Ganaini, M.A., El Sayed, W.R., and Iskander, M.M. (2004). The role of fringing coral reef in beach protection of Hurghada, Gulf of Suez, Red Sea of Egypt. *Ecological Engineering* 22 (1): 17–25.

42 Monismith, S.G. (2007). Hydrodynamics of coral reefs. *Annual Review of Fluid Mechanics* 39: 37–55.

43 Lowe, R.J., Falter, J.L., Bandet, M.D. et al. (2005). Spectral wave dissipation over a barrier reef. *Journal of Geophysical Research, Oceans* 110 (C4): 1–16.

44 Cuttler, M.V.W., Hansen, J.E., Lowe, R.J., and Drost, E.J.F. (2018). Response of a fringing reef coastline to the direct impact of a tropical cyclone. *Limnology and Oceanography Letters* 3 (2): 31–38.

45 Monismith, S.G., Rogers, J.S., Koweek, D., and Dunbar, R.B. (2015). Frictional wave dissipation on a remarkably rough reef. *Geophysical Research Letters* 42 (10): 4063–4071.

46 Rogers, J.S., Monismith, S.G., Koweek, D.A., and Dunbar, R.B. (2016). Wave dynamics of a Pacific atoll with high frictional effects. *Journal of Geophysical Research, Oceans* 121 (1): 350–367.

47 Quataert, E., Storlazzi, C., Van Rooijen, A. et al. (2015). The influence of coral reefs and climate change on wave-driven flooding of tropical coastlines. *Geophysical Research Letters* 42 (15): 6407–6415.

48 Kench, P.S. and Brander, R.W. (2006). Wave processes on coral reef flats: implications for reef geomorphology using Australian case studies. *Journal of Coastal Research* 2006 (221): 209–223.

49 Sheppard, C., Dixon, D.J., Gourlay, M. et al. (2005). Coral mortality increases wave energy reaching shores protected by reef flats: examples from the Seychelles. *Estuarine, Coastal and Shelf Science* 64 (2–3): 223–234.

50 Beck, M.W. and Lange, G.M. (2016). Managing coasts with natural solutions: Guidelines for measuring and valuing the coastal protection services of mangroves and coral reefs. Wealth Accounting and the Valuation of Ecosystem Services Partnership (WAVES) technical report. World Bank: Washington, DC.

51 Mahabot, M.M., Pennober, G., Suanez, S. et al. (2017). Effect of tropical cyclones on short-term evolution of carbonate sandy beaches on Reunion Island, Indian Ocean. *Journal of Coastal Research* 33 (4): 839–853.

52 Villanoy, C., David, L., Cabrera, O. et al. (2012). Coral reef ecosystems protect shore from high-energy waves under climate change scenarios. *Climatic Change* 112 (2): 493–505.

53 Martins, K.A., de Souza Pereira, P., Esteves, L.S., and Williams, J. (2019). The role of coral reefs in coastal protection: analysis of beach morphology. *Journal of Coastal Research* 92 (S1): 157–164.

54 Beck, M.W., Losada, I.J., Menéndez, P. et al. (2018). The global flood protection savings provided by coral reefs. *Nature Communications* 9 (1): 1–9.

55 Storlazzi, C.D., Reguero, B.G., Cole, A.D. et al. (2019). Rigorously valuing the role of US coral reefs in coastal hazard risk reduction. US Geological Survey open-file report (No. 2019-1027).

56 Hamylton, S.M., Pescud, A., Leon, J.X., and Callaghan, D.P. (2013). A geospatial assessment of the relationship between reef flat community calcium carbonate production and wave energy. *Coral Reefs* 32 (4): 1025–1039.

57 Perry, C.T., Murphy, G.N., Kench, P.S. et al. (2013). Caribbean-wide decline in carbonate production threatens coral reef growth. *Nature Communications* 4 (1402): 1–7.

58 TNC (2012). The nature conservancy. *The Economics of Oyster Reef Restoration in the Gulf of Mexico: A Case Study in Mobile Bay, Alabama.*

59 Scyphers, S.B., Powers, S.P., Heck, K.L. Jr., and Byron, D. (2011). Oyster reefs as natural breakwaters mitigate shoreline loss and facilitate fisheries. *PLoS One* 6 (8): 1–12.

60 Smith, K.A., North, E.W., Shi, F. et al. (2009). Modeling the effects of oyster reefs and breakwaters on seagrass growth. *Estuaries and Coasts* 32 (4): 748–757.

61 Bible, J. and Sanford, E. (2015). Local adaptation in an estuarine foundation species: implications for restoration. *Biological Conservation* 193: 95–102.

62 Rodriguez, A.B., Fodrie, F.J., Ridge, J.T. et al. (2014). Oyster reefs can outpace sea-level rise. *Nature Climate Change* 4: 493–497.

63 Casas, S.M., La Peyre, J.F., and La Peyre, M.K. (2015). Restoration of oyster reefs in an estuarine lake: population dynamics and shell accretion. *Marine Ecology Progress Series* 524: 171–184.

64 Morris, R.L., Bilkovic, D.M., Boswell, M.K. et al. (2019). The application of oyster reefs in shoreline protection: are we over-engineering for an ecosystem engineer? *Journal of Applied Ecology* 56 (7): 1703–1711.

65 Harris, L.E. (2009). Artificial reefs for ecosystem restoration and coastal erosion protection with aquaculture and recreational amenities. *Reef Journal* 1 (1): 235–246.

66 Defeo, O., McLachlan, A., Schoeman, D.S. et al. (2009). Threats to sandy beach ecosystems: a review. *Estuarine, Coastal and Shelf Science* 81 (1): 1–12.

67 Powell, E.J., Tyrrell, M.C., Milliken, A. et al. (2019). A review of coastal management approaches to support the integration of ecological and human community planning for climate change. *Journal of Coastal Conservation* 23 (1): 1–18.

68 Galbraith, H., Jones, R., Park, R. et al. (2002). Global climate change and sea level rise: potential losses of intertidal habitat for shorebirds. *Waterbirds* 25 (2): 173–183.

69 Precht, W.F. (2006). *Coral Reef Restoration Handbook*. USA: CRC press.

70 Lokesha, S.V. and Sannasiraj, S.A. (2011). Artificial reefs: a review. *International Journal of Ocean and Climate Systems* 4 (2): 117–124.

71 Goreau, T.J. and Trench, R.K. (2012). *Innovative Methods of Marine Ecosystem Restoration*. CRC Press.

72 Hoeke, R., Storlazzi, C., and Ridd, P. (2011). Hydrodynamics of a bathymetrically complex fringing coral reef embayment: wave climate, in situ observations, and wave prediction. *Journal of Geophysical Research, Oceans* 116 (C4).

73 Taebi, S. and Pattiaratchi, C. (2014). Hydrodynamic response of a fringing coral reef to a rise in mean sea level. *Ocean Dynamics* 64 (7): 975–987.

74 Pilarczyk, K. W. (2005). Coastal stabilization and alternative solutions in international perspective. Arabian Coast, keynote address, Verhagen.

75 Ranasinghe, R., Turner, I.L., and Symonds, G. (2006). Shoreline response to multi-functional artificial surfing reefs: a numerical and physical modelling study. *Coastal Engineering* 53 (7): 589–611.

76 Shin, E.C. and Oh, Y.I. (2007). Coastal erosion prevention by geotextile tube technology. *Geotextiles and Geomembranes* 25 (4–5): 264–277.

77 Alvarez, E., Rubio, R., and Ricalde, H. (2007). Beach restoration with geotextile tubes as submerged breakwaters in Yucatan, Mexico. *Geotextiles and Geomembranes* 25 (4–5): 233–241.

78 Blenkinsopp, C.E. and Chaplin, J.R. (2008). The effect of relative crest submergence on wave breaking over submerged slopes. *Coastal Engineering* 55 (12): 967–974.

79 Strusińska, A. and Oumeraci, H. (2008). Geotextile reef as a coastal protection against tsunami. In: *Geotechnical Engineering for Disaster Mitigation and Rehabilitation* (eds. H. Liu, A. Deng and J. Chu), 742–747. Berlin, Heidelberg: Springer.

80 Boc, S.J. and Burg, E.C. (2010). Innovative shore protection for island communities. *WIT Transactions on Ecology and the Environment* 130: 209–220.

81 Epstein, N., Bak, R.P.M., and Rinkevich, B. (2001). Strategies for gardening denuded coral reef areas: the applicability of using different types of coral material for reef restoration. *Restoration Ecology* 9 (4): 432–442.

82 Jaap, W.C. (2000). Coral reef restoration. *Ecological Engineering* 15 (3–4): 345–364.

83 Bayraktarov, E., Saunders, M.I., Abdullah, S. et al. (2016). The cost and feasibility of marine coastal restoration. *Ecological Applications* 26 (4): 1055–1074.

84 Grabowski, J.H., Brumbaugh, R.D., Conrad, R.F. et al. (2012). Economic valuation of ecosystem services provided by oyster reefs. *Bioscience* 62 (10): 900–909.

85 Piazza, B.P., Banks, P.D., and La Peyre, M.K. (2005). The potential for created oyster shell reefs as a sustainable shoreline protection strategy in Louisiana. *Restoration Ecology* 13 (3): 499–506.

86 Borsje, B.W., van Wesenbeeck, B.K., Dekker, F. et al. (2011). How ecological engineering can serve in coastal protection. *Ecological Engineering* 37 (2): 113–122.

87 Meyer, D.L., Townsend, E.C., and Thayer, G.W. (1997). Stabilization and erosion control value of oyster cultch for intertidal marsh. *Restoration Ecology* 5 (1): 93–99.

88 Hinkel, J., Lincke, D., Vafeidis, A.T. et al. (2014). Coastal flood damage and adaptation costs under 21st century sea-level rise. *Proceedings of the National Academy of Sciences* 111: 3292–3297.

89 Fuchs, H.L. and Reidenbach, M.A. (2013). Biophysical constraints on optimal patch lengths for settlement of a reef-building bivalve. *PLoS One* 8: e71506.

90 La Peyre, M.K., Serra, K., Joyner, T.A., and Humphries, A. (2015). Assessing shoreline exposure and oyster habitat suitability maximizes potential success for sustainable shoreline protection using restored oyster reefs. *PeerJ* 3: e1317.

91 La Peyre, M. K., Schwarting, L., &Miller, S. (2013). *Preliminary assessment of bioengineered fringing shoreline reefs in Grand Isle and Breton Sound, Louisiana*. US Department of the Interior, US Geological Survey Open-File Report 2013, 1040, 34.

92 Spiering, D.W., Kibler, K.M., and Kitskoudis, V. (2018). Hydrodynamic change following living shoreline restoration based on a before-after-control- impact experiment. In: *World Environmental and Water Resources Congress, American Society of Civil Engineers, Hydraulics and Waterways, Water Distribution Systems Analysis, and Smart Water* (ed. V.A. Reston), 54–64. American Society of Civil Engineers.

93 Salvador de Paiva, J.N., Walles, B., Ysebaert, T., and Bouma, T.J. (2018). Understanding the conditionality of ecosystem services: the effect of tidal flat morphology and oyster reef characteristics on sediment stabilization by oyster reefs. *Ecological Engineering* 112: 89–95.

94 Kitsikoudis, V., Kibler, K.M., and Walters, L.J. (2020). In-situ measurements of turbulent flow over intertidal natural and degraded oyster reefs in an estuarine lagoon. *Ecological Engineering* 143: 105688.

95 Chowdhury, M. S. N. (2019). Ecological engineering with oysters for coastal resilience: Habitat suitability, bioenergetics, and ecosystem services. Doctoral dissertation, Wageningen University.

96 Temmerman, S. and Kirwan, M.L. (2015). Building land with a rising sea: cost efficient nature-based solutions can help to sustain coastal societies. *Science* 349 (6248): 588–589.

97 Temmerman, S., Meire, P., Bouma, T.J. et al. (2013). Ecosystem-based coastal defence in the face of global change. *Nature* 504 (7478): 79–83.

98 McAfee, D., Bishop, M.J., Yu, T.N., and Williams, G.A. (2018). Structural traits dictate abiotic stress amelioration by intertidal oysters. *Functional Ecology* 32: 2666–2677.

99 Whitman, E.R. and Reidenbach, M.A. (2012). Benthic flow environments affect recruitment of Crassostreavirginica larvae to an intertidal oyster reef. *Marine Ecology Progress Series* 463: 177–191.

17

Mangroves, as Shore Engineers, Are Nature-Based Solutions for Ensuring Coastal Protection

Ajanta Dey[1], J.R.B. Alfred[1], Biswajit Roy Chowdhury[1], and Udo Censkowsky[2]

[1] *Nature Environment & Wildlife Society (NEWS), Kolkata, West Bengal, India*
[2] *Bluesensus – Sustainability & Seafood Consulting, Weilheim, Germany*

17.1 Introduction

Living along the interface between land and sea, the mangrove forest is a type of vegetation community comprising a variety of salt-resistant species having special adaptations and characteristics for growing in intertidal areas and estuary mouths. Mangroves are found in diverse habitats such as core forests, water bodies like rivers, crisscrossed creeks and channels, bays and backwater, mudflats, littered forest floors, and seagrass ecosystems where each is intricately interwoven into complicated relationships that form a dynamic, diverse ecosystem.

Mangrove species exist in groups and are often referred to as *Mangals* or communities. They can exist and flourish under a wide range of salinity conditions, tidal amplitudes, winds, temperatures, and muddy anaerobic situations. The highly variable conditions make mangrove forests immensely rich in biodiversity: they support a vast number of plant and animal communities, including many rare, endangered, and threatened species. The adaptive characteristics of mangroves in their root system, stems, leaves, and canopy formation have made them the most important biodiversity component of coastal areas, as they significantly contribute to building the resilience of coastal regions.

The mangrove ecosystem provides various functions:

- *Protection against storms and cyclones.* Mangroves have repeatedly been found to act as a bio-shield by dissuading wind and wave energy before it strikes the landmass. Many of the studies done after the tsunami in 2004, in particular, documented the role of mangroves [1, 2].
- They *retain terrestrial sediments and recycle nutrients,* supporting clear offshore water, which in turn favors the photosynthetic activity of phytoplankton and growth and robustness of coral reefs, seagrass beds, and reef fish communities [3].
- *Nurseries for fish, finfish, crabs, etc.* The detritus food chain supports the growth of phytoplankton, zooplankton, and peri plankton, which in turn are consumed by spawning and juvenile fish. Also, muddy waters and branching root systems prevent big fish from predating on small fish, so the ecosystem acts as a fish nursery. Thousands of fishermen depend on this productive ecosystem for their livelihoods.

Handbook of Ecological and Ecosystem Engineering, First Edition. Edited by Majeti Narasimha Vara Prasad.

- Mangroves also acts as *a carbon-sequestering pool*. They are one of the most carbon-rich biomes, with more carbon below ground, and have great below-to-above-ground carbon mass ratios than terrestrial trees. In the context of a greenhouse gas reduction strategy, mangroves are a mitigating factor for climate change [4].
- The mangrove ecosystem also acts as an *absorbent* of toxic pollutants and sediments before they reach the coastal ocean, thus providing natural treatment [5].
- Lastly, the mangroves' aesthetics and recreational facilities have immense potential to encourage *tourism* and related income for marginalized communities.

17.2 Sundarban: A Case Study

Sundarban is the largest contiguous mangrove stretch in the world, covering an area of approximately 10 000 km^2, of which 62% lies in Bangladesh and 40% in India [6]. The Indian part of Sundarban is located in the western part of the Ganges-Brahmaputra-Meghna delta, where it is flanked by the Bhagirathi Hooghly River on the west and the Raimongol-Kalindi River on the east. This is an area with crisscrossed channels, creeks, rivers, and swampy land with mangrove forests, and much of it has been reclaimed for habitation over the years.

The Sundarban Biosphere Reserve (SBR), part of the "Man and Biosphere Programme," was created in 1989 with 9630 sq.km defined by the Dampier Hodges line to the north, extending between 21°32′ to 22°40′ N and 88°05′ to 89°51′ E. The protected area consists of 4264 sq.km, while the human-inhabited area constitutes 5366 sq.km [3, 7, 8].

Tidal amplitude within the estuaries varies between 3.5 and 4 m, while seasonal variations range from 1 to 6 m [9]. The average annual rainfall varies between 1500 and 2000 mm. November to mid-February is the dry winter season, followed by the hot and humid summer months between March and May; a monsoonal weather pattern dominates the rest of the year. The average humidity is 82% and is almost the same throughout the year.

The Administrative blocks of Sundarban are spread over 13 blocks of South 24 Parganas district, 6 blocks of North 24 Parganas district, and 3 Municipalities in West Bengal. In Sundarban, there are 140 flowering plant species under 101 genera and 59 families comprising true mangroves, mangrove associates, back mangrove trees and shrubs, non-halophytic non-mangrove associates present in mangrove habitats, halophytic herbs, shrubs, weeds, epiphytes, and parasitic plants that grow among the Mangals [10].

Mangrove forests have entranced and intrigued naturalists, botanists, zoologists, and ecologists for millennia. They are home to resident and migratory birds, snakes and crocodiles, and mammals and simultaneously support incredible diversity and biomass of zooplankton, sponges, ascidians, tunicates, epibenthos, infauna, meiofauna, prawns, shrimp, crabs, insects, mollusks, fish, and amphibians [11, 12].

The SBR is distinguished from all other mangrove sites in the world and is unique to this region for the following reasons:

- It is the only marshy mangrove tiger habitat in a World Heritage Site.
- It houses very high biodiversity with unique flora and fauna.
- It provides a home for many endangered species like the royal Bengal tiger (*Panthera tigris*), fishing cat (*Prionailurus viverrinus*), Indian marsh mangoose (*Herpestes palustris*), Asian small-clawed otter, (*Aonyx cinereus*), smooth-coated otter (*Lutrogale perspicillata*), Ganges river dolphin (*Platanista gangetica*), Baer's pochard (*Aythya baeri*), masked finfoot (*Heliopais personata*), spoon-billed sandpiper (*Calidris pygmaea*), osprey (*Pandion haliaetus*), white-bellied sea

eagle (*Haliaeetus leucogaster*), slender-billed vulture (*Gyps tenuirostris*), white-rumped vulture (*Gyps bengalensis*), estuarine crocodile (*Crocodilus porosus*), Olive Ridley sea turtle (*Lepidochelys olivacea),* hawksbill sea turtle *(Eretmochelys imbricata),* river terrapin (*Batagur baska*), three-striped roofed turtle (*Batagur dhongoka*), Asian giant softshell turtle (*Pelochelys cantorii*), Pondicherry shark (*Carcharhinus hemiodon*), Ganges shark (*Glyphis gangeticus*), broadfin shark (*Lamiopsis temminckii*), and knifetooth sawfish (*Anoxypristis cuspidata*) [11, 12].

As per the 2011 census report, the population is 4.44 million, and the density of population is 1075 people per sq.km.

Human habitation is possible in this region due to the reclamation of mangrove areas. This has happened in phases since the East India Company headquarters were established in 1757 in Calcutta. The headquarters were on the edge of the forest, so British Collector General Russell proposed to divide Sundarban into plots for timber extraction and honey collection. In the late eighteenth century, the work initiated by Tillman Henkel saw marginalized and tribal populations from other states and areas engaged as labourers, and they gradually settled in Sundarban. Much of the mangrove forest that extended to Kolkata in the late eighteenth century was destroyed by the middle of the nineteenth century [3].

People in Sundarban are mostly engaged in agriculture and mainly dependent on monsoons. Many are engaged in fishing in country boats and are also dependent on honey and crab collection from the mangrove ecosystem. With the Forest Department's support through Joint Forest Management Committees, other developmental activities under the Sundarban Development Board, Panchayat and Rural Development, alternative livelihood augmentation projects and plans have been undertaken to reduce dependence on the forest. Thus, beekeeping, freshwater fishery development in ponds and canals, horticulture, tailoring, rabi-crop cultivation, etc. have been encouraged.

In this deltaic land crisscrossed by numerous rivers, rivulets, channels, and creeks, habitation on islands is possible due to dykes and embankments built around the villages to protect them from being inundated by salt water twice a day during high tides. At high tide, the water level outside the dyke on the river side is higher than the village level inside. Hence, these embankments are lifelines for the people of Sundarban.

The embankments are earthen, and many efforts have been made to concretize them, but the complex geo-hydro morphology of the region has not made this a sustainable solution. When very high tidal fluctuation occurs in spring and autumn every year, the people of Sundarban live in fear, wondering whether the embankments will be strong enough to protect them. The bricks may be destroyed or displaced or the concrete slabs may slide into the river, being scoured from below. The experiences of the tsunami in December 2004, cyclone Aila in May 2009, and Amphan in May 2020 have revealed that wherever there were mangroves in front of the dykes, they remained intact. the mangroves acted as the first line of defense against storms and cyclones, dissuaded wave and wind energy, and protected the fragile embankments.

This chapter highlights mangroves as shoreline engineers to raise and protect the landmass, thus playing a significant role in the transition from the marine ecosystem to a sustained terrestrial ecosystem.

17.3 Restoration Models

The aim of restoration should be to minimize the gap between the undisturbed (naturally disturbed) and restored ecosystem states. Ecological restoration aims to recover ecosystem integrity and resilience while delivering ecosystem services and ensuring human well-being. The conservation and restoration of biological diversity is usually a primary goal that can ensure a functioning ecosystem, securing the ecosystem services that emanate from it [13].

An ecosystem-based, situation-specific approach with community participation has been adopted for a large-scale mangrove restoration model in Sundarban, India (https://registry.verra.org/app/projectDetail/VCS/1463) to create replicable, upscale models with the help of carbon finance. A total of 4404 ha were afforested and restored, and the sites were distributed from the Raimangal River in the east to Sagar island in the west on the Bhagirathi-Hooghly River, spread over 59 *Gram Panchayats* (GPs) in 13 blocks out of 19 blocks in Sundarban.

Various experiences and results have been documented, of which two important cases of mangroves as essential contributors to coastal protection for human well-being are documented in this chapter. The chapter highlights the efficacy of the restoration models, with site selection and species selection based on the ecological variables such as salinity, inundation, soil texture, tidal amplitude, land-use and land-cover change (LULC) maps, erosion-accretion patterns, natural colonization of other mangrove species, and the indigenous technical knowledge (ITK) of local communities in establishing mangroves as functional elements for shoreline building and protection, which are especially important in the context of global warming, climate change, and sea-level rise. Case-specific illustrations for the two sites, representing systems under the same hydrological regime of the Hooghly-Padma-Brahmaputra estuary but differing in salinity, tidal wave impacts, socio-environmental conditions, etc., are presented.

It is important to note that beyond the scientific imperatives for mangrove restoration, community perception and involvement are key factors for long-term protection: ultimately, local communities that benefit from the multiple ecosystem functions that emanate from the mangroves must realize the benefits and actively engage in reducing the risks of anthropogenic pressure. However, it may be noted that activities beyond the Hooghly estuary, like building dams and barrages upstream, influence the critical minimal inflow of freshwater responsible for the most diverse mangrove vegetation in the world. A study done on the projected spatial distribution of mangrove species in the Indian Sundarbans in 2015 suggests that with progressive salinization, there has been a shift of mangrove vegetation to more saline-preferring species in the Indian Sundarban, especially in the Thakuran-Matla-Bidya-Gomor riverine areas [14]. This can be explained by the greater salinity profile in the central region of Sundarban [15]. However, the extent to which this critical minimal inflow of freshwater is required to continue with the diverse mangrove vegetation, especially the distribution of *Heritiera fomes, Nypa fruticans, Sonneratia apetala*, etc., has yet to be studied.

This chapter uses two models of restoration at Madhusudanpur Chor and Gopalganj to illustrate and document how mangroves have helped accrete land and stabilize dykes during a period, and species selection has been critical here.

17.4 Methodology

Mangrove restoration planning and related data sets were collected. Google Earth Pro high-resolution images were used to identify the mangrove cover for analysis in Sundarban. For better accuracy, all demarcations were drawn in a polygon from 500 to 800 m of ocular altitude. The files were exported in .KML file format and imported into the ArcGIS application using layers to .KML tools. After importing the maps, the boundary overlap was addressed using polygon reshape tools with tracing. Cut polygon tools were used to cut a particular mangrove cover area from the polygon. All map areas were calculated using geometry tools with a projected coordinate system of

WGS_1984_UTM_Zone_45N. In the image, the mangrove cover area is demarcated in yellow (Cat A), and silted riverine areas and accreted muddy banks (Cat B) are shown in red. The areas for Madhusudanpurchor were marked progressively for core mangrove areas as A1, A2, A3, and A4 and for the muddy banks as B1, B2, B3, and B4 for the years 2010, 2013, 2016, and 2019, respectively. Similarly, the areas for Gopalganj were marked progressively for core mangrove areas as P1, P2, P3, and P4 and for muddy banks as Q1, Q2, Q3, and Q4 for the years 2010, 2013, 2016, and 2019, respectively.

Area estimation was done with core mangrove areas and adjacent muddy banks over 10 years to understand the spatial and temporal changes in shore stability to protect the dykes, which are crucial to continued life and livelihood in this fragile delta.

Firsthand experience of working in the area for the last 15 years on mangrove restoration has also provided insight into social perspectives: the relationship between people, mangroves, and the ever-changing course of the rivers in this prograding delta.

Figure 17.1 shows an island near Lakshmipur commonly referred to as Madhusudanpurchor (muddy banks of Madhusudanpur) in Madhusudanpur GP, Kakdwip block in the South 24 Parganas district of West Bengal.

The mighty Bhagirathi-Hooghly river at its confluence with the Bay of Bengal branches into Muriganga river, and the island is located there. It is interesting to note that a 1995 landsat image shows no sign of this island (Figure 17.4). There was no natural barrier to protect the island, and extreme tidal waves prevented young saplings from growing because they had no time to anchor their roots. The site's mean elevation was 1.4 m asl, with an upper boundary elevation at 3 asl and a lower boundary at 0 asl. The slope was gradual; salinity and pH were measured with the help of a refractometer and pH meter, respectively. The salinity and pH profile were conducive for planting mangroves.

Figures 17.1 and 17.2 shows a graphical representation of the water and soil parameters in Madhusudanpur chor in the Madhusudanpur GP, Kakdwip block in the South 24 Parganas district of West Bengal.

The soil texture of the substrate is an essential factor for mangrove stabilization. Soil with a higher percentage of clay supports the growth of mangroves as it provides (i) good grip for the plants to hold on, (ii) more nutrient content, and (iii) high moisture capacity. Simple laboratory analysis of soil samples from the site confirmed the soil as clayey (sand = 14%, silt = 12%, and clay = 74%), which is favorable for planting mangroves. The soil character was derived by mapping into the Soil texture triangle format as per the United States Department of Agriculture, shown in Figure 17.3.

Keeping in mind the area's physical and chemical parameters, the species-selection strategy included *Rhizophora* sp, *Sonneratia* sp, *Bruguiera* sp, and *Avicennia* sp. It appeared that *Rhizophora mucronata* grew very well at the edge, thus anchoring the soil substrate and dissuading waves with its silt roots. Within a year, beautiful stilt root structures were visible, where the root part constituted 70% of the plant and the shoot part 30%. *Sonneratia* sp was suitable to the salinity profile and flourished. *Bruguiera gymnorrhiza* did not colonize well. *Avicennia officinalis* also survived in the Madhusudanpur chor area, as well as in the muddy banks adjacent to the dykes.

Due to the multi-species combination of the mangroves, the area benefitted from the good anchorage of *Rhizophora*, biodiversity components of *Sonneratia,* and adaptability of *Avicennia.* As a result, sediment capture progressed rapidly: the chronology shown in the figures illustrates how the small biomass of 112.77 ha in 2010 increased substantially, with a significant increase in accretion of 162% over 10 years (Figure 17.9). The gradual increase is represented in Figures 17.5–17.9. This reduced the intensity of the waves in that region; today, during low tide, one can cross from the village of Lakshmipur to the chor island.

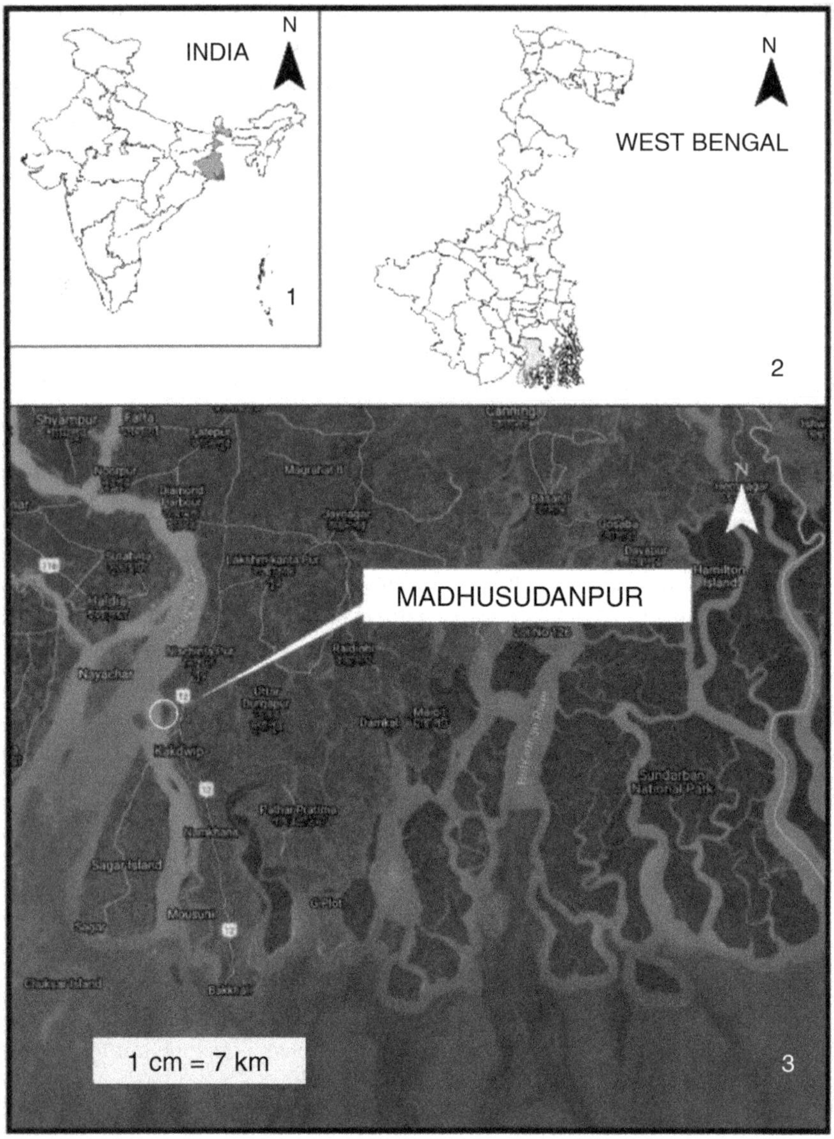

Figure 17.1 Geographical position of the island Madhusudanpurchor. Latitude: 21° 56′ 30.82″ N; longitude: 88° 10′ 26.37″ E. *See colour plate section for colour representation of this figure.*

Table 17.1 Water and soil parameters, Madhusudanpur chor.

Indicators	Pre-monsoon	Monsoon	Post-monsoon
Water salinity (ppt)	11	5	7.5
Soil salinity (ppt)	1	0	0
pH	8	8.5	8

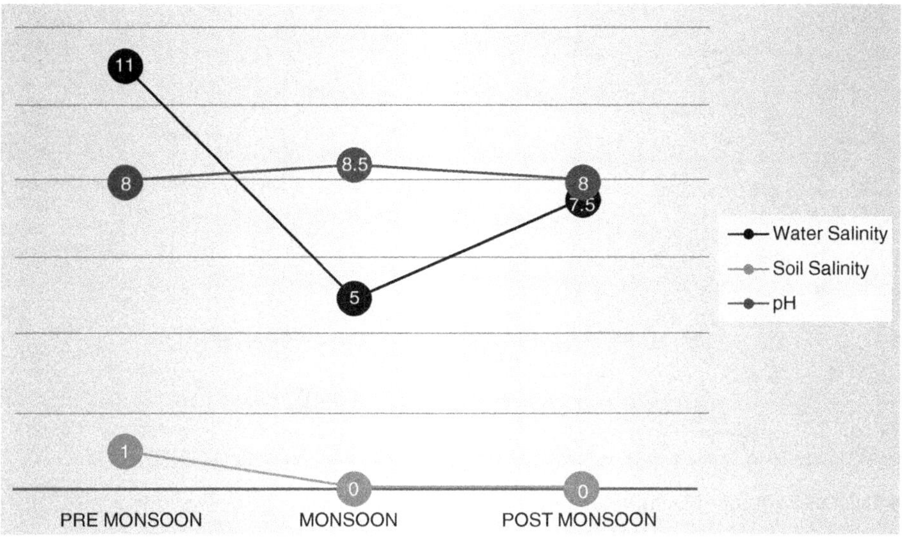

Figure 17.2 Water and soil parameters – Madhusudanpurchor – graphical representation.

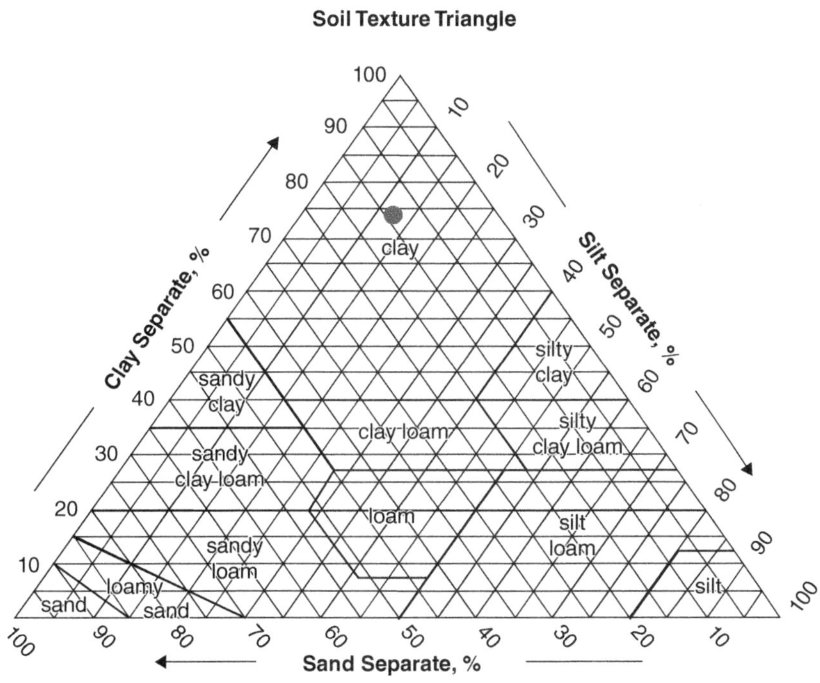

Figure 17.3 Soil texture triangle.

The local communities are happy that they took part in this conservation exercise because the roaring river is no longer a reason for sleepless nights: they were often afraid of embankment breaches and saline inundation due to the Muriganga river knocking on their doors. Cyclone Amphan proved that the Lakshmipur embankment, once a threat to locals, is now protected by layers of mangroves, which act as the first line of defense.

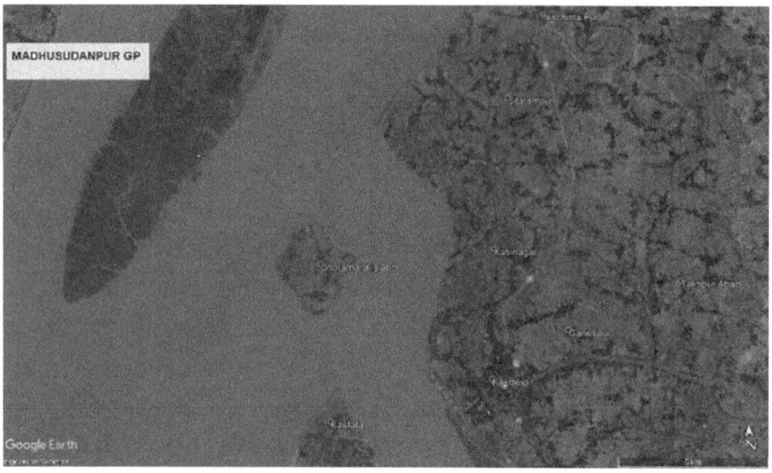

Figure 17.4 Kakdwip area without Madhusudanpurchor, 1995.

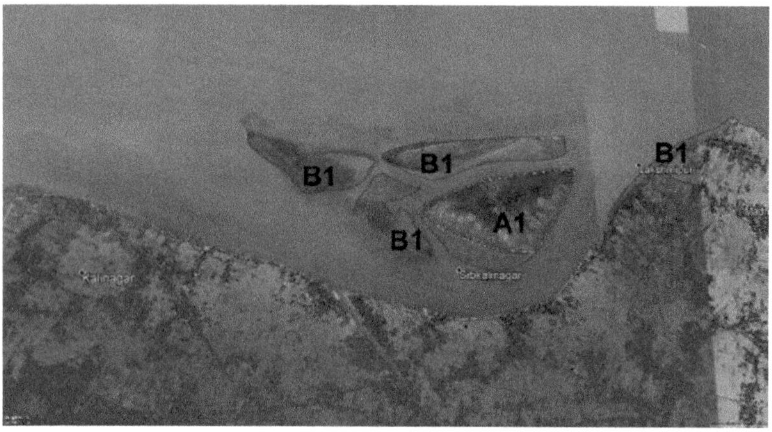

Figure 17.5 Madhusudanpur chor – 2010.

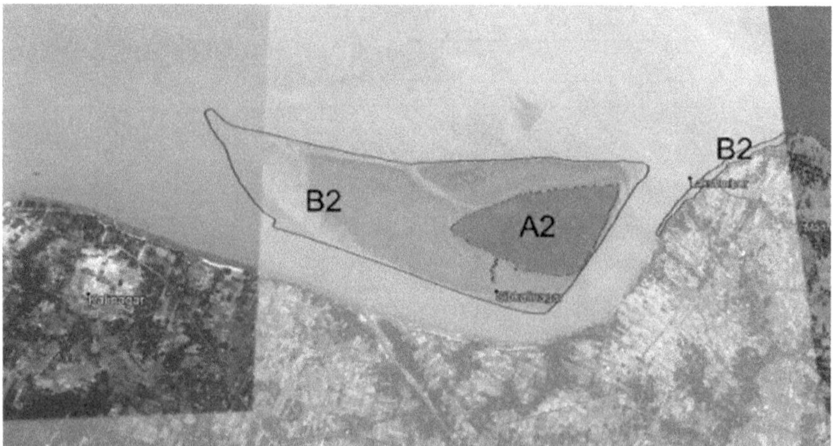

Figure 17.6 Madhusudanpur chor – 2013.

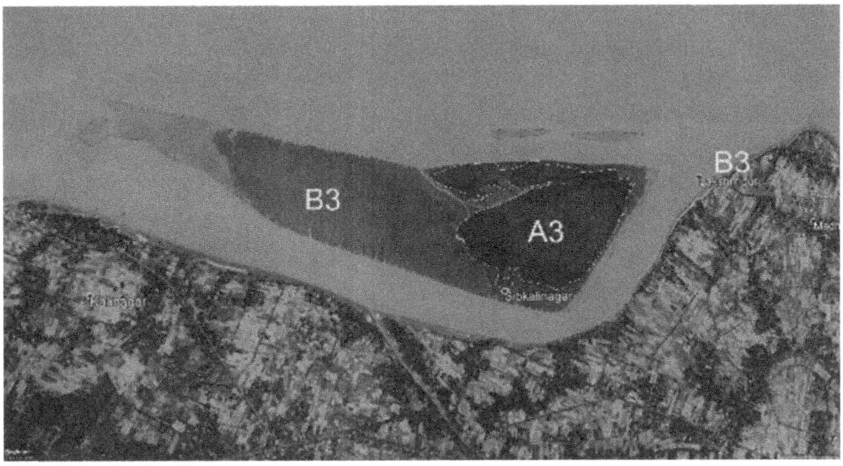

Figure 17.7 Madhusudanpurchor – 2016.

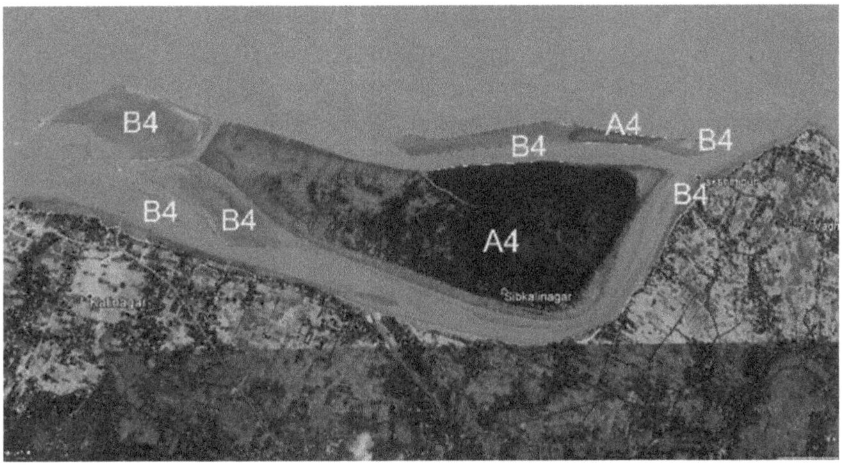

Figure 17.8 Madhusudanpurchor – 2019.

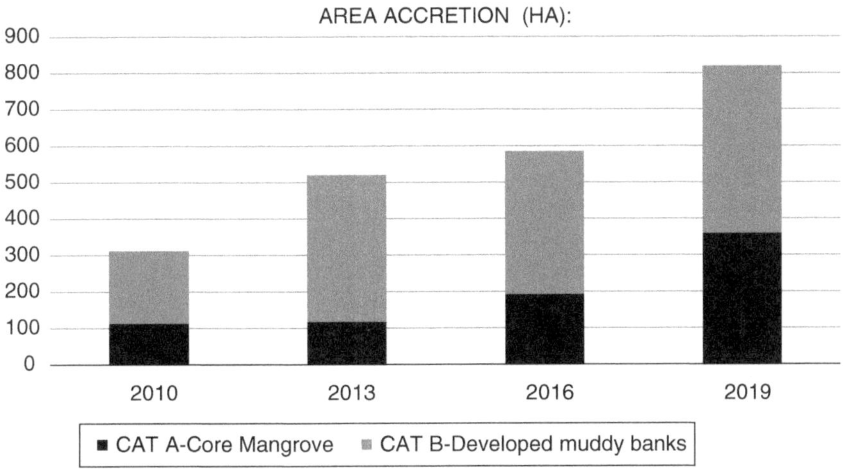

Figure 17.9 Representation of accretion in Madhusudanpur chor.

The gradual accretion status is documented in the sequential images shown in Figures 17.4–17.8 at the Madhusudanpur chor in Madhusudanpur GP, Kakdwip block in the South 24 Parganas district of West Bengal.

The pattern of accretion is illustrated in Figure 17.9.

The second shore-building process was in the central part of Sundarban, which is mainly tidal fed and receives minimal freshwater flows from upstream. In the upper part of the muddy banks adjacent to DakhinGorankathi-Gopalganj, the mangrove cover was missing. In 2009, during Aila, the embankment was breached, and knee-deep saline waters overran the village, bringing a saline curse to the area. However, significantly, Amphan in May 2020 did little damage to this particular area in Kultali block. The change is remarkable to observe.

The geographical position of the second site Gopalganj is shown in Figure 17.10.

The upstream area of the Matla River, which has become a tidal-fed river, contrasts sharply with the hydrological parameters, only 50 km from the Madhusudanpur chor, and exhibits a high salinity profile.

At Gopalganj, water samples were collected during high and low tides, along with surface soil, during the pre-monsoon, monsoon, and post-monsoon seasons. Salinity and pH were measured with a refractometer and pH meter, respectively; see Table 17.2.

Figure 17.11 shows a graphical representation of the water and soil parameters in Gopalganj in the Gopalganj GP, Kultali block in the South 24 Parganas district of West Bengal.

Beginning in 2011, a rigorous planting program was carried out, assisted by the local communities. Because salinity was high, three species of *Avicennia* were planted – *Avicennia alba, Avicennia marina,* and *A. officinalis* – along with a combination of *Sonneratia, Bruguiera*, and *Rhizophora*. Interestingly, mangrove surveillance reveals steady growth and distribution of all three species, with very few *Sonneratia* and *Rhizophora*, but no distribution of *Bruguiera. Excoecaria* sp near the embankments have been seen in good numbers, mostly due to natural regeneration.

Mangroves have played the role of shore consolidation and protect the fragile embankment from the mighty waves of the Matla River. The formation of newly silted areas has ceased over the years, and the bank has matured.

The gradual accretion is documented in the sequential images shown in Figures 17.12–17.16 at the Gopalganj site, Gopalganj GP, Kultali block, South 24 Parganas district, West Bengal.

17.5 Results and Analysis

The examples of salinity profiles and hydrological systems clearly indicate two different functions of mangroves as shore engineers. The Madhusudanpur site is fed with a freshwater flow from the Hooghly River's upper stream, which brings sediment that is crucial for the land-building function of the mangroves. The site has witnessed rapid, robust mangrove colonization with multi-species regeneration and rapid noticeable land formation. It validates the need for sediment flow for healthy and abundant mangrove colonies [16]. The soil texture triangle (Figure 17.3) and Table 17.1 indicate the presence of freshwater and sediment flow in the region and the mangroves' opportunistic role in quickly adapting and building the shoreline, consolidating the transition from marine to terrestrial ecosystem.

The diverse species of mangroves with various functions including anchorage, laterally branched roots for trapping sediments, and pneumatophores that thrive in clayey soil are natural solutions for engineering this landscape. The biodiversity is also significant, and sightings of

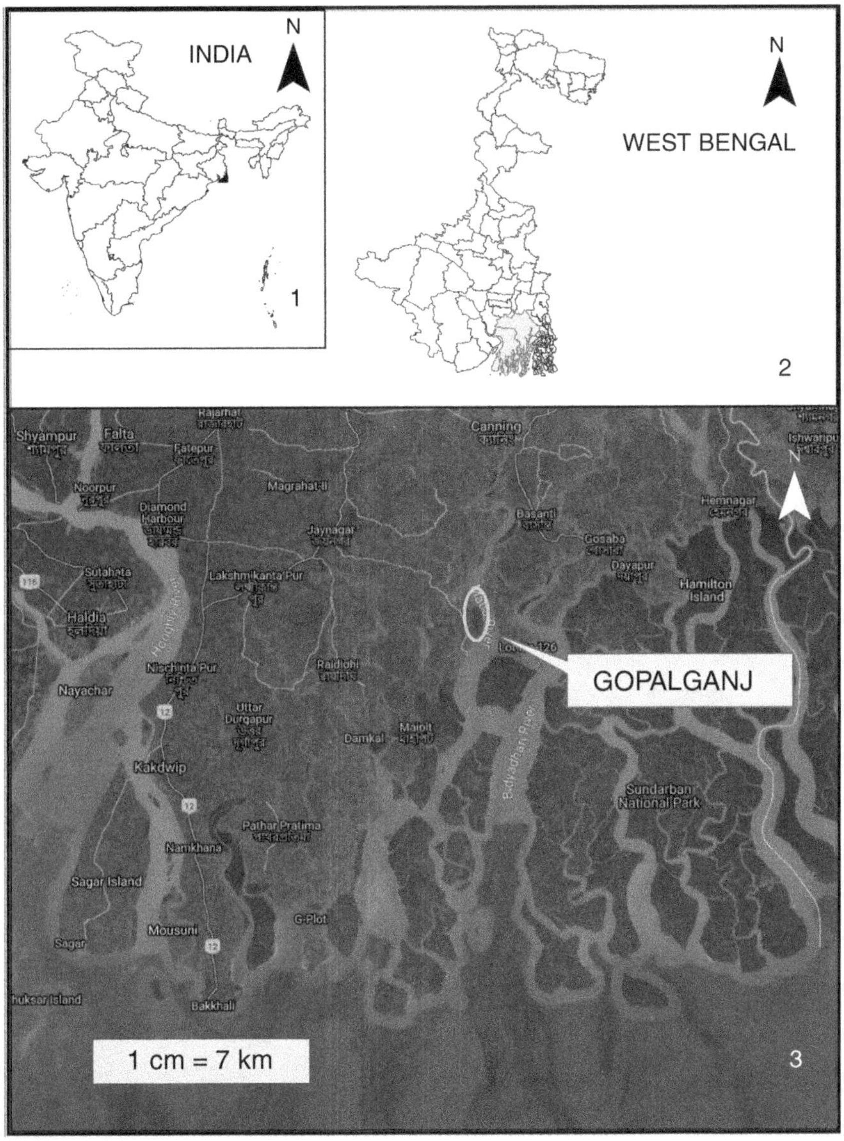

Figure 17.10 Geographical position of Gopalganj. Latitude: 22° 05′ 56.51″ N, longitude: 88° 38′ 01.96″ E. Gopalganj is a village in Gopalganj GP, in Kultali Block in the South 24 Parganas district in the Hooghly Matla estuary in Sundarban.

Table 17.2 Water and soil parameters, Gopalganj.

Indicators	Pre-monsoon	Monsoon	Post-monsoon
Water salinity (ppt)	20	15	19
Soil salinity (ppt)	6	3	6
pH	7.5	7	7

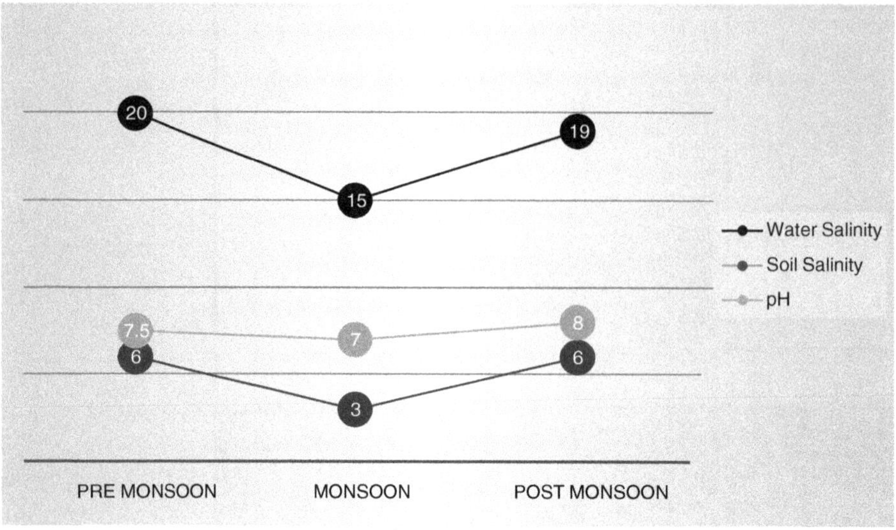

Figure 17.11 Gopalganj water and soil parameters: graphical representation.

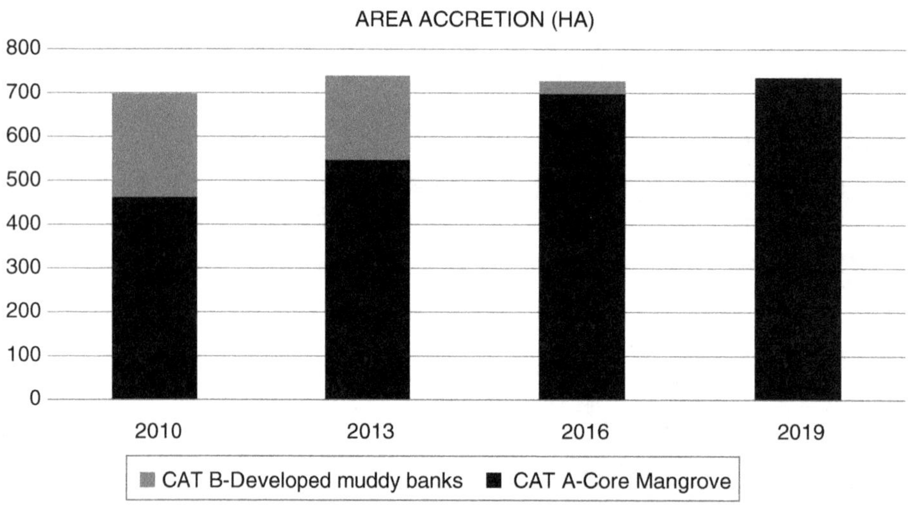

Figure 17.12 Accreted area in Gopalganj.

short-tailed shearwaters [17] and migratory birds like black-tailed godwits, bar-headed geese, curlew sandpipers, sand plovers, etc. (https://avibase.bsc-eoc.org/checklist.jsp?region=INggwb03&list=howardmoore and http://www.kolkatabirds.com/sunderchecklist.htm) have been reported. Resident wader bird colonies of purple herons and open-billed storks are seen on the island.

Interestingly, the mangroves have also contributed to the livelihoods of the local communities. Whereas in 2011, around 50 crab collectors depended on the island for sustenance, in 2019, more than 500 crab collectors are dependent on the island mangrove ecosystem for their livelihoods, as ascertained from stakeholder consultations.

The Gopalganj site is an example of mangroves providing coastline protection: the vulnerability of the dikes had been reduced significantly with the colonization of the mangroves. The mangrove area acts as the first line of defense, dissuading wind and wave energy. The experiences during Amphan, in

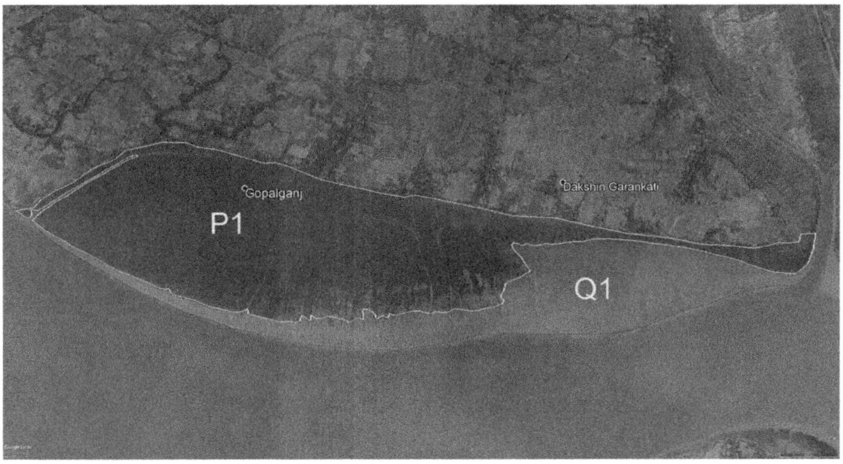

Figure 17.13 Gopalganj – 2010.

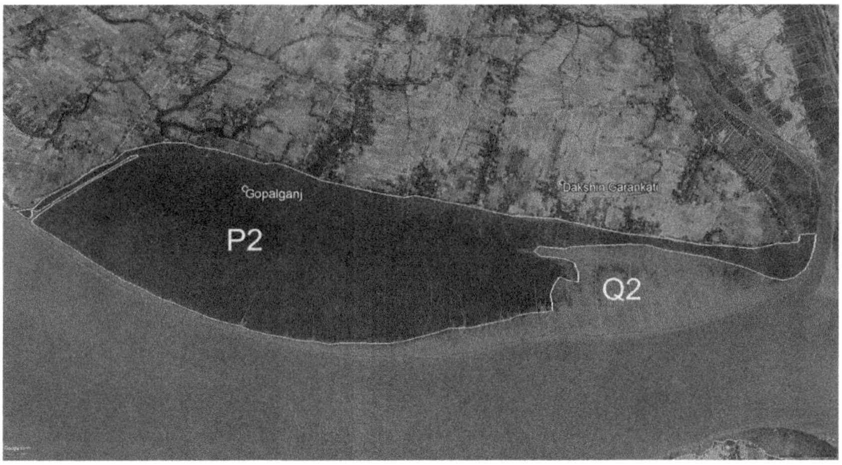

Figure 17.14 Gopalganj – 2016.

sharp contrast to Aila in 2009, bear testimony to this fact. This central part of Sundarban is known to lack species diversity, and the land accretion process is not pronounced, as shown in Figure 17.16. This may be because the freshwater connection to the Matla River is minimal, and there is reduced sediment flow. The moderate to high salinity profile also reduces species diversity. Biodiversity is also less pronounced in comparison to the Madhusudanpur site. Nevertheless, shoreline protection is apparent.

17.6 Conclusion

The two sites demonstrate the role of mangroves in building and stabilizing shorelines, contributing significantly to land building and coastal protection in the southern part of West Bengal, where land readily transients into the sea. Mangroves play a vital role as shore engineers by protecting the entire densely populated southern landmass, including the megapolis of Kolkata, against the

Figure 17.15 Gopalganj – 2013.

Figure 17.16 Gopalganj – 2019.

increasing number of tropical cyclones and storms that are a result of global warming. Thus the mangrove communities – the Mangals – with their diverse species are instrumental in providing the ecosystem services of building and protecting the shoreline as nature's engineers. However, many research gaps exist, especially regarding the role of mangroves in combatting sea level rise, carbon sequestration, impacts on fishing and livelihood, etc., and these need to be pursued.

Acknowledgments

The mangrove afforestation was carried with the support of Livelihoods Fund, Paris, and implemented by the Nature Environment & Wildlife Society (NEWS). We are grateful to the field team – ParitoshGiri, Keshab Ghosh, and Prabir Mondal – for providing all relevant information. We acknowledge the contribution of Sourav Bera in preparing the KML files for the study sites over 10 years. We fondly remember the late Dr. L.K. Banerjee, with whom we worked in these areas in

2010 and from whom we learned a great deal. Lastly, we are indebted to the communities living adjacent to the Madhusudanpur and Gopalganj areas, who have been patient with us in responding to our questions and enthusiastically accompanying us during our visits.

References

1 Dahdouh-Guebas, F., Jayatissa, L.P., di Nitto, D. et al. (2005). How effective were mangroves as a defence against the recent tsunami? *Current Biology* 15: 443–447.

2 Kathiresan, K. and Rajendran, N. (2005). Coastal mangrove forests mitigated tsunami. *Estuarine Coastal Shelf Science* 65: 610–606.

3 Ghosh, A., Schmidt, S., Flickert, T., and Nüsser, M. (2015). The Indian Sundarban mangrove forests: history, utilisation, conservation strategies and local perception. *Diversity* 7: 149–169.

4 Alongi, D.M. (2012). Carbon sequestration in mangrove forests. *Carbon Management* 3 (3): 313–322.

5 Suratman, M.N. (2008). Carbon sequestration potential of mangroves in Southeast Asia. In: *Managing Forest Ecosystems: The Challenge of Climate Change (eds.* Felipe Bravo, Valerie LeMay, Robert Jandl, Klaus Gadow*)*, MAFE, vol. 17, 297–315. Springer.

6 Spalding, M., Kainanuma, M., and Collins, L. (2010). *World Atlas of Mangroves*, 310. London, UK: Earthscan.

7 Copyright @ 2006, by Sundarban Biosphere Reserve- Department of Forests, Government of West Bengal. http://www.sundarbanbiosphere.org/html_files/sunderban_biosphere_reserve.htm

8 Government of West Bengal, Directorate of Forests, Wildlife Wing- Sundarban NP, 27th Dec 2020, last updated. http://www.wildbengal.com/sundarban-np.php

9 Islam, M.T. (2014). Vegetation changes of Sundarbans based on Landsat imagery analysis between 1975 and 2006. *Landscience and Environment* 8: 1–9.

10 Naskar, K.R., Sarkar, N.S., Ghosh, A. et al. (2004). Status of the mangroves and mangrove ecosystem of Sundarbans in West Bengal: Its impact on estuarine wetland fisheries. CIFRI-ICAR, Bulletin No. 134.

11 Sampath, V. (2003). National report on the status and development potential of the coastal and marine environment of the east coast of India and its living resources. GEF PDF block B Phase of FAO/BOBLME Programme.

12 Kailash, C., Alfred, J.R.B., Mitra, B., and Chowdhury, B.R. (2017). *Fauna of Sundarban Biosphere Reserve*, 1–240. Zoological Survey of India.

13 CBD Secretariat and Society for Ecological Restoration. (2019). A companion to the Short-Term Action Plan on Ecosystem Restoration – Resources, cases studies, and biodiversity considerations in the context of restoration science and practice. Montreal, Canada.

14 Mukhopadhyay, A., Wheeler, D., Dasgupta, S. et al. (2019). Mangrove spatial distribution in the Indian Sundarbans: predicting salinity-induced migration in a changing climate. *Journal of Management and Sustainability* 9 (1) https://doi.org/10.5539/jms.v9n1p1.

15 Mitra, A. (2016). Dynamic salinity feature of Sundarban. *Indian Journal of Geo-Marine Sciences* 45 (3): 410–415.

16 Anthony, E. and Goichot, M. (2020). Sediment flow in the context of mangrove restoration and conservation. A BMZ, IUCN and WWF project. A rapid assessment guidance manual.

17 Giri, P., Dey, A., and Sen, S.K. (2012). Short-tailed shearwater *Ardennatenuirostris* from Namkhana, West Bengal: a first record for India: Indian birds. *Journal of South Asian Ornithology* 8 (5): 131.

18

Forest Degradation Prevention Through Nature-Based Solutions

An Indian Perspective

Purabi Saikia[1], Akash Nag[1], Rima Kumari[2], Amit Kumar[3], and M.L. Khan[4]

[1] Department of Environmental Sciences, Central University of Jharkhand, Ranchi, Jharkhand, India
[2] Department of Environmental Science and Engineering, Indian Institute of Technology (Indian School of Mines), Dhanbad, Jharkhand, India
[3] Department of Geoinformatics, Central University of Jharkhand, Ranchi, Jharkhand, India
[4] Department of Botany, Dr. Harisingh Gour Vishwavidyalaya (A Central University), Sagar, Madhya Pradesh, India

18.1 Introduction

Forest and land degradation have become a major concern for the entire world, particularly for developing countries, as it poses an obstacle to eliminating poverty and hunger by means of biodiversity loss and the inability of farmers and communities to adapt to the changing climate [1]. The same is evident in the low level of growing stock and the declining trend of dense forests in India. The degradation of forests is generally measured in terms of their reduced capacity to provide ecosystem goods and services and the alterations in the forest structure and composition [2]. Woody plants, mainly trees, shrubs, and lianas, influence forests' overall composition by serving as the foundation for most food-webs [3]. They also play an important role in the economic development and human welfare by determining the life histories of millions of invertebrates. Additionally, they deliver various critical services and environmental benefits apart from the direct values that support human well-being [4, 5]. In India, millions of people who live adjacent to or within the protected areas (PAs) of forests are dependent on forests and forest-based products for their sustenance and livelihood as they share an inextricable relationship with forests [6]. However, the increasing pressure from humans on forests is not sustainable and has resulted in forest loss and degradation [7, 8]. Land-use changes have largely resulted in desertification, deforestation [9, 10], habitat fragmentation [11, 12], biodiversity loss, and eventually global warming [13], which ultimately lead to loss and degradation of the world's forests. Although India's total forest cover has been increased [14], forest quality is diminishing day by day due to overexploitation of forest resources. In India, tropical and sub-tropical forests are the most prone to forest cover loss due to agricultural expansion and overgrazing of livestock [15, 16]. The highly diverse tropical evergreen, semi-evergreen, montane wet temperate, and subtropical pine forests are replaced by relatively low diversity tropical dry deciduous and tropical thorn forests resulting from human interferences [17]. More than 40% of the country's

Handbook of Ecological and Ecosystem Engineering, First Edition. Edited by Majeti Narasimha Vara Prasad.

forests have already been degraded and understocked [18] due to overexploitation of forests [19]. The demand for food and land has increased with rapid population growth and urbanization [20, 21]. In addition, the invasion and proliferation of exotic species in India's forests have also resulted in the loss of biodiversity and a decline in other ecological services [22]. All these factors degraded forest quality and generated a series of socio-economic and ecological problems [23, 24].

Anthropogenic and natural disturbances connected to global climate change have created unprecedented pressure on forests, thereby necessitating novel ecological engineering and landscape-level approaches focused on cost-effective restoration ecosystem goods and services [25]. Successful long-term and large-scale restoration faces many challenges, including the high cost of achieving restoration, a shortage of regeneration materials [26], and to make the restored forest as a convenient socio-economic resource for communities. The new generation plantation (NGP) Platform launched in 2007 aims to complement timber production with effective maintenance and enhancement of ecosystems and simultaneously contribute to socio-economic development and climate change mitigation [27]. Integrating exotic commercial tree species with native ones in plantation programs can support biodiversity and the environment and local economies [25]. There is a direct relationship between increased biodiversity and increased primary productivity by improving soil fertility, partitioning resources, and enhanced nutrient retention [28]. Although not all species contribute equally within ecosystems, many species play significant functional, or keystone roles as 13% of forest tree species can contribute to 90% of the carbon storage [29].

In contrast, the ecological engineering approach combines both basic and applied sciences to restore, design, and construct ecosystems to recreate sustainable ecosystems that have value to both humans and nature [30]. It can be defined as "environmental manipulation by man using small amounts of supplementary energy to control systems in which the main energy drives are still coming from natural sources" [31]. The primary tool used in ecological engineering is the self-designing ecosystem, and it involves the design of the natural environment through quantitative approaches that rely on basic science [32]. Ecological engineering provides direct inputs to the restoration or conservation of altered components of natural capital, thereby restoring ecosystem functions and services [33]. Restoring the ecosystem based on ecological engineering principle inculcates the idea of designing a sustainable ecosystem through the integration of both human societies and the environment for their mutual benefit [34]. It includes a broader aspect of tools, techniques, and strategies for environmental intervention. The inclusion of engineering principles within the ecosystem framework plays an important role in predicting the dynamics of the ecosystem and providing robust solutions to the quandary associated with the ecosystem recovery [35]. Ecological engineering principles for restoring degraded ecosystems are designed to mimic the natural functioning of the previously existing ecosystem [32]. It helps manage natural resources and ensure the proper monitoring of environmental impacts [36].

Forest and landscape restoration (FLR) is an emerging concept and an essential component of conservation strategies, particularly in large scale environmental conservation [37]. It involves all the activities necessary to initiate or accelerate the recovery of a degraded ecosystem to re-establish the ecological integrity that could, in turn, support human well-being. It is an active process that involves various stakeholders from various affected land-use sectors to undertake a participatory decision-making approach to implement the practices of restoration to agreed ideal stability of the socio-ecological and economic benefits [38]. The rehabilitation and restoration of terrestrial ecosystems is a key strategy to recover ecosystem goods and services for the welfare of humankind [39], and there is a need to account for the long-term successional development of forests [40].

The broad objective of forest restoration is to return the land to its productive capacity with an ecosystem comprising of native species [41, 42]. The inclusion of indigenous knowledge about tree characteristics, planting of diverse species of ecological and economic importance, and integrating rehabilitation programs with regional development strategies are the essential elements to achieve restoration success [43]. The sustainable management and protection of ecosystem services are the most cost-effective way to sustain their benefits to human well-being [44].

18.2 Causes of Forests Degradation and Present Status Forests in India

The rate at which forests are degrading due to unsustainable forest management and over-exploitation of forest resources has become a global challenge. Natural and human-made activities affect species diversity, population structure, and natural regeneration of forest ecosystems. Degradation of tropical forests and destruction of habitat due to anthropogenic activities are the major causes of global biodiversity loss [45]. The highest rates of land-use change in the world are found in the tropics because of the rapidly growing human population [46]. Agricultural expansion; litter removal from the forest floor for organic fertilizer production, cooking, and heating of rural households; uncontrolled grazing and lopping pressure on forests; and over-stocking livestock on pasture land are the major causes of forest degradation [47–49]. Flood, fire, storms, and invasive species are natural causes, while illegal felling, over-exploitation of forest resources, encroachment, and domestic grazing are some man-induced causes of forest degradation [50]. In addition, several other factors, including the conversion of forest land for agriculture, irrigation, power projects, industrial establishments, roads, tribal colonies, settlement of evicted persons, and indiscriminate private encroachments, are the primary reasons behind the depletion of forests [51].

The depletion of forest resources and the decline of forest productivity are directly linked with forest degradation [52]. Agricultural operations, including shifting cultivation, are another major group of activities that have accelerated forest degradation and deforestation in India over a long period [53, 54]. Forest degradation, particularly in the tribal-dominated areas and hilly regions of northeast India and the Himalayas, can be attributed to shifting cultivation, where the slash-and-burn technique is used to convert forests into mosaics of crops and degraded scrub [55]. Overexploitation of soils because of forest clearing, cropping on steep slopes, deep plowing, leaving fields with low or no ground vegetation as well as overgrazing have resulted in subsequent land degradation [56].

The frequency and severity of land degradation exacerbated by erosion can have disastrous effects on downslope agricultural production and human settlements [57]. Transportation has also been found to be involved in deforestation, either directly or indirectly. Very dense and old teak trees are being selectively felled for railway sleepers and other modes of transportation, thereby reducing forest quality [58]. Urban expansion, agricultural land conversions, logging, and forest fires resulted in the loss of some 1.5–1.7 million km^2 of tree cover, or about 3.2% of the global forest cover, during 2000–2012 [59–61], thereby degrading the quality and productivity of forests.

Habitat destruction, overexploitation, environmental pollution, and anthropogenic pressure are significant disturbances to India's forest ecosystems [62]. Dry deciduous tropical forests in India are among the most threatened ecosystems, continuously experiencing disturbances such as lopping, burning, overgrazing, and clearing for cultivation [63]. These activities alter the primary forest structure, resulting in their conversion into grassland, secondary forest, savanna, or agricultural land [64]. Natural forests of the western Ghats and eastern Ghats of peninsular India have also experienced similar kinds of stress due to anthropogenic activities, where they are being replaced by forests consisting of inferior species [65]. Livestock has also been held responsible for forest degradation, resulting in the loss of fodder biomass [66]. Overgrazing has resulted in the loss of species diversity, depressed growth of seedlings, and soil erosion, thereby influencing the overall plant community structure [47, 67]. Human livelihood activities of livestock grazing, agriculture, hunting, fishing, and collection of timber, fuelwood, and non-timber forest products (NTFPs) have altered land cover and introduced new land uses in many Indian PAs [68, 69].

Indian forests have also been found to be affected by fires [70], where tropical thorn (75.27%), tropical dry deciduous (66.40%), and subtropical broad leaf hills (89.13%) are highly fire-affected [71]. Forest fires ultimately affect the quality and productivity of the forests, thereby resulting in their degradation.

India is one of 17 mega-biodiversity nations globally, with 63.7 million ha of forest cover [72]. India's forest cover is experiencing significant alterations, as evident through the biannual state of forests report by the FSI. There was an increase of 2.51% of India's total forest cover from 690 889 to 708 273 km^2 from 2007 to 2017. On the other hand, the JAXA global Forest Non-Forest (FNF) study reported a reduction in forest cover from 568 249 to 534 958 km^2 in India from 2007 to 2017 [73]. This reduction in forest cover was more evident in the eastern Himalayas followed by the western Himalayas, whereas an increase in forest cover was observed in western and eastern Ghats (Figure 18.1). The variation in forest cover status as estimated by FSI and JAXA global FNF datasets may be attributed to differences in forest cover mapping definitions and the use of different satellite datasets with different sensitivities when mapping vegetation.

The present study recorded that a significant proportion of forest cover of Madhya Pradesh and Chhattisgarh (18.69% of total forest cover) as well as Arunachal Pradesh (9.36%) was reduced during 1991–2019 (Figures 18.2 and 18.3). The percentage change in forest cover during 1991–2019 exhibited a significant increase in Delhi, followed by Kerala, West Bengal, Chandigarh, and Haryana (Figure 18.4). In contrast, the four northeastern states (Arunachal Pradesh, Nagaland, Manipur, and Mizoram) and two central Indian states (Madhya Pradesh and Chhattisgarh) recorded a loss in forest cover, with a moderate increase in the remaining states.

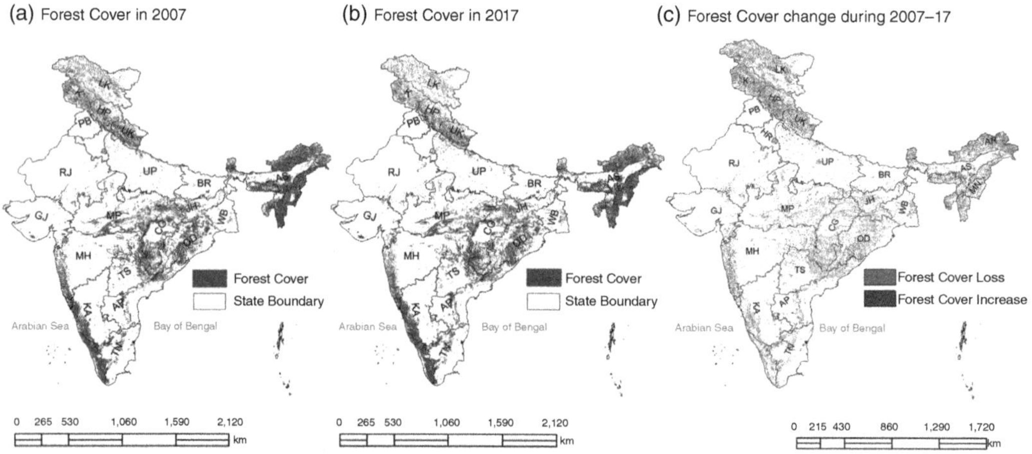

Figure 18.1 ALOS PALSAR satellite image. (a) Forest cover in 2007; (b) forest cover in 2017; (c) forest cover change during 2007–2017. *See colour plate section for colour representation of this figure.*

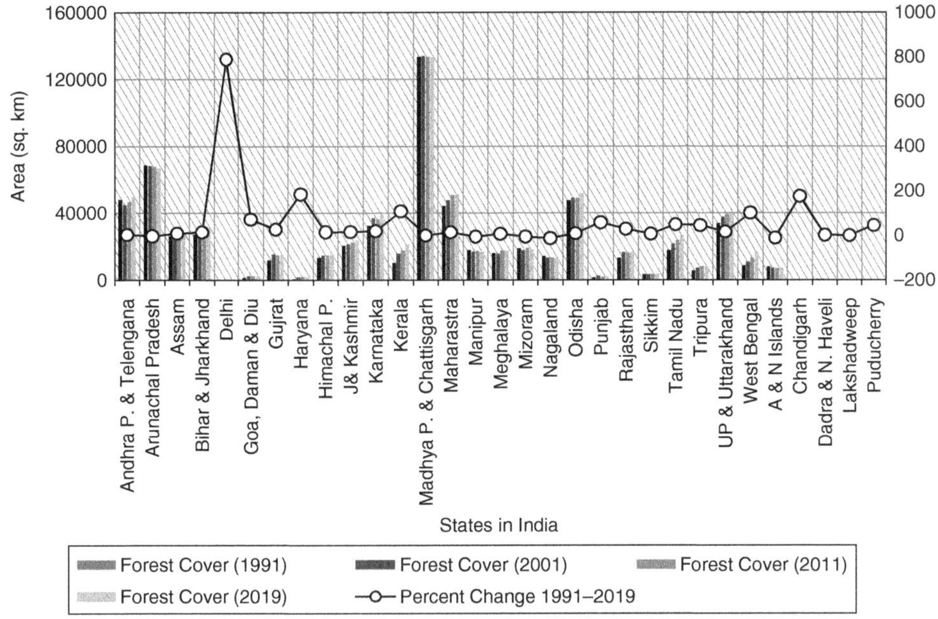

Figure 18.2 Status of forest cover in the states of India during 1991–2019.

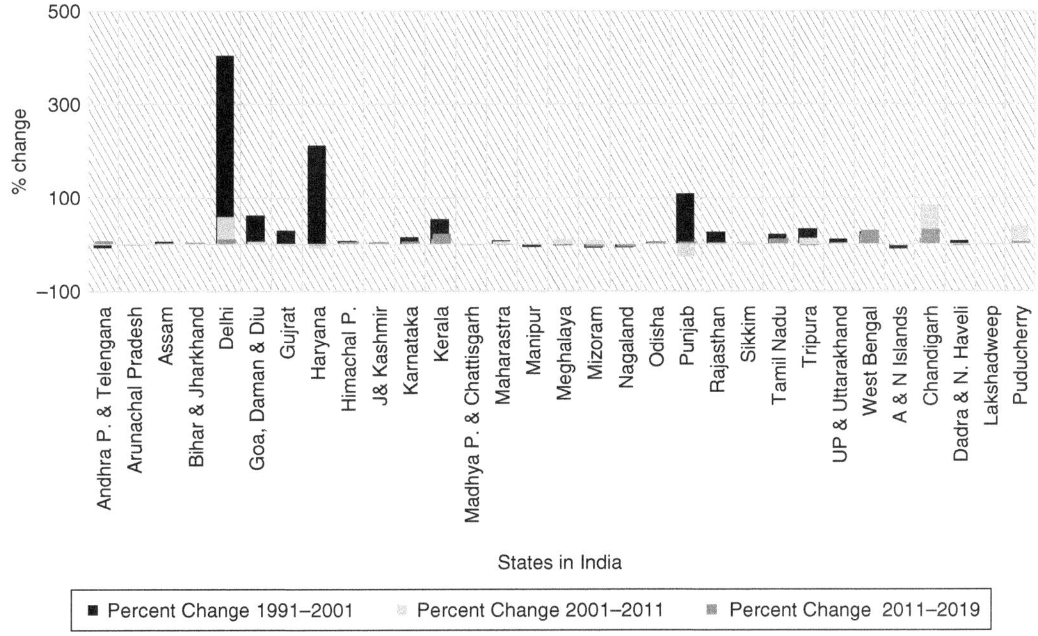

Figure 18.3 Percentage change in forest cover during different periods.

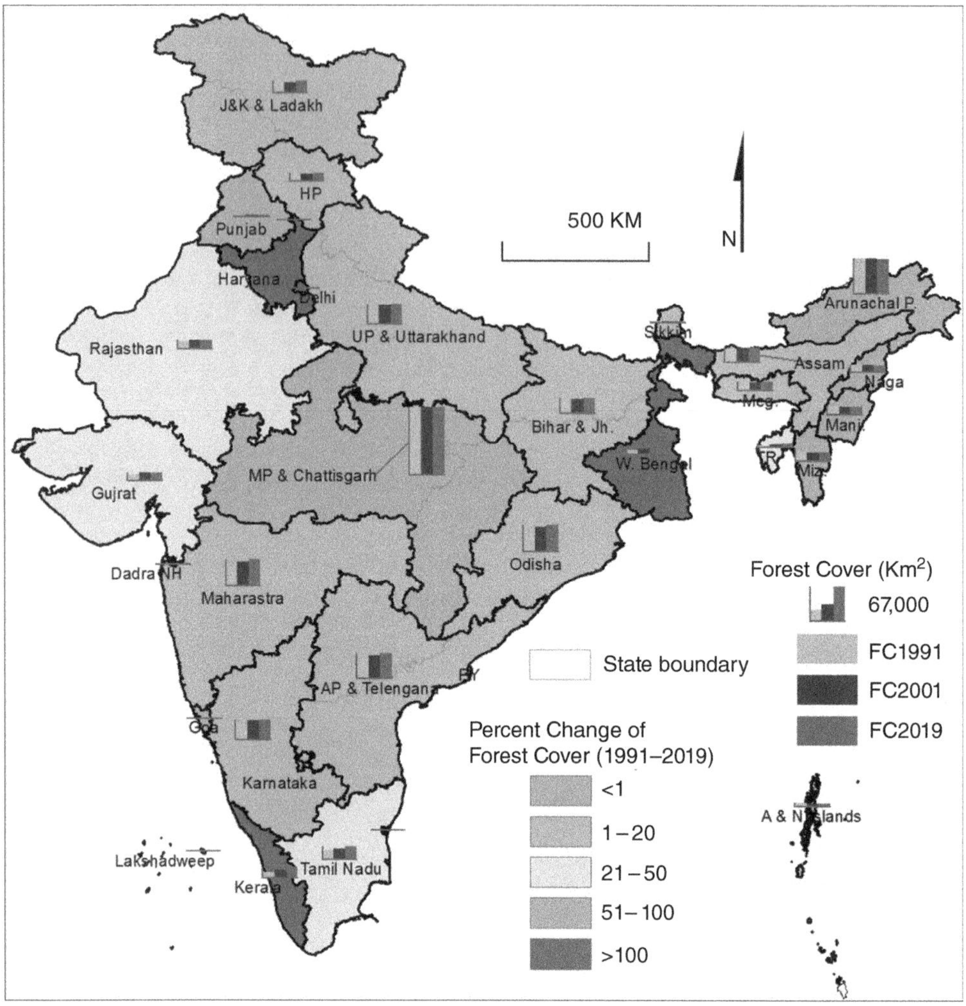

Figure 18.4 Forest cover (1991, 2001, 2019) and forest cover change during 1991–2019 in the states of India. *See colour plate section for colour representation of this figure.*

18.3 Effects of Forest Degradation

The degradation and loss of forests directly or indirectly affect forest-dependent communities, threaten the global climate and survival of many wild species, and reduce the ability of forests to provide essential ecosystem services [74]. The degradation of tropical forest quality has become a global concern because of forests' reduced potential for carbon sequestration [75]. The further destruction of tropical forests may become a catastrophe involving severe, perhaps irreparable damage to ecosystems.

The most significant effects of forest degradation are loss of habitat leading to species loss as well as a reduced supply of food resources for wild herbivores (increased crop-raiding incidences), alteration in nutrient dynamics (increased soil erosion and compaction), and changes in overall plant composition [76]. The direct effects of forest degradation on several forest-dependent communities can be observed in the form of reduced food availability and indirect effects in the form of reduced plant survival rates, regeneration, and recruitment patterns [69].

Several other concomitant effects on the ecological interactions and ecosystem functioning arise due to forest degradation [77]. The Amazon basin, which contains over half of the world's tropical rain forests, is also experiencing the world's highest absolute rate of forest destruction [78]. The recent increased rate of forest fires has created concern about the fate of the Amazon rainforests, which affect a plethora of species and large numbers of indigenous people living predominantly within remote regions of the Amazon [79]. Forest fires often follow unsustainable logging that results in poor-quality secondary forests [80].

Habitat fragmentation alters the diversity and composition of fragmented biota, thereby influencing ecological processes, including nutrient-cycling and pollination [81]. It also alters forest dynamics, causing a sharp increase in seedling mortality, damage, and canopy-gap formation due to greater desiccation and wind turbulence near forest edges [82].

The world's forests absorb about 2.4 billion tons of CO_2/year, but the destruction and degradation of forests further emits about 4.3–5.5 Gt CO_2 eq./year into the atmosphere [74]: a significant source of land-related greenhouse gas emissions. The degradation of forests also disrupts the water cycle, thereby creating drier conditions as the result of less rainfall [83].

18.4 Forest Degradation Management Strategies

Knowledge of plant diversity and the regeneration status of a particular forest is the baseline information required for a management and conservation strategy for forest biodiversity [50]. Effective monitoring techniques can be used to formulate effective forest management and land use policies [84]. Sound forest restoration strategy requires an integrated, multi-disciplinary approach of conservation and ecosystem restoration that must include preserving and protecting intact landscapes by allowing the land to heal itself [85]. Most tropical forests are highly resilient and recover their structural features, functional characteristics, and biodiversity rapidly after facing natural or anthropogenic disturbances [86]. Rehabilitation is the preferred strategy for restoring damaged tropical forests as it is the most realistic approach for dealing with complex tropical systems [46]. Sustainable forest management (SFM) is an evolving process to balance social, economic, and environmental concerns like biodiversity, productivity, and maintenance and enhance resilience [87]. Landscape management must include intact forest patches, sacred groves, ancient agro-forests, and patches regenerating after being abandoned [88].

The efficacy of forest restoration approaches depend on several factors, including the proximity of natural forests, selection of planted species or species mixtures, and degree of protection from the stressors of the recovering forest [89, 90]. The resilience of forests must be strengthened by applying integrated fire and pest management strategies and introducing resilient tree species in planted forests [91].

The degradation of forests adversely affects rare species, as their populations are reduced to such low numbers that they may not be viable [92]. The population of such species can be preserved mainly through in situ and ex situ conservation strategies that form the best defense against the erosion and degradation of forest genetic resources [93].

18.5 Policies for Preventing Forest Degradation

Restoration is mentioned in the Paris Agreement on climate change [94] as a valuable tool for addressing climate change and in the Convention on Biological Diversity [95] as a means for reducing biodiversity loss. The Millennium Ecosystem Assessment (MEA) emphasized the interdependence of people and nature to improved conservation, restoration, and sustainable ecosystem management to reverse the forests' decline [4]. As part of the Bonn Challenge, a global effort launched in 2011, 34 countries and subnational units have committed to ambitious restoration targets of restoring 350 million ha of damaged, deforested, and degraded land by 2030 (www.bonnchallenge.org). The New York Declaration on Forests agreed in 2014 to increase this target to 350 million ha by 2030 [96].

Many countries are working toward integrating their restoration strategies into sustainable development and conservation agendas through national and subnational restoration plans and policies [97–99]. The UN REDD+ program is attempting to encourage the restoration of forests by creating a market value for the carbon stored in forests [100]. In addition, the period 2021–2030 has been designated the UN Decade of Ecosystem Restoration by the United Nations to recognize the critical role of ecosystem restoration as a tool for improving environmental conditions to fight climate change and enhance food security, water supply, and biodiversity [101].

Landscape restoration also contributes to the achievement of the Sustainable Development Goals (SDGs), especially SDG15, which is to protect, restore, and promote the sustainable use of terrestrial ecosystems, sustainably manage forests, combat desertification, halt and reverse land degradation, and biodiversity loss; as well as SDG1 (no poverty), SDG2 (zero hunger), SDG6 (clean water and sanitation), and SDG13 (climate action) [57]. The United Nations Convention to Combat Desertification [102] established the Land Degradation Neutrality Fund to enable the private sector to support policies and make investments in a land-degradation-neutral world by 2030.

India's national interest in forest policy has changed from producing industrial wood to environmental stability and decentralized management to meet local subsistence requirements. The state government has played a central role in forest management in India since the establishment of the State Forest Department by the British colonial administration in 1864. After Independence in 1947, India's government maintained a state monopoly over forests and made no substantive changes to the Forest Department's administrative structure [103]. The First Five-Year Plan emphasized producing goods and services to create an industrial infrastructure in India, and during the 1960s, many existing forests were cleared and replaced with high-yielding commercially valuable species [104].

The Forest Conservation Act (FCA) was passed in 1980, which made it mandatory to obtain central government approval to convert forest land of 10 ha or more to non-forest land. The 1988 Forest Policy stated that the principal aim must be to ensure environmental stability, and the derivation of direct economic benefit must be subordinate [105]. The policy also stated that the domestic requirements of fuelwood, fodder, and minor forest produce should be the first charge of forest produce, subject to forests' carrying capacity, and is also referred to as a "massive needs-based and time-bound program of afforestation" [105]. The Eighth Five-Year Plan recognized the "need to decentralize control over nature and natural resources" [106] and to involve voluntary organizations and NGOs in this process.

In 1990, the central government sent a circular to state governments recommending the adoption of Joint Forest Management (JFM), in which local resource users would be involved in collaboration with the Forest Department and NGOs in the protection of forests [107, 108]. In India, the concept of participatory forest management evolved in the late 1970s and 1980s at Sukhomajri (Haryana), Arabari (West Bengal), and Harda (Madhya Pradesh) [109]. Experimentation in participatory forest management was observed in Asian countries including Nepal, Pakistan, Indonesia, Philippines, Cambodia, and Vietnam in parallel with the Indian experience [110] and equivalent initiatives evident in several African countries including Uganda, Congo, Gambia, Ghana, Kenya, Niger, South Africa, Mali, and Mozambique [111].

Self-Initiated Forest Protection Groups (SIFPGs) are an informal initiative by local people for protecting forests in different parts of India voluntarily without the support of any project or any help from the Forest Department [109]. In Odisha alone, more than 4000 SIFPGs were involved in protecting and managing forests [112]. Gradually, information about such community efforts was reported from West Bengal (Bankura), Jharkhand (Kudada, Bero), and many other places [113]. Community participation in forest management emerged in the hilly areas of western Himalayas in the form of "Van Panchayats" in Uttarakhand and Himachal Pradesh [114].

18.6 Ecological Engineering: A Tool for Restoration of Degraded Forests

Restoring large-scale degraded forests often poses challenges in the form of capital deprivation and a shortage of regenerative resources [25]. The integration of engineering principles within the ecosystem aids in overcoming such challenges, and the amalgamation of these diverse principles has proven to be a useful and successful ecological management approach [34, 35]. It also helped develop a bioengineered structure that further accelerated the restoration process of eroded gullies within the ecosystem [115]. The approach demonstrates a sustainable form of management practices with limited cost input and fewer negative ecological impacts [35]. The costs of restoration increase with the degree of forest degradation [116, 117] and may vary depending on methods used. Natural regeneration with native tree species may represent a cost-effective option if appropriate seed sources are available.

On the other hand, active restoration using site preparation and planting (sometimes with non-natives) represents a cost-intensive option to establish forests when sites are harsh [100]. The successful application of engineering principles for restoring degraded ecosystems depends on the ecological characteristics and ecosystem services offered by the restored ecosystem, which are directly dependent on the ecosystem's health and quality. Thus, the ecosystem's health is one of the bases for judging the sustainability and success rate of any ecological engineering approach [118].

The suitability of any eco-technology for restoration is site-specific and thus requires proper understanding and screening of the species to develop the new ecosystem. This requires identifying species with desired traits to address the specific issues for the successful design and implementation of ecological engineering principles in restoring degraded forests [119]. The ecological role of plants in reinforcing and fixing soils on degraded lands and river banks as well as on hilly terrains and slopes acts as an alternative to engineering solutions [120] as plants protect against instability in the soil structures. These bio-engineered slopes further bring dynamicity in the life of slope design [121]. Thus, the functionality associated with ecology sometimes acts as basic engineering principles that help maintain the stability of an ecosystem and prevent further degradation.

Establishing vegetation on degraded land is sometimes difficult due to poor soil conditions and strong erosive actions of wind and water, which are further aggravated in mountainous regions due to gravity's influence. Applying engineering principles and bio-engineering techniques helps to establish vegetation in degraded ecosystems [122, 123]. The techniques employed through bio-engineering principles accelerate the pace of the restoration in a degraded ecosystem by developing compatibility between vegetation and microclimatic conditions at a more advanced stage [124].

18.7 Forest Landscape Restoration: A Nature-Based Solution

NBS is one of the most sustainable and successful strategies to restore and rehabilitate any degraded ecosystem. It includes all the possible actions to protect, sustainably manage, and restore natural or modified ecosystems and effectively address societal challenges while simultaneously providing human well-being and biodiversity benefits [125]. The ecosystem approach is considered the foundation for the NBS framework, where FLR, ecosystem-based adaptation (EbA), ecological restoration (ER), and PAs are the four ecosystem-based concepts that fall under the NBS umbrella [126].

Successful implementation of NBS is possible with a deep understanding of nature's functioning and processes, and it provides multiple societal benefits, including biodiversity conservation, carbon sequestration, restoring degraded lands, food and water security [127]. The primary purpose of FLR is to regain ecological functionality and enhance human well-being across degraded landscapes at a larger scale [128, 129], and it aids in recuperating the socio-ecological, climatic, and economic welfare of forests [74, 130]. It undertakes multiple agroforestry activities, erosion control, and natural forest regeneration to implement the most appropriate restoration interventions at a landscape level.

FLR addresses the underlying causes of forest degradation and seeks to accommodate all land users [74]. It can be successful only when it reverses environmental degradation, strengthening the resilience of landscapes, securing forest-based livelihoods, and optimizing ecosystem goods and services [131]. Site-level interventions and effective planning at the landscape level become very important to accomplish and implement the concept of FLR [132].

A successful restoration strategy to be adopted at a degraded site must be based on the objectives of restoration and resources available in that site, which are welcomed by people to increase the degraded land's perceived value [133]. The rapidly changing climate is considered the fundamental rationale for managing forests toward resilience and sustainability [134]. Various types of planted and naturally regenerating forests, including agro-forests, can be established and managed in different degraded landscapes based on environmental suitability, land managers' needs, management goals, and financing arrangements [116].

Agroforestry includes a wide range of approaches for restoring degraded forests and agricultural lands, thereby contributing to landscape restoration [57]. It provides ecosystem services including increased soil productivity, control of erosion, and regulation of water availability in degraded or less-productive lands. Properly designed and managed agroforestry systems can help restore ecosystems, biodiversity conservation, and climate change adaptation and mitigation [135]. It is essential to practice agroforestry to maintain both agricultural productivity and crop diversity [136].

Another NBS, i.e., assisted natural regeneration, is a slow and gradual process of recovering the structure, function, and composition of a disturbed and degraded ecosystem [128]. It accelerates the establishment, growth, and survival of naturally colonized native tree species in abandoned fields [137]. Natural forest regrowth is promoted by high local resource availability (e.g., soil nutrients, soil moisture, microbial communities, and mineral soil properties) and high propagule (e.g., seeds and sprouts) availability. Naturally regenerating forests are thus more complex adaptive systems that possess heterogeneity, hierarchy, self-organization, openness, adaptation, memory, nonlinearity, and uncertainty [138].

18.8 Success Stories of ER from India

Forests are subjected to various natural and human-made ecological disturbances for their invaluable resources, which degrades the quality and productivity of forests [63]. Principles of FLR need to be promoted to overcome the adversities of deforestation and forest degradation. India's government has already been taken an effort to tackle the issues arising from forest degradation in the form of the Bonn Challenge, which aims to bring 21 million ha by 2030 [139]. The government has also restored an area of 9.8 million ha since 2011, of which 94.4% was contributed by government agencies, while 3.6% and 2% were contributed by the surveyed NGOs and private companies, respectively [139]. These figures can be appraised more appreciably by analyzing the success stories of restored degraded lands.

18.9 Yamuna Biodiversity Park

Yamuna Biodiversity Park (YBP) represents a remarkable success story of the ER of degraded land and serves as an example for environmental conservation in India. The park is situated in the Burari area of Jagatpur village in North Delhi, with 457 acres of nature reserve (Figure 18.5). The concept of developing a biodiversity park was established solely on the principle of ER, and the park was founded by Prof. C.R. Babu, University of Delhi, New Delhi. The park was developed to conserve the natural heritage within the periphery of the National Capital Region to enhance the quality of the environment [140]. It was created to replicate the Yamuna River's degraded ecosystem and create a self-sustaining ecosystem that would serve as a nature preserve.

The restored YBP serves as an ex situ conservation site to achieve self-sustainability like an in situ conservation site [141]. It was developed in two phases, considering inactive floodplains of the river in phase I and active floodplains in phase II. The development of YBP was initiated in 2002 by reintroducing native species to the area and continuous maintenance over the years [142, 143]. An area of about 157 acres near the village of Wazirabad was involved in phase I through adopting various restoration initiatives, including soil profiling, physico-chemical analyses, and amelioration of the soil quality through plantations [144].

The fully established park consists of two distinct zones: the *visitor zone* consists of a conservatory of medicinal plants, a butterfly garden, rangelands, sacred groves, etc., and the *nature reserve zone* harbors ~30 forest communities with multiple micro-niches and habitats for a range of birds and different animals to live and breed [145].

The work for phase II of YBP commenced in 2011 over an area of 300 acres comprising wetlands, grasslands, and floodplain forests. Impounding floodwaters and replenishing aquifers were implemented during this phase to minimize the impacts of floodwaters at the Wazirabad barrage [144].

From an ecological perspective, the YBD is recognized as a home to diverse resident and migratory birds, wild animals, and an assemblage of plant species that harbors a wide range of ecosystems native to the area.

18.10 Ecological Restoration in Corbett National Park

Corbett National Park (CNP) was established in 1936 as the first national park in the Asian continent to conserve various rare endangered animal and plant species by providing them a safe environment [146]. CNP has experienced various disturbances primarily in the form of habitat fragmentation, human-wildlife conflicts, poaching, forest fires, and invasive species resulting in habitat degradation. An invasion of *Lantana camara* in the peripheral buffer zones and forest gap zones in the core areas is one of the major management challenges [147].

A two-year restoration program was initiated in 2005 to prevent further invasions and improve the habitat by developing control measures, eradicating weeds, and restoring the weed-free

Figure 18.5 (a) Before ecological restoration (22 December 2000); (b) after ecological restoration (20 November 2018). *Source:* Google Images.

landscape. The Cut Root Stock (CRS) method was initially adopted to eradicate *L. camara* in CNP but was unable to eradicate it completely; and afterward, the land became more prone to secondary invasion by other weeds and invasive. Therefore, another suitable restoration technique was conceived to identify reference ecosystems followed by identifying targeted habitat conditions, developing field nurseries, introducing selected species sequentially into *L. camara*-free areas, and monitoring all these activities [147]. The strategy undertaken to restore the *L. camara*-invaded landscape exhibited the dominance of grassland communities and legumes in CNP, which further proves the effectiveness of NBS as a method of restoring degraded lands.

18.11 Conclusion and Recommendations

Current land-use practices and an imbalance in the inextricable relationship between forests and humans have resulted in large-scale forest loss and degradation. Information on the present scenario of forests and the extent of forest degradation helps in prioritizing and formulating conservation measures to prevent further degradation.

The best way to assess the extent of forest degradation of any region is to determine a decline in canopy cover between two times. Identifying the drivers of degradation is the foremost step to be undertaken before embarking on any restoration work. Restoration can minimize and sometimes reverse biodiversity losses, mitigate climate change, and improve local economies. It is necessary to limit climate change and other environmental risks to improve the resilience of landscapes.

The economic valuation of forest ecosystem services and biodiversity can improve ecosystem management to clarify trade-offs among conflicting environmental, social, and economic goals in the development and implementation of forest management plans and policies. NBSs including FLR, assisted natural regeneration, agroforestry plantations, ecological engineering approaches, etc. are some of the most sustainable and successful strategies to restore and rehabilitate any degraded ecosystem. FLR must be incorporated into national, regional, and local land-use planning and action. Site-level interventions are also crucial for effective planning and implementation of restoration activities at the landscape level. Efforts should also be made toward conserving dense forests and improving moderately dense forests to very dense forests, while restoring degraded and open forest areas into moderately dense forests to enhance the resilience of restored forest landscapes.

References

1 Sabogal, C., Besacier, C., and McGuire, D. (2015). Forest and landscape restoration: concepts, approaches and challenges for implementation. *Unasylva* 66: 3–10.

2 FAO (2001). Global ecological zoning for the global forest resources assessment 2000. Forestry Working Paper 56, Rome.

3 Naidu, M.T. and Kumar, O.A. (2016). Tree diversity, stand structure, and community composition of tropical forests in Eastern Ghats of Andhra Pradesh, India. *Journal of Asia-Pacific Biodiversity* 9 (1): 328–334.

4 MEA (Millennium Ecosystem Assessment) (2005). *Ecosystems and Human Well-Being: Synthesis.* Washington, DC: Island Press.

5 TEEB. (2010). Mainstreaming the economics of nature: a synthesis of the approach, conclusions and recommendations of TEEB. The Economics of Ecosystems and Biodiversity.

6 Kothari, A., Pande, P., Singh, S.D., and Variava, D. (1989). *Management of National Parks and Sanctuaries in India: A Status Report.* New Delhi: Indian Institute of Public Administration.

7 Arjunan, M., Puyravaud, J.P., and Davidar, P. (2005). The impact of resource collection by local communities on the dry forests of the Kalakad-Mundanthurai Tiger Reserve. *International Society for Tropical Ecology* 46 (2): 135–143.

8 Sagar, R. and Singh, J.S. (2004). Local plant species depletion in a tropical dry deciduous forest of northern India. *Biodiversity and Conservation* 31: 55–62.

9 Foster, D.R. (1992). Land-use history (1730–1900) and vegetation dynamics in Central New England. *Journal of Ecology* 80 (4): 753–771.

10 Sherbinin, A., Carr, D., Cassels, S., and Jiang, L. (2007). Population and environment. *Annual Review of Environment and Resources* 32: 345–373.

11 Armenteras, D., Gast, F., and Villareal, H. (2003). Andean forest fragmentation and the representativeness of protected natural areas in the eastern Andes, Colombia. *Biological Conservation* 113 (2): 245–256.

12 Noss, R.F. (2001). Beyond Kyoto: Forest management in a time of rapid climate change. *Conservation Biology* 15 (3): 578–590.

13 Fearnside, P.M. (2001). Saving tropical forests as a global warming countermeasure: an issue that divides the environmental movement. *Ecological Economics* 39 (2): 167–184.

14 FSI. (2019). State of forest report. Forest Survey of India, Ministry of Environment, Forest & Climate Change (MoEFCC), Govt. of India., Dehradun, India.

15 Anitha, K., Joseph, S., Chandran, R.J. et al. (2010). Tree species diversity and community composition in a human-dominated tropical forest of Western Ghats biodiversity hotspot, India. *Ecological Complexity* 7: 217–224.

16 Reddy, V.R. and Bahera, B. (2001). Forest degradation in India: extent and determinants. *Indian Journal of Agricultural Economics* 56 (4): 631–651.

17 Bhat, D.M., Murali, K.S., and Ravindranath, N.H. (2001). Formation and recovery of secondary forests in India: a particular reference to Western Ghats in Southern India. *Journal of Tropical Forest Science* 13 (4): 601–620.

18 Singh, U.S. (2018). Deforestation in India and climate change. *Journal of Bio Innovation* 7 (4): 631–659.

19 Mani, S. and Parthasarathy, N. (2006). Tree diversity and stand structure in inland and coastal tropical dry evergreen forests of peninsular India. *Current Science* 90: 1238–1246.

20 Kumar, A. and Kumar, A. (2019). Assessing human and carbon footprint of Ranchi urban environment using remote sensing technology. *Journal of Urban and Environmental Engineering* 13 (2): 257–264.

21 Tripathy, P. and Kumar, A. (2019). Monitoring and modelling of spatio-temporal urban growth of Delhi, India using cellular automata and geoinformatics. *Cities* 90: 52–63.

22 Saikia, P., Deka, J., Bharali, S. et al. (2017). Plant diversity patterns and conservation status of Eastern Himalayan forests in Arunachal Pradesh, Northeast India. *Forest Ecosystems* 4: 28. https://doi.org/10.1186/s40663-017-0117-8.

23 Babu, C.R. (2006). *Annual Report. Centre for Environmental Management of Degraded Ecosystems.* Delhi: School of Environmental Studies, University of Delhi.

24 Chomitz, K.M. (2007). *At Loggerheads? Agricultural Expansion, Poverty Reduction and Environment in the Tropical Forests*, 18–285. Washington, D.C.: The World Bank.

25 Löf, M., Madsen, P., Metslaid, M. et al. (2019). Restoring forests: regeneration and ecosystem function for the future. *New Forests* 50: 139–151.

26 Cernansky, R. (2018). How to rebuild a forest. *Nature* 560: 542–544.

27 Silva, L.N., Freer-Smith, P., and Madsen, P. (2019). Production, restoration, mitigation: a new generation of plantations. *New Forests* 50: 153–168.

28 Thompson, I.D., Okabe, K., Tylianakis, J.M. et al. (2011). Forest biodiversity and the delivery of ecosystem goods and services: translating science into policy. *BioScience* 61 (12): 972–981.

29 Balvanera, P., Pfisterer, A.B., Buchmann, N. et al. (2006). Quantifying the evidence for biodiversity effects on ecosystem functioning and services. *Ecology Letters* 9: 1146–1156.

30 Mitsch, W.J. (1993). Ecological engineering-a cooperative role with the planetary life-support systems. *Environmental Science and Technology* 27: 438–445.

31 Odum, H.T., Siler, W.L., Beyers, R.J., and Armstrong, N. (1963). Experiments with engineering of marine ecosystems. *Publication of the Institute of Marine Science of the University of Texas* 9: 374–403.

32 Mitsch, W.J. and Jorgensen, S.E. (2003). Ecological engineering: a field whose time has come. *Ecological Engineering* 20 (5): 363–377.

33 Rey, F., Cécillon, L., Cordonnier, T. et al. (2015). Integrating ecological engineering and ecological intensification from management practices to ecosystem services into a generic framework: a review. *Agronomy for Sustainable Development* 35: 1335–1345.

34 Mitsch, W.J. (2012). What is ecological engineering? *Ecological Engineering* 4 (5): 5–12.

35 Barot, S., Lata, J.C., and Lacorix, G. (2012). Meeting the relational challenge of ecological engineering within ecological sciences. *Ecological Engineering* 45: 13–23.

36 Xu, J. and Li, Z. (2012). A review on ecological engineering based engineering management. *Omega* 40 (3): 368–378.

37 Mansourian, S. and Vallauri, D. (2014). Restoring forest landscapes: important lessons learnt. *Environmental Management* 53 (2): 241–251.

38 Besseau, P., Graham, S., and Christophersen, T. (eds.) (2018). *Restoring Forests and Landscapes: The Key to a Sustainable Future*. Vienna, Austria: Global Partnership on Forest and Landscape Restoration.

39 Galati, A., Crescimanno, M., Gristina, L. et al. (2016). Actual provision as an alternative criterion to improve the efficiency of payments for ecosystem services for C sequestration in semiarid vineyards. *Agricultural Systems* 144: 58–64.

40 Macdonald, S.E., Simon, M., Landhausser, S.J. et al. (2015). Forest restoration following surface mining disturbance: challenges and solutions. *New Forests* 46 (5–6): 703–732.

41 Grant, C.D. and Koch, J. (2007). Decommissioning Western Australia's first bauxite mine: co-evolving vegetation restoration techniques and targets. *Ecological Management & Restoration* 8: 92–105.

42 Zipper, C.E., Burger, J.A., Skousen, J.G. et al. (2011). Restoring forest and associated ecosystem services on Appalachian coal surface mines. *Environmental Management* 47: 751–765.

43 Mansourian, S., Vallauri, D., and Dudley, N. (eds.) (2005). *Forest Restoration in Landscapes: Beyond Planting Trees*. New York: Springer.

44 Ekins, P., Simon, S., Deutsch, L. et al. (2003). A framework for the practical application of the concepts of the critical natural capital and strong sustainability. *Ecological Engineering* 41: 375–392.

45 Mishra, B.P., Tripathi, O.P., Tripathi, R.S., and Pandey, H.N. (2004). Effect of anthropogenic disturbance on plant diversity and community structure of a sacred grove in Meghalaya, Northeast India. *Biodiversity and Conservation* 13: 421–436.

46 Lugo, A.E., Parrotta, J.A., and Brown, S. (1993). Loss in species caused by tropical deforestation and their recovery through management. *Ambio* 22 (2/3): 106–109.

47 Kumar, R. and Shahabuddin, G. (2005). Effects of biomass extraction on vegetation structure, diversity and composition of forest in Sariska Tiger Reserve, India. *Environment Conservation* 32: 248–259.

48 Prabhakar, R., Somanathan, E., and Mehta, B.S. (2006). How degraded are Himalayan forests? *Current Science* 91 (1): 61–67.

49 Singh, J.S. (2006). Sustainable development of the Indian Himalayan region: linking ecological and economic concerns. *Current Science* 90 (6): 784–788.

50 Dutta, G. and Devi, A. (2013). Plant diversity, population structure, and regeneration status in disturbed tropical forests in Assam, northeast India. *Journal of Forestry Research* 24: 715–720.

51 Contreras-Hermosilla, A. (2000). underlying causes of forest decline. Occasional Paper no. 30, Centre for International Forestry Research, Indonesia. Available at: https://www.cifor.org/publications/pdf_files/OccPapers/OP-030.pdf.

52 Chirwa, P.W. and Mahamane, L. (2017). Overview of restoration and management practices in the degraded landscapes of the Sahelian and dryland forests and woodlands of East and Southern Africa. *Southern Forests: A Journal of Forest Science* 79 (2): 87–94.

53 Sati, V.P. and Rinawma, P. (2014). Practices of shifting cultivation and its implications in Mizoram, North-East India: a review of existing research. *Nature & Environment* 19 (2): 179–187.

54 Tomar, J., Das, A., Puni, L., and Chaturvedi, O.P. (2012). Shifting cultivation in Northeastern region of India: status and strategies for sustainable development. *Indian Forester* 138 (1): 52–62.

55 FSI. (2017). An assessment report on forest cover status of India. Ministry of Environment and Forest (MoEF), Government of India.

56 Gruver, J.B. (2013). Prediction, prevention and remediation of soil degradation by water erosion. *Nature Education Knowledge* 4 (12): 2.

57 FAO. (2017). Agroforestry for landscape restoration, exploring the potential of agroforestry to enhance the sustainability and resilience of degraded landscapes. Food and Agriculture Organization of the United Nations, Rome. http://www.fao.org/documents/card/en/c/fa7b6a1e-f781-4a9e-9da0-7cf02366c6f5.

58 Bowonder, B. (1982). Deforestation in India. *International Journal of Environmental Studies* 18 (3–4): 223–236.

59 DeFries, R.S., Rudel, T., Uriarte, M., and Hansen, M. (2010). Deforestation driven by urban population growth and agricultural trade in the twenty-first century. *Nature Geoscience* 3: 178–181.

60 Hansen, M.C., Potapov, P.V., Moore, R. et al. (2013). High-resolution global maps of 21st century forest cover change. *Science* 342 (6160): 850–853.

61 Riitters, K., Wickham, J., Costanza, J.K., and Vogt, P. (2016). A global evaluation of forest interior area dynamics using tree cover data from 2000 to 2012. *Landscape Ecology* 31: 137–148.

62 UNEP (2001). *India: State of the Environment-2001*. Nairobi: United Nations Environment Programme.

63 Kumari, R., Banerjee, A., Kumar, R. et al. (2019). Deforestation in India: consequences and sustainable solutions. In: *Forest Degradation around the World* (eds. M.N. Suratman, Z.A. Latif, G.D. Oliveira, et al.), 1–18. United Kingdom: IntechOpen https://doi.org/10.5772/intechopen.85804.

64 Khurana, E. and Singh, J.S. (2001). Ecology of seed and seedling growth for conservation and restoration of tropical dry forest: a review. *Environmental Conservation* 28 (1): 39–52.

65 Parthasarathy, N. (1999). Tree diversity and distribution in undisturbed and human impacted sites of tropical wet evergreen forest in Southern Western Ghats, India. *Biodiversity and Conservation* 8: 1365–1381.

66 Pandey, R., Gupta, H.S., and Kumar, A. (2014). Resource availability versus resource extraction in forests: analysis of forest fodder system in forest density classes in lower Himalayas, India. *Small-Scale Forestry* 13 (3): 267–279.

67 Sharma, C.M., Gairola, S., Ghildiyal, S.K., and Suyal, S. (2009). Forest resource use patterns in relation to socio-economic status: a case study in four temperate villages of Garhwal Himalaya, India. *Mountain Research and Development* 29 (4): 308–319.

68 Rahmani, A. (2003). Conservation outside protected areas. In: *Battles Over Nature: Science and Politics of Conservation* (eds. V. Saberwal and M. Rangarajan), 117–138. New Delhi: Permanent Black.

69 Somanathan, H. and Borges, R.M. (2000). Influence of exploitation on population structure, plant spacing and reproductive success in dioecious tree species within a fragmented cloud forest in India. *Biological Conservation* 94: 243–256.

70 Bahuguna, V.K. and Upadhay, A. (2002). Forest fires in India: Policy initiatives for community participation. *International Forestry Review* 4 (2): 122–127.

71 FSI. (2015). State of forest report. Forest Survey of India, Govt. of India, Dehradun, India.

72 FSI. (1999). State of forest report. Forest Survey of India, Govt. of India, Dehradun, India.

73 Lal, P., Dubey, A.K., Kumar, A. et al. (2019). SAR-Optical remote sensing based forest cover and greenness estimation over India. ISPRS Annals of the Photogrammetry, Remote Sensing and Spatial Information Sciences, Volume IV-5/W2. Capacity Building and Education Outreach in Advance Geospatial Technologies and Land Management, Dhulikhel, Nepal,10–11 December.

74 IUCN. (2017). Issues brief on deforestation and forest degradation. https://www.iucn.org/resources/issues-briefs/deforestation-and-forest-degradation.

75 Kale, M.P., Ravan, S.A., Roy, P.S., and Singh, S. (2009). Patterns of carbon sequestration in forests of Western Ghats and study of applicability of remote sensing in generating carbon credits through afforestation/reforestation. *Journal of the Indian Society of Remote Sensing* 37 (3): 457–471.

76 Middleton, B. (2003). Ecology and objective based management: case study of the Keoladeo National Park, Bharatpur, Rajasthan. In: *Battles Over Nature: Science and Politics of Conservation* (eds. V. Saberwal and M. Rangarajan), 86–116. New Delhi: Permanent Black.

77 Robinson, J.G. (1993). The limits to caring: sustainable living and the loss of biodiversity. *Conservation Biology* 7: 20–28.

78 Nunez, C. (2019). Rainforests, explained. *National Geographic*. https://www.nationalgeographic.com/environment/habitats/rain-forests.

79 Stewart, M. (2019). An analysis of Amazonian forest fires. Towards Data Science. https://towardsdatascience.com/an-analysis-of-amazonian-forest-fires-8facca63ba69.

80 Asner, G.P., Broadbent, E.N., Oliveira, P.J.C. et al. (2006). Condition and fate of logged forests in the Brazilian Amazon. *Proceedings of the National Academy of Sciences* 103 (34): 12947–12950.

81 Laurance, W.F., Delamônica, P., Laurance, S.G. et al. (2000). Rainforest fragmentation kills big trees. *Nature* 404 (6780): 836.

82 Laurance, W.F., Camargo, J.L.C., Luizão, R.C.C. et al. (2011). The fate of Amazonian forest fragments: a 32-year investigation. *Biological Conservation* 114 (1): 56–67.

83 Johnson, S. (2019). The effects of forest degradation on ecosystems. Sciencing. https://sciencing.com/effects-forest-degradation-ecosystems-8333738.html.

84 Ahmad, W. (1992). The use of remotely sensed data in the context of a GIS for monitoring temporal changes in a forested region of Australia. *Asia-Pacific Remote Sensing Journal* 5 (1): 133–144.

85 Dellasala, D.A., Martin, A., Spivak, R. et al. (2003). A citizen's call for ecological forest restoration: forest restoration principles and criteria. *Ecological Restoration* 21 (1): 14–23.

86 Parrotta, J.A., Knowles, O.H., and Wunderle, J.M. Jr. (1997). Development of floristic diversity in 10yearold restoration forests on a bauxite mined site in Amazonia. *Forest Ecology and Management* 99: 21–42.

87 Holvoet, B. and Muys, B. (2004). Sustainable forest management worldwide: a comparative assessment of standards. *International Forestry Review* 6: 99–122.

88 Frost, P., Campbell, B.M., Medina, G., and Usongo, L. (2006). Landscape-scale approaches for integrated natural resource management in tropical forest landscapes. *Ecology and Society* 11 (2): 30.

89 Brockerhoff, E.G., Jactel, H., Parrotta, J.A. et al. (2008). Plantation forests and biodiversity: oxymoron or opportunity? *Biodiversity and Conservation* 17: 925–951.

90 Sayer, J., Chokkalingam, U., and Poulsen, J. (2004). The restoration of forest biodiversity and ecological values. *Forest Ecology and Management* 201 (1): 3–11.

91 FAO. (2020). Reducing forest degradation. Food and Agriculture Organization of the United Nations, Rome. http://www.fao.org/sustainable-forest-management/toolbox/modules/reducing-forest-degradation/cases/en/?type=111.

92 Phartyal, S.S., Thapliyal, R.C., Koedam, N., and Godefroid, S. (2002). *Ex situ* conservation of rare and valuable forest tree species through seed-gene bank. *Current Science* 83 (11): 1351–1357.

93 Thomson, L., Graudal, L., and Kjaer, E. (2001). Conservation of genetic resources in their natural environment. In: *Forest Genetic Resources Conservation and Management (Vol. 2): In Managed Natural Forests and Protected Areas (in situ)* (eds. FAO, DFSC and IPGRI), 1–3. Rome, Italy: International Plant Genetic Resources Institute.

94 UNFCCC. (2015). The Paris Agreement. Available at: https://unfccc.int/sites/default/files/english_paris_agreement.pdf.

95 CBD (2015). *Conservation on Biological Diversity*. Paris: The Paris Agreement.

96 Climate Summit (2014). Forests: action statements and action plans. New York, USA, United Nations. http://www.un.org/climatechange/summit/wp-content/uploads/sites/2/2014/07/New-York-Declaration-on-Forest-%E2%80%93-Action-Statement- andAction-Plan.pdf; Accessed on: 22/12/2019.

97 Aguilar, M., Sierra, J., Ramirez, W. et al. (2015). Toward a post-conflict Colombia: restoring to the future. *Restoration Ecology* 23: 4–6.

98 Murcia, C., Guariguata, M.R., Andrade, A. et al. (2016). Challenges and prospects for scaling-up ecological restoration to meet international commitments: Colombia as a case study. *Conservation Letters* 9 (3): 213–220.

99 Pinto, S.R., Melo, F., Tabarelli, M. et al. (2014). Governing and delivering a biome-wide restoration initiative: the case of Atlantic forest restoration pact in Brazil. *Forests* 5 (9): 2212–2229.

100 Jacobs, D.F., Oliet, J.A., Aronsson, J. et al. (2015). Restoring forests: what constitutes success in the twenty-first century? *New Forests* 46: 601–614.

101 UN. (2019) New UN decade on ecosystem restoration to inspire bold UN environment assembly decisions. UN Environment Programme. https://www.unenvironment.org/news-and-stories/story/new-un-decade-ecosystem-restoration-inspire-bold-un-environment-assembly.

102 UNCCD (2016). *United Nation Convention to Combat Desertification*. Paris: The Paris Agreement.

103 Desai, V. (1991). *Forest Management in India: Issues and Problems*. Delhi: Himalaya Publishing House.

104 FAO (1961). *Report to the Government of India on Integration of Forests and Forest Industries*. Rome: Food and Agricultural Organisation.

105 MoEF (1988). *National Forest Policy*. New Delhi: Ministry of Environment and Forests.

106 GoI (1992). *The Eighth Five Year Plan*, vol. 2. New Delhi: Planning Commission.

107 MoEF (1990). *Involvement of Village Communities in Regeneration of Degraded Forests*. New Delhi: Ministry of Environment and Forests.

108 SPWD (1992). *Joint Forest Management: Regulations Update*. New Delhi: Society for the Promotion of Wastelands Development.

109 Bhattacharya, P., Pradhan, L., and Yadav, G. (2010). Joint forest management in India: experiences of two decades. *Resources Conservation and Recycling* 54: 469–480.

110 Colfer, C.J.P., Dahal, G.R., and Capistrano, D. (2008). *Lessons from Forest Decentralization. Money, Justice and the Quest for Good Governance in Asia Pacific*, 2–3. London: Earthscan.

111 Potters, J., Reeb, D., and Crollius, M.R. (2002). The progress of participatory forestry in Africa. Second International Workshop of Participatory Forestry in Africa. Defining the Way Forward: Sustainable Livelihoods and Sustainable Forest Management Through Participatory Forestry, Arusha, United Republic of Tanzania, 18–22 February.

112 Mishra, A. and Shete, G. (1995). *Joint Forest Management in North-West Orissa a Status Report on ISO/SWED, Consultancy in Forestry*. Bhubaneshwar, Bhopal: IIFM.

113 Bhattacharya, P. and Mittra, B. (2005). Lessons from self-initiated forest protection systems-case studies from India. In: *Contemporary Studies in Natural Management in India* (ed. S.B. Roy), 168–198. New Delhi: Inter India Publication.

114 Ballabh, V., Balooni, K., and Dave, S. (2002). Why local resource management institution decline: a comparative analysis of Van (forest) Panchayats and forest protection committees in India. *World Development* 30 (12): 2153–2167.

115 Rey, F. and Burylo, M. (2014). Can bioengineering structures made of willow cuttings trap sediment in eroded marly gullies in a Mediterranean mountainous climate? *Geomorphology* 204: 564–572.

116 Chazdon, R.L. (2008). Beyond deforestation: restoring forests and ecosystem services on degraded lands. *Science* 320: 1458–1460.

117 Stanturf, J.A., Schoenholtz, S.H., Schweitzer, C.J., and Shepard, J.P. (2001). Achieving restoration success: myths in bottomland hardwood forests. *Restoration Ecology* 9: 189–200.

118 Costanza, R. (2012). Ecosystem health and ecological engineering. *Ecological Engineering* 45: 24–29.

119 Kuzovkina, Y.A. and Volk, T.A. (2009). The characterization of willow (Salix L.) varieties for use in ecological engineering applications: co-ordination of structure, function and autecology. *Ecological Engineering* 35 (8): 1178–1189.

120 Stokes, A., Douglas, G.B., Fourcaud, T. et al. (2014). Ecological mitigation of hill slope instability: ten key issues facing researchers and practitioners. *Plant and Soil* 377: 1–23.

121 Tardío, G. and Mickovski, S.B. (2016). Implementation of eco-engineering design into existing slope stability design practices. *Ecological Engineering* 92: 138–147.

122 He, S., Wang, D., Fang, Y., and Lan, H. (2017). Guidelines for integrating ecological and biological engineering technologies for control of severe erosion in mountainous areas – a case study of the Xiaojiang River Basin, China. *International Soil and Water Conservation Research* 5 (4): 335–344.

123 Norris, J.E., Stokes, A., Mickovski, S.B. et al. (eds.) (2008). *Slope Stability and Erosion Control: Ecotechnological Solutions*. Dordrecht, The Netherlands: Springer.

124 Rey, F., Bifulco, C., Bischetti, G.B. et al. (2019). Soil and water bioengineering: practice and research needs for reconciling natural hazard control and ecological restoration. *Science of the Total Environment* 648: 1210–1218.

125 IUCN (2016). *Nature-Based Solutions to Address Global Societal Challenges* (eds. E. Cohen-Shacham, G. Walters, C. Janzen and S. Maginnis). Gland, Switzerland, pp. xiii + 97: IUCN, Gland, Switzerland. http://portals.iucn.org/library/sites/library/files/documents/2016-036.pdf.

126 Emmanuelle, C.S., Andradea, A., Daltond, J. et al. (2019). Core principles for successfully implementing and upscaling nature-based solutions. *Environmental Science and Policy* 98: 20–29.

127 Keesstra, S., Nunes, J., Novara, A. et al. (2018). The superior effect of nature based solutions in land management for enhancing ecosystem services. *Science of the Total Environment* 610–611: 997–1009.

128 Chazdon, R.L. and Guariguata, M.R. (2016). Natural regeneration as a tool for large-scale forest restoration in the tropics: prospects and challenges. *Biotropica* 48 (6): 716–730.

129 Lamb, D., Stanturf, J., and Madsen, P. (2012). What is Forest landscape restoration? In: *Forest Landscape Restoration* (eds. J. Stanturf, D. Lamb and P. Madsen). New York, London: Springer Dordrecht Heidelberg.

130 SER (Society for Ecological Restoration International Science & Policy Working Group) (2004). SER international primer on ecological restoration. Society for Ecological Restoration. http://www.ser.org/resources/resources-detail-view/ser-international-primer-on ecological-restoration.

131 IUCN and WRI. (2014). A guide to the restoration opportunities assessment methodology (ROAM): Assessing forest landscape restoration opportunities at the national or subnational level. Working paper (road-test edition).

132 Cowie, A.L., Orr, B.J., Sanchez, V.M.C. et al. (2018). Land in balance: the scientific conceptual framework for land degradation neutrality. *Environmental Science and Policy* 79: 25–35.

133 Nassauer, J.I. and Opdam, P. (2008). Design in science: extending the landscape ecology paradigm. *Landscape Ecology* 23: 633–644.

134 Crow, T.R. (2008). A framework for restoration in the national forests. In: Proceedings of the 2007 national silviculture workshop on integrated restoration of forested ecosystems to achieve multi resource benefits, Deal, R.L. (ed.), USDA Forest Service, General Technical Report. PNW-GTR-733, Portland, OR, pp. 13–21.

135 Nair, P.K.R. (1993). *An Introduction to Agroforestry*. Springer Science & Business Media.

136 Saikia, P. and Khan, M.L. (2012). Agar (*Aquilaria malaccensis* Lam.): a promising crop in the homegardens of Upper Assam, northeastern India. *Journal of Tropical Agriculture* 50 (1–2): 8–14.

137 Shono, K., Cadaweng, E.A., and Durst, P.B. (2007). Application of assisted natural regeneration to restore degraded tropical forestlands. *Restoration Ecology* 15 (4): 620–626.

138 Chazdon, R.L. and Arroyo-Mora, J.P. (2013). Tropical forests as complex adaptive systems. In: *Managing World Forests as Complex Adaptive Systems in the Face of Global Change* (eds. C. Messier, K.J. Puettmann and K.D. Coates), 35–59. New York: Routledge.

139 IUCN. (2018). Bonn challenge and India: Progress on restoration efforts across states and landscapes. https://landportal.org/fr/node/83912.

140 Patel, S. (2019). From barren land to rich ecosystem: Story behind Yamuna Biodiversity Park. *The Indian Express*. https://indianexpress.com/article/cities/delhi/from-barren-land-to-rich-ecosystem-story-behind-yamuna-biodiversity-park-5743228.

141 Jain, P. and Srivastava, D.P. (2016). The wetlands of Yamuna Biodiversity Park. Sahapedia. https://www.sahapedia.org/the-wetlands-of-yamuna-biodiversity-park.

142 Thomas, E., Jalonen, R., Loo, J. et al. (2014). Genetic considerations in ecosystem restoration using native tree species. *Forest Ecology and Management* 333: 66–75.

143 Wang, J. and Meng, J. (2018). Identifying indigenous tree species for land reforestation, forest restoration, and plantation transformation on Hainan Island, China. *Journal of Mountain Science* 15 (11): 2433–2444.

144 Khurana, E. and Tiwari, N. (2016). Yamuna Biodiversity Park: an overview. Sahapedia. https://www.sahapedia.org/yamuna-biodiversity-park-overview.

145 Harjani, A. (2013). Biodiversity Park revives Yamuna's pristine glory. *The Pioneer*. https://www.dailypioneer.com/2013/delhi/biodiversity-park-revives-yamunas-pristine-glory.html.

146 Sharma, T. (2014).Wildlife conservation projects in Corbett – saving the wilderness. Corbett National Park. https://www.corbettnationalparkindia.in/blog/wildlife-conservation-projects-in-corbett.

147 Babu, S., Love, A., and Babu, C.R. (2009). Ecological restoration of *Lantana* invaded landscapes in Corbett Tiger Reserve, India. *Ecological Restoration* 27 (4): 467–477.

19

Restoring Ecosystem Services of Degraded Forests in a Changing Climate

Smita Chaudhry[1], Gagan Preet Singh Sidhu[2], and Rashmi Paliwal[1]

[1] *Institute of Environmental Studies, Kurukshetra University, Kurukshetra, Haryana, India*
[2] *Centre for Applied Biology in Environment Sciences, Kurukshetra University, Kurukshetra, Haryana, India*

19.1 Introduction

Forests play a vital role in providing ecosystem services and help sustain humanity's well-being [1]. They are the most diverse ecosystems on earth and a storehouse of living carbon [2]. Forests are gaining attention due to their social, cultural, ecological, and economic advantages [3]. Nearly 30% of the world's land area is under forest cover that nurtures rich terrestrial flora and fauna [4]. According to the Food and Agricultural Organization (FAO)[3], approximately one-third of the earth's surface is covered by forests. Estimates of the "State of Forest Report 2017" published by the Forest Survey of India are that the country's total forest cover is 70.8 million hectares, 21.54% of the country's geographic area. Tree cover outside recorded forests is 9.38 million ha [5]. According to Kishwan et al. [6], approximately 173,000 villages in India are categorized as forest villages, indicating the dependence of communities and populations on forests.

Forests are a significant component of terrestrial ecosystems and provide food, fuel, fodder, fiber, medicines, and pulpwood, in addition to supporting commercial activities that support the livelihoods and employment of millions of people [7]. They also help sustain the ecological balance that is crucial for maintaining the overall growth of humankind. Critical functions of forests include primary ecosystem services such as climate regulation, water supply, air quality, soil enrichment, biomass production, and support for various natural resources [8, 9]. Direct and indirect benefits of ecosystem services provided by forests play a major role in determining their function in supporting rural livelihood [2].

However, in recent years, forests worldwide have shown significant deterioration [10] and transformation [11] due to society's enhanced demands for their goods and services. Human activities such as industrialization and urbanization, influence by different demographic and technological factors, have resulted in the degradation of forests [12]. Natural disturbances such as wildfires, high winds and storms, overgrazing, the spread of invasive alien species, and insect outbreaks can abruptly alter the dynamics and succession patterns in forests [13, 14]. Anthropogenic activities lead to forest loss, fragmentation, and degradation, which cause biodiversity loss and homogenization [15]. Increasing urbanization, forest fires, and conversion of

Handbook of Ecological and Ecosystem Engineering, First Edition. Edited by Majeti Narasimha Vara Prasad.

agricultural land have resulted in nearly 3.2% loss of forest cover globally [16]. Deforestation and anthropogenic activities pose serious implications for ecosystems, climate, and survival of species and can damage earth's ecosystem functions [17]. Disturbances in forests disrupt the ecosystem's structure and function, thus leading to reorganization of the ecosystem. Greater levels of disturbance strongly alter or permanently damage the composition and productivity of forests [18]. Global environmental change has led to variation in forest disturbance patterns [19], which can further affect forest ecosystem services. Forest ecosystems require a long time to adapt to disturbances, especially migration and regrowth [20]. Further, a long gestation period is required to develop and implement adaptation strategies in the forest sector [21].

Climatic conditions are major catalysts of vegetation regimes that appear globally and significantly impact the structure and distribution of forests [22]. Changing climate poses a serious threat to forests that could cause irreversible damage to unique forest ecosystems and biodiversity, rendering species extinct locally and globally [23]. Climate is an important factor that shapes the forest environment and thus has a significant impact on forest ecosystems [24]. Forests play a major role in governing the carbon cycle and climate change since they are reservoirs for a huge amount of carbon in soil and growing vegetation [25]. Abiotic and biotic disturbances due to natural processes exchange a large amount of carbon dioxide (CO_2) and non-CO_2 gases with the atmosphere. Forests function as carbon sinks and help in removing CO_2 from the atmosphere [26]. Climate change not only disturbs the forests but also threatens millions of people who depend upon forests for environmental services essential for their livelihood [27]. Changing climatic conditions result in alterations in temperature and precipitation, leading to flooding, drought, or frost [28]. The mega-fires reported in recent years are also linked to severe climatic extremes [29].

Several studies have shown that future climate change is likely to have significant impacts on forest ecosystems. In India, under the climate projection for the year 2085, 77% and 68% of the forested grids are likely to experience a shift in forest types under scenarios A2 and B2, respectively [30].

This chapter outlines the different types of forests in India and their role in maintaining an ecological balance and providing ecosystem services. The factors responsible for the degradation of forests and their impact on ecological services, nutritional status of soils, and carbon sequestration are discussed in detail. Moreover, various strategies for restoring forest ecosystems in changing climatic conditions are also briefly explored in the chapter.

19.2 Role of Forests in Maintaining Ecological Balance and Providing Services

Forests are an important driving force in ecosystem functions [31]. They regulate critical ecological functions such as the carbon cycle, the nutrient cycle, and soil retention [25]. Forests provide environmental services, e.g. remove pollution, improve water and soil quality, provide a natural habitat for organisms, conserve biodiversity, support pollination, regulate climate, and provide carbon capture and sequestration [32]. Due to the vast biodiversity and distribution of forests, they play a major role in providing ecosystem services [33].

The direct and indirect benefits that people obtain from forest ecosystems are called *ecosystem services*, which are human-centered processes [34]. They can be defined as the benefits obtained by people from nature (e.g. food) [35]. Ecosystem services are classified as provisioning (clean water), regulating (flood and drought mitigation), cultural (recreational, swimming), and supporting (biodiversity, hydrology) by the Millennium Ecosystem Assessment (MEA) [36]. Some of the essential ecosystem services provided by forests are briefly discussed below.

19.2.1 Forests and Rainfall

Forests regulate the changing patterns of atmospheric moisture and rainfall over the earth's surface. Water is released into the atmosphere as vapor from land and water surfaces by the process of evaporation and from plants by transpiration. Approximately 40% of rainfall globally occurs through the evapo-transpiration (ET) process, which involves evaporation and transpiration [17]. Transpiration causes a large portion of terrestrial ET, thereby generating vapor responsible for rainfall [37]. Spracklen et al. [38] observed that in the tropics, air passing over forests for 10 days produces twice as much rainfall as air that passes over areas with sparse vegetation.

Through transpiration and condensation, forests form low-pressure areas that attract moist air from oceans that is carried by the wind and leads to rainfall [39]. Further, increased relative humidity enhances the probability of precipitation in an area. Forests intensify rainfall by releasing biological particles such as pollen, fungal spores, and bacterial cells into the atmosphere [17]. They regulate the global water cycle and climate regulation through cloud formation, thereby cooling the climate. Natural forests support high evaporation fluxes that maintain increased air motion over forests [40].

Forest degradation and loss can decrease ET, which harms rainfall and temperature patterns in the affected area. Further, fluctuations in rainfall may alter the moisture and heat fluxes between the land and the atmosphere. Deforestation can result in a 30% decline in rainfall as predicted by various climate models [41], leading to drought and fires in some regions [42]. Moreover, aerosol particles released during forest fires lead to the scattering of solar radiation, thereby disturbing the upward movement and circulation of water vapor and altering rainfall [43].

19.2.2 Forests and Carbon Sequestration

Elevated CO_2 concentrations caused by anthropogenic activities such as industrialization, deforestation, and urbanization have caused serious environmental concerns globally [44]. The concentration of CO_2 has increased by 43% since the Industrial Revolution, and this value is expected to reach 60% by the year 2100 [45]. Through the process of carbon sequestration, CO_2 is removed from the atmosphere and stored in various biological and physical forms. It is a nature-based solution [46] wherein forests sequester carbon in tree biomass. Forests are an essential carbon sink and play an important role in mitigating global climate change [47]. Carbon is stored within the cell walls of the plant cell made up of cellulose or/and lignin. Plants perform photosynthesis that requires CO_2 to form glucose and O_2 [48]. Thus, increased production of glucose enhances the sequestration of carbon within the plant's tissues.

Carbon storage is an essential function of forests since decreased atmospheric carbon reduces the magnitude of future climate change. Forests are considered as the most important sink for atmospheric CO_2. Worldwide, approximately 30% of annual anthropogenic CO_2 emissions are absorbed by forests [49]. The world's forests absorbed 2.4 ± 0.4 petagrams of carbon per year (Pg C Yr^{-1}) (1 Pg C = 10^{15} g C) from 1990 to 2007 [49]. Saatchi et al. [50] reported the total biomass carbon stock of tropical forests (Latin America, sub-Saharan Africa, and Southeast Asia) to be 247 Gt C, with 193 Gt C and 54 Gt C stored in aboveground and belowground biomass, respectively. Latin America's forests hold a maximum of 49% of the total carbon biomass in the tropics [50]. Pan et al. [49] reported the global carbon stock in the world's forests to be 42% in live biomass (above- and belowground), 8% in deadwood, 5% in litter, and 44% in soil.

The forestry sector, including tree plantations, plays a dual role in sequestering carbon by sustaining its own carbon and absorbing carbon from the atmosphere. The significant aspects that affect the potential of tree plantations in carbon sequestration are soil organic and inorganic carbon pools, rates of CO_2 evolution from the soil surface, carbon sequestration in tree biomass, and carbon fluxes through net primary productivity [51]. The carbon sequestration potential of different tree plantations has been reported in India's social forestry sector [52].

However, emissions from industry, agriculture, and burning fossil fuels have increased the level of CO_2 in the atmosphere at an alarming rate [53]. Carbon storage and sequestration are affected by tree size, density, and health and forest growth rate [54]. Thus, maintaining biodiversity in forests helps regulate climate fluctuations that ultimately affect the environment. Forests act as effective carbon sinks and as an adaptation option due to their role in moderating the impacts of climate extremes. Climate change, along with socio-economic and land-use pressures, is likely to adversely impact forest biodiversity, carbon sinks, biomass productivity, and hence the livelihoods of forest-dependent communities [55].

19.2.3 Forests and Climate

The distribution, structure, and composition of forests are influenced by climate and indirectly through their impact on disturbances due to fire, insects, diseases, and windstorms. Forests also impact the global climate through their influence on surface roughness, albedo, hydrological cycles, and the carbon cycle [56]. Forests monitor the global temperature and heat flow of an area. They provide a cooling effect in summer and warming in winter, thereby regulating climate, which is one aspect of forest ecosystem services [57]. Evaporation and transpiration processes in forests help reduce heat and maintain a cooler environment during the day at the local level [58, 59]. Trees transpire nearly 100 l of water daily and maintain their cooling function during heat waves [17]. Thus, urban areas with dense tree growth and vegetation cover exhibit lower temperatures than areas with less vegetation [60]. Forests significantly affect air quality by directly removing air pollutants from the atmosphere. Aerosols and particulate matter, especially sulfur and nitrogen, have a cooling effect in the atmosphere [23]. Climate change alters the effect of air pollutants on forest ecosystems and tree performance [34].

Urban forests (assemblages of urban trees, shrubs, lawns) mitigate pollution and maintain environmental quality by combining ecosystem services and urban forest management [61]. They take up gaseous pollutants via stomatal apertures present on the leaf [62]. Gaseous air pollutants diffuse into the leaf's intercellular spaces, are absorbed by water films, and form acids. Forests reduce wind speed, which leads to air dispersion, and hence limit pollution [63]. However, increased urbanization and industrialization have led to the removal of trees, which hurts environmental quality [61]. The United Nations Framework Convention on Climate Change (UNFCCC) has framed two major goals for forests: (i) using forests for capturing and storing carbon to help reduce greenhouse gases (GHGs) and play a vital role in mitigation strategies; and (ii) using forests to cope with the impact of climate change and as a vital adaptation strategy for climate change [64].

19.2.4 Forests and Soil Erosion

Erosion is a physical process of removing topsoil that reduces soil quality and decreases the productivity of the ecosystem [65]. Many studies have reported that plant diversity augments soil stability and reduces soil erosion in different climatic conditions [66–68]. Preventing soil erosion and protecting forest land is an essential ecological function of forests [69]. External factors such as water and wind can erode the soil in areas with less green cover. The formation of different strata of trees in the forest prevents high-intensity rainfall from reaching the soil and causing soil erosion [70].

Trees' deep, long roots hold together the soil layers, preventing them from being carried away [70] and enhancing the strength of the soil [71]. Therefore, forests can improve the soil's water-holding capacity, prevent soil erosion, mitigate floods, and increase the level of the water table. The diversity of trees affects soil erosion in two ways: by improving the distribution of vegetation that could sustain the interception of rainfall [72] and enhancing soil properties that lead to improved soil water-holding capacity and infiltration [67, 68]. Monoculture plantations cause increased ecological insecurity due to exposure to severe environmental conditions, thereby harming soil properties [67, 73]. Thus, preventing soil erosion is a vital forest ecosystem service that prevents the degradation of soil quality and ecosystem productivity.

19.2.5 Forest and Water Quality

Water quality is a serious concern worldwide due to increased anthropogenic pollution and urban development [74]. Clean water is one of the significant ecological services provided by forests and is essential for global ecosystems [75]. Forested areas are a source of drinking water to one-third of the world's largest cities [75]. Forests regulate the hydrological cycle and strongly influence the availability of water in soil [76]. They play a major role in regulating floods and drought and thus preventing desertification and soil salinization [75].

Forest ecosystems act as buffering agents and protect water quality by reducing soil erosion, sedimentation, and pollution. They regulate flow patterns, provide water for irrigation and hydropower, control runoff, modulate water temperatures, and maintain groundwater [77]. The interaction between forest and water cycle is very complex and must include both blue water (liquid form) and green water (vapor form) [40]. Therefore, the gradual reduction of forest area negatively affects ecosystem services, leading to compaction or hardening of soils, soil erosion, loss of transpiration, and enhanced runoff, thus increasing the risk of floods [17].

19.3 Types of Forests in India

Based on satellite remote sensing and geographic information system (GIS), Reddy et al. [78] have prepared an integrated and schematic classification of 16 forest types with vast distribution throughout various parts of India. The mean annual rainfall in these forests ranges from <100 to 4000 mm annually with varied floral diversity. Table 19.1 describes the forest types in India.

19.4 Forest Degradation

Two terms are most frequently used to describe the depletion of forests: *forest degradation* and *deforestation*. Forest degradation is the impairment of a forest's quality, whereas deforestation reduces forest area due to the removal of trees in large numbers. It includes the thinning of the canopy and carbon loss due to direct and indirect drivers of degradation. It hampers ecological functioning and services to humanity.

Degradation of forests occurs due to various natural and human activities under the influence of many factors such as demography, technology, environment, and politics [79]. According to Hosonuma et al. [80], forest degradation is related to decreased forest production over time. The main causes of forest degradation include overgrazing, forest fires, uncontrolled exploitation of forest products, the spread of invasive alien species, and pest attacks. Being extremely dynamic, forests naturally undergo slight changes in their structure and composition that help maintain balance in the ecosystem.

Table 19.1 Different forest types in India.

	Forest type	Geographical distribution	Mean annual rainfall (mm)	Elevation zone (m)	Floral diversity
1)	Tropical wet evergreen forest	Eastern Himalayas, northeast and western Ghats	3000–4000	<100 to 2000–3000	*Michelia montana*, *Dipterocarpus* spp., *Ailanthus integrifolia*, *Artocarpus chaplasha*, *Vateria indica*
2)	Tropical dry evergreen forest	Deccan and coastal areas	1000–2000	<100 to 100–500	*Atalantia monophylla*, *Albizia amara*, *Prosopis juliflora*, *Mundulea sericea*
3)	Subtropical broad-leaved hill forest	Lower Himalayan slopes, Aravallis, northeast and western Ghats	2000–3000	100–2000	*Quercus*, *Mangifera indica*, *Rhododendronarboretum*, *Litsea* spp.
4)	Subtropical dry evergreen forest	Himalayan foothills and Shivalik hills	500–1000	Up to 1500	*Acacia modesta*, *Canthium dicoccum*, *Holoptelea integrifolia*
5)	Montane wet temperate forest	Eastern Himalayas, northeast region, and hills in the western Ghats	2000–3000	2000–3000	*Magnolia campbelli*, *Michelia champaca*, *Rhododendron* spp., *Acer campbelli*
6)	Tropical semi-evergreen forest	Western and eastern Himalayas, Gangetic plains, western Ghats, Deccan region and islands	3000–4000	<100–2000	*Shorea robusta*, *Terminalia myriocarpa*, *Terminalia bellirica*, *Artocarpus hirsutus*, *Syzygium nervosum*, *Dipterocarpus gracilis*
7)	Tropical moist deciduous forest	Western and eastern Himalayas, Gangetic plains, semi-arid region, northeast, Deccan coastal area, western Ghats and islands	1000–2000	<100–2000	*Shorea robusta*, *Terminalia alata*, *Dalbergia latifolia*, *Lagerstroemia parviflora*, *Tectona grandis*
8)	Tropical dry deciduous forest	Western Himalayas, Gangetic plains, semi-arid region, northeast, Deccan region and western Ghats	1000–2000	<100–1500	*Anogeissus latifolia*, *Tectona grandis*, *Shorea robusta*, *Terminalia alata* and *Hardwickia binata*
9)	Tropical thorn forest	Plains, semi-arid region, desert area and Deccan region	<100–2000	<100–1000	*Acacia* spp., besides *Butea monosperma*, *Bauhinia racemosa*, *Prosopis cineraria* and *Capparis decidua*

10)	Subtropical pine forest	Western and eastern Himalayas and northeast region	500 to >4000	500–4200	*Pinus roxburghii*, *Cedrus deodara*, *Abies densa*, *Picea spinulosa* and *Quercus serrata*
11)	Himalayan moist temperate forest	Trans Himalayas, eastern and western Himalayas	1000–2000	1500–4200	*Abies pindrow*, *Quercus incana*, *Rhododendron arboretum*, *Betula utilis*
12)	Himalayan dry temperate forest	Eastern and western Himalayas	<500–4000	1000–4200	*Cedrus deodara*, *Pinus gerardiana*, *Pinus wallichiana*, *Juniperus macropoda*, *Rhododendron arboretum*
13)	Sub-alpine forest	Trans-eastern and western Himalayan	<500–4000	2000–4200	*Betula utilis*, *Pinus wallichaiana*, *Rhododendron arboretum*, *Juniperus wallichiana* and *Quercus semecarpifolia*
14)	Littoral and swamp forest	Coastal	<500–4000	<100	*Rhizophora mucronata*, *Avicennia marina*, *Excoecaria agallocha*, *Bruguiera gymnorrhiza*
15)	Moist alpine scrub	Himalayas and higher hills along the Myanmar border			*Rhododendron*, *Betula*, *Quercus* and *Berberis*
16)	Dry alpine scrub	Himalayan region	800	3000–5200	*Juniperus* sp., *Hippophae* and herbs such as *Sedum* sp., *Primulaminutissima*, *Potentilla fruticosa*

Source: Reddy et al. [78].

Forest degradation generally refers to the depletion of the forest's capacity to provide goods, articles, ecosystems, and socio cultural services. Forests provide natural habitat to various species and timber and non-timber products for human use and play an important ecological role in cycling nutrients and gases in the environment, sequestering carbon from the atmosphere, binding the soil, and much more. Changes or disturbances in environmental conditions and processes damage forest productivity. These changes are caused by natural disturbances as well as human activities. Natural disturbances include natural fires, drought, pests, and diseases.

Human-induced changes are endless, such as intentional fires, overexploitation of forest resources for timber and non-timber forest products, pollution, alien species invasion, conversion of forest area for urban development and agricultural practices, etc. Forests as an ecosystem have always tended to evolve; therefore, they remain in a continuous state of successional and natural development. Forest degradation is a significant threat to the natural world and the survival of species.

Forest degradation often leads to full-scale removal of trees or deforestation, as it is easy to convert degraded forest land for other land uses. Forest degradation and deforestation both affect the survival of the world's population, whose livelihoods are forest-dependent. According to the FAO, *deforestation* is the conversion of forest to another land use or the long-term reduction of tree canopy cover below the 10% threshold. According to an analysis of "The State of the World's Forest" by the FAO, global forest areas were reduced from 31.6% of total land to 30.6% between 1990 and 2015 [81, 82]. According to Chaturvedi et al.[83], in northeastern India, more than 80% of land area classified as forests is under severe pressure of deforestation. This region has the highest rate of deforestation in India (65% of total deforestation in India over the period of 2005–2007), mainly due to encroachment and shifting cultivation [84].

A continuous decline in forest cover poses a severe threat to the environment and the entire human race. Some of the factors responsible for forest degradation are discussed below.

19.4.1 Invasive Alien Species

Among the various drivers of forest degradation, invasive alien species cause huge harm to the forest ecosystem and are a significant threat to biodiversity worldwide [85]. Invasive alien species harm the native biota and can cause the extinction of native plant species [86]. They cause a reduction in ecosystem services [87] of the forest, thus leading to substantial economic losses [88]. Invasive alien species compete with native species for sunlight, soil, nutrients, and water. These alien plant species negatively affect the resident communities and ecosystem functions such as water loss, nutrient enrichment, and changed fire patterns [86]. Massive plant invasions lead to homogenization of floras wherein native species are suppressed, leading to the dominance of invasive alien species [89]. These species invade many forests and change their growth patterns.

For example, Campagnaro et al. [90] observed the effect of five major invasive alien tree species (*Acer negundo*, *Ailanthus altissima*, *Prunus serotina*, *Quercus rubra*, and *Robinia pseudoacacia*) in European Union forests. They reported that these tree species pose detrimental effects on the forest habitat structure and functions. According to Qiu [91], invasive plants, animals, and pests caused substantial economic losses in China that amount to $15 billion annually. According to Straube et al. [92], invasive species such as ants and earthworms resulted in severe harm to large areas by replacing endemic species due to competition. Invasive alien pests have been a major threat to the country's agroforestry, economic stability, and ecosystem services [93].

19.4.2 Forest Fires

Forest fires caused by increased human activities are a significant cause of forest degradation. Injudicious anthropogenic input leads to global warming that alters the climatic patterns and causes forest fires [94]. Uncontrolled fires can cause huge losses to ecosystem services. Forest fires cause significant ecological and economic problems in many countries, with long-term consequences to the natural environment [95].

The spreading pattern of forest fires depends on the type of forest. For example, a deciduous forest has more humidity and does not catch fire easily; however, a coniferous forest with high resin content is more prone to forest fires [95]. Excessive forest fires can cause the loss of natural habitats of plant communities and damage the forest ecosystem's trophic structures. Anthropogenic activities such as increasing tourism, smoking in forests, and burning agricultural waste are significant reasons for forest fires. Forest fires lead to soil erosion due to the destruction of vegetation, which damages the soil's upper fertile layer [96]. Moreover, a large amount of CO_2 is liberated in the atmosphere due to wood fires that increase greenhouse gas concentrations, leading to global warming [97]. Therefore, knowledge of forest structure, causes of fires, and their impact on the environment can help with monitoring forest fires.

19.4.3 Overpopulation and Exploitation of Forest Resources

The rapid growth of the human population and industrialization are the main reasons for the overexploitation of forest resources worldwide [98]. Humans use forestry resources to meet their requirements for food, fodder, medicines, etc. The harvesting of wood for fuel in developing countries is another reason for the degradation of forests. Many people from rural areas depend on forested areas and their ecosystem services, thus damaging the environment [99]. Both authorities and environmentalists have neglected the harmful effect of fuelwood consumption on biodiversity since they consider disturbances caused by deforestation due to shifting land use to be more harmful [100]. Barrett et al. [101] reported that most biodiversity hotspots are located in developing countries, where most of the population depends on forest biomass for cooking; this can cause significant loss of biodiversity and forest degradation.

Many authors have used mathematical models and reported the effect of increased globalization and industrialization on forest resources [102–104]. For example, Singh et al. [104] opined that the collection of firewood, leaf fodder, and leaf litter from the forest floor is the leading cause of degradation in the oak and pine elevation belt of the western and central Himalayas. They observed a 40–50% reduction in soil carbon and nitrogen, a 25% reduction in soil water holding capacity, and forest biomass degradation that disintegrates ecosystem structure and function [104]. Therefore, many more studies are required to increase food production and forest resources and, simultaneously, protect biodiversity.

19.4.4 Overgrazing

During the last century, climate change and the unprecedented increase in mean global temperatures have affected forested regions. Climate change causes prolonged windstorms, drought, and flooding in forests, leading to forest degradation [105]. Moreover, the combined effect of climate change and human activities such as forest fires, deforestation, and overexploitation of forest products affects the structure and services of forest ecosystems [106]. Overgrazing is a primary driver of forest degradation worldwide since it leads to water runoff, sediment loss, soil moisture, precipitation, and plant loss [107]. For example, Li et al. [108] reported desertification of grassland in Inner Mongolia (China) due to increased herd sizes that enhanced pressures on the remaining pastures in the area. Hilker et al. [109] have reported widespread decline of vegetation across the Steppes and~70% degradation of the ecosystem due to overgrazing. Further, this can lead to a loss in plant communities' structure and composition due to the consumption of plants by livestock. Overgrazing is known to decrease soil organic matter, which can harm soil nutrients [110]. Therefore, enhanced grazing can lead to a huge loss of biodiversity and further augment climate change.

19.5 Impacts of Forest Degradation

The impacts of degradation are critical for the various ecosystem functions and ecological services provided by forests.

19.5.1 Carbon Sequestration

Forest destruction releases absorbed carbon dioxide (CO_2) into the atmosphere during biomass degradation. Forest degradation and deforestation were responsible for 12–20% of global anthropogenic GHG emissions in the 1990s and 2000s. Deforestation and degradation of forests further affect the potential for forests to remove global CO_2 from the atmosphere [50]. Reduction or unavailability of CO_2 sinks leads to the continuous increase in CO_2 concentration in the atmosphere. The increase in atmospheric CO_2 concentration leads to a change in global climatic conditions by trapping more and more heat in the earth's atmosphere. Forest degradation also affects hydrological and meteorological processes, leading to changes in local and global climates. Research suggests that the change in global climatic conditions is highly correlated with forest degradation. It has been reported that more than 2 billion forests have been degraded globally [79, 111, 112]; therefore, there is an urgent need to develop mitigation strategies for the conservation and restoration of forest resources.

19.6 Nutritional Status of Soil

Forest cover keeps the soil intact, thereby reducing erosion and loss of soil nutrients. Deforestation reduces the compactness of soil, thereby degrading the soil structure. Removing vegetation cover reduces the soil's organic carbon and other soil nutrients, which causes further loss of soil productivity. Degradation of forests removes vegetation cover from forest land, exposing the soil to extreme solar heat and heavy rainfall. Extreme solar heat reduces the biological activities of soil, and heavy rainfall erodes the soil layers. Increased soil erosion leads to more siltation in rivers and, hence, flooding events in major rivers and low-lying areas. Forest cover is necessary to maintain the soil's nutritional status and quality of water resources.

19.7 Hydrological Regimes

Forest cover is vital for water availability, maintaining the global water cycle, conserving water flow, reducing floods, and preventing soil erosion. Forest cover maintains the hydrological cycle through evapo-transpiration. Deforestation increases flooding events by reducing the water-holding capacity of the soil, which promotes excess runoff, streamflow, and soil erosion. Forests are often considered sponges, capable of absorbing water during rainfall and releasing it slowly in the dry season [113].

19.8 Ecological Services

Degradation of forests adversely affects ecosystem integrity by impairing the provisioning, regulating, supporting, and cultural services of ecosystems. Forest degradation and deforestation reduce ecological services such as the production of goods and food, nutrient cycling, water regulation, carbon sequestration, biodiversity, and much more. Habitat destruction due to forest removal leads to a reduction in species richness, changes in species distribution, and even extinction. Potapov et al. [114] reported that only 23.5% of the world's old and original forest cover, i.e. an unbroken extension of the natural ecosystem within the present forest area, remains. According to the FAO [8], forest land covering 30% (~4 billion hectares) of the earth's total land area is decreasing at a rate of 13 million hectares per year (Mha/yr) due to changes in land use. Deforestation and degradation make the conditions worse as these destructive activities adversely affect biodiversity. This further affects the four broadly classified ecosystem services:

- *Provisioning services of a forest ecosystem*: Food and goods, genetic resources, raw materials, etc.
- *Regulatory services*: Carbon sequestration, climate regulation, waste decomposition and detoxification; water and air purification, pollination, pest and disease control; and flood protection
- *Supporting services*: Nutrient cycling, soil formation, and primary productivity
- *Cultural services*: Spiritual and recreation benefits

19.9 Social Implications

Forest degradation negatively impacts many species' survival and maintenance and hinders local forests' ability to provide essential services to humanity and nature. The degradation of forests affects every sphere of human life: sociocultural, economic, political, and even psychological. Humans depend on forests for various reasons, and their degradation may result in the loss of essential services, thus negatively impacting sustainable livelihoods. A blow to socioeconomic benefits is observed, more so in the poorest areas, as billions of people directly depend on forests for their livelihood. Various tribal and forest-dwelling communities are the direct beneficiaries and are primarily dependent on forests for their lives. Forest degradation may impact their cultural traditions and their way of living. Cultural harmony may be disturbed, inducing social and health disturbances.

Agricultural and forest produce is affected because of the loss of soil fertility and retarded secondary succession. Societies are disturbed, particularly on the periphery of forest regions. And forest land conversion for human purposes may impact vulnerable communities by increasing man-animal conflicts on the forest edge.

19.10 Methods for Restoring and Rehabilitating Forests

Restoration of the forest ecosystem is a long-term process that involves strategy development, implementation, and monitoring of the restored site. Restoration of an ecosystem aims to return the degraded ecosystem's original state and repair its functioning. Similarly, forest restoration activities involve the reestablishment of pre-existing conditions of a forest, such as its structure, species diversity, and productivity. Some restoration exercises include controlling fires, grazing, and erosion; weed control; planting native species; and introducing trees in single- and mixed-species plantations.

Forest restoration programs include strategies such as rehabilitation, reclamation, replacement, and reconstruction [79, 115, 116]. Several works have studied different strategies to restore degraded forest landscapes with specific objectives and methods (Table 19.2).

Some policies and strategies suggested by Ravindranath et al. (2006) for proper forest management include incorporating climate concerns in long-term forest policy-making processes; conserving forests and reducing forest fragmentation; expanding protected areas and linking them to promote migration; promoting mixed species forestry to reduce vulnerability; undertaking anticipatory planting and assisting natural migration through transplanting plant species; promoting in situ and ex situ gene pool conservation; and initiating forest fire management strategies.

Forest landscape restoration (FLR) is an emerging concept within forest restoration programs. It is a natural method that significantly reduces forest degradation impacts and effectively restores the ecological balance. It is a large-scale approach involving the restoration of multiple functions of forests for local people. Hence, FLR differs from small-scale and forest stand-management approaches by regaining of multiple forest functions and enhancing well-being through a landscape-approach [135, 136].

The FLR process is an ongoing approach incorporating multiple action areas for restoring landscapes. The FLR landscapes are broadly classified into three land-use situations: forest land, agricultural land, and protective lands and buffers (Figure 19.1). These three land-use situations include seven general categories of FLR options:

1) Forest land: Restoration through
   a) Planting (*Category 1*)
   b) Natural regeneration (*Category 2*)
   c) Rehabilitation and silvicultural treatments (*Category 3*)
2) Agricultural land: Restored through
   a) Agroforestry (*Category 4*)
   b) Improved fallow (*Category 5*)
3) Protective lands and buffers: FLR interventions can involve
   a) Mangrove restoration (*Category 6*)
   b) Watershed protection and erosion control (*Category 7*)

As the comprehensive approach of FLR proceeds, it ensures that actions are linked together (Figure 19.2). Hence, according to the International Union for Conservation of Nature (IUCN) and World Resources Institute (WRI) [137], the guiding principles of FLR include:

- Focus on landscapes: The approach must cover restoring the entire ecosystem. It uses a mosaic interaction of lands and management practices to balance the social, ecological, and economic integrity of land use.
- Maintenance of natural ecosystems: It enhances the conservation and recovery of forest production sustainably and avoids the conversion and/or destruction of the natural ecosystem.

Table 19.2 Strategies of forest restoration.

Strategies	Processes/Methods	Objective	References
Rehabilitation	Restoring desired species composition, structure, and processes to an existing but degraded ecosystem. Methods involved in rehabilitation can include conversion, transformation, and reforestation.	Ecosystem function restoration (carbon sequestration, coastal protection, etc.), enhancing biodiversity, restoring provisioning services	[117], [118], [119], [120], [121], [79], [122]
Reconstruction	Restoring native plant species/communities on converted land through afforestation, recolonization, planting, and seeding.	Restoring provisioning services, enhancing biodiversity	[123], [124], [11], [125], [126]
Reclamation	Restoring severely degraded land after extensive destruction activities such as mining, oil or gas drilling, etc. Methods of restoration include revegetation of sites through natural recolonization, improvement of soil properties (physical, chemical, and biological), seeding or out-planting seedlings, and weed control.	Restoring provisioning services, biodiversity conservation	[127], [128], [129], [130], [131]
Replacement	Restoration involves replacing species or locally adapted genotypes that are displaced due to climate change with new species or new genotypes of that species that have been historically absent from the site. Example: Restoration of mined or polluted lands. The method involves assisted migration or managed relocation.	Restoring provisioning services, biodiversity conservation	[132], [133], [134]

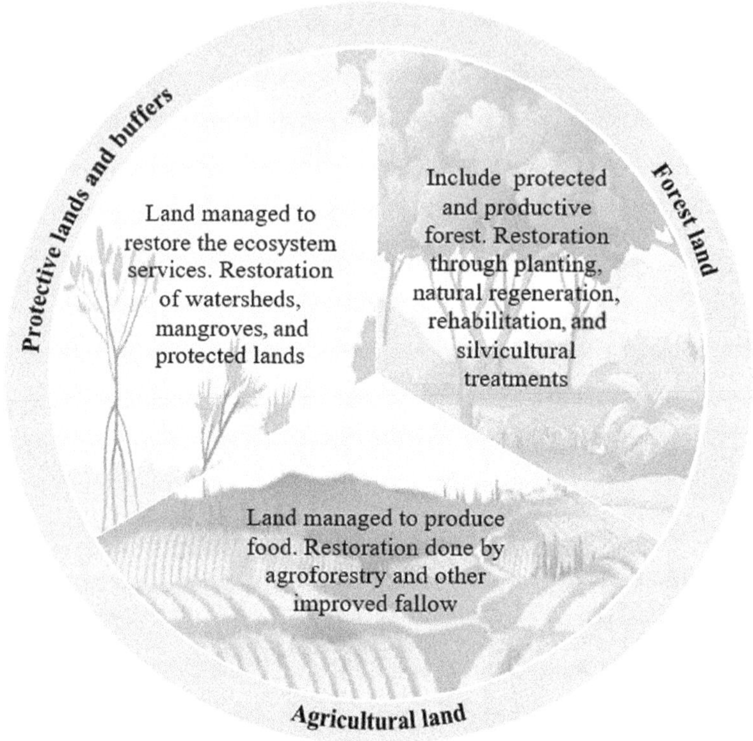

Figure 19.1 Illustration of land-use situations and FLR options.

- Engaging stakeholders and supporting participatory governance: It involves the active participation of stakeholders of all land uses and a local population whose livelihood directly depends on the forest in the planning and decision-making process for conservation and management.
- Planning FLR with different flexible approaches based on local context: FLR provides a flexible approach and uses various conservation and management approaches based on local contexts such as landscape history, economics, and the social, cultural, and ecological needs of locals. Thus, it include both scientific and traditional approaches for restoring the ecosystem.
- Multiple benefits: Restoration should benefit all stakeholders and locals.
- Successfully manage the long-term resilience of the forest/ecosystem: FLR approaches must enhance the landscape's biodiversity and resilience over time. Restoration progress must be assessed and monitored to review the successful implementation of FLR.

India's REDD+ strategy plan includes mechanisms for addressing direct benefits of biodiversity and benefit-sharing for native and local communities. The incentives received from the implementation of REDD+ are passed to local communities that are actively involved in forest protection and management against deforestation. In the Indian context, forests are not to be managed for "carbon services" alone, but for all the ecosystem services that flow from forests to the local community [138]. Incentives for carbon services are added to the benefits that local communities are already receiving from forest ecosystems.

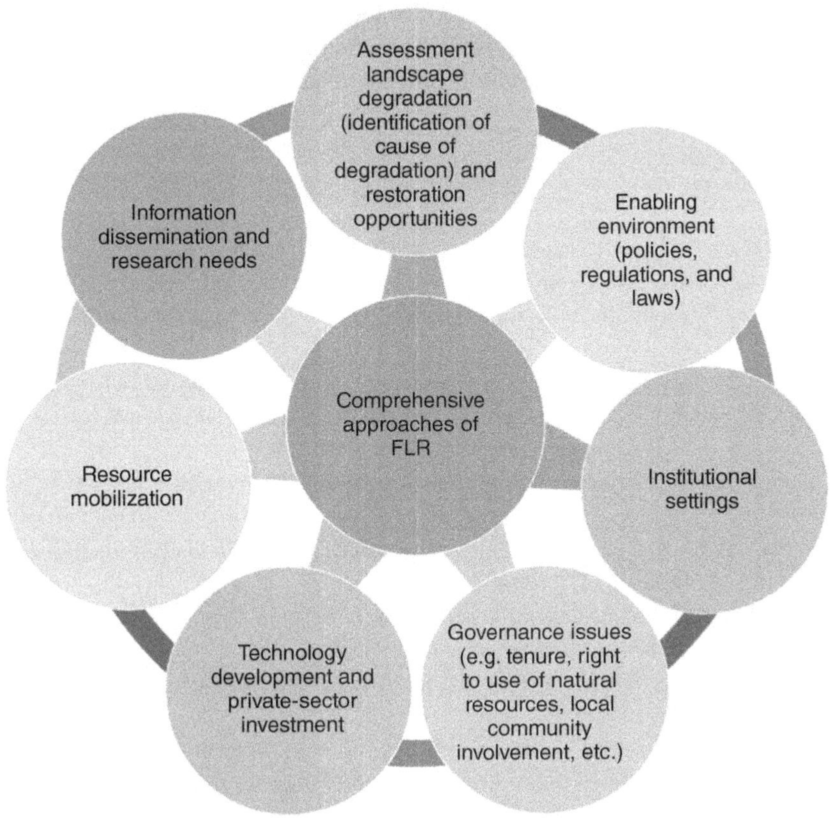

Figure 19.2 Illustration of a comprehensive approach of FLR.

19.11 Conclusion

Climate change poses substantial risks to the survival of many species. Forest ecosystems capture and store CO_2 and play a significant role in mitigating extreme climate variations. The potential for forest ecosystems to sequester carbon is considered an important step in mitigating climate change impacts and is also essential for adapting to present and future global climate changes. However, human-induced impacts and natural disturbances impair the forest ecosystem's productivity and functioning and, hence, lead to forest degradation. This has severe consequences for the physical, biological, and social environment of forest-dependent human populations. Thus, forest restoration strategies aim to strengthen forest services to sustain them for future generations. FLR is an emerging participatory decision-making approach, involving the stakeholders in different divisions of affected land use during restoration. It encourages people to work together to identify the problem and implement restoration strategies.

India is one of the few countries where deforestation rates have been reduced and regulated, and forest cover has nearly stabilized due to legislation, forest conservation and afforestation programs, community awareness, and participation. The implementation of the REDD+ mechanism has also played a vital role in forest conservation. Restoring degraded forest ecosystems is crucial for maintaining ecological services at the local, regional, and global scales. Successful restoration is possible by taking appropriate steps that involve the maintenance and integration of all environmental components of forest ecosystems so they function properly. There is also a need to develop and implement adaptation strategies for adjustments in natural systems in response to projected climate change.

References

1 Matthews, S.N., Iverson, L.R., Peters, M.P. et al. (2014). Assessing and comparing risk to climate changes among forested locations: implications for ecosystem services. *Landscape Ecology* 29 (2): 213–228.

2 Ahammad, R., Stacey, N., and Sunderland, T.C. (2019). Use and perceived importance of forest ecosystem services in rural livelihoods of Chittagong Hill tracts, Bangladesh. *Ecosystem Services* 35: 87–98.

3 FAO (2016). *Global Forest Resources Assessment 2015: How Are the World's Forests Changing?* Rome: FAO.

4 Gustafsson, L., Baker, S.C., Bauhus, J. et al. (2012). Retention forestry to maintain multifunctional forests: a world perspective. *Bioscience* 62 (7): 633–645.

5 Bahuguna, V.K. (2018). Changing dimensions of forests in India: impact of climate change and deforestation. *MOJ Ecology and Environmental Sciences* 3 (5): 319–320.

6 Kishwan, J., Pandey, R., and Dadhwal, V.K. (2009). India's forest and tree cover: contribution as a carbon sink. *Technical paper* 130: 1–12.

7 HLPE, 2017. Sustainable Forestry for Food Security and Nutrition. A report by the high level panel of experts on food security and nutrition of the committee on World Food Security. Rome

8 FAO, (2010). Global Forest Resources Assessment 2010. FAO Forestry Paper No. 163. Food and Agriculture Organisation of the United Nations, Rome, Italy.

9 Reed, J., van Vianen, J., Foli, S. et al. (2017). Trees for life: the ecosystem service contribution of trees to food production and livelihoods in the tropics. *Forest Policy and Economics* 84: 62–71.

10 Liu, Y.Y., van Dijk, A.I., McCabe, M.F. et al. (2013). Global vegetation biomass change (1988–2008) and attribution to environmental and human drivers. *Global Ecology and Biogeography* 22 (6): 692–705.

11 Ciccarese, L., Mattsson, A., and Pettenella, D. (2012). Ecosystem services from forest restoration: thinking ahead. *New Forests* 43 (5–6): 543–560.

12 Vergara, D.G., Coladilla, J., Alcantara, E. et al. (2019). Conservation under regional industrialization: fragmentation and cover change in a forest reserve. *Journal of Environmental Science and Management* 22 (1): 36–53.

13 Carpenter, S.R. and Turner, M.G. (2000). Hares and tortoises: interactions of fast and slow variables in ecosystems. *Ecosystems* 3: 495–497.

14 Seidl, R., Honkaniemi, J., Aakala, T. et al. (2020). Globally consistent climate sensitivity of natural disturbances across boreal and temperate forest ecosystems. *Ecography* 43: 967–978.

15 Van Der Plas, F., Manning, P., Soliveres, S. et al. (2016). Biotic homogenization can decrease landscape-scale forest multifunctionality. *Proceedings of the National Academy of Sciences* 113 (13): 3557–3562.

16 Riitters, K., Wickham, J., Costanza, J.K., and Vogt, P. (2016). A global evaluation of forest interior area dynamics using tree cover data from 2000 to 2012. *Landscape Ecology* 31 (1): 137–148.

17 Ellison, D., Morris, C.E., Locatelli, B. et al. (2017). Trees, forests and water: cool insights for a hot world. *Global Environmental Change* 43: 51–61.

18 Schmitt, S., Maréchaux, I., Chave, J. et al. (2019). Functional diversity improves tropical forest resilience: insights from a long-term virtual experiment. *Journal of Ecology* 108 (3): 831–843.
19 Seidl, R., Rammer, W., and Spies, T.A. (2014). Disturbance legacies increase the resilience of forest ecosystem structure, composition, and functioning. *Ecological Applications* 24 (8): 2063–2077.
20 Leemans, R. and Eickhout, B. (2004). Another reason for concern: regional and global impacts on ecosystems for different levels of climate change. *Global Environmental Change* 14 (3): 219–228.
21 Ravindranath, N.H. and Sathaye, J.A. (2002). *Climate Change and Developing Countries (Advances in Global Change Research)*. Netherlands: Springer.
22 Wieczynski, D.J., Boyle, B., Buzzard, V. et al. (2019). Climate shapes and shifts functional biodiversity in forests worldwide. *Proceedings of the National Academy of Sciences* 116 (2): 587–592.
23 IPCC, Climate Change (2001). *Synthesis Report, Intergovernmental Panel on Climate Change*. Geneva: Switzerland.
24 Yu, D., Liu, Y., Shi, P., and Wu, J. (2019). Projecting impacts of climate change on global terrestrial ecoregions. *Ecological Indicators* 103: 114–123.
25 Nunes, L.J., Meireles, C.I., Pinto Gomes, C.J., and Almeida Ribeiro, N. (2019). Forest management and climate change mitigation: A review on carbon cycle flow models for the sustainability of resources. *Sustainability* 11 (19): 5276.
26 Boegelsack, N., Withey, J., O'Sullivan, G., and McMartin, D. (2018). A critical examination of the relationship between wildfires and climate change with consideration of the human impact. *Journal of Environmental Protection* 9 (5): 461–467.
27 Dawson, I.K., Leakey, R., Clement, C.R. et al. (2014). The management of tree genetic resources and the livelihoods of rural communities in the tropics: non-timber forest products, smallholder agroforestry practices and tree commodity crops. *Forest Ecology and Management* 333: 9–21.
28 IPCC, (2011). Summary for Policymakers: Intergovernmental Panel on Climate Change Special Report on Managing the Risks of Extreme Events and Disasters to Advance Climate Change Adaptation. Cambridge: Cambridge University Press.
29 Stephens, S.L., Agee, J.K., Fule, P.Z. et al. (2013). Managing forests and fire in changing climates. *Science* 342 (6154): 41–42.
30 Ravindranath, N.H., Joshi, N.V., Sukumar, R., and Saxena, A. (2006). Impact of climate change on forests in India. *Current Science*: 354–361.
31 Schulze, E.D. and Mooney, H.A. (eds.) (2012). *Biodiversity and Ecosystem Function*. Heidelberg: Springer.
32 Müller, F., Burkhard, B., Hou, Y. et al. (2016). Indicators for ecosystem services. In: *Routledge Handbook of Ecosystem Services* (eds. M. Potschin, R. Haines-Young, R. Fish and R.K. Turner), 157–169. Routledge.
33 Brockerhoff, E.G., Barbaro, L., Castagneyrol, B. et al. (2017). Forest biodiversity, ecosystem functioning and the provision of ecosystem services. *Biodiversity and Conservation* 26: 3005–3035.
34 Cudlín, P., Seják, J., Pokorný, J. et al. (2013). Forest ecosystem services under climate change and air pollution. In: *Developments in Environmental Science*, vol. 13 (eds. R. Matyssek, N. Clarke, P. Cudlin, et al.), 521–546. Elsevier.
35 Binder, S., Haight, R. G., Polasky, S., Warziniack, T., Mockrin, M. H., Deal, R. L., and Arthaud, G. (2017). Assessment and valuation of forest ecosystem services: state of the science review. *Gen. Tech. Rep. NRS-170. Newtown Square, PA: US Department of Agriculture, Forest Service, Northern Research Station. 47 p., 170*, 1–47.
36 Millennium Ecosystem Assessment (MEA). Ecosystems and Human Well-Being: Synthesis; Island Press: Washington, DC, USA, 2005.

37 Jasechko, S., Sharp, Z.D., Gibson, J.J. et al. (2013). Terrestrial water fluxes dominated by transpiration. *Nature* 496 (7445): 347–350.

38 Spracklen, D.V., Arnold, S.R., and Taylor, C.M. (2012). Observations of increased tropical rainfall preceded by air passage over forests. *Nature* 489 (7415): 282.

39 Makarieva, A. M., Gorshkov, V. G., Sheil, D., Nobre, A. D., & Li, B. L. (2010). Where do winds come from? A new theory on how water vapor condensation influences atmospheric pressure and dynamics. *arXiv preprint arXiv:1004.0355.*

40 Birot, Y., Gracia, C., Matteucci, G. et al. (2011). Water for forests and people in the Mediterranean region. *Forêt Mediterranéenne* 32: 363–366.

41 Spracklen, D.V. and Garcia-Carreras, L. (2015). The impact of Amazonian deforestation on Amazon basin rainfall. *Geophysical Research Letters* 42 (21): 9546–9552.

42 Duffy, P.B., Brando, P., Asner, G.P., and Field, C.B. (2015). Projections of future meteorological drought and wet periods in the Amazon. *Proceedings of the National Academy of Sciences* 112 (43): 13172–13177.

43 Tokinaga, H., Xie, S.P., Deser, C. et al. (2012). Slowdown of the Walker circulation driven by tropical indo-Pacific warming. *Nature* 491 (7424): 439–443.

44 Kumar, M., Sundaram, S., Gnansounou, E. et al. (2018). Carbon dioxide capture, storage and production of biofuel and biomaterials by bacteria: A review. *Bioresource Technology* 247: 1059–1068.

45 Kumar, M., Gazara, R.K., Verma, S. et al. (2016). Genome sequence of carbon dioxide-sequestering Serratia sp. strain ISTD04 isolated from marble mining rocks. *Genome Announcements* 4 (5): e01141–e01116.

46 Jin, L., Yi, Y., and Xu, J. (2020). Forest carbon sequestration and China's potential: the rise of a nature-based solution for climate change mitigation. *China Economic Journal* 13: 1–23.

47 Favero, A., Daigneault, A., and Sohngen, B. (2020). Forests: carbon sequestration, biomass energy, or both? *Science Advances* 6 (13): eaay6792.

48 Sadatshojaie, A. and Rahimpour, M.R. (2020). CO_2 emission and air pollution (volatile organic compounds, etc.)–related problems causing climate change. In: *Current Trends and Future Developments on (Bio-) Membranes* (eds. A. Figoli, Y. Li and A. Basile), 1–30. Elsevier.

49 Pan, Y., Birdsey, R.A., Fang, J. et al. (2011). A large and persistent carbon sink in the world's forests. *Science* 333 (6045): 988–993.

50 Saatchi, S.S., Harris, N.L., Brown, S. et al. (2011). Benchmark map of forest carbon stocks in tropical regions across three continents. *Proceedings of the National Academy of Sciences* 108 (24): 9899–9904.

51 Arora, P. and Chaudhry, S. (2017). Vegetation and soil carbon pools of mixed plantation of Acacia nilotica and Dalbergia sissoo under social forestry scheme in Kurukshetra, India. *Journal of Materials and Environmental Science* 8: 4565–4572.

52 Arora, P. and Chaudhry, S. (2015). Carbon sequestration potential of populus deltoides plantation under social forestry scheme in Kurukshetra, Haryana in Northern India. *Journal of Materials and Environmental Science* 6 (3): 703–720.

53 Sheikh, A.Q., Skinder, B.M., Ashok, K.P., and Ganai, B.A. (2014). Terrestrial carbon sequestration as a climate change mitigation activity. *Journal of Pollution Effects & Control* 2: 1–8.

54 Fares, S., Paoletti, E., Calfapietra, C. et al. (2017). Carbon sequestration by urban trees. In: *The Urban Forest* (eds. D. Pearlmutter, C. Calfapietra, R. Samson, et al.), 31–39. Cham: Springer.

55 Chaturvedi, R. K. (2011) Impact of Climate Change on Forests and Biodiversity of Madhya Pradesh – An Assessment Report. https://www.devalt.org/images/L2_ProjectPdfs/(15)ForestsDiversity.pdf?Oid=150

56 Bhatti, J. S., Van Kooten, G. C., Apps, M. J., Laird, L. D., Campbell, I. D., Campbell, C.,Turetsky, M.R., Yu, Z., & Banfield, E. (2003). Carbon balance and climate change in boreal forests. *Towards Sustainable Management of the Boreal Forest. NRC Research Press, Ottawa, Ontario*, 799–855.

57 Feng, H., & Zhang, L. (2010, October). Forest ecological services of regulating temperature based on remote sensing technology. 3rd International Congress on Image and Signal Processing (Vol. 5, pp. 2180–2183). IEEE.

58 Hesslerová, P., Pokorný, J., Brom, J., and Rejšková-Procházková, A. (2013). Daily dynamics of radiation surface temperature of different land cover types in a temperate cultural landscape: consequences for the local climate. *Ecological Engineering* 54: 145–154.

59 Maes, W.H., Pashuysen, T., Trabucco, A. et al. (2011). Does energy dissipation increase with ecosystem succession? Testing the ecosystem exergy theory combining theoretical simulations and thermal remote sensing observations. *Ecological Modelling* 222 (23–24): 3917–3941.

60 Bounoua, L., Zhang, P., Mostovoy, G. et al. (2015). Impact of urbanization on US surface climate. *Environmental Research Letters* 10 (8): 084010.

61 Escobedo, F.J., Kroeger, T., and Wagner, J.E. (2011). Urban forests and pollution mitigation: analyzing ecosystem services and disservices. *Environmental Pollution* 159 (8–9): 2078–2087.

62 Nowak, D.J., Hirabayashi, S., Bodine, A., and Greenfield, E. (2014). Tree and forest effects on air quality and human health in the United States. *Environmental Pollution* 193: 119–129.

63 Nowak, D.J., Crane, D.E., and Stevens, J.C. (2006). Air pollution removal by urban trees and shrubs in the United States. *Urban Forestry & Urban Greening* 4: 115–123.

64 Ussiri, D.A. and Lal, R. (2017). Global forests management for climate change mitigation. In: *Carbon Sequestration for Climate Change Mitigation and Adaptation* (eds. D.A.N. Ussiri and R. Lal), 395–432. Cham: Springer.

65 Flores, B.M., Staal, A., Jakovac, C.C. et al. (2019). Soil erosion as a resilience drain in disturbed tropical forests. *Plant and Soil* 450: 1–15.

66 Berendse, F., van Ruijven, J., Jongejans, E., and Keesstra, S. (2015). Loss of plant species diversity reduces soil erosion resistance. *Ecosystems* 18 (5): 881–888.

67 Song, Z., Seitz, S., Li, J. et al. (2019). Tree diversity reduced soil erosion by affecting tree canopy and biological soil crust development in a subtropical forest experiment. *Forest Ecology and Management* 444: 69–77.

68 Wang, Z., Hou, Y., Fang, H. et al. (2012). Effects of plant species diversity on soil conservation and stability in the secondary succession phases of a semihumid evergreen broadleaf forest in China. *Journal of Soil and Water Conservation* 67 (4): 311–320.

69 Hamilton, L.S., Dudley, N., Greminger, G. et al. (2008). *Forests and Water: A Thematic Study Prepared in the Framework of the Global Forests Resources Assessment 2005*. Roma (Italia): FAO.

70 Bakirci, M. (2010). Negative impacts of forest fires on ecological balance and environmental sustainability: case of Turkey. *Revija za geografijo* 5 (1): 15–32.

71 Reubens, B., Poesen, J., Danjon, F. et al. (2007). The role of fine and coarse roots in shallow slope stability and soil erosion control with a focus on root system architecture: a review. *Trees* 21 (4): 385–402.

72 Liu, J., Gao, G., Wang, S. et al. (2018). The effects of vegetation on runoff and soil loss: multidimensional structure analysis and scale characteristics. *Journal of Geographical Sciences* 28 (1): 59–78.

73 Li, Y., Härdtle, W., Bruelheide, H. et al. (2014). Site and neighborhood effects on growth of tree saplings in subtropical plantations (China). *Forest Ecology and Management* 327: 118–127.

74 Kreye, M.M., Adams, D.C., and Escobedo, F.J. (2014). The value of forest conservation for water quality protection. *Forests* 5 (5): 862–884.

75 FAO (2013). *Forests and Water – International Momentum and Action*, 76. Italy: Rome.

76 Miura, S., Amacher, M., Hofer, T. et al. (2015). Protective functions and ecosystem services of global forests in the past quarter-century. *Forest Ecology and Management* 352: 35–46.

77 Creed, I.F., Weber, M., Accatino, F., and Kreutzweiser, D.P. (2016). Managing forests for water in the Anthropocene – the best kept secret services of forest ecosystems. *Forests* 7 (3): 60.

78 Reddy, C.S., Jha, C.S., Diwakar, P.G., and Dadhwal, V.K. (2015). Nationwide classification of forest types of India using remote sensing and GIS. *Environmental Monitoring and Assessment* 187 (12): 777.

79 Stanturf, J.A., Palik, B.J., and Dumroese, R.K. (2014a). Contemporary forest restoration: a review emphasizing function. *Forest Ecology and Management* 331: 292–323.

80 Hosonuma, N., M.Herold, D. S.Veronique, D. F. R. S, M.Brockhaus, L.Verchot, A.Angelsen, and E.Romijn (2012). "An Assessment of Deforestation and Forest Degradation Drivers in Developing Countries." In: Environ. Res. Lett. 7.4, p. 044009. doi: 10.1088/1748-9326/7/4/044009.

81 FAO (2015). Global Forest Resources Assessment. FAO Forestry Paper No. 1. Rome: UN Food and Agriculture Organization.

82 Kumari, R., Banerjee, A., Kumar, R., Kumar, A., Saikia, P., & Khan, M. L. (2019). Deforestation in India: Consequences and Sustainable Solutions. In Deforestation Around the World. IntechOpen. DOI: http://dx.doi.org/10.5772/intechopen.85804

83 Chaturvedi, R.K., Gopalakrishnan, R., Jayaraman, M. et al. (2011). Impact of climate change on Indian forests: a dynamic vegetation modeling approach. *Mitigation and Adaptation Strategies for Global Change* 16 (2): 119–142.

84 FSI (2009) The state of forest report, Forest Survey of India, Ministry of Environment and Forests, Government of India

85 Baider, C. and Florens, F.V. (2011). Control of invasive alien weeds averts imminent plant extinction. *Biological Invasions* 13 (12): 2641–2646.

86 Hejda, M., Pyšek, P., and Jarošík, V. (2009). Impact of invasive plants on the species richness, diversity and composition of invaded communities. *Journal of Ecology* 97 (3): 393–403.

87 Pejchar, L. and Mooney, H.A. (2009). Invasive species, ecosystem services and human well-being. *Trends in Ecology & Evolution* 24 (9): 497–504.

88 Mack, R.N., Simberloff, D., Mark Lonsdale, W. et al. (2000). Biotic invasions: causes, epidemiology, global consequences, and control. *Ecological Applications* 10 (3): 689–710.

89 Daehler, C.C. (2003). Performance comparisons of co-occurring native and alien invasive plants: implications for conservation and restoration. *Annual Review of Ecology, Evolution, and Systematics* 34 (1): 183–211.

90 Campagnaro, T., Brundu, G., and Sitzia, T. (2018). Five major invasive alien tree species in European Union forest habitat types of the alpine and continental biogeographical regions. *Journal for Nature Conservation* 43: 227–238.

91 Qiu, J. (2013). China battles army of invaders. *Nature* 503 (7477): 450.

92 Straube, D., Johnson, E.A., Parkinson, D. et al. (2009). Nonlinearity of effects of invasive ecosystem engineers on abiotic soil properties and soil biota. *Oikos* 118 (6): 885–896.

93 Pan, X., Zhang, J., Xu, H. et al. (2015). Spatial similarity in the distribution of invasive alien plants and animals in China. *Natural Hazards* 77 (3): 1751–1764.

94 De Faria, B.L., Brando, P.M., Macedo, M.N. et al. (2017). Current and future patterns of fire-induced forest degradation in Amazonia. *Environmental Research Letters* 12 (9): 095005.

95 Földi, L. and Kuti, R. (2016). Characteristics of forest fires and their impact on the environment. *Academic and Applied Research in Military and Public Mgmt. Sc* 1: 5–17.

96 Fonseca, F., de Figueiredo, T., Nogueira, C., and Queirós, A. (2017). Effect of prescribed fire on soil properties and soil erosion in a Mediterranean mountain area. *Geoderma* 307: 172–180.

97 Köster, E., Köster, K., Berninger, F. et al. (2018). Changes in fluxes of carbon dioxide and methane caused by fire in Siberian boreal forest with continuous permafrost. *Journal of Environmental Management* 228: 405–415.

98 Tandon, A., Jyotsna, K., and Dey, S. (2018). A mathematical model to investigate the impact of overgrowing population-induced mining on forest resources. *Environment, Development and Sustainability* 20 (4): 1499–1516.

99 Specht, M.J., Pinto, S.R.R., Albuquerque, U.P. et al. (2015). Burning biodiversity: Fuelwood harvesting causes forest degradation in human-dominated tropical landscapes. *Global Ecology and Conservation* 3: 200–209.

100 Puyravaud, J.P., Davidar, P., and Laurance, W.F. (2010). Cryptic destruction of India's native forests. *Conservation Letters* 3 (6): 390–394.

101 Barrett, C.B., Travis, A.J., and Dasgupta, P. (2011). On biodiversity conservation and poverty traps. *Proceedings of the National Academy of Sciences* 108 (34): 13907–13912.

102 Lewis, S.L., Edwards, D.P., and Galbraith, D. (2015). Increasing human dominance of tropical forests. *Science* 349 (6250): 827–832.

103 Ramdhani, V. and Nugrahani, E.H. (2015). Dynamical system of modelling the depletion of forestry resources due to crowding by industrialization. *Applied Mathematical Sciences* 9 (82): 4067–4079.

104 Singh, S.P., Pandey, A., and Singh, V. (2019). Nature and extent of Forest degradation in Central Himalayas. In: *Tropical Ecosystems: Structure, Functions and Challenges in the Face of Global Change* (eds. S.C. Garkoti, S.J. Van Bloem, P.Z. Fulé and R.L. Semwal), 27–43. Singapore: Springer.

105 Scarascia-Mugnozza, G., Oswald, H., Piussi, P., and Radoglou, K. (2000). Forests of the Mediterranean region: gaps in knowledge and research needs. *Forest Ecology and Management* 132 (1): 97–109.

106 Doblas Miranda, E., Attorre, F., Azevedo, J. et al. (2018). Drivers of degradation and other threats. *State of Mediterranean Forests* 2018: 72–89.

107 Kairis, O., Karavitis, C., Salvati, L. et al. (2015). Exploring the impact of overgrazing on soil erosion and land degradation in a dry Mediterranean agro-forest landscape (Crete, Greece). *Arid Land Research and Management* 29 (3): 360–374.

108 Li, S.G., Harazono, Y., Oikawa, T. et al. (2000). Grassland desertification by grazing and the resulting micrometeorological changes in Inner Mongolia. *Agricultural and Forest Meteorology* 102 (2–3): 125–137.

109 Hilker, T., Natsagdorj, E., Waring, R.H. et al. (2014). Satellite observed widespread decline in Mongolian grasslands largely due to overgrazing. *Global Change Biology* 20 (2): 418–428.

110 Deng, L., Shangguan, Z.P., Wu, G.L., and Chang, X.F. (2017). Effects of grazing exclusion on carbon sequestration in China's grassland. *Earth-Science Reviews* 173: 84–95.

111 Stanturf, J.A., Palik, B.J., Williams, M.I. et al. (2014b). Forest restoration paradigms. *Journal of Sustainable Forestry* 33 (sup1): S161–S194.

112 Vásquez-Grandón, A., Donoso, P.J., and Gerding, V. (2018). Forest degradation: when is a forest degraded? *Forests* 9 (11): 726.

113 Qazi, N.Q., Bruijnzeel, L.A., Rai, S.P., and Ghimire, C.P. (2017). Impact of forest degradation on streamflow regime and runoff response to rainfall in the Garhwal Himalaya, Northwest India. *Hydrological Sciences Journal* 62 (7): 1114–1130.

114 Potapov, P., Yaroshenko, A., Turubanova, S. et al. (2008). Mapping the world's intact forest landscapes by remote sensing. *Ecology and Society* 13 (2): 51.

115 Stanturf, J. and van Oosten, C. (2014). Operational poplar and willow culture. In: *Poplars and Willows – Trees for Society and the Environment* (eds. J.G. Isebrands and J. Richardson), 200–257. CABI.

116 Stanturf, J.A. and Madsen, P. (2002). Restoration concepts for temperate and boreal forests of North America and Western Europe. *Plant Biosystems-An International Journal Dealing with all Aspects of Plant Biology* 136 (2): 143–158.

117 Brunner, A., Hahn, K., Biber, P., and Skovsgaard, J.P. (2006). Conversion of Norway spruce: a case study in Denmark based on silvicultural scenario modelling. In: *Sustainable Forest Management: Growth Models for Europe* (ed. H. Hasenauer), 343–371. Heidelberg: Springer.

118 Drouineau, S., Laroussinie, O., Birot, Y. et al. (2000). *Joint Evaluation of Storms, Forest Vulnerability and Their Restoration*. Joensuu, Finland: European Forest Institute.

119 Hansen, J. and Spiecker, H. (2005). Conversion of Norway spruce (*Picea abies* [L.] Karst.) forests in Europe. In: *Restoration of Boreal and Temperate Forests* (eds. J.A. Stanturf and P. Madsen), 339–347. Boca Raton: CRC Press.

120 Harmer, R. and Morgan, G. (2009). Storm damage and the conversion of conifer plantations to native broadleaved woodland. *Forest Ecology and Management* 258: 879–886.

121 Spiecker, H., Hansen, J., Klimo, E., Skovsgaard, J.P., Sterba, H., & von Teuffel, K. (Eds.). (2004). Norway spruce conversion: options and consequences.EFI Research Report 18. Brill: Leiden, Boston, Köln.

122 Zerbe, S. (2002). Restoration of natural broad-leaved woodland in Central Europe on sites with coniferous forest plantations. *Forest Ecology and Management* 167: 27–42.

123 Blay, D. (2012). Restoration of deforested and degraded areas in Africa. In: *A Goal-Oriented Approach to Forest Landscape Restoration* (eds. J. Stanturf, P. Madsen and D. Lamb), 267–319. Dordrecht: Springer.

124 Ciccarese, L., Brown, S., and Schlamadinger, B. (2005). Carbon sequestration through restoration of temperate and boreal forests. In: *Restoration of Boreal and Temperate Forests* (eds. J.A. Stanturf and P. Madsen), 111–120. Boca Raton: CRC Press.

125 Lamb, D. (2011). *Regreening the Bare Hills: Tropical Forest Restoration in the Asia– Pacific Region*. Dordrecht: Springer Science + Business Media.

126 Lithgow, D., Martinez, M., Gallego-Fernandez, J. et al. (2013). Linking restoration ecology with coastal dune restoration. *Geomorphology* 199: 214–224.

127 Evans, D.M., Zipper, C.E., Burger, J.A. et al. (2013). Reforestation practice for enhancement of ecosystem services on a compacted surface mine: path toward ecosystem recovery. *Ecological Engineering* 51: 16–23.

128 Fields-Johnson, C., Zipper, C., Burger, J., and Evans, D. (2012). Forest restoration on steep slopes after coal surface mining in Appalachian USA: soil grading and seeding effects. *Forest Ecology and Management* 270: 126–134.

129 Prach, K. and Hobbs, R.J. (2008). Spontaneous succession versus technical reclamation in the restoration of disturbed sites. *Restoration Ecology* 16: 363–366.

130 Prach, K. and Pysek, P. (2001). Using spontaneous succession for restoration of human disturbed habitats: experience from Central Europe. *Ecological Engineering* 17: 55–62.

131 Zipper, C.E., Burger, J.A., Barton, C.D., and Skousen, J.G. (2013). Rebuilding soils on mined land for native forests in Appalachia. *Soil Science Society of America Journal* 77: 337–349.

132 McLachlan, J.S., Hellmann, J.J., and Schwartz, M.W. (2007). A framework for debate of assisted migration in an era of climate change. *Conservation Biology* 21: 297–302.

133 Pedlar, J.H., McKenney, D.W., Aubin, I. et al. (2012). Placing forestry in the assisted migration debate. *Bioscience* 62: 835–842.

134 Williams, M.I. and Dumroese, R.K. (2013). Preparing for climate change: forestry and assisted migration. *Journal of Forestry* 114: 287–297.
135 Chazdon, R.L., Brancalion, P.H., Laestadius, L. et al. (2016). When is a forest a forest? Forest concepts and definitions in the era of forest and landscape restoration. *Ambio* 45 (5): 538–550.
136 Laestadius, L., Buckingham, K., Maginnis, S., and Saint-Laurent, C. (2015). Before Bonn and beyond: the history and future of forest landscape restoration. *Unasylva* 66 (245): 11–18.
137 IUCN and WRI, (2014). A guide to the Restoration Opportunities Assessment Methodology (ROAM): Assessing forest landscape restoration opportunities at the national or sub-national level. Working Paper (Road-test edition). Gland, Switzerland: IUCN. pp. 125.
138 Sharma, V., & Chaudhry, S. (2013). An overview of Indian forestry sector with REDD. *International Scholarly Research Notices, 2013.*

20

Forest Degradation Prevention

Marta Jaskulak[1,2] and Anna Grobelak[1]

[1] *Faculty of Infrastructure and Environment, Institute of Environmental Engineering, Czestochowa University of Technology, Czestochowa, Poland*
[2] *LGCgE, Laboratoire de Génie Civil et géo-Environnement, University of Lille, IMT Lille Douai, Univ. Artois, Yncrea Hauts-de-France, Lille, France*

20.1 Introduction

Forest degradation is the decline in a forest's ability to provide resources, products, and sociocultural and environmental services [33]. It entails an alteration that negatively impacts a forest's attributes: for instance, stock growth and biomass, carbon reservoir, biodiversity, soils, and attraction values, leading to a reduction in the provision of products and services [34]. The degradation of tropical forests in Sub-Saharan Africa alone emits over ~0.5 $Pg\,y^{-1}$ of carbon and severely reduces the biodiversity of the region [30].

Various environmental disturbances can cause adverse changes in forest health, all of which have different degrees, extremity, standards, sources, and prevalence [35]. There may be natural disturbances (e.g. fires, storms, wind, droughts, the presence of pests, and invasive species) or disturbances caused by human activities (e.g. unfeasible logging, the spread of industrial contamination, the introduction of invasive non-native "new" species, constructing roads, mining, shifting cultivation, hunting, and grazing), or a mix of both natural and human-made [23]. Anthropogenic disturbances may be deliberate, such as logging or grazing, or it may be accidental, such as the spread of invasive species [16]. Moreover, there are underlying or hidden causes of forest degradation, including poverty, adverse policies, and ambiguous ownership rights [8].

Forests lands often have an internal scope of natural changes and succession phases, and natural or anthropogenic disruptions do not necessarily result in degeneration. Degradation happens when the production of a particular forest product or service is continuously below the anticipated capacity and is not within the variation range that would be naturally expected [12]. While it is difficult to define and quantify, forest deterioration is a significant problem worldwide. It adversely affects forest ecosystems and the products and services obtained from forests; for instance, it is deemed an essential source of land-associated greenhouse gas (GHG) production [37].

In 2010, it was estimated that 27% of worldwide forest landscapes (approximately 1.5 billion hectares) suffered degradation [36]. Often, there is confusion between the concepts of forest degradation and deforestation. *Deforestation* is an extended period of forest loss, with no promise that the forest will regenerate through natural growth or reforestation activities; it leads to forest area

Handbook of Ecological and Ecosystem Engineering, First Edition. Edited by Majeti Narasimha Vara Prasad.

decrease. Forest *degradation* does not entail a decrease in the size of the forest but a qualitative reduction in the state of the forest. Forest degradation often results in full-blown deforestation because degraded forests can be easily changed to agricultural lands [15]. Figures 20.1 and 20.2 show the current extent of this problem from 2001 to 2018.

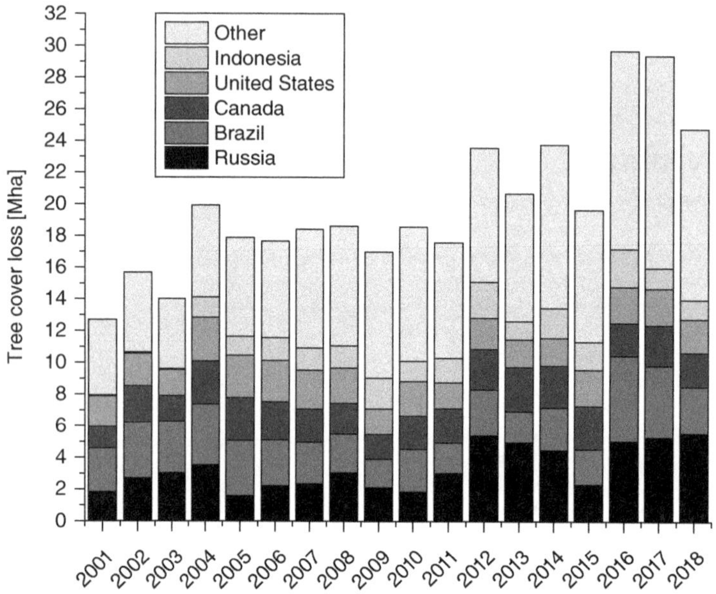

Figure 20.1 Tree cover loss from 2001 to 2018. *Source:* Based on open data from The Sustainability Consortium, World Resources Institute, and University of Maryland. "Tree Cover Loss by Driver."

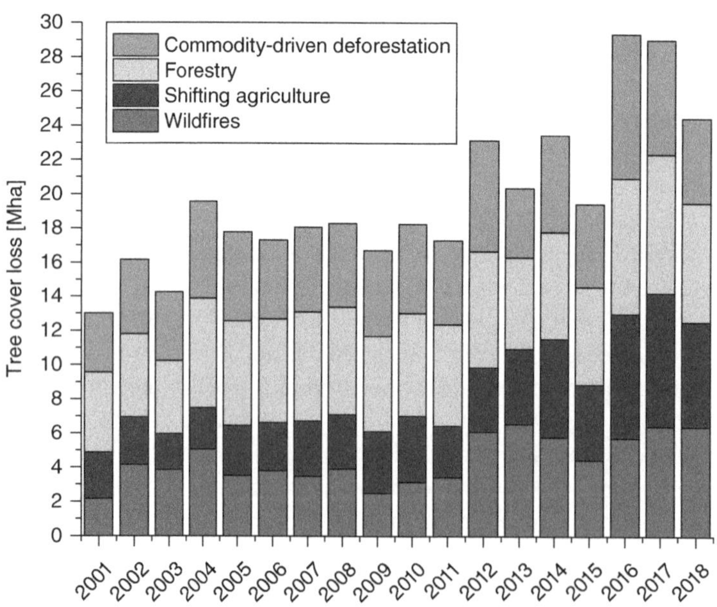

Figure 20.2 Global tree cover loss by its dominant driver from 2001 to 2018. *Source:* Based on the open data of The Sustainability Consortium, World Resources Institute, and University of Maryland. "Tree Cover Loss by Driver."

20.2 The Problem of Forest Degradation

Forests offer numerous services. For instance, they avert soil erosion, control the water cycle, hold and store carbon, release oxygen, offer freshwater and habitats, minimize the dangers of fire (in the tropics), and provide wood and non-wood forest goods [15]. Therefore, forest degradation negatively impacts millions of people who wholly or partly depend on forest products or services at the local level and billions of individuals who benefit from the services of forests at the regional or worldwide scale [1]. In particular, for owners of forests and managers, forest degradation directly threatens their livelihoods because it reduces the productivity and profitability of a forest, indicating unsustainable activities in the forests they supervise [43]. Between 2001 and 2018, a total of 361 Mha of tree cover was lost globally, which is the equivalent of a 9.0% decrease in tree cover since 2001 and a total of 98.7 Gt of CO_2 emissions (Figure 20.1).

In recent years, forest fires have become the most significant environmental impact in many countries globally, including Latin America, Sub-Saharan Africa, Australia, and the Mediterranean region [2]. Moreover, Sub-Saharan Africa is currently experiencing the fastest rate of urban migration in the world and, at the same time, is strictly dependent on forest-based resources; forest fires cause the severe deterioration of land and result in losses in soil fertility and productivity [19].

Currently, the rate of forest fires worldwide is continually increasing. A substantial area is burned each year, but little has been done to change and improve soil and water management strategies and techniques in burned areas, despite their significant impact on future soil erosion and hydrological processes [45]. Furthermore, given enough time without reclamation attempts, past forest fires cause complete erosion of the soil, leaving it exposed to the atmosphere. They also bring significant secondary issues in the burned areas, including a high rate of soil and water degradation that occurs after the fire due to the lack of vegetative cover and microbiome. This severely limits the possibilities to mitigate the deterioration of the land.

Most nutrients are lost in the first months after a fire [41]. Hence, strategies and technologies applied *ex-ante* are crucial in forest management. Such techniques include forest planning and preventive forestry with mulching, hillslope barriers, proper seeding, creating infiltration pathways and constant monitoring, and, if necessary, the restoration of the ecosystem (Table 20.1) [9]. These *ex-post* strategies potentially can mitigate future degradation but need to be applied as soon as possible to be effective, which adds to the short-term costs [26]. Moreover, the post-fire restoration needs to be implemented in all essential parts of the burned land, driven by an understanding of erosion patterns. In addition, to be truly implementable on a large burned area, such strategies must be cost-effective and include integrating various techniques to reduce the output of water, nutrients, and sediments [48].

In many areas that have been affected by forest degradation in recent years, few or no efforts have been taken to minimize its detrimental impact, which has led to a loss of forest-related recourses and biodiversity [32]. Hence, a significantly more assertive attitude toward the conservation of land and water, especially in areas prone to droughts and wildfires, should be sought [4]. According to estimates, over 2 billion hectares of deforested and degraded forest globally are potentially appropriate and ready for reforestation. Sustainably managing degraded areas would change the trend of degradation, help forest landscape recovery, and restore related forest products and services. Besides replenishing the forest's reproductive capacity and diversity, it would provide economic, social, and environmental gains, including alleviating and adapting to climate change [47].

Table 20.1 Main advantages and disadvantages of *ex-ante* strategies in forest management.

Technique/strategies	Advantages	Disadvantages	References
Mulching and hydro mulching	Increase in soil organic matter Increase in foil fertility Increase in soil water-holding capacity and erosion reduction, promoting plant growth	–	Chen et al. [12]
Seeding	Improved fertility and infiltration if applied with other techniques	The seeds can be easily washed out or eaten by animals. Low survival of seeds on burned areas	Zarin et al. [34]
Erosion barriers	Increases soil fertility and vegetative cover	Possibility of rill erosion High cost	Zeb et al. [48]
Natural check dams	Creates areas of nutrient accumulation; increases soil quality and fertility	Fast decomposition, not long-lasting	Boucher et al. [7]
Scarifying	Increased seed survival Increased content of nutrients	Increased seed survival Increased content of nutrients	Vlasov et al. [47]
Plowing	Increased seed survival Increased content of nutrients	Loss of soil moisture Soil erosion Damages the soil structure	Fry [22]

Source: Based on Brewer et al. [9].

Goal 15 of the Sustainable Development Goals to be attained by 2020 mentions degradation in the Global Objectives on Forests of the United Nations Forest Instrument and in the Aichi Biodiversity Targets of the Convention on Biological Diversity. The United Nations Framework Convention on Climate Change (UNFCCC) agreement to reduce greenhouse gas emissions from deforestation and forest degradation (known as the Reducing Emissions from Deforestation and Forest Degradation program, or REDD+) potentially incentivizes third-world countries to reduce forest degradation (and emissions of associated greenhouse gases), and reestablish or promote forest management (thus increasing forest-related carbon sequestration) [41, 46].

20.3 Assessing Levels of Forest Degradation

It is crucial to determine whether forests are undergoing degradation and the causes of the degradation to implement strategies that can combat and reverse the process [44]. Collection and analysis of information about the condition of a forest and the extent of degradation helps prioritize resources and strategies to prevent further degradation, solve the root causes, and restore and rehabilitate the forest landscape that has been degraded [31]. Assessing forest degradation enables countries to meet their promises and commitments for international reporting [17].

It is challenging to quantify the scale of the problem as different stakeholders perceive degradation in different ways. One person may perceive the issue as degradation, while others may perceive it as a source of livelihood, making it difficult to have a universal approach to define degradation (Table 20.2) [22]. As a result, various criteria apply, such as health and vitality, production capacity, species diversity, and aesthetic value [27]. However, more weight is given to criteria that influence the stakeholder's perception of degradation. For instance, a planted forest may be perceived as degraded if the consideration is biodiversity. The definition of degradation

Table 20.2 Benefits of sustainable forest management on the community, environmental, national, and global scale.

Type of benefit	Community level	Environmental/landscape level	National/global level
Forest productivity	• Increased wood production • Increased production of non-wood forest products • Increase in the long-term productivity of forests • Diversification of production	• Limiting risks with regard to losses in forest production • Increase in the availability of clean water	• Food and water security
Ecological	• Improved biodiversity • Improved micro-climate • Protection of hydrological recourses and increase of water availability • Reclamation of natural forests • Protection of forest species • Reduced soil erosion • Reduced wind velocity • Increase in soil fertility and content of organic matter • Limiting the occurrence of wildfires	• Intact ecosystem • Water availability • Water purity • Reduced degradation	• Reduced desertification • Reduced land degradation • Increase in resistance to climate change • Increased carbon sequestration • Decreased carbon emissions • Increased and protected biodiversity
Economical	• Vast range of products from forest resources • Income diversification	• Limitation of infrastructure damage (caused by desertification and wildfires) • Diversification of places of employment (including tourism) • Stimulation of economic growth	• Increased livelihood and well-being
Sociological and cultural	• Increase in local micro-economy, including the availability of forest recourses and possibilities of income • Improved local landscape • Protection of natural heritage	• Attractive landscape • Increase in awareness with regard to environmental protection	• Protection of environmental heritage

Sources: Beddington et al. [6]; Brewer et al. [9]; Lambin and Meyfroidt [29]; Yemiru et al. [36].

should be associated with forest management objectives and social goals. Therefore, *degradation* can be defined as the forest's capacity to produce the products and services desired by stakeholders [19].

20.4 Drivers of Forest Degradation

Worldwide, the most significant actions that cause adverse disturbances and ultimately forest degradation include unfeasible logging and extracting timber, unsustainably collecting fuelwood, unsustainably producing charcoal, overgrazing within forests, uncontrolled bushfires, and defaunation (i.e. life loss or wildlife reduction, which may provide significant ecological repercussions: for instance, regarding tree pollination, pest control, and seed dispersal) (Figure 20.2) [23]. Hidden drivers are the sophisticated associations of social, economic, political, and technological issues that influence direct drivers. These include inappropriate policies, weak leadership, insufficient law enforcement, landlessness, ambiguous rights allocation, rural deprivation, failure to invest and lack of financial assets, population growth and movement, and unrest (Figure 20.3) [7]. The difference between direct and hidden drivers, and between anthropogenic and natural alterations, is not often as precise as it may initially seem. Anthropogenic and natural drivers are often related, and degradation results from long, sophisticated causation chains [3, 45].

20.4.1 Strategies to Address Causes of Forest Degradation

Degradation may or may not be a predecessor of deforestation as forests may be degraded for a significant time but may not result in deforestation [9]. Some instances where degradation is a predecessor to deforestation are the deforestation of selectively logged areas with a few years of

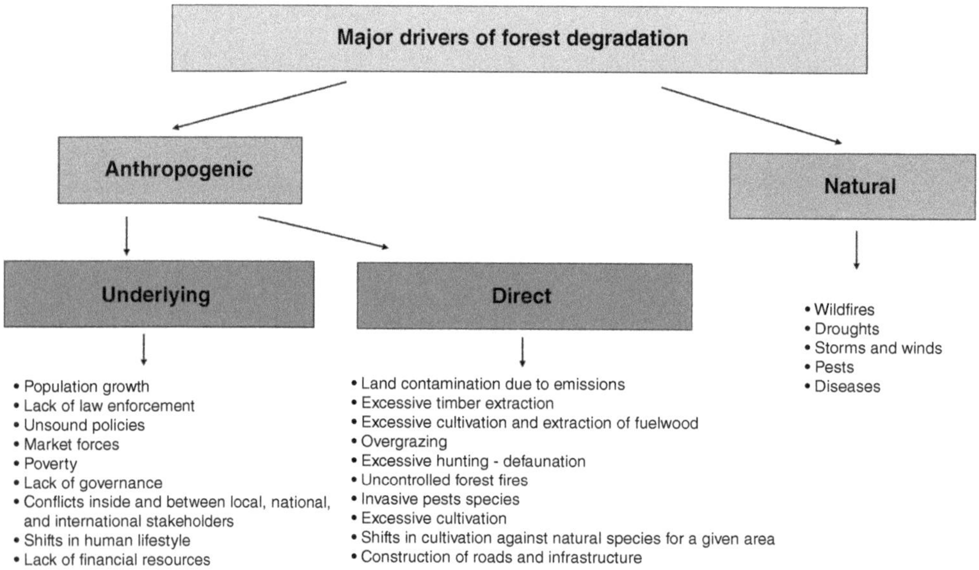

Figure 20.3 Overview and classification of the main drivers of forest degradation. *Source:* Based on Boucher et al. [7].

logging in the absence of sustainable forest management and the existence of deforestation pressures [33]. Thus, understanding the drivers and patterns of degradation is crucial in forest management, limiting the loss of biodiversity, and mitigating climate change [42]. New modeling studies, especially those able to predict the spatiotemporal patterns of forest degradation, can be used by stakeholders to extract the necessary recourses without the unsustainable loss of biodiversity and carbon storage [24, 25].

Strategies that address factors that lead to forest degradation should consider the potential impact on climate-change mitigation and adaptation, food security, and local livelihood. These measures should also be more focused on governance improvement and dealing with illegal activities. The intervention should consider direct and indirect drivers and levels, as well as a mixture of measures [18]. The primary responsibilities of forest managers are to develop actions and strategies at local levels and view forest degradation in the context of larger scales. Managers should not underestimate the time required to design programs to address these factors and reconcile all stakeholder interests [28].

20.4.2 The Hierarchy of Land Degradation Responses

Avoid-reduce-reverse is the hierarchy considered in land degradation neutrality (LDN) and is an overarching principle for the implementation of sustainable forest use (Figure 20.4) [38]. It is critical for interventions within the hierarchy to be prioritized based on the potential to enhance conservation of land-based natural capital while acknowledging that measures to reduce or avoid land degradation are cost-effective compared to reversing degradation.

LDN interventions apply to a single forest unit. However, at the landscape level, the application of the response hierarchy involves a mixture of protective strategies and implementation of sustainable forest management to reduce or prevent degradation alongside interventions that focus on localized restoration and rehabilitation to reverse the degradation process [28]. Degradation reversal is a counterbalance aimed at losses from new gradation in other land areas within one land type. This is evident from land-use and land-management interventions applied to every response hierarchy level, which focus on agricultural land [8].

The response hierarchy emphasizes the integration of planning of forest use at the environmental level as part of the LDN implementation process. It also focuses on identifying and selecting the

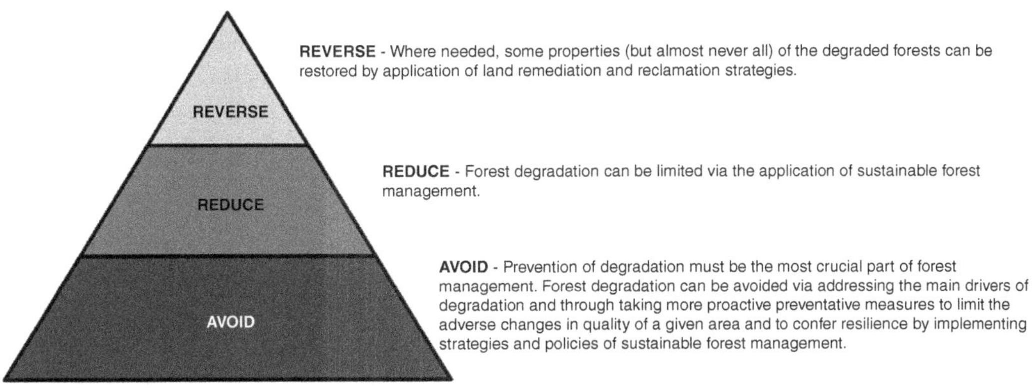

Figure 20.4 The hierarchy of the forest degradation neutrality response. *Source:* Based on Parrotta, Yeo-Chang, and Camacho [38].

most suitable combination of interventions to achieve LDN. Preliminary assessment within LDN planning involves assessing potential forest degradation, resilience, and relevant economic and social factors that decision-makers should consider when determining the most appropriate intervention [38].

Overall, the basic principles related to achieving forest production neutrality can be described as follows: [10, 12, 28]

1) Balance environmental sustainability with economic and social demand for forest-related products. Maintain or improve the quality of ecosystem services. Increase the efficiency of processes to minimize trade-offs between environmental protection and economic growth. Encourage LDN practices by integrating economic and social development and environmental protection within the limits of forest capital while avoiding spreading burdens to other regions or future generations.
2) Base all decisions about forest management on multi-variable assessments: forest production potential, its current condition, resilience to chosen factors, and social, cultural, and economic factors in a given region. Test management strategies at a limited local level before any large-scale intervention to ensure their evidence-based benefits and reduce possible risks.
3) Avoidance, prevention, and reduction need to have priority over reclamation of past degradation so that future actions are identified and implemented with a key aim of achieving no environmental degradation in the future.
4) Include local governance, stakeholders, and especially land users to search, design, implement, and monitor all actions and interventions in a given region. Applied strategies need to consider not only scientific knowledge but also local and traditional circumstances.
5) Encourage and implement good governance by removing or improving policies that lead to poor forest management.
6) Consider the availability of resources (both human and economic) in a given region for good practices to combat forest degradation.
7) Implement upward and downward accountability and transparency.

When the land for LDN intervention is determined at the planning stage, possible losses can be identified and counterbalanced to achieve greater gains through either restoration or rehabilitation [21].

20.5 The Role of Forest Management in Degradation Prevention

Forest degradation assessment at every site ensures the implementation of effective measures at the local scale by forest owners and managers [39]. Degradation indicators measured during regular forest assessment should be determined first. Early identification of local problems enhances the revision of the forest management plan, which is essential in preventing further degradation, dealing with the causes of degradation, enabling actions to be taken to restore the damage, and initiating a rehabilitation process [5]. Forest managers should acknowledge that indicators of a forest's capacity to provide goods and services may change over time and may not mean the forest is undergoing degradation [14].

Overall, forest degradation at the subnational, national, or global level is often approximated using larger scales to help in program design and policy implementation and payment mechanisms and incentive strategies to prevent degradation, such as paying for environmental services [10]. Larger-scale monitoring is required for national reporting and international processes like those associated with biodiversity and greenhouse gas emissions [28]. In addition, it remains a challenge to measure and monitor forest degradation at various levels. It is also time-consuming and expensive compared to deforestation. Therefore, the use of on-the-ground forest inventories and remotely sensed data can provide reliable approximations [40]. Remote sensing is a cost-effective option due to proxies, such as canopy cover percentage, where a decreasing trend means degradation. Focused field surveys such as rapid rural assessments and biodiversity assessments can be carried out in localities where remote sensing reports possible solutions [11].

The most appropriate monitoring approach is determined from parameters such as climate, type of vegetation in the forest, and degradation dynamic, including whether the degradation occurs on a large or small scale and whether it is diffuse or concentrated [20]. In some cases, degradation may directly result from forest management as different forests may be observed on different managerial boundaries. Also, degradation may become visible after many time-series of observations [13]. Some countries have a suitable national forest monitoring system to monitor forest degradation. However, achieving high certainty levels when monitoring forest degradation is expensive and time-consuming [34].

It should be considered in the REDD+ context whether degradation is likely to be a significant greenhouse gas emissions contributor. When monitoring is deemed necessary, it should emphasize areas that are susceptible to degradation [41]. The approximation of degradation is likely to be accurate due to the inclusion of a large number of variables and difficulties experienced when measuring the variables. However, even with the most efficient monitoring and measurement systems, it is difficult to estimate the yearly degradation rate. Therefore, a long-term approach is more suitable [40]. Sources of remotely sensed data used in monitoring forest degradation include active sensors such as radar and lidar and high-spatial-resolution optical methods such as RapidEye [18].

Thus, identifying and analyzing drivers of forest degradation is a complex, highly location-specific process that requires the analysis of direct and indirect causes [9]. Identifying the scale and areas of forest degradation involves an in-depth assessment of the degradation drivers that should be considered for every area. Moreover, stakeholders should participate in these assessments and help with data analysis to understand change dynamics [39].

20.5.1 Sustainable Forest Management (SFM) for Prevention of Degradation and the Restoration of Degraded Areas

The degradation of forests is a global problem, and it should be handled at every level (from local to regional, national, and finally, international). It is often the result of previous unsustainable forest management, overexploitation of forest resources, and factors outside the particular forest area under supervision [4]. Forest degradation may be deemed a warning sign that sustainable forest management (SFM) has not been attained. Implementing the strategies of SFM is ambiguous when its facilitating conditions – i.e. appropriate policies, leadership regimes, organizations, motivation, ownership, rights, honesty, and engagement of stakeholders – do not exist [5]. Forest and land managers and stakeholders can fight for an SFM-enabling environment, and they can play an essential role in stopping and changing forest degradation through their actions at the local level [30].

The most crucial work for forest managers is their adherence to the SFM principles in their approach to management, with the goals of ensuring that current use and practices maintain and promote the economic, social, and environmental benefits of forests while providing livelihoods (Table 20.3) [15]. Forest managers can explain and use precise management techniques, explain and respect the limits of harvesting, seek and determine forest degradation signals and environmental impacts, and take prompt actions to stop or change degradation – including asking for help and technical assistance when necessary [20]. In SFM forests, forest managers should continuously supervise the effects of management practices and evaluate whether they promote degradation. Overall, SFM is a flexible process in which plans and methods for managing forests should be adapted to management and evaluation and changing economic, social, and environmental circumstances. Forest degradation can be handled effectively using a broad approach that targets its direct and hidden drivers (Figure 20.3) [14].

When forest degradation results from natural causes, such as, storms, winds, pests, droughts, or wildfires, the aim of forest managers should be promoting forest resilience to facilitate the preparedness of stands for future occurrences [13]. This can be achieved, for instance, by maintaining biodiversity at different forest scales (e.g. stand, landscape, and region); utilizing integrated fire management (significantly reducing risk and recovery); using combined pest management; and (in planted forests) choosing tree species and varieties with possible resilience to the conditions anticipated in the future (Table 20.3). Forest and land managers can prevent degeneration by reestablishing and renovating forest landscapes and through the right planning of land utilization [8].

Table 20.3 Example of strategies and actions to prevent and halt forest degradation.

Problem/cause of degradation	Most commonly used solution	References
Unsustainable logging and excessive timber extraction	Implementation of sustainable forest management practices, which include: Certification Reinforcement and extension of protected areas Encourage agroforestation Encourage reforestation as a way to address the growing demand for forest-related products Strategies to reduce the impact of logging Changes in local forestry laws Open dialogue with local stakeholders Increase forest governance and strengthen law enforcement against illegal harvest, import, and export	Defries et al. [17] Yemiru et al. [36] Malik et al. [30]
Reduction in land productivity	Modify annual harvest patterns	Fry [22]
Lack of/reduced regeneration of forests after logging	Implement active practices to help with regeneration – enrichment planting, assisted regeneration with soil supplementation Shift logging polices to avoid future degradation Avoid limiting the population of tree species to the point where self-replacement becomes impossible Increase the biodiversity of the vegetative layer to avoid diseases and improve soil quality	Zeb et al. [48]
Unsustainable extraction of fuelwood and excessive charcoal production	Implement sustainable strategies of fuelwood extraction and charcoal production – increase awareness, local polices, law enforcement, and efficient stoves and heaters Increase agroforestry and reforestation practices as a response to growing demand for the production of fuelwood and charcoal Promote alternative fuels as energy sources (solar, wind, biogas) Support local communities in their transition to alternative sources of energy	Houghton [23] Zarin et al. [34]

Table 20.3 (Continued)

Problem/cause of degradation	Most commonly used solution	References
Natural disasters	Implement strategies to minimize the impact and strengthen resilience	Lambin and Meyfroidt [29]
Overgrazing	Increase silvo-pastoral systems Promote more sustainable grazing	Fahay et al. [20]
Wildfires	Fire prevention and management policies	Faria et al. [21]
Pests and diseases	Pest management – ensure sustainable practices – promote biodiversity, control entry pathways	Southworth et al. [44]
Soil erosion	Maintain and improve the growth of the vegetative layer Mulch to improve soil properties Reduce the impact of logging	Bustamante et al. (2016)

Source: Based on Das [15].

20.6 Conclusions – Prioritization and Implementation

Measures and actions of forest degradation prevention should be based on agreed criteria such as estimated costs, prevailing implementation capacities, and objectives by first consulting with local stakeholders. The consultation process can help determine the drivers to be given priority, the most effective action, and choice rationale. To create and choose the most beneficial cost-effect relations, the strategies and potential solutions to deal with forest degradation prevention and management of degraded areas need to be rooted in an understanding of bio-geo-chemical properties specific to a given area.

References

1 Ahenkan, A. and Boo, E. (2011). Improving the supply chain of non-timber forest products in Ghana. In: *Supply Chain Management: New Perspectives* (ed. S. Renko), 443–458. Rijeka: InTech Publisher https://doi.org/10.5772/19253.

2 Aragão, L.E.O.C., Anderson, L.O., Fonseca, M.G. et al. (2018). 21st century drought-related fires counteract the decline of Amazon deforestation carbon emissions. *Nature Communications* 9 (1) https://doi.org/10.1038/s41467-017-02771-y.

3 Arima, E.Y., Richards, P., Walker, R., and Caldas, M.M. (2011). Statistical confirmation of indirect land use change in the Brazilian Amazon. *Environmental Research Letters* 6 https://doi.org/10.1088/1748-9326/6/2/024010.

4 Averina, L.V., Voronin, V.V., Vlasov, A.G., and Vasilieva, D.I. (2014). Lands with special legal regime of use – problems and ways of their solution. *Problems of Regional Ecology* 5: 223–228.

5 Baldassini, P. and Paruelo, J.M. (2020). Deforestation and current management practices reduce soil organic carbon in the semi-arid Chaco, Argentina. *Agricultural Systems* 178: 102749. https://doi.org/10.1016/j.agsy.2019.102749.

6 Beddington, J., Asaduzzaman, M., Clark, M. et al. (2012). Achieving food security in the face of climate change: Final report from the Commission on Sustainable Agriculture and Climate Change. CGIAR Research Program on Climate Change, Agriculture and Food Security (CCAFS). Copenhagen, Denmark.

7 Boucher, D., Elias, P., Lininger, K. et al. (2011). *The Root of the Problem: What's Driving Tropical Deforestation Today?* Cambridge, Massachusetts: Union of Concerned Scientists.

8 Bustamante, M.M., Roitman, I., Aide, T.M. et al. (2015). Toward an integrated monitoring framework to assess the effects of tropical forest degradation and recovery on carbon stocks and biodiversity. *Global Change Biology* 22 (1): 92–109. https://doi.org/10.1111/gcb.13087.

9 Brewer, J., Langston, J.D., Ferretti-Gallon, K. et al. (2020). Alleviating forest degradation in the Lancang-Mekong region requires closing management – measurement gaps. *Journal of Forestry Research* 31: 2033–2051. https://doi.org/10.1007/s11676-020-01111-z.

10 Bullock, E.L., Woodcock, C.E., and Olofsson, P. (2020). Monitoring tropical forest degradation using spectral unmixing and Landsat time series analysis. *Remote Sensing of Environment* 238: 110968. https://doi.org/10.1016/j.rse.2018.11.011.

11 Charnley, S., Spies, T.A., Barros, A.M. et al. (2017). Diversity in forest management to reduce wildfire losses: implications for resilience. *Ecology and Society* 22 (1) https://doi.org/10.5751/es-08753-220122.

12 Chen, H., Zhu, T., Krott, M. et al. (2013). Measurement and evaluation of livelihood assets in sustainable forest commons governance. *Land Use Policy* 30 (1): 908–914. https://doi.org/10.1016/j.landusepol.2012.06.009.

13 Creutzburg, M.K., Scheller, R.M., Lucash, M.S. et al. (2017). Forest management scenarios in a changing climate: trade-offs between carbon, timber, and old forest. *Ecological Applications* 27 (2): 503–518. https://doi.org/10.1002/eap.1460.

14 Daniel, C.J., Ter-Mikaelian, M.T., Wotton, B.M. et al. (2017). Incorporating uncertainty into forest management planning: timber harvest, wildfire and climate change in the boreal forest. *Forest Ecology and Management* 400: 542–554. https://doi.org/10.1016/j.foreco.2017.06.039.

15 Das, N. (2010). Incidence of forest income on reduction of inequality: evidence from forest dependent households in milieu of joint forest management. *Ecological Economics* 69 (8): 1617–1625.

16 DeFries, R. and Pandey, D. (2010). Urbanization, the energy ladder and forest transitions in India's emerging economy. *Land Use Policy* 27: 130–138.

17 Defries, R.S., Rudel, T., Uriarte, M., and Hansen, M. (2010). Deforestation driven by urban population growth and agricultural trade in the twenty-first century. *Nature Geoscience* 3: 178–181.

18 Dupuis, C., Lejeune, P., Michez, A., and Fayolle, A. (2020). How can remote sensing help monitor tropical moist forest degradation? – a systematic review. *Remote Sensing* 12 (7): 1087. https://doi.org/10.3390/rs120.

19 Dwomoh, F.K., Wimberly, M.C., Cochrane, M.A., and Numata, I. (2019). Forest degradation promotes fire during drought in moist tropical forests of Ghana. *Forest Ecology and Management* 440: 158–168. https://doi.org/10.1016/j.foreco.2019.03.014.

20 Fahey, R.T., Alveshere, B.C., Burton, J.I. et al. (2018). Shifting conceptions of complexity in forest management and silviculture. *Forest Ecology and Management* 421: 59–71. https://doi.org/10.1016/j.foreco.2018.01.011.

21 Faria, B.L., Brando, P.M., Macedo, M.N. et al. (2017). Current and future patterns of fire-induced forest degradation in Amazonia. *Environmental Research Letters* 12 (9): 095005. https://doi.org/10.1088/1748-9326/aa69ce71087.

22 Fry, I. (2008). Reducing emissions from deforestation and Forest degradation: opportunities and pitfalls in developing a new legal regime. *Review of European Community & International Environmental Law* 17 (2): 166–182. https://doi.org/10.1111/j.1467-9388.2008.00597.x.

23 Houghton, R.A. (2012). Carbon emissions and the drivers of deforestation and forest degradation in the tropics. *Current Opinion in Environment Sustainability* 4: 597–603.

24 Jahani, A., Feghhi, J., Makhdoum, M.F., and Omid, M. (2015). Optimized forest degradation model (OFDM): an environmental decision support system for environmental impact assessment using an artificial neural network. *Journal of Environmental Planning and Management* 59 (2): 222–244. https://doi.org/10.1080/0964056.

25 Jaskulak, M., Grobelak, A., and Vandenbulcke, F. (2020). Modelling assisted phytoremediation of soils contaminated with heavy metals – Main opportunities, limitations, decision making and future prospects. *Chemosphere* 249: 126196. https://doi.org/10.1016/j.chemospher.

26 Jaskulak, M. and Grobelak, A. (2019). Cadmium phytotoxicity – biomarkers. In: *Cadmium Tolerance in Plants* (eds. M. Hasanuzzaman, M.N.V. Prasad and K. Nahar), 177–191. Elsevier Inc https://doi.org/10.1016/b978-0-12-815794-7.00006-0.

27 Khasaev, G., Vlasov, A., Vasilieva, D., and Parsova, V. (2018). Assessment of the role of forest land in Samara region as environmental protection and land degradation prevention factor. *Baltic Surveying* 8: 38–45. https://doi.org/10.22616/j.balticsurveying.2018.005.

28 Kim, Y., Bae, J.S., Fisher, L.A. et al. (2016). Indonesia's forest management units: effective intermediaries in REDD+ implementation? *Forest Policy and Economics* 62: 69–77. https://doi.org/10.1016/j.forpol.2015.09.004.

29 Lambin, E. and Meyfroidt, P. (2011). Global land use change, economic globalization, and the looming land scarcity. *Proceedings of the National Academy of Sciences of the United States of America* 108 (9): 3465–3472.

30 Malik, A., Mertz, O., and Fensholt, R. (2017). Mangrove forest decline: consequences for livelihoods and environment in South Sulawesi. *Regional Environmental Change* 17: 157–169. https://doi.org/10.1007/s10113-016-0989-0.

31 Masda, Y., Magi, M., Nanao, H. et al. (2002). Coastal erosion due to long–term human impact on mangrove forests. *Wetlands Ecology and Management* 10: 1–9.

32 Muttaqin, M.Z., Alviya, I., Lugina, M. et al. (2019). Developing community-based forest ecosystem service management to reduce emissions from deforestation and forest degradation. *Forest Policy and Economics* 108: 101938. https://doi.org/10.1016/j.forpol.2019.05.024.

33 Pearson, T.R.H., Brown, S., Murray, L. et al. (2017). Greenhouse gas emissions from tropical forest degradation: an underestimated source. *Carbon Balance and Management* 12: 3. https://doi.org/10.1186/s13021-017-0072-2.

34 Zarin, D.J., Harris, N.L., Baccini, A. et al. (2015). Can carbon emissions from tropical deforestation drop by 50% in five years? *Global Change Biology* 22: 1336–1347. https://doi.org/10.1111/gcb.13153.

35 van der Werf, G.R., Randerson, J.T., Giglio, L. et al. (2010). Global fire emissions and the contribution of deforestation, savannah, forest, agriculture, and peat fires (1997–2009). *Atmospheric Chemistry and Physics* 10: 11707–11735.

36 Yemiru, T., Roos, A., Campbell, B.M., and Bohlin, F. (2010). Forest incomes and poverty alleviation under participatory forest management in the bale highlands, Southern Ethiopia. *International Forestry Review* 12 (1): 6677.

37 Kamanga, P., Vedeld, P., and Sjaastad, E. (2009). Forest incomes and rural livelihoods in Chiradzulu District, Malawi. *Ecological Economics* 68: 613–624.

38 Parrotta, J., Yeo-Chang, Y., and Camacho, L.D. (2016). Traditional knowledge for sustainable forest management and provision of ecosystem services. *International Journal of Biodiversity*

Science, Ecosystem Services and Management 12 (1–2): 1–4. https://doi.org/10.1080/21513732.2016.1169580.

39 Santika, T., Meijaard, E., Budiharta, S. et al. (2017). Community forest management in Indonesia: avoided deforestation in the context of anthropogenic and climate complexities. *Global Environmental Change* 46: 60–71. https://doi.org/10.1016/j.gloenvcha.2017.08.0028.2015.1005732.

40 Schall, P., Schulze, E., Fischer, M. et al. (2018). Relations between forest management, stand structure and productivity across different types of central European forests. *Basic and Applied Ecology* 32: 39–52. https://doi.org/10.1016/j.baae.2018.02.007.

41 Sharma, P., Thapa, R.B., and Matin, M.A. (2019). Examining forest cover change and deforestation drivers in Taunggyi district, Shan State, Myanmar. *Environment, Development and Sustainability* 22: 5521–5538. https://doi.org/10.1007/s10668-019-00436-y.

42 Silvério, D.V., Brando, P.M., Bustamante, M.M. et al. (2018). Fire, fragmentation, and windstorms: a recipe for tropical forest degradation. *Journal of Ecology* 107 (2): 656–667. https://doi.org/10.1111/1365-2745.13076.

43 Singh, M., Bhojvaid, P., Jong, W.D. et al. (2017). Forest transition and socio-economic development in India and their implications for forest transition theory. *Forest Policy and Economics* 76: 65–71. https://doi.org/10.1016/j.forpol.2015.10.013.

44 Southworth, J., Nagendra, H., and Cassidy, L. (2010). Forest transition pathways in Asia: studies from Nepal, India, Thailand and Cambodia. *Journal of Land Use Science* 7 (1): 51–65.

45 Tacconi, L., Rodrigues, R.J., and Maryudi, A. (2019). Law enforcement and deforestation: lessons for Indonesia from Brazil. *Forest Policy and Economics* 108: 101943. https://doi.org/10.1016/j.forpol.2019.05.029.

46 Thangata, P. and Hildebrand, P. (2012). Carbon stock and sequestration potential of agroforestry systems in smallholder agroecosystems of sub-Saharan Africa: mechanisms for 'reducing emissions from deforestation and forest degradation' (REDD). *Agriculture, Ecosystems & Environment* 158: 172–183. https://doi.org/10.1016/j.agee.2012.06.007.

47 Vlasov, A.G., Voronin, V.V., and Vasilieva, D.I. (2013). Legislative base of land and property complex. *Problems of regional ecology* 4: 117–121.

48 Zeb, A., Hamann, A., Armstrong, G.W., and Acuna-Castellanos, D. (2019). Identifying local actors of deforestation and forest degradation in the Kalasha valleys of Pakistan. *Forest Policy and Economics* 104: 56–64. https://doi.org/10.1016/j.forpol.2019.04.005.

21

Use of Plants for Air Quality Improvement

Richa Rai[1], Madhoolika Agrawal[2], and S.B. Agrawal[2]

[1] *Plant Ecology and Climate Change Science Division, CSIR National Botanical Research Institute, Lucknow, Uttar Pradesh, India*
[2] *Department of Botany, Institute of Science, Banaras Hindu University, Varanasi, Uttar Pradesh, India*

21.1 Introduction

Air pollution has emerged as a significant problem in urban areas of developing countries. A significant cause of air pollution is unchecked, uncontrolled, and unprecedented urbanization, which has led to an increase in construction sites, motor vehicles, and industrialization. Air pollutants cause the degradation of air quality. Deteriorated air quality affects the health of the ecosystem and vegetation and contributes significantly to the global disease burden [1, 2]. The World Health Organization (WHO) has reported that air pollutants cause adverse human health issues like cardiovascular, respiratory, and pulmonary disease [3].

The health effects of air pollution remain a public health concern worldwide (Figure 21.1). Detrimental effects of air pollution are confirmed to be associated with morbidity and mortality. PM with a diameter of less than 2.5 μm ($PM_{2.5}$) and O_3 are the primary air pollutants of health concern [6]. Air pollutants such as O_3 can increase the risk of appendicitis, leading to the risk of a perforated appendix. A higher concentration of ambient air pollutants is harmful to people suffering from chronic pulmonary disease. Air pollution causes a decrease in pulmonary function, increased infections, increased respiratory symptoms, acute exacerbations of chronic obstructive pulmonary disease (COPD), the onset of asthma, and the frequency of childhood asthma. Long-term exposure to ambient air pollution can damage lung function in children. Air pollution also leads to lung cancer and a risk of childhood asthma after exposure to NO_2. The prevalence of chronic phlegm, a major symptom of chronic respiratory disease, indicates long-term exposure to inhaled irritants primarily related to exposure to combustion-source air pollution.

Particulate pollution is mostly associated with respiratory ailments [7, 8] as well as low birth rate, fetal growth characteristics and preterm birth [9], DNA damage, mutagenic activity, congenital heart defects [10], ischemic heart disease [11], inflammatory responses [12], infant mortality [13], oxidative stress [14], and atherosclerosis [15]. People with asthma are more sensitive to SO_2, and brief exposure triggers shortness of breath [16]. Particles less than 10 μm in diameter may get deep into the lungs, and some may get into the bloodstream.

Handbook of Ecological and Ecosystem Engineering, First Edition. Edited by Majeti Narasimha Vara Prasad.

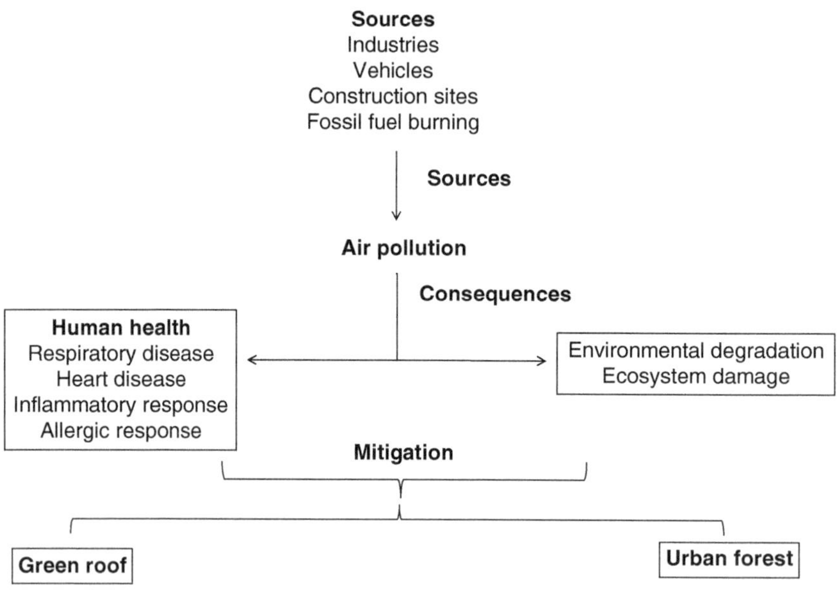

Figure 21.1 Sources, consequences, and mitigation of air pollution in urban areas. *Sources:* Abhijith et al. [4]; Janhall [5].

The harmful impacts on human health have motivated research into methods to reduce pollutant emission by control techniques and using urban vegetation; developing green barriers using trees, hedges, green walls, and green roofs on open roads and streets may help improve air quality. Vegetation is preferably used to mitigate the harmful impact of air pollutant as it can clean the air by filtering out pollutants. Plants remove air pollutants by absorbing gaseous pollutants through their stomata, deposition of particulate and aerosols over leaf surfaces, and fallout of particulates on the vegetation's leeward side [5] (Figure 21.1).

This chapter highlights different air pollutants' impact on human health and measures to improve air quality through urban greening or developing green barriers and plant traits to be used in selecting plant species for vegetation.

21.2 Current Status of Air Pollutants

Anthropogenic emissions have resulted in increased sulfur deposition with high atmospheric sulfate loadings near most industrialized areas. Sulfur dioxide (SO_2) forms sulfate aerosols that have a significant effect on global and regional climate. Sulfur is ubiquitous in the biosphere and mostly found in fossil fuels and with coal and crude oil deposits. The primary source of SO_2 emissions in the atmosphere is fossil fuel combustion [17]. Industrial sources, thermal power plants, and transport sectors are identified as the sole contributors to SO_2 emissions. SO_2 sulfate aerosols were the principal component of the historic "London Smog" that caused over 1 million premature deaths each year [18, 19].

SO_2 emissions have been reduced in most the developed and developing countries due to stringent pollution control measures. India and China are major contributors of SO_2 emissions and current haze problems [20, 21]. The main reason is that China and India are the top two global coal consumers [22]. Despite the large reductions in SO_2, Li et al. [23] estimated the annual SO_2 emissions from China and India during 2005–2016 and showed that emissions from China were highest, at 36.6 Mt (106 tons)/year in 2007 and gradually decreased to around 8.4 Mt/year in 2016.

For India, estimated emissions are 9.5–12.6 Mt/year (2016), and it is predicted that in the future, India will become the top emitter of SO_2. In both countries, 70–90% of SO_2 emissions are from coal. Mukherjee et al. [24] reported 21 μg/m^3 SO_2 concentrations in the city site of Varanasi, which was below the threshold of the Central Pollution Control Board (CPCB). SO_2 concentrations in metro cities in India like Delhi, Mumbai, Chennai, and Kolkata were measured at 2.3, 2.3, 3.4, and 6.1 ppb, respectively [25].

Nitrogen dioxide (NO_2), a highly reactive gas, is formed in the ambient air through the oxidation of nitric oxide (NO). Nitrogen oxides occur in the atmosphere naturally and as a result of human activities. NOx is emitted from both anthropogenic sources (transportation, power plants, and fertilizers) and natural sources (biomass burning, lightning, and soil) [26]. This results from a chemical reaction between atmospheric N_2 and O_2 in the presence of heat to form NO, which then reacts again with O_2 to form NO_2. The rate of reaction is determined by the temperature of combustion. Thus, unlike SO_2, NO_2 is not a component of fossil fuel but results from a catalytic reaction of heat with atmospheric N_2 and O_2 in the combustion process. NOx reacts with volatile organic compounds (VOCs) under sunlight to form tropospheric ozone [27] and particulate nitrate, an important component of $PM_{2.5}$.

Anthropogenic emissions of NOx lead to the majority of all nitrogen inputs to the environment. The US EPA showed that annual NO_2 concentrations decreased by 30% from 1990 to 2006 [28]. In China, the number of vehicles has increased by sixfold since 1980 [29], resulting in an increase of NOx concentrations to 12 Tg N/year [30].

During the last few decades, tropospheric ozone has been identified as an important secondary pollutant of concern worldwide that affects humans, animals, and plants [31–33]. Tropospheric O_3 is formed by a chain of photochemical reactions that involve various precursors such as methane (CH_4), carbon monoxide (CO), nitrogen oxides (NOx), and VOCs in the presence of bright sunlight [31].

Background concentrations of ground-level O_3 have increased from an estimated preindustrial concentration of 10 ppb to average summer concentrations varying from 30 to 50 ppb in mid-latitudes of the northern hemisphere and O_3 episodes reaching 100 ppb [34]. According to the current emission scenario, it is predicted that global surface O_3 will increase by 1.5 ± 1.2 ppb on average by 2030; and for the IPCC SRES A2 scenario, by 4.3 ± 2.2 ppb, with the most substantial increases in South Asia, Southeast Asia, and the Middle East [35].

Monitoring reports of O_3 in suburban and rural areas of Asia also showed that mean monthly O_3 concentration now commonly reaches to 50 ppb [36]. [37] reported an increase of 0.4 ppb/year of O_3 concentration for the north tropical Atlantic during 1977–2002 through in situ measurements in the marine boundary layer. For the tropical Pacific, a positive trend of 0.14 ppb/year was recorded for Hawaii [38]. Ebojie et al. [39] showed a significant increase in tropospheric O_3 at a rate of 2%/year in South America, 1–3%/year in South Asia (1–3%/year), 2%/year in parts of Africa, and 2%/year over the marine region. Sharma et al. [40], using the WRF-Chem model, simulated O_3 concentration over the South Asian region in the range of 30–70 ppbv, with the highest levels (>50 ppbv) seen over the eastern regions of India, the Indo-Gangetic Plain (IGP), and parts of southern India. The mean ozone (O_3) concentration recorded at the Doon Valley varied from 13.7 to 71.3 ppb, with the peak O_3 concentration reaching 101.3 ppb [41].

21.3 Green Roofs, Urban Forests, and Air Pollution

Air quality today is a primary health concern as 54% of the world's population lives in urban areas – and this number is predicted to increase to 66% in 2050 [42]. Traffic emissions are the primary source of air pollution in cities around the globe. To improve air quality, various measures are taken, like low-emission zones and stringent use of technology to check air pollution. In addition to these policies, there is a requirement to develop green roofs and green space to mitigate urban air pollution (Table 21.1).

A range of passive methods have been identified to reduce personal exposure to primary pollutant concentrations in the built environment. Passive barriers are used for air quality improvement, including porous or solid barriers, and partially or fully act as baffles between a pollutant source and porous (trees and vegetation) or solid (noise barriers, low boundary walls, and parked cars) barriers [53]. These mechanisms can improve air quality and provide healthier conditions for urban dwellers [53, 54].

They include streets, trees, vegetation barriers (including hedges), green (or living) walls, and green (or living) roofs. A green roof, i.e. a roof with a vegetation cover, is a passive technique to mitigate the harmful effect of air pollution in an urban setting. Green roofs come in three forms: extensive, intensive, and semi-intensive. An extensive green roof has 50–150 mm of growing medium to grow plants, and it limits the range of plants based on its size and weight. Due to the shallow and fragile root system of the vegetation, foot traffic is not allowed. Intensive green roofs generally have from 150 to 1200 mm of growing medium, which helps for growing bushes and trees. These green roofs can withstand foot traffic and give aesthetic value to the building. Semi-intensive green roofs combine both intensive and extensive green roofs, and 5% of the growing medium is under or over 15 cm [47]. Semi-intensive roofs allow for more types of plants than intensive green roof systems but are not as diverse as extensive green roofs.

Currie and Bass [55] recorded the role of vegetation, i.e. trees, shrubs, and grass, in removing air pollutants in the greater Toronto area's midtown region using the UFORE model (Urban Forest Effects). The modeled data showed that intensive green roofs with shrubs removed significant air pollutants compared to extensive grass. Using shrubs as baseline and increasing green roofs by 20% could remove 4.48, 1.74, 1.24, and 0.41 metric tons/years of PM_{10}, O_3, NO_2, and SO_2, respectively.

Table 21.1 Air pollutants removed by green roofs and urban forests.

References	City	Green roof/urban forest	Types of pollutant removed
Speak et al. [46]	Manchester, UK	Red fescue	PM_{10}
Speak et al. [46]	Manchester, UK	Creeping bentgrass	PM_{10}
Speak et al. [46]	Manchester, UK	Ribwort, Plantin	PM_{10}
Yang et al. [47]	Chicago, US	Grass, herbaceous plants, trees/shrubs	Total particulate, PM_{10}, O_3, NO_2, SO_2
Gourdji [48]	Montreal, Canada	Pinus strobes, *Acer* palmatus, Magnolia	PM, O_3, NO_2
Sicard et al. [49]	Mediterranean forest near Rome, Italy	*Acer* sps., *Ailanthus* sps., *Casurina*, *Eucalyptus*, *Populus*, *Quercus*	O_3, NO_2, PM
Weber et al. [50]	Berlin, Germany	*Sisymbrium loeselii*, *Polygonum aviculare*, *Convolvolus arvensis*, *Chenopodium album*	PM
Bottalico et al. [51]	Florence, Italy	Deciduous trees, Coniferous trees	O_3 and PM
Morikawa et al. [52]	Ontario, Canada	Kalanchoe blossefeldiana	NO_2

Source: Clark et al. [43]; Davies et al. [44]; Jayasooriya et al. [45].

As a part of green roof structures, trees effectively reduce air pollutants to the greatest degree due to their higher surface area. Research in Los Angeles measured the effects of tree planting and re-roofing in lighter colors on ambient temperatures and air pollution. The results confirmed that combining trees with cool roofs could lower the ambient temperature in Los Angeles by 3 °C and cool the air around buildings [56]. Smog formation occurs when nitrogen oxides (NOx) react with relative organic compounds released from the incomplete combustion of fossil fuels and accelerated under higher temperatures. Cooler air reduces the reaction rate, and hence the formation of smog is reduced. Akbari et al. [56] calculated that daytime temperature reductions would decrease reliance on air conditioning and reduce NOx emissions from coal-fired electricity plants, resulting in an estimated 10% reduction in smog precursors or a reduction of 350 tons of NOx/day.

Johnson and Newton [57] calculated that uncut grass on a green roof in an area of 2000 m^2 can remove 4000 kg of PM. Clark et al. [43] showed that covering approximately 20% of all industrial and commercial roof surfaces in Detroit, Michigan with sedum green roofs would lead to the removal of 800 000 kg/year of NO_2; and in Singapore, it would lead to a reduction of 37 and 21% of sulfur dioxide and nitrous acid [58]. In Chicago, 19.8 ha of green roof, equivalent to 71% of the total green roof area, consisting of 67.42% intensive or semi-intensive, removed 1675 kg of pollutants, including 52, 27, 14, and 7%, respectively, of O_3 NO_2, PM_{10}, and SO_2 [47]. The study showed that intensive green roofs consisting of a 50 : 50 ratio of tall herbaceous plants to shrubs with small deciduous trees would remove 2046.89 metric tons of air pollutants.

Speak et al. [46] reported that extensive green roofs in Manchester, England, of 50 ha with red fescue, creeping bentgrass, ribwort plantin, and sedum could remove 1.61 ± 0.05, 0.9 ± 0.03, 0.25 ± 0.01, and 0.21 ± 0.01 metric tons/year of PM_{10}, respectively; maximum removal was observed with red fescue.

Maximum reductions in PM, O_3, NOx, and SO_2 occur when plants are actively growing and in leaf. Individual plant species also exhibit vast differences in their ability to take up pollutants [52, 59]. Morikawa et al. [52] analyzed 217 species against pollutant uptake, including *Kalanchoe blossfeldiana* and sedum, a succulent belonging to the Crassulaceae family that has the maximum potential to remove NO_2 (Table 21.1). Clark et al. [43] compared the annual uptake of NO_2 for the succulent *K. blossfeldiana*, herbaceous species (*Nicotiana tabaccum*), and trees (*Eucalyptus viminalis*). The tree had more uptake capacity than succulent and herbs. Tobacco had 30 times more uptake capacity of NO_2 than sedum, suggesting that it may not be the ideal plant for air pollution mitigation, which is the primary objective of a green roof. Deciduous broadleaf species tend to remove O_3 most effectively [60, 61]. Deciduous trees are found to take up 90% of the total O_3 [60, 61]. In Montreal, deciduous broadleaved vegetation on green roof structures reduced O_3 pollution levels [48]. Benjamin and Winer [62] found that *Magnolia* emits only 0.6 g monoterpene/tree.day and has an O_3-forming potential of 2 g O_3/tree.day, due to which *Magnolia* could be preferred for O_3 mitigation in a green roof.

Another approach to mitigating urban air pollutants is the development of urban green space or forests. Urban green space is defined as a network of natural and semi-natural green spaces such as forest, parks, green roofs, and walls, and can provide nature-based and cost-effective solutions [63] and contribute to ecosystem resilience and human well-being through ecosystem services from urban centers to peri-urban areas [44, 45] (Table 21.1). Urban vegetation removes gaseous pollutants by absorption through leaf stomata or plant surfaces ([64]). Studies have shown that pollutant removal depends on the deposition velocity, pollutant concentration, tree cover, length of leaf season, precipitation, and other meteorological variables that affect tree transpiration [65–67]. Air pollutants and PM are deposited on the surface when they pass close to it. Most plants have a large surface area per unit volume, increasing the probability of deposition.

Nowak and Crane [68] have developed a deposition model that can estimate the pollutant removal capacity of an urban forest. Urban forests can affect air pollution by (i) directly removing pollutants, (ii) emitting atmospheric chemicals directly from vegetation or indirectly through vegetation maintenance practices, (iii) altering urban microclimates (e.g. reducing air temperature), and (iv) altering building energy use and consequently emissions from power plants.

Planning and management of urban green space is a crucial issue in the compact city concept, as these spaces provide essential benefits to urban dwellers [69, 70]. Urban forests – the stock of trees in urban areas – can improve environmental conditions and quality of life in cities by providing multiple ecosystem services. They have important aesthetic and cultural values, and urban trees can, for instance, increase biodiversity, mitigate the heat island effect, reduce stormwater runoff, sequester carbon, and filter pollutants from the air (e.g. [68, 71]). Urban forests and trees can also have detriments, though, such as blocking sunlight, generating leaf litter, being hosts to pests, producing allergenic pollen and VOCs as precursors to ozone, and being hazardous during storms. The benefits and disadvantages of the urban forest are influenced by its planning and management, such as location and configuration of tree plantings, the density of tree stands, species selection, and more.

Hourly air-quality improvement in Brooklyn due to pollution removal by trees during the day when the trees were in leaf averaged 0.26% for O_3 and SO_2, 0.25% for PM_{10}, 0.17% for NO_2, and 0.001% for CO. Air pollution removal by trees at night is minimal due to stomatal closure. Air quality improves with increased tree cover and decreased boundary-layer heights. In urban areas with 100% tree cover (i.e. contiguous forest stands), short-term improvements in air quality (one hour) due to pollution removal by trees were as high as 14.7% for SO_2, 14.2% for O_3, 10.5% for PM_{10}, 6.8% for NO_2, and 0.05% for CO. Individual large trees removed up to 65 times more pollution than small trees; large trees remove up to 1.95 kg/year versus 0.03 kg/year for small trees.

Removal of gaseous pollutants by trees can be permanent, while trees typically serve as a temporary retention site for particles. The removed particles can be resuspended in the atmosphere during turbulent winds, washed off by precipitation, or dropped to the ground with leaf and twig fall [72]. Trees can also act as barriers between sources and populations, although vegetation is inherently more complicated to study than solid structures, and the effectiveness of vegetative barriers for reducing ultrafine particle (UFP) concentration has been shown to be variable [73]. This variability is likely due to a number of confounding factors. The complex and porous structures of trees and bushes can modify near-road concentrations via pollutant capture or through altering airflow, which can result in either diminished dispersion through the reduction of wind speed and boundary layer heights [72, 74] or enhanced dispersion due to increased air turbulence and mixing. Recirculation zones have also been observed immediately downwind of forested areas with a flow structure consistent with an intermittent recirculation pattern (Detto et al. 2008; Frank and Ruck 2008).

Using dry deposition models (UFORE/i-Tree), a number of studies showed that annual O_3 removal by green roofs from 1.2 to 4.4 g/m^2/year [45, 75]. Deutsch et al. [75] reported that 109 ha of green roofs removed 3.1 tons of O_3 in Toronto [55], and 202 ha of green roofs removed 6 tons of O_3 in Washington, D.C., equivalent to approximately 25 000 street trees.

Sicard et al. [49], using a dry deposition model, reported average annual O_3 removal of 5.4 g/m^2 of tree cover/year in 55 US cities, 3.7 g/m^2/year in 86 Canadian cities, and 3.3 g/m^2/year in 10 Italian cities in 2003. The study also found that the O_3 removal rate varied worldwide depending on the amount of air pollution, length of in-leaf season, precipitation and other meteorological conditions. The rates of removal were recorded in Melbourne (0.9 g/m^2/year), Stockholm (1.3/g^2/year), Montreal (3.9 g/m^2/year), Beijing (5.5 g/m^2/year), Florence (7.3 g/m^2/year), and San Diego (7.6 g/m^2/year).

Coniferous trees tend to be among the most effective in capturing PM due to their intricate needle structure [76] and foliage structure, which is associated with greater deposition velocity of PM. Small conifers like *Pinus strobus* "Nana," *Pinus mugho* var. pumilio, *P. mugho* "Slowmound," and *Pinus pumila* "Dwarf Blue" may mitigate particulate pollution due to their intricate needle structures [48].

21.4 Traits for Phytoremediation of Air Pollution

Urban vegetation removes gaseous pollutants by absorption through leaf stomata or plant surfaces [64, 77]. Studies have shown that pollutant removal depends on the deposition velocity, pollutant concentration, tree cover, length of in-leaf season, amount of precipitation, and other meteorological variables that affect tree transpiration [65–67] (Figure 21.2).

Plants play an essential role in removing air pollutants from the atmosphere in urban areas, depending mainly on their leaves' surface area to collect PM (Beckett et al. 1998; [66]). The capacity of plants to accumulate PM indicates considerable species-specific differences that depend primarily on the properties of the leaf surfaces. Leaf surfaces are composed of cuticles and superimposed epicuticular waxes with considerable ultrastructural and chemical diversity [79] (Figure 21.2).

Tree parameters such as crown size, porosity and leaf density, height, and spacing are responsible for dispersing pollutants. Trees also remove pollution by intercepting airborne particles. Some particles can be absorbed into the tree, although most intercepted particles are retained on the plant surface [80, 81]. Vegetation is a temporary retention site for atmospheric particles as the intercepted particles may be resuspended into the atmosphere, washed off by rain, or dropped to the ground with leaf and twig fall [82]. Factors that affect pollution removal by trees include the amount of healthy leaf-surface area, concentrations of local pollutants, and local meteorology.

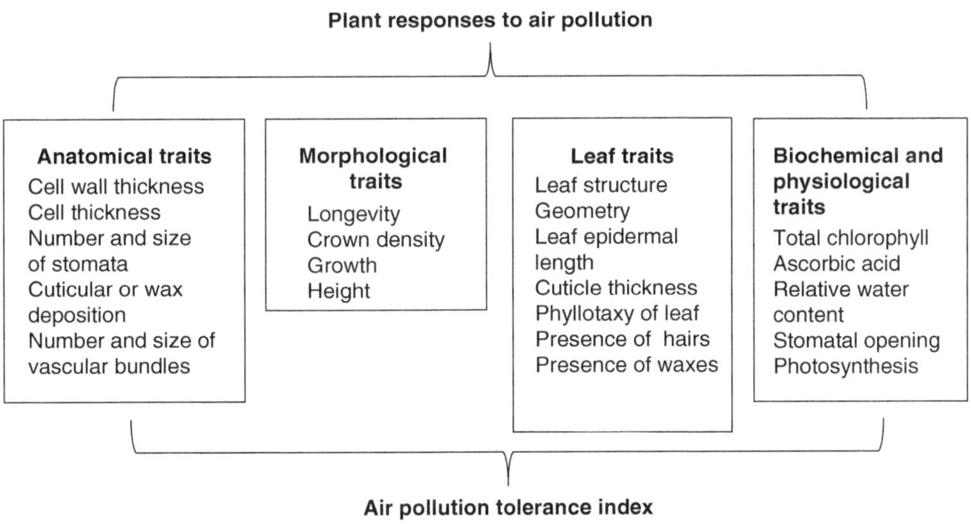

Figure 21.2 Traits contributing to identifying the Air Pollution Tolerance Index (APTI). *Sources:* Nowak [65]; Nowak et al. [66]; Nowak et al. [67]; Prusty, Mishra, and Azeez [78].

Modeling the effects of increased urban tree cover on O_3 concentrations from Washington, DC to central Massachusetts showed that urban trees generally reduce O_3 in cities, though average concentrations tend to increase slightly in the overall modeling domain. Interactions of the effects of trees on the physical and chemical environment demonstrate that trees can cause changes in pollution removal rates and meteorology, particularly air temperature, wind fields, and boundary layer heights, which may influence O_3 concentrations [72].

Traits important for phytoremediation of air pollution include the tree crown projection area and crown volume, which can be used as proxy variables for leaf area and leaf biomass [83]. Tree crown size is dependent on tree crown projection area and crown volume. Reductions in plant height, basal diameter, canopy area, root-shoot ratio, and plant biomass were also recorded by Pandey and Agrawal [84] in a tropical urban environment, with maximum reduction in *Delonix regia* followed by *Cassia Fistula* and *Carissa carandas*. Ajuru and Friday [85] studied the effects of PM on the anatomy of three tropical plants, *Alchornea cordifolia*, *Musa paradisiaca*, and *Manihot esculenta*, in Port Harcourt, Nigeria, and found significant effects of pollution load, with an increase in the number of vascular bundles, although no effect was observed in cortex, epidermis, and pith tissues. To avoid air pollution effects from traffic, plants develop different strategies including reductions in leaf area, specific leaf area, leaf dry matter content, and leaf water content [86, 87]. Prusty et al. [78] reported that the dust accumulation potential was higher in the leaves of *Ipomoea carnea*, *Tabernaemontana divaricata,* and *Pongamia pinnata* compared to *Ficus religiosa* and *Quisqualis indica* near the national highway at Sambalpur, Orissa, India, which suggests that the roughness of the leaf surface, the small size of the petiole, and the smaller height of plants contribute more to dust accumulation.

Leaf morphological traits such as structure, geometry, epidermal length, cuticle thickness, phyllotaxy, plant height, canopy size, and nature of the tree determine plant responses at sites with greater dust pollution [78, 88]. Dust accumulation was greater in shorter plants than taller plants [78] (Figure 21.2). Nighat et al. [88] recorded reductions in leaf area, length and width of stomata, length of the stomatal pore, and stomatal density in *Ruellia tuberosa* at the site of thermal power plant emissions in Aligarh, India, whereas the stomatal index was unaffected.

Leaf functional traits such as leaf mass per area or specific leaf area, tissue thickness, and leaf area are considered environmental conditions. Therefore, these characteristics provide the opportunity to correlate variations in relation to changing environment under natural and modified systems. The dust-capturing capacity of plants depends on plant height, density, size, age and area of leaves, canopy structure, leaf inclination, and external characteristics such as cuticle hair, cuticle, and meteorological factors [89]. A more comprehensive index called the Anticipated Pollution Index (API) also indicates the suitability of plants for green belt development.

21.4.1 Physiological and Biochemical Traits

Biochemical traits include ascorbic acid, total chlorophyll, leaf extract, pH, and relative water content (RWC), and these parameters are used to calculate Air Pollution Tolerance Index (APTI) (Figure 21.2). Species that score higher APTI values are considered the most tolerant species in polluted environments and are preferred for the development of green belts. Increases in antioxidants and defense-related metabolites are the primary response of plants against air pollution stress. Ninave et al. [90] also observed an increase in ascorbic acid content in plants in a polluted environment.

The Air Pollution Index (API) is examined by evaluating the various biochemical, socioeconomic, and biological characteristics of the species, such as canopy structure, APTI plant habit, and economic value [86]. Roy et al. [91] identified *Magnifera indica* L., *Azadirachta indica* L., and *Ficus religosa* L. as the most tolerant species, as they had high APTI, while *Ficus bengalensis* and *Alstonia scholaris* were intermediate toward air pollution due to low their APTI at the Jharkhand industrial site.

Verma and Singh [87] recorded reductions in Chl a, Chl b, and carotenoid contents by 6–64% at higher pollution load in both *Thevetia nerifolia* and *F. religiosa* in Lucknow, India, indicating a decrease in productivity with a higher load of air pollutants. A similar trend was also observed for foliar protein content in both plants, indicating inhibition in protein synthesis or damage to existing protein, although the percent reduction in protein content was lower than the photosynthetic pigments, indicating more sensitivity of pigments to air pollutants [87]. In contrast, Ninave et al. [90] found an increase in Chl content in all the studied plants except *Polyalthia longifolia* in polluted environments. Based on the plants' responses, *A. indica* was identified as a tolerant species in the urban environment of Nagpur, India.

Due to foliar dust deposition, a decline in both chlorophyll and carotenoid contents was observed in most plants. Chl and total chlorophyll showed maximum reductions of 46.5 and 45.4% in *Ficus benghalensis* due to dust deposition [78]. Dust contains many organic and inorganic compounds, which on entry into plant tissues cause increases in reactive oxygen species (ROS) and primarily damage to the membrane and pigments. Moraes et al. [92] found a decline in ascorbate concentration in *Titanophora pulchra,* while chlorophyll a, Chl a/b ratio, and foliar F, N, and S concentrations increased at the polluted site compared to a control in Cubatão, Brazil. The increase in pigment concentration is attributed to increasing leaf N content, indicating atmospheric NOx and NH_4^+ in N assimilation. Pandey and Agrawal [84], however, recorded reductions in chlorophyll, ascorbic acid, and nitrogen contents in leaves, whereas sulfate content was incremented with increasing pollution load in three tropical tree species.

Baek and Woo [93] assessed the responses of *Penaeus indicus* and *Eleutherodactylus orientalis* growing in four sites in the Philippines with differential pollution loads. In both test plants, chlorophyll content and Chl a/b ratio declined at sites with higher pollution loads, whereas antioxidative enzymes showed increments. A decrease in the size of starch grains in plant leaves indicated consumption of stored carbon at the expense of defense response [93]. Moraes et al. [92] found greater variations in water-soluble thiols and peroxidase activity at higher pollution loads compared to ascorbic acid in *Psidium guajava* and *Psidium cattleyanum* around the industrial complex of Cubatão, SE Brazil. Mukherjee and Agrawal [86] also observed an increase in ascorbic acid content in four tropical tree species (*M. indica*, *P. longifolia*, *F. benghalensis*, and *P. guajava*) growing at higher pollution loads in different parks in Varanasi City, India. Baek and Woo [93] observed an increase in cell wall thickness in response to the pollution load in P. indicus, which may be due to increased defense-related metabolites to protect against membrane damage.

In the urban area of Varanasi City, Mukherjee and Agrawal [86] found an increase in photosynthetic pigments with an increase in air pollution load in *M. indica* and *P. longifolia,* whereas *F. benghalensis* and *P. guajava* showed marked reductions. Carotene content showed a significant increase with a higher pollution load in all the studied plant species, with a maximum increase of 88% in *P. longifolia*. Chl a/b ratio and TChl/Car ratio showed reductions in most plants with increasing pollution loads [84, 86]. Most studies have identified an increase or maintenance of leaf biomass allocation and cellular turgidity at higher pollution loads in tolerant plant species [86].

Plants can accumulate, detoxify, and concentrate pollutants ([94]; Mukherjee and Agrawal 2016), which enables them to tolerate harsh environments. Agrawal and Singh [95] recorded increases in total S content of 41%, 25%, 35%, 27%, 33%, and 24%, respectively, in *M. indica*, *P. guajava*, *Cassia siamea*, *D. regia*, *Eucalyptus hybrid*, and *Bougainvillea spectabilis* around two thermal power plants in the Sonbhadra district of Uttar Pradesh, India, and found a direct correlation with average SO_2 concentrations in the ambient air.

Sicard et al. [49] suggested criteria for selecting trees for removal of O_3: air pollutant removal capacity and ozone-forming potential (OFP) are the most important factors. Secondary criteria are pollen allergenicity, O_3 sensitivity, and tolerance to drought and pest and diseases. Based on these criteria, a numerical scale using the range of 5–15 to 1–10 facilitates interpretation. This species-specific air quality index (S-AQI) is used for categorizing and recommending urban tree and shrub species for air quality planning (>8).

21.5 Conclusions

This chapter concludes that green roof development and urban forestation may help to mitigate urban air pollution. Intensive green roofs are more efficient in absorbing air pollutants and PM; however, green roofs are more costly than urban vegetation. Urban vegetation is also an efficient remover of airborne pollutants. Urban vegetation and urban forests are more cost-effective in removing air pollutants and dust particles than green roofs. Removal of pollutants is dependent on various leaf traits, deposition velocity, biomass, leaf area index, and biochemical traits. Plants with higher APTI and API are preferred for urban vegetation along streets.

References

1 Cohen, J., Brauer, M., Burnett, R. et al. (2017). Estimates and 25-year trends of the global burden of disease attributable to ambient air pollution: an analysis of data from the Global Burden of Diseases Study 2015. *Lancet* 389 (10082): 1907–1918.

2 WHO. (2016). Database: Outdoor Air Pollution in Cities; 2003–2010.

3 Liu, M.X., Zhang, Q., Huang, J. et al. (2018). Effect of leukotriene receptor antagonists (LTRAs) on airway inflammation in patients with asthma-COPD overlap (ACO) and its mechanism. *J. Med. Sci.* 45: 852.

4 Abhijith, K.V., Kumar, P., Gallagher, J. et al. (2017). Air pollution abatement performances of green infrastructure in open road and built-up street canyon environments – a review. *Atmos. Environ.* 162: 71–86.

5 Janhall, S. (2015). Review on urban vegetation and particle air pollution: deposition and dispersion. *Atmos. Environ.* 105: 130–137.

6 Huang, Y. (2014). Outdoor air pollution: a global perspective. *J. Occup. Environ. Med.* 56 (Suppl. 10): S3–S7.

7 Anenberg, S.C., Horowitz, L.W., Tong, D.Q., and West, J.J. (2010). An estimate of the global burden of anthropogenic ozone and fine particulate matter on premature human mortality using atmospheric modeling. *Environ. Health Perspect.* 118: 1189–1195.

8 Gehring, U., Gruzieva, O., Agius, R.M. et al. (2013). Air pollution exposure and lung function in children: the ESCAPE project. *Environ. Health Perspect.* 121: 1357–1364.

9 van den Hooven, E.H., Pierik, F.H., de Kluizenaar, Y. et al. (2012). Air pollution exposure during pregnancy, ultrasound measures of fetal growth, and adverse birth outcomes: a prospective cohort study. *Environ. Health Perspect.* 120 (1): 150–156.

10 Agay Shay, K., Friger, M., Linn, S. et al. (2013). Air pollution and congenital heart defects. *Environ. Res.* 124 https://doi.org/10.1016/j.envres.2013.03.005.

11 Zhang, L.W., Chen, X., Xue, X.D. et al. (2014). Long-term exposure to high particulate matter pollution and cardiovascular mortality: a 12-year cohort study in four cities in northern China. *Environ. Int.* 62: 41–47.

12 Silbajoris, R., Alvaro, R., Osornio-Vargas, A.R. et al. (2011). Ambient particulate matter induces interleukin 8 expression through an alternative NF-κB (Nuclear Factor-Kappa B) mechanism in human airway epithelial cells. *Environ. Health Perspect.* 119 (10): 1379–1383.

13 Son, J.Y., Bell, M.L., and Lee, J.T. (2011). Survival analysis of long-term exposure to different sizes of airborne particulate matter and risk of infant mortality using a birth cohort in Seoul, Korea. *Environ. Health Perspect.* 119: 725–730.

14 Kim, J.B., Kim, C., Choi, E. et al. (2012). Particulate air pollution induce arrhythmia via oxidative stress and calcium calmodulin kinase II activation. *Toxicol. Appl. Pharmacol.* 259: 66–73.

15 Tonne, C., Yanosky, J.D., Beevers, S. et al. (2012). PM mass concentration and PM oxidative potential in relation to carotid intima-media thickness. *Epidemiology* 23: 486–494.

16 Jiang, X.Q., Mei, X.D., and Feng, D. (2016). Air pollution and chronic away diseases: what should people know and do? *J. Thorac. Dis.* 8: E31–E40.

17 Smith, S.J., Pitcher, H., and Wigley, T.M.L. (2001). Global and regional anthropogenic sulfur dioxide emissions. *Global Planet. Change* 29 (1–2): 99–119.

18 Lelieveld, J., Evans, J.S., Fnais, M. et al. (2015). Contribution of outdoor air pollution sources to premature mortality on a global scale. *Nature* 525: 367–371.

19 Zhang, Q. et al. (2017). Transboundary health impacts of transported global air pollution and international trade. *Nature* 543: 705–709.

20 Chakraborty, A., Gupta, T., and Tripathi, S.N. (2016). Chemical composition and characteristics of ambient aerosols and rainwater residues during Indian summer monsoon: insight from aerosol mass spectrometry. *Atmos. Environ.* 136: 144–155.

21 Huang, R., Zhang, Y., Bozetti, C. et al. (2014). High secondary aerosol contribution to particulate pollution during haze events in China. *Nature* 514: 218–222.

22 BP P.L.C. (2016). BP statistical review of world energy 2016. https://www.bp.com/content/dam/bp/pdf/energyeconomics/statistical-review-2016/bp-statistical-review-of-world-energy-2016-full-report.pdf.

23 Li, C., McLinden, C., Fioletov, V. et al. (2017). India is overtaking China as the world's largest emitter of anthropogenic Sulphur dioxide. *Sci. Rep.* 7: 14304.

24 Mukherjee, A., Pandey, B., Agrawal, S.B., and Agrawal, M. (2018). Responses of tropical and subtropical plants to air pollution. In: *Tropical Ecosystems: Structure, Function and Challenges in the Face of Global Change* (eds. S.C. Garkoti, S.J. Van Bloem, P.Z. Fule and R.L. Semwal), 129–162. Singapore: Spinger Nature.

25 ENVIS. (2010). ENVIS newsletter. Centre for Control of Pollution.

26 Warneck, P. (2000). *Chemistry of the Natural Atmosphere.* 0-12-735632-0. Academic Press.

27 Liu, S.C., Trainer, M., Fehsenfeld, F. et al. (1987). Ozone production in the rural troposphere and the implications for regional and global ozone distributions. *J. Geophys. Res.* 92: 4191–4207.

28 US EPA, 2007. Latest finding on National air quality: Status and trends through 2006. www.gov.epa/air/airtrends/2008

29 Liu, J. and Diamond, J. (2005). China's environment in a globalizing world. *Nature* 435: 1179–1186.

30 Streets, D.G. and Waldhoff, S.T. (2000). Present and future emissions of air pollutants in China: SO_2, NOx, and CO. *Atmos. Environ.* 34: 363–374.

31 Karlsson, P.E., Klingberg, J., Engardt, M. et al. (2017). Past, present and future concentrations of ground –level ozone and potential impacts on ecosystems and human health in Northern Europe. *Sci. Total Environ.* 576: 22–35.

32 Monks, P.S. et al. (2015). Tropospheric ozone and its precursors from the urban to the global scale from air quality to short-lived climate forcer. *Atmos. Chem. Phys.* 15: 8889–8973.

33 van Zelm, R., Preiss, P., van Goethem, T. et al. (2016). Regionalized life cycle impact assessment of air pollution on the global scale: damage to human health and vegetation. *Atmos. Environ.* 134: 129–137.

34 Royal Society. (2008). Ground-level ozone in the 21st century: future trends, impacts and policy implications. Royal Society Policy Document 15/08. The Royal Society, London.

35 Dentener, F., Stevenson, D., Ellingsen, K. et al. (2006). The global atmospheric environment for the next generation. *Environ. Sci. Technol.* 40: 3586–3594.

36 EANET. (2006). Data report on acid deposition on East Asia region 2005. Network Centre of EANET, Japan. http://www.eanet.cc.

37 Lelieveld, J., van Aardenne, J., Fischer, H. et al. (2004). Increasing ozone over the Atlantic ocean. *Science* 304: 1483–1487.

38 Oltmans, S.J., Johnson, B.K., and Harris, J.M. (2012). Springtime boundary layer ozone depletion at Barrow, Alaska: meteorological influence, year-to-year variation, and long-term change. *J. Geophys. Res.* 117: D00R18.

39 Ebojie, F., Burrows, J.P., Gebhardt, C. et al. (2016). Global tropospheric ozone variations from 2003 to 2011 as seen by SCIAMACHY. *Atmos. Chem. Phys.* 16: 417–436.

40 Sharma, A., Ojha, N., Ansari, T.U. et al. (2020). Effects of dry deposition on surface ozone over South Asia inferred from a regional chemical transport model. *ACS Earth Space Chem.* 4: 2.

41 Ojha, N., Girach, I., Sharma, K. et al. (2019). Surface ozone in the Doon valley of the Himalayan foothills during spring. *Environ. Sci. Pollut. Res.* 26 https://doi.org/10.1007/s11356-019-05085-2.

42 United Nations. (2014). World urbanization prospects 2014. Demographic research report 32.

43 Clark, C., Talbot, B., Bulkley, J., and Adriaens, P. (2005). Optimization of green roofs for air pollution mitigation. 3rd North American Green Roof Conference: Greening Rooftops for Sustainable Communities, Washington, DC, 4–6 May. The Cardinal Group, Toronto.

44 Davies, H.J., Doick, K.J., Hudson, M.D., and Schreckenberg, K. (2017). Challenges for tree officers to enhance the provision of regulating ecosystem services from urban forests. *Environ. Res.* 156: 97–107.

45 Jayasooriya, V.M., Ng, A.W.M., Muthukumaran, S., and Perera, B.J.C. (2017). Green infrastructure practices for improvement of urban air quality. *Urban For. Urban Greening* 21: 34–47.

46 Speak, A.F., Rothwell, J.J., Lindley, S.J., and Smith, C.L. (2012). Urban particulate pollution reduction by four species of green roof vegetation in a UK city. *Atmos. Environ.* 61: 283–293.

47 Yang, J., Yu, Q., and Gong, P. (2008). Quantifying air pollution removal by green roofs in Chicago. *Atmos. Environ.* 42: 7266–7273.

48 Gourdji, S. (2018). Review of plants to mitigate particulate matter, ozone as well as nitrogen dioxide air pollutants and applicable recommendations for green roofs in Montreal, Quebec. *Environ. Pollut.* 241: 378–387.

49 Sicard, P., Agatholelous, E., Araminiene, V. et al. (2018). Should we see urban trees as effective solutions to reduce increasing ozone levels in cities? *Environ. Pollut.* 243: 163–176.

50 Weber, F., Kowarik, I., and Saumel, I. (2014). Herbaceous plants as filters: Immobilization of particulates along urban street corridors. *Environmental Pollution* 186: 234–240.

51 Bottalico, F., Travaglini, D., Chirci, G. et al. (2017). A spatially explicit method- modified to assess the dry deposition of air pollution by urban forests in the city of Florence, Italy. *Urban Forestry and Urban Greening* 27: 221–234.

52 Morikawa, H., Higaki, A., Nohno, M. et al. (1998). More than a 600-fold variation in nitrogen dioxide assimilation among 217 plant taxa. *Plant Cell Environ.* 21: 180–190.

53 McNabola, A. (2010). New directions: passive control of personal air pollution exposure from traffic emissions in urban street canyons. *Atmos. Environ.* 44: 2940–2941.

54 Amorim, J.H., Rodrigues, V., Tavares, R. et al. (2013). CFD modelling of the aerodynamic effect of trees on urban air pollution dispersion. *Sci. Total Environ.* 461–462: 541–551.

55 Currie, B. and Bass, B. (2008). Estimates of air pollution mitigation with green plants and green roofs using the UFORE model. *Urban Ecosyst.* 11: 409–422.

56 Akbari, H., Pomerantz, M., and Taha, H. (2001). Cool surfaces and shade trees to reduce energy use and improve air quality in urban areas. *Solar Energy* 70 (3): 295–310.

57 Johnson, J. and Newton, J. (1996). *Building Green, a Guide for Using Plants on Roofs and Pavement.* London: The London Ecology Unit.

58 Tan, P. and Sia, A. (2005). A pilot green roof research project in Singapore. 3rd North American Green Roof Conference: Greening Rooftops for Sustainable Communities, Washington, DC. 4–6 May. The Cardinal Group, Toronto.

59 Takahasi, M., Kondo, K., and Morikawa, M. (2003). Assimilation of nitrogen dioxide in selected plant taxa. *Acta Biotechnol.* 23: 241–247.

60 Alonso, R., Vivanco, M.G., González-Fernández, I. et al. (2011). Modelling the influence of peri-urban trees in the air quality of Madrid region (Spain). *Environ. Pollut.* 159: 2138–2147.

61 Manes, F., Marando, F., Capotorti, G. et al. (2016). Regulating ecosystem services of forests in ten Italian metropolitan cities: air quality improvement by PM10 and O3 removal. *Ecol. Indic.* 67: 425–440.

62 Benjamin, M.T. and Winer, A.M. (1998). Estimating the ozone-forming potential of urban trees and shrubs. *Atmos. Environ.* 32: 53–68.

63 Connop, S., Vandergert, P., Eisenberg, B. et al. (2016). Renaturing cities using a regionality- focused biodiversity- led multifunctional benefits approach to urban green infrastructure. *Environ. Sci. Policy* 62: 99–111.

64 Salmond, J.A., Williams, D.E., Laing, G. et al. (2013). The influence of vegetation on the horizontal and vertical distribution of pollutants in a street canyon. *Sci. Total Environ.* 443: 287–298.

65 Nowak, D.J. (2006). Air pollution removal by urban trees and shrubs in the United States. *Urban Forestry and Urban Greening* 4: 115–123.

66 Nowak, D.J., Hoehn, R.E. III, Bodine, A.R. et al. (2013). "Assessing urban forest effects and values: Toronto's urban forest". U.S. Department of Agriculture, Forest Service, Northern Research Station. *Resour. Bull.* NNRS-79, 59 p.

67 Nowak, D.J., Hirabayashi, S., Bodine, A., and Greenfield, E. (2014). Tree and forest effects on air quality and human health in the United States. *Environ. Pollut.* 193: 119–129.

68 Nowak, D.J. and Crane, D.E. (2000). The Urban Forest Effects (UFORE) model: quantifying urban forest structure and functions. Integrated tools for natural resources inventories in the 21st century. USDA Forest Service General Technical Report NC-212.

69 James, P., Tzoulas, K., Adams, M.D. et al. (2009). Towards an integrated understanding of green space in the European built environment. *Urban Forestry and Urban Greening* 8 (2): 65–75.

70 Pauleit, S. (2003). Perspectives on urban greenspace in Europe. *Built Environment* 29: 89–93.

71 McPherson, E.G., Nowak, D., Heisler, G. et al. (1997). Quantifying urban forest structure, function, and value: the Chicago urban forest climate project. *Urban Ecosystems* 1: 49–61.

72 Nowak, D.J., Civerolo, K.L., Rao, S.T. et al. (2000). A modelling study of the impact of urban trees on ozone. *Atmos. Environ.* 34: 1601–1613.

73 Hagler, G.S.W., Lin, M.-Y., Khlystov, A. et al. (2012). Field investigation of roadside vegetative and structural barrier impact on near-road ultrafine particle concentrations under a variety of wind conditions. *Science of The Total Environment* 419: 7–15.

74 Wania, A., Bruse, M., Blond, N., and Weber, C. (2012). Analysing the influence of different street vegetation on traffic-induced particle dispersion using microscale simulations. *J. Environ. Manage.* 94: 91–101.

75 Deutsch, B., Whitlow, H., Sullivan, M., and Savineau, A. (2005). Re-greening Washington, DC: a green roof vision based on environmental benefits for air quality and storm water management. 3rd North American Green Roof Conference: Greening Rooftops for Sustainable Communities, Washington, DC, 4–6 May. The Cardinal Group, Toronto.

76 Freer-Smith, P.H., El-Khatib, A.A., and Taylor, G. (2004). Capture of particulate pollution by trees: a comparison of species typical of semi-arid areas (*Ficus nitida* and *Eucalyptus globulus*) with European and North American species. *Water Air Soil Pollut.* 155: 173–187.

77 Escobedo, F.J., Kroeger, T., and Wagner, J.E. (2011). Urban forests and pollution mitigation: analyzing ecosystem services and disservices. *Environ. Pollut.* 159: 2078–2087.

78 Prusty, B.A.K., Mishra, P.C., and Azeez, P.A. (2005). Dust accumulation and leaf pigment content in vegetation near the national highway at Sambalpur, Orissa, India. *Ecotoxicol. Environ. Saf.* 60: 228–235.

79 Koch, K., Bhushan, B., and Barthlott, W. (2009). Multifunctional surface structures of plants: an inspiration for biomimetics. *Prog. Mater Sci.* 54: 137–178.

80 Ziegler, I. (1973). The effect of air-polluting gases on plant metabolism. In: *Environmental quality and safety*, vol. 2, 182–208. New York: Academic Press.

81 Rolfe, G.L. (1974). Lead distribution in tree rings. *Forest Science.* 20 (3): 283–286.

82 Smith, W.H. (1990). *Air Pollution and Forests*. New York: Springer.

83 Binkley, D., Campoe, O.C., Gspaltl, M., and Forrester, D.I. (2013). Light absorption and use efficiency in forests: why patterns differ for trees and stands. *For. Ecol. Manage.* 288: 5–13.

84 Pandey, J. and Agrawal, M. (1994). Evaluation of air pollution phytotoxicity in a seasonally dry tropical urban environment using three woody perennials. *New Phytol.* 126: 53–61.

85 Ajuru, M.G. and Friday, U. (2014). Effects of particulate matter on the anatomy of some tropical plants (*Alchonea cordifolia*, *Musa paradisiaca* and *Manihot esculenta*). *Int. J. Sci. Technol. Res.* 3: 304–308.

86 Mukherjee, A. and Agrawal, M. (2016). Pollution response score of tree species in relation to ambient air quality in an urban area. *Bull. Environ. Contam. Toxicol.* 96: 1–6.

87 Verma, A. and Singh, S.N. (2006). Biochemical and ultrastructural changes in plant foliage exposed to auto-pollution. *Environ. Monit. Assess.* 120: 585–602.

88 Nighat, F., Mahmooduzzafar, and Iqbal, M. (2000). Stomatal conductance, photosynthetic rate, and pigment content in Ruellia tuberosa leaves as affected by coal-smoke pollution. *Biol. Plant.* 43: 263–267.

89 Wang, H., Li, Y., and Shi, H. (2011). Leaf dust capturing of urban greening plant species in relation to leaf micromorphology. IEEE 978-1-61284-340-7/11.

90 Ninave, S.Y., Chaudhari, P.R., Gajghate, D.G., and Tarar, J.L. (2001). Foliar biochemical features of plants as indicators of air pollution. *Bull. Environ. Contam. Toxicol.* 67: 133–140.

91 Roy, A., Bhattacharya, T., and Kumari, M. (2020). Air pollution tolerance, metal accumulation and dust capturing capacity of common tropical trees in commercial and industrial sites. *Sci. Total Environ.* 722: 137622.

92 Moraes, R.M., Delitti, W.B.C., and Moraes, J.A.P.V. (2003). Gas exchange, growth, and chemical parameters in a native Atlantic forest tree species in polluted areas of Cubatão, Brazil. *Ecotoxicol. Environ. Saf.* 54: 339–345.

93 Baek, S.G. and Woo, S.Y. (2010). Physiological and biochemical responses of two tree species in urban areas to different air pollution levels. *Photosynthetica* 48: 23–29.

94 De França, E.J., De Nadai Fernades, E.A., Bacchi, M.A., and Saiki, M. (2004). Native trees as biomonitors of chemical elements in the biodiversity conservation of the Atlantic Forest. *J. Atmos. Chem.* 49: 579–592.

95 Agrawal, M. and Singh, J. (2000). Impact of coal power plant emission on the foliar elemental concentrations in plants in a low rainfall tropical region. *Environ. Monit. Assess.* 60: 261–282.

22

Phylloremediation for Mitigating Air Pollution

Majeti Narasimha Vara Prasad

School of Life Sciences, University of Hyderabad, Hyderabad, Telangana, India

22.1 Introduction

More than half of the world's population lives in cities, a trend that is rapidly accelerating, especially in developing countries. Urban and suburban areas are merging into huge megalopolitan areas, resulting in megacities. Megacities require and use large quantities of energy and materials, metabolizing them and generating large quantities of waste products and pollutants. The results are unsustainable environments that adversely affect ecological integrity and diversity and human health and well-being. Many city dwellers are disconnected from forests and the natural world, resulting in the emergence of a new species: "homo urbanus." To manage such a situation in a climate-constrained world, plants are valuable tools for monitoring and mitigating air pollution. Several remediation technologies for air pollution are suggested (Table 22.1), which are often cost-prohibitive. Plants provide a natural and inexpensive way to cleanse the atmosphere by absorbing gaseous and particulate matter by virtue of their architecture and tolerance. Plant leaves are efficient pollutant-trapping devices. In India, field investigations have been carried out in various places during the last three decades.

Urban and periurban ecosystems must maintain green belts to achieve for ecological balance, particularly with perennials that have high Air Pollution Tolerance Indexes (APTIs). Tolerant plants act as an effective sink for pollution, and therefore, species with a high APTI are recommended for green belts in polluted areas. The APTI is an empirical value suggesting the tolerance potential of the plant to air pollution. It can be calculated using this formula:

$$\text{APTI} = \left[\text{A}\left(\text{T}+\text{P}\right)+\text{R}\right]/10$$

where A = ascorbic acid content, T = total chlorophyll, P = pH of leaf extract, and R = relative water content of the leaf. The sum is divided by 10 to reduce it to a small number.

Particulate pollutants are as follows:

- Coal dust
- Fly-ash
- Petrocoke
- Suspended particulate matter

Handbook of Ecological and Ecosystem Engineering, First Edition. Edited by Majeti Narasimha Vara Prasad.

Table 22.1 Remediation techniques for air pollution.

Technique	Process
Adsorption	Air pollutants are adsorbed onto activated carbon or zeolites.
Filtration	Polluted air is passed through fibrous material coated with a viscous substance.
Ozonation	Pollutants are oxidized by the generation of ozone.
Photolysis	Pollutants are oxidized by the application of UV alone or in combination with a photocatalyst.
Biofiltration	Pollutants are degraded by passing the contaminated air through a microbe-colonized packed bed of solid support.
Membrane separation	Membranes are used to separate pollutants.
Enzyme oxidation	Pollutants are degraded by catalysts or enzymes through the transfer of air emissions into an aqueous phase.
Botanical purification	Pollutants are degraded using enzymes in plants by passing the air through vegetation.
Nanotechnology	Carbon nanotubes for air pollutant control via adsorption.

Source: Dhanabalan et al. [1], Fahri et al. [2], Gabal et al. [3], Gangupomu et al. [4], Hamadani et al. [5], Hussain [6], Lee et al. [7], Subhramanyam et al. [8], and Vani Padmaja et al. [9].

These are the primary gaseous pollutants:

Sulfur dioxide
Nitrogen dioxide
Carbon monoxide

And the following are other miscellaneous air pollutants:

Toxic metals
Polyaromatic hydrocarbons

SO_2 is a reducing gas and is 2.7 times heavier than air. The phytotoxicity of SO_2 is ascribed to its oxidation-reduction potential. It undergoes rapid hydration, and one volume of H_2O can dissolve 80 volumes of SO_2.

Nitrogen oxides (mostly nitric oxide) are formed in high-temperature combustion sources from reactions of nitrogen in the air with oxygen, and additional nitrogen oxides may be formed from nitrogen in fuel (important for coal). Although estimates of naturally formed nitrogen oxides are less certain than sulfur emissions, natural nitrogen oxide emissions are thought to be more important to total emissions of nitrogen oxides than natural emissions of sulfur are to total sulfur emissions. On a global scale, natural nitrogen oxide emissions are thought to be comparable in magnitude to the amount from human sources. Nitrogen oxide emissions from human sources primarily come from motor vehicles and stationary combustion sources, mainly power plants.

Nitric oxide is rapidly oxidized by ozone to form nitrogen dioxide. In the photochemical relationship between nitric oxide, ozone, and nitrogen dioxide, the concentration of these chemical species is directly affected by the intensity of sunlight. These chemicals are also known to react photochemically with hydrocarbons and other atmospheric chemicals to form photochemical smog. Although the rates and mechanisms of the main sequence of reactions are fairly well known, many details relating to acid formation are unknown. For example, the sources and classes of hydrocarbons that most effectively promote the reactions leading to nitrogen acids have not been determined. Also unknown is the quantity of acid formed that survives to form acid precipitant compared to that which is destroyed by further photochemical reactions. Even less is known about the reactions of nitrogen species in cloud droplets or on particles.

Primary sources of NO_x are power plants that burn fossil fuels, and automobile systems. The most common type of nitrogenous air pollutant is nitrogen dioxide (NO_2). In the atmosphere, NO_2 is reduced by UV light to nitrogen monoxide and atomic oxygen. Atomic oxygen reacts with O_2 to form O_3. O_3 reacts with NO to form NO_2 and O_2,, thus closing the cycle. Thus O_3, an oxidizing agent, is formed secondarily through photochemical reactions in the presence of primary pollutants such as NO_2 and HC.

In the presence of sunlight, atomic oxygen from the photochemical reduction of NO_2 reacts with reactive C to form reactive intermediates called *radicals*. These radicals then take part in a series of reactions to form still more radicals that combine with O_2, HC, and NO_2. As a result, NO_2 is regenerated. O_3 accumulates, and a number of secondary pollutants form, such as formaldehyde, aldehydes, and PAN (peroxy acetyl nitrate). All of these collectively form a *photochemical group*.

These, along with other pollutants such as SO_2, CO, and CO_2, act synergistically. The total effect is much greater than the individual effects. A number of vegetables are sensitive to PAN (= photochemical smog), which has driven vegetable farms several kilometers away from the United States' metropolitan centers.

22.2 Significance of Tree Canopy Architecture and Types of Canopies for Mitigating Air Pollution

Different species of trees have different architectures and leaf shapes as the result of genetically fixed growth programs for space occupancy and gaining exposure to sunlight. While animal species adopt different strategies for obtaining food and orienting their relationship with predators, plants adopt growth habits, leaf shapes, and branching orientations so that each plant species occupying a particular ecological niche can use sunlight to its maximum advantage.

The development of trees with a particular canopy and growth habit is genetically fixed, and it is suggested that the growth habits of plants can be judiciously used to buffer the deleterious effects of atmospheric pollutants. The micro-environments created by tree canopies with a large system of exposed leaf surfaces in constant exchange with the atmosphere are evolutionary mechanisms for maximum sunlight utilization. The tree's total leaf area index is many times greater than the area occupied by the tree. These vast exposed areas of living systems with different canopies, leaf orientations, phyllotaxy, and leaf movements to meet the trees' sunlight requirements function as a filtering labyrinth for pollutant gases and a sink for pollutant particles. These living systems also have the advantage of renewable leaves that are replaced by new ones in every growing season.

Closely knit shelter belts around industrial complexes also act as a buffer for noise pollution. Different types of tree canopies help purify the air by supplying oxygen and absorbing carbon dioxide during the day. Each plant canopy functions as a micro-environment, creating its own dynamics of airflow and gas exchange (Figure 22.5).

It is well known that a tree is a photosynthetic device consisting of assimilating units: leaves. The orientation of trees in a definite pattern is based on their form and physiology, which maximizes the net rate of carbon gain from photosynthesis; net carbon gain is of primary evolutionary importance. The primary elements of trees are the crown, trunk, and roots. For tree canopy studies, the nature of the crown and trunk are taken into consideration (Figure 22.5).

It is generally considered that the *canopy* is defined as the volume occupied by leaves and associated stems. Leaf area is the main component of canopies, and the leaf area index gives quantitative data about the canopy depth. The *leaf area index* is defined as the total area per unit ground area and is used to define the canopy. The *foliage area index* gives both leaf and stem area indexes. Thus the canopy is defined in terms of the height of supporting systems besides leaf and stem areas. The leaves' nature, size, and deciduous or evergreen nature are important in understanding the size and depth of tree canopies. Canopies orient themselves to the particular architecture of trees and can be classified distinctly, as shown in Figure 22.1.

Ambient air contains various sizes of solid particles commonly recognized as particulates or dust, which are continuously agglomerated and deposited on surfaces. The deposited particulate matter is a conglomerate of chemically heterogeneous substances of many different types, which act on plants in a variety of ways.

The biological activity of the particles varies based on pollutant source profiles. These variations are expressions of quantitative and qualitative differences such as the relative presence of sulfuric acid mist, sulfates, or other reactive substances in the particulate mix or the relative amounts of specific carcinogenic compounds in the organic fraction of airborne particulates. It therefore can be seen that the evaluation of biological activity ascribable to "particulates" is extremely complex and depends not only on their total quantity, size range, and intrinsic physical or chemical properties but also on their chance for interaction in the polluted air.

Research has shown that plant leaves can act as biological filters, removing large quantities of particles from the urban atmosphere. *Stomata* are microscopic pores on the underside (abaxial) of the leaf. These allow air into and out of the leaf, through which the plant takes in CO_2 and lets out O_2, and allow water vapor out in the process of transpiration. As air passes through the stomata, most airborne particles do not pass through but are instead eliminated on the leaf's outer surface due to available moisture.

A certain amount of force is needed for particles to stick to the surface. This amount is greater depending on the size of the particle. The airflow through the stomata is not very powerful, so only smaller particles stick to the bottom surface. The particles on the top surface of leaves are mainly

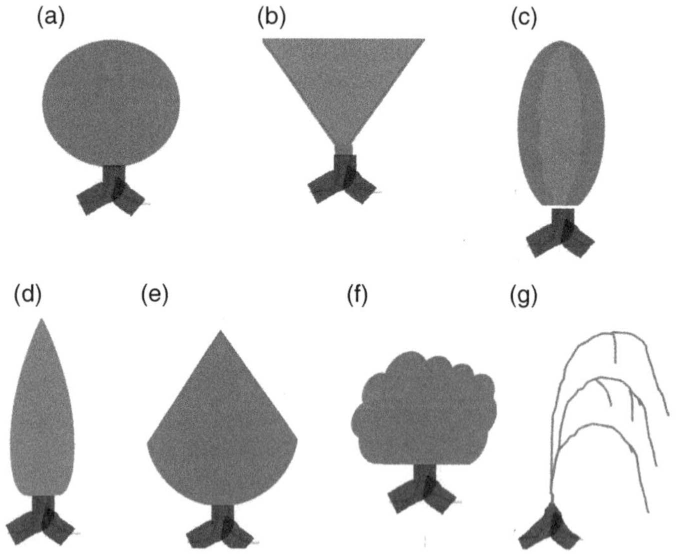

Figure 22.1 Different types of canopies: (a) rotund: *Mangifera indica, Lagerstroemia flost-regina, Mimusops elengi*; (b) flat-topped: *Samanea saman, Delonix regia*; (c) cylindrical: *Dalbergia sisso, Heterophragma roxburghii*; (d) chimney: *Eucalyptus* sp.; (e) conical: *Thuja, Cryptomeria*; (f) broken: *Ceiba pentandra*; (g) drooping *Salix*. (Redrawn from Subrahmanyam et al 1985).

from dust settling. The leaf surface captures both coarse and fine particles because of favorable morphological features of leaves. These particulates remain adhered to the plant leaves until they are washed down through rain or artificially.

Because the particles deposited on the leaves are from the air, particle analysis also provides information about the source of particulate air pollution in the area. The higher the concentration of particulate matter in the area, the higher the particle concentration on the leaf surface will be. The particles' sizes and chemical composition on the leaf surface are representative of the airborne particles in the sampled area.

Different types of leaves tend to vary in several aspects of their surfaces. Some leaves have greater surface rigidity or roughness than others, affecting their stickiness or particle solubility. Stickier leaves are better for collecting particles because more particles stick to their surface; and once the deposited particles are washed away, the leaves are ready for dust capture again. Therefore, certain morphological features of leaves are more favorable for particulate capture from the surrounding environment (Table 22.2).

Air particles are derived from innumerable natural and human-made sources, each contributing different physical and chemical properties. While portions of the particulate milieu consist of relatively non-reactive substances such as carbon and calcium, they differ in their biological activity by virtue of the presence of adsorbed chemicals that may be added through condensation of volatile chemicals in a stack as flue gases cool or through interactions with other pollutants in the atmosphere. Therefore, the potential for biological activity of given particle is governed by its opportunistic ability to interact with other environmental pollutant substances capable of chemical or physical interaction.

The biological activities of particles necessarily vary due to differing pollutant source profiles. It therefore can be seen that the evaluation of biological activity ascribable to "particulates" is complex and depends not only on their total quantity, size range, and intrinsic physical or chemical properties but also on their chance for interaction in the polluted air. The opportunity for variation in biological activity is enormous, and at this time, it would seem unwise to attempt to relate health effects to the total amount of respirable particulate material in the atmosphere without regard for their qualitative differences. Leaf margin type also plays a prominent role in air pollution abatement (Figure 22.2).

Plants differ in the composition of endophytic fungi and bacteria in the tissue of phyllosphere [66–68]. Leaf-associated endophytes may play positive and negative roles (Figure 22.3). Leaf tissue in association with endophytes converts toxic gaseous pollutants into nutrients required by plants (Figure 22.4) [27, 66–74].

Air quality improvement by different species of plants depends on various parameters such as the following:

Particulate matter (10 mm size)
Dust-capturing mechanism by plants
Morphological features of plant leaves for dust-capture efficiency
Anatomical features of plant leaves for efficient dust capture
Phyllotaxi
External structure of leaves' epidermis
Physiological features of plant leaves for efficient dust capture
Transpiration process
Environmental factors for efficient dust capture by plants
Light intensity
Moisture
Wind velocity

Table 22.2 Bibliographic resume of air quality improvement by plant leaves.

Research contribution	Reference
Plant physiological mechanisms of air	[11]
Air quality in different indoor and outdoor settings in Africa	[12]
Asian tree species for refinement of air quality by estimating air pollution tolerance index	[13]
Dynamics of plant responses to combinations of air pollutants	[14]
Canopy position has a stronger effect than tree species on phyllosphere bacterial diversity	[15]
Assessment of indoor fine and ultra-fine particulates	[16]
Infection with *Pseudomonas citronellolis*	[17]
Carpinus betulus L. phyllomicrobiome in urban and forest areas	[18]
Potential of plant-microbe interactions in environmental decontamination	[19]
Variation in tree species' ability to capture and retain airborne fine particulate matter	[20]
The molecular basis of glandular trichome development and secondary metabolism in plants	[21]
PM deposited on *Platanus acerfolia* leaves change across different cities in Europe	[22]
Plants in air phytoremediation	[23]
Phylloremediation of air pollutants: exploiting the potential of plant leaves and leaf-associated microbes	[24]
Leaf bacterial diversity mediates plant diversity and ecosystem function relationships	[25]
Impact of air pollution-induced climate change	[26]
Draft genome sequence of *Rhodococcus erythropolis* VSD3 *snd Bacillus licheniformis* VSD4, a diesel fuel-degrading and plant growth-promoting bacterium isolated from *Hedera helix* leaves	[27]
Diversity and hydrocarbon-degrading epiphytic microbial communities on Platanus x acerifolia leaves in an urban area	[28]
Anthropogenic emissions in air pollution	[29]
A mixed-integer programming model for national ambient air quality standards (NAAQS) attainment strategy analysis	[30]
The physiological role of the ribulose monophosphate pathway in bacteria and archaea	[31]
Removal of indoor volatile organic compounds via photocatalytic oxidation	[32]
Removal of ethylbenzene from contaminated air by *Zamioculcas Zamiifolia* and associated microorganisms	[33]
Plant surface reactions	[34]
Impacts of particulate matter pollution on plants: implications for environmental biomonitoring	[35]
Host species identity, site, and time drive temperate tree phyllosphere bacterial community	[36]
Exploitation of endophytic bacteria to enhance the phytoremediation potential of the wetland helophyte *Juncus acutus*	[37]
Endophytic *Bacillus cereus* ERBP-*Clitoria ternatea* interactions: potentials for the enhancement of gaseous formaldehyde removal	[38]
Effect of endophytic *Bacillus cereus* ERBP inoculation into non-native host: potentials and challenges for airborne formaldehyde removal	[39]

Table 22.2 (Continued)

Research contribution	Reference
Diversity, abundance, and activity of ammonia-oxidizing microorganisms in fine particulate matter	[40]
Biochemical and molecular mechanisms of plant-microbe-metal interactions: relevance for phytoremediation	[41]
Air pollution trends over Indian megacities and their local-to-global implications	[42]
Air pollution and risk of respiratory and cardiovascular hospitalizations in the most populous city in Vietnam	[43]
Air pollutant emissions from Chinese households: a major and underappreciated ambient pollutant source	[44]
The role of plant-microbe interactions and their exploitation for phytoremediation of air pollutants	[45]
Structure and functional metabolism of bacterial communities on leaves of typical urban greening tree species	[46]
Relative and contextual contribution of different sources to the composition and abundance of indoor air bacteria in residences	[47]
Formaldehyde degradation by *Aspergillus* sp.	[48]
Bacteria naturally associated with leaves of higher plants degrade hydrocarbons	[49]
Air pollution and public health: emerging hazards and improved understanding of risk	[50]
Urban microbiomes and urban ecology	[51]
Uptake of toluene and ethylbenzene by plants: removal of volatile indoor air contaminants	[52]
The air quality of a Mediterranean urban environment area and its relation to major meteorological parameters	[53]
Relationships between phyllosphere bacterial communities and plant functional traits in a neotropical forest	[54]
Plant traits and taxonomy drive host associations in tropical phyllosphere fungal communities	[55]
Isolation, plant colonization potential, and phenanthrene degradation performance of the endophytic bacterium	[56]
Endophytic bacteria degrade organic pollutants	[57]
Degradation, phytoprotection, and phytoremediation of phenanthrene by endophyte *Pseudomonas putida*	[58]
Community structures of N 2-fixing bacteria associated with the phyllosphere of a Holm oak forest and their response to drought	[59]
The influences of diesel particulate filter installation on air pollutant emissions for used vehicles	[60]
Seasonally dynamic fungal communities in the *Quercus macrocarp* phyllosphere differ between urban and nonurban environments	[61]
Human health effects of air pollution	[62]
Degradation of phenanthrene on plant leaves by phyllosphere bacteria	[63]
Degradation of benzene, toluene, and xylene by *Pseudomonas aeruginosa*	[64]
Study of ultrafine particles near a major highway with heavy-duty diesel traffic	[65]

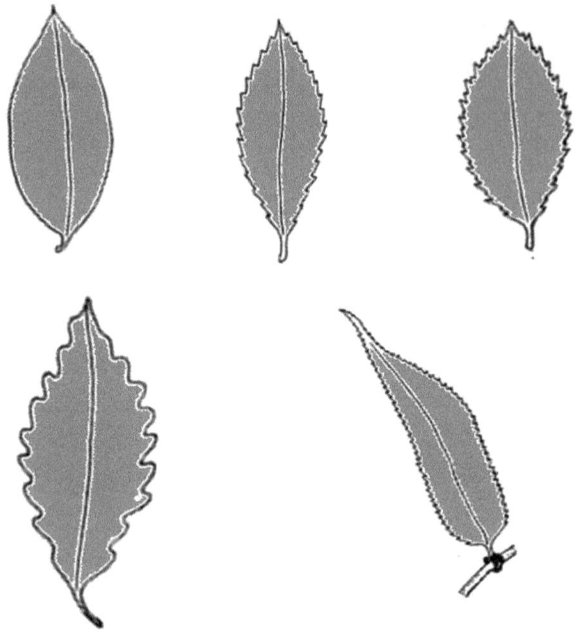

Figure 22.2 Different types of margins of leaves. (Redrawn based on Shrivastava et al. [10], Central Pollution Control Board, MOEF&CC Govt of India New Delhi).

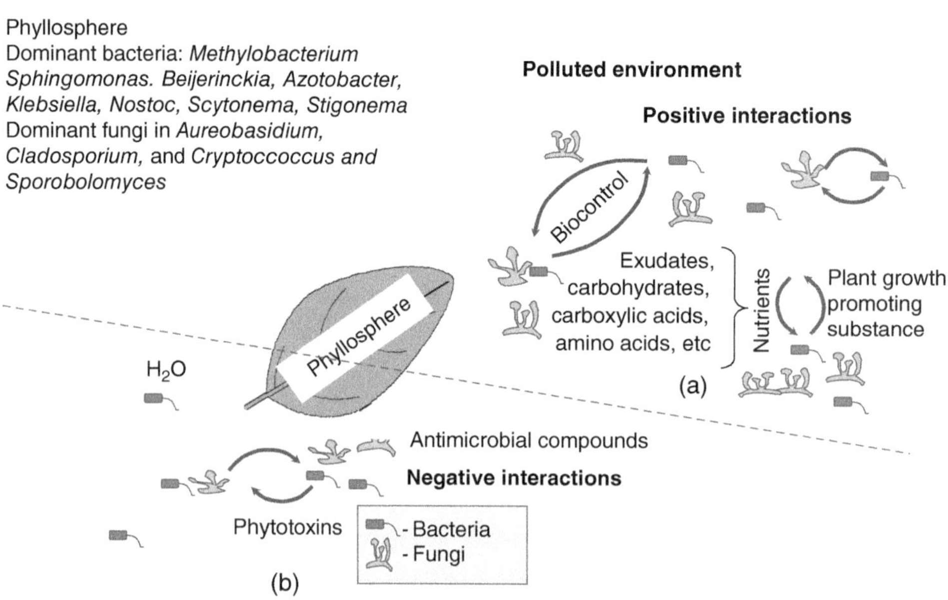

Figure 22.3 Leaf fungal and bacterial endophytes with positive and negative functions.

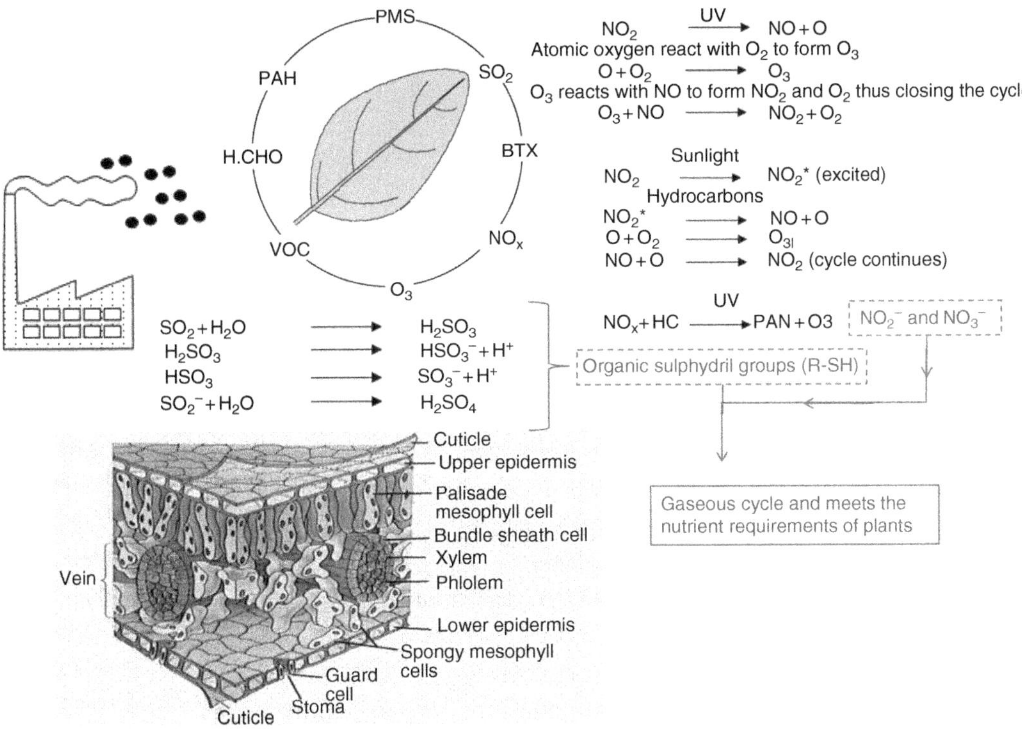

Figure 22.4 Leaf tissue involvement and chemistry in mitigating gaseous pollutants and their conversion to nutrients required by plants. *See colour plate section for colour representation of this figure.*

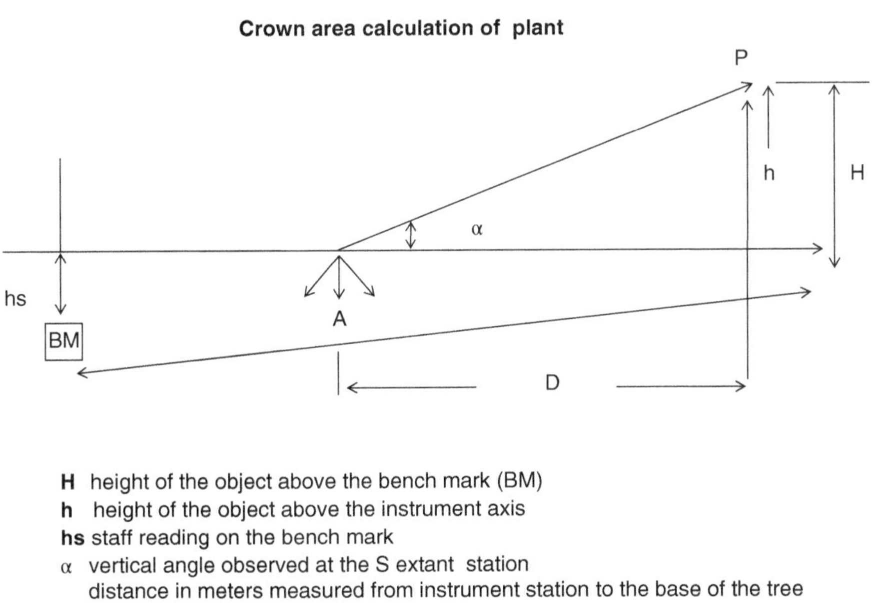

Figure 22.5 Canopy/Crown area calculation of plant species. *Source:* Redrawn based on Shrivastava et al. [10], Central Pollution Control Board, MOEF&CC, Govt. of India, New Delhi.

Plant height, crown area, and density of plants play crucial roles in capturing dust particles and abating air pollution.

22.3 Air-Improving Qualities of Plants

22.3.1 Dust-Capturing Mechanisms Using Plants

Dust capture by leaves/plants depends on the moisture conditions on and around the leaf surface (Figure 22.6). The moisture condition on the leaf surface depends on the rate of plant transpiration, and the underlying processes are diagrammatically represented in Figure 22.6.

22.3.2 Environmental Factors for Efficient Dust Capture by Plants

The following environmental factors influence the dust-capturing efficiency of plants:

22.3.2.1 Light Intensity

The photosynthesis and transpiration processes of plants depend on light intensity. The leaves may be categorized based on light intensity: sun leaves (smaller and thicker, with smaller and more numerous chloroplasts) and shade leaves (larger and thinner, with fewer chloroplasts and fewer well-defined mesophyll layers).

22.3.2.2 Moisture

Mesophytes grow in environments with adequate amounts of water and have an extensive role in capturing dust from the environment (Figure 22.6).

Xerophytes grow in habitats characterized by seasonal or persistent drought. They have small, thick leaves with well-developed spongy and palisade layers, a thick cuticle, and often sunken stomata overlaid with trichomes (hairy coverings). Some species have a multiseriate epidermis or hypodermis, and some species have succulent or no leaves.

Hydrophytes grow in habitually wet, aquatic environments. They have large, thin leaves with poorly developed spongy and palisade layers (large aerenchyma), a thin cuticle, stomata on the upper surface of floating leaves, and relatively little xylem and supporting tissue.

22.3.2.3 Wind Velocity

The wind velocity also influences dust deposition and dust capture by the leaves/plants.

22.4 Effects of Vegetation on Urban Air Quality

Following are the primary mechanisms by which urban plants influence the local air quality:

- Intercept particulate pollution and absorb gaseous pollutants
- Lower air temperatures
- Alter pollutant emissions from power plants

22.4.1 Interception and Absorption of Pollution

When plant leaves are transpiring water and taking up carbon dioxide during daylight, other gases, including pollutants, are simultaneously taken up by the plant leaves. Once inside the leaf, these gases are absorbed and removed from the atmosphere. Some pollution removal can

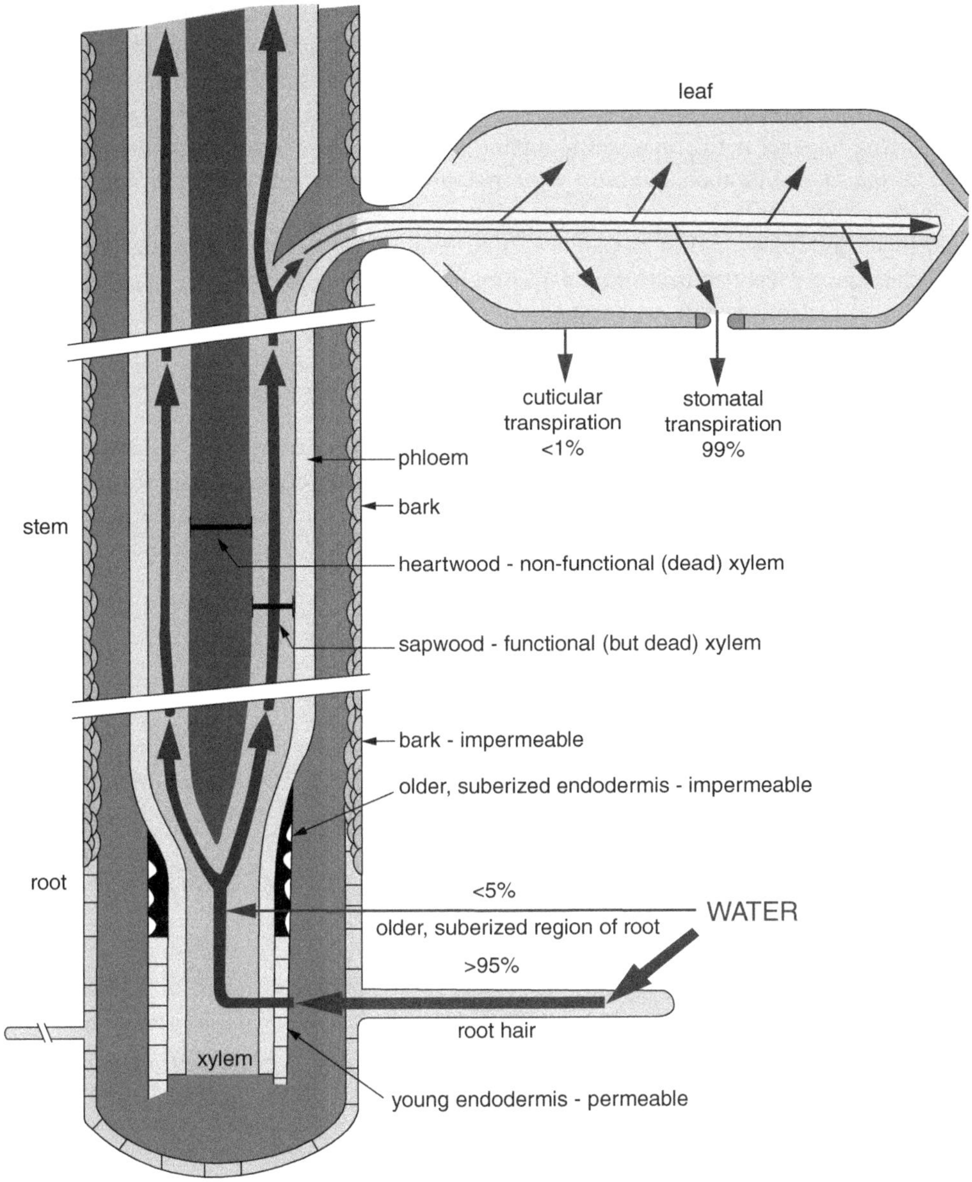

Figure 22.6 Schematic representation of moisture condition surrounding the phyllosphere. *Source:* Shrivastava et al. [10], Central Pollution Control Board, MOEF&CC, Govt. of India, New Delhi. *See colour plate section for colour representation of this figure.*

also occur on the plant surfaces. Particles can be deposited on plant surfaces through sedimentation under the influence of gravity or through impaction under the influence of wind. Particles hitting the plant surfaces may be retained on the surface, may rebound off the surface, or may be temporarily retained and subsequently removed (re-suspended into the air or transported to soil or another surface). Thus, vegetation acts as a temporary retention site for atmospheric particles, as many particles can be re-suspended to the atmosphere, be washed off by rain, or drop to the ground through leaf and twig fall. However, trees can store various trace metals (e.g. lead) in their tissue.

Pollutant uptake by plants is highly variable as it is regulated by numerous plant, pollutant, and environmental forces (e.g. plant water deficit, light intensity, wind speed, gas solubility in water, leaf size and geometry, etc.). Pollution removal by vegetation is dependent on local meteorological conditions, the plant's inherent ability to remove the pollutant, and the concentration of pollution in the atmosphere. In general, the more pollution in the air, the more the plant can remove, up to a point where the plant becomes affected by the pollutant and the stomata shut, limiting the removal of gaseous pollutants.

The majority of pollution removal by plants occurs inside the leaves during the day, as this is when leaf surfaces are actively transpiring, and maximum pollution concentration can be absorbed. Individual plant size affects the total removal of pollutants per tree.

22.4.2 Temperature Effects

As water vapor is moved out of open leaf stomata (transpiration), it changes from a liquid to a gas. This change requires a large amount of energy and reduces air temperatures mainly within and near the plant canopy. This mechanism lowers air temperatures and can significantly affect local air pollution. By lowering air temperatures, plants may also affect ozone photochemistry and ozone precursor emission rates, thus influencing ozone formation.

22.4.3 Impact on Energy Use

Trees and plants can increase or decrease building energy use by shading buildings, altering air flows, and lowering air temperatures through transpiration. Shading residences can affect building energy use: summer shade can reduce air conditioning use and reduce energy consumption.

22.5 Urban Air Quality Improvement through Dust-Capturing Plant Species

Urban environments have moderate to high particulate levels. In certain residential areas, SPM levels are at a critical level. Particulate emissions from industrial areas may be controlled through various pollution-control devices; however, in a residential area, there is no cost-effective technology for controlling particulates/dust buildup except green belt development. The SPM level may reach critical levels due to high traffic movement, vehicular emissions, and other anthropogenic activities.

Despite concerted efforts to control air pollution at sources, the particulate problem persists in urban areas. The growth of population, number of vehicles, and development of industries further enhances particulate pollution problems. Public awareness, proper industrial siting, land-use planning, and other measures to combat pollution such as proper traffic and transportation planning, etc. are still needed.

Green belt development with dense plantation of effective dust-capturing plant species such as plants with medium and high dust-capturing efficiency may help with the particulate problem in residential and industrial areas. A green belt with efficient dust-capturing plants can act as an efficient biological filter, removing significant amounts of particulates from the urban atmosphere; it is a cost-effective technology and also enhances the aesthetic value in urban locations.

Acknowledgments

Thanks are due to the Central Pollution Control Board, Ministry of Environment, and Forests, Govt. of India, New Delhi for providing access to the report from which Figures 22.5 and 22.6 are redrawn.

References

1 Dhanabalan, S.S., Avaninathan, S.R., Rajendran, S., and Carrasco, M.F. (2020). Photocatalysts for indoor air pollution: A brief review. In: *Green Photocatalysts for Energy and Environmental Process*, 247–274. Cham: Springer.
2 Fahri, F., Bacha, K., Chiki, F.F. et al. (2020 Apr 23). Air pollution: new bio-based ionic liquids absorb both hydrophobic and hydrophilic volatile organic compounds with high efficiency. *Environmental Chemistry Letters*: 1–9.
3 Gabal, E., Chatterjee, S., Ahmed, F.K., and Abd-Elsalam, K.A. (2020 Jan 1). Carbon nanomaterial applications in air pollution remediation. In: *Carbon Nanomaterials for Agri-Food and Environmental Applications*, 133–153. Elsevier.
4 Gangupomu, R.H., Sattler, M.L., and Ramirez, D. (2014 Jun 1). Carbon nanotubes for air pollutant control via adsorption: A review. *Reviews in Nanoscience and Nanotechnology* 3 (2): 149–160.
5 Hamadani, H., Rashid, S.M., Rehman, M.U. et al. (2020). Global Scenario of Remediation Techniques to Combat Environmental Pollution. In: *Bioremediation and Biotechnology*, 93–106. Cham: Springer.
6 Hussain, C.M. (ed.) (2020 Mar 19). *The handbook of environmental remediation: classic and modern techniques*. Royal Society of Chemistry.
7 Lee, B.X., Hadibarata, T., and Yuniarto, A. (2020 Aug). Phytoremediation mechanisms in air pollution control: a review. *Water, Air, & Soil Pollution*. 231 (8): 1–3.
8 Subhramanyam, G.V., Rao, D.N., Varshney, C.K., and Biswas, D.K. (eds.) (1985). *Air pollution and plants: A state-of-the-art report*, 1–193. New Delhi: Ministry of Envir. and Forests, GOI.
9 Vani Padmaja, G. and Shruti, S. (2020 May). Engineered nanoparticles in air pollution remediation technologies. *Research Journal of Chemistry and Environment* 24: 5.
10 Shrivastava, N.G., Sharma, C.S., and Bhatnagar, S. (2007). *Phytoremediation of Particulate Matter from Ambient Environment Through Dust Capturing Plant Species*, 123. Delhi: Central Pollution Control Board.
11 Kim, K.J., Shagol, C.C., Torpy, F.R. et al. (2020). Plant physiological mechanisms of air treatment. In: *From Biofiltration to Promising Options in G Aseous Fluxes Biotreatment*, 219–244. Elsevier.
12 Osunmakinde, C.O., Selvarajan, R., Ogola, H.J. et al. (2020). Microbiological air quality in different indoor and outdoor settings in Africa and beyond: challenges and prospects. In: *Current Microbiological Research in Africa Selected Applications for Sustainable Environmental Management* (eds. A.L.K. Abia and G.R. Lanza), 137–174. Cham: Springer.
13 Kwak, M.J., Lee, J.K., Park, S. et al. (2020 Jan). 2020 evaluation of the importance of some east Asian tree species for refinement of air quality by estimating air pollution tolerance index, anticipated performance index, and air pollutant uptake. *Sustainability*. 12 (7): 3067.
14 Papazian, S. and Blande, J.D. (2020). Dynamics of plant responses to combinations of air pollutants. *Plant Biol*. 22: 68–83.

15 Herrmann, M., Geesink, P., Richter, R., and Kuesel, K. (2020). *Canopy Position Has a Stronger Effect than Tree Species Identity on Phyllosphere Bacterial Diversity in a Floodplain Hardwood Forest.* BioRxiv.

16 Mishra, A.K., Mishra, P., Gulia, S., and Goyal, S.K. (2020). Assessment of indoor fine and ultra-fine particulate matter in a research laboratory. In: *Indoor Environmental Quality* (eds. A. Sharma, R. Goyal and R. Mittal), 19–26. Singapore: Springer.

17 Williams, G. (2019). First report of infection with Pseudomonas citronellolis: a case of urosepsis. *New Microbes New Infect.* 30: 100531.

18 Imperato, V., Kowalkowski, L., Portillo-Estrada, M. et al. (2019). Characterisation of the *Carpinus betulus* L. phyllomicrobiome in urban and forest areas. *Front. Microbiol.* 10: 1110.

19 Ma, Y. (2019). Biotechnological potential of plant-microbe interactions in environmental decontamination. *Front. Plant Sci.* 10: 1519.

20 Chen, L., Liu, C., Zhang, L. et al. (2017). Variation in tree species ability to capture and retain airborne fine particulate matter (PM2.5). *Sci. Rep.* 7: 3206. https://doi.org/10.1038/s41598-017-03360-1.

21 Tian, N., Liu, F., Wang, P. et al. (2017). The molecular basis of glandular trichome development and secondary metabolism in plants. *Plant Gene.* 12: 1–12. https://doi.org/10.1016/j.plgene.2017.05.010.

22 Baldacchini, C., Castanheiro, A., Maghakyan, N. et al. (2017). How does the amount and composition of PM deposited on *Platanus acerfolia* leaves change across different cities in Europe. *Environ. Sci. Technol.* 51: 1147–1156. https://doi.org/10.1021/acs.est.6b04052.

23 Gawronski, S.W., Gawronska, H., Lomnicki, S. et al. (2017). Plants in air phytoremediation. *Adv. Bot. Res.* 83: 319–346. https://doi.org/10.1016/bs.abr.2016.12.008.

24 Wei, X., Lyu, S., Yu, Y. et al. (2017). Phylloremediation of air pollutants: exploiting the potential of plant leaves and leaf-associated microbes. *Front. Plant Sci.* 8: 1318. https://doi.org/10.3389/fpls.2017.01318.

25 Laforest-Lapointe, I., Paquette, A., Messier, C., and Kembel, S.W. (2017). Leaf bacterial diversity mediates plant diversity and ecosystem function relationships. *Nature* 546: 145–147. https://doi.org/10.1038/nature2239.

26 Duan, K., Sun, G., Zhang, Y. et al. (2017). Impact of air pollution induced climate change on water availability and ecosystem productivity in the conterminous United States. *Clim. Chang.* 140: 259–272. https://doi.org/10.1007/s10584-016-1850-7.

27 Stevens, V., Thijs, S., McAmmond, B. et al. (2017b). Draft genome sequence of *Rhodococcus erythropolis* VSD3, a diesel fuel-degrading and plant growth-promoting bacterium isolated from *Hedera helix* leaves. *Genome Announc.* 5: e01680–e01616. https://doi.org/10.1128/genomeA.01680-16.

28 Gandolfi, I., Canedoli, C., Imperato, V. et al. (2017). Diversity and hydrocarbon-degrading potential of epiphytic microbial communities on *Platanus* × *acerifolia* leaves in an urban area. *Environ. Pollut.* 220: 650–658. https://doi.org/10.1016/j.envpol.2016.10.022.

29 Archibald, A.T., Folberth, G., Wade, D.C., and Scott, D. (2017). A world avoided: impacts of changes in anthropogenic emissions on the burden and effects of air pollutants in Europe and North America. *Faraday Discuss.* https://doi.org/10.1039/C7FD00004A.

30 Macpherson, A.J., Simon, H., Langdon, R., and Misenheimer, D. (2017). A mixed integer programming model for national ambient air quality standards (NAAQS) attainment strategy analysis. *Environ. Model. Softw.* 91: 13–27. https://doi.org/10.1016/j.envsoft.2017.01.008.

31 Kato, N., Yurimoto, H., and Thauer, R.K. (2006). The physiological role of the ribulose monophosphate pathway in bacteria and archaea. *Biosci. Biotechnol. Biochem.* 70: 10–21. https://doi.org/10.1271/bbb.70.10.

32 Huang, Y., Ho, S.S., Lu, Y. et al. (2016). Removal of indoor volatile organic compounds via photocatalytic oxidation: a short review and prospect. *Molecules* 21: 56. https://doi.org/10.3390/molecules21010056.

33 Toabaita, M., Vangnai, A.S., and Thiravetyan, P. (2016). Removal of ethylbenzene from contaminated air by *Zamioculcas Zamiifolia* and microorganisms associated on *Z. Zamiifolia* leaves. *Water Air Soil Pollut.* 227: 1–11. https://doi.org/10.1007/s11270-016-2817-z.

34 Jud, W., Fischer, L., Canaval, E. et al. (2016). Plant surface reactions: an opportunistic ozone defence mechanism impacting atmospheric chemistry. *Atmos. Chem. Phys.* 16: 277–292. https://doi.org/10.5194/acp-16-277-2016.

35 Rai, P.K. (2016). Impacts of particulate matter pollution on plants: implications for environmental biomonitoring. *Ecotoxicol. Environ. Saf.* 129: 120–136. https://doi.org/10.1016/j.ecoenv.2016.03.012.

36 Laforest-lapointe, I., Messier, C., and Kembel, S.W. (2016). Host species identity, site and time drive temperate tree phyllosphere bacterial community structure. *Microbiome* 4: 27. https://doi.org/10.1186/s40168-016-0174-1.

37 Syranidou, E., Christofilopoulos, S., Gkavrou, G. et al. (2016). Exploitation of endophytic bacteria to enhance the phytoremediation potential of the wetland helophyte Juncus acutus. *Front. Microbiol.* 7: 1016. https://doi.org/10.3389/fmicb.2016.01016.

38 Khaksar, G., Treesubsuntorn, C., and Thiravetyan, P. (2016b). Endophytic Bacillus cereus ERBP-Clitoria ternatea interactions: potentials for the enhancement of gaseous formaldehyde removal. *Environ. Exp. Bot.* 126: 10–20. https://doi.org/10.1016/j.envexpbot.2016.02.009.

39 Khaksar, G., Treesubsuntorn, C., and Thiravetyan, P. (2016a). Effect of endophytic Bacillus cereus ERBP inoculation into non-native host: potentials and challenges for airborne formaldehyde removal. *Plant Physiol. Biochem.* 107: 326–336. https://doi.org/10.1016/j.plaphy.2016.06.020.

40 Gao, J., Fan, X., Pan, K. et al. (2016). Diversity, abundance and activity of ammonia-oxidizing microorganisms in fine particulate matter. *Sci. Rep.* 6: 38785. https://doi.org/10.1038/srep38785.

41 Ma, Y., Oliveira, R.S., Freitas, H., and Zhang, C. (2016). Biochemical and molecular mechanisms of plant-microbe-metal interactions: revelance for phytoremediation. *Front. Plant Sci.* 7: 918. https://doi.org/10.3389/fpls.2016.00918.

42 Gurjar, B.R., Ravindra, K., and Nagpure, A.S. (2016). Air pollution trends over India megacities and their local-to-global implications. *Atmos. Environ.* 142: 475–495. https://doi.org/10.1016/j.atmosenv.2016.06.030.

43 Phung, D., Hien, T.T., Linh, H.N. et al. (2016). Air pollution and risk of respiratory and cardiovascular hospitalizations in the most populous city in Vietnam. *Sci. Total Environ.* 557: 322–330. https://doi.org/10.1016/j.scitotenv.2016.03.070.

44 Liu, J., Mauzerall, D.L., Chen, Q. et al. (2016). Air polluant emissions from Chinese households: a major and underappreciated ambient pollutant source. *Proc. Natl. Acad. Sci. U. S. A.* 113: 7756–7761. https://doi.org/10.1073/pnas.1604537113.

45 Weyens, N., Thijs, S., Popek, R. et al. (2015). The role of plant–microbe interactions and their exploitation for phytoremediation of air pollutants. *Int. J. Mol. Sci.* 16: 25576–25604. https://doi.org/10.3390/ijms161025576.

46 Zhang, H.H., Chen, S.N., and Huang, T.L. (2015). Structure and functional metabolism of bacterial communities on leaves of typical urban greening tree species. *Pol. J. Environ. Stud.* 24: 823–828. https://doi.org/10.15244/pjoes/30592.

47 Miletto, M. and Lindow, S.E. (2015). Relative and contextual contribution of different sources to the composition and abundance of indoor air bacteria in residences. *Microbiome* 3: 61. https://doi.org/10.1186/s40168-015-0128-z.

48 Yu, D.S., Song, G., Song, L.L. et al. (2015). Formaldehyde degradation by a newly isolate fungus Aspergillus sp. *Hua. Intl. J. Environ. Sci. Technol.* 12: 247–254. https://doi.org/10.1007/s13762-013-0411-0.

49 Ali, N., Al-Awadhi, H., Dashti, N. et al. (2015). Bioremediation of atmospheric hydrocarbons via bacteria naturally associated with leaves of higher plants. *Int. J. Phytoremed.* 17: 1160–1170. https://doi.org/10.1080/15226514.2015.1045125.

50 Kelly, F.J. and Fussell, J.C. (2015). Air pollution and public health: emerging hazards and improved understanding of risk. *Environ. Geochem. Health* 37: 631–649. https://doi.org/10.1007/s10653-015-9720-1.

51 King, G.M. (2014). Urban microbiomes and urban ecology: how do microbes in the built environment affect human sustainability in cities? *J. Microbiol.* 52: 721–728. https://doi.org/10.1007/s12275-014-4364-x.

52 Sriprapat, W., Suksabye, P., Areephak, S. et al. (2014). Uptake of toluene and ethylbenzene by plants: removal of volatile indoor air contaminants. *Ecotoxicol. Environ. Saf.* 102: 147–151. https://doi.org/10.1016/j.ecoenv.2014.01.032.

53 Karagiannidis, A., Poupkou, A., Giannaros, T. et al. (2014). The air quality of a mediterranean urban environment area and its relation to major meteorological parameters. *Water Air Soil Pollut.* 226: 2239. https://doi.org/10.1007/s11270-014-2239-8.

54 Kembel, S.W., O'connor, T.K., Arnold, H.K. et al. (2014). Relationships between phyllosphere bacterial communities and plant functional traits in a neotropical forest. *Proc. Natl. Acad. Sci. U. S. A.* 111: 13715–13720.

55 Kembel, S.W. and Mueller, R.C. (2014). Plant traits and taxonomy drive host associations in tropical phyllosphere fungal communities. *Botany* 92: 303–311. https://doi.org/10.1139/cjb-2013-0194.

56 Sun, K., Liu, J., Gao, Y. et al. (2014). Isolation, plant colonization potential, and phenanthrene degradation performance of the endophytic bacterium Pseudomonas sp. Ph6-gfp. *Sci. Rep.* 4: 5462. https://doi.org/10.1038/srep05462.

57 Afzal, M., Khan, Q.M., and Sessitsch, A. (2014). Endophytic bacteria: prospects and applications for the phytoremediation of organic pollutants. *Chemosphere* 117: 232–242. https://doi.org/10.1016/j.chemosphere.2014.06.078.

58 Khan, Z., Roman, D., Kintz, T. et al. (2014). Degradation, phytoprotection and phytoremediation of phenanthrene by endophyte pseudomonas putida, PD1. *Environ. Sci. Technol.* 48: 12221–12228. https://doi.org/10.1021/es503880t.

59 Rico, L., Ogaya, R., Terradas, J., and Penuelas, J. (2014). Community structures of N 2-fixing bacteria associated with the phyllosphere of a holm oak forest and their response to drought. *Plant Biol.* 16: 586–593. https://doi.org/10.1111/plb.12082.

60 Tsai, Y.I., Yang, H.H., Wang, L.C. et al. (2011). The influences of diesel particulate filter installation on air pollutant emissions for used vehicles. *Aerosol Air Qual. Res.* 11: 578–583. https://doi.org/10.4209/aaqr.2011.05.0066.

61 Jumpponen, A. and Jones, K.L. (2010). Seasonally dynamic fungal communities in the *Quercus macrocarp* phyllosphere differ between urban and nonurban environments. *New Phytol.* 186: 496–513. https://doi.org/10.1111/j.1469-8137.2010.03197.x.

62 Kampa, M. and Castanas, E. (2008). Human health effects of air pollution. *Environ. Pollut.* 151: 362–367. https://doi.org/10.1016/j.envpol.2007.06.012.

63 Waight, K., Pinyakong, O., and Luepromchai, E. (2007). Degradation of phenanthrene on plant leaves by phyllosphere bacteria. *J. Gen. Appl. Microbiol.* 53: 265–272. http://www.ncbi.nlm.nih.gov/pubmed/18057816.

64 Kahraman, H. and Geckil, H. (2005). Degradation of benzene, toluene and xylene by *Pseudomonas aeruginosa* engineered with the Vitreoscilla Hemoglobin gene. *Eng. Life Sci.* 5: 363–368. https://doi.org/10.1002/elsc.200520088.

65 Zhu, Y., Hinds, W.C., Kim, S. et al. (2002). Study of ultrafine particles near a major highway with heavy-duty diesel traffic. *Atmos. Environ.* 36: 4323–4335. https://doi.org/10.1016/S1352-2310(02)00354-0.

66 Chaudhary, D., Kumar, R., Sihag, K., and Kumari, A. (2017). Phyllospheric microflora and its impact on plant growth: a review. *Agric. Rev.* 38 (1): 51–59.

67 Kapley, A., Purohit, H.J., Chhatre, S. et al. (1999). Osmotolerance and hydrocarbon degradation by a genetically engineered microbial consortium. *Bioresour. Technol.* 67: 241–245. https://doi.org/10.1016/S0960-8524(98)00121-7.

68 Ramos, J.L., Molina, L., and Segura, A. (2009). Removal of organic toxic chemicals in the rhizosphere and phyllosphere of plants. *Microb. Biotechnol.* 2 (2): 144–146.

69 Kumar, S., Prasad, S., and Yadav, K.K. (2019). Utilization of air pollutants by plants: need for present and future scrutiny. *J. Agric. Food Chem.* 2019 (67): 2741–2742. https://doi.org/10.1021/acs.jafc.9b00921.

70 Sandhu, A., Halverson, L.J., and Beattie, G.A. (2007). Bacterial degradation of airborne phenol in the phyllosphere. *Environ. Microbiol.* 9 (2): 383–392.

71 Stevens, V., Thijs, S., McAmmond, B. et al. (2017a). Draft genome sequence of *Bacillus licheniformis* VSD4, a diesel fuel–degrading and plant growth–promoting phyllospheric bacterium. *Genome Announc.* 5: e00027–e00017. https://doi.org/10.1128/genomeA.00027-17.

72 Ugrekhelidze, D., Korte, F., and Kveitdze, G. (1997). Uptake and transformation of benzene and toluene by plant leaves. *Ecotoxicol. Environ. Saf.* 37: 24–29. https://doi.org/10.1006/eesa.1996.1512.

73 Undugoda, L.J., Kannangara, S., and Sirisena, D.M. (2016). Aromatic hydrocarbon degrading fungi inhabiting the phyllosphere of ornamental plants on roadsides of urban areas in Sri Lanka. *J Bioremed Biodeg.* 7 (328): 2.

74 UNFPA. (2004). Migration and urbanization. State of World Population 2004. The Cairo Consensus at Ten: Population, Reproductive Health and the Global Effort To End Poverty. New York, NY: United Nations Population Fund.

23

Green Belts for Sustainable Improvement of Air Quality

S.B. Chaphekar[1], R.P. Madav[2], and Seemaa S. Ghate[3,4]

[1] *Homi Bhabha State University, Mumbai, Maharashtra, India*
[2] *Terracon Ecotech Pvt. Ltd., Mumbai, Maharashtra, India*
[3] *Women Scientist 'B', Department of Science and Technology, New Delhi, India*
[4] *Knowhow Foundation, Pune, Maharashtra, India*

23.1 Introduction

Planting trees along roads to provide relief to travelers from the hot sun has been practiced for centuries. The practice continues and is now supplemented by more systematically organized planting in cities and towns and along roads and highways. Scientific evidence to confirm this age-old wisdom is evident in observation of moderation of micro-environments in built-up urban areas, reducing temperatures by 2–3 °C and increasing relative humidity by 3–4% under canopies of trees when compared to the same parameters in nearby open spaces [1]. Characteristics of plants for noise pollution abatement are listed by Rameshkumar et al. [2]. Extensive observations have been made of reduction in noise levels by 6–7 dB below tree canopies compared to open areas [3]. Plants are also found to be effective in reducing indoor pollution caused by volatile organic compounds (VOCs) [4].

The capacity of plants to act as a sink for air contaminants has been addressed by a number of reviews [5–13]. These studies indicate that vegetation provides a significant filtration and reaction surface for removing pollutants from the atmosphere. Rameshkumar et al. [2] have reviewed the role of plant species in controlling environmental pollution.

Trees are also important particulate sinks. Studies dealing with diverse types of particulates, including heavy metals, pollen, spores, fungal hyphae, salts, radio-nuclides, etc. are numerous. All trees are reported to capture large quantities of particulates, from 10 to <1 μm size classes [14]. The quality of air in local built-up environments changes in the presence of vegetation, as reported by Abhijith et al. [15].

Systematic measurement of air pollution in urban-industrial areas in India started in the 1960s. Frequent episodes of health-affecting emissions from a number of industries in suburban Chembur in Mumbai triggered repeated surveys of pollution in the city by scientists from national and regional agencies like BARC, NEERI, MPCB, and CLI, some NGOs, and educational institutions like IIT and The Institute of Science, for identification and quantification of pollutant gases in several parts of the urban environment and their possible effects on human health. Effects of pollution on plants were also observed [16, 17].

Handbook of Ecological and Ecosystem Engineering, First Edition. Edited by Majeti Narasimha Vara Prasad.

23.2 Tolerance of Plants to Air Pollutants

Repeated surveys in industrial and urban areas in different parts of the world helped identify and describe injury symptoms exhibited by plants. Some of the data was compared with observational and experimental data collected from research centers ([18, 19], and many others) (Table 23.1).

Dust deposition on leaf surfaces was observed mostly in field studies. Chemical analysis of leaf washes helped identify the chemical nature of particulate emissions from sources like vehicular exhaust and industries in the neighborhood. The collected information helped distinguish sensitive and tolerant plant species (Table 23.2).

Apart from field observations and experimental exposure of plants to regulated pollutant dosages in fumigation chambers, some researchers have been using the concept of the Air Pollution

Table 23.1 Symptoms of pollution-induced foliar injury in plants.

Pollutants	Plant injury symptoms
SO_2	Interveinal necrotic blotches; red-brown banding in pines
NO_x	Interveinal necrotic blotches; red-brown distal necrosis in pines
HF	Tip and margin necrosis
NH_3	Interveinal necrotic blotches; distal necrosis in pines
Cl_2	Chlorosis, upper surface fleck; distal necrosis in pines
Ethylene	Chlorosis, necrosis, abscission, dwarfing, premature defoliation
H_2S	Interveinal necrotic blotches
O_3	Upper surface flecks
PAN	Lower surface bronzing, chlorosis, early senescence

Source: Smith [20].

Table 23.2 Some plant species identified as sensitive/tolerant to air pollution.

Plant species	Sensitive	Tolerant	Source
Citrus lemon		Coal dust	Rao [21]
Syzygium cuminii	Cement dust		Jafri et al. [22]
Psidium guyava	Cement dust		Yunus and Ahmad [23]
Calotropis procera		Polluted areas	Yunus and Ahmad [24]
Thespesia populnea		Dust	Shetye and Chaphekar [25]
Ficus benghalensis		Dust	Das [26]
Banana crop	SO_2, dust		Bedi et al. [27]
Mangifera indica	SO_2		Rao [28]
Croton, Nerium		SO_2	Chaphekar [16]
Terminalia tomentosa	HF		Rao and Pal [29]
Tectona grandis		Dust	Ahmad et al. [30]
Polyalthia longifolia	SO_2		Raza et al. [31]
Pithecolobium dulce		Dust	Raza et al. [31]

Note: Only a small number of publications are quoted here to represent more than 90 published papers referred to and listed in the original document. For a comprehensive list through 1994, refer to [32].

Tolerance Index (APTI) to assess the sensitiveness/tolerance of species. The APTI is a product of (i) ascorbic acid content (mg/g fresh weight), (ii) leaf extract pH, (iii) total chlorophyll content (mg/g dry weight) of leaves, and (iv) relative water content (%) of leaves [33]. Agrawal [34], Rameshkumar et al. [2], have used this index to identify a variety of evergreen, deciduous, and herbaceous plants for their air pollution tolerance capacity.

Morphological characteristics of leaves of plants [35] that play an important role in responding to air pollutants are summarized in Table 23.3.

Table 23.4 summarizes leaf properties conducive to capturing airborne dust particles.

Shapes and sizes of tree canopies play an essential role in forming an interface between trees' foliage and atmospheric pollutants – both gaseous and particulate. It is observed that spacing between canopies of trees facilitates a larger interface than when the canopies are closely packed [2].

23.2.1 Agro-Climates in India

Plant species for plantation in green belts are selected primarily according to their capacity to remove pollutants from the air. However, the success of the growth of the selected plants depends on their being adapted to the local climatic and soil conditions. For this purpose, the agro-climatic conditions of the region where the belt is to be cultivated and the compatibility of the selected species must match. Knowledge of local vegetation and reference to regional flora helps in short-listing the species before launching a project.

The Planning Commission of India recognizes 15 agro-climatic zones (Figure 23.1). These zones are further divided into several subzones based on rainfall, climate, and soils [37]. For the

Table 23.3 Morphological properties of leaves of plants showing high efficiency for removal of gaseous pollutants.

1. Crowns well exposed, open, spreading
2. Leaves in mosaics, exposed freely
3. Long internodes
4. Leaves freely suspended in air, swinging easily on air currents
5. Many stomata, in level with the epidermis
6. Protective coatings like thick cuticle, wax arches, hairs, scales, etc. absent on the leaf epidermis
7. Mesophyll cells arranged loosely, thin-walled, and parenchymatous
8. Metabolism rates high

Source: Chaphekar [36] ©1995, Surya International Publications.

Table 23.4 Morphological properties of plants showing high efficiency for removal of particulate pollutants from the atmosphere.

1. Crowns well exposed, open, spreading
2. Leaves well exposed, dorsi-ventral, on strong, short, and stiff petioles
3. Venation prominent, closely netted
4. Leaf epidermis rough, hairy, or scaly, and/or with protective cuticle or wax
5. Stomata sunken below the epidermal surface, protected with wax arches or rings, dead hairs, trichomes, etc.

Source: Chaphekar [36] ©1995, Surya International Publications.

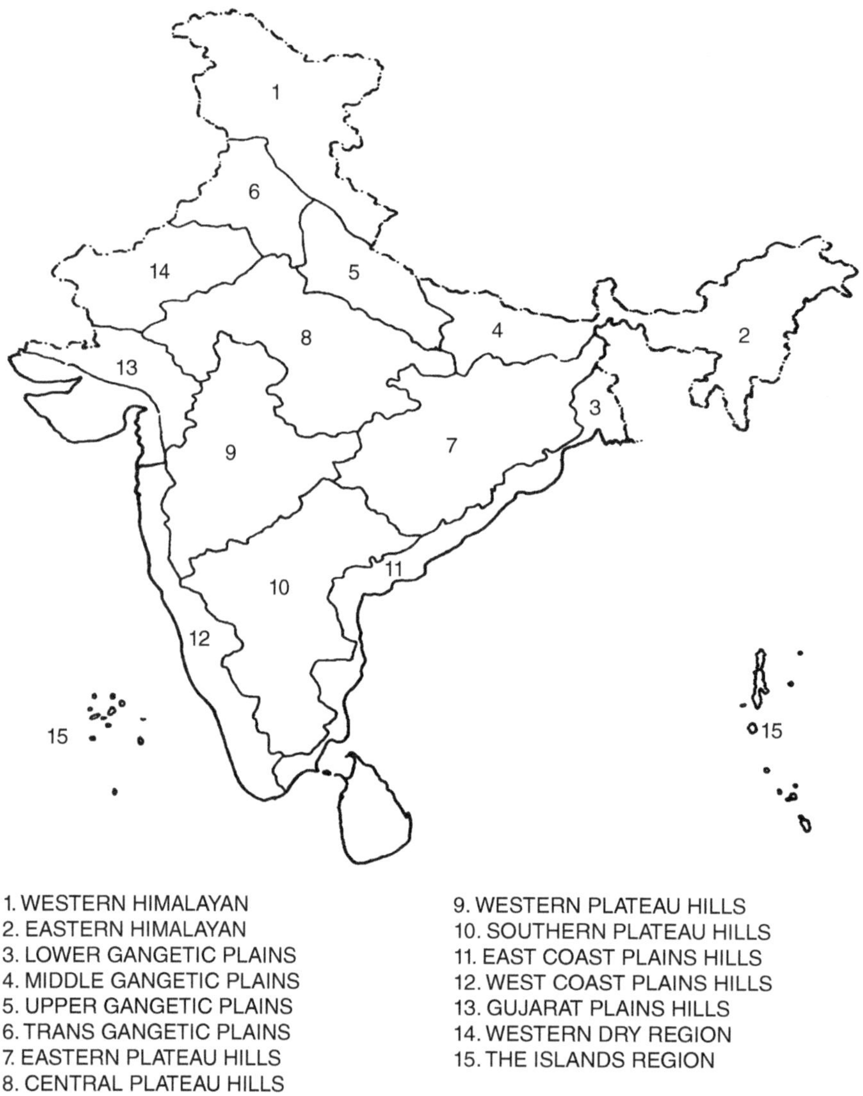

Figure 23.1 Agro-climatic zones of India.

convenience of project operators, revenue districts falling in each subzone are also listed. An example of one subzone is given here.

Zone IX – Western Plateau and Hills – Subzone No. 1. This hilly subzone receives, on average, 988 mm rainfall and has medium to deep black or shallow red loamy soils. The subzone includes the districts of Kolhapur, Satara, Pune, and Nasik of Maharashtra state and Jhabua in Madhya Pradesh. This information helps select suitable plant species for green belts after referring to information about local vegetation and flora.

23.2.2 Green Belts

A popular slogan from the 1970s and continuing to this day is, "Grow more trees and reduce pollution." People ask, "Pollution injures plants, even sometimes causes total defoliation; can their plantation provide a reliable solution to pollution?" In response, it is emphasized that plants absorb pollutants within their limits of tolerance and, to that extent, remove pollutants from ambient air. It is also necessary to choose suitable plant species and locate them suitably, in numbers determined using available models, to scavenge air pollutants. The need for more greenery, including in urban areas, is further emphasized by plants' capacity for carbon sequestration, which can help mitigate the effects of the emerging climate crisis.

Plant uptake of air pollutants appears to follow the order HF > SO_2 > Cl > NO_2 > O_3 > PAN > NO > CO, commensurate with the water-solubility of these gases.

The rate of pollutant removal is found to increase linearly as the concentration of the pollutant increases. This happens until the concentration causes closure of the stomata [5, 9, 38]. Smith [20] guesstimated that a hectare consisting of five-year-old vegetation constituting 365 broad-leaved trees and 700 pines would remove gaseous pollutants annually in quantities as O_3– 9.6×10^4, SO_2–748, CO –2.2, NO_x–0.38, PAN –0.17 (all values in tons).

23.2.3 Choosing Plant Species

Selection of plant species is based on specific criteria: (i) indigenous species adapted to local climatic, edaphic conditions – for this, the country's map based on agro-climatic conditions is refered; (ii) reference to regional flora; (iii) canopy shapes and sizes of trees at maturity to help determine architectural and landscaping aspects of the belt; (iv) number of trees and their location with reference to the source of pollution; (v) local wind roses; and (vi) spacing of trees to ensure the free movement of air within the belt to facilitate efficient pollutant sorption by foliage.

In addition to field observations and experimental exposure of plants to controlled dosages of pollutants for selecting tolerant species, the concept of APTI can be employed for a relatively quick check for pollutant-tolerant species. Based on the criteria for tolerant species of plants discussed so far, a comprehensive list of suitable species for planting in green belts in the districts of Subzone No. 1 of Zone IX of India is compiled as an example. A total of 213 species are listed, from which only 12 are given in Table 23.5 for space reasons. The wide range of species available and effort involved in preparation for green belt cultivation is evident from the fact that the list contains the following number of species per letter of the alphabet: A – 44, B – 17, C – 22 D – 8, E – 6, F –9, G – 9, H–5, I – 5, J – 2, K – 1, L – 4, M – 14, N – 2, O – 1, P- 21, Q – 3, R – 1, S – 20, T – 14, U – 1, and Z – 4. Thus a total of 213 species of trees are found suitable for the development of green belts in Zone IX, Subzone No. 1.

23.2.4 Designing Green Belts

23.2.4.1 Ground-Level Concentration (GLC) of Emitted Pollutants

Conventionally, practitioners in the field of environmental management followed a rule of thumb that the middle portion of a green belt should coincide with the location where the maximum concentration of pollutants land after being emitted from a stack. For this purpose, the assessment was that ground-level concentration (GLC) of the pollutant was highest at a point 10 times the stack's height (in meters).

A more scientific approach to decide the GLC is proposed in the Gaussian formula, which takes into account (i) concentration of the effluent at the emission point, i.e. top of the stack; (ii) emission factor from the source; (iii) lateral and vertical standard deviations of the diffusing plume; (iv) average wind speed; and (v) height of the plume.

In the case of non-stack "fugitive" emissions near ground level, the green belt is mostly of an ad hoc nature around the factory. Taking into account that wind directions change during different parts of the day and by season, Dr. Ganguli [39] suggested that a scientifically planned green belt

Table 23.5 A partial list of plant species recommended for planting in green belts in Zone IX, Subzone No. 1.

Name of plant	Approx. height in meters	Growth rate	Reproduction	Canopy shape	Leaf duration	Remarks
Abutilon indicum	3	Fast	Seeds	Oblong	Evergreen	—
Acacia catechu	3	Fast	Seeds	Oblong	Evergreen	—
Achras sapota	10	Slow	Seeds, grafts	Spreading	Evergreen	Fruits
Actinodaphne hookenakeri	10	Slow	Seeds	Oblong	Evergreen	Medicinal
Albizzia lebbeck	20	Fast	Seeds	Round/spreading	Deciduous	—
Annona squamosa	10	Fast	Seeds, grafts	Round	Evergreen	Fruit
Ailanthus excelsa	20	Fast	Seeds, cuttings	Round	Deciduous	—
Alstonia scholaris	15	Fast	Seeds	Round	Evergreen	—
Anogeissus latifolia	12	Slow	Budding	Round/Oblong	Evergreen	—
Anthocephalus chinensis	20	Fast	Seeds	Spreading	Evergreen	—
Artocarpus heterophyllus	10	Slow	Seeds	Oblong	Evergreen	Fruits
Azardirachta indica	20	Fast	Seeds	Spreading	Evergreen	Medicinal

has to be present in all directions around the source of emissions. A diagram of a sample green belt suggested by him is shown in Figure 23.2.

23.2.4.2 Mathematical Model

A mathematical model for calculating the dispersal and landing of gaseous pollutants was presented by Gupta and Kapoor [40], which aimed at perfection since it was meant for abating the consequences of pollution by radionuclides escaping from (a rare possibility as per Atomic Energy Commission [AEC] claim) an accident in a nuclear power plant. The model is applicable to all types of pollution-generating industries [41]. They devised the concept of an Attenuation Coefficient (Af) [42], which is based on (i) deposition velocity of pollutant, (ii) its mass flux with distance traveled up to the tree canopy, (iii) average wind speed through a green belt with adequate spacing between trees, (iv) average foliar surface area density of a single tree and the green belt, and (v) stable (A) to very unstable (F) atmospheric conditions.

A significant advantage of this model is that the Af gives information about the effectiveness of a green belt in attenuating pollution. Af is defined as the ratio of the mass flux of a pollutant reaching a given distance in the absence of green belt to the mass flux of the pollutant reaching the same distance in the presence of a green belt.

Significant values of Af occur for ground-level pollution sources (fugitive emissions), especially in stable (A) atmospheric conditions. The results of the model's application also suggest that a green belt is more effective in attenuating pollution in stable atmospheric conditions and with less

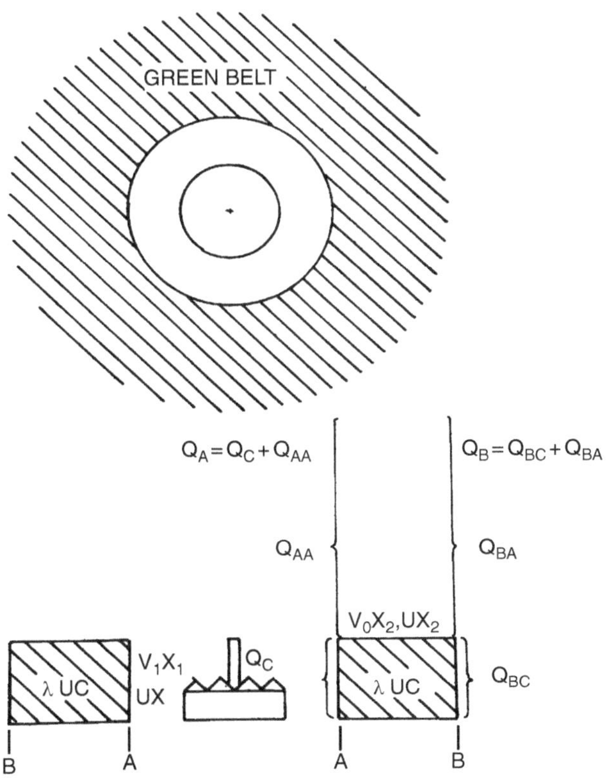

Figure 23.2 Simplified diagram of model green belt around a factory with ground-level emissions. A, inner edge of the green belt; B, outer edge.

distance between the source and the belt. Taller trees are advantageous at smaller distance between the source and the belt, and increasing the width of the green belt beyond 250–300 m does not give an added advantage (Figure 23.3).

23.2.4.3 Two Approaches

The model is valid for both pollution-source and pollution-receptor oriented approaches for determining green-belt dimensions. In the first case, the green belt is planted around the pollution source, as shown in Figure 23.2; whereas in the second case, the receptor is protected against diffused or multiple pollution sources by cultivating a green belt around the receptor (e.g. protecting the Taj Mahal from multiple emission sources in and near Agra; monuments and even residences in or near industrial estates).

23.2.4.4 Planting Along Roadsides

Many cities are recognized as suffering mainly from pollution from automobiles. Indian cities are experiencing the same problem [43]. An automobile is a point source of pollution that is mobile when emissions occur. The ever-increasing number of automobiles and their movement on the network of roads converts those roads into an area source. This type of pollution affects a large number of citizens residing in urban areas. It is estimated that almost 67% of the air pollution in Mumbai city is caused by automobiles. On a national scale, the figure stands near 40% and is increasing rapidly [43]. Hence, it is logical that citizens should be protected from air pollution on and along roads by all available means, including with green belts.

Vehicular pollution includes gaseous and particulate emissions. Of these, particulate pollution has been reported to be beyond permissible limits of 140 μg/m^3 in almost all cities in India [43].

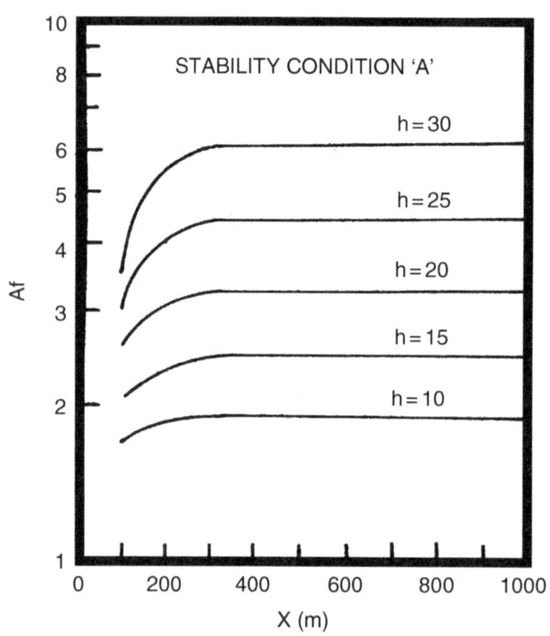

Figure 23.3 Attenuation factor (Af) is a function of the width of the green belt, tree height 50 m, and atmospheric stability condition A. Increasing the value of X more than 300 m does not serve any useful purpose for tree height 10–30 m. *Source:* Chaphekar et al. [32].

Suspended particulate matter (SPM) is caused by wear and tear of road surfaces and car tires and remains airborne for long durations, flying up about 10 m vertically and 25–30 m horizontally. SPM functions as nuclei for deposition of airborne air pollutants and intensifies pollution hazards and is suspected to be a significant contributor to respiratory problems, especially among children and senior citizens.

Roads are of different types, from super-express and express highways connecting major cities to streets, roads, and lanes in cities and towns. Trees in residential premises form green belts along lanes, and those planted by local authorities along roads and streets, mainly for esthetics and shade, form green belts. Studies have shown that low-level hedges, rather than tall trees, improve air quality in street canyons, i.e. the lanes and small roads between buildings [15]. However, scope for planting is limited by lack of space along roadsides, which is needed for free movement of pedestrians.

Intercity highways carry heavy traffic of all types of vehicles. The potential for pollution is high and likely to affect agricultural farms and forest biota along these highways. It has been reported that agricultural production in the country is reduced by 10–80% due to air pollution [34]. When properly planned, highways are clearly marked with Right of Way (RoW) reserved, and declared as No Development Zone (NDZ) where green belts are to be developed.

23.2.4.5 Choice of Plants for Roadsides

Tree and shrub species for planting are primarily selected from the list prepared for the respective zone and subzone (refer to Table 23.5). Further short-listing needs to be done on the basis of the following:

1) Pollution tolerance and dust removal efficiency of foliage
2) Space required above and below ground, especially with reference to buttress roots
3) Littering of roads due to leaf fall (deciduousness), flowering profusion (like *Peltophorum*), fruit litter (figs, causing slippery surface), etc.
4) Straightness of the main trunk for 3–4 m from the ground (natural or manipulated)
5) Water requirements in terms of quantity and frequency of watering

In addition to roadside plantings, plants are also needed on road dividers, medians, and traffic islands, for aesthetic and other purposes (Figure 23.4). They have to be tall enough to protect the vision of vehicle drivers from the headlight glare of vehicles moving in the opposite direction at night and short enough to maintain visibility on curves and bends in the road. Shrub species 1–1.5 m tall are recommended to satisfy these requirements (Table 23.6).

23.2.4.6 Nurturing Green Belts

Green belts in public areas frequently suffer from neglect. A remedial measure suggested here is based on personal experience. An average household is fond of nurturing plants. Once they are convinced that green belts provide them with usufruct benefits like beautiful, fragrant flowers, fruits, vegetables, and cleaner air and healthy ambience, they will be interested in deriving these benefits.

In today's world, people spend more than 90% of their time inside close spaces, where the levels of many common organic pollutants can be 12 times higher than outside. Indoor air pollution ranks among the top five major hazardous health issues and is a global concern. Significant impacts of indoor air pollution are seen in hospitals, offices, factories, malls, residential complexes, IT parks, etc. Indoor pollution is now considered by many experts to be a major threat to human health [44]. An interesting research project was carried out in an elementary school in Seoul, South Korea that showed the importance of indoor greenery (IG). The comparative study of two groups

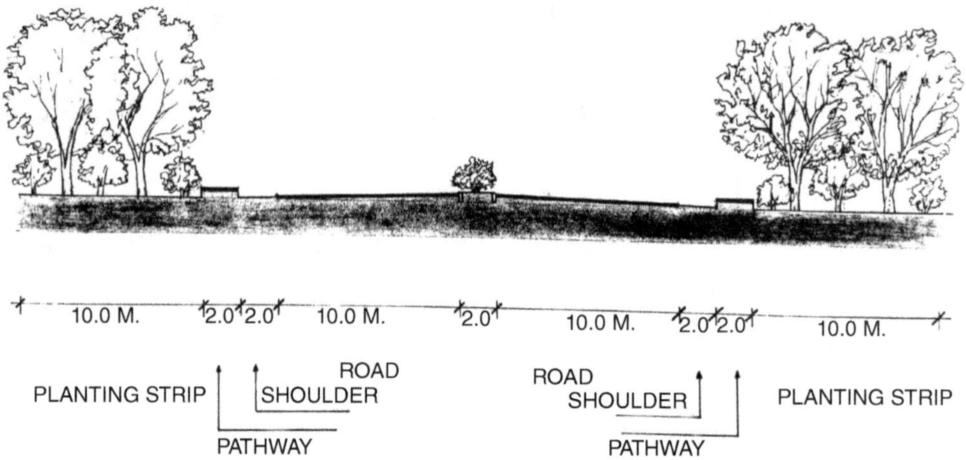

Figure 23.4 Cross section of a road showing a planting scheme.

Table 23.6 A partial list of shrub species recommended for planting on medians of express highways in Zone IX, Subzone 1.

1. *Duranta plumieri*
2. *Hamelia patens*
3. *Hibiscus rosa-sinensis*
4. *Ixora coccinia*
5. *Lantana camara*
6. *Lowsonia inermis*
7. *Nerium indicum*
8. *Ricinus communis*
9. *Melaleuca leucadendron*
10. *Poincettia pulcherrima*
11. *Tecoma stans*
12. *Thevetia peruviana*
13. *Tabernemontana divaricata*
14. *Dracaena sp.*
15. *Croton sp.*

of students revealed a significant improvement in the attention capacity of pupils in classrooms with indoor plants [45].

Removing indoor pollution with potted ornamentals is achievable as some studies have shown [4, 7, 46], Ghate tested house plants such as *Syngonium podophyllum* (maza red) [46], *Chlorophytum comosum* (green spider plant), *Aglaonema commutatum* (Chinese evergreen), *Polyscia fructicosa* (ming aralia), *Epipremnum aureum* (money plant), *Rhapis excelsa* (lady palm), *Spathiphyllum wallichi* (peace lily), and *Sansevieria trifasciata* (mother-in-law's tongue) for their pollution absorption potential. These plants were tested against common indoor pollutants released through the smoke of mosquito coils, heavy smelling incense sticks, cigarette smoke, paints, and varnishes. *Sanseviera trifasciate* and *C. comosum* showed the highest capacity for absorption of nicotine, a dangerous chemical released by cigarette smoke. The absorption of nicotine was in the range of 0.01 to 0.8 ppm [6, 7].

Similar experiments carried out at some households in the neighborhood inspired people to plant indoor ornamentals and adopt trees and bushes in their vicinity in public places and nurture them. This act not only added aesthetics to their locality but also improved their 'feel-good factor'. Encouraging a similar strategy for nurturing roadside green belts is the need of the day.

23.3 Conclusion

Setting up green belts in urban and industrial areas and along roadsides needs a revised approach. Information is now available to select suitable plant species that are effective for sorption of pollutants. Meticulously designed green belts, using a wide diversity of indigenously available plant species, supplemented when useful with known beneficial but harmless exotics, and planted as per developed norms for optimized efficiency, ensure efficient functioning. Improving aesthetics of the operational regions is an added bonus. Active participation of local residents and stakeholders can help ensure the success of the effort to increase green cover, conserve biodiversity in areas dominated by human activity, and delay the predicted unfavorable consequences of global warming.

References

1 Chaphekar, S.B. and Kulkarni, S.S. (1973). Light, Heat, and Roadside Plants. *Indian Forester* 99 (9): 579–581.

2 Rameshkumar, S., Arumugam, T., Anandakumar, C.R. (2013). Use of plant species in controlling environmental pollution – a review. *Bull. Environ., Pharmacol. Life Sci.* 2 (2): 52–63.

3 Rane, P.P. (2010). Studies on relationship between plants and noise levels. PhD thesis, Bombay University.

4 Ghate, S. (2016). Assessment of phytoremediating potential of *Aglaonema commutatum* Schott for indoor pollutants. *Int. J. Plant Environ.* 2 (1–2): 87–92.

5 Bennett, J.H. and Hill, A.C. (1975). Interaction of air pollutants with canopy of vegetation. In: *Response of Plants to Air Pollution* (eds. J.B. Mudd and T.T. Kozlowski), 273–306. New York: Academic Press.

6 Ghate, S. (2019). Phytoremediation of indoor organic air pollution using house plants. *Environews* 25 (2): 6–7.

7 Ghate, S. (2019). Management and remediation of indoor air quality with phytoremediation technology using house plants. *Int. J. Sci. Res* 10 (12(D)): 36315–36319.

8 Ghate, S. (2019). Screening house plants for indoor pollution removal capacity, specifically allethrins, a poisonous pesticide released from smoke of mosquito coil. *Int. J. Appl. Sci. Technol.* 4 (4): 114–127.

9 Hill, A.C. (1971). A sink for atmospheric pollutants. *J. Air Pollut. Control Assoc.* 21: 341–346.

10 Keller, T. (1978). How effective are forests in improving air quality? Eighth World Forestry Conference, Jakarta, Indonesia, 1978, pp 9.

11 Rao, D.N., Agrawal, M., and Singh, J. (1990). Study of pollution sink efficiency. Growth response and productivity patterns of plants with respect to fly-ash and SO_2. Final Tech. Report of MAB project DOE/14/256/85-MAB/RE.

12 Smith, W.H. and Dochinger, L.S. (1976). Capabilities of metropolitan trees to reduce atmospheric contaminants. Better Trees for Metropolition Landscapes, U.S.D.A., Forest Service, Gen. Tech. Report No. NE-22.
13 U.S. Environmental Protection Agency (1976). *Diagnosing Vegetation Injury Caused by Air Pollution*. Washington, DC: U.S.E.P.A.
14 Beckett, K.P., Freer-Smith, P., and Taylor, G. (2000). Effective tree species for local air quality management. *J. Arboric.* 26 (1): 12–19.
15 Abhijith, K.V., Kumar, P., Gallaghar, J. (2017). Air pollution abatement performances of green infrastructure in open road and built-up street canyon environments – a review. *Atmos. Environ.* 162: 71–86.
16 Chaphekar, S.B. (1972). Effect of atmospheric pollutants on plants in Bombay. *J. Biol. Sci.* 15: 1–6.
17 Chaphekar, S.B. (1979). An evidence of high pollution level in Chembur. *Scavenger* 9: 22–25.
18 Agrawal, M. and Agrawal, S.B. (1989). Phytomonitoring of air pollution around a thermal power plant. *Atmos. Environ.* 23: 763–769.
19 Treshaw, M. (1970). *Environment and Plant Response*. New York: Mc Graw Hill.
20 Smith, W.H. (1981). *Air Pollution and Forests. Interactions between Air Contaminants and Forest Ecosystems*. New York: Springer-Verlag.
21 Rao, D.N. (1971). A study of air pollution problem due to coal unloading in Varanasi, India. In: *Proceedings of Second International Air Congress* (eds. M.M. Englend and G.T. Beny), 273–276. New York: Academic Press.
22 Jafri, S., Srivastava, K., and Ahmed, K.J. (1979). Environmental pollution and epidermal structure in *Syzygium cumini* L. Skeel. *Indian J. Air Pollut. Control* 2: 74–77.
23 Yunus, M. and Ahmad, K.J. (1980). Effect of air pollution on leaf epidermis of *Psidium guyava* L. *Indian J. Air. Pollut. Control* 3: 62–67.
24 Yunus, M. and Ahmad, K.J. (1981). Changes in cuticular and epidermal features of *Calotropis procrea* (Asclepiadaceae) due to air pollution. *J. Biol.* 10 (2 and 3): 275–282.
25 Shetye, R.P. and Chaphekar, S.B. (1981). On measurement of dustfall with plants. Sem. Management of Environment, BARC, Bombay.
26 Das, T.M. (1981). Plants and pollution. Presidential Address in Section of Agricultural Sciences. Indian Science Congress Association Meeting, Varanasi.
27 Bedi, S.J., Patel, M.D., and Shah, S.J. (1982). Ecomorphological and phytochemical studies to determine the pollutant and its source, causing damage to vegetation. A case study of damage to banana plantation. *Ind. J. Ecol.* 2: 335–338.
28 Rao, D.N. (1972). *Mangifera indica*, a bioindicator of pollution in the tropics. 22nd Int. Geog. Cong. Montreal.
29 Rao, D.N. and Pal, O. (1978). Effect of hydrogen fluoride on growth and yield of corn (*Zea mays*). In: *Glimpses of Ecology* (eds. J.S. Singh and B. Gopal), 461–470. Jaipur, India: International Scientific Publications.
30 Ahmad, K.J., Yunus, M., Singh, S.N. (1991). Air pollution and plants. *CSIR News* 41 (15).
31 Raza, S.H., Murthy, M.S.R., Bhagyalakshmi, O., and Shylaja, G. (1990–1991). Effect of vegetation on urban climate and healthy urban colonies. *Energy Build.* 15–16: 487–491.
32 Chaphekar, S.B., Kapoor, R.K., Gupta, V.K. (1994). *Guidelines for Development of Green Belts*. New Delhi: Project sponsored by C.P.C.B.. Published by the Sponsors.
33 Singh, S.K., Rao, D.N., Agrawal, M. (1991). Air pollution tolerance index of plants. *Environ. Manage.* 31: 45–56.
34 Agrawal, M. (2005). Effects of air pollution on agriculture, an issue of national concern. *Technol. Dev. Policy Issues* 28 (3 and 4): 93–104.

35 Yunus, M. and Ahmad, K.J. (1979). Use of epidermal traits of plants in pollution monitoring. Nat. Sem. On Environ. Pollution and Its Control. A Status Review. National Productivity Council, Bombay.

36 Chaphekar, S.B. (1995). Green belts. In: *Landmarks of Botany in India* (eds. P. Soni and A.K. Srivastava), 17–24. Dehradun: Surya International Publications.

37 Alagh, Y.K. and Wadia, F.K. (1990). Agro-climatic regional planning: Overview of a new approach. *J. Ind. Sch. Political Econ.* 2 (1): 81–111.

38 Bennett, J.H. and Hill, A.C. (1973). Absorption of gaseous air pollutants by a standardized plant canopy. *J.Air. Pollut. Control Assoc.* 23: 203–206.

39 Ganguli, A.K. (1976). *A Proposal for a National Programme of Development of Perennial Green Envelop and Large Scale Afforestation*, Bombay: Scavenger 6: 8–17.

40 Gupta, V.K. and Kapoor, R.K. (1985). Reducing the consequences of reactor accidents, with a green belt. *Nucl. Technol.* 70 (2): 204–214.

41 Gupta, V.K. and Kapoor, R.K. (1992). Attenuation of air pollution by green belt: Optimization of density of tree plantation. Conference on Precipitation, Scavenging and Atmosphere Surface Exchange, Washington D.C.

42 Kapoor, R.K. and Gupta, V.K. (1984). A pollution attenuation coefficient concept for optimization of green belt. *Atmos. Environ.* 18: 1107–1113.

43 National Environmental Engineering Research Institute (NEERI). (2005). Periodic reports and lecture notes.

44 Hewitt, C.N., Ashworth, K., and MacKenzie, A.R. (2020). Using green infrastructure to improve urban air quality (GI4AQ). *Ambio* 49 (1): 62–73.

45 Kim, H.H., Yeo, I.Y., and Lee, J.Y. (2020). Higher attention capacity after improving indoor air quality by indoor plant placement in elementary school classrooms. *Hortic. J.* 89 https://doi.org/10.2503/hortj.UTD-110.

46 Ghate, S. (2015). *Syngonium podophyllum* 'Maza red' as indoor phytoremediator. *Environews* 21 (3): 6–7.

24

Air Quality Improvement Using Phytodiversity and Plant Architecture

D.N. Magana-Arachchi[1] *and R.P. Wanigatunge*[2]

[1] *National Institute of Fundamental Studies, Kandy, Sri Lanka*
[2] *Department of Plant and Molecular Biology, Faculty of Science, University of Kelaniya, Kelaniya, Sri Lanka*

24.1 Introduction

Air quality can be defined as the degree to which air is suitable for humans, animals, and plants – that is, any living organism – to remain healthy. Good air quality means the air is clean, clear, and free from pollutants. The air quality in an area is not static and can be changed within hours.

Recent reports predict that by 2050, almost 9.7 billion people will inhabit the earth [1], and two-thirds of them are expected to live in urban areas [2]. Globally, many countries monitor their air quality by maintaining an Air Quality Index (AQI). However, air quality maintenance cannot be achieved by a single country: it requires universal cooperation since air pollution does not follow geopolitical boundaries. Air pollution is defined as the earth's atmosphere being polluted by anthropogenic or naturally occurring substances at higher concentrations than their normal ambient levels, causing adverse effects on human health and the ecosystem. It is a universal problem, and people who live in cities are more vulnerable than those who live in rural areas or villages.

The World Health Organization (WHO) estimates that globally, 3 million people die each year due to exposure to air pollution [3]. With the current global coronavirus pandemic, more attention has been directed to air pollution. It is assumed that the causes of death in COVID-19 patients and their conditions have been or can be aggravated by long-term exposure to air pollutants.

The concentrations of pollutants, the rates at which they are released into the atmosphere, and the length of time they stay trapped in a particular location are the three main factors that affect air quality in a specific area. Major air pollutants are particulate matter (PM10 and PM2.5), volatile organic compounds (VOCs) like benzene, toluene, ethylbenzene, and xylenes (BTEX), and inorganic pollutants (CO_2, CO, SO_2, NO, NO_2). Many of these outdoor air pollutants are also found indoors, but concentrations can be higher. The role of trees in removing air pollutants has been well documented. This chapter summarizes the influence of plants in improving air quality with particular emphasis on phytodiversity and plant architecture.

Handbook of Ecological and Ecosystem Engineering, First Edition. Edited by Majeti Narasimha Vara Prasad.

24.2 Phytodiversity

Phyto means plant or plant-related, and *diversity* means variety. Thus, the term *phytodiversity* is generally used to describe the number, variety, and variability of plants. There are an estimated 500 000 land plant species comprising approximately 450 000 species of angiosperm, 1000 gymnosperms, 10 000 ferns, 1300 lycophytes, 9000 mosses, 200–250 hornworts, and 7500 liverworts [4]. Patterns of local diversity of trees on a regional scale can be obtained from plot inventories. The highest tree species diversity is shown in lowland rainforests in Borneo and Sumatra (>210 tree species >10 cm diameter in 1 ha of forest), while high diversities are also shown in lowland rainforests from Sulawesi to southern China (>100 tree species) in tropical East Asia including southeast Asia, southern China, and northeast India [5]. Within the higher altitudes (>1200 m) in the northern tropics, the tree species diversity is lower due to extreme soil types and long dry seasons [4]. Diversity of trees between tropical and temperate zone are different. High local plant diversities in tropical forests can be explained in numerous ways, such as differences between species, resource usage (water, nutrients, light), and/or their pests and diseases [5].

Plant biodiversity in an area is influenced by a mixture of natural and anthropogenic factors that interact to produce the vegetation structure. Globally, much is known about plant distribution, but less is known about factors that affect the distribution of plant diversity and human influences on plant biodiversity [6]. Variations in urban tree cover across regions and within cities indicate the types of factors that affect urban tree structure and consequently biodiversity. Land use can also affect species composition, as land without human interference tends to be dominated by natural regeneration of native and exotic species, while within areas of managed land use, species diversity and composition are decided by people. Plant diversity can be improved by plant species richness and is governed by changes in plant communities' functional composition and diversity [7].

24.3 Plant Architecture

Plant architecture is the three-dimensional organization of the plant body. Most parts, whether roots, shoots, stems, leaves, etc., need to withstand environmental stresses during their growth phases, and the pattern of growth is species-specific. Plants grow systematically according to a particular strategy. This architectural model governs the different phases of plant growth. This architectural concept has provided powerful tools for studying different plant forms.

The identification of the architectural model of any given plant is based on the observation of the simple morphological features presented by plants: (i) the growth pattern, i.e. determinate vs. indeterminate growth and rhythmic vs. continuous growth; (ii) the branching pattern, i.e. terminal vs. lateral branching vs. no branching, monopodial vs. sympodial branching, rhythmic vs. continuous vs. diffuse branching, immediate vs. delayed branching; (iii) the morphological differentiation of axes, i.e. orthotropic vs. plagiotropic vs. axes with mixed morphological and/or geometrical features (with plagiotropic and orthotropic portions); and (iv) lateral vs. terminal flowering [8]. In nature, although certain plants obey their architectural unit during their entire life, others change their architectural unit during their development, late in ontogeny, or under particular conditions.

The primary growth of a plant results from several processes that can be grouped into two distinct harmonized morphogenetic events. The undifferentiated cells of the apical meristem function to form new organs, which is called *organogenesis*. The insertion zone of a leaf on the stem is called a *node*, and the stem portion between two adjacent nodes is called the *internode* [8]. Shapes and sizes of leaves and flowers are major determinants of plant architecture [9]. Further, the development of plant branching plays a pivotal role in plant architecture.

We will discuss the leaf architecture as leaves, or the amount of leaf area plays a significant role in assisting plants to remove air pollutants from the atmosphere; leaves are a highly structured, multi-layered plant organ.

24.3.1 Leaf Architecture – Regulation of Leaf Position

Axillary buds are positioned in relation to leaves, so phyllotaxis is an essential determinant of overall shoot architecture. Expansin, which is a family of extracellular proteins, induces local changes in cell wall extensibility, which is a key event in leaf initiation. Expression patterns of several transcriptional regulators provide molecular markers for the site of leaf initiation. Stems and roots are radially symmetrical, but leaves have dorsoventral symmetry from inception [10]. The adaxial side of the primordium arises from the central part of the shoot apical meristem (SAM), while the abaxial side arises from the more peripheral part of the SAM. Leaves typically display at least a subtle asymmetry along the medial-lateral axis as well, and in some species with horizontally oriented shoots, the right and left halves of the blade may be strongly unequal – an adaptation thought to reduce self-shading in low-light environments. Some molecular mechanisms that determine leaf dorsoventrality are conserved between monocots and dicots. Despite sharing common architectural attributes, lateral position, internal tissue patterning, etc. evolutionary diversification has resulted in dramatic variation in leaf architecture. Several families of transcriptional regulators play a central role.

24.3.2 Development of Internal Leaf Architecture

Internal leaf anatomy is more or less the same for most leaf shapes and sizes. In general, tissues in leaves are organized into three systems: dermal, ground, and vascular. Within each tissue system, cells differentiate spatially as appropriate for the context provided by the external leaf architecture. Surface differences of the upper and lower leaf can be observed in dicot plants by more stomatal density seen in the abaxial and more trichome distribution on the adaxial epidermal layer. Above the upper epidermis is a waxy cover called the *cuticle*, with thickness ranging <0.1–$10\,\mu m$. The wax contents and compositions vary among plant species. Palisade parenchyma cells (involved in photosynthesis) are specialized for light capture. They are arranged on the leaf's adaxial side, and the spongy parenchyma cells, specialized for gas diffusion, are on the abaxial side.

When considering the internal leaf architecture, how the mesophyll and dermal tissues are layered and the arrangement of leaf vein patterns (vein size, continuity between veins, and regular spacing) are remarkable. The stomatal spacing pattern depends on a series of stereotypical cell divisions that result in clonally guard cells and neighboring epidermal cells [11]. Stomatal patterning responds to environmental cues, including ambient CO_2 concentration. Wild types plants generally have reduced stomatal density under elevated CO_2 levels, because loss-of-function mutations in the high carbon dioxide (HIC) gene have increased stomatal density [12].

The leaf surface is filled with trichomes, which are epidermal outgrowths, and their positioning is also of nonrandom spacing. It is not lineage-based but is derived from a lateral inhibition mechanism. Over 20 genes are needed for trichome morphogenesis [10]. Analysis of the cuticular water permeability of leaves with regard to life form and climate of origin shows that the highest permeances are found within the group of mesomorphic leaves of deciduous species growing in temperate climates. Cuticles from evergreen leaves from epiphytic plants growing in tropical climates have the lowest permeance. However, it is not definitive. The group of xeromorphic leaves of evergreen species growing in mostly Mediterranean non-tropical climates has cuticular permeances ranging from 10^{-4} to 10^{-6}/ms, covering nearly the whole spectrum of leaf cuticular permeances measured so far [13].

In this chapter, we have not discussed phytodiversity and plant architecture in detail, and readers are encouraged to refer to the reviews of Barthélémy and Caraglio [8] and Wei et al. [14].

24.4 Phytoremediation

Phytoremediation is defined as using plants and associated microbes to reduce the concentrations of pollutants in the air, soil, and water. It is widely accepted as a cost-effective, environmentally friendly natural cleanup process to reduce and detoxify/degrade indoor and outdoor air pollutants.

24.4.1 Role of Plants During Particulate Matter and Gaseous Phytoremediation

Particulate matter (PM) has become the most dangerous pollutant in some countries [15], and studies have revealed that PM is also composed of microorganisms [16]. Plants can scavenge significant amounts of PM, especially in urban areas [17], by dry deposition, particle impaction, interception, and Brownian diffusion [18].

Plant species differ in particle retention and encapsulation due to differences in leaf characteristics, including size, shape, hairs/trichomes, stomatal length and pore size, cuticle chemical composition, and structure. Multiple vegetation characteristics influence plants' overall effectiveness and determine whether PM is retained on plant surfaces, accumulated within plant wax, re-suspended into the atmosphere, or washed off by rain [19].

There are two distinct trends in PM removal by different plant species [20]. *Ficus aurata, Muntingia calabura, Grewia laevigata*, and *Bauhinia semibifida,* which have high, hairy, dense leaves, show higher PM removal efficiency; while less efficiency is seen in *Syzygium myrtifolium* and *Adinandra dumosa,* as they have fewer or no hairs on their leaf surfaces. However, *Gnetum gnemon*, with a smooth leaf surface, shows relatively high deposition velocities for PM10. This is due to the presence of microscopic cup-shaped structures or grooves on leaf surfaces. Greater upper-surface hairiness and a low specific leaf area (SLA) can achieve a higher deposition velocity for both PM2.5 and PM10 size fractions. Hairs on leaves serve as obstacles, while low SLA coupled with smaller and thicker leaves favors the deposition and retention of PM [20]. Smaller leaves and complex leaf shapes (Palmately lobed) have a significant impact on PM removal [21]. Needle shape leaves in conifers are more efficient in removing PM2.5 particles than the broadleaved plants [22]. A larger number of grooves and trichomes in broadleaved plants is more effective at particle removal. Figure 24.1 summarizes the role of leaf architecture in air pollutant removal, including PM. The stomatal cavity allows PM up to 2 μm to enter, and the highest PM deposition is reported in furrowed areas in the leaf surfaces [23]. *Juniperus formosana* was most effective in PM removal at leaf scale, while *Pinus bungeana* was effective at tree scale [23].

Trees remove gaseous air pollution primarily by uptake via leaf stomata, but some gases are removed by the plant surface itself. The aerial surfaces of plants, particularly leaves, are estimated to total 4×10^8 km^2 on the earth. Plant leaves can adsorb or absorb air pollutants, and microorganisms colonized on the leaf surface and endophytes aid in biodegrading pollutants into less toxic or nontoxic molecules [14]. When gas molecules are inside the leaf, they spread into intercellular spaces and can be absorbed by water to form acids or can react with inner leaf surfaces.

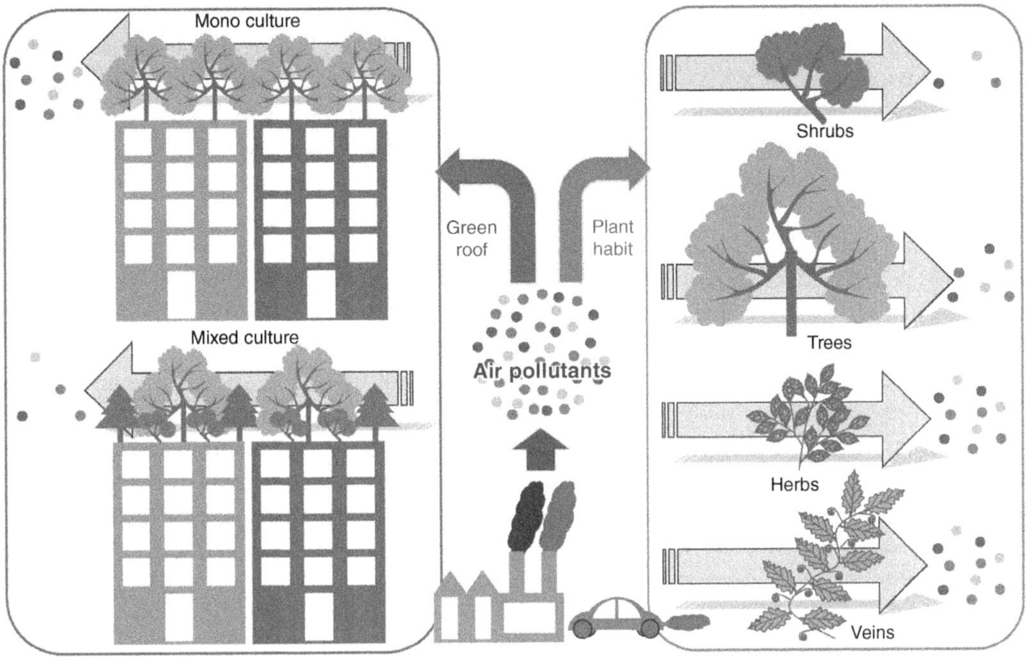

Figure 24.1 A schematic illustration of the role of leaf architecture in removing air pollutants. *See colour plate section for colour representation of this figure.*

Some particles are absorbed into the tree, while most particles are held on the plant surface. Captured particles are often re-suspended into the atmosphere; some are washed off by rain or dropped to the ground with the leaf.

Large healthy trees greater than 77 cm in diameter remove approximately 70 times more air pollutants annually (1.4 kg/year) than small healthy trees less than 8 cm in diameter (0.02 kg/year). Modeling studies by Nowak et al. [24–26] have displayed how effective urban trees are in removing gaseous air pollutants.

Increased urban tree cover reduces ozone in cities but slightly increases the overall regional ozone concentration [27]. Moreover, ozone studies that integrate temperature, deposition, and emission effects of trees reveal that trees can cause significant reductions in ozone concentrations in urban areas. Air quality improvement by trees in cities during the daytime of the in-leaf season averages around 0.51% for particulate matter, 0.45% for O_3, 0.44% for SO_2, 0.33% for NO_2, and 0.002% for CO. However, in areas with 100% tree cover (i.e. contiguous forest stands), air pollution improvement is on average around four times higher than city averages, with short-term improvements in air quality (1 hour) as high as 16% for O_3 and SO_2, 13% for particulate matter, 8% for NO_2, and 0.05% for CO [24]. It has been shown that the difference in O_3 concentrations above and below forest canopies exceeded 50 ppb, meaning a 40% improvement. In a study in Shanghai, China, SO_2 and NO_2 concentrations decreased by 5.3 and 2.6%, respectively, when comparing concentrations in an external urban woodland and at a distance of 50–100 m into the forest.

Plants can remove VOCs from the air. VOCs can be removed by up-taking via leaf stomata or by the cuticle. After entering the leaf, a pollutant undergoes degradation, storage, or excretion, either at the site of uptake or after translocation to other parts of the plant. During the daytime, with the presence of light, a higher removal rate is observed. Exceptions are so-called Crassulacean acid metabolism (CAM) and facultative CAM plants, which either constitutively or after drought stress exposure (facultative) close their stomata during the day and open them during the night; they are thereby involved in air phytoremediation under drought conditions, taking up pollutants from the air during the night, along with CO_2 absorption. Species belonging to genera *Sedum* are used in green rooftops because of their ability to switch to CAM photosynthesis. After uptake by leaves, the way VOCs are translocated is discussed in detail by Weyens et al. [28]. However, certain trees emit VOCs. VOC emissions are temperature-dependent, and trees are capable of lowering air temperatures. Hence increased tree cover can lower overall VOC emissions and thereby lowering ozone levels. Trees and shrubs filter PM flowing through urban greenery and the urban forest structure [25, 29]. Studies have shown that VOC removal rates depend on plant species (even between cultivars), which means phytodiversity [30].

Urban vegetation has a significant impact on improving local and regional air quality by *T*emperature reduction, *R*emoval of air pollutants, *E*mission of VOCs and tree maintenance emissions, and *E*nergy effects on buildings. Temperature reduction and climate change can happen due to tree transpiration rates and tree canopies affecting air temperature, radiation absorption and heat storage, wind speed, relative humidity, turbulence, surface albedo, surface roughness, and consequently the evolution of the mixing-layer height. Those local meteorological changes are directly related to the concentrations of air pollutants in urban areas. Further, the emissions of many pollutants, including ozone-forming chemicals, are temperature-dependent. Hence a reduction in temperature by trees can improve air quality.

It has to be understood that vegetation is only a temporary retention site for many atmospheric particles. Removal of gaseous pollutants is simple as they are easily absorbed and diffused within the leaf interior. However, as different plant species have different effects based on their species characteristics, the overall impact on air quality is not constant. In general, wide species diversity with greater growth forms and size at maturity affects air quality either positively or negatively. Plant species that can be used in the mitigation of air pollution are summarized in Table 24.1.

24.4.2 Ways of Improving Air Quality

24.4.2.1 Outdoor Air Pollutants

The air in traffic-congested cities is highly polluted, and green infrastructure (GI), including street canyons, roadside hedges, crossings, and vegetation barriers adjacent to traffic sources, can effectively remove these harmful toxicants. Recent findings highlighted the efficiency of GI in improving near-road air quality and thereby minimizing people's exposure to local air pollutants [46–48]. Plant species with broad, rough leaves can be used as dust barriers close to the point of origin (e.g. traffic-congested roads). Trees in parking areas can minimize evaporative emissions from vehicles through their shade. A study revealed that increasing tree cover at vehicle parking from 8 to 50% can reduce light-duty vehicle VOC evaporative emission rates by 2% and NO_X start emissions by less than 1%.

The current state of the use of vegetation to mitigate traffic-related air pollutants (TRAP) in both modeling and field measurements can be found in the given reviews [47, 48]. Factors that govern roadside vegetation's effectiveness to reduce pollutants are the properties of the plants used, such as density and shape, particle size, and wind speed. Vegetation can reduce downwind pollutant concentrations [46, 49] but depends on the nature of the greenery. The effectiveness of roadside vegetation to mitigate TRAP is inconclusive, due to increased downwind pollutant concentrations in some locations as a result of thin vegetation and tree spacings that allow for gaps in the barrier [50]. Highly porous vegetation results in negligible particle reduction and elevated gaseous concentrations downwind compared to air quality measurements in clearings [49]. This implies

Table 24.1 Major air pollutants and viable plant species for their removal.

Pollutant	Plant species	References
PM	*Chlorophytum comosum* L.	31
	Catalpa speciosa, Ulmus pumila, Amygdalus triloba, Broussonetta papyrifera, coniferous species	22
	Sorbus aria, Pinus spp.	32
	Pinus mugo, Pinus sylvestris, Taxus media, Taxus baccata, Stephanandra incisa, Betula pendula	33
	Syringa meyeri, Sorbaria sorbifolia, Fraxinus pennsylvanica	34
	Pinus pinea, Cornus mas, Acer pseudoplatanus	35
	C. speciosa, Broussonetia papyrifera, U. pumila	22
	Juniperus formosana, Pinus bungeana	23
	Fraxinus chinensis, Sophora japonica, Ailanthus altissima, Syringa oblata and *Prunus persica*	36
	P. sylvestris	37
	Hedera helix	38
	Aloe vera, Epipremnum aureum	39
Benzene	*C. comosum, E. aureum*	40
	Syngonium podophyllum	41
	H. helix, Spathiphyllum wallisii, S. podophyllum	42
	E. aureum, H. helix, and *C. comosum*	43
	Howea forsteriana, S. wallisii, Dracaena deremensis	44
	Hemigraphis alternata, H. helix, Hoya carnosa, Asparagus densiflorus, Tradescantia pallida	30
	Ruscus hyrcanus, Danae racemosa	45
Toluene	*H. helix, S. wallisii, S. podophyllum*	42
	H. alternata, H. helix, H. carnosa, A. densiflorus, T. pallida	30
	R. hyrcanus, D. racemosa	45
n-hexane	*H. forsteriana, S. wallisii, D. deremensis*	44
Ethylbenzene	*R. hyrcanus, D. racemosa*	45
Xylene	*R. hyrcanus, D. racemosa*	45

that plant density in vegetation barriers can strongly influence the reduction or increase of pollutants. There are many reports on how to design these vegetative roadside barriers in a cost-effective manner.

The quantification of effects of vegetation on the dispersion and deposition of pollutants and their effects on ambient concentrations are described in a detailed report of the impacts of vegetation on urban air pollution by the Air Quality Expert Group [51]. The effects of trees on airflow and turbulence are discussed, taking a single tree and a group of trees as examples. When a single tree is considered, the upwind speed is reduced and distorted because the wind is blocked by the tree. Close to the tree, the downwind is highly disturbed; but moving away from the tree, the air stream gradually gathers speed with a reduction in turbulence. For a group of trees, the effect can be calculated as the total of the sum of each tree. Configurations of the trees (branching, stem diameter, leaf sizes, structure, and shape, etc.) can modify the wind patterns, and as a result, the dispersion and deposition of PM vary.

In cities, ventilation can be inhibited by tall buildings, reducing wind speed and trapping heat in air pockets. Vegetated roofs (either intensive or extensive) can be used as biofilters to remove these air pollutants. In green roofs, diverse communities are more productive than monocultures [52]. Structural complexity is more important than biomass in roof insulation, rainwater retention [53], and plant survival [54], meaning that variation in plant height, branching, and leafiness must be considered when selecting green roof species. Increasing plant diversity without limiting plants to one species can increase structural diversity if functionally and morphologically different species are tested together. The amount of biomass suitable for green roofs is constrained by wind shear, limited nutrient variability, stress tolerance, and ease of maintenance [55]. Hence, when constructing vegetated roofs, attention must be paid to both phytodiversity and plant architecture (Figure 24.2).

24.4.2.2 Indoor Air Pollutants

People spend in excess of 90% of their lives indoors, and as a consequence, the quality of the ambient indoor environment can have serious implications. The effectiveness of plants in removing indoor air pollutants has been well documented [56, 57]. Plants can reduce indoor PM concentrations because they increase the surface area of the available space. Further, a relatively slow airflow speed helps create a stable indoor environment, thereby increasing plants' efficiency. Research in China recorded removal efficiencies in plants of approximately 0.2–0.36 for PM2.5 and 0.24–0.39 for PM10, respectively [58].

Installation of indoor "plant walls" (i.e. green walls, living walls, bio-walls, or vertical gardens) is another way of improving indoor air quality, assuming that plant walls can provide a greater density of plants per unit floor area than potted plants and greater atmospheric exposure of their substrate, expecting better air quality than potted plants [57]. However, pollution reduction in such green walls is likely to be very slow [59] or to vary with air-conditioning flows. Green walls can be used as indoor biofilters, thereby improving air quality, and the plant species that can be tested are *Chlorophytum comosum* and *Epipremnum aureum*. These species can significantly reduce increasing doses of PM, with or without augmented substrate airflow; however, benzene removal rates decrease with increasing aeration [40].

24.4.2.3 Phyllosphere Microorganisms

The aerial part or the above-ground organs of plants are known as the phyllosphere. According to the literature, it is predicted that globally, the phyllosphere harbors around 10^{26} bacterial cells. The composition and diversity of phyllosphere microorganisms varies according to the host plant species [60]. This was confirmed by studying trees in temperate and tropical climates and Mediterranean perennials. Additionally, microbial compositions within plant species may differ due to geographic locations [61], and these differences can be due to climatic variation, limited dispersal of the colonizing taxa [61], locality (urban vs. rural) [62], or seasonal variations [63]. It is assumed that phyllosphere microorganisms play a major role in removing pollutants from air. However, data in such studies is scarce [14].

Bacterial acquisition and degradation of phenol on leaves by the introduced phenol-degrading bacterium *Pseudomonas fluorescens* and by natural phyllosphere communities have been demonstrated by Sandhu et al. [64]. Benzene, toluene, and n-hexane removal by rhizosphere bacteria have also been discussed [44, 65].

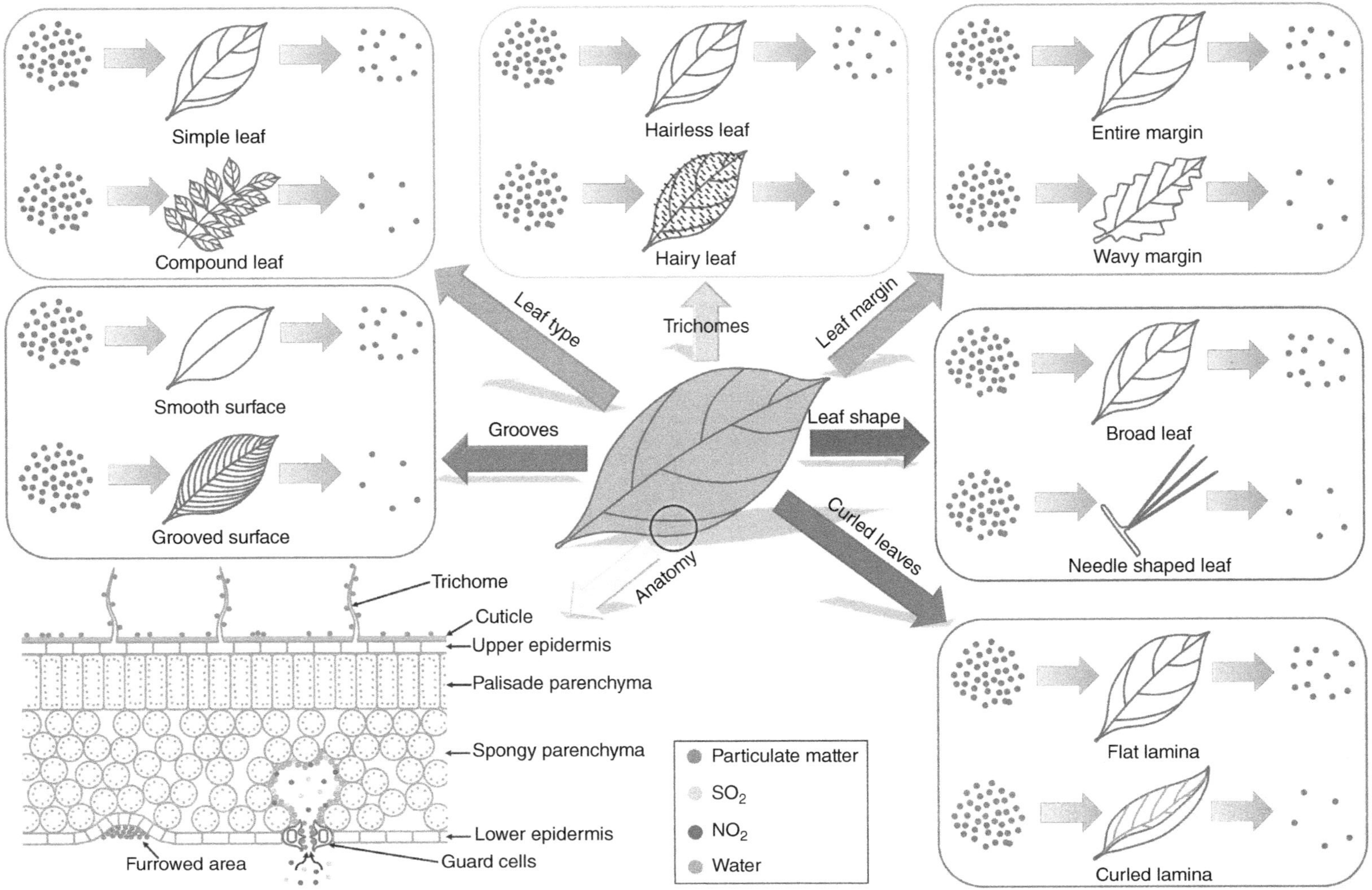

Figure 24.2 Air quality improvement in cities with green infrastructure. *See colour plate section for colour representation of this figure.*

24.5 Conclusion

Air pollution is serious global issue, and trees are a natural way to trap the pollutants in the air. In this chapter, we have summarized how phytodiversity and plant architecture can be utilized in improving air quality in both outdoor and indoor environments. We emphasized leaves as they play a major role in this process.

Acknowledgment

We thank Mr. Thenuka Deshan Kodituwakku for his assistance in preparing the figures.

References

1 United Nations (2015). *World Population Prospects*. New York: United Nations.
2 United Nations (2018). *2018 Revision of World Urbanization Prospects*. Washington, DC: Department of Economic and Social Affairs of the United Nations.
3 World Health Organization (2016). *Ambient Air Pollution: A Global Assessment of Exposure and Burden of Disease*. Geneva: WHO.
4 Corlett, R.T. (2016). Plant diversity in a changing world: status, trends, and conservation needs. *Plant Diversity* 38 (1): 10–16.
5 Corlett, R. (2014). *The Ecology of Tropical East Asia*. USA: Oxford University Press.
6 Kreft, H. and Jetz, W. (2007). Global patterns and determinants of vascular plant diversity. *Proc. Natl. Acad. Sci.* 104 (14): 5925–5930.
7 Reiss, J., Bridle, J.R., Montoya, J.M., and Woodward, G. (2009). Emerging horizons in biodiversity and ecosystem functioning research. *Trends Ecol. Evol.* 24 (9): 505–514.
8 Barthélémy, D. and Caraglio, Y. (2007). Plant architecture: a dynamic, multilevel and comprehensive approach to plant form, structure and ontogeny. *Ann. Bot.* 99 (3): 375–407.
9 Reinhardt, D. and Kuhlemeier, C. (2002). Plant architecture. *Eur. Mol. Biol. Organ. Rep.* 3 (9): 846–851.
10 Turnbull, C.G. (2005). *Plant Architecture and its Manipulation*. Blackwell Publishing.
11 Bergmann, D.C., Lukowitz, W., and Somerville, C.R. (2004). Stomatal development and pattern controlled by a MAPKK kinase. *Science* 304 (5676): 1494–1497.
12 Gray, J.E., Holroyd, G.H., Van Der Lee, F.M. et al. (2000). The HIC signaling pathway links CO_2 perception to stomatal development. *Nature* 408 (6813): 713–716.
13 Burghardt, M. and Riederer, M. (2008). Cuticular transpiration. In: *In Annual Plant Reviews, Biology of the Plant Cuticle*, vol. 23 (eds. M. Riederer and C. Müller), 292. Blackwell Publishing.
14 Wei, X., Lyu, S., Yu, Y. et al. (2017). Phylloremediation of air pollutants: exploiting the potential of plant leaves and leaf-associated microbes. *Front. Plant Sci.* 8: 1318.
15 Zhou, X., Cao, Z., Ma, Y. et al. (2016). Concentrations, correlations and chemical species of PM2.5/PM10 based on published data in China: potential implications for the revised particulate standard. *Chemosphere* 144: 518–526.
16 Weerasundara, L., Amarasekara, R.W.K., Magana-Arachchi, D.N. et al. (2017). Microorganisms and heavy metals associated with atmospheric deposition in a congested urban environment of a developing country: Sri Lanka. *Sci. Total Environ.* 584: 803–812.

17 Yli-Pelkonen, V., Scott, A.A., Viippola, V., and Setälä, H. (2017). Trees in urban parks and forests reduce O_3, but not NO_2 concentrations in Baltimore, MD, USA. *Atmos. Environ.* 167: 73–80.

18 Maro, D., Connan, O., Flori, J.P. et al. (2014). Aerosol dry deposition in the urban environment: assessment of deposition velocity on building facades. *J. Aerosol Sci.* 69: 113–131.

19 Przybysz, A., Sæbø, A., Hanslin, H.M., and Gawroński, S.W. (2014). Accumulation of particulate matter and trace elements on vegetation as affected by pollution level, rainfall and the passage of time. *Sci. Total Environ.* 481: 360–369.

20 Chiam, Z., Song, X.P., Lai, H.R., and Tan, H.T.W. (2019). Particulate matter mitigation via plants: understanding complex relationships with leaf traits. *Sci. Total Environ.* 688: 398–408.

21 Weerakkody, U., Dover, J.W., Mitchell, P., and Reiling, K. (2018). Evaluating the impact of individual leaf traits on atmospheric particulate matter accumulation using natural and synthetic leaves. *Urban For. Urban Greening* 30: 98–107.

22 Chen, L., Liu, C., Zhang, L. et al. (2017). Variation in tree species ability to capture and retain airborne fine particulate matter (PM 2.5). *Sci. Rep.* 7 (1): 1–11.

23 Song, Y., Maher, B.A., Li, F. et al. (2015). Particulate matter deposited on leaf of five evergreen species in Beijing, China: source identification and size distribution. *Atmos. Environ.* 105: 53–60.

24 Nowak, D.J., Crane, D.E., and Stevens, J.C. (2006). Air pollution removal by urban trees and shrubs in the United States. *Urban For. Urban Greening* 4 (3-4): 115–123.

25 Nowak, D.J., Crane, D.E., Stevens, J.C. et al. (2008). A ground-based method of assessing urban forest structure and ecosystem services. *Arboric. Urban For.* 34 (6): 347–358.

26 Nowak, D.J., Hirabayashi, S., Bodine, A., and Greenfield, E. (2014). Tree and forest effects on air quality and human health in the United States. *Environ. Pollut.* 193: 119–129.

27 Nowak, D.J. and Crane, D.E. (2000). The Urban Forest Effects (UFORE) Model: quantifying urban forest structure and functions. Integrated Tools For Natural Resources Inventories in the 21st Century. Gen. Tech. Rep. NC-212. US Dept. of Agriculture, Forest Service, North Central Forest Experiment Station. 714-720.

28 Weyens, N., Thijs, S., Popek, R. et al. (2015). The role of plant–microbe interactions and their exploitation for phytoremediation of air pollutants. *Int. J. Mol. Sci.* 16 (10): 25576–25604.

29 Popek, R., Gawrońska, H., and Gawroński, S.W. (2015). The level of particulate matter on foliage depends on the distance from the source of emission. *Int. J. Phytorem.* 17 (12): 1262–1268.

30 Yang, D.S., Pennisi, S.V., Son, K.C., and Kays, S.J. (2009). Screening indoor plants for volatile organic pollutant removal efficiency. *HortScience* 44 (5): 1377–1381.

31 Gawrońska, H. and Bakera, B. (2015). Phytoremediation of particulate matter from indoor air by *Chlorophytum comosum* L. plants. *Air Q. Atmos. Health* 8 (3): 265–272.

32 Beckett, K.P., Freer Smith, P.H., and Taylor, G. (2000). Effective tree species for local air quality management. *Arboric. J.* 26 (1): 12–19.

33 Sæbø, A., Popek, R., Nawrot, B. et al. (2012). Plant species differences in particulate matter accumulation on leaf surfaces. *Sci. Total Environ.* 427: 347–354.

34 Popek, R., Gawrońska, H., Wrochna, M. et al. (2013). Particulate matter on foliage of 13 woody species: deposition on surfaces and phytostabilisation in waxes – a 3-year study. *Int. J. Phytorem.* 15 (3): 245–256.

35 Terzaghi, E., Wild, E., Zacchello, G. et al. (2013). Forest filter effect: role of leaves in capturing/releasing air particulate matter and its associated PAHs. *Atmos. Environ.* 74: 378–384.

36 Wang, L., Liu, L.Y., Gao, S.Y. et al. (2006). Physicochemical characteristics of ambient particles settling upon leaf surfaces of urban plants in Beijing. *J. Environ. Sci. (China)* 18 (5): 921–926.

37 Teper, E. (2009). Dust-particle migration around flotation tailings ponds: pine needles as passive samplers. *Environ. Monit. Assess.* 154 (1-4): 383–391.

38 Ottelé, M., van Bohemen, H.D., and Fraaij, A.L. (2010). Quantifying the deposition of particulate matter on climber vegetation on living walls. *Ecol. Eng.* 36 (2): 154–162.
39 Cao, Y., Li, F., Wang, Y. et al. (2019). Assisted deposition of PM2.5 from indoor air by ornamental potted plants. *Sustainability* 11 (9): 2546.
40 Torpy, F. and Zavattaro, M. (2018). Bench-study of green-wall plants for indoor air pollution reduction. *J. Living Archit.* 5 (1): 1–15.
41 Irga, P.J., Torpy, F.R., and Burchett, M.D. (2013). Can hydroculture be used to enhance the performance of indoor plants for the removal of air pollutants? *Atmos. Environ.* 77: 267–271.
42 Yoo, M.H., Kwon, Y.J., Son, K.C., and Kays, S.J. (2006). Efficacy of indoor plants for the removal of single and mixed volatile organic pollutants and physiological effects of the volatiles on the plants. *J. Am.Soc. Hortic. Sci.* 131 (4): 452–458.
43 Gong, Y., Zhou, T., Wang, P. et al. (2019). Fundamentals of ornamental plants in removing benzene in indoor air. *Atmosphere* 10 (4): 221.
44 Wood, R.A., Orwell, R.L., Tarran, J. et al. (2002). Potted-plant/growth media interactions and capacities for removal of volatiles from indoor air. *J. Hortic. Sci. Biotechnol.* 77 (1): 120–129.
45 Fooladi, M., Moogouei, R., Jozi, S.A. et al. (2019). Phytoremediation of BTEX from indoor air by Hyrcanian plants. *Environ. Health Eng. Manage. J.* 6 (4): 233–240.
46 Abhijith, K.V. and Kumar, P. (2019). Field investigations for evaluating green infrastructure effects on air quality in open-road conditions. *Atmos. Environ.* 201: 132–147.
47 Abhijith, K.V., Kumar, P., Gallagher, J. et al. (2017). Air pollution abatement performances of green infrastructure in open road and built-up street canyon environments – a review. *Atmos. Environ.* 162: 71–86.
48 Baldauf, R. (2017). Roadside vegetation design characteristics that can improve local, near-road air quality. *Transp. Res. Part D: Transp. Environ.* 52: 354–361.
49 Deshmukh, P., Isakov, V., Venkatram, A. et al. (2019). The effects of roadside vegetation characteristics on local, near-road air quality. *Air Q. Atmos. Health* 12 (3): 259–270.
50 Hagler, G.S., Lin, M.Y., Khlystov, A. et al. (2012). Field investigation of roadside vegetative and structural barrier impact on near-road ultrafine particle concentrations under a variety of wind conditions. *Sci. Total Environ.* 419: 7–15.
51 Lewis, A., Moller, S.J., and Carslaw, D. (2018). Report: impacts of vegetation on urban air pollution. Air Quality Expert Group.
52 Lundholm, J., MacIvor, J.S., MacDougall, Z., and Ranalli, M. (2010). Plant species and functional group combinations affect green roof ecosystem functions. *PLoS One* 5 (3).
53 Dunnett, N., Nagase, A., Booth, R., and Grime, P. (2008). Influence of vegetation composition on runoff in two simulated green roof experiments. *Urban Ecosyst.* 11 (4): 385–398.
54 Mulder, C.P.H., Uliassi, D.D., and Doak, D.F. (2001). Physical stress and diversity-productivity relationships: the role of positive interactions. *Proc. Natl. Acad. Sci.* 98 (12): 6704–6708.
55 Snodgrass, E.C. and Snodgrass, L.L. (2006). *Green Roof Plants: A Resource and Planting Guide* (No. 04; SB419. 5, S5.). Portland, OR: Timber Press.
56 Pérez-Urrestarazu, L., Fernández-Cañero, R., Franco-Salas, A., and Egea, G. (2015). Vertical greening systems and sustainable cities. *J. Urban Technol.* 22 (4): 65–85.
57 Soreanu, G., Dixon, M., and Darlington, A. (2013). Botanical biofiltration of indoor gaseous pollutants – a mini-review. *Chem. Eng. J.* 229: 585–594.
58 Peng, Z., Deng, W., Hong, Y., and Chen, Y. (2020). An experimental work to investigate the capabilities of plants to remove particulate matters in an enclosed greenhouse. *Air Q. Atmos. Health* 13: 477–488.

59 Llewellyn, D. and Dixon, M. (2011). Can plants really improve indoor air quality. In: *Comprehensive Biotechnology*, 2e (ed. M. Moo-Young), 331–338. Burlington: Academic Press.

60 Whipps, J., Hand, P., Pink, D., and Bending, G.D. (2008). Phyllosphere microbiology with special reference to diversity and plant genotype. *J. Appl. Microbiol.* 105 (6): 1744–1755.

61 Qvit-Raz, N., Finkel, O.M., Al-Deeb, T.M. et al. (2012). Biogeographical diversity of leaf associated microbial communities from salt-secreting Tamarix trees of the Dead Sea region. *Res. Microbiol.* 163: 142–150.

62 Jumpponen, A. and Jones, K.L. (2010). Seasonally dynamic fungal communities in the *Quercusmacrocarpa* phyllosphere differ between urban and nonurban environments. *New Phytol.* 186: 496–513.

63 Redford, A.J. and Fierer, N. (2009). Bacterial succession on the leaf surface: a novel system for studying successional dynamics. *Microb. Ecol.* 58: 189–198.

64 Sandhu, A., Halverson, L.J., and Beattie, G.A. (2007). Bacterial degradation of airborne phenol in the phyllosphere. *Environ. Microbiol.* 9: 383–392.

65 Chun, S.C., Yoo, M.H., Moon, Y.S. et al. (2010). Effect of bacterial population from rhizosphere of various foliage plants on removal of indoor volatile organic compounds. *Hortic. Sci. Technol.* 28 (3): 476–483.

25

Information Explosion in Digital Ecosystems and Their Management

Chanchal Kumar Mitra[1] *and Majeti Narasimha Vara Prasad*[2]

[1] *Department of Biochemistry (Retd.), University of Hyderabad, Hyderabad, Telangana, India*
[2] *School of Life Sciences, University of Hyderabad, Hyderabad, Telangana, India*

25.1 Introduction

The word *digital* has connotations in counting: an online etymology says, "digital (adj.) mid-15c., 'pertaining to numbers below ten;' 1650s, 'pertaining to fingers,' from Latin *digitalis*, from *digitus,* 'finger or toe.'" The numerical sense is because numerals under 10 were counted on fingers. The meaning "using numerical digits" is from 1938, especially for computers that begun to use data in the form of digits (as opposed to analogue) after about 1945. The meaning in reference to recording or broadcasting is from 1960 (https://www.etymonline.com/word/digital). We use the modern meaning related to computers that run on, use, and produce discrete data in the form of (binary) digits. So it will be good to get some basics of computers: they have become wonderfully complex, but the fundamentals can be described briefly in simple terms.

By *ecosystem*, we mean a set of living systems (organisms) that mutually interact among themselves (group or colony) and also with the environment (consisting of both living and non-living things). The complex set of interactions forms a dynamic system that evolves slowly with time (evolution) and is a characteristic of the system in the sense that living systems change with interactions. An ecosystem usually has a biological component (living system) as an integral part. In general use, we often mean a complex network or interconnected system (https://www.lexico.com/en/definition/ecosystem). When we speak about a *digital ecosystem*, we mean a network of interconnected computers and the information (data) flow between them and humans' peripheral interaction with this network at an individual level or as a group. Modern computers can process, analyze, and transfer an incredible amount of information (data) very quickly. They are also spread out more or less uniformly across the globe, and their access to information is extensive. This is particularly true because computers are no longer instruments confined to science laboratories but are mass devices that cater to news, entertainment, business, personal communication, and education (among others). In recent times, smartphones are used for all these purposes, and the user does not have to understand the operation or function of the device. Although the digital ecosystem's frontend remains the smartphone (or a laptop or desktop computer to a lesser extent), the backends are very powerful computers (server farms) spread over multiple continents. The digital ecosystem has finally arrived (https://www.gstic.org/expert-story/building-strong-foundations-for-a-globally-connected-circular-economy).

Handbook of Ecological and Ecosystem Engineering, First Edition. Edited by Majeti Narasimha Vara Prasad.

25.1.1 Digital Computers

Computers are a modern invention: Charles Babbage (b: 26 December 1791, d: 18 October 1871) (Figure 25.1) is credited with the invention of modern computers (https://science.howstuffworks.com/innovation/inventions/who-invented-the-computer1.htm). He worked out the details of two machines – the difference engine and the analytic engine – that could solve mathematical problems (https://commons.wikimedia.org/w/index.php?curid=7665728). Although both were extremely complex to make or operate (no functional machines were built), the idea was important (https://commons.wikimedia.org/w/index.php?curid=4807331). Von Neuman (b. 28 December 1903, d. 8 February 1957) was a mathematician and computer scientist who contributed to computer architecture, programming, and computer viruses. Alan Turing (b. 23 June 1912 and d. 7 June 1954) was a mathematician and computer scientist who developed the theoretical aspects of the modern computing machine. The first functional electronic computer, nicknamed ENIAC (Electronic Numerical Integrator And Computer), used vacuum tubes and became operational at the University of Pennsylvania on 15 February 1946 (https://en.wikipedia.org/wiki/ENIAC). A smartphone today has far greater computing power than this first computer.

25.1.2 Modern Architectures for Computer Systems

At a broad level, all computers are functionally very similar: they have one or more CPUs (central processing unit; often called the "brain"), a primary storage or memory unit (in modern jargon, RAM, or random access memory), and suitable interfaces for connecting with external devices (often called *ports*). The task to be performed is coded in a software program and put into a

Figure 25.1 (Left) A photo of Charles Babbage by an unknown wood engraver, after an intermediary drawing by Thomas Dewell Scott (British, 1828–1911) for *Illustrated London News*. The newsprint from the reverse side can be seen in the background. *Source:* Thomas Dewell Scott(1871). Licensed under CC BY SA 3.0. (Right) The analytical engine. *Source:* Bruno Barral (ByB) (2009). Licensed under CC BY SA 2.5.

memory location (code section). Memory locations are also used for storing temporary results (stack section). The CPU fetches small parts of the code (instructions), executes them, and repeats the process. A subset of the instructions are arithmetic operations (addition, subtraction, multiplication, and division). Instructions also control the flow of the program. Some machines have more than one CPU, and work is divided between them. All user interactions (data input and output) are conducted via ports and are in direct control of the software program. Usually, computers run one operating system (OS) by default, and the OS controls the user interactions and software management (for example, running several programs in the background to monitor various ports). We show in Figure 25.2 the three giants who shaped modern computers: Alan Turing (https://en.wikipedia.org/wiki/Alan_Turing), John von Neumann (https://en.wikipedia.org/wiki/John_von_Neumann), and Claude Shannon (https://en.wikipedia.org/wiki/Claude_Shannon).

The first general-purpose programmable, digital electronic computer was the ENIAC, which used vacuum tube technology (https://www.phy.ornl.gov/csep/ov/node10.html). They were the first-generation computers (1937–1953) and were used mostly for computational purposes. There was no proper programming language, and the software was written directly in machine code and did not permit a very complex program to be developed.

The second generation of computers followed rapidly (1954–1962) and was based on semiconductor technology. Memory storage also improved and was based on magnetic beads (similar to modern RAM). However, computers were still expensive and bulky because they used discrete components. But software development progressed rapidly, and several high-level programming languages (including FORTRAN, developed by IBM [FORmula TRANslation], which is still used widely by the scientific community) arose in this era.

In the third generation (1963–1972), integrated circuits (ICs) replaced several discrete components, and that trend has continued today. During this period, ICs were designated as SSI (small-scale integration, equivalent to about 10 discrete components), whereas we see today LSI devices

Figure 25.2 Three giants who shaped modern computers. (Left) Alan Turing (b: 23 June 1912, d: 7 June 1954): mathematician, computer scientist, logician, cryptanalyst, philosopher, and theoretical biologist. *Source:* Alan Turing at age 16 (1928). Licensed under CC BY SA 3.0. (Center) John von Neumann (b: 28 December 1903, d: 8 February 1957): mathematician, physicist, computer scientist, engineer, and polymath. *Source:* By LANL (2008). Licensed under CC BY SA 3.0. (Right) Claude Shannon (b: 30 April 1916; d: 24 February 2001): mathematician, electrical engineer, and cryptographer, known as "the father of information theory." *Source:* Jacobs, Konrad (2015). Licensed under CC BY-SA 2.0.

(large-scale integration, equivalent to about 1000 discrete components). Modern CPUs can have more than 10 billion equivalent transistors made using VLSI (very large scale integration) technology (https://commons.wikimedia.org/wiki/File:Moore's_Law_Transistor_Count_1971-2018.png).

Magnetic core memory was replaced with cheaper and faster dynamic RAM. Internal architecture and peripheral devices (card readers, printers, hard disk, and others, some of which are no longer in use) became more common. Payroll processing became the most important commercial application during this period, but inventory management also became automated with cheaper processing and storage.

Today (2020), the most powerful computer is considered to be the Summit model IBM Power System AC922, using a POWER9 Tesla V100 series processor, with a peak processing power of 200.795 pflops (pflops = 10^{15} floating-point operations per second, a common measure of computing power or speed) (https://www.datacenterknowledge.com/supercomputers/world-s-10-fastest-supercomputers-pictures also see, https://top500.org/resources/top-systems). It has 2 414 592 cores and consumes 10.096 MW of power.

25.1.3 Microprocessors

While large and powerful computers were being made for highly specific tasks (e.g. complex engineering problems, weather predictions, simulations, etc.), another group of applications was coming to the forefront. Relatively simple but repetitive tasks can be performed by computers very efficiently. ICs containing a CPU and some memory were being developed for use in smart instruments. This process was accelerated due to the development of large-scale IC technology (LSI technology). In 1971, Intel presented the LSI device 4004 as the first four-bit single-chip microprocessor. The next year (1972), Intel announced the first eight-bit single-chip microprocessor 8008, and two years later (1974), a better and versatile eight-bit microprocessor 8080, which started the revolution in personal computers (https://www.intel.com/content/www/us/en/history/museum-story-of-intel-4004.html).

Today, microprocessors (also called microcontrollers or single-chip computers) can be found in many consumer devices. Some are quite powerful (32- and 64-bit microprocessors are now common) and can perform complex tasks (for example, microcontrollers found in mobile phones perform digital signal processing). Most modern personal computers use 64-bit processors but with considerable peripheral support (both memory and external devices). Microprocessors can be found in consumer items like toasters, washing machines, mobile phones, and televisions. The current trend is to enable the communication of these embedded microprocessors with external systems (software). Usually, a desktop (or a laptop) computer can be used to run software under the supervision of the OS (Windows, Linux, and Mac are the most popular OSs) with multiprocessing, multithreading, and multitasking support.

25.1.4 Networks of Computers

It was realized that two computers could be connected and share resources. For example, several computers may share one printer, file server, scanner, and other common peripherals – this provides optimum utilization of expensive resources. But if two computers are connected so that they can share information – for example, two different programs running on two different machines can communicate with each other and share information – we can have a vastly superior system (compared to two standalone machines). The information exchange can be quite simple: for example, one machine can start (execute) one new program (existing code) to be executed on a remote machine, and the execution results are made available to the first machine. These concepts first crystallized in 1961, and the ARPANET (Advanced Research Projects Agency Network) was born. The first computers were connected in 1969, packet switching (data broken into small packets with additional information) was implemented, and a network protocol was established [1].

Using suitable software, email, remote login, and file transfer were possible within a network (1971). Soon the internet, the worldwide network of computer networks, followed. It consists of private, public, academic, business, and government networks from local to global scope (https://en.wikipedia.org/wiki/Internet). Today it also includes telephony, internet TV and radio, music, newspapers, and a host of diverse information and entertainment media. Sir Timothy John Berners-Lee is an English engineer and computer scientist best known as the inventor of the World Wide Web (WWW). In simple terms, the internet provides the backend hardware, whereas the WWW is the software component (the browser provides the user interface) useful to access other resources on the internet.

25.1.5 Development of Databases

In the early days of computers (1960–1970), memory (medium of information storage) was relatively expensive, and offline storage of information (magnetic tapes, floppy and hard disks) was widely practiced. RAM and offline storage prices are now low, and computers with large memory and storage are now very common. Businesses embraced computers early on (COBOL, Common Business Oriented Language, was one of the early business-friendly programming languages, developed c. 1959) and wanted more storage and lower computational power. Databases first appeared in the form of spreadsheets. One of the early spreadsheets was VisiCalc (Figure 25.3), but it was rapidly superseded by Lotus 1-2-3, which offered more features. Along with Lotus 1-2-3, dBase was also hugely popular (1980–1990) because of its business-friendly features. SQL (Structured Query Language) was introduced by IBM (1974) and worked in relational databases.

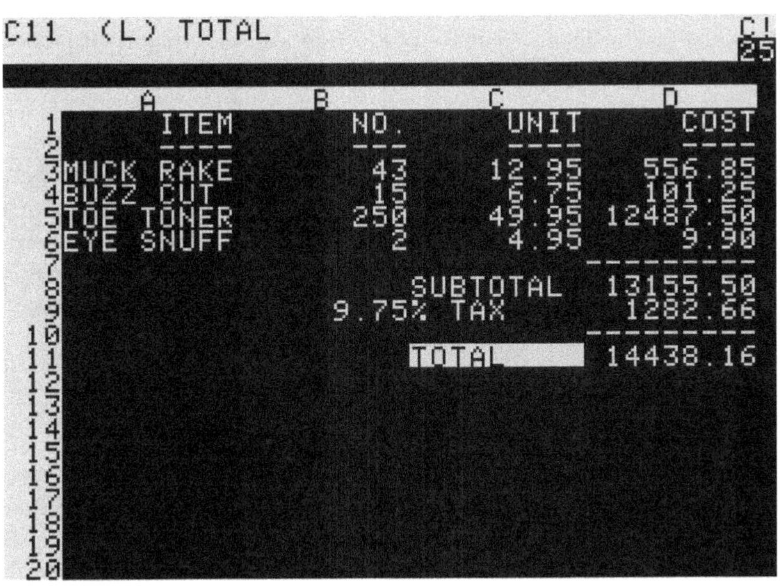

Figure 25.3 A VisiCalc spreadsheet running on an Apple II computer. *Source:* Screenshot of VisiCalc running on an Apple II computer; September 2005.

Although the software has evolved considerably, the basic features are present in all modern database systems (e.g. Oracle and MySQL). Modern software can handle very large databases without any difficulty. Databases containing pictures, movies, songs, etc. (objects in general) are quite common these days.

25.1.6 Data as Knowledge

In 1948, Claude Shannon described the theory of communication that started the Information Age [2]. Basically, information present in a given message can be quantified and represented as bits (==binary digit; a system with only two well-defined states) (https://en.wikipedia.org/wiki/Shannon_(unit)). If a system has N possible (equally likely) states, the maximum information it can contain is given by $\log_2 N$, and the relevant unit is called a shannon ($=\log_2 2$; information content of 1 bit). However, the unit is sparingly used, and bits are more common.

This information content has a phenomenological relation with entropy, a measure of disorder in a thermodynamic system. Higher entropy equals greater randomness in a thermodynamic system. In a similar way, information entropy was defined as the expected value of a random variable's self-information. Information entropy is similar to the entropy defined by Boltzmann's H-theorem. Just as randomness increases in a thermodynamic system because of spontaneous or irreversible processes, in a communication channel, information degrades over time due to noise. These concepts have been widely used in simple English texts and also in protein and nucleic acid sequences. However, decoding the information requires additional information (software). Thus it is possible to extract additional information from existing static data.

25.2 Growth

One of the essential visible parameters of a living system is growth. This is a change over time and can rightly be called an evolutionary parameter in the short term. In a biological system, growth is usually limited and is often controlled by resources and reproduction. Another important factor that controls growth is death, whether from natural or unnatural causes; it has a serious effect if it takes place before successful reproduction.

For a very simple system such as the growth of bacteria in a nutrient culture medium, the initial growth rate is directly proportional to the number of active bacteria. This gives rise to an exponential rate of growth. However, this does not hold true in later phases because when the concentration of bacteria is high, the nutrient concentration becomes limiting, and the accumulated waste products become inhibitory. While extremely simple in form and application, this model, with minor modifications, has been used in other branches of science, e.g. economics, social sciences, and commerce. We will focus on computers and digital commerce.

25.2.1 Traditional Models for Growth

Rapid industrialization had a characteristic property: it catalyzed its own development. Since the beginning of the Industrial Revolution, technology has progressed very fast and has affected the life (quality of life) of common people in many ways. However, this is particularly remarkable in the case of computers, as we see in Moore's law.

In 1965, Gordon Moore, one of the founders of Intel Corp, predicted that the number of transistors in a computer microchip would double every two years. This has held for about 50 years, although it cannot continue indefinitely (see https://newsroom.intel.com/wp-content/uploads/sites/11/2018/05/moores-law-electronics.pdf, https://en.wikipedia.org/wiki/Moore%27s_law, and http://news.bbc.co.uk/2/hi/technology/4472549.stm). Like biochemical growth (say, bacterial growth), the increase is fueled by research and development, and the widespread use of computers has kept down user costs. As Figure 25.4 shows, it is quite crowded at the top. As processors become more complex, raw computing power does not increase the same way, but the processors can now handle more complex tasks (multiprocessing, multitasking, and multithreading).

25.2.2 Growth Curves

Although processors have become extremely complex, operating a desktop or a laptop has become easier for the typical user. A computer user (a modern desktop or laptop) is unlikely to be an expert in computers. The user is more likely to use the device to exchange common information: an email, popular music or a video, or simply some news. However, a very large amount of information is systematically stored in databases. The concept of databases is not new; since ancient times, governments have stored information about subjects (mostly tax matters), and business people have kept records of lending and borrowing on semi-permanent media. With the advent of affordable computers, databases have become very important, and information that needs to be searched is invariably stored in some type of database. Today, the top three database companies are Microsoft, IBM, and Oracle (MySQL is now part of Oracle) [3]. Microsoft Excel is the most popular spreadsheet application for MSME (micro, small, and medium enterprises) businesses.

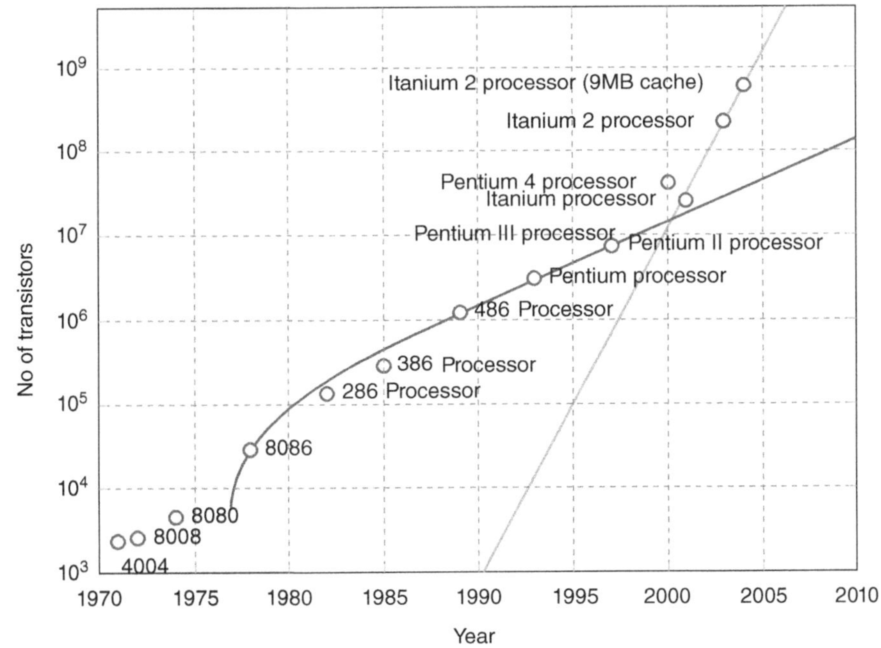

Figure 25.4 Graphical demonstration of Moore's law. The relationship is exponential, and the plot is therefore made on semi-log graph paper. For clarity, we have used only Intel CPU data in this plot. The numbers were taken from https://www.intel.com/pressroom/kits/events/moores_law_40th. The numbers suggest two exponentials in the overall results: growth has accelerated in recent times. *Source:* Based on Moore's law outline, graphical demonstration of Moore's law.

Although exact numbers are not available, we can consider the social media user base to be indicative of the number of active users of social media. One report (https://datareportal.com/social-media-users) says that the number of active social media users (as of April 2020) stands at 3.81 billion, i.e. approximately 49% of the world population (based on all age groups). This number is growing at the rate of 8.7% annually, and 3.76 billion users (about 99% of all social media users) access the internet via their mobile phones (https://www.statista.com/topics/1164/social-networks). Facebook is the most popular social media platform (2.498 billion active users vs. 0.800 billion for TikTok and 0.516 billion for Sina Weibo), and India has the largest user base (280 million). An average user spends about 2.5 hours daily on social media. The last point is relevant because mobile phones are considered personal devices, and people store personal and confidential information on their phones (e.g. personal videos, passwords, and financial information). One only needs to provide basic and minimal information while registering for social media, but often the software has access to several important files on the phone. If security is compromised, confidential data can be leaked to the server or a third party (in case of a malware attack). Security is therefore critical in the case of mobile phones (which are lost, stolen, and misplaced quite often).

Information held by governments is very likely to be equal to or greater than the volume of information held by social media. Although there may be considerable duplication, an average user is well represented in social media databases (if we consider the individual contributions or *posts*).

25.2.3 Limits of Growth

Moore's law suggests computer performance (represented by processing speed) or the equivalent cost for a given task will continue to decrease indefinitely, but there may be physical limitations. Transistors have shrunk in physical size on circuit layouts, but we may hit the limit as we approach atomic dimensions. Perhaps there may be a new technological breakthrough (for example, quantum computing), but we are more concerned about the software or computational limit.

Adding more people to a given (intrinsically simple) task does not get the desired result (getting the job done faster) beyond a certain level. However, some tasks are intrinsically repetitive, and an army of workers can do them efficiently. Modern computers excel at the second class of tasks (e.g. searching a database, ordering a sequence, multiplying large numbers, working on grids or meshes, etc.), and today's faster and faster machines are geared for business. Here they manage huge amounts of user data and equally large inventories.

For example, a modern search engine caches (copies and stores locally) many internet sites in a searchable form. It also stores (more often than not) the user's search preferences (explicit or deduced) and history to be used for targeted advertisements. It also tracks the user's current location so that information can be locally relevant (e.g. local news is given preference). Also, information delivered needs to comply with local rules and customs. In addition, search results are customized as per the search engine's internal algorithms and user preferences. Naturally, modern search engines are very large and complex. For example, Google indexes hundreds of terabytes of information from web pages and runs on hundreds of physically distributed servers (https://en.wikipedia.org/wiki/Google_Search). The seven most popular search engines in 2018 were (i) Google, (ii) YouTube, (iii) Amazon, (iv) Facebook, (v) Bing, (vi) Baidu, and (vii) Yandex. It is interesting to note that Amazon is a shopping site (not designed as a search engine), and Facebook is a social networking site (originally intended for students) (https://www.searchenginejournal.com/seo-101/meet-search-engines). Also, note that Baidu is a search engine for Chinese users, and Russian-speaking nations widely use Yandex.

25.2.4 Growth vs. Life

All the search engines (and social media giants) have accumulated vast amounts of user data (information) used for internal research and development. Initially, basic data is obtained from the user at the registration level, but this is often not useful for business applications. Therefore, a model is used to extract additional information based on shopping history, travel preferences, search preferences, social media posts, and other online activities. Artificial Intelligence (AI) is also used to get a reliable classification. Thus every user gets a profile online that is a very specific description useful for several applications.

Because of the large number of users involved in the process, this classification/ systematization is invariably automated without any direct user intervention. In 2018, several hundred thousand user data profiles were taken from Facebook by Cambridge Analytica without user consent and used for political purposes, ending in a scandal (https://en.wikipedia.org/wiki/Facebook%E2%80%93Cambridge_Analytica_data_scandal). That the ecosystem has grown so large is not the real challenge; the real concern is that only a few have control over the system.

25.3 Sustainability

That a vast amount of private user information is in the hands of a very few select individuals is a real concern for all. Cambridge Analytica (Cambridge Analytical Limited, an offshoot of the SCL group, formerly Strategic Communications Laboratories, a private British behavioral research, and strategic communication company) demonstrated very effectively how bulk user data can be used to undermine democratic foundations. The company no longer exists, but its former employees have set up new initiatives (Auspex International, which proposed working in the Middle East and Africa) to focus on geopolitics, health, and extremism (https://www.bbc.com/news/technology-44807093).

Data mining tools are widely used (and are quite powerful) in scientific research (using scientific literature as the base), but their application in politics and elections is new. Although the attention was on Facebook (mainly because the primary data was sourced from it without explicit consent), the current focus is to produce user bases independently. In this regard, the Aarogya Setu mobile app developed by the Government of India (GoI) (https://play.google.com/store/apps/details?id=nic.goi.aarogyasetu) for contact tracing for Covid-19 has generated considerable privacy concerns (https://www.bbc.com/news/world-asia-india-52659520). A digital ecosystem should have built-in data protection, but data protection is applied as an outer layer that offers less protection in today's software model.

25.3.1 Production vs. Consumption

Like a disease, information has no potential unless it circulates. As we mentioned earlier, information is associated with entropy or randomness in a sense similar to thermodynamics. Information that is widely circulated or easily available has low usable potential. Similarly, information that is not readily available is considered valuable, much like any commercial item (e.g. price is determined by demand and not by the cost of production). When such information changes hands, it is used and acquires entropy, and the usable potential decreases (it loses its surprise factor). Thus, the wide distribution of any information causes a loss of value. Now we know why a customer's bank account password has high value compared to the customer's name!

In the same way, as soon as a new password is created, the information content of the earlier one is automatically reduced. In this sense, information has a sense of direction: it tends to flow from high information (low circulation) to low information (fully randomized) in a spontaneous fashion. Thus, we can have the concept of *source*, where information is produced by individuals, to *sinks*, where the information is uniformly spread and has very low value. During this transition from source to sink (or production to consumption), the information is used for some purpose (consolidating other information, somewhat like a heat engine) and finally discarded (simply because it has lost its potential).

In earlier days (circa 1980), computer networks were considered very useful for defense and scientific research, but today the most useful computer network applications lie in e-commerce and social networks. Today we measure "usefulness" with a very simple parameter: generation of actionable information. In that sense, information has some value that can be quantified with currency. Thus e-commerce and social networking (sites) are related, or even connected, with the economy or the circulation economy. This is relevant because a static system does not allow for growth; a dynamic and expanding system, both in the e-economy and e-commerce, can develop sustainability. For many researchers, sustainability is a more important parameter than development.

25.4 Knowledge vs. Information

Although knowledge and information are related concepts, they are distinct. Most researchers agree that knowledge refers to an interpreted version of reality from a subject's given perspective in a given space and time [4]. This cannot be quantified in absolute terms, but relative transfers are quantifiable in principle (much like information, entropy, etc.). Using the thermodynamics analogy, information is related to entropy, whereas knowledge is related to free energy [5]. Knowledge transfers (associated with the change in knowledge) are essential concerns in information technology. Knowledge transfers are functions of the knowledge givers, knowledge receivers, and channel capacity (also associated with noise). The current information explosion is also accompanied by a concurrent knowledge explosion. If we assume that both knowledge givers and receivers are humans, this knowledge explosion is very likely to lead to considerable societal changes.

Hofkirchner has developed an epistemological foundation for knowledge, information, and information exchange. He has proposed a heuristic model of social information research that can be constructed in a human-centric fashion [6]. In essence, knowledge is associated with an intelligent system (knowledge givers or knowledge takers), but information can exist all by itself. In a thermodynamic case, we can say that seawater has a large reserve of heat energy but can only be converted into free energy when the heat can be made to flow from a higher temperature to a lower temperature (with the intervention of a heat engine). Although less precise, similar considerations can be applied in economics: the flow of money contributes to the dynamics (economic well-being); and the beginning and endpoints (money givers and money takers) determine the quality of conversion.

25.5 Circulation of Information

Information stored in the hard disk of a computer cannot achieve any purpose. For this, the information must be circulated. During this flow, information can perform useful tasks. Fake news, most commonly propagated via social media, has only a short life (it is likely to be noticed as fake and pulled down), but it can do considerable damage even in this short time. Both protests and riots are also organized via closed groups, and the information circulates only within the group. Even if the same information is released in a public space, it has minimal work potential because the damage (the intended objective) has already been done (the value has expired). Today, WhatsApp (a subsidiary of Facebook) and Twitter are the most common vehicles of misinformation and fake news. It is important to note that algorithms exist to detect fake news and misinformation using conventional tools (AI, machine learning, and deep learning) [7].

Fake news (and misinformation; hate speech included) has reached epidemic proportions because of several factors. Political figures and entertainers have taken considerable interest in social media, and they love to have a huge fan base. Periodically they make statements, primarily intended for their fan base, which are expected to be informational. Government functionaries, as well as political leaders, regularly announce policy matters to the public via social media. Over time, people have developed a trust in the medium: information coming via a known medium is trusted to be true. If the format (encapsulation) appears right, the message (content) must be true. Some additional reliability factors (e.g. photos or references) may be provided to make it more convincing. For example, if I see a tweet from NASA that claims the moon is made of green cheese, I will be tempted to accept it as truth (https://en.wikipedia.org/wiki/The_Moon_is_made_of_green_cheese).

Fake news, to be acceptable, must be designed well. The originator (sender) must be reliable (to the recipient), and the medium should also be good (low noise; reliable against tampering). The receiver (intended recipient) must accept and not question (if the message is challenged, the quality of the medium and the originator suffers). Most of the time, a genuine message tends to be informative only, and the recipient is not expected to act in any way. Fake news prepares the recipient for any subsequent action. In the US, approximately 40% of the population receives their daily news via social media. Traditional news media (both print and online) offer a more balanced view and do not directly promote fake news or hate speech. In addition, the channel is open, and people can question the information being distributed.

25.6 Quality vs. Quantity

In both the examples discussed in this section, large-scale data mining takes place. The program's efficacy depends on numbers: i.e. unless a significant number of people are covered, the desired objective will not be achieved. However, in the Cambridge Analytica scandal, the initial group of people were persuaded to voluntarily part with the desired information. In the Aarogya Setu case, people were forced to install the app on their mobile phones. In the first case, people were unaware of how the harvested information was going to be used. In the second case, users have been told that the government can use the individual information in any way it deems fit. In the first example, the user data was used to undermine election results (the democratic process), whereas in the latter case, the user data can be misused against the public (privacy issues). Finally, in the former case, the desired objective was attained (Donald Trump won the 2016 US election); we are still uncertain about the latter.

25.6.1 Case Study 1: Facebook and Cambridge Analytica Scandal

Very briefly, in 2018 (when the news broke), Cambridge Analytica Ltd., a British political consultancy firm, used millions of Facebook users' personal information, without proper user consent, in political advertisements in US elections. The data collection took place in 2013 via an app (software running on a mobile phone) in the form of a series of questions that built a psychological profile of the individual. The data corresponding to US voters was sold to political campaigns for targeted advertisements. The Facebook CEO finally apologized to the public via newspaper advertisements (https://en.wikipedia.org/wiki/Facebook%E2%80%93Cambridge_Analytica_data_scandal, https://www.icmrindia.org/casestudies/catalogue/Business%20Ethics/BECG160.htm).

Although only 270 000 users participated in the original survey, approximately 87 million users' data was downloaded by Cambridge Analytica. In other words, not only the participants' data was used, but their friends' data was also harvested. The information harvested included the user's public profile, page likes, current location, and birthday. For some, private messages were also collected. The information was detailed enough to build a psychological profile of the user. The profile was used to create a targeted voting advertisement for each individual.

From an individual perspective, the information harvested is perhaps insignificant; but when considered in totality for millions of similar users, the application of AI tools (to build a psychological profile), and the potential consequences to an institution (democratic election), the results can be staggering. The company (Cambridge Analytica) has used similar tools and techniques in other democratic processes. This goes for fake news, hate speech, and foreign interference in democratic election processes.

Facebook was fined $5 billion for the data breach by the US Federal Trade Commission. Cambridge Analytica Ltd. closed down (1 May 2018, filed for insolvency), but the major players have joined new companies (https://www.icmrindia.org/casestudies/catalogue/Business%20Ethics/BECG160.htm).

25.6.2 Case Study 2: Aarogya Setu Mobile App by National Informatics Centre (NIC) of the GoI

Aarogya Setu ("a bridge to cure") is a mobile application launched by the GoI to check the propagation of the Covid-19 pandemic (https://play.google.com/store/apps/details?id=nic.goi.aarogyasetu). In essence, this is a contact-tracing app; it checks if two users, one of them infected with the virus (SARS-CoV-2), happen to be close together (measured in terms of Bluetooth signal strength). The description says that "Aarogya Setu is a mobile application developed by the GoI to connect essential health services with the people of India in our combined fight against COVID-19." However, since the beginning, the app has been plagued by privacy issues.

The basic problem is that India does not have a data protection law. The Aadhar database (the UIDAI – Unique IDentification Authority of India – repository of a biometric database of Indian residents) has been breached in the past (https://www.businessinsider.com.au/cambridge-analytica-executives-and-mercer-family-launch-emerdata-2018-3). The user location data in Aarogya Setu is periodically uploaded to a server, but the government can share it with anyone. Experts also claim that there are flaws in the code, and security has been compromised. However, the app's use is mandatory for government and private employees and all travelers (both air and rail). The controversies, at this point, are yet to die down (https://www.bbc.com/news/world-asia-india-42575443).

25.7 How Does the Digital Ecosystem Work?

Today, most users of the internet need not understand the internet. Most active users of the internet use it for information, education, and entertainment. Most users access the internet on their smartphones (mobile phones with advanced features) for news, games, weather, movies, songs,

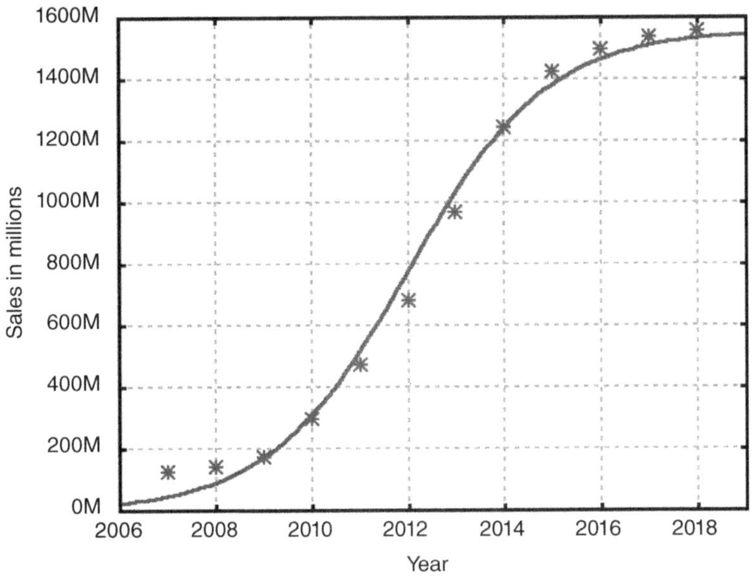

Figure 25.5 Cell phone sales over the last decade. Low-end cell phones (basic features only) are not included in the count. The solid curve is a logistic curve (sigmoidal) fitted to the existing data points. The curve suggests that the market is close to saturation, and sales continue because people replace their handsets every 18 months on average. *Source:* Data taken from Clement J., Global smartphone sales to end users 2007-2021, 2020; https://www.statista.com/statistics/263437/global-smartphone-sales-to-end-users-since-2007.

and travel. Most smartphones (88% in 2018) use the Android OS (mobile OS, based on Linux and customized by Google), and the sale of smartphones has already reached a saturation point (Figure 25.5).

The total number of active smartphones is around 4 billion. The average lifetime (in 2019) of a smartphone was close to 18 months (in other words, 50% of new phone owners discard their phones after 12.5 months; the half-life of a smartphone is, therefore, ~12.5 months). Thus the total number of active smartphones in the year 2019 was approximately given by (new phones sold in 2019) + 0.5*(new phones sold in 2018) + 0.25*(new phones sold in 2017) + 0.12*(new phones sold in 2016): about 2.860 billion.

In 2020, people spent 144 minutes/day on social media interactions. This is good news for online sales (and other marketers) and allows fake news to circulate more intensively.

25.7.1 Digital Ecosystem and Sustainable Development

The origin of inequality is rooted in deep societal structures [7]. Jean-Jacques Rousseau, in 1755, wrote in his second discourse that modern societal inequality originates from a process of natural evolution that has included social laws and the right to property, and this is unjustified based on moral considerations (https://aub.edu.lb/fas/cvsp/Documents/DiscourseonInequality.pdf879500092.pdf). In modern times, social laws have become more rigid, economic disparity has generally increased, and inequality between people has grown.

With rapid industrialization, the demand for cheap labor increased, and local inequality rapidly increased. Administrators and planners realized that inequality urgently needed to be addressed because it destabilizes society. This was a difficult task because the people who benefitted from the system were reluctant to give up their privileges.

With the development of mass transport, the problem is no longer confined to a city, state, or single country; it has become an international problem, although individual countries need to address it. It has now become essential to address the problem from the outside because the inside forces resisted change. Inequality has grown in part due to the advancement of technology; many believe that a technological solution to inequality, in particular economic inequality, can be addressed by technology itself. The United Nations Development Program (UND, founded in 1965) has tasked itself with eradicating poverty and reducing inequalities and exclusion (https://www.undp.org/content/undp/en/home/about-us.html).

In 2015, all 193 United Nations member states resolved to adopt a universal call to action to end poverty, protect the planet, and ensure that all people enjoy peace and prosperity by 2030. The call is in the form of 17 sustainable development goals (SDGs), also known as Global Goals, for developments that balance social, economic, and environmental sustainability. The 17 SDGs are shown in Figure 25.6. All 17 goals are interconnected and interrelated (https://open.undp.org). In other words, no poverty (SDG 1) automatically means zero hunger (SDG 2).

UNDP has a very modest budget ($5790 million in 2020), and it has supported 4188 projects in 151 countries. The largest number of projects has been focused on SDG 16 (Peace, Justice, and Strong Institutions; 1178 projects with 23.5% of the total budget), followed by SDG 1 (No Poverty; 1064 projects with 21.6% of the total budget) (https://open.undp.org). The focus is clear: without peace, justice, and strong institutions, development and poverty eradication cannot succeed. SDGs 2, 6, and 14 each received <1% (of the total budget) of support.

It is implicit that nations needed prodding to improve on SDGs 16 and 1, but they should move on their own on other matters. Afghanistan is the largest recipient under SDG 16 (followed by Sudan), whereas, under SDG 1 (no poverty), Iraq is the biggest recipient (followed by Yemen). It is understandable (the focus) that Afghanistan and Sudan are war-torn countries and lack democratic modern institutions. However, both Iraq and Yemen are considered oil-rich countries, and it is rather surprising that they are suffering from poverty and inequality.

Figure 25.6 Sustainable Development Goals (SDG) relevant to digital science. SDGs 4, 8, 9, 11, and 12 are intimately connected with digital ecosystems. *Source:* Sustainable Development Goals UNDP, 2020. United Nations Development Programme, https://www.undp.org/content/undp/en/home/sustainable-development-goals.html.

In developed countries, technology, automation, and modernization are perhaps significant causes of poverty and inequality. And it is equally likely that the same technology can be used to reduce human suffering and inequality. Digital technology can be an enabler by improving connectivity. SDGs 4, 8, 9, 11, and 12 primarily depend on digital science and technology.

25.7.2 SDG 4: Quality Education

SDG 4 seeks to ensure inclusive and equitable quality education and promote life-long learning opportunities for all. First, the bad news about education: poverty, armed conflict, and other emergencies keep many children around the world out of school.

In developing regions, children from the poorest households are four times more likely to be out of school than those from the richest households. Now for some good news. Since 2000, there has been enormous progress to provide primary education to all children worldwide. The primary school enrolment rate in developing regions has reached 91%. By measures in any school, that's a good grade. Now, let's get an even better grade for all children and achieve the goal of universal primary and secondary education, affordable vocational training, access to higher education, and more.

25.7.3 SDG 8: Decent Work and Economic Growth

This SDG promotes sustained, inclusive, and sustainable economic growth, full and productive employment, and decent work for all. An essential part of economic growth is that people have jobs that pay enough to support themselves and their families. The good news is that the middle class is growing worldwide – almost tripling in size in developing countries in the last 25 years, to more than a third of the population. But in 2015, we also had widening inequalities, and job growth was not keeping pace with the growing labor force – over 200 million people didn't have jobs. That's the equivalent of the entire population of Brazil. Things don't have to be that way. We can promote policies that encourage entrepreneurship and job creation. We can eradicate forced labor, slavery, and human trafficking. And in the end, we can achieve the goal of decent work for all women and men by 2030.

25.7.4 SDG 9: Industry, Innovation, and Infrastructure

Industrialization produced concentrated regions of high demands for manual work. In the pre-industrialization era, the population was more uniformly distributed over the countryside, and density was thin. With high demand for manual work, cities emerged around industrial areas, and villages had to feed a labor force that was earning money but not practicing agriculture. Also, local high concentrations of people in cities put demands on the local infrastructure (health and sanitation). Today, more than 50% of the world's people live in cities, but the mega cities are straining to maintain the infrastructure (including transport, education, and health). With modern mega cities, poverty is also concentrated in mega slums.

Poverty is not new, but cities have now created large, visible inequalities that urgently need to be addressed. The inequality is a social force that can cause irreversible breakdown in order (societal, political, and economic) that needs to be corrected. In other words, the current model of growth is not sustainable. We need a more equitable system and social order for growth that benefits all. Industrial growth must accompany growth in social harmony and physical infrastructure.

25.7.5 SDG 11: Sustainable Cities and Communities

More than half the world's population now lives in cities, and that figure will reach about two-thirds of humanity by 2050. Cities are getting bigger. In 1990, there were 10 "mega-cities" with 10 million inhabitants or more. In 2014, there were 28 mega-cities, home to 453 million people. Incredible, isn't it?

Many people love cities; they're centers of culture, business, and life. But they're also often centers of extreme poverty. To make cities sustainable for all, we can create good, affordable public housing. We can upgrade slum settlements. We can invest in public transport, create green spaces, and involve a broader range of people in urban planning decisions. That way, we can keep the things we love about cities and change the things we don't.

25.7.6 SDG 12: Responsible Consumption and Production

Ensure sustainable consumption and production patterns. Some people use a lot of stuff, and some people use very little; a significant share of the world population is consuming too little to meet even their basic needs. It doesn't have to be this way. We can have a world where everybody gets what they need to survive and thrive. And we can consume in a way that preserves our natural resources so that our children can enjoy them, and their children and their children after that.

The hard part is how to achieve that goal. We can manage our natural resources more efficiently and dispose of toxic waste better. Cut per capita food waste in half globally. Get businesses and consumers to reduce and recycle waste. And help countries that have typically not consumed a lot to move toward more responsible consumption patterns [8].

25.8 Conclusions

At the current time, the digital ecosystem all over the world has taken control of people via social networking and e-commerce sites. Close to this comes popular entertainment in the form of games, movies, and music (multimedia entertainments). Mobile phones are the preferred vehicle for accessing the internet. Although mobile phone sales have reached a saturation point, about half of the world's population owns a mobile phone (about 2.8 billion people owned a smartphone in 2019). Internet coverage, however, remains an issue in several parts of the world. Privacy issues will remain a concern. Fake news, hate speech, and misinformation may perhaps continue to increase. People will perhaps continue to be vulnerable in this Information Age.

References

1 Bidgoli, H. (2004). *The Internet Encyclopedia,* vol. 2 (G–O), 39. Wiley. ISBN: 978-0-471-68996-6.
2 Shannon, C.E. (1948). A mathematical theory of communication. *Bell System Technical Journal* 27: 379–423 and 623–656.
3 Berg, K., Seymour, T.J., and Goel, R. (2013). History of databases December 2012. *International Journal of Management and Information Systems* 17 (1) https://doi.org/10.19030/ijmis.v17i1.7587.
4 Sudhindra, S., Ganesh, L.S., and Arshinder, K. (2017). Knowledge transfer: an information theory perspective. *Knowledge Management Research and Practice* 15: 400–412. https://doi.org/10.1057/s41275-017-0060-z.

5 Kakihara, M. and Sørensen, C. (2002). Exploring knowledge emergence: from chaos to organizational knowledge. *Journal of Global Information Technology Management* 5 (3): 48–66. https://doi.org/10.1080/1097198X.2002.10856331.

6 Hofkirchner, W. (2014). Epistemology and the Study of Social Information Within the Perspective of a Unified Theory of Information. In: *Theories of Information, Communication and Knowledge*, Studies in History and Philosophy of Science, vol. 34 (eds. F. Ibekwe-SanJuan and T. Dousa). Dordrecht: Springer https://doi.org/10.1007/978-94-007-6973-1_3.

7 Abdullah-All-Tanvir, E., Mahir, M., Akhter, S. and Huq, M.R., Detecting Fake News using Machine Learning and Deep Learning Algorithms, *2019 7th International Conference on Smart Computing & Communications (ICSCC)*, Sarawak, Malaysia, Malaysia, 2019, pp. 1-5, doi: https://doi.org/10.1109/ICSCC.2019.8843612.

8 World Commission on Environment and Development (1987). *Our Common Future,* 27. Oxford: Oxford University Press. ISBN 019282080X.

26

Nanotechnology in Ecological and Ecosystem Engineering

L.R. Sendanayake[1], *H.A.D.B. Amarasiri*[2], *and Nadeesh M. Adassooriya*[3]

[1] *Postgraduate Institute of Science, University of Peradeniya, Peradeniya, Sri Lanka*
[2] *Department of Pharmacy, The Open University of Sri Lanka, Nugegoda, Sri Lanka*
[3] *Department of Food Science & Technology, Wayamba University of Sri Lanka, Makandura, Sri Lanka*

26.1 Ecology, Ecosystem, and Ecosystem Engineering

Careful observation of the environment reveals many interactions among living things and the physical environment. The depth and the complexity of these interactions are surprising. Animal–animal interactions for food, reproduction, and defense can be easily observed. More in-depth observations and research reveal a diverse range of interactions, such as atomic-level connections within the physical environment. This pool of interactions among living organisms and the non-living components in the surrounding environment is called the *ecosystem*. *Ecology* is a subdiscipline of biology that studies these interactions.

The ecosystem is an essential component of the environment to maintain an eco-balance. Ecologists have revealed that organisms in the environment exert a great effect on the environment's physical and chemical processes. These effects are not just simple one-step reactions; some are directly related to energy recycling and have more than one step. The environment's physical structures are altered by organisms that facilitate changes in chemical reactions. These physical changes can decrease the activation energy of a particular reaction. Aforesaid organisms are called *ecosystem engineers*, and they can create, modify, maintain, and destroy habitats. They heavily impact the habitat's landscape heterogeneity and species richness.

All micro and macro organisms and plants are potential ecosystem engineers, including human beings. Engineering procedures vary from micro to macro. Ecology and ecosystems have a broad scope, but in this chapter, our primary focus is on nanotechnology and nanoscience in ecological and ecosystem engineering [1, 2].

26.2 Nanomaterials, Nanotechnology, and Nanoscience

Nature is a unique scientist with many creations in all the science disciplines, including nanotechnology. Humans have applied this technology in every branch of science, including ecology. Therefore, it can be concluded that nanotechnology has a universal role in science. We will focus

Handbook of Ecological and Ecosystem Engineering, First Edition. Edited by Majeti Narasimha Vara Prasad.

on a few essential terms and definitions before discussing the nanotechnological concerns of ecological and ecosystem engineering.

Nanotechnology (NT) is the manipulation of nanomaterials. *Nanomaterials* (NM) can be defined as materials with at least one dimension in the nano range, typically 1–100 nm. There are four basic classes of nanomaterials: 0D, with all dimensions in the nano range; 1D, with two dimensions in the nano range; 2D, with one dimension in the nano range; and 3D, with building blocks consisting of 0D–2D nanomaterials. The study of these nanomaterials is known as *nanoscience* [2]. Nanomaterials have different physical and chemical properties than their parent material due to their enormous surface area–to-mass ratio. Nanomaterials' unique properties made them potentially useful in fields including medicine, engineering, and environmental remediation [3, 4].

26.3 Nanotechnology in Ecological and Ecosystem-Engineering

The expansion and application of nanotechnology during the last few decades has made it a potent factor in the environment; advantages and disadvantages are widely discussed. Some nanotechnology products are added to the environment to enhance the quality of the environment, and others are accidentally or purposely added and become pollutants. Environmental cleaning can be discussed under advantages, and addition of waste products of nanomaterials and by-products to the environment can be discussed under disadvantages.

Sometimes, nanomaterials used for environmental cleaning can have a long-term impact as pollutants. When the environment is polluted due to nanotechnology products and procedures, it is called *nanopollution*. The most common cause is poor environmental impact evaluation [5]. This emphasizes that the ecosystem and its engineers are directly and indirectly affected by nanotechnology, positively and negatively. As many nanotechnology products are foreign and novel to the environment, it is difficult to predict future outcomes. Let's discuss the role of nanotechnology in ecological and ecosystem engineering in detail.

26.4 Nanotechnology to Remediate Environmental Pollution

Pollution can be defined in many ways, but some recognized institutes have given good definitions. According to the United States Environmental Protection Agency, pollution is "The presence of a substance in the environment whose chemical composition or quantity prevents the functioning of natural processes and produces undesirable environmental and health effects." Today's number-one global burden is environmental pollution. Each country has tried to minimize it to a different extent based on its capacity.

Manufacturing and other essential processes add thousands of pollutants to the environment each minute. Urbanization and population growth are the key factors that govern these processes. As in the definition, natural processes are disturbed by pollution. This implies that ecosystem balance and ecosystem engineering, including engineers, are affected by pollution. Therefore, it is crucial to take necessary measures to prevent or minimize pollution, such as by introducing and implementing rules and regulations.

When new technologies are discussed, they have the potential to add pollutants to the environment. Nanotechnology is booming and is the youngest technology in modern science; it is widely used in many fields, including environmental pollution remediation. This utilization can either prevent or reduce environmental pollutants [6].

26.5 Environmental Remediation

Environmental remediation can be defined as reducing or removing pollutants from the environment using chemical, physical, and biological methods. The application can be categorized based on the target where it is applied: soil, air, or water. Remediation methods can also be classified into two broad categories: conventional and modern methods.

Nanotechnological approaches are modern methods, and they are used to remediate all areas: soil, air, and water. There are pros and cons for both conventional and modern methods. When balanced, modern methods are the front runners, with low cost, high efficacy and effectiveness, and relatively low onset time [7]. However, during the cleanup process, undesirable by-products and end products can be formed. The best approach to maintaining ecological and ecosystem engineering is removing pollutants that humans have added. Doing so directly and indirectly facilitates natural engineering and cyclic procedures.

26.6 Surface Water Remediation

Water is vital for all animals and plants. Around 70% of the earth's surface is covered by water, but only 1% is available as clean water for human consumption, and many people are suffering due to an insufficient water supply. Furthermore, each year, 18 million children die from diarrhea caused by drinking polluted water [8]. These numbers are for human beings – it is hard to estimate the total adverse effect on the environment, including other animals and plants. Rapid urbanization, industrialization, and hazardous chemicals used for day-to-day activities add tons of pollutants to water and water sources. It is a significant threat to the ecosystem balance, because all living things depend on water.

It is crucial to clean polluted water or minimize water pollution to ensure ecosystem balance and the lives of ecosystem engineers. As previously discussed, there are both traditional and new methods to remediate water. Due to drawbacks such as high energy consumption and low efficiency and effectiveness of conventional methods, new technologies such as nanotechnology are used more often than traditional methods [9].

Due to nanomaterials' specific surface areas, they have remarkable adsorbing properties. The surface can be modified with surface engineering technics, and surface reactivity can be optimized/maximized. The large surface area enables the catalytic activity of reactions (for example, oxidation reactions) when NMs are used as catalysts. In addition to surface area, the surface charge density, chemical affinity, and ability to transfer electrons should be considered when developing NMs for a particular purification process. High mobility can be observed due to the small size and high surface reactivity of NMs. This property is very useful when liquids are purified, because a small quantity of NM is sufficient to quickly treat a large volume. These qualities enable testing both inorganic and organic NMs as water-purification agents, disinfectants, and heavy-metal removers. Gravitational and magnetic forces are used in addition to well-known chemical bonds when removing pollutants.

The discovery of nanotechnology has opened a new door to handle wastewater more effectively, but NMs also have drawbacks. The use of NMs to remediate polluted water can be divided into four categories based on the mode of action: adsorption, photocatalysis, disinfection, and membrane filtration. For any nanotechnological approach, one or more of these modes of action can be used to clean the environment. In addition, if an NM is developed to remediate one primary part of the environment, such as soil, often it is also used to remediate the other parts, such as water and air. Now we will focus on each mode of action in detail for water remediation and briefly for soil and air.

26.6.1 Adsorption

Adsorption is a promising method for removing pollutants from water. The *adsorbent* is the NM, and the *adsorbate* is the pollutant. Physiosorption and chemisorption forces are involved in this process. The large active surface area and relatively small diffusion distance among the particles makes NMs efficient substrates to carry out the process by creating a high surface area normalized adsorption capacity [10]. The ability to modify the surface of NMs enables the selective adsorption of pollutants instead of essential components in the water.

Based on environmental conditions such as temperature and pH, NMs can be developed to release (desorption) the adsorbed components [11–13]. Design and development of NMs with adsorption and desorption ability are crucial in the reusability/recycling of NMs, leading to a reduced economic burden and less secondary pollution. In addition, they last a long time.

The porosity of NMs is another property that can be modified. Electrospun activated carbon nanofibers can be used to control adsorption kinetics as their pore morphologies can be tuned [14]. NMs used as adsorbents can be classified as magnetic nanoparticles (NPs), metallic oxide NPs, metallic NPs, and nanostructured mixed oxides, based on their function. In addition to these categories, carbon-based NMs such as carbon nanotubes (CNTs), carbon NPs, carbon nanosheets, and silicon-based NMs such as silicon NPs, silicon nanosheets, silicon NPs and nanoclays, polymer-based NMs, nanofibers, and aerogels are utilized [15, 16]. Some of these NMs are available as commercial preparations resulting from continuous research and development. Metal and metal oxide NMs have been highly developed to clean hazardous pollutants from water.

Iron is often used due to its low cost and natural abundance [17]. Iron-based NPs are mainly used to remove heavy metals and dyes as they have high surface charge, redox potential, and reusability [13, 18, 19]. In addition to an NM's properties, the nature of the environment in which it is applied also plays an essential role in adsorption. Incubation time, pH, temperature, surrounding living organisms, and adsorbent dose are a few factors. There are various approaches and modifications to enhance the adsorption of iron NMs, such as additives (for example, aminopropyltrimethoxysilane) [20]. Nanoscale zero-valent iron (nZVI) is the newest modification of iron NPs: nZVI can catalyze the reduction of heavy metals into insoluble forms. Since nZVI has magnetic properties, it tends to aggregate and then oxidize. This reduces the activity of the nZVI by limiting the active surface and reducing the reaction rate. Hence, it is important to do further studies to control the unwanted aggregation and oxidation of nZVI.

On the other hand, when iron is oxidized in water, the water becomes more polluted. To address this issue, a coating can be used. For example, Cr in oxidation state VI can be reduced and precipitated with chitosan-coated nZVI as Cr(III). Another form of iron NP is magnetic Fe_2O_3, which is highly reactive and reusable; but the high reactivity is sometimes a disadvantage as it readily corrodes in an aqueous environment. Again, a coating can be used with available resources such as porous carbon and mesoporous silica.

Other than iron, magnesium oxide (MgO), zinc oxide (Zn), and manganese oxide (MnO) also showed promising qualities in reducing heavy metals in polluted water. All these NMs were designed with different nanostructures and high Brunauer–Emmett–Teller (BET) surface areas. MnO is usually developed as hydrous MnO or nanotunnel materials to treat polluted water. In addition, MnO and other metal oxides are developed as nanoplates, nano assemblies, nanorods, and microspheres for water remediation.

Next, let's move on to CNTs, as they are highly effective as adsorbents, primarily for organic pollutant adsorption. There are four possible sites to interact with pollutants: inner sites, external surface, peripheral groove, and interstitial channel. Hydrophobic, covalent, π-π, electro-statistics interactions and hydrogen bonding help form bonds among pollutants and CNTs [21]. Organic compounds with carboxylic acid groups, hydroxyl groups, and amine groups are adsorbed to CNTs due to these interactions, especially due to hydrogen bonding [22]. When positively charged species come into contact, due to the electron-rich nature of CNTs, pollutants are adsorbed via electrostatics bonds. Environmental factors may affect these bindings, such as pH.

Small particle size, separation difficulties, and poor dispersion in water are limitations of CNTs. CNTs have been modified by researchers to achieve specific properties and qualities. Multi-wall CNTs (MWCNTs) are one approach, and magnetic properties have been introduced into them. In one study, polyacrylamide functionalized MWCNTs were synthesized to selectively adsorb humic acid and Pb from polluted water [23]. Phenol water pollutants such as 2,4-dichlorophenolcan are adsorbed when CNTs are modified with cyano groups. Common inorganic acids such as HCl are used to modify CNTs by introducing –OH groups to their surface, and they are highly effective in removing heavy metals like Cd from polluted water.

These examples clearly show that the CNTs' surface can be modified by introducing functional groups that enhance the selectivity of pollutant adsorption. When CNTs are treated with acids with the purpose of surface modifications, the process removes impurities from the CNTs. Modifying CNTs with magnetic properties also enhances their dispersion ability in water, compared to regular CNTs, as each NT tend to repulse each other. Whenever CNTs need to be removed from treatment procedures, an external magnetic field can be used.

When cost is compared, graphene oxide nanosheets (NSs) are more effective than CNTs. They can be an alternative to CNTs to adsorb metallic and organic pollutants. Graphene is an allotrope of carbon, and two-dimensional graphene oxide NSs are produced via chemical oxidation of graphene. Hydrophilic groups such as carboxylic acid and hydroxyl groups can be introduced to the graphene oxide to facilitate the adsorption of heavy metals. Inorganic pollutants are adsorbed via π-bonding, electrostatic bonding, and chelating. Hydrogen bonding also involves adsorbing organic pollutants. Graphene can adsorb both cationic pollutants (for example, metal ions and dyes) and anionic pollutants such as perchlorate and fluoride.

The next nanostructured adsorbent is nanopolymers. Nano cellulose and dendrimers are used in wastewater treatment as they are biocompatible, cost-effective, and biodegradable. Both organic and inorganic and hydrophobic and hydrophilic pollutants can be removed using dendrimers as their structure includes both hydrophobic and hydrophilic parts. The main advantage of dendrimers is that they need only a pH change to desorb captured pollutants.

Nanostructured materials that can be used as nanoadsorbents are relatively easy to manufacture, and as most of them are eco-friendly as they are biodegradable and biocompatible. They have significant potential for use in treating polluted water.

26.6.2 Photocatalysis

Many human activities add hazardous materials to the environment, including microorganisms, viruses, pesticides, crude oils, dyes, and hormonal products such as estrogen. There are naturally occurring microorganisms and viruses in the environment, but in a balanced number. Extensive human activities disturb this eco-balance, and it is vital to correct it. When hormonal products are added to the environment, ecosystem engineers and their breeding patterns are disturbed, ultimately disturbing ecosystem engineering. For example, it was proven that adding estrogen to water alters the gender of fish. This impact can decrease the fish population in water and the role of fish in the environment. Thus, via the disturbance of food chains and networks, the entire ecological system can be disturbed.

Photocatalysis is the breakdown of pollutants using the energy of light, usually UV. The primary scenario of photocatalysis is to support oxidation of pollutants by providing energy. This procedure creates highly reactive transitory species such as H_2O_2 and O_3, converts organics into readily biodegradable products, and finally produces CO_2 and water. It is an advanced oxidation process that generates hydroxyl radicals as the main species, which degrade organic pollutants in water [24].

From a nanotechnological perspective, semiconductors of metal oxides are often utilized in water treatment due to their achievable bandgap properties. Irradiation with light generates electron-hole pairs, which initiate the degradation process. Their low or zero toxicity, ability to tune, reusability, and doping ability make them effective in water treatment. ZnO and TiO_2 were extensively studied for photocatalytic activity and proved to be promising. Rapid recombination of electrons and holes reduces the efficiency of TiO_2, but size reduction leads to an increase in the interfacial charge carrier transfer [25]. It is crucial to adjust the size in the nano range (the optimal size is in this range) to achieve the best efficiency. This step is a little tricky as the extensive size reduction facilitates the recombination of electrons and holes. Even for ZnO, incrementing the particle size discourages photocatalytic activity [26]. The shape also plays an important role, as TiO_2 nanorods are more efficient than NPs due to their rapid reactant transfer to the surface and short carrier diffusion path length [27].

Shape and size are not only the factors to consider when NMs are designed for photocatalytic action. Band-gap also plays a significant role. Each individual NM has its optimal wavelength absorption. For TiO_2, it is 3.0–3.2 eV. This energy is at the UV range, and UV is only 3–5% of the solar spectrum. Therefore, NMs should be efficient enough to capture the available UV energy. Hybrid NPs, narrow-bandgap semiconductors, doping metal impurities, doping with anions, doping with noble metals, and dye sensitization are few approaches to increase photocatalytic efficiency [28–33].

Photocatalysis with NMs has become a commercially available method to treat polluted water. Artificial UV light sources are used to catalyze the degradation of organic pollutants. As we all know, they reduce the activation energy of the degradation pathway.

26.6.3 Disinfection

Disinfection of polluted water was practiced in the past, but due to this procedure's disadvantages, such as harmful disinfection by-product (DBP) formation, it requires modifications to be more effective [34]. Pathogenic microorganisms are mainly targeted by using ozone and chlorine-like disinfectants, but they can produce harmful DBPs by reacting with the other available components in the targeted water. NMs bring a new dimension to disinfecting water with no or minimal DBP production as they are relatively inert and mild oxidants [35].

NM-based disinfectants' mode of action is different than traditional techniques. They can penetrate cells and disrupt electron transfer pathways or produce reactive oxygen species that eventually kill the cell [36]. Chitosan-based NPs can interact with negatively charged cell membranes (as they are positively charged) and increase cell permeability, ultimately rupturing it and causing cell death [37]. Silk polymeric peptide NPs can create nanoscale channels in the cell wall of microorganisms, leading to osmotic pressure collapse and, finally, cell death [38].

Low toxicity to human beings, broad spectrum, and strong antimicrobial action made Ag NPs a promising agent in disinfection [39]. As in many cases, the size and shape of these Ag NPs are important in determining the antimicrobial properties. Compared to rod-shaped or spherical NPs, triangular NPs demonstrated a stronger biocidal activity [40, 41]. These Ag NPs can bind with various microbial enzymes and cause fatal DNA damage [42–44].

Again, TiO_2 can be discussed here as it produces reactive oxygen species and causes bacterial inactivation and, finally, death [45]. ZnO also shows antibacterial properties, and when it is joined with Pd, the bactericidal property is increased [46–49]. Fullerene and CNTs also demonstrated significant antibacterial activity, but their solubility in water is poor. The addition of surfactants can be one approach to increase the solubility of these NMs [50, 51].

26.6.4 Nanomembranes

Used/polluted water can be filtered using selectively permeable (semipermeable) membranes before release into the environment, to minimize or prevent damage. Membranes can keep the pollutants and release water and other components. Nanofiltration can be used under low pressure like available well-established filtration techniques such as microfiltration, reverse osmosis, and ultrafiltration [52]. Surface polluted water has relatively low osmotic pressure, and nanofiltration is highly effective there. The pores allow only water and monovalent salts like Na^+ to to pass through. Nanofiltration traps large pollutant molecules such as heavy metals and other organic molecules, including microbes [14]. As reported in the literature, in addition to the targeting molecules' size, charge of the targeting molecules will help in interacting with the charged membrane in permeation. Therefore, divalent sulfate ions are successfully rejected by the membrane. High-quality nanofiltration membranes are useful in removing solutes from polluted water [53, 54].

Some nanomembranes are developed using inorganic materials, and sometimes mixtures are used to enhance properties: for example, antimicrobial properties. For example, Ag NPs consist of nanomembranes to enhance their antimicrobial property, and TiO_2-embedded nanomembranes can enhance photocatalytic activity [52, 55]. Hydrophilic metal oxides such as Al_2O_3 are used to increase the hydrophilicity of the membranes to reduce fouling. The incorporation of antimicrobial NPs reduces biofouling.

26.7 Groundwater Remediation and Soil Remediation

Special consideration should be given to groundwater in ecology. Water present under the surface of the earth is known as *groundwater*. It can be in pores of the soil, rock fractures, etc. The location where these soil spaces and rock fractures are completely saturated with water is known as the *water table*. Water at the surface eventually moves through the soil to replace groundwater that is extracted for agricultural and human use via extraction wells. Groundwater also moves naturally to the earth's surface: for example, through springs.

Animals that live in groundwater are special and are not found on the earth's surface or in surface water. These living organisms do a unique job as ecosystem engineers, which cannot be replaced by any other living things on earth. When polluted water moves into the water table and mixes with groundwater, the water's ecosystem balance is disturbed. Therefore it is crucial to clean polluted groundwater and maintain the delicate ecosystem balance. When water pollution happens in any location, the consequences will simultaneously be experienced everywhere else, as water moves cyclically in the environment [56].

Nanotechnological approaches can be used to clean up polluted groundwater. The approach is different and difficult as the solution needs to reach water in the soil. Advancement of knowledge has revealed that human activities and accidental scenarios add pollutants, let pollutants enter the water table, and exert toxicological effects on ecological health [6]. It is important to discuss soil remediation parallel to groundwater remediation because, in most cases, they take place simultaneously.

As discussed above, water from the surface goes through the soil to the water table beneath the earth's surface. Even when water is moving through polluted soil, water-soluble and partially water-soluble pollutants mix with the water and ultimately pollute the groundwater. Therefore, when the soil is remediated, it indirectly remediates water and vice versa. Many conventional and new technologies can remediate soil and groundwater, but new technological approaches are more efficient when the pros and cons are balanced.

In situ and ex situ remediation methods are used. During in situ remediation, polluted soil is treated in its original environment; and in ex situ remediation, the soil is transferred to a separate place, and necessary treatments are carried out [57]. Physical, biological, and chemical methods are used to treat pollution, and combining these methods gives very good results. The contaminated area/site to be remediated (the type of soil) is the primary factor determining the selection of remediation methods and necessary adjustments [58].

In nanotechnological approaches, reactive NMs are applied to the field to destroy organic and inorganic pollutants, including heavy metals and microorganisms [59]. In situ remediation is possible with NMs due to their small size and wider distribution. Distribution can be further facilitated by coating the NMs [6]. In situ remediation with NMs such as nanoscale zeolite and CNTs provides more cost-effective and rapid remediation. Among these NMs, nZVI is widely used in groundwater treatment [59]. Non-aqueous liquids such as oil (pollutants) are removed/decomposed by using nano oxides. nZVI is utilized for groundwater remediation due to its small size, high abundance, low by-product formation, and adsorption and reduction properties [60].

During the last decade, nZVI was studied for its activity in polluted groundwater and soil. During these studies, many pollutants such as phenolic pollutants and chlorinated organic solvents were tested for detoxification, and nZVI was found to be highly effective [61]. In another study, nZVI was more effective in removing As from contaminated soil and groundwater; the tested NMs are around five times more effective than micro-sized particles [62]. Introducing magnetic properties helps remove heavy metals trapped in nZVI from contaminated fields using an external magnetic source, just as in surface water remediation. It was demonstrated and proven at the laboratory level that around 12 nm iron oxide NPs are efficient for removing As [63].

NMs can be mixed with useful microorganisms to enhance the degradation of pollutants. For example, Pd NPs are deposited on the cell wall and plasma of *Shewanella oneidensesis,* and when these modified microorganisms come in contact with chlorinated pollutants/compounds, they are degraded catalytically [64]. Some microorganism-conjugated NPs can be used to immobilize target groups/species of microorganisms that can degrade specific pollutants. Their movement can be externally controlled using a magnetic field when metal NMs are used to immobilize them. For example, the surface of *Pseudomonas delafieldii* was coated with magnetic nanoparticles (Fe_2O_4), which were functionalized with ammonium oleate and by using an external magnetic field. Subsurface uranium can be bioremediated with biogenic uraninite, which has a biological origin [65]. A significant environmental impact can be achieved with polycyclic aromatic hydrocarbons (PAHs) as they are hydrophobic and remain in the soil for a long time. Tungittiplakorn et al. developed poly(ethylene) glycol-modified urethane acrylate (PMUA) NPs to bioremediate PAH. These polymeric NPs were able to efficiently trap and release targeting pollutants from the soil. These prepared PMUA NPs were more stable in a heterogeneous bacteria-rich environment [66, 67].

Natural enzymes have a relatively short life span, and hence their usage is less cost-effective than synthetic enzymes that are developed with resistance to oxidation [68–70]. These enzymes can be used in bioremediation, and the lifetime can be extended by conjugating with NPs such as magnetic iron NPs. At the end of the reaction, the enzymes can be easily separated from the mixture by applying an external magnetic field, enabling reuse [38, 41].

Zeolite is the next NM to be discussed and it has a unique cage-like structure with a network of open channels that have a diameter range from 0.3 to 0.7 nm. This porous structure enables them to capture ions and exchange them with the surroundings based on environmental conditions. This porosity also makes them less dense than other minerals. Currently, natural as well as synthetic zeolites are available [71, 72]. They can be used to remediate contaminated soil as they are capable of adsorbing materials, ion exchange, and being used as sieves [73, 74]. Zeolite is used for many purposes to adjust pH, absorb H_2S and SO_2 in the gaseous state from soil, etc. [75–77].

When soil is polluted, its N and P composition is significantly lowered, which greatly affects plant growth. As plants are also ecosystem engineers, it is essential to encourage their growth. Therefore, N and P can be added to the soil artificially for this purpose, but excess addition leads to polluting the soil. To address this issue, N- and P-loaded zeolites can be used [73]. Fertilizers are heavily used in agricultural fields to increase crops and to polluted soil to encourage plant growth. In addition to zeolite-containing fertilizers, CNTs are also utilized by loading P and N into their channels. They can slow-release the minerals into the soil and hence prevent secondary accidental pollution [78].

Some research studies have shown secondary eco-friendly impacts. When heavy metals in polluted fields were treated with vivianite and apatite NPs, it was found that P constituents in the tested NPs were able to release into the soil [79, 80]. Zeolite has a broad working capacity as it can adsorb and absorb pollutants (especially heavy metals), adjust the soil's pH, and work as fertilizer to improve soil quality. It was proven that the soil leachate after zeolite addition has a low heavy metal concentration and low plant Hg absorption [81–85].

In situ remediation with trichloroethylene makes vinyl chloride and dichloroethylenes as undesirable by-products. These types of chlorinated hydrocarbons can migrate into the water table and cannot be removed with existing simple procedures [86]. Many NPs are readily transported with water, and due to this property, they can be used in treating polluted underground water without evacuation. Their extremely small size facilitates injection into small areas, and they last a long time [59, 61]. Their movements within the soil are affected by gravitational sedimentation. NPs can be designed to suspend throughout the targeting area without or with minimal coagulation [61].

As mentioned previously, a diverse range of NMs such as CNTs, nanoscale dendrimers, bimetallic particles, zeolites, and metal oxides are utilized in environmental remediation. Among them, iron NPs are the most popular and common [87–90]. Iron plays a role as a reducing agent through donating electrons. Pollutants/species to remediate are reduced and become less active or inactive. When iron NPs degrade heavy metals and polyhalogenated organic compounds, the iron NPs are oxidized into ferric or ferrous ions [59]. The remediation activity of iron NPs covers a wide range, including pesticides, as well as 2,4-Dichlorophenoxyacetic acid with magnetic NPs [91]. It was proven that iron NPs could remedy surface contaminants and subsurface contaminants such as fertilizers, heavy metals, and organic solvents.

In certain conditions, iron NPs agglomerate due to attractive forces among the particles (for example, high concentrations). Aggregation of particles adversely affects the activity because targeted reactivity is lost as the aspect ratio of NPs decreases. Various approaches were reported in the literature to overcome this issue. In one study, calcium alginate was used to entrap iron NPs. It reduced the mobility of the NPs and prevented aggregation. Release studies were also done, and it was found that coated NPs were slowly released compared to bare NPs, but statistical analysis clearly showed that the difference between the control and the test was not significant. Therefore, biopolymer coating can be recommended to minimize the mobility of the NPs and prevent aggregation [92].

Nanoclays play an essential role in polluted soil remediation as they have good porosity, less toxicity, relatively low cost, and good adsorption properties. They can adsorb organic and inorganic pollutants, including ionic forms. The cation exchange capacity of these layered materials also facilitates the adsorption of pollutants [93, 94]. Clay is a natural constituent of the soil, and natural clay particles are transporters/carriers of soil constituents, including nutrients and pollutants within the soil. Frequently, positively charged pollutants are bound to clay particles, which immobilizes the pollutants. This sorption happens to the external planar surface or the interlayer spaces.

The amount or extent of adsorption and the selectivity are highly dependent on the type of clay. In a study, montmorillonite clay and kaolin clay were tested for binding pyrene, and good adsorption of kaolin was observed compared to the other [95]. Groundwater contamination can be minimized by covering landfills with nanoclays. This is achieved via the stabilization and isolation of pollutants due to adsorbed immobilization [96]. When clays are utilized for remediation purposes, the nature of the targeting soil can affect efficiency. Therefore, the nature of the soil must be studied before application [97].

Recently, biochar was combined with clays to produce another NM. Biochar is produced by heating biomass with or without limited air to 250 °C or more. Biochar is nearly somewhat similar to activated charcoal but not completely equal. Biochar's porosity, surface functional groups, and high surface area facilitate adsorption of organic and inorganic pollutants from water and soil [98, 99]. Among various tested pollutants, antibiotic adsorption from polluted sources was promising. It is crucial to ensure the lives of useful microorganisms in the soil and water to maintain ecosystem balance. In 2016, Li et al. tested potato stem biochar using clay composite for norfloxacin adsorption and found its adsorption capacity to be 5.24 mg/g [100]. In two Sri Lankan studies, tetracycline and ciprofloxacin were tested for their adsorption to municipal solid waste biochar. Montmorillonite and red earth were combined with biochar for tetracycline and bentonite for ciprofloxacin. Two studies showed significant adsorption of antibiotics from aqueous solutions and variability in adsorption based on the clay type [101, 102]. This is a good approach to protect microbes in soil and water.

26.8 Air Remediation

We next turn our attention to air remediation. Air is essential for all living thing. Animals, including human beings, depend on O_2. Some living things, such as microorganisms, need different gases to maintain their lives and metabolic reactions. Air is composed of a variety of gases in different percentages. To maintain healthy air, these percentages should be nearly a constant. Whenever these values fluctuate, the air is said to be polluted. Humans and other mechanisms are usually involved in disturbing the balance. Air is a crucial factor to ecosystem engineers and ecosystems directly and indirectly to ensure life on earth. Research revealed that air pollution with sulfur compounds added excess acids to water sources (lakes, rivers). This directly damages trees and soil. Nitrogenous compounds in the air mix with surface water, damage aquatic life, and eventually enter the groundwater. Due to fuel combustion and manufacturing processes, heavy metal vapors mix with air in a cyclic manner, are transferred into soil and water, and damage living things.

Various pollutants can be present in the air, such as gases, organic materials, and microbes. Oxidation, adsorption, biofiltration, and simple ventilation methods are used in cleaning polluted air [103–108]. Most of the time, oxidation is cost-effective, safe, and reliable. This process directly converts volatile organic compounds (VOCs) into harmless or less harmful substances without additional treatments [109]. NMs can be used as catalysts to save energy and increase efficiency. The catalytic role and behavior of NMs is the same as discussed early.

Filters can be used: specific membranes are installed to purify the air, which requires relatively low energy, and this approach is highly recommended in industries that expel smoke/air during manufacturing and processing [110–112]. Filtration techniques include ultrafiltration and gas separation, but nanofiltration is interesting for the nanotechnological aspect [113–117]. Many development methods were used such as track-etching, sintering, and electrospinning [118]. As previously discussed, NMs can be joined with various membranes to enhance their selectivity and efficiency.

Evidence from the literature shows that methods to clean the air can be broadly categorized as physical and biological processes. Physical methods require energy at high temperatures, but biological methods can be carried out around 40 °C maximum. Microbes such as bacteria are employed for this method, and at the end of the cleaning procedure, gases such as CO_2, H_2O vapor, and nitrogenous and sulfur-containing products are released to the environment [119, 120]. Active living cells may be poor in efficiency as they can capture less of the targeted pollutants. Therefore, with the aid of biotechnology, NMs were introduced and bound with the microorganisms, which enhanced their efficiency for capturing and degrading pollutants. Sometimes microbes are immobilized on the working substrate using forces such as magnetic NPs and external magnetic fields. Based on research results, combined methods are more effective and efficient in the remediation of contaminated air. Photocatalytic NMs combined with cold plasma and biological methods combined with NMs are two examples [121]. In addition, nanosensors can check the air quality in the environment or check the expelled air after treatment to ensure that the expecting cleaning levels are achieved.

26.9 Future Scope of Nanotechnology and Nanoscience in Ecological and Ecosystem Engineering

As mentioned at the beginning of this chapter, ecology and ecosystem engineering is a vast area. Human activities disturb these engineering processes purposely or accidentally. But these disturbed processes can be put on the right path by applying various techniques, as it is essential to maintain the ecological balance and ultimately life on earth. This is not an easy task, as each traditional and new method has various drawbacks, but the methods can be analyzed and the best method selected with relatively more advantages than disadvantages. It was proven that new technologies are far better than existing traditional methods to heal the environment.

There are natural mechanisms to remediate the environment, but these mechanisms are not enough as environmental pollution keeps increasing daily. As humans, we need to find methods to reduce the burden on the environment, and we need to create mechanisms to support nature and modify the existing natural mechanisms to remediate pollution. Environmental pollution remediation is the primary thing we can do to support nature as many other things are beyond our limits to maintain ecology and ecosystem engineering. Nanotechnology is a new approach to find solutions for these issues. Broadly, nanotechnological applications in environmental remediation can be divided into four classes: adsorption, photocatalysis, disinfection, and membrane filtration. Directly or indirectly, these methods are supported by researchers. NMs also have drawbacks when used in environmental remediation. As this technology is new and many pieces of research are in the initial stages (in laboratories), there are many areas to be studied, particularly the long-term effect of NMs on the environment. Sometimes NM end products and by-products can be potential pollutants after years in the environment, including NMs used to remediate the environment. In the future, it is crucial to produce NMs, as they are promising agents for rapidly remediating environments with high efficacy; but at the same time, they should be studied for their long-term impact on nature.

References

1 Wolanski, E. and Elliott, M. (2016). Ecohydrology solutions. Estuarine Ecohydrology Elsevier; https://linkinghub.elsevier.com/retrieve/pii/B9780444633989000076.

2 Wright, J.P. and Jones, C.G. (2006). The concept of organisms as ecosystem engineers ten years on: progress, limitations, and challenges. *BioScience* 56 (3): 203.

3 Mehndiratta, P., Jain, A., Srivastava, S., and Gupta, N. (2013 Mar 13). Environmental pollution and nanotechnology. *Environment and Pollution* 2 (2): 49.

4 Mihranyan, A., Ferraz, N., and Strømme, M. (2012 Jun). Current status and future prospects of nanotechnology in cosmetics. *Progress in Materials Science* 57 (5): 875–910.

5 Alvarez, P.J. (2006 Oct). Nanotechnology in the environment—the good, the bad, and the ugly. *Journal of Environmental Engineering* 132 (10): 1233–1233.

6 Tratnyek, P.G. and Johnson, R.L. (2006 May). Nanotechnologies for environmental cleanup. *Nano Today* 1 (2): 44–48.

7 Salipira, K.L., Mamba, B.B., Krause, R.W. et al. (2007 Jan 29). Carbon nanotubes and cyclodextrin polymers for removing organic pollutants from water. *Environmental Chemistry Letters* 5 (1): 13–17.

8 Weltgesundheitsorganisation, UNICEF. (2010). Progress on sanitation and drinking-water. World Health Organization.

9 Sanchez, F. and Sobolev, K. (2010 Nov). Nanotechnology in concrete – a review. *Construction and Building Materials* 24 (11): 2060–2071.

10 Kalfa, O.M., Yalçınkaya, Ö., and Türker, A.R. (2009 Jul). Synthesis of nano B2O3/TiO2 composite material as a new solid phase extractor and its application to preconcentration and separation of cadmium. *Journal of Hazardous Materials* 166 (1): 455–461.

11 Saikia, J., Saha, B., and Das, G. (2011 Feb). Efficient removal of chromate and arsenate from individual and mixed system by malachite nanoparticles. *Journal of Hazardous Materials* 186 (1): 575–582.

12 Saikia, J., Sikdar, Y., Saha, B., and Das, G. (2013 Dec). Malachite nanoparticle: a potent surface for the adsorption of xanthene dyes. *Journal of Environmental Chemical Engineering* 1 (4): 1166–1173.

13 Saha, B., Das, S., Saikia, J., and Das, G. (2011 Apr 28). Preferential and enhanced adsorption of different dyes on iron oxide nanoparticles: a comparative study. *Journal of Physical Chemistry C* 115 (16): 8024–8033.

14 Shi, Q., Liang, H., Feng, D. et al. (2008 Apr). Porous carbon and carbon/metal oxide microfibers with well-controlled pore structure and Interface. *Journal of the American Chemical Society* 130 (15): 5034–5035.

15 Gupta, V.K., Tyagi, I., Sadegh, H. et al. (2015 Mar 1). Nanoparticles as adsorbent; a positive approach for removal of noxious metal ions: a review. *Science, Technology and Development* 34 (3): 195–214.

16 Anjum, M., Miandad, R., Waqas, M. et al. (2019 Dec). Remediation of wastewater using various nanomaterials. *Arabian Journal of Chemistry* 12 (8): 4897–4919.

17 Li, L., Fan, M., Brown, R.C. et al. (2006 Oct). Synthesis, properties, and environmental applications of nanoscale iron-based materials: a review. *Critical Reviews in Environmental Science and Technology* 36 (5): 405–431.

18 Li, Y.-H., Ding, J., Luan, Z. et al. (2003). Competitive adsorption of Pb2+, Cu2+ and Cd2+ ions from aqueous solutions by multiwalled carbon nanotubes. *Carbon* 41 (14): 2787–2792.

19 Istratie, R., Stoia, M., Păcurariu, C., and Locovei, C. (2019 Dec). Single and simultaneous adsorption of methyl orange and phenol onto magnetic iron oxide/carbon nanocomposites. *Arabian Journal of Chemistry* 12 (8): 3704–3722.

20 Palimi, M.J., Rostami, M., Mahdavian, M., and Ramezanzadeh, B. (2014 Nov). Surface modification of Cr2O3 nanoparticles with 3-amino propyl trimethoxy silane (APTMS). Part 1: studying the mechanical properties of polyurethane/Cr2O3 nanocomposites. *Progress in Organic Coatings* 77 (11): 1663–1673.

21 Das, R., Vecitis, C.D., Schulze, A. et al. (2017). Recent advances in nanomaterials for water protection and monitoring. *Chemical Society Reviews* 46 (22): 6946–7020.

22 Yang, K., Wu, W., Jing, Q., and Zhu, L. (2008 Nov). Aqueous adsorption of aniline, phenol, and their substitutes by multi-walled carbon nanotubes. *Environmental Science & Technology* 42 (21): 7931–7936.

23 Yang, K. and Xing, B. (2010 Oct 13). Adsorption of organic compounds by carbon nanomaterials in aqueous phase: Polanyi theory and its application. *Chemical Reviews* 110 (10): 5989–6008.

24 Kuwahara, Y., Kamegawa, T., Mori, K., and Yamashita, H. (2010 Apr 1). Design of new functional titanium oxide-based photocatalysts for degradation of organics diluted in water and air. *Current Organic Chemistry* 14 (7): 616–629.

25 Zhang, Z., Wang, C.-C., Zakaria, R., and Ying, J.Y. (1998 Dec 9). Role of Particle Size in Nanocrystalline TiO2-Based Photocatalysts. *J. Phys. Chem. B* 102 (52): 10871–10878.

26 Hayat, K., Gondal, M.A., Khaled, M.M. et al. (2011 Feb). Nano ZnO synthesis by modified sol gel method and its application in heterogeneous photocatalytic removal of phenol from water. *Applied Catalysis A: General* 393 (1–2): 122–129.

27 Macak, J.M., Zlamal, M., Krysa, J., and Schmuki, P. (2007 Feb 5). Self-organized TiO2 nanotube layers as highly efficient photocatalysts. *Small* 3 (2): 300–304.

28 Ni, M., Leung, M.K.H., Leung, D.Y.C., and Sumathy, K. (2007 Apr). A review and recent developments in photocatalytic water-splitting using TiO2 for hydrogen production. *Renewable and Sustainable Energy Reviews* 11 (3): 401–425.

29 Fujishima, A., Zhang, X., and Tryk, D. (2008 Dec 15). TiO2 photocatalysis and related surface phenomena. *Surface Science Reports* 63 (12): 515–582.

30 Asahi, R. (2001 Jul 13). Visible-light photocatalysis in nitrogen-doped titanium oxides. *Science* 293 (5528): 269–271.

31 Torres, G.R., Lindgren, T., Lu, J. et al. (2004 May). Photoelectrochemical study of nitrogen-doped titanium dioxide for water oxidation. *The Journal of Physical Chemistry. B* 108 (19): 5995–6003.

32 Hattori, A., Yamamoto, M., Tada, H., and Ito, S. (1998 Aug). A promoting effect of NH_4F addition on the photocatalytic activity of sol-gel TiO_2 films. *Chemistry Letters* 27 (8): 707–708.

33 Ohno, T., Akiyoshi, M., Umebayashi, T. et al. (2004 Jun). Preparation of S-doped TiO2 photocatalysts and their photocatalytic activities under visible light. *Applied Catalysis A: General* 265 (1): 115–121.

34 Rienzie, R., Sendanayake, L.R., and Adassooriya, N.M. (2020). Nanotechnology applications for removal of disinfection by-products from water. Disinfection By-products in Drinking Water. Elsevier. https://linkinghub.elsevier.com/retrieve/pii/B9780081029770000123.

35 Adhikari, M.D., Mukherjee, S., Saikia, J. et al. (2014). Magnetic nanoparticles for selective capture and purification of an antimicrobial peptide secreted by food-grade lactic acid bacteria. *Journal of Materials Chemistry B* 2 (10): 1432.

36 Li, Q., Mahendra, S., Lyon, D.Y. et al. (2008 Nov). Antimicrobial nanomaterials for water disinfection and microbial control: potential applications and implications. *Water Research* 42 (18): 4591–4602.

37 Qi, L., Xu, Z., Jiang, X. et al. (2004 Nov). Preparation and antibacterial activity of chitosan nanoparticles. *Carbohydrate Research* 339 (16): 2693–2700.

38 Gazit, E. (2007). Self-assembled peptide nanostructures: the design of molecular building blocks and their technological utilization. *Chemical Society Reviews* 36 (8): 1263.

39 Xiu, Z., Zhang, Q., Puppala, H.L. et al. (2012 Aug 8). Negligible particle-specific antibacterial activity of silver nanoparticles. *Nano Letters* 12 (8): 4271–4275.

40 Morones, J.R., Elechiguerra, J.L., Camacho, A. et al. (2005 Oct 1). The bactericidal effect of silver nanoparticles. *Nanotechnology* 16 (10): 2346–2353.

41 Pal, S., Tak, Y.K., and Song, J.M. (2007 Mar 15). Does the antibacterial activity of silver nanoparticles depend on the shape of the nanoparticle? A study of the gram-negative bacterium Escherichia coli. *Applied and Environmental Microbiology* 73 (6): 1712–1720.

42 Kim, J.S., Kuk, E., Yu, K.N. et al. (2007 Mar). Antimicrobial effects of silver nanoparticles. *Nanomedicine: Nanotechnology, Biology and Medicine* 3 (1): 95–101.

43 Liau, S.Y., Read, D.C., Pugh, W.J. et al. (1997 Sep). Interaction of silver nitrate with readily identifiable groups: relationship to the antibacterialaction of silver ions. *Letters in Applied Microbiology* 25 (4): 279–283.

44 Danilczuk, M., Lund, A., Sadlo, J. et al. (2006 Jan). Conduction electron spin resonance of small silver particles. *Spectrochimica Acta Part A: Molecular and Biomolecular Spectroscopy* 63 (1): 189–191.

45 Kikuchi, Y., Sunada, K., Iyoda, T. et al. (1997 Jun). Photocatalytic bactericidal effect of TiO2 thin films: dynamic view of the active oxygen species responsible for the effect. *Journal of Photochemistry and Photobiology A: Chemistry* 106 (1–3): 51–56.

46 Jones, N., Ray, B., Ranjit, K.T., and Manna, A.C. (2008 Feb). Antibacterial activity of ZnO nanoparticle suspensions on a broad spectrum of microorganisms. *FEMS Microbiology Letters* 279 (1): 71–76.

47 Khalil, A., Gondal, M.A., and Dastageer, M.A. (2011 Jul). Augmented photocatalytic activity of palladium incorporated ZnO nanoparticles in the disinfection of Escherichia coli microorganism from water. *Applied Catalysis A: General* 402 (1–2): 162–167.

48 Sawai, J. (2003 Aug). Quantitative evaluation of antibacterial activities of metallic oxide powders (ZnO, MgO and CaO) by conductimetric assay. *Journal of Microbiological Methods* 54 (2): 177–182.

49 Huang, Z., Zheng, X., Yan, D. et al. (2008 Apr). Toxicological effect of ZnO nanoparticles based on bacteria. *Langmuir* 24 (8): 4140–4144.

50 Lyon, D.Y., Brown, D.A., and Alvarez, P.J.J. (2008 May 1). Implications and potential applications of bactericidal fullerene water suspensions: effect of nC60 concentration, exposure conditions and shelf life. *Water Science and Technology* 57 (10): 1533–1538.

51 Vecitis, C.D., Zodrow, K.R., Kang, S., and Elimelech, M. (2010 Sep 28). Electronic-structure-dependent bacterial cytotoxicity of single-walled carbon nanotubes. *ACS Nano* 4 (9): 5471–5479.

52 Gehrke, I., Geiser, A., and Somborn-Schulz, A. (2015 Jan). Innovations in nanotechnology for water treatment. *Nanotechnology, Science and Applications* 1.

53 Qu, X., Alvarez, P.J.J., and Li, Q. (2013 Aug). Applications of nanotechnology in water and wastewater treatment. *Water Research* 47 (12): 3931–3946.

54 Jagadevan, S., Jayamurthy, M., Dobson, P., and Thompson, I.P. (2012 May). A novel hybrid nano zerovalent iron initiated oxidation – biological degradation approach for remediation of recalcitrant waste metalworking fluids. *Water Research* 46 (7): 2395–2404.

55 Arancibia-Miranda, N., Baltazar, S.E., García, A. et al. (2016 Jan). Nanoscale zero valent supported by Zeolite and Montmorillonite: template effect of the removal of lead ion from an aqueous solution. *Journal of Hazardous Materials* 301: 371–380.

56 Maurice, L. (2009). Groundwater ecology literature review. *British Geological Survey* 09: 29.

57 Lu, C. and Su, F. (2007 Dec 1). Adsorption of natural organic matter by carbon nanotubes. *Separation and Purification Technology* 58 (1): 113–121.

58 Khan, F.I., Husain, T., and Hejazi, R. (2004 Jun). An overview and analysis of site remediation technologies. *Journal of Environmental Management* 71 (2): 95–122.

59 Karn, B., Kuiken, T., and Otto, M. (2009). Nanotechnology and in situ remediation: a review of the benefits and potential risks. *Environmental Health Perspectives* 117 (12): 9.

60 Tratnyek, P.G. and Johnson, R.L. (2006 May). Nanotechnologies for environmental cleanup. *Nano Today* 1 (2): 44–48.

61 Zhang, W. (2003). Nanoscale iron particles for environmental remediation: an overview. *Journal of Nanoparticle Research* 5 (3/4): 323–332.

62 Yean, S., Cong, L., Yavuz, C.T. et al. (2005 Dec 1). Effect of magnetite particle size on adsorption and desorption of arsenite and arsenate. *Journal of Materials Research* 20 (12): 3255–3264.

63 Rickerby, D.G. and Morrison, M. (2007 Jan). Nanotechnology and the environment: a European perspective. *Science and Technology of Advanced Materials* 8 (1–2): 19–24.

64 Windt, W.D., Aelterman, P., and Verstraete, W. (2005 Mar). Bioreductive deposition of palladium (0) nanoparticles on Shewanella oneidensis with catalytic activity towards reductive dechlorination of polychlorinated biphenyls. *Environmental Microbiology* 7 (3): 314–325.

65 Bargar, J.R., Bernier-Latmani, R., Giammar, D.E., and Tebo, B.M. (2008 Dec 1). Biogenic uraninite nanoparticles and their importance for uranium remediation. *Elements* 4 (6): 407–412.

66 Tungittiplakorn, W., Cohen, C., and Lion, L.W. (2005 Mar). Engineered polymeric nanoparticles for bioremediation of hydrophobic contaminants. *Environmental Science & Technology* 39 (5): 1354–1358.

67 Tungittiplakorn, W., Lion, L.W., Cohen, C., and Kim, J.-Y. (2004 Mar). Engineered polymeric nanoparticles for soil remediation. *Environmental Science & Technology* 38 (5): 1605–1610.

68 Qiang, Y., Sharma, A., Paszczynski, A., and Meyer, D. (2007). Conjugates of magnetic nanoparticle-enzyme for bioremediation. Cleantech Conference and Trade Show. https://www.taylorfrancis.com/books/9781498721837/chapters/10.1201/9780429187469-25.

69 Hegedüs, I. and Nagy, E. (2009). Comparison of the structure and the stability of single enzyme nanoparticles. *Hungarian Journal of Industry and Chemistry* 37 (2): 123–130. doi:10.1515/239.

70 Kim, J., Jia, H., Lee, C. et al. (2006 Jul). Single enzyme nanoparticles in nanoporous silica: a hierarchical approach to enzyme stabilization and immobilization. *Enzyme and Microbial Technology* 39 (3): 474–480.

71 Pabalan, R.T. and Bertetti, F.P. (2001 Jan 1). Cation-exchange properties of natural zeolites. *Reviews in Mineralogy and Geochemistry* 45 (1): 453–518.

72 Mumpton, F.A. (1985). Using Zeolites in Agriculture. Innovative Biological Technologies for Lesser Developed Countries.

73 Ming, D.W. and Allen, E.R. (2001 Jan 1). Use of natural zeolites in agronomy, horticulture and environmental soil remediation. *Reviews in Mineralogy and Geochemistry* 45 (1): 619–654.

74 Ming, D.W. and Boettinger, J.L. (2001 Jan 1). Zeolites in soil environments. *Reviews in Mineralogy and Geochemistry* 45 (1): 323–345.

75 Githinji, L.J.M., Dane, J.H., and Walker, R.H. (2011 Jan). Physical and hydraulic properties of inorganic amendments and modeling their effects on water movement in sand-based root zones. *Irrigation Science* 29 (1): 65–77.

76 Shi, W., Shao, H., Li, H. et al. (2009 Oct). Progress in the remediation of hazardous heavy metal-polluted soils by natural zeolite. *Journal of Hazardous Materials* 170 (1): 1–6.

77 Liu, R. and Lal, R. (2013 Sep). A laboratory study on amending mine soil quality. *Water, Air, and Soil Pollution* 224 (9): 1679.

78 DeRosa, M.C., Monreal, C., Schnitzer, M. et al. (2010 Feb). Nanotechnology in fertilizers. *Nature Nanotechnology* 5 (2): 91–91.

79 Liu R. In-situ lead remediation in a shoot-range soil using stabilized apatite nanoparticles. 2011.

80 Liu, R. and Zhao, D. (2007 Jun). Reducing leachability and bioaccessibility of lead in soils using a new class of stabilized iron phosphate nanoparticles. *Water Research* 41 (12): 2491–2502.

81 Edwards, R., Rebedea, I., Lepp, N.W., and Lovell, A.J. (1999 June). An Investigation Into the Mechanism by Which Synthetic Zeolites Reduce Labile Metal Concentrations in Soils. *Environmental Geochemistry and Health* 21: 157–173. https://doi.org/10.1023/A:1006677115486.

82 Lin, C.-F., Lo, S.-S., Lin, H.-Y., and Lee, Y. (1998 Jul). Stabilization of cadmium contaminated soils using synthesized zeolite. *Journal of Hazardous Materials* 60 (3): 217–226.

83 Shanableh, A. and Kharabsheh, A. (1996 Feb). Stabilization of Cd, Ni and Pb in soil using natural zeolite. *Journal of Hazardous Materials* 45 (2–3): 207–217.

84 Moirou, A., Xenidis, A., and Paspaliaris, I. (2001 May). Stabilization Pb, Zn, and Cd-contaminated soil by means of natural zeolite. *Soil and Sediment Contamination: An International Journal* 10 (3): 251–267.

85 Haidouti, C. (1997 Dec). Inactivation of mercury in contaminated soils using natural zeolites. *Science of The Total Environment* 208 (1–2): 105–109.

86 Masciangioli, T. and Zhang, W.-X. (2003 Mar). Peer reviewed: environmental technologies at the nanoscale. *Environmental Science & Technology* 37 (5): 102A–108A.

87 Zhang, W., Wang, C.-B., and Lien, H.-L. (1998 May). Treatment of chlorinated organic contaminants with nanoscale bimetallic particles. *Catalysis Today* 40 (4): 387–395.

88 Ponder, S.M., Darab, J.G., and Mallouk, T.E. (2000 Jun). Remediation of Cr(VI) and Pb(II) aqueous solutions using supported, nanoscale zero-valent iron. *Environmental Science & Technology* 34 (12): 2564–2569.

89 Kanel, S.R., Manning, B., Charlet, L., and Choi, H. (2005 Mar). Removal of arsenic(III) from groundwater by nanoscale zero-Valent iron. *Environmental Science & Technology* 39 (5): 1291–1298.

90 Shipley, H.J., Engates, K.E., and Guettner, A.M. (2011 Jun). Study of iron oxide nanoparticles in soil for remediation of arsenic. *Journal of Nanoparticle Research* 13 (6): 2387–2397.

91 Fang, G., Si, Y., Tian, C. et al. (2012 Mar). Degradation of 2,4-D in soils by Fe3O4 nanoparticles combined with stimulating indigenous microbes. *Environmental Science and Pollution Research* 19 (3): 784–793.

92 Bezbaruah, A.N., Krajangpan, S., Chisholm, B.J. et al. (2009 Jul 30). Entrapment of iron nanoparticles in calcium alginate beads for groundwater remediation applications. *Journal of Hazardous Materials* 166 (2–3): 1339–1343.

93 Churchman, G.J., Gates, W.P., Theng, B.K.G., and Yuan, G. (2006). Clays and clay minerals for pollution control. Developments in Clay Science. Elsevier. https://linkinghub.elsevier.com/retrieve/pii/S1572435205010202.

94 Undabeytia, T., Nir, S., Rytwo, G. et al. (2002 Jun). Modeling adsorption–desorption processes of cu on edge and planar sites of Montmorillonite. *Environmental Science & Technology* 36 (12): 2677–2683.

95 Wei, Y., Liang, X., Lin, W. et al. (2015 Mar). Clay mineral dependent desorption of pyrene from soils by single and mixed anionic–nonionic surfactants. *Chemical Engineering Journal* 264: 807–814.

96 Konta, J. and Kühnel, R.A. (1997 May). Integrated exploration of clay deposits: some changes of strategy. *Applied Clay Science* 11 (5–6): 273–283.

97 Mobasser, S. and Taha, M.R. (2013 May 12). Adsorption of PCB from contaminated soil using nano clay particles. *Journal of Industrial Pollution Control* 29: 145–148.

98 Lehmann, J. and Joseph, S. (eds.) (2015). Biochar for Environmental Management: Science, Technology and Implementation, 2e, 928. London; New York: Routledge, Taylor & Francis Group.

99 Tan, X., Liu, Y., Gu, Y. et al. (2016 Jul). Biochar-based nano-composites for the decontamination of wastewater: a review. *Bioresource Technology* 212: 318–333.

100 Li, Y., Wang, Z., Xie, X. et al. (2017 Feb). Removal of Norfloxacin from aqueous solution by clay-biochar composite prepared from potato stem and natural attapulgite. *Colloids and Surfaces A: Physicochemical and Engineering Aspects* 514: 126–136.

101 Premarathna, K.S.D., Rajapaksha, A.U., Adassoriya, N. et al. (2019 May). Clay-biochar composites for sorptive removal of tetracycline antibiotic in aqueous media. *Journal of Environmental Management* 238: 315–322.

102 Ashiq, A., Adassooriya, N.M., Sarkar, B. et al. (2019 Apr). Municipal solid waste biochar-bentonite composite for the removal of antibiotic ciprofloxacin from aqueous media. *Journal of Environmental Management* 236: 428–435.

103 Wargocki, P., Wyon, D.P., Baik, Y.K. et al. (1999 Sep). Perceived air quality, sick building syndrome (SBS) symptoms and productivity in an office with two different pollution loads. *Indoor Air* 9 (3): 165–179.

104 Hyttinen, M., Pasanen, P., Björkroth, M., and Kalliokoski, P. (2007 Jun). Odors and volatile organic compounds released from ventilation filters. *Atmospheric Environment* 41 (19): 4029–4039.

105 Zhang, X., Gao, B., Creamer, A.E. et al. (2017 Sep). Adsorption of VOCs onto engineered carbon materials: a review. *Journal of Hazardous Materials* 338: 102–123.

106 Waring, M.S., Siegel, J.A., and Corsi, R.L. (2008 Jun). Ultrafine particle removal and generation by portable air cleaners. *Atmospheric Environment* 42 (20): 5003–5014.

107 Iranpour, R., Cox, H.H.J., Deshusses, M.A., and Schroeder, E.D. (2005 Oct). Literature review of air pollution control biofilters and biotrickling filters for odor and volatile organic compound removal. *Environmental Progress* 24 (3): 254–267.

108 Johnson, M.S., Nilsson, E.J.K., Svensson, E.A., and Langer, S. (2014 Aug 5). Gas-phase advanced oxidation for effective, efficient in situ control of pollution. *Environmental Science & Technology* 48 (15): 8768–8776.

109 Huang, H., Xu, Y., Feng, Q., and Leung, D.Y.C. (2015). Low temperature catalytic oxidation of volatile organic compounds: a review. *Catalysis Science & Technology* 5 (5): 2649–2669.

110 Judd, S. (2008 Feb). The status of membrane bioreactor technology. *Trends in Biotechnology* 26 (2): 109–116.

111 Balamurugan, R., Sundarrajan, S., and Ramakrishna, S. (2011 Aug 22). Recent trends in nanofibrous membranes and their suitability for air and water filtrations. *Membranes* 1 (3): 232–248.

112 Fillaudeau, L., Blanpain-Avet, P., and Daufin, G. (2006 Jan). Water, wastewater and waste management in brewing industries. *Journal of Cleaner Production* 14 (5): 463–471.

113 Petersen, R.J. (1993 Aug). Composite reverse osmosis and nanofiltration membranes. *Journal of Membrane Science* 83 (1): 81–150.

114 Radjenović, J., Petrović, M., Ventura, F., and Barceló, D. (2008 Aug). Rejection of pharmaceuticals in nanofiltration and reverse osmosis membrane drinking water treatment. *Water Research* 42 (14): 3601–3610.

115 Cho, J., Amy, G., and Pellegrino, J. (2000 Jan). Membrane filtration of natural organic matter: factors and mechanisms affecting rejection and flux decline with charged ultrafiltration (UF) membrane. *Journal of Membrane Science* 164 (1–2): 89–110.

116 Cui, Z.F., Bellara, S.R., and Homewood, P. (1997 May). Airlift crossflow membrane filtration — a feasibility study with dextran ultrafiltration. *Journal of Membrane Science* 128 (1): 83–91.

117 Zodrow, K., Brunet, L., Mahendra, S. et al. (2009 Feb). Polysulfone ultrafiltration membranes impregnated with silver nanoparticles show improved biofouling resistance and virus removal. *Water Research* 43 (3): 715–723.

118 Agarwal, S., Greiner, A., and Wendorff, J.H. (2013 Jun). Functional materials by electrospinning of polymers. *Progress in Polymer Science* 38 (6): 963–991.

119 Revah, S. and Morgan-Sagastume, J.M. Methods of Odor and VOC Control. In: Shareefdeen Z, Singh A, editors. Biotechnology for Odor and Air Pollution Control [Internet]. Berlin/Heidelberg: Springer-Verlag; 2005 [cited 2020 May 7]. p. 29–63. Available from: http://link.springer.com/10.1007/3-540-27007-8_3

120 Muñoz, R., Malhautier, L., Fanlo, J.-L., and Quijano, G. (2015 Aug 9). Biological technologies for the treatment of atmospheric pollutants. *International Journal of Environmental Analytical Chemistry* 95 (10): 950–967.

121 Assadi, A.A., Loganathan, S., Tri, P.N. et al. (2018 Sep). Pilot scale degradation of mono and multi volatile organic compounds by surface discharge plasma/TiO2 reactor: investigation of competition and synergism. *Journal of Hazardous Materials* 357: 305–313.

Index

Handbook of Ecological and Ecosystem Engineering, First Edition. Edited by Majeti Narasimha Vara Prasad.

n

o

S